SIGNAL PROCESSING FOR COGNITIVE RADIOS

SIGNAL PROCESSING FOR COGNITIVE RADIOS

SUDHARMAN K. JAYAWEERA
University of New Mexico

WILEY

Published by John Wiley & Sons, Inc., Hoboken, New Jersey
Published simultaneously in Canada

Library of Congress Cataloging-in-Publication Data:

Jayaweera, Sudharman K., 1972–
Signal processing for cognitive radios / Sudharman K. Jayaweera.
pages cm
Includes index.
ISBN 978-1-118-82493-1 (hardback)
1. Cognitive radio networks. 2. Signal processing. I. Title.
TK5103.4815.J39 2015
621.382′2–dc23

2015020555

Printed in the United States of America

10 9 8 7 6 5 4 3 2 1

To my mother (Amma) Mulin
and my son
Milidu

CONTENTS

PREFACE

Arguably, it is signal processing that makes a cognitive radio cognitive. Its predecessor, software-defined radio (SDR) technology, already provides a software-reconfigurable device platform by implementing most baseband radio operations in software instead of in hardware. Cognitive radios are meant to be SDRs that are cognitive and intelligent. Acquisition of knowledge through learning is a common aspect of both cognition and intelligence, while self-awareness and reasoning (application of acquired knowledge) are perhaps distinctive features of cognition and intelligence, respectively. A cognitive radio, thus, is supposed to possess all these features: self-awareness, learning, and reasoning. Clearly, these are attributes that a radio can possess mostly through signal processing. It is the signal processing algorithms, implemented on an SDR platform, that will endow a radio with self-awareness, learning, and reasoning abilities.

There are many books devoted to cognitive radios. However, none are devoted to signal processing in cognitive radios. This book is an attempt to highlight the fundamental role of signal processing in cognitive radios. One may identify two types of signal processing within a cognitive radio: signal processing for gaining spectrum awareness and signal processing for achieving efficient communications. Many processing algorithms that fall under the latter are already present in all wireless communications systems and devices. However, signal processing for spectrum awareness is unique to cognitive radios. These are aimed at providing the cognitive radio with self-awareness, a necessary ingredient of cognition. Attempting to comprehensively cover signal processing of both these types in a single book is perhaps an unrealistic goal. Moreover, it is unnecessary given the fact that there are many excellent books devoted to signal processing in wireless communications systems. Hence, the focus of this book is on signal processing that is unique to a cognitive radio. Not surprisingly, signal processing for gaining spectrum awareness constitutes a bulk of this focus. Still though, there are certain advanced signal processing techniques aimed at achieving efficient communications that are realistically well suited for implementations on sophisticated devices such as cognitive radios, thus deserving to be included in a book on signal processing for cognitive radios. An example is the advanced cooperative communications and processing techniques.

Any radio that is simply cognitive and intelligent can be considered as a cognitive radio. Hence, the notion of cognitive radios is a broad concept. Depending on the application context and the type of communications network in question, cognition and intelligence in a cognitive radio may be directed at achieving different objectives. One such widely pursued objective is the development of radios that may coexist with licensed spectrum users through dynamic spectrum sharing (DSS). However, there are other useful objectives as well, including, among others, multiband/multimode operation and antijamming. In general, such cognitive radios may be taken to be wideband radios in the sense that they may be able to operate over a wide span of non-contiguous spectrum. By default, the cognitive radios in this book are such wideband radios whose cognition and intelligence are aimed at arbitrary performance objectives. The DSS cognitive radios are treated as an important special case.

I would like to thank my former graduate students Mario Bkassiny, Yang Li, and Tianming Li whose pursuit of graduate research on the topic of cognitive radios led to much of the new results in this book. In particular, this book would not have been possible without the help of Mario, Yang, and Ding Li (another former graduate student of mine), who helped me in preparing many diagrams and figures for the book. I cannot thank them enough. I would also like to thank Prof. Christos Christodoulou, my colleague at the ECE Department of UNM, for the opportunity to collaborate with him on cognitive radio research as well as his encouragement during this book writing project. I am grateful for Prof. Carlos Mosquera of the University of Vigo, Spain, and his wife Vicky for their friendship and support during the times that I spent in Vigo. You gave me much needed distraction to get through the tough times while still keeping the interest in working on CR research. I must mention Prof. Hyuck Kwon's help with some of the proof-reading and my friend Thusitha Liyanage for checking on me time-to-time when I was too absorbed in work to keep in touch with any friends.

I would like to thank Simone Taylor and Kari Capone at Wiley for getting me on board to do this book with Wiley in the first place. From there, I am thankful to Brett Kurzman and Ho Kin Yunn at Wiley USA and Singapore, respectively, for handling the book project in such a speedy manner while making the whole process painless to me. I must also mention Alex Castro at Wiley. Finally, I would like to thank Jayapriya Purushothaman at SPi Global for her timely management of the copy-editing, typesetting and proof reading stages.

Albuquerque, NM, United States SUDHARMAN K. JAYAWEERA
05/12/2014

PART I

INTRODUCTION TO COGNITIVE RADIOS

1

INTRODUCTION

1.1 INTRODUCTION

Cognitive radios (CRs) are one of the most researched topics since the introduction of the concept in 1999 [1]. The original definition of the CR concept was ambitious and visionary [1]: CRs have the ability for "... automated reasoning about the needs of the user," and they are "... radio-domain-aware intelligent agents that search out ways to deliver the services the user wants even if that user does not know how to obtain them" [1]. May be this definition and our intuitive understanding of the word *cognitive* are enough for now. We can postpone defining the CRs more formally till the next chapter.

In recent years, there have been many efforts toward developing CRs that are, being true to the above definition, user-, and environment-aware. Intuitively, cognitive abilities of a radio must emerge from being able to interpret and react to its RF environment and user's needs (the performance goals of the radio). Moreover, these cognitive abilities should lead to effective learning. But, of course, we are talking about a radio device. Just as human perception of the external world is gained through the five sensory organs (nose, smell; ears, hearing; eyes, sight; skin, touch; and mouth, taste), a radio's perception of its external world has to be gained through the single sensory organ a radio is equipped with, that is, its antenna. The sensory input is essentially *signals* or, in general, some form of electromagnetic radiation, as in Figure 1.1. Hence, the ability to interpret signals to comprehend its environment and acquire knowledge, followed by reasoning, decision-making, and learning from

Signal Processing for Cognitive Radios, First Edition. Sudharman K. Jayaweera.

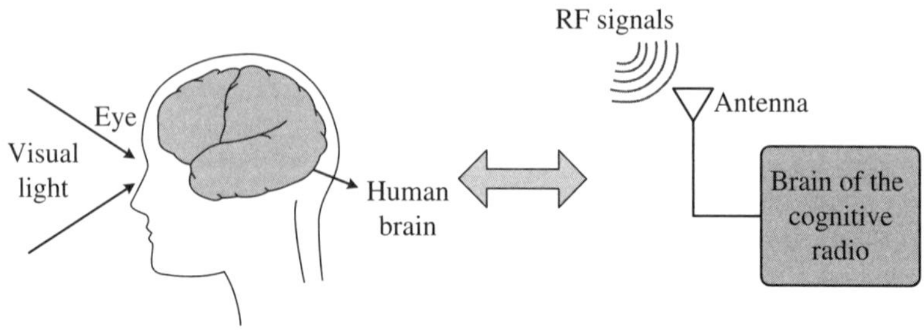

FIGURE 1.1 RF signals and the antenna are the sensations and sensory organ of a cognitive radio.

such acquired knowledge, is at the heart of a radio's cognition. In that sense, arguably, all cognitive abilities of a radio must emerge from advanced *signal processing* techniques it possesses: be they for interpreting RF environment, making decisions on how to communicate, or learning new knowledge. The focus of this book, thus, is on signal processing techniques for CRs.

1.2 SIGNAL PROCESSING AND COGNITIVE RADIOS

Mathematically, a signal is simply a function. Depending on in what domain we are looking at the signal, the independent variable of the function can be time, frequency, space, or even a combination of them. Since a signal as a function of time is the most familiar, perhaps, let us think of it as such, for now. Signals are, of course, fundamental to all communications and can be of various types physically. But for our purposes, it may be thought of as an electrical current/voltage or an RF waveform.

Generating, processing (or manipulating), transforming, transmitting, and detecting signals are involved in all communications systems and devices. By signal processing, in a (digital) communications receiver, we essentially mean the processing involved from the reception at the antenna until the detection of binary bits, as in Figure 1.2. In a communications receiver, information bits are extracted from a received RF signal by converting it into several different forms in several steps: analog RF waveform to an intermediate frequency (IF) waveform, then to a discrete-time sequence, then to symbol values, and finally to bits. We may consider processing up to this point as signal processing. There may be further decoding of error control coding (ECC) based on these bits that we may not necessarily treat as signal processing. Detected binary bits may need to be processed again before they can be in a form that is useful for a human user, but we may skip treating such processing as part of the communications receiver.

Similarly, signal processing in a transmitter can include processing involved from the output of an error control coder till the generation of an analog waveform that is ready to be radiated from an antenna as in Figure 1.3. We may loosely treat processing of bits as coding (or ECC) and not as part of signal processing. Of course, this is not completely precise. For example, both trellis coding and space–time coding can

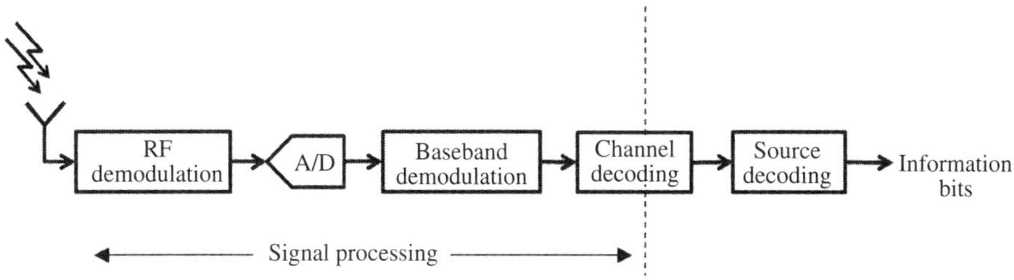

FIGURE 1.2 Signal processing in a receiver.

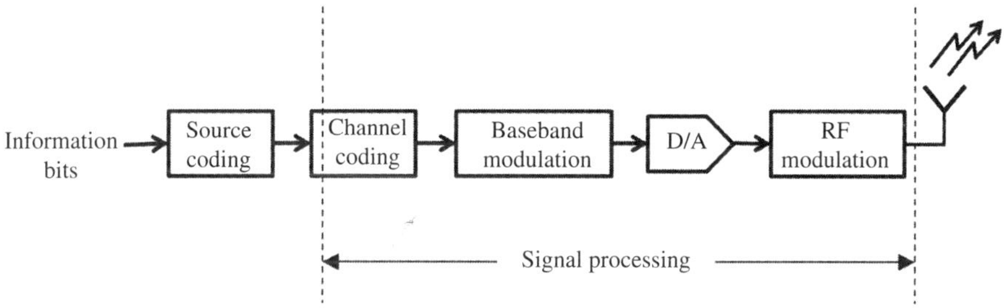

FIGURE 1.3 Signal processing in a transmitter.

make the coding and signal processing to overlap. In general, this is true with *coded modulation* techniques that combine coding and modulation together.

Signal processing can be either analog or digital. Early communications systems and devices mainly used analog signal processing implemented in hardware. Thus, all processing of signals were performed directly in the analog domain on the analog waveforms using analog circuits. For example, in an old AM or FM radio, the message waveform $m(t)$ would be filtered using an analog filtering circuit and modulated using a mixer circuit. On the other hand, in digital communications systems, part of the system (or the radio device) holds the signals in digital format. The interface between the analog and digital worlds is usually a pair of an *analog-to-digital* (A/D) and a *digital-to-analog* (D/A) converters. Processing a signal in digital form is called digital signal processing (DSP). DSP can, however, be performed also in either hardware or software. But usually, even if the processing is in hardware, it is essentially a software algorithm specifically implemented on a hardware circuit for speed and efficiency. Hence, in a certain sense, we may call all DSP essentially as implemented in software. Of course, to implement signal processing algorithms as software, one needs a computer or a processor. In the case of a communications device, this could be a microprocessor, a general-purpose processor (GPP), a DSP chip, an application-specific integrated circuit (ASIC), or a field-programmable gate array (FPGA). All these essentially provide digital hardware platforms to implement and run the software programs that implement DSP algorithms.

The importance of signal processing in modern digital communications cannot be overemphasized. In fact, without the advanced signal processing techniques, many of today's sophisticated and high-performance communications systems may not have been possible. For instance, over the last two decades, cellular communications systems have seen a tremendous worldwide growth and expansion. This was mostly due to the fact that each generation of new cellular systems rapidly evolved into providing breathtaking quality improvements compared to the previous generation. Many of these gains, in part, were due to advanced signal processing techniques that facilitated overcoming various channel impairments and interference issues. For example, the cellular communications concept relies on dividing the geographical area into a set of cells and reusing the same set of carrier frequencies in nonadjacent cells. Frequency reuse pattern and the cell sizes are carefully optimized so that signals sufficiently attenuate beyond its own cell. It is, however, still unavoidable that the same frequency signals from one cell will spill over and interfere with that of another cell. This is called the *cochannel interference* (CCI). Earlier cellular systems used the so-called *matched-filter*-based single-user detection at the receiver. Later generations of cellular systems, on the other hand, improve the performance of the simple matched-filter receiver by incorporating sophisticated CCI suppression algorithms. There are many such examples of the use of advanced signal processing algorithms in modern communications systems.

For a CR, signal processing is even more important because it is fundamental to the radio's identity as a cognitive device. If the processing in the brain (brain functions) is the source of cognition and intelligence of a biological entity, then signal processing algorithms are the source of cognition and intelligence in a CR. In other words, signal processing algorithms are the brain functions of a CR.

1.3 SOFTWARE-DEFINED RADIOS

The theory of evolution states that the intelligence in biological entities develops and evolves out of necessity. Cognition and intelligence in radios seem to be no different. The concept of CRs evolved from its primitive ancestors, early hardware defined radios, as the application domains of radios expanded, just like life on earth moved from oceans to land followed by various land migrations.

Early wireless devices were hardware defined in the sense that they were developed to communicate using a single waveform format as is permitted by their hardware. By the early 1990s, it has been noted that keeping up with rapid technology advancements in wireless communications is a challenge with these traditional hardware-defined radios. To take advantage of new communications technologies, one has to redesign the hardware.

Moreover, different wireless systems were designed to achieve different performance objectives. Hence, they were designed to operate in various spectrum bands available, or best suited, for the application in mind. The RF characteristics, air interfaces, and signaling waveforms/protocols used were different from one system to another. A hardware-defined radio designed to operate according to one scheme could not communicate with a device using another scheme. Final result was that one system could not interact with another system. As wireless systems proliferated, this inability of interoperability of hardware-defined radios emerged as a clear limitation.

These were particularly significant limitations in situations in which it is important for users on different systems to be able to easily interact with each other. A particularly interested group of interoperability and rapid technology advancement was the US military [2]. In many cases, different service branches within the US military had developed their own systems to meet specific demands and performance goals. In fact, even within the same service, there can be multiple radio systems based on multiple legacy standards. In addition, it is also of importance to be able to communicate with friendly forces while denying interception by enemies. For the US military, limitations of hardware-defined radios were in full display in 1991 during the First Gulf War (Operation Desert Storm). One of the major lessons learned from the war was the inadequacy of interoperability of the joint forces communications technologies [3]: The systems were mostly inconsistent among each other, and thus, communications among different services, or with friendly forces, that were using different radio systems were almost impossible. In addition, taking advantage of rapid technological advancements made by the commercial industry was not possible in a cost-effective way when one has to replace or redesign the system hardware.

It was against this backdrop that the notion of *software radios* emerged in the early 1990s, which later in 1995 became known as *software-defined radios* (SDRs) [4]. According to Wikipedia,[1] however, the origin of the term software radio goes back to 1984. A team at the Garland Texas Division of E-Systems Inc. (now part of Raytheon) has used the term software radio to refer to a digital baseband receiver in a company newsletter in 1985 [5]. However, it was [4] that used the term in the context of a radio that also included a transmitter and is widely credited for coining the term in the sense it is used today.

Software radios are radios that perform some, or all, baseband operations in software. Fixed, or hardware-defined, radios are limited in their operational capabilities by their hardware configurations. If, however, some or all of these hardware can be controlled by software, then at least to some degree these limitations can be overcome. In this sense, the concept of SDR promised a path toward interoperability of radio systems. Indeed, if all systems were to follow a common SDR architecture, then, at least in theory, the interoperability issue can be resolved. Moreover, with software radios, it is much easier and cost-effective to incorporate new waveforms and technology advancements to existing radios without having to redesign or replace the hardware components.

SDR was also motivated by the popularity and proliferation of consumer wireless systems, in particular cellular and WLAN networks. Indeed, during the 1980s and the early 1990s, the telecommunications industry players were aware of the pending potential rapid worldwide growth in demand and popularity of wireless systems. The competitive pressure to develop better, faster, and improved wireless technologies

[1] http://en.wikipedia.org/wiki/Software-radio#History

was enormous. There were a lot of money to be made. As a result, there were a large amount of new signal processing and coding algorithms being pumped out of an ever-increasing number of research groups spread all over the world. The SDR technology provided a flexible approach to integrate these algorithms to wireless systems and then to switch or choose among them in real time.

Moreover, in the same geographical area, one may usually find several different wireless networks. A wireless device that is hardwired to operate on one particular network (following a certain wireless standard) can only transmit and receive signals from that network. For example, an early analog Advanced Mobile Phone System (AMPS) cell phone could only transmit and receive analog AMPS signals. As wireless networks proliferate, users needed to carry several different devices, each designed to communicate over a different network, to take advantage of different networks and services available at different locations. The reason is that since signal generation and processing is done in hardware, the same circuit cannot generate or process a different signaling waveform. With the rapid growth of computers and software, it was clear that if, instead of hardware, software programs could be used to generate and process signals in these communications devices, then things could be much simpler: Instead of having to have several different circuits to generate and process different waveforms, all we need is several different software program instructions residing in a microprocessor's memory. When a certain type of signaling waveform is needed, all one needs to do is to give a command to activate that portion of the program. Everything can be done in software, provided signals are represented in digital domain. This transition from A/D and D/A can be achieved by using A/D and D/A converters. Hence, all we need to do is once the signal needs to be transmitted (say), construct an analog waveform corresponding to the software-defined digital signal and feed it to an antenna.

Example 1.1 (Analog, Digital, and Discrete-Time Signals) An *analog signal* is a function of the continuous-time variable t that can take a continuum of values as in Figure 1.4(a). Hence, when we say $x(t)$ is an analog signal, we mean that as a function both its domain and range are the set of real values: That is, if $x(t)$ is an analog signal, then $x : \mathbb{R} \longrightarrow \mathbb{R}$. On the other hand, a *discrete-time signal* is a function whose domain is the set of integers but the range is still the set of reals: That is, if $\tilde{x}[n]$ is a discrete-time signal, then $\tilde{x} : \mathbb{N} \longrightarrow \mathbb{R}$. In other words, a discrete-time signal can be considered as a signal that is temporally quantized (or sampled) as shown in Figure 1.4(b).

A *digital signal* is a discrete-time signal whose range is constrained to only a finite set of finite values $\mathcal{A}$. Hence, if $x[n]$ is a digital signal, then $x : \mathbb{N} \longrightarrow \mathcal{A}$. For example, a digital signal whose amplitude is only allowed to take the two values A and $-A$ will have $\mathcal{A} = \{A, -A\}$ as in Figure 1.4(c). Comparing to a discrete-time signal, we see that a digital signal is a signal whose amplitude and time are both quantized. ■

Hence, an ideal SDR (receiver) architecture may directly feed the antenna output to an A/D converter, which then interfaces with what we may call the software-defined radio SDR platform. The SDR platform is essentially a software-based digital radio

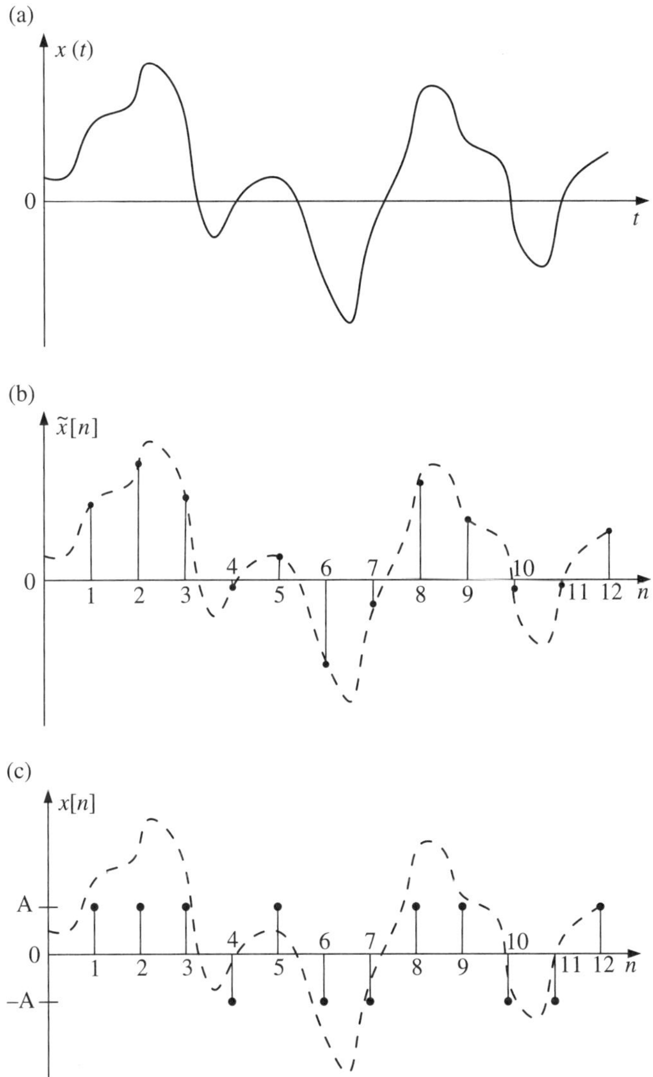

FIGURE 1.4 Signals involved in the process of analog-to-digital conversion. (a) An analog signal. (b) A sampled discrete-time signal. (c) A digital signal.

platform as shown in Figure 1.5. Sometimes, this is also called the reconfigurable digital radio [6].

The ideal SDR architecture of Figure 1.5, of course, is not realistic. To start with, at least one needs a BPF/LPF before feeding the antenna output to the A/D converter to avoid aliasing (since all A/D converters have to have a finite sampling rate). We will also at least need some amplification of the RF output of the antenna since it might be too weak for any further processing. Hence, some filtering and power amplification

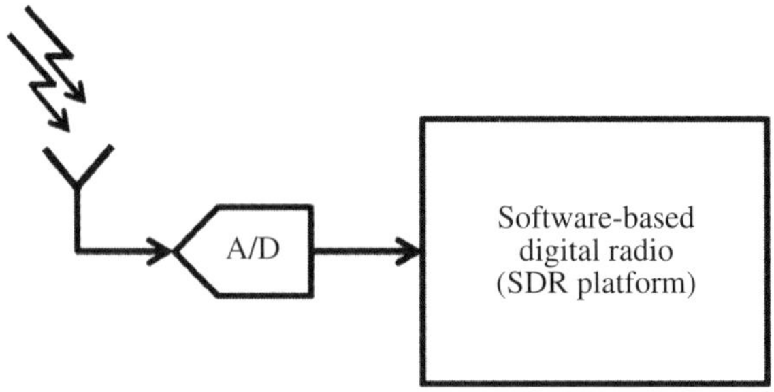

FIGURE 1.5 The ideal software-defined radio concept.

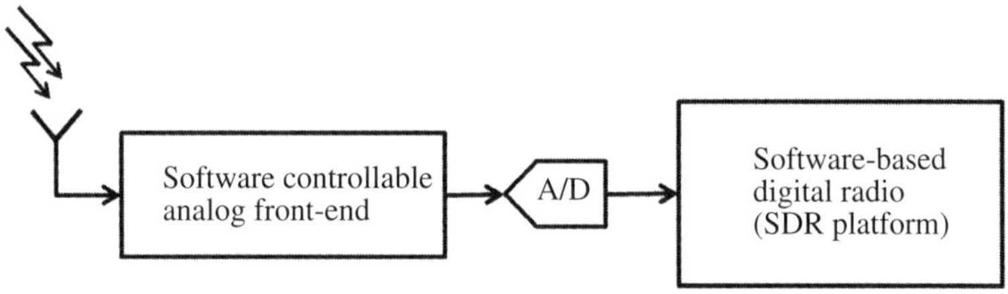

FIGURE 1.6 A feasible software-defined radio architecture.

may be needed to get a clean and sufficiently strong signal for later digital processing. These components must then be in an analog RF stage. This forces us down to the somewhat more realistic SDR architecture of Figure 1.6 that consists of a software-based digital radio platform connected to a software-controllable analog RF front-end and an (RF) antenna.

The A/D converter plays the key role of gateway between digital and analog domains. The sampling theory, which is the underlying theory assuring the equivalence between the analog signal and its A/D converted digital signal, places a fundamental demand on the A/D converter: The sampling rate must be at least twice that of the analog signal's bandwidth (or the highest frequency) to avoid aliasing and to ensure accurate equivalence between the two signals (analog and discrete time). Of course, output from the antenna may not have a clear cutoff frequency. Hence, it is customary to use an *antialiasing filter* before the A/D converter to avoid aliasing.

It is, however, not possible to have arbitrarily large sampling rates and arbitrarily fine resolutions in an A/D converter for physical reasons such as power consumption, electronics, and memory requirements. These essentially boil down to a maximum possible bit rate for a given A/D converter output. As Example 1.2 shows, in an A/D converter, each sample is quantized and represented by a binary string of a certain number of bits. Hence, the higher the sampling rate, the lower the resolution will be to keep the final (maximum) bit rate fixed, since we have to reduce the number of bits used to represent each sample. The lower the number of bits per sample, the smaller the number of levels that can be used in the quantizer, leading to crude quantization of analog sample values. This distortion is called the *quantization noise*.

Example 1.2 (A/D conversion) A/D conversion takes an analog (i.e., continuous-valued, continuous-time) signal, or a waveform, into a digital signal. Hence, A/D conversion involves following steps:

1. *Time quantization* (*sampling*): Given an analog signal $x(t)$, time quantization step produces a discrete-time signal. Of course, this is the familiar process called sampling. Since the objective is to obtain a digital signal that is equivalent to the original analog signal $x(t)$, we would like to ensure that the discrete-time signal produced by the sampling process, say, $\tilde{x}[n]$, is equivalent to the original signal $x(t)$. We know from *the sampling theorem* (see Appendix A for details) that this is guaranteed by sampling at a rate not less than twice the bandwidth (or the maximum frequency) of the signal. Hence, if the analog signal is a baseband signal with highest frequency f_m to ensure that the time quantized signal is equivalent to the original signal, we need a sampling rate of at least $f_s = 2f_m$ since the signal bandwidth is $W = f_m$. As long as we satisfy this condition, time quantization step is fully reversible. Indeed, as shown in Appendix A, it is possible to reconstruct the original analog signal $x(t)$ from the discrete-time sampled sequence $\tilde{x}[n]$. However, when the signal of interest is centered at a certain (high) frequency f_c with a much narrower bandwidth W than the center frequency, then it is sufficient to have a sampling rate of $f_s = 2W$ rather than $f_s = 2\left(f_c + \frac{W}{2}\right)$. In this case, we do have frequency aliasing. However, as long as we use a suitable band-pass antialiasing filter before sampling, it is possible to recover the signal from the sampled sequence. In practice, it is customary to choose an IF that is larger than $f_s = 2W$ and perform the conversion to the baseband directly in digital domain.
2. *Amplitude quantization* (*quantization*): Output from the sampling stage is a discrete-time signal $\tilde{x}[n]$. These sample values, however, can take any arbitrary real value (assuming real-valued signals). The amplitude quantization step, often simply referred to as the quantization step, limits the amplitude values to a finite set of finite values. Once the number of allowed amplitude levels and their values (called the quantizer levels) are determined, the simplest form of quantization can be achieved by rounding off the arbitrary amplitude values of the discrete-time signal $\tilde{x}[n]$ to the nearest allowed quantizer levels. For a chosen number of allowed quantizer levels L, if the quantizer levels (amplitude values) are chosen by equally dividing the range of the original signal into L levels and then rounding off to the nearest level as aforementioned, then we obtain what is called a *uniform quantizer*. One may choose more elaborate quantization schemes than the uniform quantization in order to obtain better performance in specific situations. Unlike sampling, the amplitude quantization is not reversible.
3. *Binary encoding*: The output from the quantization step is a digital signal that assumes L number of possible amplitude levels. These quantizer amplitude levels are each assigned a unique bit-sequence (bit-string) label. Then, rather than the actual quantizer amplitude level, the A/D converter outputs the bit-sequence

label associated with each amplitude level. If the quantizer is to have an L number of amplitude levels, in order to assign a unique bit-sequence label to each of them, the minimum length of the bit-sequence label must be $\log_2 L$ number of bits.

From the earlier discussion, we can determine the output bit rate of an A/D converter that employs a sampling rate of f_s (samples/second). If the quantizer has L levels, each sample must be represented by at least $\log_2 L$ bits. Hence, the output bit rate of the A/D converter is

$$R = f_s \log_2 L \text{ bps.} \tag{1.1}$$

This points out the relationship between the signal bandwidth, the quantization noise, and the output bit rate of the A/D converter. If the original analog signal has a large bandwidth, then f_s has to be chosen to be large, as required by the sampling theorem. Hence, we need an A/D converter with a higher output bit rate. Also, if smaller quantization noise (distortion) is desired, then a quantizer with finer levels is needed. This implies a larger value of L, leading again to an A/D converter with a higher output bit rate. ■

Example 1.3 (Digital Voice) Voice signals have a bandwidth of about 4 kHz. Hence, voice is digitized by sampling at a rate of $f_s = 8$ kHz. Each sample is quantized with an $L = 256$-level quantizer resulting in an output bit rate of

$$R = 8 \log_2 256 \text{ kbps} = 64 \text{ kbps.} \tag{1.2}$$

In practice, this is too large a bit rate for supporting voice communications, especially with wireless systems. However, it is quite common that the output bit rates of A/D converters can be reduced by removing some of the bits without losing information or with very small distortion. This is called *compression* or *source coding*, which relies on the fact that information-bearing signals, such as voice signals, have a significant amount of redundancies that can be removed without losing information. Many cellular phones, for example, use a technique called *linear predictive coding* (LPC) to compress the 64 kbps toll-quality voice by a factor somewhere between 4 and 8, resulting in effective bit rates of 8–16 kbps.

The redundancies in an information signal imply that successive signal samples are not independent of each other. They are correlated. The idea of LPC is to exploit this correlation to linearly predict the next signal sample value as a linear combination of some of the past signal sample values. Then, instead of the exact next sample value, an LPC speech coder quantizes the difference between the exact and the predicted sample values. Since this prediction error can have a much smaller dynamic range than the actual voice sample value itself, the quantizer only needs to have a smaller number of levels to maintain the same resolution. ■

The required preconditioning of the signal before sampling (e.g., filtering by a suitable antialiasing filter) also shows why it is not practical to have an SDR that directly attaches an A/D converter (or a D/A converter) to the (RF) antenna itself as implied in Figure 1.5. Instead, as shown in the feasible SDR architecture of Figure 1.6, there will be a certain amount of processing still performed in analog hardware before the A/D converter (in an SDR receiver) and after the D/A converter (in an SDR transmitter). The analog circuitry that performs this processing is called the (RF) front-end in Figure 1.6.

In practice, it is not just an RF antialiasing filter that may make up the front-end. It is advantageous to perform the A/D conversion at an IF rather than directly in baseband. Sampling at an IF does require the sampling rate to be higher than what it would have been in the baseband. However, it makes the design of frequency-tunable components more manageable as well as makes all hardware to be essentially fixed after the IF stage. The signal that is sampled at the IF, for example, is brought to baseband by direct digital down-conversion that is essentially achieved in software. Hence, the analog front-end in a practical SDR may include both RF and IF stages.

The A/D and D/A converters form the interface between the software-controllable analog RF and IF front-end and the software-based digital radio platform as in Figure 1.7. A practical SDR goal is to shrink this RF/IF front-end as much as possible so that the A/D converter and the D/A converter become as close to the RF antenna as possible.

In the SDR architecture of Figure 1.7, the digital radio platform operates in digital baseband. The software-controllable analog front-end may be either a homodyne or a superheterodyne receiver but with components that are controllable/tunable by software commands from the digital radio platform. A simple homodyne receiver design in which sampling is performed in baseband may look as in Figure 1.8.

As mentioned earlier, early communications systems were designed in such a way that all components of the radio were hardwired. On the other hand, in a software-defined radio as in Figure 1.7, the waveforms to be communicated are generated in the

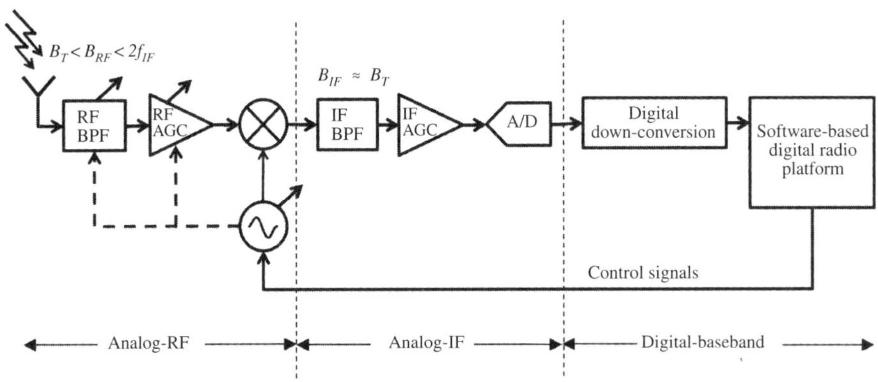

FIGURE 1.7 A realistic software-defined radio architecture.

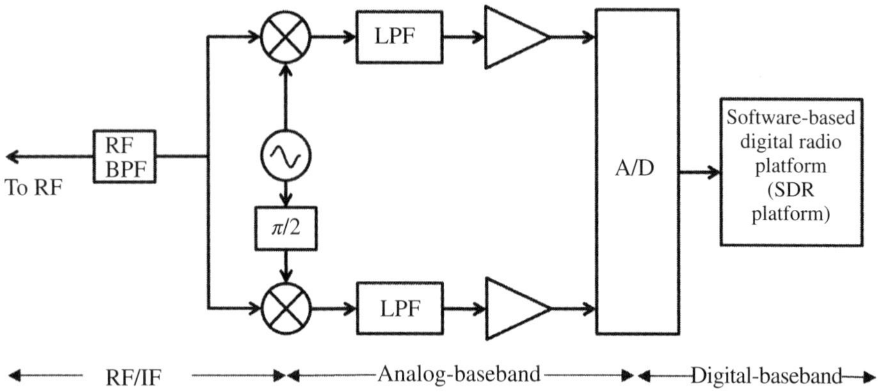

FIGURE 1.8 A realistic software-defined radio (SDR) receiver (homodyne) with baseband I/Q channels.

software-based digital radio platform. Essentially, the type of waveform to be generated is written into a software program. If there is a need to change the waveform, all one has to do is to change the software program residing in the software-based digital radio platform. Or we may actually have definitions (or codes) to produce several different types of waveforms in the software program residing in the software-based digital radio platform at the same time and just switch among them as needed. Of course, depending on the type of waveform (modulated signal) one wants to transmit, there may be a need to modify some of the analog front-end components. The software-based digital radio platform is also responsible for generating these control signals as can be seen in Figure 1.7.

1.3.1 Software-Defined Radio Platforms

Implementing an SDR as described earlier requires analog RF components that are controllable by software commands. These include RF and IF stage amplifiers and filters. Over the last two decades, there have been significant advances in developing software-controllable tunable RF front-end components including filters, power amplifiers, reconfigurable antennas, and duplexers, among others. A discussion on some of these can be found in [6].

On the other hand, software-based digital radio platforms are nothing but digital computers, perhaps especially designed to handle specific tasks required in a communications device. Indeed, in early SDRs, the software-based digital radio platform was based on common microprocessors or GPPs. Digitally processing a signal, however, involves certain computations that are very much specialized (such as computing the FFT) and commonly needed. This led to the digital signal processors, which are processors with certain built-in capabilities that facilitate efficient computation of certain specific signal processing steps. Although DSPs provided a significant improvement over the GPPs, they alone were not sufficient as research led to complex signal processing algorithms that required faster and complex computations. This led to another class of platforms called ASICs that especially addressed the issue of processing speed of DSPs. ASICs offered better performance compared to DSPs in terms of speed, but as their name suggests, they were designed and programmed for specific applications and thus were not easily reconfigurable like DSPs. Many software-based digital radio platforms of SDRs may thus use a combination of DSPs and ASICs to implement different parts. For example, tasks that are so specialized for a particular radio, and need to run fast, may be implemented in ASICs since they may not be needed to be changed much, while algorithms that may need a certain level of modifications later may be implemented on DSPs. Currently emerging platforms, however, tend to use FPGAs that offer high performance of ASICs with the additional advantage of high reconfigurability, even better than that of DSPs. The drawback of FPGAs, however, is their high power consumption. As a result, they tend to be larger in size and more expensive compared to both DSPs and ASICs.

In this book, we may use the terms software-based digital radio platform and SDR platform interchangeably. An SDR system contains not only the SDR platform but also the front-end hardware as in Figure 1.6.

1.3.2 Software-Defined Radio Systems

Over the years, there have been several SDR systems that have been developed by both commercial industry and by the US military. In the following, we briefly review some of these systems.

1.3.2.1 Military-Use SDR Systems

1. *Speakeasy I*: In the early 1990s, especially after the First Gulf War, the US military realized the limitations of its then-existed communications systems. In addition to perhaps obvious desirables such as improvements on communications speed, jammer/interference resistance, and low probability of interception/detection (LPI/LPD), the lack of interoperability among radios of joint forces was exposed as a critical issue [3]. The Speakeasy program, managed by the Air Force Research Laboratory (AFRL), was a response to these challenges that aimed at developing a communications system with waveform reconfigurability by implementing waveforms and their associated processing in software on programmable signal processors [2]. The primary objective of Speakeasy I (1992–1995) SDR project was to develop and demonstrate in a field test a modular, reprogrammable modem with an open architecture. The secondary objective was to develop a generic software architecture that will facilitate addition of new waveforms [7]. The initial goal was to be able to demonstrate a multiband multimode radio (MBMMR) that can support any waveform in 2 MHz–2 GHz range.

   In 1995, the Speakeasy I project demonstrated the feasibility of a military SDR modem in which rapid waveform reconfigurability was achieved by implementing waveforms in software rather than in analog circuits. The DSPs were used to develop the required modular and programmable architecture for software implementation of waveforms [2]. However, the phase I demonstration only covered the waveforms in frequency range from 30 to 400 MHz. Indeed, it had been determined that the 2 MHz–2 GHz range cannot be covered by using a single RF front-end with then state-of-the-art technology. Hence, the radio was to have three RF sections covering ranges from 2 to 30 MHz, 30 to 400 MHz, and 400 to 2 GHz [7]. Although the basic goal was achieved, the modem software, the user interface, and the waveform development platform of Speakeasy I did not have ease of use, which was also one of the objectives of the program.

2. *Speakeasy II*: The 4-year Speakeasy II program began in 1995 immediately following the success of Speakeasy I project. The main objective of Speakeasy II was to expand the software-based, open, modular, and reprogrammable architectural implementation of the modem, which was achieved during Speakeasy I, to include the entire radio system or the network [7]. Hence, Speakeasy II was to be an open architecture, networked, secure, and software-programmable radio system with simultaneous multichannel, multiband and multimode (MBMM) capabilities to operate from 2 MHz to 2 GHz [8]. An additional goal was to encourage the commercial off-the-shelf (COTS) modules and commercial standards in the development [7].

   The Speakeasy II program successfully demonstrated the ability to implement certain radio waveform functions in software modules that can be developed to be independent from the underlying hardware [8]. It also demonstrated that interoperability among radios can be achieved by using such SDRs. Indeed, in a US Army field demonstration in March 1997, the Speakeasy II was able to successfully bridge voice channels between an Air Force aircraft using HAVE QUICK UHF systems and soldiers using commercial handheld land mobile radios (LMRs) as well as Army "Single Channel Ground and Airborne Radio System" (SINCGARS) VHF radios [7]. The implementation included an especially developed RF homodyne transceiver and DSPs augmented with FPGAs. Although the waveforms involved were limited to AM and FM and the frequency range covered by the especially developed transceiver was 4–400 MHz, the performance of this field demonstration was so impressive that it was determined to be ready to enter production without even having to wait till all original objectives are achieved [7].

   The portable SDRs produced from the Speakeasy program were not small or flexible enough for handheld use. Indeed, the initial Speakeasy II platforms were about the size of old PCs [8]. Still, the hardware-independent software module implementation of waveform functions and programmable radios based on such modules achieved by the Speakeasy program have to be considered a great success.

3. *Joint Tactical Radio System* (*JTRS*): The success of the Speakeasy program paved the way for the JTRS program, arguably the most ambitious SDR development project undertaken to date. Until 2011, it was managed by the Joint Program Executive Office (JPEO) for the Joint Tactical Radio System (JPEO JTRS). In 2012, it was restructured under the Joint Tactical Networking Center

(JTNC). The goal of the JTRS project was to achieve real-time interoperable multimedia (voice, data, video) communications among radios of different branches of the US military as well as those of friendly forces. Additionally, compatibility with newer systems, reduced time to adapt technology advancements, and affordability are expected.

The JTRS program aims to achieve these objectives by ensuring software reuse and portability [9]. It recognizes the fact that while SDRs provide a means to achieve the aforementioned goals of interoperability, technology enhancement, and affordability, unless the codes used in implementing originally analog circuit-based functions in software are portable and reusable, they themselves may become the limiting factor to achieving the stated goals. As a result, all JTRS software are required to be compliant with the open SDR reference architecture called the Software Communications Architecture (SCA).

The SCA was developed collaboratively by a broad coalition of stakeholders in developing SDRs, including the US, Japan, the UK, Sweden, and Canada, in response to the need for a common standard for the software development for SDRs. As different groups pursued SDR platform and system development, it had quickly become clear that without a common standard for the software development, SDR systems will run into the same interoperability issues that plagued the earlier hardwired systems. The SCA specifies a set of baseline requirements for any software developed for SDRs in order to achieve interoperability, reconfigurability, and portability [10]. According to the 2006 report from the JPEO of the JTRS [10], SCA is essentially a set of rules that guides software developers to achieve the following goals, independent of the particular implementation:

- Portability of application software
- Reduced developmental cost
- Reusing of design modules
- Leveraging of commercial architectures

Adhering to the SCA can obviously lead to compatibility among different manufacturers and suppliers, a desirable objective for the military but also important for manufacturers. It also helps promote competition among developers, which may ultimately lead to reduced costs for the users.

Thus, the JTRS is not a single system, but a family of software-programmable, modular, MBMM communications systems that maximizes the software and hardware commonality and reusability. The commonality and the compliance with the SCA lead to interoperability, while reusability ensures both affordability and compatibility with legacy systems as well as newer systems that may become available due to technological advances.

Since 1998, it is mandatory that all new radio procurements by the US Department of Defense (DoD) agencies be JTRS compliant with the hope that ultimately JTRS communications devices will replace the old single-function, hardware-intensive communications systems with software-centric systems that satisfy the requirements of the user domains while being interoperable with legacy communications systems and supporting growth for new waveforms [11]. In addition, JTRS system was to be scalable to match the communications requirements of different users, extendible to support growth and change, affordable, and usable through open systems standards and technology that will serve as the baseline for evolution [11].

It is an ambitious project that aims to develop an open radio architecture that will allow affordable interoperable operation among perhaps hundreds of different systems used by various branches of military services. While JTRS itself has perhaps turned out to be more and more complex and inflexible, one of its lasting impacts seems to be the development of the SCA that essentially provides a set of rules for software developers for SDRs to ensure compatibility among each other.

1.3.2.2 Commercial-Use SDR Systems The commercial telecommunications industry has been a driving force of the advancement of wireless technologies including the SDR. In fact, one of the objectives of the initial military SDR programs was to be able to take advantage of the rapid technological advancements made by the commercial wireless industry by being able to quickly, and cost-effectively, incorporate them into military systems. Similarly, the commercial telecommunications industry was also quick to realize the potential of SDR technology. Indeed, by the end of the 1990s, there were already several successful commercial-use SDR systems by various industry players (for more details, see [12]):

1. *Motorola Horizon 3G*: Motorola's Horizon 3G SDR was aimed at supporting then-existed GSM- and CDMA-based cellular systems.
2. *Vanu Inc.'s Anywave*: One of the successful commercial SDR products was the Vanu Inc.'s Anywave SDR cellular base station that supported several cellular standards such as GSM and cdma2000 [13]. Compared to other competing cellular base stations that used specialized protocols on the wire-line part of the network, the Anywave SDR used VoIP for both internal and external traffic.
3. *AirNet Communications Corp.'s AdaptaCell*: In 1997, AirNet Communications Corp. deployed the first commercial SDR cellular base station. This broadband multicarrier system was designed to support several standards including GSM, GPRS, and EDGE. Fully adaptive later versions of AirNet SDR base transceiver subsystems, named AdaptaCell BTS, can be configured to support the emerging 802.16e mobile Worldwide Interoperability for Microwave Access (WiMAX) standard, using single carrier, orthogonal frequency division multiplexing (OFDM), and/or orthogonal frequency division multiple access (OFDMA) waveforms.

It is noteworthy that many commercial SDR platforms were SDR base stations rather than SDR handsets. While one reason for this was the increased power consumption of SDR platforms, another reason was that perhaps the objective was to achieve base station interoperability among many different standards.

1.4 FROM SOFTWARE-DEFINED RADIOS TO COGNITIVE RADIOS

SDRs were concerned with a radio device implementation that provided flexibility in changing the signal waveforms to be received or transmitted. How and when to switch from one signaling waveform to another was not a focus of SDR. When interoperability is desired, perhaps an operator may switch the radio to the desired mode. The ability to change the radio's physical communications mode by flipping a switch, or giving a command, was still a great achievement. Toward the end of the 1990s, however, the SDRs were ready to evolve into its logical next level: Rather than having an external entity making the decisions of when and how to change the communications mode and configure operating characteristics, maybe the radio itself can make these decisions based on its RF environment conditions and performance objectives.

This naturally implies that the radio needs to have some sort of self-awareness, or a consciousness, built into it. Hence, such self-aware SDRs were termed the CRs in [1]. This evolution of wireless communications from analog hardware-based implementations to SDRs and then from there to CRs is shown in Figure 1.9.

Arguably, CR concept is one of the most important innovations in wireless communications during the last two decades. CRs offer the potential to radically change the future of wireless and mobile communications. However, that has to wait due to another development.

1.4.1 The Spectrum Scarcity Problem

The primary objective of a communications system designer is to be efficient in power budget and spectrum usage. This is motivated by the commonly accepted fact that these two resources, power and bandwidth, are both scarce. While transmit power at least seems somewhat less fundamental a concern, it is usually observed that almost all spectrum that can possibly be useful, or suitable, for communications seems to be already assigned for particular services or systems. Moreover, according to the traditional way of assigning spectrum if a certain frequency band is assigned for a particular type of service, or licensed for a particular system, then no other system or service can operate in that frequency band. Thus, while in the 1980s and 1990s mobile wireless communications were gaining wide popularity all over the world and the possible applications seem to be limitless, the only spoiler seemed to be the radio spectrum: There was not much unassigned spectrum to go around.

During the late 1990s several research groups, as well as the Federal Communications Commission (FCC) in the US, measured the utilization of spectrum in the highly desirable range for mobile wireless communications (roughly from a few tens of hundreds of MHz to a few GHz). The results were eye-opening: Most of the allocated radio spectrum was heavily underutilized. Specifically, most of the frequencies traditionally allocated for certain services (e.g., broadcast TV) were not used most of the times in many places. It appeared that, after all, the perception of overcrowded spectrum was not exactly accurate. The spectrum was overcrowded on the paper. But in practice, the license holders of certain spectrum bands were not really using it in many

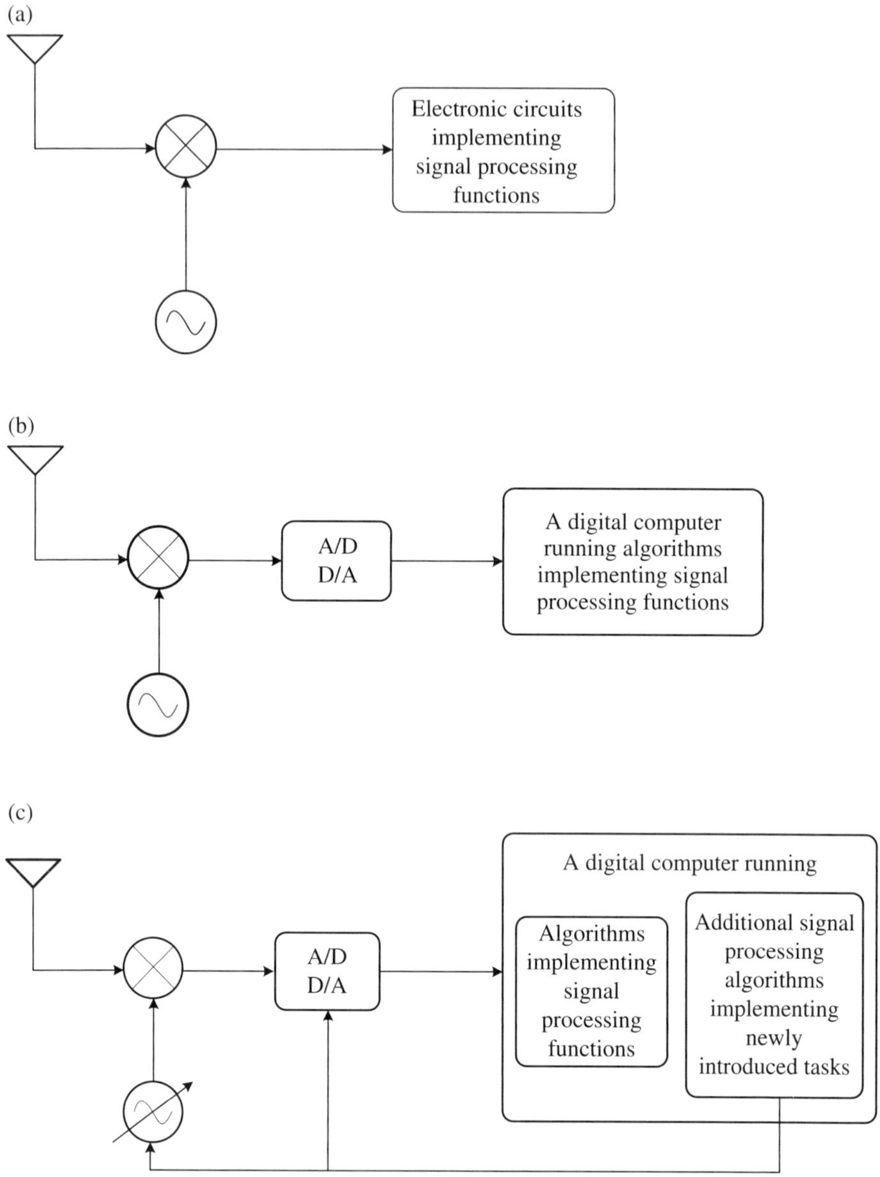

FIGURE 1.9 The evolution of wireless devices from conventional to SDR to cognitive radio. (a) A conventional radio. (b) A software-defined radio. (c) A cognitive radio.

places and/or much of the time. Another finding from these measurement campaigns was that compared to exclusively allocated frequency bands, the so-called industrial, scientific and medical (ISM) bands were highly crowded and better utilized. The 2.4–2.4835 GHz ISM band in the US, for example, was found to be highly crowded with WiFi WLANs, Bluetooth, and other wireless signals.

These observations led to a rethinking of spectrum allocation policies practiced by governmental agencies. The realization that the perceived scarcity of radio spectrum is mainly due to the inefficiency of traditional spectrum allocation policies led the FCC to recommend several broad solutions to improve the spectrum utilization in its 2002 Spectrum Policy Task Force Report [14]. One important theme of these recommendations was permitting unlicensed users and services to access the already allocated spectrum bands if and when they are not used by the license holder. Although this was a reasonable approach to better utilize the spectrum, achieving this would require new technological breakthroughs.

The concept of CRs emerged against the backdrop of these new observations on the inefficient spectrum utilization and the growing demand for various wireless services. It did not take too long for many to realize that the concept of CRs can provide a promising solution to the problem of underutilization of the allocated spectrum: If a radio can be aware of its RF environment, as was envisioned with CRs, then it may be able to take advantage of unused spectrum opportunities. Self-awareness of a CR combined with the reconfigurability of an SDR can possibly facilitate dynamically sharing the spectrum with a legacy licensed system while not harming its performance.

Hence, a key driver of such dynamic spectrum sharing (DSS) approaches for improving spectrum utilization was the CRs. This in turn made DSS to be a key driver for the development of CR technology. These two became so much intertwined that it was almost difficult to separate them. As a consequence, during its initial phase of development, CRs were interpreted as synonymous with DSS or dynamic spectrum access [15], almost replacing the original vision and definition of [1] of CRs as autonomous, intelligent, and self-aware radio devices with much broader scope.

1.4.2 Emergence of CRs

A CR is an evolution of the concept of SDRs in the sense that the radio has built-in intelligence and self-awareness to manage its own operations in response to the perceived state of the RF environment and the radio's performance objectives. Hence, obviously, an SDR is the host platform of a CR. Indeed, it was as an evolution of SDR that will make SDRs more personal that the CR was initially introduced in [1]. Since the motivation for SDR was interoperability, affordability, and compatibility with future systems, then it follows that these much broader objectives are also the motivations for CRs. Indeed, it appears that the objective of MBMMR of Speakeasy II SDR program, for example, had been retained as an objective of CRs. This is clear from the way the successor program to the Speakeasy II, the JTRS, has naturally embraced the CR technology. A CR, defined as a self-aware SDR with decision-making and learning abilities, promised great potential in realizing the goals of interoperability, affordability, and compatibility with future systems. Naturally, one may expect such CRs to be wideband, multichannel, and MBMM.

On the other hand, CRs pursued for achieving DSS did not have the interoperability or compatibility with future systems as a primary motivation. There, the key driver was the spectrum coexistence, in particular with license holders to the spectrum, so as to improve the spectrum utilization. It is, however, clear that the objective of spectrum coexistence can easily be met by being cognitive, wideband, and MBMM. Hence, while reinterpreting CRs as a technology specifically aimed at achieving DSS may have helped the progress of CR technology in that direction faster, the same objectives might have been still achieved if the focus stayed as a general evolution of SDR with the original SDR objectives intact as envisioned in [1]. The latter has more broad implications for the future of wireless communications: Self-aware SDRs with autonomous decision-making ability are a robotic radio, or a Radiobot [16]. Such robotic radios promise the potential to transform today's wireless telecommunications in almost limitless exciting ways. Ultimately, CRs may lead to autonomous and universal communications devices that attend to human needs without us even having to be aware of them.

CR is still a developing technology. While CRs explicitly aimed at achieving DSS seem more mature in their development, certainly, CRs with broader application scope are still only at a very young stage. It is, however, conceivable that the full promise of CRs will only be realized when the technology evolves into CRs as just intelligent and self-aware autonomous radios ready to take up any application, not just DSS. For this reason, in this book, our main focus is on emphasizing this more general vision of CRs. However, given the fact the CRs for DSS are more developed already and in many cases provide the foundation for the more general CRs, we will treat many aspects of them specifically as well.

1.5 WHAT THIS BOOK IS ABOUT

This book is about signal processing techniques for CRs. While implementation of traditionally analog circuit-based signal processing functions in open standard, modular software is the key idea behind SDR technology, it is the addition of powerful advanced signal processing algorithms that perform newly defined cognitive and intelligent tasks that forms the key idea of CR technology. As it will be clear from the next chapter, by a CR, we necessarily mean a wideband MBMMR with self-awareness and intelligence. CRs specifically aimed at DSS are treated as a special case.

Cognition implies that the radio has the ability to make its own decisions based on its interpretation of the RF environment. Hence, we may organize signal processing within a CR under two broad themes: signal processing for spectrum awareness and signal processing for communications. Signal processing for communications is already present in all wireless communications systems and devices. It refers to processing of information-bearing signals to achieve communications. Signal processing for spectrum awareness is, however, unique to CRs. This is what indeed makes CRs cognitive, or self-aware. The purpose of signal processing algorithms for spectrum awareness is to allow the radio to gain an understanding of what is present and happening in the radio's RF environment as well as user needs. In the context of DSS networks, the unique signal processing task that is aimed at gaining spectrum awareness is called the *spectrum sensing*. As we will see later in this book, the spectrum sensing problem defined in the context of DSS CRs does not go far enough to allow acquiring complete spectrum knowledge that may be required by a wideband MBMM CR. To make the distinction clear, we may term signal processing involved in gaining spectrum awareness in such a CR as *spectrum knowledge acquisition*.

Signal processing algorithms for communications and spectrum awareness are not completely isolated from one another within a CR. Indeed, spectrum knowledge gained through the spectrum awareness signal processing algorithms provides additional side information to the signal processing algorithms for communication, and vice versa. Still, this broad classification allows us to focus on signal processing that is unique to a CR in this book without delving too much into signal processing tasks that are common to traditional radios. This is necessary not only to emphasize and highlight the unique aspects of signal processing in CRs but also to keep the scope of the book at a manageable level. Note that signal processing for communications is covered in much more detail in many other books dedicated to signal processing for wireless communications including [17–19]. Almost all signal processing techniques that have been developed for achieving better wireless communications can conceivably be integrated also into CRs with little modifications.

The remainder of Part I of the book is dedicated to introduce the concept of CR. First, in Chapter 2, we will formally define the concept of CRs. As already mentioned, our emphasis will be on CRs as self-aware, intelligent, autonomous radio devices with broad applications. In Chapter 3, we will treat in detail the special type of CRs that are aimed at achieving DSS. Thus, in Chapter 3, we will also discuss the DSS problem and various different approaches for achieving DSS.

All signal processing encountered in CRs, as well as in general wireless communications, are applications of theories of detection, estimation, and inference. Thus, these underlying theories form the basis for understanding all signal processing algorithms we discuss throughout this book. Hence, Part II of the book is devoted to introduce theoretical foundations on detection, estimation, and inference to make the book self-contained. In Chapter 4, we will start by first giving a brief introduction to the theory of signal detection in discrete time. Note that throughout this book we will mostly work in discrete time since all signal processing in CRs will be performed in discrete time on sampled and digitized signals. Chapter 4 will present both parametric and nonparametric detection theories. This will be followed by an introduction to the theory of parameter estimation in Chapter 5. We will first discuss Bayesian parameter estimation with special attention to the minimum mean-squared error (MMSE) parameter estimation. Next, we will introduce several approaches for estimating parameters that are not modeled as being random. We will briefly discuss the theory of minimum variance unbiased estimation (MVUE) followed by the widely used maximum likelihood (ML) estimation.

A special estimation problem that is of interest in many communications and signal processing applications, and in particular in wideband CRs, is the power spectrum estimation. This refers to estimating the power spectral density (PSD) of a signal from a finite number of signal samples. Broadly, these techniques can be classified as nonparametric and parametric spectrum estimation approaches. Parametric spectrum estimation approaches employ knowledge of the specific structure of signals whose spectrum is to be estimated, whereas nonparametric approaches do not assume such knowledge. Taken together, the literature on PSD estimation is broad and it is perhaps not possible to cover in detail in a single chapter. Hence, in Chapter 6, we limit our discussion to nonparametric spectrum estimation approaches. This is justified because in wideband spectrum knowledge acquisition, a CR must mostly rely on nonparametric spectrum estimation approaches. Chapter 6 will also discuss spectrum estimation when the signals of interest are cyclostationary.

An important feature of CRs is their ability to make decisions based on its knowledge of its own state, the environmental conditions, and user needs. In Chapter 7, we introduce one approach to making optimal decisions. Specifically, Chapter 7 deals with decision-making in Markov environments. We will introduce the theory of Markov decision processes (MDPs) followed by the so-called partially observable MDPs (POMDPs). MDPs can be solved by using well-known dynamic programming principles. However, solving POMDPs can be considerably more difficult due to computational complexity. In Chapter 7, we will discuss several algorithms that have been proposed to compute optimal decision policies for POMDPs.

We will conclude our discussion of theoretical foundations by discussing Bayesian nonparametric classification in Chapter 8. Signal processing in CRs can involve various classification and clustering problems. One specifically important situation is in spectrum knowledge acquisition in which the radio may need to classify detected signals. As we will see later in the book, this problem is better modeled as a Bayesian nonparametric classification problem. Unlike many commonly used classification and clustering algorithms, Bayesian nonparametric approach to classification allows the number of clusters to be unknown and unbounded. Chapter 8 will introduce the theory behind Bayesian nonparametric classification and specifically discuss an approach based on the so-called Dirichlet process mixture model (DPMM).

Part III of the book is devoted to signal processing problems and techniques in CR. In Chapter 9, we start our discussion by introducing the wideband spectrum knowledge acquisition problem in a wideband MBMM CR. As we will see, the general wideband spectrum knowledge acquisition problem consists of several subproblems, some of which are not encountered in the spectrum sensing problem addressed in the context of DSS CRs. After developing a basic framework to achieve wideband spectrum sensing with realistic hardware and SDR platforms, the rest of Chapter 9 will delve into one of the most important decision-making problems that is unique to wideband CRs, termed the subband selection problem. Simply stated, the CR needs to determine which part of the spectrum it should sense at any given time in order to best achieve its performance objectives. This problem is formulated as a POMDP, and both optimal and reduced complexity suboptimal solution approaches are discussed. Closed-form solutions however presume perfect knowledge of the state of the wireless spectrum that in many cases cannot be justified. One of the most powerful tools available to a CR is the ability to learn. Hence, in Chapter 9, we will also introduce machine learning approaches and specifically discuss machine learning-aided subband selection algorithms.

Chapters 10 and 11 will be focused on signal processing techniques for extracting spectrum knowledge from sensed observations corresponding to a subband. This involves both detection of spectral (signal) activity in the sensed subband and classifying and identifying these detected signals. Chapter 10 will deal with spectral activity detection. First, a general framework will be introduced based on classical power spectrum estimation and detection theory. When usual Gaussian noise assumptions cannot be justified, the performance of this approach may degrade. Hence, Chapter 10 will further discuss improved spectral activity detection approaches based on non-Gaussian and robust detectors. A more powerful wavelet-based approach that can improve the detection performance of signals that are scattered over a wide spectrum range will also be discussed. Finally, Chapter 10 will conclude by briefly discussing compressive sampling-based wideband spectrum sensing that may ultimately reduce the need to have extremely high-rate A/D converters. In Chapter 11, we will discuss how these detected signals are to be identified. After discussing the general classification and identification architecture in a wideband CR, Chapter 11 will develop a Bayesian nonparametric classification framework to identify detected signals without knowing how many active systems are there and how many signals belong to each of these systems.

As we mentioned earlier as well, CR technology for DSS is perhaps the most matured due to the wide attention they received over the past decade. Thus, in Chapters 12–14, we will specifically discuss signal processing related to DSS CRs. First, in Chapter 12, we will define the so-called primary signal detection problem faced by a DSS CR. This is a somewhat simpler special case of the more general spectral activity detection problem discussed in Chapter 10. Similarly, in Chapter 13, we will discuss the spectrum decision-making problem faced by a DSS CR. While, as we saw in Chapter 9, a wideband CR needs to make decisions on which subband to be sensed, a DSS CR's decisions have to be concerned with picking a particular primary channel to be sensed. This problem will also be formulated as a POMDP problem, and various solution approaches will be discussed. In Chapters 12 and 13, we will also discuss how CRs may cooperate among each other to make these decisions more accurate and efficient in the presence of wireless channel dynamism and limited sensing abilities. Finally, in Chapter 14, we will introduce a different approach for DSS called the dynamic spectrum leasing in which spectrum owners are provided with incentives to make the spectrum sharing with CRs more efficient.

While almost all advanced signal processing approaches that have been developed for traditional wireless communications systems can possibly be also integrated with CRs, one powerful approach that stands out as being particularly more justified in CRs than in traditional radios is the cooperative communications. The reason is that many cooperative communications protocols inherently presume advanced processing and computing abilities from the cooperating wireless nodes. CRs possess these abilities by their design and thus are well placed to take advantage of these powerful cooperative communications techniques. Certainly, the cooperative communications itself is a topic that can fill a textbook, and indeed, there already exist several such books. In Chapter 15, we will briefly introduce the topic of cooperative communications and discuss how cooperative protocols can be integrated into CR networks.

Finally, in Chapter 16, we will discuss learning functionality in a CR. After looking at the need for learning in a CR, we will introduce various machine learning techniques that have either already been applied in CRs or have the potential to be useful in the future as they evolve.

1.6 SUMMARY

The concept of cognitive radios emerged as a natural evolution of wireless telecommunications. In particular, limitations of traditional hardware-defined radio systems were broadly acknowledged in the early 1990s, leading to the interest in the so-called software-defined radios (SDRs). The waveforms in SDRs were defined in software, rather than in hardware, allowing the radio to easily switch among different waveforms. For the same reason, these radios were easily reprogrammable to introduce new waveforms as well. An ideal SDR architecture may thus be to perform all signal processing operations in software on the software-defined digital waveform before D/A converting to feed into a transmitting antenna. However, in practice, there may still be a need for some operations to be in RF/IF hardware. Thus, normally, an SDR architecture will consist of an RF front-end between the antenna and the A/D (and D/A) converter. The digital signal is processed in the software-radio platform that is made of DSPs, ASICs, or FPGAs.

Cognitive radios are the natural next step in the evolution of SDRs. The first proposal of the cognitive radio concept viewed it as an SDR with built-in intelligence that will allow the radio to autonomously manage its operations in response to the perceived state of the RF environment and the user needs. Such a radio was quickly recognized as an ideal device to take advantage of unused spectrum opportunities in licensed spectrum bands based on a technique called dynamic spectrum sharing. This gave a strong motivation for the investments in the development of cognitive radio technology, and indeed, cognitive radios and dynamic spectrum sharing became almost synonymous. However, the concept of cognitive radios does not have to be limited to dynamic spectrum sharing. In general, a cognitive radio can be a self-aware, intelligent, wideband, MBMM wireless device that is capable of optimal autonomous operation.

2

THE COGNITIVE RADIO

2.1 INTRODUCTION

The Merriam-Webster Dictionary defines the word *cognition* as conscious mental activities.[1] The mental activities include thinking, learning, understanding, and remembering. Hence, cognition necessarily means performing these mental activities consciously or while being aware of doing them. A little more detailed is the definition given in the Oxford English Dictionary[2]: The word cognition means the mental action or process of acquiring knowledge and understanding through thought, experience, and the senses. Again, it is a mental process but is defined to be involved specifically in the acquisition of knowledge and comprehension through thought, experience, and the senses. Cognition, however, only goes so far as to acquire knowledge and comprehend. In addition, importantly, both are achieved by the entity itself through its own thought, senses, and experience. On the other hand, according to the Merriam-Webster Dictionary, the word *intelligence* means the ability to learn or understand or to deal with new or trying situations or the ability of skilled use of reason.[3] Similarly, the Oxford English Dictionary defines it as the ability to acquire and apply knowledge and skills.[4] Thus, intelligence means not only the ability to acquire knowledge

[1] http://www.merriam-webster.com/dictionary/cognition
[2] http://www.oxforddictionaries.com/definition/english/cognition
[3] http://www.merriam-webster.com/dictionary/cognition
[4] http://www.oxforddictionaries.com/definition/english/cognition

Signal Processing for Cognitive Radios, First Edition. Sudharman K. Jayaweera.

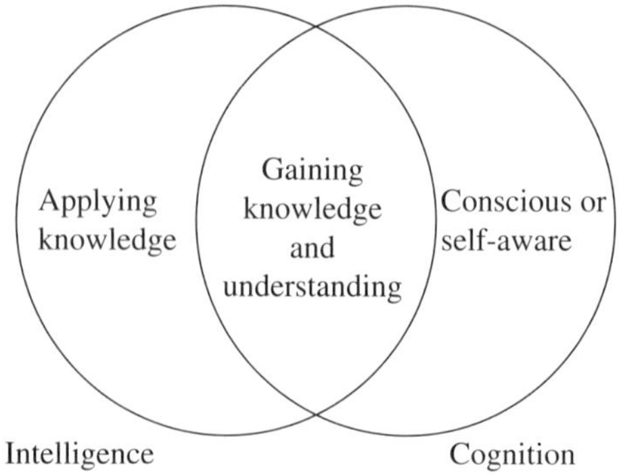

FIGURE 2.1 Cognition and intelligence have certain overlap. But they each also possess certain attributes that are unique to each one of them.

and understand but also the ability to apply the acquired knowledge and understanding through proper reasoning. Cognition, on the other hand, goes so far as gaining knowledge and understanding but consciously.

While some believe that intelligence is part of cognition and others believe cognition is part of intelligence, from the above definitions, it seems that neither is strictly true. Cognition and intelligence are two concepts with certain aspects that overlap with each other. But also, there are some aspects implied by each concept that seem to be unique to each one as Figure 2.1 attempts to show. Consciousness is uniquely required in cognition. The ability to apply and reason with knowledge is required for intelligence. Both deals with gaining knowledge and comprehension.

Surely, the cognitive radio implied in [1,20] is not just a radio with self-awareness and the ability to gain (spectrum) knowledge. It also is expected to have the ability to apply the acquired knowledge to better fulfill the radio user's needs (in fact, even without the user having to be aware of them). In other words, the cognitive radio first envisioned by [1,20] must also be intelligent. Hence, the cognitive radios are not *just* cognitive radios. They are intelligent cognitive radios. Indeed, this is what we will mean by a cognitive radio throughout this book.

Definition 2.1.1 (Cognitive radio (1)) A cognitive radio is a software-defined radio with built-in cognition and intelligence.

As this definition emphasizes, the term cognitive radio is perhaps incomplete. It leaves out an important aspect of these radios that is not implied by the word cognitive. Yet, the term cognitive radio has already taken root in the wireless communications literature so deeply, and it is almost always implicitly assumed to be an intelligent device. Hence, the above explicit definition that emphasizes the term cognitive radio as shorthand for intelligent cognitive radio seems sufficient.

The connection between cognition and intelligence can also be seen through their relationship to *learning*. Recall that cognition is the conscious mental process of acquiring knowledge and understanding through one's own thought, experience, and senses. However, such acquisition of knowledge and skills is self-learning (or cognitive/passive learning as it is called in the literature on educational psychology). Thus, cognition can be thought of as the process involved in, or that enables, self-learning. It does this through the so-called cognitive, or brain, abilities that include perception, attention, memory, motor skills, language/auditory, and spatial and visual processing and the so-called executive (logic and reasoning) functions that are the skills that enable goal-oriented behavior [21]. Example brain executive functions include planning, problem-solving, decision-making, and strategizing [21, 22]. It must be pointed out that by self-learning we mean a process of executing deliberate actions, observing the outcomes, and learning from them. Of course, the same attributes of cognition that enable self-learning can also facilitate learning in general. Indeed, it is known that cognitive abilities are critical to learning. Strong cognitive abilities imply easy and fast learning.

On the other hand, as shown in Figure 2.2, intelligence consists of the following attributes [23]:

1. *Perception*: The ability of sensing the surrounding environment and the internal states to acquire information
2. *Learning*: The ability of transforming the acquired information into knowledge and keeping them for future use
3. *Reasoning*: Using knowledge to achieve certain goals

Hence, learning is the key attribute of intelligence that converts information gathered through the sensory organs into knowledge. It is what links perception with reasoning. In the absence of cognition, the learning ability assumed in intelligence may not constitute self-learning. However, if the entity is both cognitive and intelligent, then not only it will have the ability for self-learning, but also other forms of learning abilities will get a boost due to the presence of cognitive abilities that enable self-learning.

Similarly, a cognitive radio as in the above definition is an intelligent SDR that not only constitutes attributes of intelligence but also the cognitive abilities that enable self-learning and self-awareness. In the case of a radio, being conscious, or self-aware, means knowing the internal state and capabilities of the radio, user needs, and the state of its RF environment. Being cognitive, it is endowed with abilities

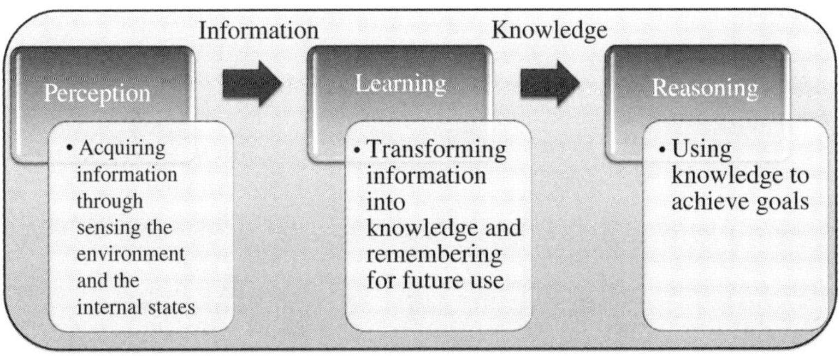

FIGURE 2.2 Attributes of intelligence.

such as perception, attention, memory, reasoning, and decision-making that enable self-learning. However, being intelligent, its learning does not necessarily have to be limited to self-learning. Other forms of learning can also be present.

The purpose of self-awareness required in a cognitive radio is to ensure that the radio has the ability to autonomously response to its RF environment and user needs given its internal state. Hence, a more explicit secondary definition of a cognitive radio is as follows.

Definition 2.1.2 (Cognitive radio (2)) A cognitive radio is a multiband, multimode, wideband SDR with autonomous decision-making and learning abilities that can optimally reconfigure its operating mode in response to the perceived state of its RF environment and user needs.

This definition emphasizes that a cognitive radio is capable of multimode operation over multiple spectrum bands, implying that it is also inherently wideband. While it is built on an SDR platform so that the baseband reconfigurability is already assumed, in general, it may also consist of real-time reconfigurable RF antenna and front-end components as well. Hence, reconfigurability does not necessarily have to be limited to baseband, but can extend to RF portion of the radio front-end. This is important for agile wideband operation of the radio. Apart from these, everything else in the definition is related to signal processing. It is the advanced signal processing algorithms that allow a cognitive radio to interpret its RF environment and user needs and self-determine the optimal mode of operation to meet user performance objectives under the perceived state of the RF environment and its own internal state.

2.2 A FUNCTIONAL MODEL OF A COGNITIVE RADIO

As is clear from Definition 2.1.2, a cognitive radio must have the ability for autonomous decision-making/reasoning and learning and the ability to modify radio's behavior based on such self-learning. Thus, a cognitive radio must possess the following attributes: self-management, self-learning, and self-reconfigurability. It should be able to learn from experience and optimally self-reconfigure to adapt to the observed RF environment in order to operate in the best mode to achieve user performance objectives. For this, a cognitive radio must necessarily support the following functionalities:

1. Spectrum knowledge acquisition (spectrum awareness)
2. Communications decision-making
3. Learning

This leads to the abstract functional model of a cognitive radio shown in Figure 2.3.

2.2.1 Spectrum Knowledge Acquisition (Spectrum Awareness)

Spectrum awareness is at the heart of a cognitive radio's identity as an intelligent and cognitive entity. It involves gaining knowledge and understanding of the radio's

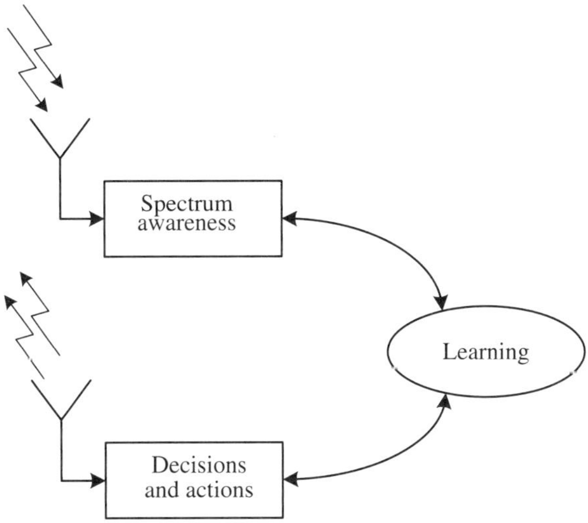

FIGURE 2.3 An abstract functional model of a cognitive radio.

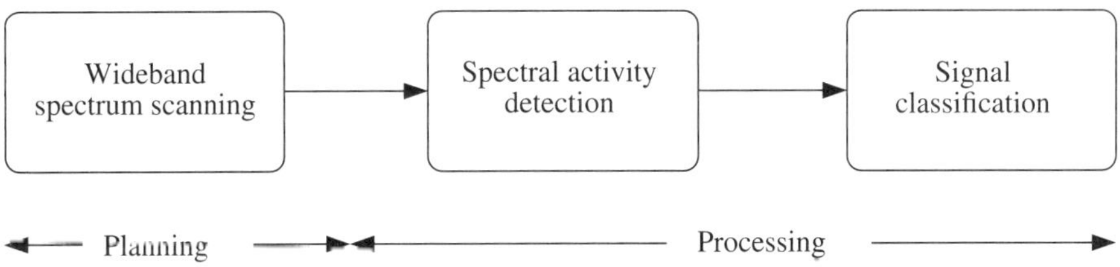

FIGURE 2.4 Spectrum knowledge acquisition consists of a planning stage and a processing stage.

RF environment. For this, the radio needs to sense its surrounding environment to identify RF activities. This is called spectrum knowledge acquisition, which involves both observing (sensing) the spectrum and comprehending what is observed. Sometimes, this is also called the (wideband) spectrum sensing although in many cases the comprehension step in spectrum sensing is limited to detecting signal activities. In general, a cognitive radio can be interested in determining the state of RF spectrum over a range of frequencies of interest to it possibly containing several active systems. Hence, spectrum comprehension also requires the ability to classify and identify the detected, possibly heterogeneous, signals. Thus, the spectrum knowledge acquisition consists of three steps (see Figure 2.4):

1. Wideband spectrum scanning
2. Spectral activity detection
3. Signal classification and identification

The first of these, the wideband spectrum scanning, addresses the issue of how a cognitive radio may handle real-time scanning of a wide spectrum range under realistic hardware constraints. No matter how sophisticated the technology, there will always be a limit to what is possible in terms of processing speed, storage memory, and dynamic range of hardware. Hence, without careful planning, it is not possible to scan and process a wide spectrum band of interest in real time. This planning is needed to determine which portion of the spectrum of interest to scan at a given time so that the knowledge required to achieve the radio's objectives is duly acquired. In Chapter 9, we will discuss in detail the wideband spectrum scanning problem in a cognitive radio.

Processing stage of Figure 2.4 is aimed at knowledge acquisition and comprehension. The first step in processing is the spectral activity detection. Given the observed spectrum data, the cognitive radio needs to detect any and all active signals in the scanned spectrum. Unlike in a conventional communications receiver, a cognitive radio may not be able to presume channelization information to limit its search to signal energy in a particular frequency band/channel. Hence, processing at this stage needs to be based on the signal over the whole scanned spectrum range, which can be computationally demanding. Detection of valid spectral activities, that is, signals, from spurious glitches due to noise is the main challenge. Adding to the problem will be the difficulty in modeling noise over a wide spectrum range. Chapter 10 will discuss the spectral activity detection problem in a wideband cognitive radio. In Chapter 12, on the other hand, we will discuss the problem in the special case of a dynamic spectrum sharing cognitive radio.

The second step in processing is to identify the detected spectral activities. There can be many signals belonging to several different active systems in different portions of the scanned spectrum. Hence, a cognitive radio may detect many active signals at various carrier frequencies in wideband spectrum sensing data. Such detection of signals itself does not provide the radio with a complete picture of the state of its RF environment. For better comprehension, it is important that the radio can classify detected signals into different groups and associate these groups with known systems expected to be active in the frequencies scanned. It is also possible that some of the detected signals may belong to systems that may be new to the radio. The cognitive radio must be able to create classification classes for such signals and flag them as being new. A priori, then, the cognitive radio may not know the number of classes to which the set of detected signals need to be classified.

Classification works by identifying common and distinct aspects among a set of entities. In many cases, rather than the direct entities themselves, if we extract these important aspects from the entity and provide them to a classification algorithm, it can make the job of the classification algorithm much easier and efficient. Such aspects are called *features*, and the classification based on features is called the feature-space classification. In the case of signal classification, the original entities are the detected signals. A cognitive radio must be able to extract relevant features from detected signals to help in such classification. Signal classification problem faced by a cognitive radio will be discussed in Chapter 11.

The importance of efficient spectrum knowledge acquisition to a cognitive radio cannot be overemphasized. The difference between an SDR and a cognitive radio is the cognitive radio's ability to optimally respond to its RF environment. For this, the cognition radio must first know the state of its RF environment.

2.2.2 Communications Decision-Making

Ultimately, the goal of a cognitive radio is to achieve efficient and effective communications. We go from SDR to cognitive radios so that the radio itself can make decisions on the optimal mode of operation under given circumstances. Hence, the purpose of being aware of its spectrum environment is so that the radio can select to operate in the most appropriate way under the given state of the environment and the performance objectives of the user.

Hence, a cognitive radio must be able to make its own decisions on the way actual communications to be achieved. Current wireless devices such as smart phones can point out to a user which networks are available in its vicinity. A user may either manually switch the device to an available network or may have given prior instructions on what to do. A cognitive radio can be different from this in several ways. First, the radio may not have to necessarily suggest to the user which networks are available and wait for the user to select the network. The radio may use historical data associated with user location, time, behavior, and the state of the RF environment to predict what the user's needs might be and select the best communications mode to operate in. It can automatically perform the switching in many cases. Moreover, the best mode of operation may not have to be necessarily selecting one particular network or a protocol. The radio may simultaneously operate in multimode, meaning that it may transmit and receive over a set of selected protocols at the same time in order to best meet its owner's performance objectives. This optimization and reconfiguration of the communications mode is what we call the communications decision-making.

Communications decision-making functionality involves solving optimization and inferencing problems. A key input to this module is the learned knowledge from the spectrum knowledge acquisition step, as shown in Figure 2.3. The processing algorithms take these inputs and compute the best mode of operation and associated system parameter settings to achieve the implementation of the selected mode. Finally, the cognitive radio must be able to implement this particular mode by reconfiguring itself. In general, this reconfiguration may involve not only the baseband reconfiguration in the SDR platform but also some RF front-end modules.

In Chapters 13–15, we will discuss communications decision-making in cognitive radios.

2.2.3 Learning in Cognitive Radios

Learning is the acquisition of knowledge, comprehension, and skills and remembering them for future use. As we saw earlier, it is a key component of intelligence. Learning in nonbiological entities is normally termed *machine learning*. Hence, by learning in a cognitive radio, what we mean is machine learning.

Learning can be of different forms. Perhaps the most commonly known are the *supervised learning* and *unsupervised learning*. Supervised learning is the learning with the aid of a supervisor or a teacher. This implies that there are correct examples of what to learn from certain given information that are available to the learner. On the other hand, in unsupervised learning, there is no teacher. This is similar to what we earlier called self-learning. In the case of a cognitive radio, learning functionality should specifically support self-learning (when needed), in addition to supervised learning. This means that a cognitive radio must have the ability to observe the resulting outcomes of its chosen actions and learn from them to make better operating decisions in the future.

In signal processing terminology, learning problem is essentially an (adaptive) estimation problem. Thus, learning functionalities are already present in most signal processing modules used in modern wireless communications devices and systems. Recall that supervised learning is the learning with the aid of a teacher who provides example learning problems with the associated correct answers. In signal processing terminology, this is called training. Many signal processing algorithms contain parameters that need to be adjusted to the specific scenario. Hence, quite commonly, a sequence of training bits or symbols is used to adapt these parameters of a signal processing algorithm. For example, in LPC of speech, the coefficients of the linear predictor are recomputed and updated regularly. Training symbols are preagreed between the transmitter and receiver. Hence, having them is equivalent to having a teacher that provides the correct answers. One may see supervised learning, as learning by instruction in which the radio is taught to learn something. On the other hand, in the case of unsupervised learning, there are no solved examples to get trained. In the context of signal processing, this means that there are no training bits or symbols. In signal processing terminology, such a scenario is called blind signal processing. The cognitive radio must learn something on its own. To emphasize this, we may thus call unsupervised learning in a cognitive radio as autonomous learning or self-learning.

Whether it is training based (supervised) or blind (unsupervised), the learning (estimation) can be based on either batch or adaptive processing. In adaptive signal processing, the new value of the learned quantity (estimate) will be obtained by updating, or modifying, the currently available value in light of new evidence or data. Hence, in real-time applications under dynamic conditions, adaptive signal processing makes particular sense. In wireless communications, adaptive signal processing is particularly useful due to inherent dynamism of the wireless medium and channels.

In light of adaptive processing, one may recognize a third paradigm in signal processing between training based and blind called the decision-directed. Initially, in decision-directed mode of signal processing, training-based learning is available. Thus, we have an estimate of a quantity based on training data. However, going forward, rather than keeping this quantity fixed at the learned value, the algorithm can adapt it as if the decisions made using the same algorithm were correct. In other words, the radio acts as its own teacher and assumes each decision it makes as being correct. So it uses them to learn necessary parameter values. Obviously, there is the danger of catastrophic error propagation in this type of systems. This can partly be avoided if we have access to new training data at regular intervals to recalibrate the estimated values.

2.3 THE COGNITIVE RADIO ARCHITECTURE

Architecturally, a cognitive radio is an SDR platform with a *cognitive engine* that controls it. If we draw a parallel between a cognitive radio and a biological entity, then the SDR is its body and the cognitive engine is its brain and the central nervous system. Of course, as we have already seen, an SDR itself constitutes a software-controllable RF front-end and a software-reconfigurable digital radio platform (see Figure 1.6). Hence, an ideal cognitive radio architecture may look like Figure 2.5.

The essential difference between an SDR and a cognitive radio is that in an SDR the decisions on when and how to change the operating mode are either pre-programmed or user supplied, while in a cognitive radio these decisions are made by the cognitive engine of the radio itself.

Figure 2.6 shows a detailed architecture of a realistic cognitive radio. Note that the basic cognitive functionalities identified in the previous section (spectrum knowledge acquisition, communications decision-making, and learning) are all embedded in the cognitive engine. It is responsible for being aware of the RF environment and the user

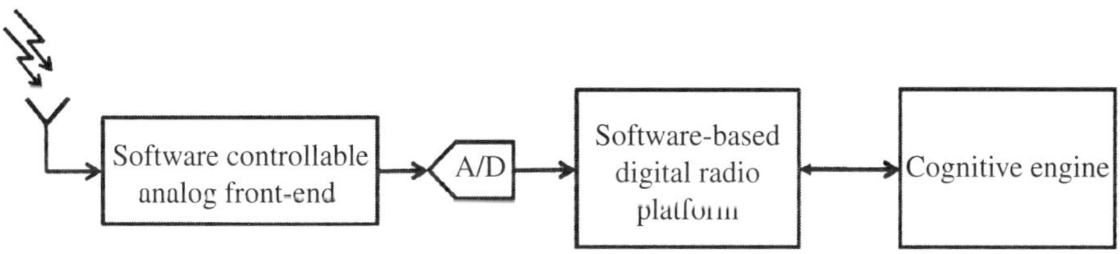

FIGURE 2.5 An ideal cognitive radio architecture.

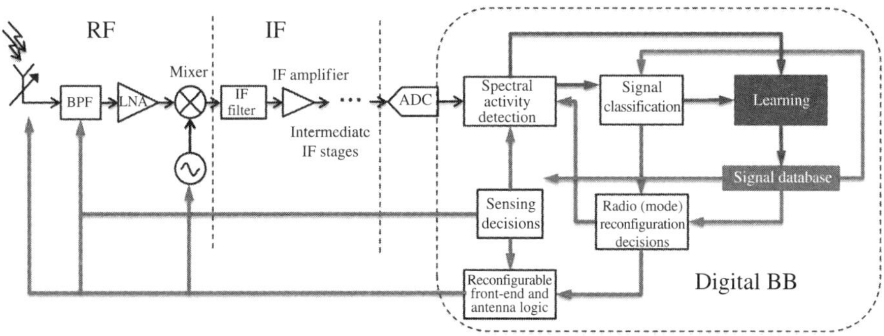

FIGURE 2.6 Basic architecture of a cognitive radio.

needs (spectrum awareness) and determining the optimal mode of communications in response to them (communications decision-making). The communications decisions in a cognitive radio may specify not only the baseband reconfiguration in software but also how to reconfigure the software-controllable front-end to achieve the desired communications mode. The cognitive engine also consists of a learning module that specifically supports learning from its own experience. Hence, we can subdivide the cognitive engine to identify several regions based on their responsibilities and functions:

1. Spectrum sensing region
2. Radio reconfiguration region
3. Learning region
4. Memory region

The cognitive engine is essentially a computer. It is a set of software programs with associated memory, interfaces, and I/O devices. Hence, for example, it can be implemented on a suitable digital processing platform such as DSP or FPGA. The software programs embedded in the cognitive radio are the signal processing algorithms.

2.3.1 Spectrum Sensing Region of a Cognitive Engine

The spectrum sensing region is responsible for acquiring spectrum awareness. It is responsible for making decisions on which spectrum portions to scan and process sensed measurements to detect signal activities and using them to interpret the RF environment. The results of this analysis are passed on to radio reconfiguration and learning units, as shown in Figure 2.6. The sensing region also makes sensing decisions. These decisions generate commands to reconfigure the radio's sensing front-end, in real time, for spectrum scanning. Arguably, spectrum sensing module is the unique feature of a cognitive radio compared to any other form of radio, including SDR. The signal processing algorithms for spectrum knowledge acquisition that reside within the spectrum sensing region of the cognitive engine are thus at the heart of all signal processing unique to cognitive radios. In Chapters 9–12, we will discuss signal processing for spectrum knowledge acquisition in detail. Chapter 12 will specifically be devoted to the spectrum sensing in a dynamic spectrum sharing cognitive radio network.

2.3.2 Radio Reconfiguration Region of a Cognitive Engine

The primary objective of any radio is to achieve successful communications. In the case of a cognitive radio, success means achieving user communications objectives by operating in the optimal configuration for the given RF conditions. Hence, all spectrum sensing and analysis efforts are aimed at facilitating this objective. The outputs of the spectrum knowledge acquisition algorithms are thus forwarded to the communications decision-making algorithms. These algorithms are responsible for determining the best (optimal) operating mode to achieve communications performance objectives (user needs) given the perceived state of the RF environment. This characterization is used to generate control signals that will instruct the front-end and the SDR platform on how to optimally configure the operating parameters. Hence, radio reconfiguration region of the cognitive engine consists of a set of reasoning and decision-making algorithms. The difference between these algorithms and the adaptive communications algorithms already found in many current wireless communications systems is that these algorithms explicitly take into account its own understanding of the state of the RF environment in making decisions. Chapters 13–15 will discuss various communications decision-making problem formulations encountered in cognitive radios.

2.3.3 Learning Region of a Cognitive Engine

The learning region of a cognitive radio may consist of both supervised and unsupervised learning algorithms. In signal processing terminology, learning algorithms in cognitive radios are estimation algorithms. The defining feature of cognitive intelligence is the ability to self-learn. A cognitive radio should be able to self-learn new signal classes and new operating modes and modify and update the existing ones based on its own experience. To support self-learning, the learning region will specifically support blind and decision-directed signal processing (unsupervised and semi-supervised learning) algorithms in addition to the training-based signal processing (supervised learning) procedures.

In many cases, the signal processing algorithms in a cognitive radio may need to be adaptive given the inherent dynamism of the wireless environment. These adaptive algorithms, among others, help with updating current belief of state of the environment and the user and the best communications strategies for the given situation based on this belief. The input to the learning algorithms is the radio's experience made of sensing measurements and decisions, current and past radio reconfiguration decisions, and their observed impact on the RF environment and achieving user performance goals. This is characterized by the so-called information history. The knowledge learned is fed back into the decision-making parts of the cognitive engine as well as knowledge database (long-term memory).

It is normally the case that learning is better formulated and implemented in the context of specific learning problems. This is particularly true in the case of machine learning, as in a cognitive radio platform. For this reason, in this book, we will incorporate learning, or estimation, procedures within the sensing or communications problems themselves whenever it is necessary. In Chapter 16, we will review several machine learning paradigms that can be ideally suited for cognitive radio applications.

2.3.4 Memory Region of a Cognitive Engine

Cognitive radio's memory is essentially a database. One may divide this into long-term memory and short-term memory. The long-term memory will hold spectrum regulation rules, information on known RF systems and their characteristics, and the capabilities of the radio itself. On the other hand, the short-term memory may consist of environment-dependent signal classes, parameters, and instantaneous performance objectives of the user. The quantities (knowledge) residing in short-term memory can be overwritten easily in real time. The modification or overwriting of quantities in the long-term memory may be prohibited and/or controlled by strict guidelines, which may also be a set of instructions residing in the long-term memory itself. A suitable analogy for this is the operating system of a computer.

2.4 THE IDEAL COGNITIVE RADIO

An ideal cognitive radio will presume an ideal SDR platform as its host. Recall that an ideal SDR would be able to digitize and process either directly the RF antenna output or at least after very few RF processing. Assuming such an ideal SDR is available, an ideal cognitive radio must be able to effectively scan a wide spectrum band in real time and detect all spectral activities present in this spectrum of interest. It must be able to classify these detected spectral activities into known signal classes and new signal classes accordingly. The signals classified into known classes should be associated with known systems by matching their appropriate features with a database in the radio's memory that holds known system characteristics. On the other hand, the detected signals that were classified as new signals must be associated with other such signal points the radio might have encountered in the past or might encounter in the future. After some time, the radio must learn whether to create a new signal class, or classes, or expand some of the already known system classes to absorb these new signals.

The spectrum knowledge acquired through detection, classification, and association of spectral activities must be used to generate an internal RF map of its environment. Based on this, and based on the performance objectives of the user, an ideal cognitive radio will determine the optimal communications mode and the associated parameter values in order to implement that optimal operating mode. The radio must then generate control signals to reconfigure its SDR platform as well as the front-end hardware in real time to be aligned with these specifications so that the communications can be achieved in the selected optimal operating mode.

An ideal cognitive radio will monitor how successfully its performance objectives were met by operating according to its selected optimal mode. This feedback will be used to learn how to better reason in future to determine the optimal operating mode given similar spectrum conditions and user needs. Such learning is one of the most important features of an intelligent cognitive radio.

It is likely that an ideal cognitive radio will be wideband, multimode, and multiband. It will have the ability to coexist and interoperate with other radio networks in the same spectrum range. With minimal user intervention, an ideal cognitive radio will be able to autonomously manage its own operations to meet the user performance needs. It must be pointed out, however, that autonomous does not mean that the radio will, or can, operate without regard to other radios. Clearly, for any radio to establish a communications link, at least, there must be one other radio. Hence, in general, any cognitive radio will operate within some communications network.

Moreover, autonomous cognitive radios may also pursue cooperative communications strategies with other radios when that is beneficial to achieving its own performance goals. However, it will have intelligence and cognition to make its own autonomous decisions regarding the best operating procedure and configuration subject to network capabilities, rules, and etiquettes including when to cooperate with other radios and when not to.

2.5 SIGNAL PROCESSING CHALLENGES IN COGNITIVE RADIOS

Advanced and sophisticated signal processing algorithms are quite commonplace in almost all modern wireless communications systems. Depending on the application context, they attempt to overcome, or mitigate, various challenges to achieving acceptable performance. For example, multiuser detection algorithms are used in code-division multiple-access (CDMA)-based networks to suppress multiple-access interference (MAI) caused by other wireless users within the same system. Array processing algorithms, or smart antenna algorithms, are used to process signals received from an antenna consisting of an array of elements when the SNR is insufficient with a single antenna element. Sophisticated channel equalization algorithms are used to remove, or mitigate, the intersymbol interference (ISI) caused by using transmission rates that cannot be supported by finite channel bandwidths. Many other examples are abound.

Signal processing for cognitive radios faces the same challenges met by any other wireless communications system. However, there are also certain signal processing issues that can either be new or exacerbated in the context of cognitive radios. These arise as a result of new signal processing objectives found in cognitive radios or due to the fact that certain signal processing tasks have become more complicated due to the cognitive and autonomous operating requirements of the radio. Some of these challenges include:

1. *Wideband operation*: Ideally, a cognitive radio must be able to operate over a wide spectrum range. The range of spectrum that is of interest to the radio can be noncontiguous and spread over even several GHz. Both spectrum knowledge acquisition and communications modules of the cognitive radio have to be able to handle operation over such a wide spectrum range. In the case of signal processing algorithms associated with spectrum knowledge acquisition, this leads to challenges of how to effectively scan and process the whole spectrum of interest in real-time, unknown, and non-Gaussian noise and interference characteristics, frequency-dependent noise, unknown traffic conditions, and unknown signal classes, to name a few. Similarly, the communications itself also face the same challenges of unknown, non-Gaussian, and frequency-dependent noise and interference. Moreover, wideband operation may require use of different propagation and channel fading models. Hence, signal processing algorithms need to be both adaptive and robust against these variations.

2. *Partially observable environment*: Cognitive radios are expected to make their own autonomous decisions that are optimal under the given RF environment conditions. This requires that the cognitive radio does have access to the true state of the RF environment. However, due to limited hardware capabilities, processing power, noisy observations, and sensing errors, a cognitive radio can only obtain partial observations of the state variables. Hence, both decision-making and learning in cognitive radios are problems defined in a partially observable environment. It is well known that obtaining solutions to partially observable decision processes is much more difficult compared to their completely observable counterparts.
3. *Distributed multiagent environment*: Wireless medium is an open medium. Hence, there will be more than one radio operating in a spectrum range of interest to a given cognitive radio. Some of these may be cognitive radios, while others may not. In addition, some of the radios may belong to the same system as the cognitive radio in question, and others may belong to different cognitive or legacy (noncognitive) radio networks. All these radios attempt to achieve their communications objectives within their individual capabilities. Hence, decision-making problems of all cognitive radios in this spectrum range can be considered as an interaction of multiple distributed (rational) agents, also called a *multiagent system* (MAS). Corresponding learning problems are also naturally *multiagent learning* (MAL) *processes*. Distributed multiagent decision-making and learning problems are inherently more complicated and can be very difficult to solve. Depending on the exact system details, the radios may operate either cooperatively or noncooperatively, and each cognitive radio might have either full or partial knowledge of the actions of the other radios. Hence, signal processing algorithms for learning and decision-making also need to take into account these attributes of the distributed multiagent environment.
4. *Autonomous operation*: Signal processing algorithms in cognitive radios must support autonomous decision-making and learning. This means in many cases the algorithms may not be able to rely completely on training. Sometimes, the algorithms will have to be in decision-directed mode, whereas in other cases blind adaptivity may be the only option. Guaranteeing sufficient performance under these cases is significantly more challenging than that with training-based signal processing algorithms. Moreover, autonomous decision-making can be strictly suboptimal when there are multiple radios operating in the same spectrum range.

2.6 SUMMARY

- A cognitive radio is a software-defined radio (SDR) with built-in cognition and intelligence. Hence, it is self-aware of its environment and user needs and its own internal state and is intelligent to reason with and apply the acquired knowledge to better achieve its performance objectives.

- Functionally, a cognitive radio must consist of the ability of spectrum knowledge acquisition, communications decision-making, and learning.
- The spectrum knowledge acquisition is a unique functionality to a cognitive radio. It deals with gaining knowledge and comprehending the radio's RF environment. This process consists of three steps: wideband spectrum sensing, spectral activity detection, and signal classification and identification.
- The acquired spectrum knowledge serves as an input to the communications decision-making algorithms whose responsibility is to determine the best mode of operation to achieve user needs given the state of the RF environment. This involves computing associated system parameter settings to achieve the implementation of the selected mode and instructing the SDR platform to reconfigure itself accordingly.
- Learning functionality, or machine learning, is needed in a cognitive radio to transform the gathered information into useful knowledge. The ability to observe the resulting outcomes of its chosen actions and learn from them to make better operating decisions in the future is an essential element of intelligence. Learning can be either supervised or unsupervised. In signal processing terminology, this learning is a problem of estimation, and supervised and unsupervised learning can be identified, respectively, with training-based and blind signal processing.
- An ideal cognitive radio architecture is made of an SDR platform and a cognitive engine that controls it. In reality, the SDR platform itself will be made of a software-based digital radio platform and an analog RF/IF front-end.
- The unique component that makes a radio a cognitive radio is the cognitive engine that is formed by a collection of processing algorithms that provide the previously identified functionalities. Thus, we may identify four distinct parts within the cognitive engine: spectrum sensing region, radio reconfiguration region, learning region, and memory region.
- Signal processing in cognitive radios may face certain unique challenges compared to traditional wireless systems. These challenges arise primarily as a result of new signal processing requirements imposed on cognitive radios so that they can be self-aware and intelligent. These include requirements of autonomous and wideband operation in partially observable, distributed multi-agent environments. In addition, certain signal processing tasks have also become more complicated in cognitive radios than in traditional wireless devices due to cognitive and autonomous operating requirements of the radio.

3

COGNITIVE RADIOS AND DYNAMIC SPECTRUM SHARING

3.1 INTRODUCTION

The original motivation for cognitive radios came as a natural evolution of software-defined radios (SDR): automating multimode and multiband operation. However, as mentioned in Chapter 1, the notion of cognitive radios emerged around the same time that many measurement campaigns pointed out the high underutilization of many desirable radio spectrum bands. Regulatory agencies such as the US Federal Communications Commissions (FCC) were looking to encourage new approaches to address the perceived scarcity of radio spectrum due to this wasteful usage. The question was, is it possible to improve the spectrum utilization in already assigned frequency bands without harming the existing legacy systems? Due to their inherent ability to comprehend and respond to the spectrum state, the cognitive radio concept seems to provide a solution. Thus, immediately after its initial introduction, cognitive radio concept was viewed as an ideal vehicle to improve spectrum utilization by capitalizing on the unused spectrum opportunities.

In its 2002 Spectrum Policy Task Force report referred to in Chapter 1, the FCC noted several important trends on spectrum usage below 1 GHz [14]. For instance, it observed that in many places the VHF and UHF frequencies that are historically designated for broadcast TV are not always used. Frequencies that are used by a TV station in a particular geographical area tend to have continuous use. But not all frequencies allocated for TV broadcasting are used in any given area. Even when a

Signal Processing for Cognitive Radios, First Edition. Sudharman K. Jayaweera.

particular TV channel frequency is used, it was not used efficiently. This was because analog TV transmissions were not the most spectrally efficient, perhaps because spectrum efficiency was historically not a key objective in their original design. This was highlighted by digital TV transmission systems that promised the ability of fitting more channels into a given spectrum band compared to that with analog systems. It was observed that the same was also true in frequencies allocated for cellular base stations and land mobile radios [14]. Thus, in its 2002 report, the FCC made the observation that allocated spectrum can be divided into two types:

1. *Access-limited spectrum bands*: Frequencies that are not fully used by the licensed users
2. *Throughput-limited spectrum bands*: Frequencies that are fully used by the licensed users

The only way to improve the efficiency of throughput-limited spectrum bands was to improve the transmission techniques themselves. On the other hand, [14] argued that the inefficiency in access-limited bands originated due to the traditional *command and control model* of spectrum allocation followed by regulatory agencies. The command and control model allocates and assigns frequencies to limited categories of spectrum users only for well-defined specific purposes. In this model, the licensee has sole rights to a specific spectrum band but only for a specific type of operation, and the rights are not transferable.

The FCC noted that the efficiency in overall spectrum utilization can be improved by improving the utilization in access-limited bands. It went on to identify several methods to achieve this. Some of these were obvious choices such as encouraging/discouraging efficient/inefficient spectrum usage. However, the task force report also suggested several nontraditional approaches, including permitting unlicensed users to access the spectrum unused by license holders on a time, frequency, or space basis and modifying the regulations as technology evolves. Along this rationale, the FCC went on to encourage the so-called *commons* and *exclusive-use* models of spectrum allocation, whenever possible, over the traditional command and control model of spectrum allocation. As shown in Figure 3.1, these two models seem to provide starkly different alternatives to the traditional command and control model: The commons model essentially takes away licensed rights completely by arguing for a framework in which an unlimited number of unlicensed users can share the same spectrum with usage rights that are only governed by technical standards or etiquettes, but with no right to protection from interference. The exclusive-use model, on the other hand, gives license holders flexible and transferable rights to its licensed spectrum within a defined geographic area.

The commons model was not new to wireless communications. Indeed, it was already the model used in the ISM bands and has been credited with stimulating many new technologies including WiFi and Bluetooth. The exclusive-use model, however, was a novel approach that prompted search for new methods of spectrum access. Under the exclusive-use model, a spectrum band is not allocated for a licensee for a *specific type of use*. Moreover, transferable rights implied that a licensee may actually

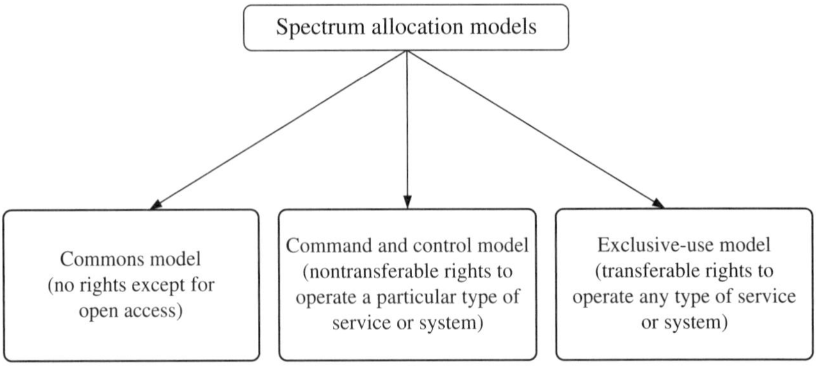

FIGURE 3.1 Different spectrum allocation models.

permit another system to operate within its own spectrum band. Hence, there is the possibility that several different types of systems can be operated within the same licensed spectrum band as long as the licensee is either happy to allow it or is forced to allow it by regulatory laws.

One possible way to achieve the transfer of spectrum rights is based on static (or off-line) agreements. However, the notion of cognitive radios provided a more flexible alternative in which such spectrum sharing agreements can be achieved dynamically in real time: A cognitive unlicensed radio can observe its RF environment to accurately determine when the license-holding user is not using its spectrum or when the license-holding user is willing to allow unlicensed users to access its spectrum. The cognitive radio can then access and use this spectrum without compromising the licensed user's quality of service (QoS) or without violating the terms of access. Since the interaction between the licensed and unlicensed cognitive users is in real time, the amount and terms of access can be easily modified as needs and capabilities change either in the licensed system or unlicensed system or both. This type of spectrum coexistence that is to be enabled by new spectrum allocation paradigms is called the dynamic spectrum sharing (DSS). Cognitive radio technology is an essential enabler of DSS because spectrum awareness is at the heart of its success.

DSS implies a hierarchical access method, in that there is a *primary system* that owns the spectrum rights and a *secondary system* that wants to access this spectrum whenever possible as needed. The secondary system users were expected to be cognitive radios, while primary system can be a legacy system.

Definition 3.1.1 (Primary user) A primary user (PU) in a DSS system is a radio that has license rights to use the particular frequency channel/band at a given time.

Definition 3.1.2 (Secondary user) A secondary user (SU) in a DSS system is a radio that may not have license rights to use the particular frequency channel/band but may access it whenever it is available for use without compromising or violating PU QoS, rights and terms.

PU belongs to a particular system, or service, that has a license to operate in the given frequency band. For example, in the US, a broadcast TV operator may hold a license to operate either in the 54–216 MHz VHF band or in the 420–890 MHz UHF band. Hence, broadcast TV signals are the PU in these spectrum bands in the US. Under the traditional command and control model of spectrum allocation rights, these were supposed to be the only users that can access this spectrum band at any time. Even if secondary radios were to access this spectrum under DSS, the PU will still retain the priority. The SUs are supposed to access a given frequency channel only when a licensed user (i.e., a PU) is either not transmitting on it or willing to tolerate a certain amount of interference. This requires that the SU be able to determine (or be informed of) when a given channel is actually free (idle) or available for its use, as well as be able to immediately vacate the channel as soon as a PU starts transmission on it or the interference to the PU exceeds a certain threshold. This is where cognitive abilities of cognitive radios come in handy.

There are two broad approaches to achieve DSS-based spectrum coexistence:

- *Spectrum underlay*: Spectrum underlay refers to a technique in which secondary cognitive radios may use the spectrum simultaneously with PU as long as they ensure interference to primary receivers is below a certain threshold, as shown in Figure 3.2(a). Usually, but not always, this implies that the primary communications are narrowband, while the secondary communications are based on some form of spread spectrum or, in general, wideband. In spectrum underlay systems, cognitive SU are expected to monitor the spectral activity in their RF environment to determine whether the interference they might cause to the PU is within the acceptable limit. If not, the SU transmit power level needs to be adjusted until the maximum tolerable interference constraint is satisfied again. Hence, the *maximum tolerable interference level* acts as the parameter that controls the amount of spectrum coexistence possible in spectrum underlay systems.
- *Spectrum interweave (or spectrum overlay)*: Spectrum interweave refers to a DSS paradigm in which SUs access the primary spectrum only when PUs are not using it. Hence, an SU accesses a primary channel only when a PU is not transmitting on it, as shown in Figure 3.2(b). SUs in spectrum interweave systems are also expected to monitor the primary transmission activity in their RF environment to first determine when a certain frequency channel is indeed idle before accessing it. This is called the *listen-before-talk* (LBT). Moreover, they need to immediately vacate the particular frequency band if and when a primary transmission starts on it. Collisions with primary transmissions may not always be avoidable since no perfect detection of primary activity may be possible by an SU. Thus, usually DSS systems based on spectrum interweave are designed subject to an allowed *maximum collision probability*, which acts as the parameter that controls the amount of spectrum coexistence possible.

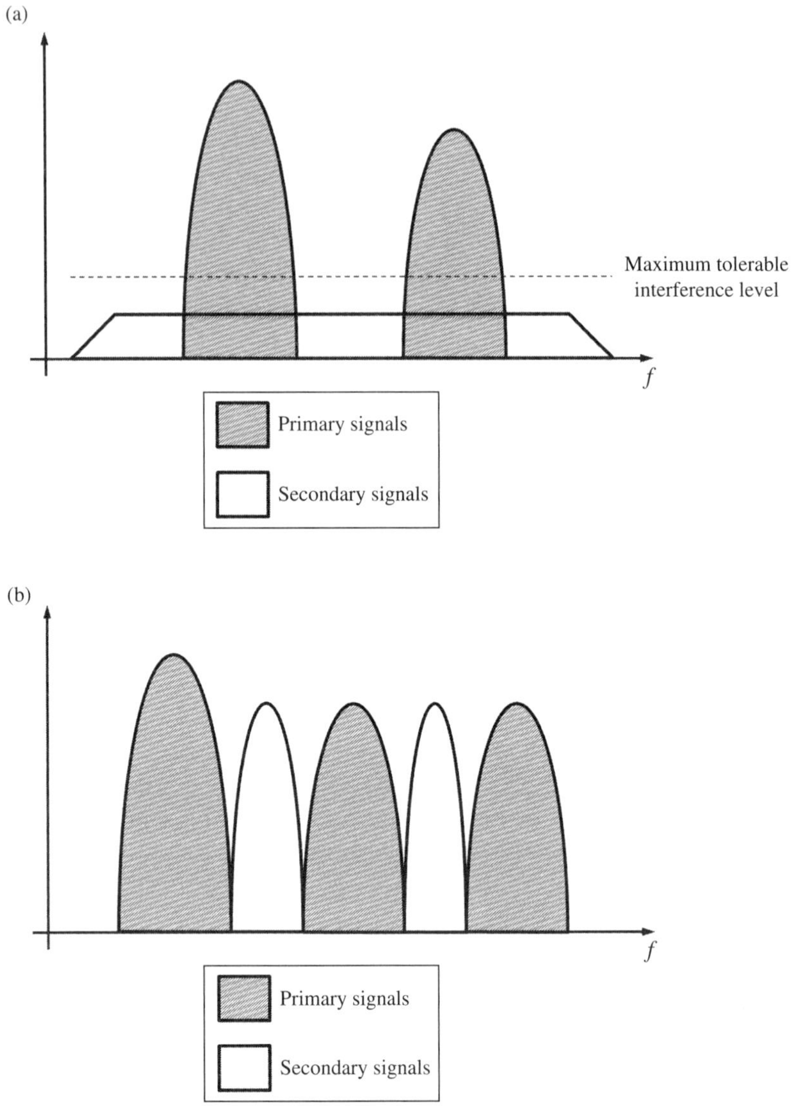

FIGURE 3.2 Spectrum coexistence under dynamic spectrum sharing. (a) Spectrum underlay. (b) Spectrum interweave (spectrum overlay).

3.2 INTERFERENCE AND SPECTRUM OPPORTUNITIES

Whether it is spectrum interweave or spectrum underlay, the key to DSS lies in proper management of signal transmission power level and interference. Signal power level is the major resource, besides spectrum, one has control over to ensure sufficient receiver performance. Interference level, on the other hand, is the limiting factor of

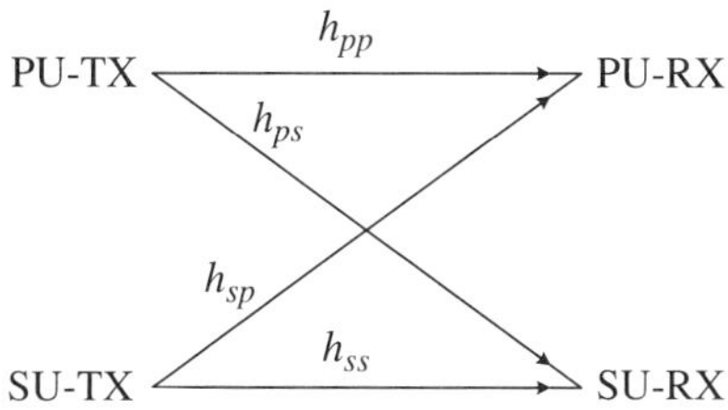

FIGURE 3.3 Dynamic spectrum sharing as an interference channel.

performance for a link operating above the noise-limited region. Ironically, power and interference are one and the same. Signal power that ensures acceptable performance for one receiver becomes the interference to another receiver. This is not a new observation. In CDMA wireless communications systems, for instance, many users transmit simultaneously over the same channel using different spreading waveforms (or codes). A receiver attempting to demodulate one particular transmission sees all other transmissions as interference.

In the simplest DSS setup of Figure 3.3, the primary user transmitter (PU-TX) transmits to its intended primary user receiver (PU-RX), while the secondary user transmitter (SU-TX) transmits to its intended secondary user receiver (SU-RX). Power unnormalized channel coefficient from primary to primary, primary to secondary, secondary to primary, and secondary to secondary are denoted, respectively, by h_{pp}, h_{ps}, h_{sp}, and h_{ss}. This is a version of the so-called *interference channel* [24–26]. The performance of any intended link may be improved in isolation by increasing the corresponding transmit power. However, this also increases the interference it causes to the other receiver degrading its performance. For example, if the primary transmit power is P_1, it causes an amount $h_{ps}P_1$ of interference power to the secondary receiver. Similarly, if the SU transmit power is P_2, it causes an amount $h_{sp}P_2$ of interference to the primary receiver. Clearly, increasing the transmit power in each individual link (primary to primary or secondary to secondary) leads to increased interference to the other link, thereby possibly degrading its performance.

The ability of a cognitive radio for spectrum awareness provides an opportunity to carefully manage the mutual interference between PU and SU. In DSS, an instant that a secondary transmitter-receiver (TX-RX) pair can successfully communicate is called a *spectrum opportunity*. A spectrum opportunity has to meet two requirements:

1. The PU QoS should not be compromised.
2. The SU performance should meet a minimum threshold.

We may translate these conditions, respectively, into as keeping the interference from secondary to primary below a threshold and ensuring the received power from secondary to secondary above a certain minimum received power requirement. Note that these requirements for a spectrum opportunity apply regardless of whether the DSS mechanism is based on spectrum interweave or spectrum underlay.

Figure 3.4 shows possible spectrum opportunities in the three-dimensional space formed by space, time, and frequency. In a spectrum interweave-based DSS system, possible spectrum opportunities are those time and frequency slots that are unused by a PU at a particular spatial location (see Figure 3.4(a)). Possible spectrum opportunities are interweaved along time and frequency dimensions at a fixed space coordinate: At a particular location in space, possible spectrum opportunities in a certain frequency channel are those time intervals during which the PU is not transmitting. Similarly, at a particular location in space, possible spectrum opportunities at a fixed time instant are found in frequency bands that are not used by any PU. The assumption is that both primary and secondary TX-RX pairs are within the same spatial area, which will cause one system's transmitter to interfere at the receiver of the other system. Hence, only one system is to transmit at a time with priority given to the primary system. In the three-dimensional space of (space–time–frequency), the possible spectrum opportunities are thus located on planes parallel to (time–frequency) plane as shown in Figure 3.4(a).

On the other hand, spectrum opportunities in a spectrum underlay system are located on planes parallel to the (space–frequency) plane. Primary and secondary transmissions can be simultaneous as long as they do not cause unacceptable interference at the opposing receivers. Of course, the primary transmissions always have the priority. Hence, as shown in Figure 3.4(b), possible spectrum opportunities in a spectrum underlay-based DSS system are those spatial locations and frequency slots that are unused by a PU at any fixed time instant. We may thus view possible spectrum opportunities in a spectrum underlay system as being interweaved along space and frequency dimensions for a fixed time coordinate: At any given time, possible spectrum opportunities in a certain frequency channel are those spatial locations at which a PU is not going to be effected by a secondary transmission on that channel. Similarly, at a given time, possible spectrum opportunities at a fixed location can be found in frequency bands that are not used by a PU. Hence, the possible spectrum opportunities are located on planes parallel to (space–frequency) plane as shown in Figure 3.4(b).

Figure 3.4 only shows *possible* spectrum opportunities. For them to be actual spectrum opportunities, the secondary TX–RX pair must also be able to communicate successfully without compromising the primary QoS. For example, let us consider a spectrum underlay-based DSS system and suppose average transmit powers of primary and secondary transmitters are P_1 and P_2, respectively. Then, we may view conditions for a spectrum opportunity in terms of their relative positions. Let us denote by R_Q and r_Q the maximum TX-RX separation in the primary and secondary systems, respectively, for acceptable receiver performance with these transmit power levels. In other words, the PU-RX must be inside the radius R_Q circle centered at PU-TX for acceptable primary performance. Similarly, the SU-RX must be within the radius r_Q circle centered at the SU-TX for acceptable secondary performance. Let us denote by r_I the *interference radius* of a secondary transmitter relative to the PU. That is, a secondary transmission will cause an unacceptable level of interference to the primary receiver as long as the secondary transmitter is within r_I distance from the primary receiver. Hence, we call the interior of the radius r_I circle centered

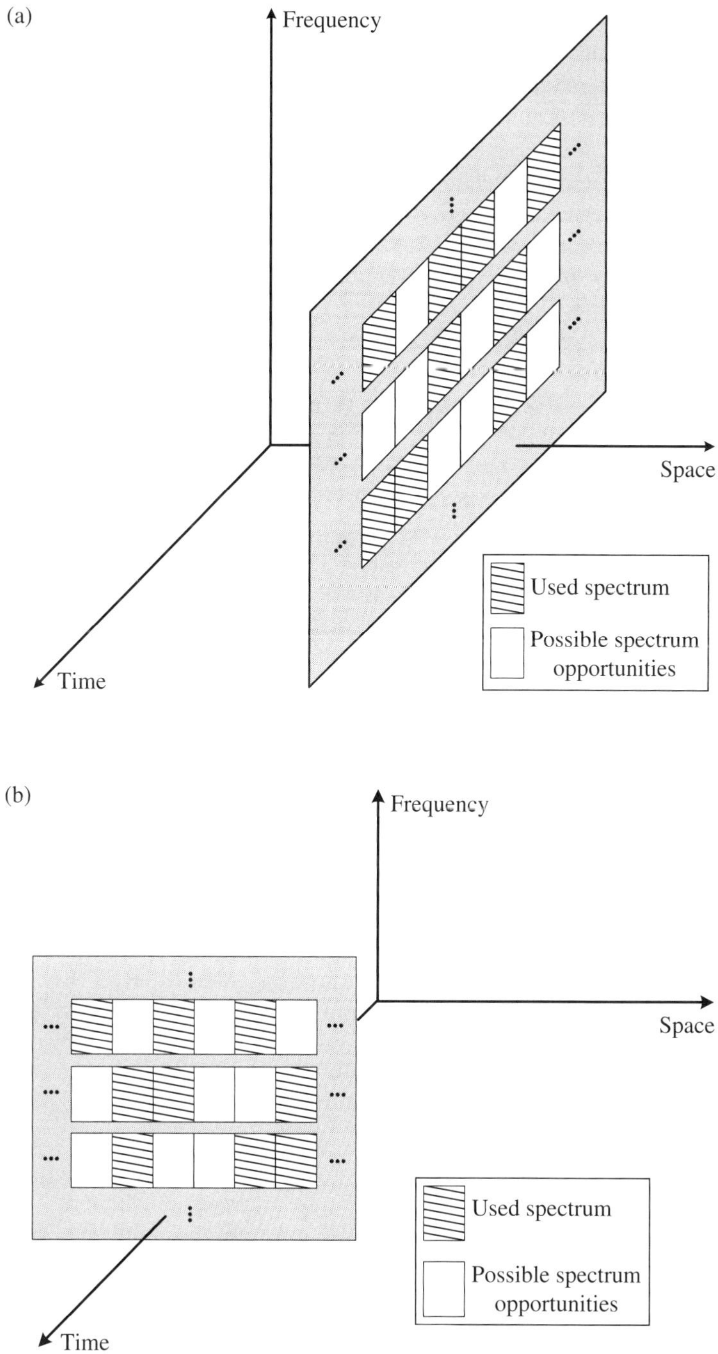

FIGURE 3.4 Possible spectrum opportunities. (a) In a spectrum interweave-based DSS system, spectrum opportunities are considered along the time and frequency dimensions. (b) In a spectrum underlay-based DSS system, spectrum opportunities are considered along the space and frequency dimensions.

at the SU-TX as its interference region with respect to the primary system. Thus, when there are multiple secondary transmitters, there should not be a secondary transmitter within a radius r_I circle centered at a primary receiver so that it is not affected by an unacceptable amount of secondary interference. Similarly, denote by R_I the interference radius of the primary transmitter relative to the secondary system. The performance of a secondary receiver within a distance of at most R_I from a primary transmitter will be so poor due to the primary interference that it may not worth attempting to communicate. Then, the radius R_I circle centered at the PU-TX is called the *primary interference region* with respect to SU. When there are multiple primary transmitters in the region, thus, there should not be any of them within a radius R_I circle centered at a secondary receiver for a secondary transmission to be successful.

Figure 3.5(a) shows a scenario in which there is no spectrum opportunity because the secondary transmission is to interfere at the primary receiver although secondary receiver is outside of the interference region of the primary transmitter. This is seen by noting that in Figure 3.5(a) the primary receiver is inside the secondary interference region. On the other hand, Figure 3.5(b) shows a situation in which there is a spectrum opportunity because both primary and secondary receivers are outside of the interference region of the opposing transmitter: The primary receiver is outside of the secondary interference region, and the secondary receiver is outside of the primary interference region.

Figure 3.5(b) only shows one possible relative geometry of PU-TX, PU-RX, SU-TX, and SU-RX so that there is a spectrum opportunity for the secondary system. There are various ways to rearrange them relative to each other to define all possible spectrum opportunities. However, it is not easy to view this in a simple way without keeping some of the nodes fixed. In Figure 3.6, we consider a DSS system with a PU-TX and a PU-RX that are at fixed locations. We have assumed that PU-RX is within R_Q of the PU-TX to ensure sufficient primary receiver performance. Given these primary locations, Figure 3.6 shows the possible locations for the secondary nodes to obtain a spectrum opportunity. Specifically, the SU-TX can be anywhere outside of the radius r_I circle centered at the PU-RX, and the SU-RX can be anywhere outside of the radius R_I circle centered at the PU-TX. We have a spectrum opportunity whenever the secondary transmitter and receiver are in allowed regions shown in Figure 3.6 with their separation being at most r_Q.

3.3 DYNAMIC SPECTRUM ACCESS

By definition, cognitive radios are radios that can observe and understand their RF environment and adjust their operating mode accordingly in response [1]. Hence, in early proposals for hierarchical DSS proposals, the burden of managing the interference and coexistence was placed squarely on the cognitive secondary system: The secondary system is expected to be vigilant of spectrum opportunities and be able to efficiently use them. In this type of DSS systems, the PU may not even need to be aware of the ongoing spectrum sharing by the SU, at least in theory. SUs find

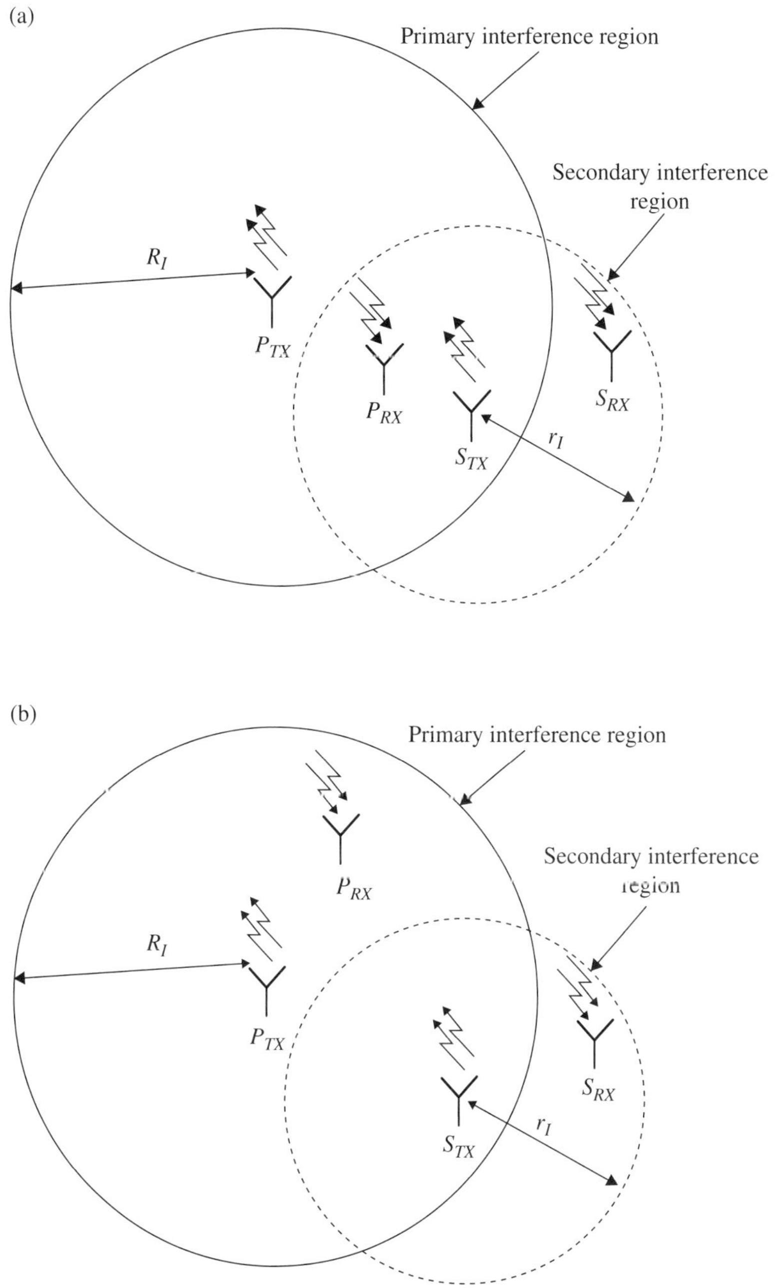

FIGURE 3.5 Interference regions and spectrum opportunities in a spectrum underlay system. (a) No spectrum opportunity because of significant interference to the primary receiver from the secondary transmitter. (b) A spectrum opportunity because both receivers are outside of the interference regions of the opposing transmitter.

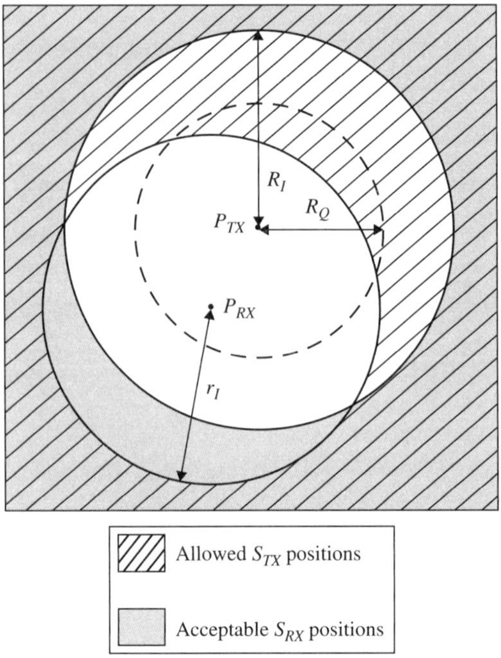

FIGURE 3.6 Allowed secondary transmitter and receiver locations for spectrum opportunities with respect to a fixed primary (transmitter, receiver) location.

and access the spectrum opportunities on their own. Hence, we term these spectrum sharing methods as the *dynamic spectrum access* (DSA) methods. In early literature, however, the term DSA was in general used to mean dynamically sharing the spectrum [27]. This made sense because the most extensively studied early approach was indeed what we defined above as the DSA: Secondary cognitive radios *access* the primary spectrum whenever it is available without explicit real-time permission from the PU (Figure 3.7). The availability here means either the spectrum band/channel is idle or the SU can transmit and not exceed the maximum tolerable interference threshold at the primary receiver.

DSA is a DSS method in which SU gain access to PU-owned spectrum without PU explicitly, or proactively, instructing them to do so [28,29]. The spectrum coexistence process is managed completely by the secondary cognitive radios, or the secondary *cognitive radio network* (CRN), to ensure that their access does not degrade the QoS of primary communications. The PUs can be oblivious to the fact that secondary cognitive radios are accessing their channels when they do not use them (in spectrum interweave DSA systems) or when they can operate without compromising their QoS requirements (in spectrum underlay DSA systems). Hence, in spectrum interweave-based DSA systems, the cognitive secondary radios search for channels that are not being used by a PU and access them. In spectrum underlay-based DSA systems, the cognitive secondary radios sense the spectrum to determine whether there is a safe

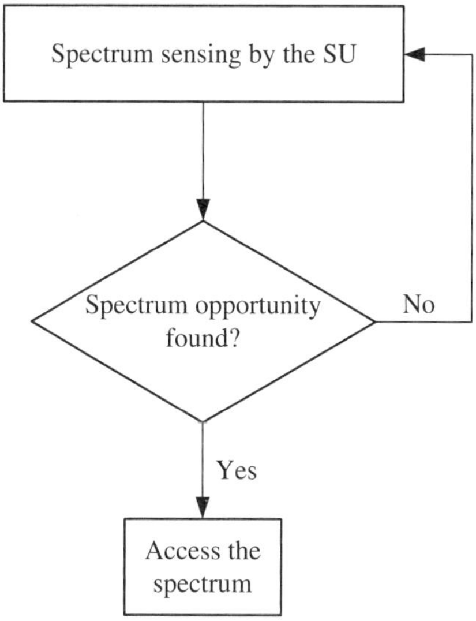

FIGURE 3.7 Dynamic spectrum access is almost entirely managed by the secondary cognitive radios.

transmit power level that will not degrade primary QoS and access the channel with that transmit power. In general, these are the spectrum opportunities in the case of DSA-based DSS systems.

The advantage of DSA approach to DSS is that, in theory, it does not require any modifications or changes to existing legacy systems. However, in reality, some small amount of modifications may be necessary. Hence, as long as cognitive secondary system can manage the spectrum coexistence process without violating the QoS requirements of the PU, DSA is feasible in already allocated spectrum bands. Following the FCC report [14], a spectrum band that was identified as a possible candidate for such cognitive secondary system operation was the spectrum historically allocated for VHF/UHF broadcast TV transmissions. The measurement campaigns had shown that spectrum allocated for broadcast TV can be largely underutilized in many places. Due to their relatively low frequencies, transmissions on these bands can cover relatively large areas compared to, for example, WiFi transmissions in 2.4 GHz ISM band. Hence, these frequencies were prime candidates for wireless regional area networks (WRAN) except that they had already been allocated for broadcast TV. By using DSA technology, cognitive radios provided a way to use this spectrum band for WRAN operation without compromising the existing TV transmissions.

The focus on achieving DSA with cognitive radios was so strong, that DSA became almost synonymous with cognitive radios. For example, [15] went so far as to directly redefine cognitive radios as purely a vehicle for improving spectrum utilization based on DSA. While such narrowing of the original cognitive radio definition to simply DSA perhaps led to temporarily halting the progress toward the full potential of an intelligent cognitive radio as originally envisioned in [1], there is no denying that many research established during this effort has provided a foundation to build toward the general cognitive radios.

3.4 DYNAMIC SPECTRUM LEASING

Recall that under the exclusive-use model of spectrum allocation, a spectrum licensee is granted with transferable rights to a particular spectrum band. Hence, if and when it is possible, a spectrum licensee may in fact proactively allow a secondary system to access its spectrum. Of course, such a situation can be anticipated only if there is a gain for the PU for allowing the secondary cognitive users to access its spectrum. When such access is granted for a limited duration, we may call it as *spectrum leasing*. The SU may have to pay a leasing fee to facilitate such spectrum sharing. Spectrum leasing can be of either static or dynamic. In *static spectrum leasing* (SSL), the PU and SU may negotiate an *off-line* arrangement for the SU to access the spectrum within certain parameters (leasing agreement). On the other hand, in *dynamic spectrum leasing* (DSL), the PU may negotiate with SU in *real time* and allow access to its spectrum (Figure 3.8) [28–33]. As far as signal processing is concerned, we are interested in such DSL approaches for DSS.

DSL is motivated by the observation that DSA approach provides no incentive for the PU to share the spectrum with SU. As pointed out earlier, in DSA it is not expected that PU be even aware of the existence of SU. Their coexistence is to be facilitated by regulatory laws and be an add-on to the existing networks. DSL, on the other hand, is an attempt to take advantage of the exclusive-use model for spectrum

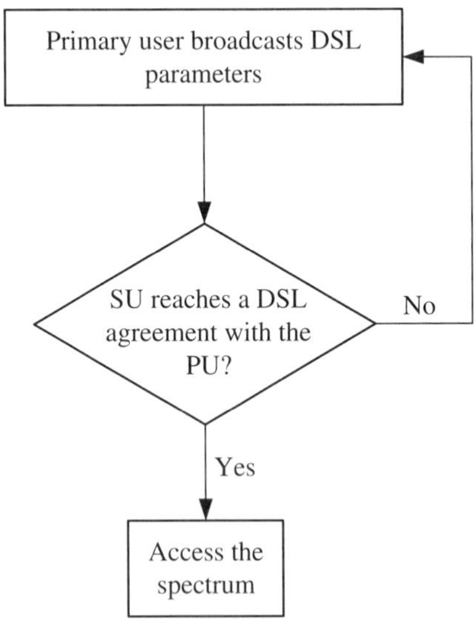

FIGURE 3.8 Dynamic spectrum leasing provides the primary users an opportunity to be proactive in managing the spectrum sharing process in order to maximize their rewards.

allocation under which the spectrum licensees have the rights to trade or lease their spectrum to third parties. In DSL-based DSS, PUs are provided with an explicit incentive to share the spectrum with cognitive SUs. This incentive can be either a monetary reward (leasing payments) or an *in-kind* payment. The advantage of such an incentive-based framework is that it gives PU a motive to dynamically adjust the extent to which they are willing to lease, or share, the spectrum with SU. Hence, in contrast to DSA, DSL makes PUs proactive in the spectrum sharing process. It encourages PUs themselves to facilitate secondary access, whenever possible, to the maximum possible extent [28, 30, 33]. Hence, arguably, DSL may lead to better overall spectrum utilization, albeit at the expense of required modifications to the existing legacy systems compared to what is needed in DSA-based DSS. Perhaps surprisingly, however, DSL can be achieved with very little effort from the part of the primary system. Indeed, it was argued in [28, 30] that it is still possible to relegate most of the interference management burden to the cognitive secondary system just as in DSA-based systems.

Similar to DSA, DSL can also be implemented in both spectrum underlay and interweave modes. In spectrum underlay DSL mode, the PU, or the primary system, may actively adapt the maximum secondary interference it is willing to tolerate, named the *interference cap*, in response to the state of the RF environment and its required QoS parameters. In spectrum interweave DSL mode, the PU may leave out a portion of its channels for the SU if it can meet its required QoS by using only the remaining channels. In both cases, the leasing payment to the PU, called the *primary reward*, can, in general, be a monotonic increasing function of the amount of access allowed to the SU. For example, in underlay DSL, the primary reward may be a monotonic increasing function of the interference cap, while in spectrum interweave DSL, it can be a monotonic increasing function of the amount of channels freed up for secondary use.

3.5 CHALLENGES IN DSS COGNITIVE RADIOS

The main challenge in DSS cognitive radios is to ensure that performance of the PU is not compromised due to the presence of SU activity. To ensure this, secondary cognitive radios should access the spectrum only when it is possible to do so without degrading primary performance. Thus, being aware of the RF environment is extremely important, and it is exactly the ability of cognitive radios for spectrum knowledge acquisition what makes cognitive radios the ideal technology to achieve DSS. This process of observing the RF environment to draw conclusions regarding its state is what is called the spectrum sensing. Spectrum sensing is essential to all cognitive radios. It allows the radio to get a feel for the state and conditions of the surrounding RF environment. Based on what is inferred from such spectrum sensing, the cognitive radio can then make intelligent decisions about how to communicate.

Having accurate spectrum knowledge is perhaps more critical to DSA-based DSS systems in which the secondary cognitive radios are responsible for managing the whole process of spectrum coexistence. Only if spectrum sensing is accurate and efficient the secondary cognitive radios will ever get successful access to the spectrum. In a DSL-based system, on the other hand, this may be somewhat less challenging due to the proactive engagement of PU that informs SU when to access the spectrum. However, even in this case, primary system may still rely on faithful reporting of spectrum sensing measurements by the SUs themselves. Thus, regardless of the coexistence mechanism, it can be argued that spectrum sensing is one of the greatest signal processing challenges in any DSS cognitive radio system. This is, in fact, true in general for any cognitive radio even if its primary objective is not DSS.

DSS requires a cognitive radio to simultaneously monitor the spectrum as well as communicate. For example, consider a DSA-based cognitive SU in spectrum interweave mode. Suppose that the secondary radio is transmitting on a particular primary channel. It needs to monitor the primary signal activity on that channel at the same time so that it can stop its transmission on that channel as soon as a PU starts transmission. We call such operation of a cognitive radio as *listen while talking* (LWT). It should be noted that sensing while transmission is different from the normal full-duplex communication. Full-duplex communication uses a designated transmission and a reception channel. Normally, this is achieved by either allocating two separate frequencies (frequency division duplexing or FDD) or two separate time slots within a time frame (time division duplexing or TDD) for transmission and reception. On the other hand, simultaneous spectrum sensing and communications under LWT may require the radio to receive and transmit at the same frequency.

In an FDD secondary system, a link in secondary system will consist of two simultaneous frequency channels, say, centered at f_1 and f_2, respectively: one for transmission (f_1) and the other for reception (f_2). In spectrum interweave mode, the cognitive SU needs to sense the channel f_1 for possible primary signals. In spectrum underlay mode, the SU needs to sense the channel f_1 to ensure that the interference to the primary transmissions are below the acceptable threshold. In either case, the cognitive radio needs to simultaneously receive and transmit signals at frequency f_1: receive primary signals (spectrum sensing) and transmit its own signals (communications). It may not necessarily need to monitor the frequency f_2 on which it is receiving, unless to help out its corresponding partner node who is actually transmitting on frequency f_2. Even if it does sense frequency f_2 for possible primary signals, it will still only be receiving signals at frequency f_2 and not transmit. The received signals will possibly be a mix of primary signals and the secondary signals intended for it.

On the other hand, in a TDD-based secondary system, a secondary communications link consists of two time slots within each time frame. Let us denote the indices of the two time slots as n_1 and n_2 where slot n_1 is assigned for transmission and slot n_2 is assigned for reception. It is during time slot n_1 that the secondary radio needs to simultaneously transmit and sense. Hence, similar to an FDD system, during the time slot n_1, the radio will need to both transmit (its own signals) and receive (possible primary) signals on the same frequency channel. During the time slot n_2 allocated for reception, the radio only receives signals. Even if it monitors primary activity during slot n_2, still, it only needs to receive a mix of communications signals intended for it and possible primary signals.

Hence, regardless of whether the duplexing is FDD or TDD, in general, there is a need for simultaneously transmitting and receiving signals at the same frequency for LWT systems. This is different from traditional FDD operation in which two separate frequencies that are sufficiently far apart are used for simultaneous transmission and reception. For LWT, it is the same frequency that the cognitive radio needs to monitor while also transmitting. This poses a significant challenge in terms of managing the self-interference caused by radio's own transmission because the self-interference is at the same frequency at the same time. For example, consider a radio communicating on a certain primary channel. If it senses this channel for possible primary activity while transmitting, the receptions will include its own transmit signal as well. Unless there is sufficient protection against self-interference, this much stronger signal could swamp the other received signals. A possible solution for achieving simultaneous sensing and communications as required for LWT is the use of a self-interference cancelation circuit.

A commonly used approach for achieving self-interference cancelation is to use a self-interference cancelation circuit. Such circuits can be either analog or digital. An analog self-interference cancelation circuit can be used closer to the antenna itself. These circuits are usually based either on a balun (balanced–unbalanced) transformer or an RF choke. Figure 3.9 shows a possible self-interference canclation circuit-based LWT radio implementation. The idea is essentially to feed a properly

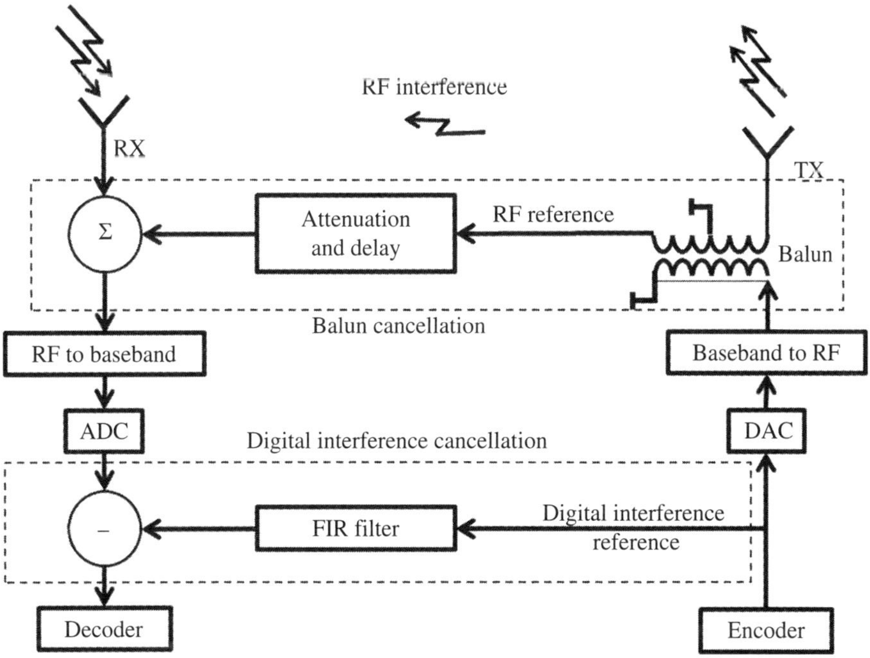

FIGURE 3.9 Simultaneous sensing and communications with self-interference cancelation.

compensated version of the transmit signal into the received signal so that the self-interference present in the received signal can be canceled out. Figure 3.9 also shows that a digital self-interference cancelation circuit can also be used, if needed, based on the same principle. The advantage of digital self-interference cancelation circuit is that it can be implemented within the digital SDR platform itself.

Many DSS proposals, however, assume that primary transmissions will be time slotted so that SU need to sense a channel only at the beginning of a time slot to determine whether a channel is free for access or not. With such structured transmissions, the spectrum sensing requirement reduces to the so-called LBT in which the cognitive radio senses the primary channels before starting any access.

LBT approach assumes that time axis is divided into time frames and transmission intervals as in Figure 3.10. It is assumed that during each time frame, or time slot, there is a designated portion of time that is used for spectrum sensing as shown, respectively, in Figures 3.10(a) and (b). It is common to take this sensing period to be at the beginning of the transmission interval. If a PU intends to transmit on a particular channel during a transmission interval, it will indicate that by starting its transmission during the corresponding sensing duration as shown in Figure 3.10. Secondary cognitive radios are supposed to monitor channels during these sensing periods and, specifically, are not supposed to transmit during these periods. This ensures that there is no self-interference problem during channel sensing for primary signals.

The use of designated sensing periods commonly assumed in many DSS proposals, is perhaps conceptually easier than the self-interference cancelation approach. However, the approach requires a certain amount of synchronicity between primary and secondary systems. First, the primary transmissions need to follow the slotted structure. Second, SUs must be able to maintain timing synchronization with primary transmissions in order to sense during the exact sensing intervals. Both these may not always be possible due to the nature of the primary network or difficulty in synchronization. In addition, secondary users themselves need to conform with the etiquette of not transmitting during the designated sensing periods. Especially when there are multiple secondary systems, maintaining such conformity can also be difficult.

There are, however, other challenges as well. One of them is that the secondary cognitive radios should be able to maintain a certain minimum level of QoS in their own communications. Again, especially in DSA, since secondary transmissions have to be terminated when a PU starts to transmit or secondary power must be lowered as soon as interference to a primary receiver exceeds a certain threshold, the secondary communications can get abruptly interrupted. For successful communications, the SU have to be able to maintain a minimum QoS in the presence of such interruptions. Hence, the secondary cognitive radio system may have to be equipped with sophisticated cross-layer algorithms. For example, the secondary radio may always maintain a list of possible backup channels that is up to date in case it needs to terminate transmission on the current channel. This, however, requires that the radio must sense several primary channels at the same time.

To maintain uninterrupted communications, the secondary system must also be equipped with protocols to ensure that the secondary receiver and transmitter can

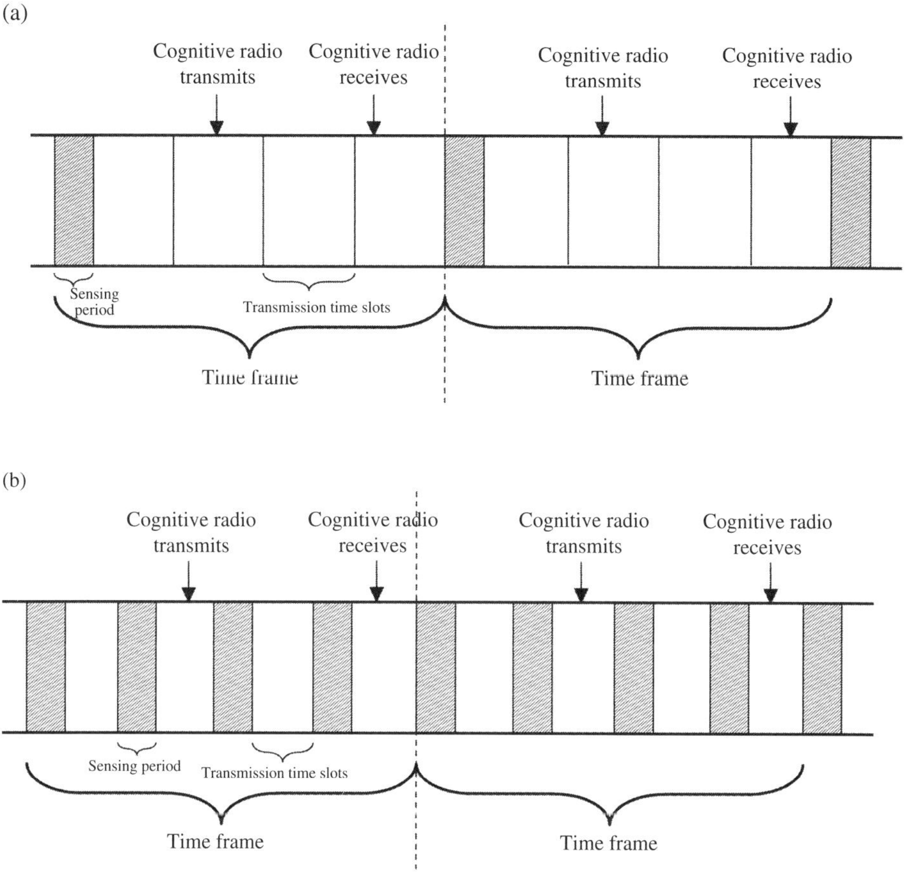

FIGURE 3.10 DSS with designated sensing time slots for simultaneous sensing and transmission. (a) One sensing slot per time frame. (b) One sensing slot per TDMA time slot.

simultaneously switch to the same new channel. Ensuring that the secondary transmitter and receiver can both switch to the same channel to establish a link may not be easy without having a central control. In systems with a central control, this is relatively easy to achieve. But if the secondary radios are autonomous, then ensuring both transmitter and receiver switch to the same frequency channel is called the *frequency rendezvous* problem [34, 35]. Note that not only to maintain uninterrupted communications but also to establish a link at the first place, one needs a frequency rendezvous method. Hence, adequate solutions to this problem is essential in DSS secondary systems without centralized channel assignment.

It can be expected that future spectrum allocation policies may evolve in such a way to allow DSS in many spectrum bands. This may lead to many secondary systems with varying capabilities trying to dynamically share a given primary spectrum. Such a scenario will require attention to various potential problems. For simplicity, suppose there is only a single centralized primary system in the spectrum band of interest. One problem is that secondary systems may compete among themselves to access the spectrum opportunities. Without proper designs, such competition may lead to either inadvertently harming the primary QoS or inefficient use of the spectrum or both. Moreover, some of the secondary systems may be centrally controlled, while others may be ad hoc networks. Again, without proper protocols and spectrum etiquettes, spectrum sharing can lead to chaos. Hence, for both DSA and DSL, there is a need for having efficient protocols for spectrum sharing. In the case of DSL, this must also include the necessary coordination and information exchange between the primary and secondary systems.

3.6 COGNITIVE RADIOS AND FUTURE OF WIRELESS COMMUNICATIONS

An ideal cognitive radio characterized by wideband, autonomous, and multimode operation has the potential to transform current wireless telecommunications technology in a radical way. In particular, such a cognitive radio may replace the plethora of wireless devices performing a multitude of different tasks with a single intelligent radio device that can perform all wireless communications needs of a user anytime, anywhere, and anyhow, be it broadband internet, mobile phone service, satellite radio, TV/multimedia, or GPS/location services: a single autonomous device that may make its own decisions regarding how to reconfigure (transform) itself to meet the user's performance objectives without the user even having to instruct it explicitly. Instead, the radio can be intelligent enough to infer user's routine objectives through *learn by experience* based on user's location, time of day, other RF devices in its environment, and the user's behavior and actions. Once it verifies its predictions on the user's communications needs and the performance objectives, it will have the ability to self-determine the optimal mode of operation in order to meet them given its internal state and the state of its RF spectrum.

Realizing the full potential of such a universal wireless device, however, will require removing the artificial barriers to spectrum access caused by current spectrum allocation policies. This seems to be already understood by the policy makers as the following excerpt from the 2012 report by the President's Council of Advisors on Science and Technology (PCAST) on Realizing the Full Potential of Government-Held Spectrum to Spur Economic Growth [36] shows: " ... spectrum should be managed not by fragmenting it into ever more finely divided exclusive frequency assignments, but by specifying large frequency bands that can accommodate a wide variety of compatible uses and new technologies that are more efficient with larger blocks of spectrum [36]."

Taking full advantage of such large frequency bands requires radio devices and systems that are wideband and can operate over many modes simultaneously. The wideband cognitive radio concept provides exactly that solution. It is already evident that the path toward such universal wideband radios will first go through DSS cognitive radios. As spectrum allocation based on exclusive-use and commons models see wider adaptation, it will encourage more DSS-based wireless communications systems enabled by cognitive radio technologies. While such DSS cognitive radios may be a stepping stone toward the truly intelligent wideband autonomous cognitive radios, they may however demonstrate the redundancy in allocating spectrum for dedicated services since these cognitive radios may access spectrum opportunities based on their location and needs.

Removing the barriers to spectrum access can stimulate development of new wireless technologies. If the experience of ISM bands is any guide, having more spectrum available for newcomers to test their technologies, systems, and services without having to spend enormous amount of capital for spectrum licenses as is the case today can be a catalyst for innovation. This is also likely to spur more competition, improved efficiencies and better technologies.

Cognitive radios will impact not just the consumer wireless communications but all forms of communications. Already there have been efforts to leverage cognitive radio capabilities to meet specific objectives of many applications including e-health, telemedicine, and Wireless Medical Telemetry Services (WMTS) [37–40], emergency and first-responder communications [41–43], and jamming and antijamming systems for military communications [44–50]. In these applications, cognitive communications techniques are used to better address some of the fundamental requirements of these systems including meeting regulatory constraints and ensuring reliability, security, and robustness. In WMTS communications, for instance, cognitive radios have been proposed as a way to seamlessly operate over multiple noncontiguous frequency bands using possibly multiple radio interfaces [38, 39]. In military satellite communications, on the other hand, cognitive radios are an ideal means to achieve jamming and antijamming capabilities [44–50].

Over the last few decades, wireless telecommunications has revolutionized the way people communicate. An ideal cognitive radio technology will complete this forward journey by not only making radios to be universal in their abilities but also giving them the power to be self-aware and autonomous.

3.7 SUMMARY

- Cognitive radio concept emerged around the same time when there was a broad consensus that the perceived scarcity of radio spectrum was, for a large part, caused by the traditional command and control model of spectrum allocation.
- Under the command and control model, spectrum is allocated for a particular type of service or system and these rights were nontransferable. Hence, in many places, the allocated spectrum was not either completely or efficiently used by the license holder but could not be used by another service or a system as it would violate the regulatory laws. This prompted encouraging alternative

forms of spectrum allocation paradigms, namely, the commons model and the exclusive-use model.

- It was also noted that the utilization of already allocated spectrum bands can be improved by modifying existing spectrum allocation rules to allow unlicensed users to access these spectrum bands whenever possible without violating or compromising the QoS of the license holders. Cognitive radios were identified as an ideal technology to enable such dynamic spectrum sharing (DSS) due to their inherent ability to comprehend and respond to the state of the spectrum.
- Most of initial cognitive radio research was geared toward DSS. In this context, the cognitive radios who were the unlicensed users were termed the secondary users, while the license holders to a particular spectrum band of interest were termed the primary users.
- Spectrum underlay and spectrum interviewing provide two broad technological paradigms to achieve DSS between primary and secondary users.
- The key to successful DSS lies in the proper interference management.
- A spectrum opportunity for a secondary cognitive radio can be defined as a situation in which the mutual interference between these two systems satisfies certain specific requirements: Secondary transmitters do not cause excessive interference at the primary receiver, and primary transmitters do not cause excessive interference at the secondary receiver.
- Depending on how interference between primary and secondary users is managed, there are two paradigms of DSS called dynamic spectrum access (DSA) and dynamic spectrum leasing (DSL). In DSA, primary users can be completely oblivious to their spectrum coexistence with secondary users, and thus, the interference management is the sole responsibility of the secondary radios. On the other hand, in DSL, there is an incentive for the primary users to share the spectrum with cognitive secondary radios, encouraging them to be proactive in facilitating the process.
- The key challenge in DSS cognitive radio networks is to ensure that the primary QoS is not compromised by the secondary access to their spectrum. In general, this may require cognitive radio to operate in LWT mode to ensure immediate backing off either when there is a primary transmission or excessive interference to the primary receiver.
- Combatting self-interference in LWT mode may need special circuitry.
- If primary transmissions are time slotted and a certain amount of synchronicity between primary and secondary systems can be justified, however, a cognitive radio may be able to operate in the LBT mode. This may avoid having to use self-interference cancelation circuits, provided all secondary users conform to the etiquette of not transmitting during the designated sensing periods.
- Other challenges faced by DSS cognitive radios include guaranteeing a minimum QoS when operating as a secondary system, frequency rendezvous of secondary transmitter and receiver, and managing competition among multiple secondary systems vying for the access to the same primary spectrum.

- Cognitive radios have the potential to radically change the future wireless telecommunications. In particular, they may render current spectrum allocation policies redundant, giving rise to universal wireless access devices that may meet all communications needs of a user with a single device. Removing access barriers to spectrum through modification to current policies may also further stimulate new radio technologies, spurring competition and economic growth.

PART II

THEORETICAL FOUNDATIONS

4

INTRODUCTION TO DETECTION THEORY

4.1 INTRODUCTION

A common problem encountered in many signal processing applications is that of having to detect a certain signal activity. For example, a scientist may use a seismometer to detect a seismic signal caused by either an earthquake or an underground nuclear explosion. In sonar signal processing, a submarine detects a reflected, usually acoustic, signal from an underwater object or another submarine using a hydrophone (see Figure 4.1). In yet another example, a ballistic missile defense (BMD) system may use a radar sensor to detect an incoming missile or a decoy. In communications system applications, signals of interest are modulated analog or digital communications signals. A communications receiver must first detect the signal of interest before proceeding to further processing. In the case of a cognitive radio, spectrum knowledge acquisition requires that the radio must first be able to detect signals in its RF environment before being able to determine what type of signals they are. While basic theory on signal detection discussed in this chapter is common to all these and other similar applications, we will present the material mostly using the communications terminology for obvious reasons.

Detection of a signal is made complicated by the inherent noise in communications receivers, anomalies in propagation media, and various interference phenomena. It is a challenge to detect a particular signal when it is embedded in noise. Moreover, if the signal is present only at certain times but not at other times, the receiver needs to distinguish whether it is seeing only the ambient noise or the combination of noise and

Signal Processing for Cognitive Radios, First Edition. Sudharman K. Jayaweera.

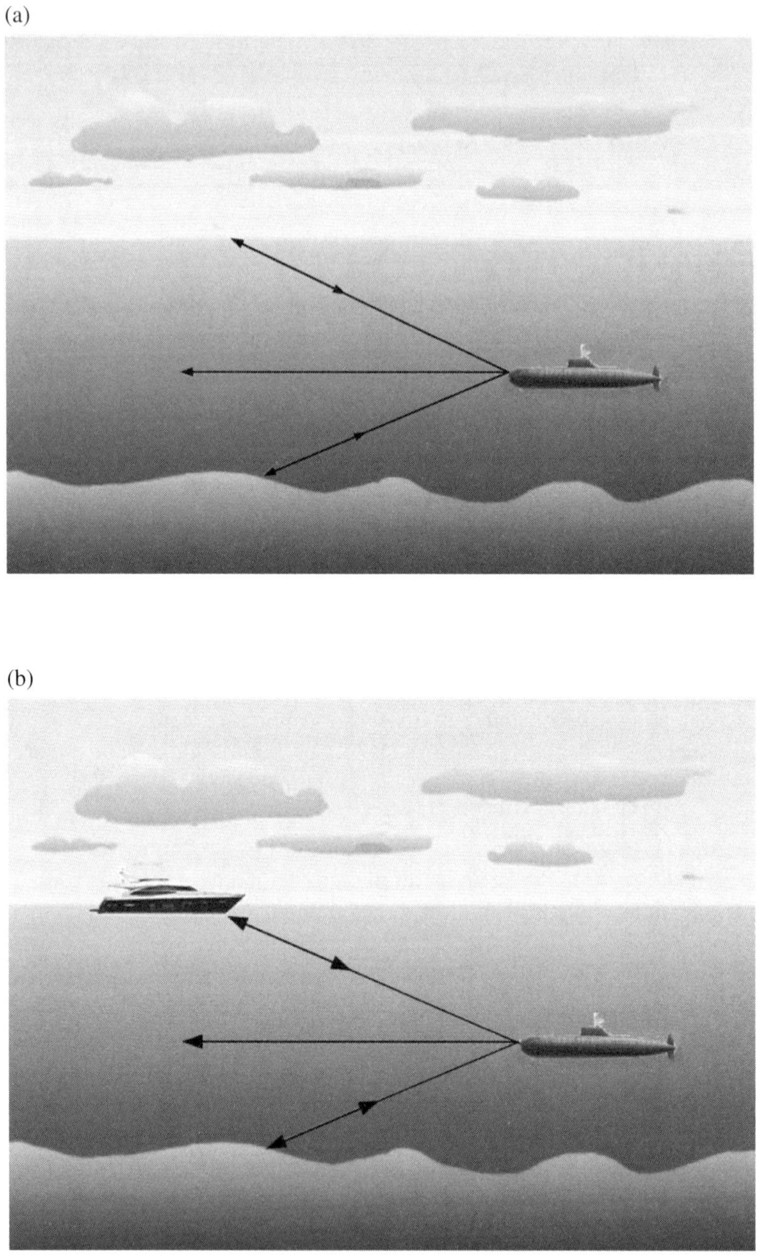

FIGURE 4.1 A submarine radar needs to detect when a signal reflected from a target is present in its received signal. (a) When there is no target, the received signal is comprised of noise and reflections from the ocean floor and surface. (b) When there is a target, the received signal contains a reflected signal from the target in addition to noise and reflections from the ocean floor and surface.

the signal of interest. This is the basic problem encountered in a radar system: A radar receiver needs to determine whether it is receiving a return signal reflected from some remote object or not. Interestingly, it is the same problem a dynamic spectrum sharing cognitive radio faces during spectrum sensing: The cognitive secondary user needs to determine whether there is a primary signal present or not in its sensed observation of a primary channel.

Noise may affect a signal in various ways depending on its origin. In wireless communications, by the time the signal arrives at the receiving antenna, it may already be corrupted by random propagation effects in the channel, including shadowing, fading, and attenuation. The effect of these corruptions is usually modeled as being of multiplicative, and it is common to refer to multiplicative noise as *fading*. On the other hand, the noise introduced in a communications receiver is usually modeled as being additive. Additive noise is present in all detectors since its origin can be traced to electronics themselves. It is the additive noise that is of interest to us here.

In general, noise is unpredictable. Hence, it is customary to model it as being a random process $n(t)$. Thus, signal detection problem is a problem of determining whether a signal $s(t)$ is present or not in an observed signal $y(t)$ that is corrupted by additive noise $n(t)$. Of course, one would like to do this in a certain optimal way. Depending on the chosen optimality criteria, we may end up with different solutions.

A problem in which we need to determine between two possible hypotheses based on an observation y is called a *binary hypothesis testing* problem. We may denote by H_0 and H_1 the two possible hypotheses of interest. For example, in a radar detection problem, H_0 may correspond to the case that there is no return signal reflected from a target in $y(t)$, as in Figure 4.1(a), while H_1 may correspond to the case that there is a return signal embedded in $y(t)$, as in Figure 4.1(b). In other words, H_1 hypothesis means that there is a target present.

The binary hypothesis testing problem forms the foundation for handling all detection problems. Clearly, detection of a signal in the presence of noise can easily be identified with two distinct hypotheses. For instance, in binary digital communications, we may transmit bits 0 and 1 using two distinct waveforms: signal $s_0(t)$ for bit 0 and signal $s_1(t)$ for bit 1. Hence, the detection problem at a digital communications receiver can be identified as a binary hypothesis testing problem in which the hypothese H_0 and H_1 correspond to the cases that the signal embedded in $y(t)$ is either $s_0(t)$ or $s_1(t)$, respectively. The radar problem corresponds to the case in which one of these signals is identically zero (i.e., no reflected signal present).

Throughout this book, we will assume discrete-time signals with the exception of a few instances. Since noise is modeled as being random, the received discrete-time signal $y[n]$ is also random which we emphasize by using the notation $Y[n]$. Depending on which hypothesis is held true, the distribution of $Y[n]$ will be different. Let us denote by P_0 and P_1 the distributions of observed signal $Y[n]$ under the two hypotheses H_0 and H_1, respectively, with corresponding densities f_0 and f_1. Hence, binary hypothesis testing problem, or the signal detection problem, is essentially equivalent to having to distinguish between these two distributions based on the observation $y[n]$.

A detector, or a decision rule, is a mapping (function) of a given observation y from the observation space to the set of hypothesis indices 1 or 0: $\delta : \mathcal{Y} \longrightarrow \{0, 1\}$, where $\mathcal{Y}$

denotes the observation space. If $\delta(y) = 0$ for a given observation y, then we take the hypothesis H_0 as being the true hypothesis corresponding to that observation. If, on the other hand, $\delta(y) = 1$ for a given observation y, then we take the hypothesis H_1 as being the true hypothesis corresponding to that observation. Because the detector output deterministically asserts which hypothesis must be taken as the true hypothesis, such a decision rule is termed a *nonrandomized*, or deterministic, *decision rule*.

We may, however, interpret a decision rule also as a mapping from the observation space to the probability with which we take hypothesis H_1 as being true: $\delta : \mathcal{Y} \longrightarrow [0, 1]$. With this interpretation, if $\delta(y) = 0$ for a given observation y, we interpret this as saying that with probability 0 we take the hypothesis H_1 to be the true hypothesis corresponding to that observation. Of course, this is equivalent to saying that with probability 1 we take hypothesis H_0 as the true hypothesis corresponding to this particular observation. Similarly, if $\delta(y) = 1$ for a given observation y, then with probability 1 we take the hypothesis H_1 to be the true hypothesis corresponding to that observation. With this probabilistic interpretation, the decision rule is called a *randomized decision rule*. Clearly, nonrandom decision rules can be expressed as a special case of the randomized detectors: Indeed, if $\delta(y)$ takes only the two extreme probability values of 0 and 1 for all observations y, then the resulting detector is simply a nonrandom detector.

Either way, the question is that what should be this mapping function under an assumed optimality condition? The answer to this question is developed in the *theory of detection*.

In detection theory terminology, correctly detecting hypothesis H_1 when it is in fact the true underlying hypothesis corresponding to an observation is what is called a *detection*. The probability that a given detector correctly declares hypothesis H_1 as being true when it is indeed true is thus called the *detection probability*, denoted by P_D, of the detector. Note that, with the interpretation $\delta(y)$ is the probability with which the hypothesis H_1 is accepted as the true hypothesis for a given observation y, the detection probability of a detector $\delta(.)$ can be written as

$$P_D = \mathbb{E}\{\delta(Y)|H_1\}, \tag{4.1}$$

where $\mathbb{E}\{.\}$ denotes the expectation operator. On the other hand, declaring hypothesis H_0 as being true when the underlying true hypothesis is in fact H_1 is called a *miss*. If we denote by P_M the *miss probability* of a detector in the case of binary hypothesis testing, clearly,

$$P_M = 1 - P_D. \tag{4.2}$$

Declaring hypothesis H_1 as being true when the hypothesis H_0 is in fact true is called a false alarm. With the randomized decision-rule interpretation of a detector δ, the *false-alarm probability* can be written as

$$P_F = \mathbb{E}\{\delta(Y)|H_0\}. \tag{4.3}$$

In this chapter, we discuss basics of detection theory in discrete time that will be used throughout this book. The continuous-time case leads to similar solutions but requires advanced measure theoretic concepts for a rigorous presentation that is not assumed here. Our self-contained concise introduction is mostly based on material drawn from [51] and [52]. For more in-depth details, the reader should indeed refer to those references.

4.2 OPTIMALITY CRITERIA: BAYESIAN VERSUS NON-BAYESIAN

Given two decision rules (detectors) δ_1 and δ_2, how can we determine whether one is better than the other? Better yet, can we determine whether one is better than all other possible detectors? Both these questions are imprecise if we do not clarify *good in what sense*. Clearly, one detector may be better than another detector in one aspect but not in another aspect. For example, detector δ_1 might be computationally more efficient in that it gives the output faster compared to δ_2, whereas decisions of the detector δ_2 might appear to be accurate more often than those of δ_1.

Thus, in order to meaningfully compare decision rules for a hypothesis testing problem, we need to specify our assumed performance criterion. Of course, we would then like to have the best decision rule in terms of that performance criterion. A decision rule that is the best among all decision rules in terms of a given performance metric is called an *optimal decision rule* according to that performance, or optimality, criterion. Hence, it is imprecise to call a decision rule optimal if we do not specify according to what optimality criteria it is optimal. In some contexts, the relevant performance metric might be obvious that the need to always remind in what sense we mean a detector is optimal might not be so great. In other cases, however, it might cause confusion if we are not explicit.

Since our objective is to distinguish between the two hypotheses H_0 and H_1 in designing a signal detector for a binary hypothesis testing problem, perhaps, a natural goal is to try to be correct as often as we can. Equivalently, we would like to be less in error. In the context of binary hypothesis testing, there are only two ways that a decision rule may be in error: declaring H_0 when H_1 is in fact true or declaring H_1 when H_0 is in fact true. Hence, a natural approach may be to minimize the probability that we declare H_0 when H_1 is in fact true as well as the probability that we declare H_1 when H_0 is in fact true. The former is the miss probability P_M and the latter is the false-alarm probability P_F. Note also that in lieu of (4.2), minimizing P_M is equivalent to maximizing the detection probability P_D. Hence, our natural inclination may be to look for a detector that simultaneously minimizes the false-alarm probability P_F and maximizes the detection probability P_D.

Clearly, either one of these objectives alone can be achieved very easily. For example, minimizing P_F alone is straightforward: set $\delta(y) = 0$ for all y. This way, regardless of the observation, we will always declare the hypothesis H_0 as being true, and so we will never make a false alarm! We have not only just minimized P_F, but we have minimized it to be $P_F = 0$. However, this detector's detection probability is also $P_D = 0$ for it never declares H_1 as true. Hence, this detector minimizes both P_F and P_D, whereas we would like to maximize P_D. Similarly, we may maximize the detection probability by setting $\delta(y)=1$ for all y. This detector will maximize the detection probability to be $P_D=1$, but it will now have the highest possible false-alarm value of $P_F=1$ as well.

Hence, minimizing false alarm and minimizing miss probabilities at the same time are conflicting objectives. Whenever we minimize one, the other shoots up to 1. But naturally, they are both error events, and thus, we would like to avoid them. A reasonable approach at this point is to adapt a trade-off between these two probabilities. Instead of trying to minimize both P_F and P_M at the same time, we may try to find a detector that minimizes a combination, a trade-off, of these two error probabilities. But what is the correct, or the best, trade-off?

4.2.1 The Bayesian Approach

In general, there is no best trade-off. It depends on what we consider as the suitable trade-off. The Bayesian approach is to treat the two hypotheses H_0 and H_1 as being random themselves. That is, with the relative frequency interpretation of the probabilities, sometimes, the underlying hypothesis may be H_0, and other times, the underlying hypothesis may be H_1. The Bayesian approach assumes that there are a priori probabilities, p_0 and p_1, associated with each of the hypotheses H_0 and H_1 being true, respectively.

If we consider an ensemble of observations, some of them will correspond to an underlying hypothesis H_0, while the others will correspond to an underlying hypothesis H_1. The probabilities of these events are indeed p_0 and p_1 a priori. A given detector will have a P_M miss probability on those observations corresponding to an underlying true hypothesis H_1, while it will have a P_F false-alarm probability on those observations corresponding to an underlying true hypothesis H_0. We know that no detector can minimize both these probabilities simultaneously. However, now, we may try to find a detector that minimizes the average (overall) error probability, denoted as P_e, over the whole ensemble of H_0 and H_1 hypotheses:

$$\begin{aligned} P_e &= \Pr(H_1|H_0)\Pr(H_0)+\Pr(H_0|H_1)\Pr(H_1) \\ &= p_0P_F+p_1P_M. \end{aligned} \tag{4.4}$$

A decision rule that minimizes this *average error probability* P_e is called a *Bayesian optimal decision rule* or, more precisely, a Bayesian minimum probability of error decision rule. In general, any detector that minimizes a cost function averaged over a priori probabilities is called a *Bayesian optimal detector*. Hence, minimum average error probability detection is just one special case of the Bayesian optimal detection.

The distinguishing feature of the Bayesian approach is the averaging of an assumed cost function with respect to the a priori probabilities. Hence, the Bayesian approach presumes that:

(a) Hypotheses can be modeled as endowed with a certain prior distribution.
(b) This prior distribution is known.

For a more general discussion on Bayesian optimal hypothesis testing, the reader is referred to classic texts on detection theory such as that in Refs. [51,53] and [54].

While Bayesian optimality with general cost functions can be more suitable in some contexts, for communications receivers considered in this book, the average error probability cost function of (4.4) is the most natural choice. Hence, throughout this book, we will only consider the average error probability cost function, and the Bayesian optimality will always mean the optimality in the sense of minimum average error probability.

4.2.2 A Non-Bayesian Approach: Neyman–Pearson Optimality Criterion

Recall the two key assumptions under which the Bayesian approach can be justified: (i) Hypotheses can be modeled as endowed with a certain prior distribution and (ii) that prior distribution is known. There are situations in which either or both these assumptions may not be valid. It is easy to imagine situations in which the latter may not be true because we may not have a good prior knowledge about the distribution of the two hypotheses. The former assumption, on the other hand, can be questionable on more philosophical grounds: Let us consider a radar that attempts to detect an enemy drone. Suppose that the enemy drone is programmed so that either it flies over the area under the surveillance once during each hour or it does not. Let us consider hour 1. If the target is the drone, can we assume that it is either present or absent with a certain probability? One may say yes, but one may object. Note that, if the drone is present during hour 1, there is no randomness about it. Every day during the first hour, the target will always be present! This is because it is programmed to do that. Similarly, if the drone is absent during the fourth hour, it is always absent during the fourth hour. There is nothing random about it. At the heart of this is the question that whether it is reasonable to model something as being random just because we do not know it. At each hour, we do not know whether the target is present or not a priori. But it is not a random choice. It is either present or absent, and that ground truth is fixed a priori. The only issue is that we do not know which one is true.

Philosophical arguments aside, it should be clear that there are situations in which either the Bayesian assumption may not be justified or that the prior distribution of the hypotheses may not be known. Then, we are faced with a situation in which Bayesian averaging (i.e., averaging with respect to the prior distribution of the hypotheses) cannot be justified or cannot be evaluated. There are many natural problems in which this is the case, including the cognitive radio spectrum sensing problem that we will explore in Chapters 10 and 12. When the Bayesian approach is not applicable or possible, we need a suitable alternative optimality criteria for binary hypothesis testing.

Recall that the two types of errors in a binary hypothesis testing problem are the misses and false alarms. As we already mentioned, simultaneous minimization of both these errors is the natural optimality criterion. The need for alternatives arises because it is not possible to simultaneously minimize these two types of errors. Reducing one type of errors increases the errors of the other type, hence the need for a trade-off. The Bayesian minimum error probability approach was to minimize the weighted sum of these two types of errors, in which weights were chosen to be equal to the a priori probabilities of the two hypotheses. Essentially, it gives the two types of errors an importance that is proportional to the relative frequencies of the two hypotheses. We have already seen that if, on the other hand, we wanted to give one type of error the primary importance, we can make that particular type of error probability go to zero (at the expense of the other type of error probability shooting up to unity).

In radar signal detection, the hypothesis H_0 corresponds to the case that the observed signal does not contain a desired signal of interest but just the noise, whereas hypothesis H_1 corresponds to the case that the observation is a signal of interest plus noise. The primary objective of the radar is to detect a desired signal of interest whenever it is present in the received signal. For example, if it is a military radar looking for enemy aircrafts, missing the H_1 hypothesis can be catastrophic. So one would naturally wants the detection probability of a detector $\delta(y)$, which is $\mathbb{E}\{\delta(Y)|H_1\}$, to be as high as possible. As we already saw, maximizing the detection probability has a simple solution that is, to set $\delta(y) = 1$ for all y, and we may achieve detection probability of one. But, of course, this detector is not very useful because its false-alarm probability is also equal to 1. Every time we declare that there is an enemy target, it might have to be followed up with contingency plans that are supposed to be put in place when there is an enemy signal. Hence, every time the detector falsely declares the presence of a target, we will be spending resources, only to find out later all that was just a waste. So, there is a price, or a cost, to pay for each false alarm. If there was no cost attached to these false alarms, then a detector that always declares hypothesis H_1 might be reasonable. But all practical situations of interest involve some cost attached to false alarms.

In such scenarios, there is a maximum limit to how much false alarms we can tolerate on average. This then gives us a natural alternative optimality criterion that does not require Bayesian prior probabilities. We may simply posed a constrained optimization problem: Find the detector that maximizes the detection probability subject to a given maximum level of false alarms. This is called the *Neyman–Pearson optimality* criterion. It is a non-Bayesian approach since it does not assume, and does not need, a prior distribution. A detector that maximizes the detection probability subject to a maximum false alarm probability constraint is thus called a *Neyman–Pearson optimal*. In the context of radar detection, the maximum allowed false-alarm level, called the *size of the detector*, is the maximum tolerable probability of erroneously declaring that there is a target when actually there is no target. Since detection probability of a detector is called the *power of the detector*, the Neyman–Pearson problem is to find the most powerful test of a required size.

The Neyman–Pearson formulation also arises naturally in the context of dynamic spectrum sharing cognitive radio systems. There, the objective of a cognitive secondary user is to detect a channel that is absent of licensed (primary) user transmissions. Let y be the cognitive radio's observed signal on a particular primary channel. If we define the H_1 hypothesis as the noise-only hypothesis and H_0 as the primary signal plus noise hypothesis, then a dynamic spectrum sharing cognitive radio's goal is to maximize the probability of correctly detecting the noise-only hypothesis (i.e., H_1) on the primary channel. The maximum false-alarm probability in this case specifies the maximum probability with which we may be comfortable in declaring that there is no primary signal when there is indeed a primary signal present. The purpose of finding a channel without an active primary transmission is for the secondary cognitive radio to transmit on that channel. Hence, a false alarm as defined in this case will lead to a secondary cognitive radio transmission on a primary channel while primary user is actually in transmission, thereby colliding with it. This is called a collision, or interfering, with a primary user. Hence, in the context of dynamic spectrum sharing cognitive radios, the false-alarm probability is called the *collision probability*. Usually, in DSS systems, there is a maximum tolerable probability of collision with primary users. Hence, the problem naturally reduces to the classical Neyman–Pearson hypothesis testing formulation: Maximize the detection probability subject to a given collision (false-alarm) probability constraint.

4.3 PARAMETRIC SIGNAL DETECTION THEORY

Let us assume that the observation probability density (distribution) functions are completely known except for a finite number of possibly unknown parameters. When the probability density of the observations can be parameterized by a finite number of parameters, and the form of the parameterized distribution is known, the detection problem is called a parametric detection problem [51].

Formally, let us assume that the two hypotheses are determined by a parameter $\theta \subset \Theta \subseteq \mathbb{R}$. Hypotheses H_0 and H_1 correspond to θ values in two nonempty disjoint sets Θ_0 and Θ_1 with $\Theta = \Theta_0 \cup \Theta_1$. When Θ contains only two elements so that both Θ_0 and Θ_1 contains a single element each, this is called a *simple binary hypothesis testing* problem. Otherwise, it is a *composite hypothesis testing* problem [51, 52]. The detection problem is to declare the true hypothesis given an observation Y whose distribution depends on the parameter θ.

We may denote the observation density for a given parameter θ as $f_{Y|\theta}(y|\theta)$, assuming it exists. Note that this could be either a probability density function (pdf) or a probability mass function (pmf). When there is no ambiguity, we may let $f_{Y|\theta}(y|\theta) = f_\theta(y)$. This will be particularly convenient when Θ_0 and Θ_1 contains a single element each so that the problem is a simple hypothesis testing problem. In that case, we might as well set $\Theta_0 = \{0\}$ and $\Theta_1 = \{1\}$ and denote the density of Y under the two hypotheses of H_0 and H_1 as $f_{Y|\theta=0}(y|\theta = 0) = f_0(y)$ and $f_{Y|\theta=1}(y|\theta = 1) = f_1(y)$, respectively.

In general, however, the binary hypothesis testing problem is

$$\begin{aligned} H_0 : Y &\sim f_{Y|\theta}(y|\theta) \quad \text{with } \theta \in \Theta_0 \\ &\text{versus} \\ H_1 : Y &\sim f_{Y|\theta}(y|\theta) \quad \text{with } \theta \in \Theta_1. \end{aligned} \tag{4.5}$$

4.3.1 Bayesian Optimal Detection

If a known prior distribution $f_\Theta(\theta)$ is available, one may easily derive a Bayesian minimum probability of error optimal decision rule. Indeed, we may express the average probability of error in this case as

$$P_e = \Pr(H_1|\theta \in \Theta_0)\, p_0 + \Pr(H_0|\theta \in \Theta_1)\, p_1, \tag{4.6}$$

where

$$p_0 = \Pr(\theta \in \Theta_0) \tag{4.7}$$

and

$$p_1 = \Pr(\theta \in \Theta_1) \tag{4.8}$$

are the prior probabilities of the hypotheses H_0 and H_1, respectively. We may express (4.6) as

$$\begin{aligned} P_e &= p_0 \mathbb{E}\{\delta(Y)|\theta \in \Theta_0\} + p_1 (1 - \mathbb{E}\{\delta(Y)|\theta \in \Theta_1\}) \\ &= p_0 \int \delta(y) f(y|\theta \in \Theta_0)\, dy - p_1 \int \delta(y) f(y|\theta \in \Theta_1)\, dy + p_1 \\ &= \int \delta(y) (p_0 f(y|\theta \in \Theta_0) - p_1 f(y|\theta \in \Theta_1))\, dy + p_1. \end{aligned} \tag{4.9}$$

Since $\delta(y)$ can only be assigned values in $[0,1]$, from (4.9), it is easily seen that the minimum probability of error optimal (Bayesian) decision rule $\delta^{(B)}$ is such that

$$\delta^{(B)}(y) = \begin{cases} 1 & \text{if } p_0 f(y|\theta \in \Theta_0) - p_1 f(y|\theta \in \Theta_1) \le 0 \\ 0 & \text{if } p_0 f(y|\theta \in \Theta_0) - p_1 f(y|\theta \in \Theta_1) > 0 \end{cases} \tag{4.10a}$$

$$= \begin{cases} 1 & \\ & \text{if } \dfrac{f(y|\theta \in \Theta_1)}{f(y|\theta \in \Theta_0)} \overset{\geq}{\underset{<}{}} \dfrac{p_0}{p_1}. \\ 0 & \end{cases} \tag{4.10b}$$

This shows that essentially the Bayesian minimum error probability decision rule $\delta^{(B)}$ is a simple threshold test on the *likelihood ratio* (LR) defined as

$$L(y) = \frac{f(y|\theta \in \Theta_1)}{f(y|\theta \in \Theta_0)}, \tag{4.11}$$

where, for $j = 0, 1$,

$$f\left(y|\theta \in \Theta_j\right) = \frac{f\left(y, \theta \in \Theta_j\right)}{f\left(\theta \in \Theta_j\right)} = \frac{1}{p_j} \int_{\Theta_j} f_{Y|\theta}(y|\theta) f_\Theta(\theta) d\theta \tag{4.12}$$

with p_j's given by (4.7) and (4.8), $f_{Y|\theta}(.)$ is called the *likelihood function* of the observation Y, and $f_\Theta(.)$ is the prior distribution.

Thus, we may rewrite the optimal minimum error probability Bayesian detection rule as

$$\delta^{(B)}(y) = \begin{cases} 1 & \geq \\ & \text{if } L(y) \quad \tau_B, \\ 0 & < \end{cases} \tag{4.13}$$

where the optimal threshold τ_B is completely determined by the prior probabilities:

$$\tau_B = \frac{p_0}{p_1}. \tag{4.14}$$

It is interesting to observe that regardless of the form of the observation (e.g., whether it is a scalar or a vector), all that is needed for Bayesian optimal decision-making is the LR (4.11), which is a real scalar.

Note also from (4.10a) that

$$\delta^{(B)}(y) = \begin{cases} 1 & \text{if } \dfrac{p_0 f(y|\theta \in \Theta_0)}{f(y)} \leq \dfrac{p_1 f(y|\theta \in \Theta_1)}{f(y)} \\ 0 & \text{if } \dfrac{p_0 f(y|\theta \in \Theta_0)}{f(y)} > \dfrac{p_1 f(y|\theta \in \Theta_1)}{f(y)} \end{cases} \tag{4.15a}$$

$$= \begin{cases} 1 & \text{if } f(\theta \in \Theta_1|y) \geq f(\theta \in \Theta_0|y) \\ 0 & \text{if } f(\theta \in \Theta_0|y) > f(\theta \in \Theta_1) \end{cases}. \tag{4.15b}$$

Since, for $j = 0, 1$, $f\left(\theta \in \Theta_j|y\right)$ is the a posteriori probability of the hypothesis H_j given the observation y, from (4.15b), we note that the minimum probability of error decision rule (4.10a) is also equivalent to declaring the hypothesis having the *maximum a posteriori probability* (MAP). For this reason, the Bayesian minimum error probability detector is also called the *MAP detector*.

4.3.1.1 Simple Hypothesis Testing: Bayesian Optimal Detector Consider the situation in which both Θ_0 and Θ_1 contains just a single element each, which we denote as $\Theta_0 = \{0\}$ and $\Theta_1 = \{1\}$. This, as we already mentioned earlier, is the case of simple hypothesis testing. The prior density f_Θ in this case is simply

$$f_\Theta(\theta) = p_0\delta_D(\theta) + p_1\delta_D(\theta - 1),$$

where $\delta_D(.)$ denotes the *Dirac-delta function* that is defined implicitly by the following integral called the *sifting integral*:

$$\int_{-\infty}^{\infty} g(t)\delta_D(t)dt = g(0),$$

for any function $g(t)$ that is continuous at $t = 0$.

As stated earlier, in this case, it is more convenient to denote the observation likelihood for a given θ as $f_{Y|\theta}(y|\theta) = f_\theta(y)$, so that the density of Y under the two hypotheses of H_0 and H_1 can be written as $f_{Y|\theta=0}(y|\theta=0) = f_0(y)$ and $f_{Y|\theta=1}(y|\theta = 1) = f_1(y)$, respectively:

$$\begin{aligned} H_0 : Y &\sim f_0(y) \\ &\text{versus} \\ H_1 : Y &\sim f_1(y). \end{aligned} \tag{4.16}$$

The LR for the simple hypothesis testing problem thus simplifies to

$$L(y) = \frac{f_1(y)}{f_0(y)}, \tag{4.17}$$

and the Bayesian minimum error probability optimal decision rule, or equivalently the MAP rule, is given by

$$\delta^{(B)}(y) = \begin{cases} 1 & \text{if } L(y) \geq \frac{p_0}{p_1} \\ 0 & \text{if } L(y) < \frac{p_0}{p_1} \end{cases}. \tag{4.18}$$

Example 4.1 (Optimal Minimum Probability of Error Detection of a Known Signal in Additive Gaussian Noise) Let us consider the simple hypothesis testing scenario of detecting a known deterministic signal $\mathbf{s}$ embedded in additive, zero-mean Gaussian noise $\mathbf{w} \sim \mathcal{N}(\mathbf{0}, \Sigma)$ where Σ is the known covariance matrix of the noise vector $\mathbf{w}$. Specifically, the observations under the two hypotheses are given by

$$\begin{aligned} H_0 &: \mathbf{y} = \mathbf{w} \\ H_1 &: \mathbf{y} = \mathbf{s} + \mathbf{w}, \end{aligned} \tag{4.19}$$

so that

$$\begin{aligned} H_0 &: \mathbf{Y} \sim \mathcal{N}(\mathbf{0}, \Sigma) \\ H_1 &: \mathbf{Y} \sim \mathcal{N}(\mathbf{s}, \Sigma), \end{aligned} \tag{4.20}$$

where $\mathbf{y}$, $\mathbf{s}$, and $\mathbf{w}$ are all assumed to be L-vectors and $\mathcal{N}(\boldsymbol{\mu}, \Sigma)$ denotes the multivariate normal distribution with mean vector $\boldsymbol{\mu}$ and covariance matrix Σ (see Appendix B for a collection of pdf's that will be useful throughout this book).

Note that (4.19) is a general signal model that is applicable in many different situations encountered in communications. For example, the L components of the observation vector $\mathbf{y}$ may correspond to time samples of a continuous-time received waveform, spatial samples of a wavefront received by a set of antenna array elements, or, more generally, the sampled output of a parallel bank of filters. An example of this last case is a bank of matched filters (MFs) assumed in a *multiuser detector* [55]. Moreover, even the hypothesis testing situations in which there are two nonzero known signals, say, $\mathbf{s}_0$ and $\mathbf{s}_1$, present under the hypotheses H_0 and H_1, respectively, can be reduced to a problem of the form of (4.19). Indeed, suppose that

$$\begin{aligned} H_0 &: \tilde{\mathbf{y}} = \mathbf{s}_0 + \mathbf{w} \\ H_1 &: \tilde{\mathbf{y}} = \mathbf{s}_1 + \mathbf{w}. \end{aligned} \tag{4.21}$$

Since $\mathbf{s}_0$ is assumed to be known, given the original observation $\tilde{\mathbf{y}}$ of (4.21), we may define a new observation $\mathbf{y} = \tilde{\mathbf{y}} - \mathbf{s}_0$. It is easy to verify that this $\mathbf{y}$ corresponds to model (4.19) with $\mathbf{s} = \mathbf{s}_1 - \mathbf{s}_0$.

The LR for (4.20) is

$$\begin{aligned} L(\mathbf{y}) &= \frac{f_1(\mathbf{y})}{f_0(\mathbf{y})} \\ &= \frac{\frac{1}{(2\pi)^{\frac{L}{2}}|\Sigma|^{\frac{1}{2}}} e^{-\frac{1}{2}(\mathbf{y}-\mathbf{s})^T\Sigma^{-1}(\mathbf{y}-\mathbf{s})}}{\frac{1}{(2\pi)^{\frac{L}{2}}|\Sigma|^{\frac{1}{2}}} e^{-\frac{1}{2}\mathbf{y}^T\Sigma^{-1}\mathbf{y}}} \\ &= e^{\mathbf{s}^T\Sigma^{-1}\mathbf{y} - \frac{1}{2}\mathbf{s}^T\Sigma^{-1}\mathbf{s}}. \end{aligned} \tag{4.22}$$

From (4.18), the Bayesian optimal decision rule is a threshold test on the LR (4.22). However, since exp(.) is monotonic in its argument, we might as well perform the threshold test on the argument of the exp(.) function appearing in the LR. This can be accomplished by taking the *LLR*

$$\mathscr{L}(\mathbf{y}) \triangleq \log L(\mathbf{y}) = \mathbf{s}^T\Sigma^{-1}\mathbf{y} - \frac{1}{2}\mathbf{s}^T\Sigma^{-1}\mathbf{s}$$

as the decision statistic and comparing it with the log(.) of the optimal threshold. Hence, the Bayesian minimum probability of error detector for problem (4.20) simplifies to

$$\delta^{(B)}(\mathbf{y}) = \begin{cases} 1 & \geq \\ & \text{if } \mathbf{s}^T\Sigma^{-1}\mathbf{y} \quad \log\tau + \frac{1}{2}\mathbf{s}^T\Sigma^{-1}\mathbf{s}\,, \\ 0 & < \end{cases} \tag{4.23}$$

where $\tau = \frac{p_0}{p_1}$.

If prior probabilities are equal so that $\tau = 1$, then the above detector simplifies to

$$\delta^{(B)}(\mathbf{y}) = \begin{cases} 1 & \geq \\ & \text{if } \mathbf{s}^T\Sigma^{-1}\mathbf{y} \quad \frac{1}{2}\mathbf{s}^T\Sigma^{-1}\mathbf{s}\,. \\ 0 & < \end{cases} \tag{4.24}$$

If we further assume the case of zero-mean, iid Gaussian noise so that $\Sigma = \sigma^2\mathbf{I}_L$, where $\mathbf{I}_L$ denotes the $L \times L$ identity matrix, the optimal decision rule becomes

$$\delta^{(B)}(\mathbf{y}) = \begin{cases} 1 & \geq \\ & \text{if } \mathbf{s}^T\mathbf{y} \quad \frac{||\mathbf{s}||^2}{2}\,, \\ 0 & < \end{cases} \tag{4.25}$$

where $||\mathbf{s}|| = \sqrt{\mathbf{s}^T\mathbf{s}} = \sqrt{\sum_{l=1}^{L} s_l^2}$ is the Euclidean norm of signal vector $\mathbf{s}$.

Figure 4.2 shows the Bayesian optimal decision rule for the problem (4.20) in the case of a scalar observation Y with $s = 1$. In this case, the decision statistic is $\mathbf{s}^T\mathbf{y} = y$, and the optimal threshold is $\tau' = \frac{||\mathbf{s}||^2}{2} = \frac{1}{2}$. Note that if noise was not present, then observations corresponding to the two hypotheses H_0 and H_1 should be $y = 0$ and $y = 1$, respectively. Addition of Gaussian noise causes the observation to take any arbitrary

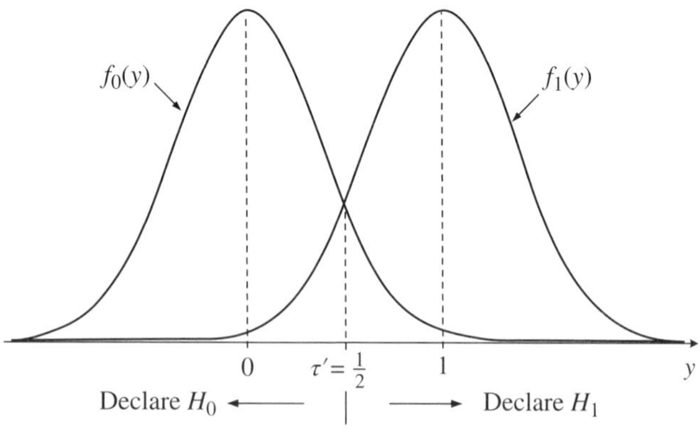

FIGURE 4.2 The structure of the Bayesian minimum probability of error optimal detector in the case of a scalar observation y.

real value. However, since noise has mean zero, the means of the observations under hypotheses H_0 and H_1 are still 0 and 1, respectively. The optimal Bayesian detector in Figure 4.2 declares the hypothesis whose mean is more closer to the observed y as the true hypothesis. This interpretation is true for the problem (4.20) even when the observation is a vector. Since we may consider the two hypothesis H_0 and H_1 in this case as being corresponding to the two signals $s = 0$ and $s = 1$, respectively, then the optimal Bayesian detector can also be viewed as declaring the signal that is closest to the observation y in the Euclidean distance sense. Hence, the Bayesian minimum error probability optimal detector in the Gaussian noise case is also called the *minimum distance detector*.

Let us denote by $T(\mathbf{Y})$ the decision statistic of the detector (4.25):

$$T = T(\mathbf{Y}) \triangleq \mathbf{s}^T \mathbf{Y}. \tag{4.26}$$

Note that under either of the hypothesis, T is a linear transformation of a Gaussian random vector $\mathbf{Y}$. Hence, T is a Gaussian random variable under both hypotheses. Thanks to this fact, in this case, we may easily compute the actual minimum probability of error achieved by this detector in closed form.

Recall that the average probability of error is the average of the false alarms and the missed detections where the averaging is with respect to the prior distribution of the two hypotheses. False alarms are caused when the decision statistic T in (4.26), is above the threshold $\frac{||\mathbf{s}||^2}{2}$ under the hypothesis H_0, and the misses are caused when T is below the threshold $\frac{||\mathbf{s}||^2}{2}$ under the hypothesis H_1. But, from (4.19), we note that, under the hypothesis H_0,

$$T|H_0 = \mathbf{s}^T \mathbf{W} \tag{4.27}$$

and, under the hypothesis H_1,

$$T|H_1 = ||\mathbf{s}||^2 + \mathbf{s}^T \mathbf{W}. \tag{4.28}$$

It is easy to see that

$$\mathbf{s}^T \mathbf{W} \sim \mathcal{N}\left(0, \sigma^2 ||\mathbf{s}||^2\right). \tag{4.29}$$

Hence, from (4.6), the minimum error probability achievable in the simple hypothesis testing in Gaussian noise scenario is

$$\begin{aligned} P_e &= p_0 \Pr(H_1|H_0) + p_1 \Pr(H_0|H_1) \\ &= \frac{1}{2}\left[\Pr\left(T \geq \frac{||\mathbf{s}||^2}{2}|H_0\right) + \Pr\left(T < \frac{||\mathbf{s}||^2}{2}|H_1\right)\right] \\ &= \frac{1}{2}\left[\Pr\left(\mathbf{s}^T\mathbf{W} \geq \frac{||\mathbf{s}||^2}{2}\right) + \Pr\left(||\mathbf{s}||^2 + \mathbf{s}^T\mathbf{W} < \frac{||\mathbf{s}||^2}{2}\right)\right] \end{aligned}$$

$$
\begin{aligned}
&= \frac{1}{2}\left[\Pr\left(\mathbf{s}^T\mathbf{W} \geq \frac{||\mathbf{s}||^2}{2}\right) + \Pr\left(\mathbf{s}^T\mathbf{W} < -\frac{||\mathbf{s}||^2}{2}\right)\right] \\
&= \Pr\left(\mathbf{s}^T\mathbf{W} \geq \frac{||\mathbf{s}||^2}{2}\right) \\
&= Q\left(\frac{\frac{||\mathbf{s}||^2}{2}}{\sqrt{\sigma^2||\mathbf{s}||^2}}\right) \\
&= Q\left(\frac{1}{2}\sqrt{\frac{||\mathbf{s}||^2}{\sigma^2}}\right) \\
&= Q\left(\frac{1}{2}\sqrt{SNR}\right),
\end{aligned} \tag{4.30}
$$

where we have defined the signal-to-noise ratio (SNR) as

$$
SNR = \frac{||\mathbf{s}||^2}{\sigma^2} \tag{4.31}
$$

and $Q(x)$ denotes the Gaussian complimentary cumulative distribution function (ccdf) defined as

$$
\begin{aligned}
Q(x) &= \Pr\left(\mathcal{N}(0,1) > x\right) \\
&= \frac{1}{\sqrt{2\pi}}\int_x^{\infty} e^{-\frac{t^2}{2}}\,dt.
\end{aligned} \tag{4.32}
$$

Figure 4.3 shows this minimum achievable average probability of error in simple hypothesis testing in the presence of zero-mean additive Gaussian noise. It is customary in communications applications to plot the error probabilities as a function of SNR in decibel scale where $SNR_{dB} = 10\log_{10}(SNR)$. ■

4.3.2 Neyman–Pearson Optimal Detection

When either the Bayesian framework is not suitable or when prior probabilities are unknown, we need alternative optimality criteria in designing reasonable signal detectors. As we saw earlier, one such alternative is to use the Neyman–Pearson optimality criterion.

Let us consider the power (detection probability) of a detector $\delta(y)$ to distinguish between the two hypotheses H_0 and H_1 in (4.5). This is the probability that $\delta(y)$ declares H_1 as true when indeed H_1 is true. But H_1 being true means that $\theta \in \Theta_1$. Missed detections imply that while actual $\theta \in \Theta_1$, our detector mistook it as in Θ_0. Certainly, this confusion can be more or less for different $\theta \in \Theta_1$. Consider two particular θ values $\theta_1, \theta_2 \in \Theta_1$. It is possible that θ_1 leads to a distribution of Y that is easily distinguishable from Y's distribution corresponding to any $\theta \in \Theta_0$, whereas this is not so with $\theta = \theta_2$. This will make the detection probability of the detector conditioned on θ to be higher if $\theta = \theta_1$ than what it is when $\theta = \theta_2$.

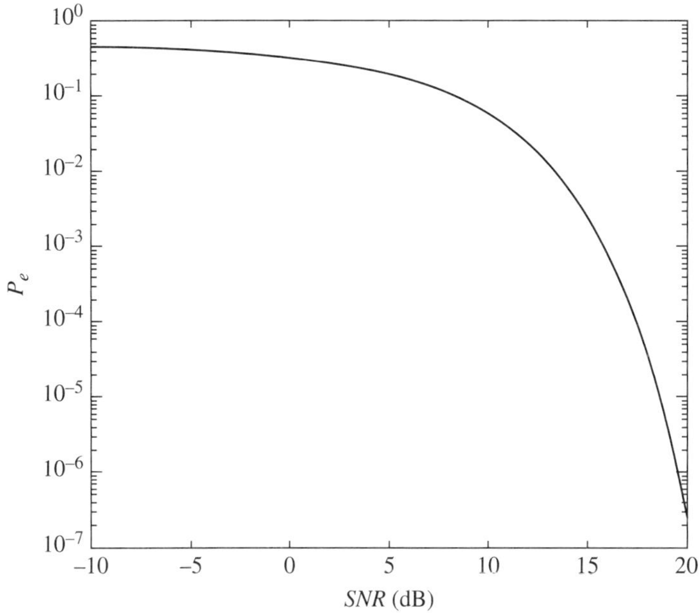

FIGURE 4.3 Minimum achievable average probability of error in simple binary hypothesis testing in the presence of zero-mean additive Gaussian noise.

In general, then the detection probability of a detector for the problem (4.5) can be a function of the exact value of $\theta \subset \Theta_1$. A little thought should show that this is also true for the false alarm probability of a detector. Hence, let us denote the detection and false-alarm probabilities of a detector $\delta(y)$ for the problem (4.5) as

$$P_D(\delta,\theta) = \mathbb{E}\{\delta(Y)|\theta\} \quad \text{for } \theta \in \Theta_1 \tag{4.33}$$

and

$$P_F(\delta,\theta) = \mathbb{E}\{\delta(Y)|\theta\} \quad \text{for } \theta \in \Theta_0. \tag{4.34}$$

Then we may formally state the Neyman–Pearson hypothesis testing problem for (4.5) as

$$\max_{\delta} \; P_D(\delta,\theta) \quad \text{subject to} \quad P_F(\delta,\theta) \le \alpha, \tag{4.35}$$

where α denotes the size of the required detector.

Before investigating the general solution to this problem, let us consider the simple hypothesis testing problem for which the solution is particularly simple.

4.3.2.1 Simple Hypothesis Testing: Neyman–Pearson Optimal Detector Consider again the simple hypothesis testing in which $\Theta_0 = \{0\}$ and $\Theta_1 = \{1\}$. As before, denote the observation density for a given θ as $f_{Y|\theta}(y|\theta) = f_\theta(y)$, so that the density of Y under the two hypotheses of H_0 and H_1 can be written as $f_{Y|\theta=0}(y|\theta = 0) = f_0(y)$ and $f_{Y|\theta=1}(y|\theta = 1) = f_1(y)$, respectively:

$$\begin{aligned} H_0 : Y &\sim f_0(y) \\ &\text{versus} \\ H_1 : Y &\sim f_1(y). \end{aligned} \tag{4.36}$$

The *Neyman–Pearson lemma* below specifies the optimal solution to the detection problem (4.35) when the hypotheses are simple as in (4.36) [51]:

Theorem 4.1 (Neyman–Pearson Lemma) Consider the simple hypotheses pair of (4.36) in which distribution of Y has an associated density $f_j(y)$ for $j = 0$ and $j = 1$. Let $\tilde{\delta}$ be a decision rule of the form

$$\tilde{\delta}(y) = \begin{cases} 1 & > \\ \gamma(y) & \text{if } L(y) = \eta \\ 0 & < \end{cases} \tag{4.37}$$

where $L(y) = \frac{f_1(y)}{f_0(y)}$ is the LR and the threshold $\eta \geq 0$ and the *randomization parameter* $0 \leq \gamma(y) \leq 1$ are chosen such that $P_F(\tilde{\delta}) = \alpha$ for $\alpha > 0$. Let δ be any decision rule satisfying $P_F(\delta) \leq \alpha$. Then

$$P_D(\tilde{\delta}) \geq P_D(\delta). \tag{4.38}$$

Moreover, for every $\alpha \in (0, 1)$, there is always a decision rule $\tilde{\delta}$ of the form of (4.37) with $\gamma(y) = \gamma_0$ (a constant) for which $P_F(\tilde{\delta}) = \alpha$.

Proof: See [51] for a proof.

It is also true that the detector (4.37) specified in the Neyman–Pearson lemma is the unique solution to the problem (4.36) up to a measure zero set. In other words, if δ^* is any α-level Neyman–Pearson optimal decision rule for H_0 versus H_1 in (4.36), then δ^* must be of the form of (4.37), except possibly on a set of $\mathbf{Y}$ values having zero probability under H_0 and H_1. A proof of the Neyman–Pearson lemma can be found in [51].

It is worth noting from (4.37) that the Neyman–Pearson optimal decision rules, which we will denote as $\delta^{(NP)}$, are also threshold tests on the LR $L(y)$ similar to the Bayesian optimal decision rules. The only difference is that now the optimal threshold η and the randomization parameter γ are chosen so that the false-alarm probability is equal to exactly α. The proof in [51] is constructive in that it also specifically gives a procedure to compute η and γ for a given α. Although we omit the detailed proof here, this procedure is illustrated in the example below and will also be used in

Chapters 10 and 12 in which we design Neyman–Pearson optimal decisions rules for spectral activity detection and primary signal detection, respectively.

Example 4.2 (Neyman–Pearson Optimal Detection of a Known Signal in Additive Gaussian Noise) Let us consider the simple hypothesis testing scenario of detecting a known signal $\mathbf{s}$ embedded in additive, zero-mean Gaussian noise $\mathbf{w} \sim \mathcal{N}(\mathbf{0}, \Sigma)$ considered earlier in Example 4.1:

$$\begin{aligned} H_0 &: \mathbf{y} = \mathbf{w} \\ H_1 &: \mathbf{y} = \mathbf{s} + \mathbf{w}, \end{aligned} \tag{4.39}$$

where $\mathbf{y}$, $\mathbf{s}$, and $\mathbf{w}$ are all assumed to be L-vectors. The LR $L(\mathbf{y})$ for (4.39) was computed in (4.22) so that from (4.37) the optimal Neyman–Pearson decision rule should be of the form of

$$\delta(\mathbf{y}) = \begin{cases} 1 & > \\ \gamma & \text{if } e^{\mathbf{s}^T\Sigma^{-1}\mathbf{y} - \frac{1}{2}\mathbf{s}^T\Sigma^{-1}\mathbf{s}} = \eta. \\ 0 & < \end{cases} \tag{4.40}$$

To obtain the optimal Neyman–Pearson test, all that we need is to compute the threshold η and the randomization parameter γ so that the false-alarm probability is exactly equal to a given threshold α. Due to the monotonicity of the exp(.) function, however, we may equivalently write (4.40) as

$$\delta(\mathbf{y}) = \begin{cases} 1 & > \\ \gamma & \text{if } \mathbf{s}^T\Sigma^{-1}\mathbf{y} = \tilde{\eta} \\ 0 & < \end{cases} \tag{4.41}$$

where $\tilde{\eta} = \log \eta + \frac{1}{2}\mathbf{s}^T\Sigma^{-1}\mathbf{s}$. For simplicity of illustration, let us consider the case of zero-mean, iid Gaussian noise so that $\Sigma = \sigma^2\mathbf{I}_L$. Then (4.41) becomes

$$\begin{aligned} \delta(\mathbf{y}) &= \begin{cases} 1 & > \\ \gamma & \text{if } \mathbf{s}^T\mathbf{y} = \eta' \\ 0 & < \end{cases} \\ &= \begin{cases} 1 & > \\ \gamma & \text{if } T(\mathbf{y}) = \eta' \\ 0 & < \end{cases} \end{aligned} \tag{4.42}$$

with $\eta' = \sigma^2 \log \eta + \frac{||\mathbf{s}||^2}{2}$ and the decision variable $T(.)$ is as defined in (4.26). For an α-level Neyman–Pearson optimal test, the false-alarm probability of (4.42) should be exactly equal to α:

$$\Pr\left(T(\mathbf{y}) > \eta' | H_0\right) + \gamma \Pr\left(T(\mathbf{y}) = \eta' | H_0\right) = \alpha. \tag{4.43}$$

As we saw in (4.27) and (4.29), $T(.)$ is a Gaussian random variable under the hypothesis H_0:

$$T|H_0 = \mathbf{s}^T\mathbf{W} \sim \mathcal{N}\left(0, \sigma^2||\mathbf{s}||^2\right). \tag{4.44}$$

Hence, $\Pr\left(T(\mathbf{y}) = \eta'|H_0\right) = 0$ for any η'. Thus, we see that randomization parameter γ does not affect the false-alarm probability in (4.43). Hence, we can arbitrarily set $\gamma = 1$. From (4.43) and (4.44),

$$\begin{aligned}\alpha &= \Pr\left(T(\mathbf{y}) \geq \eta'|H_0\right)\\ &= \Pr\left(\mathcal{N}\left(0, \sigma^2||\mathbf{s}||^2\right) \geq \eta'\right)\\ &= Q\left(\frac{\eta'}{\sigma||\mathbf{s}||}\right).\end{aligned} \tag{4.45}$$

Hence, the optimal threshold needed for test (4.42) is given by

$$\eta' = \sigma||\mathbf{s}||Q^{-1}(\alpha), \tag{4.46}$$

where $Q^{-1}(.)$ denotes the inverse Q-function. Thus, the optimal α-level Neyman–Pearson test for the simple hypothesis testing problem (4.39) is

$$\delta^{(NP)}(\mathbf{y}) = \begin{cases} 1 & \geq \\ & \text{if } \mathbf{s}^T\mathbf{y} \quad \sigma||\mathbf{s}||Q^{-1}(\alpha). \\ 0 & < \end{cases} \tag{4.47}$$

The detector (4.47) has the highest detection probability among all detectors not exceeding false-alarm probability $P_F = \alpha$. Indeed, we can easily compute this maximum detection probability achieved by the Neyman–Pearson optimal detector:

$$P_D = \Pr\left(T(\mathbf{y}) \geq \eta'|H_1\right) \tag{4.48a}$$

$$= \Pr\left(\mathcal{N}\left(||\mathbf{s}||^2, \sigma^2||\mathbf{s}||^2\right) \geq \eta'\right) \tag{4.48b}$$

$$\begin{aligned}&= Q\left(\frac{\eta' - ||\mathbf{s}||^2}{\sigma||\mathbf{s}||}\right)\\ &= Q\left(\frac{\sigma||\mathbf{s}||Q^{-1}(\alpha) - ||\mathbf{s}||^2}{\sigma||\mathbf{s}||}\right)\\ &= Q\left(Q^{-1}(\alpha) - \frac{||\mathbf{s}||}{\sigma}\right)\\ &= Q\left(Q^{-1}(\alpha) - \sqrt{SNR}\right),\end{aligned} \tag{4.48c}$$

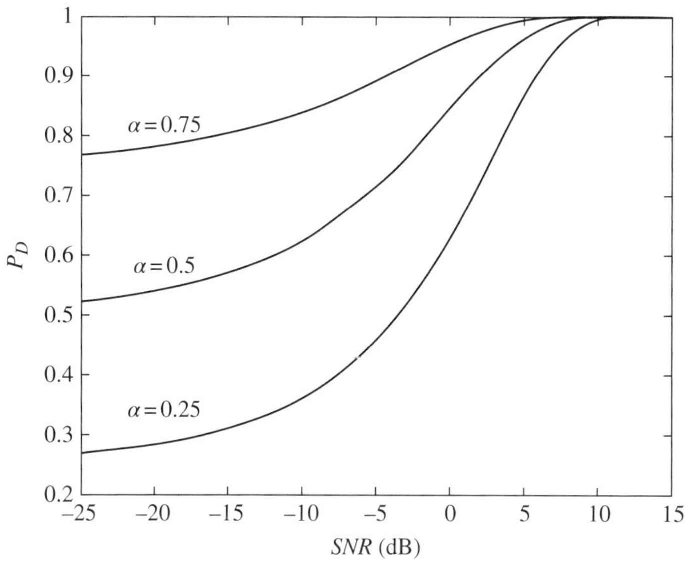

FIGURE 4.4 Probability of detection P_D of the optimal Neyman–Pearson detector as a function of SNR for a fixed false-alarm probability constraint α.

where in (4.48b) we have used the fact from (4.28) that $T(.)|H_1 \sim \mathcal{N}\left(||\mathbf{s}||^2, \sigma^2||\mathbf{s}||^2\right)$ and in (4.48c) we have used the SNR definition from (4.31).

Observe from (4.48c) that the maximum detection probability achieved by the size α Neyman–Pearson optimal detector is both a function of the false-alarm constraint α and the SNR. Figure 4.4 shows the probability of detection P_D as a function of SNR for a fixed false-alarm probability constraint α. According to (4.48c), $P_D \longrightarrow \alpha$ as $SNR \longrightarrow 0$. This is as expected, since $SNR \longrightarrow 0$ implies that $\mathbf{s} \longrightarrow \mathbf{0}$ so that essentially $H_1 \longrightarrow H_0$. On the other hand, for any fixed false-alarm probability constraint α, $P_D \longrightarrow 1$ as $SNR \longrightarrow \infty$.

Detection probability (4.48c) is shown in Figure 4.5 as a function of the false-alarm probability α for a fixed SNR. Such a figure is called the *receiver operating characteristic* (ROC) of the detector. Since $Q(x)$ is a monotonic decreasing function of x, we see from (4.48c) that for a fixed SNR, the optimal detection probability P_D monotonically increases with increasing α. Of course, this is as we expected from the basic trade-off between the false alarms and misses: If we are willing to tolerate more false alarms, we can reduce the misses leading to increased detection probability. ■

4.3.2.2 Composite Hypothesis Testing: Uniformly Most Powerful Neyman–Pearson Optimal Detection When the two hypotheses are not simple hypotheses as earlier, both detection and false-alarm probabilities can be functions of θ as noted in (4.33) and (4.34). If we want a size α Neyman–Pearson optimal detector for this general composite hypothesis testing problem, then false-alarm probability for which $\theta \in \Theta_0$ should be made not to exceed α? Naturally, it makes sense for this to be

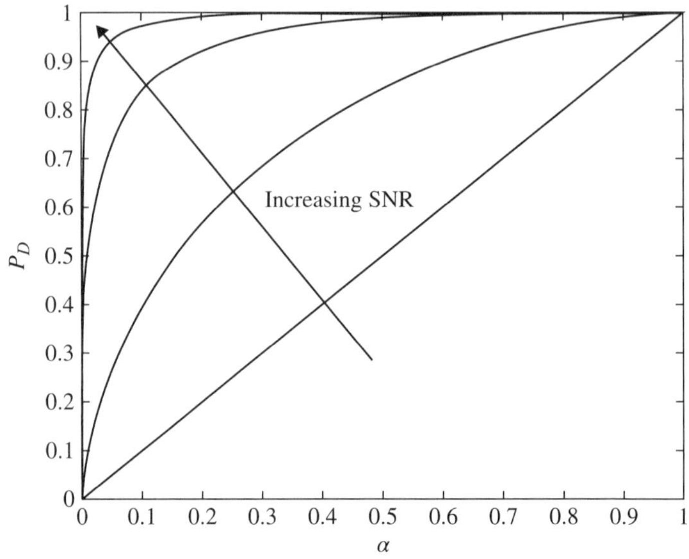

FIGURE 4.5 Receiver operating characteristic (ROC) of the optimal Neyman–Pearson detector.

the worst-case false-alarm probability in that it is the maximum allowed false-alarm probability for any $\theta \in \Theta_0$. That is, our interest is in decision rules δ that satisfy $\sup_{\theta\in\Theta_0} P_F(\delta,\theta) \leq \alpha$. Subject to this constraint, we would like to maximize the detection probability $P_D(\delta,\theta)$.

Hence, we may state the optimal Neyman–Pearson detection problem as follows:

$$\max_{\delta} \quad P_D(\delta,\theta) \quad \text{subject to} \quad \sup_{\theta\in\Theta_0} P_F(\delta,\theta) \leq \alpha. \tag{4.49}$$

Of course, since $P_D(\delta,\theta)$ is also a function of $\theta \in \Theta_1$, the solution to problem (4.49) may very well likely be a detector δ_θ that depends on the particular value of $\theta \in \Theta_1$. For example, suppose that H_0 hypothesis were to be simple as in the previous section so that $\Theta_0 = \{\theta_0\}$. Then for any fixed $\theta_1 \in \Theta_1$, the solution to (4.49) is given by the Neyman–Pearson lemma as applied for a simple hypothesis testing between $\theta = \theta_1$ and $\theta = \theta_0$. If we denote this Neyman–Pearson optimal decision rule as δ_{θ_1}, then it is the most powerful size α test for simple hypothesis H_0 versus the composite hypothesis H_1 when $\theta = \theta_1 \in \Theta_1$. If we now allow the hypothesis H_0 to be also composite and impose the worst-case false-alarm probability constraint as in (4.49), then still the optimal Neyman–Pearson detector for a fixed $\theta_1 \in \Theta_1$ can be expected to be dependent on this θ_1. For different values of $\theta \in \Theta_1$, the optimal test may be different and thus need to be denoted as δ_θ.

Since a priori we do not know the value of $\theta \in \Theta_1$ corresponding to a particular observation, then it is not clear which optimal Neyman–Pearson test δ_θ we should use. Of course, we would not have to worry about this if all these optimal tests δ_θ

happen to be the same test! That, however, would mean that there is a single size α test δ that is the *most powerful test* regardless of the value of $\theta \in \Theta_1$:

$$\max_{\delta} \; P_D(\delta,\theta) \quad \text{for all } \theta \in \Theta_1 \quad \text{subject to } \sup_{\theta\in\Theta_0} P_F(\delta,\theta) \leq \alpha. \tag{4.50}$$

If a solution δ existed for (4.50) that does not depend on $\theta \in \Theta_1$, it is called the *uniformly most powerful (UMP) test* or the UMP detector. Of course, it is to be expected that such tests may only exist under special circumstances. In general, finding a UMP detector might be difficult, and it may not even exist as in the following example.

Example 4.3 (Nonexistence of a UMP Test in the Presence of Additive Gaussian Noise) To see how a UMP test may not exist even in a seemingly simple scenario, consider the following composite hypothesis testing problem from [51]:

$$\begin{aligned} H_0 &: \mathbf{y} = \mathbf{w} \\ H_1 &: \mathbf{y} = \theta\mathbf{s} + \mathbf{w}, \end{aligned} \tag{4.51}$$

where, as before, $\mathbf{y}$, $\mathbf{s}$, and $\mathbf{w} \sim \mathcal{N}(\mathbf{0}, \Sigma)$ are all L-vectors and $\theta \neq 0$. In terms of parameter θ, we may express the hypothesis testing problem as

$$\begin{aligned} H_0 &: \theta = 0 \\ H_1 &: \theta \neq 0. \end{aligned} \tag{4.52}$$

Since, for a fixed $\theta \neq 0$,

$$\begin{aligned} H_0 &: \mathbf{Y} \sim \mathcal{N}(\mathbf{0}, \Sigma) \\ H_1 &: \mathbf{Y} \sim \mathcal{N}(\theta\mathbf{s}, \Sigma), \end{aligned}$$

the LR for (4.51) is

$$\begin{aligned} L(\mathbf{y}) &= \frac{f_1(\mathbf{y})}{f_0(\mathbf{y})} \\ &= \frac{\frac{1}{(2\pi)^{\frac{L}{2}}|\Sigma|^{\frac{1}{2}}} e^{-\frac{1}{2}(\mathbf{y}-\theta\mathbf{s})^T\Sigma^{-1}(\mathbf{y}-\theta\mathbf{s})}}{\frac{1}{(2\pi)^{\frac{L}{2}}|\Sigma|^{\frac{1}{2}}} e^{-\frac{1}{2}\mathbf{y}^T\Sigma^{-1}\mathbf{y}}} \\ &= e^{\theta\mathbf{s}^T\Sigma^{-1}\mathbf{y} - \frac{\theta^2}{2}\mathbf{s}^T\Sigma^{-1}\mathbf{s}}. \end{aligned} \tag{4.53}$$

The corresponding LLR is

$$\mathscr{L}(\mathbf{y}) = \theta\mathbf{s}^T\Sigma^{-1}\mathbf{y} - \frac{\theta^2}{2}\mathbf{s}^T\Sigma^{-1}\mathbf{s}.$$

For any fixed $\theta \neq 0$, (4.51) is a simple hypothesis testing problem. Hence, the optimal Neyman–Pearson test for any fixed $\theta \neq 0$ is of the form of

$$\delta^{(NP)}(\mathbf{y}) = \begin{cases} 1 & \\ & \text{if } \theta \mathbf{s}^T \Sigma^{-1} \mathbf{y} - \dfrac{\theta^2}{2} \mathbf{s}^T \Sigma^{-1} \mathbf{s} \quad \begin{matrix} \geq \\ \\ < \end{matrix} \quad \log \eta, \\ 0 & \end{cases} \tag{4.54}$$

where η is to be chosen to satisfy a given false-alarm constraint. For simplicity, let us consider zero-mean, iid Gaussian noise $\mathbf{w}$ so that $\Sigma = \sigma^2 \mathbf{I}_L$. Then the optimal test (4.54) reduces to

$$\delta^{(NP)}(\mathbf{y}) = \begin{cases} 1 & \\ & \text{if } \theta \dfrac{\mathbf{s}^T \mathbf{y}}{\sigma^2} - \theta^2 \dfrac{||\mathbf{s}||^2}{2\sigma^2} \quad \begin{matrix} \geq \\ \\ < \end{matrix} \quad \log \eta \\ 0 & \end{cases} \tag{4.55a}$$

$$= \begin{cases} 1 & \\ & \text{if } T_\theta(\mathbf{y}) \quad \begin{matrix} \geq \\ \\ < \end{matrix} \quad \log \eta, \\ 0 & \end{cases} \tag{4.55b}$$

where we have defined $T_\theta(\mathbf{y})$ as the decision statistic:

$$T_\theta(\mathbf{y}) = \theta \frac{\mathbf{s}^T \mathbf{y}}{\sigma^2} - \theta^2 \frac{||\mathbf{s}||^2}{2\sigma^2}. \tag{4.56}$$

For a known fixed θ, we may compute the above decision statistic T_θ and implement the optimal size α Neyman–Pearson decision rule. But if we do not know the fixed value of θ, we cannot evaluate the required decision statistic $T_\theta(\mathbf{y})$ in (4.56). Indeed, the problem (4.51) requires us to design an optimal detector that is valid for any $\theta \neq 0$.

However, observe from (4.55a) that for all $\theta > 0$, we may rewrite the optimal test as

$$\delta_1^{(NP)}(\mathbf{y}) = \begin{cases} 1 & \\ & \text{if } \mathbf{s}^T \mathbf{y} \quad \begin{matrix} \geq \\ \\ < \end{matrix} \quad \sigma^2 \dfrac{\log \eta}{\theta} + \theta \dfrac{||\mathbf{s}||^2}{2} \\ 0 & \end{cases}$$

$$= \begin{cases} 1 & \\ & \text{if } \mathbf{s}^T \mathbf{y} \quad \begin{matrix} \geq \\ \\ < \end{matrix} \quad \eta_{1,\theta}, \\ 0 & \end{cases} \tag{4.57}$$

where $\eta_{1,\theta}$ must be chosen so that the false-alarm probability of the test (4.57) is equal to α. However, note that the decision statistic in (4.57) is $T(\mathbf{y}) = \mathbf{s}^T\mathbf{y}$, and clearly, its distribution does not depend on θ under hypothesis H_0 (this is the same $T(\mathbf{y})$ in (4.27), and the hypothesis H_0 here is exactly the same as that in Example 4.2 earlier). Hence, the threshold $\eta_{1,\theta}$ required to ensure a false-alarm probability α in the test (4.57) will not depend on θ. Indeed,

$$\begin{aligned}\alpha &= \Pr(H_1|H_0)\\ &= \Pr\left(\mathbf{s}^T\mathbf{Y} \geq \eta_{1,\theta}|H_0\right)\\ &= \Pr\left(\mathbf{s}^T\mathbf{W} \geq \eta_{1,\theta}\right)\\ &= Q\left(\frac{\eta_{1,\theta}}{\sigma||\mathbf{s}||}\right).\end{aligned}$$

Hence, as long as $\theta > 0$, the optimal size α Neyman–Pearson test for problem (4.51) can be written in the form of (4.57) with

$$\eta_{1,\theta} = \sigma||\mathbf{s}||Q^{-1}(\alpha). \tag{4.58}$$

An interesting point is that $\eta_{1,\theta}$ does not depend on the specific value of θ as long as $\theta > 0$. Hence, as long as $\theta > 0$, the optimal size α Neyman–Pearson test for the problem (4.51) is the following fixed test:

$$\delta_1^{(NP)}(\mathbf{y}) = \begin{cases} 1 & \geq \\ & \text{if } \mathbf{s}^T\mathbf{y} \quad \eta_1, \\ 0 & < \end{cases} \tag{4.59}$$

where

$$\eta_1 \triangleq \sigma||\mathbf{s}||Q^{-1}(\alpha). \tag{4.60}$$

In other words, (4.59) is the most powerful size α test for (4.51) for all $\theta > 0$.

Figure 4.6 shows the decision regions of test $\delta_1^{(NP)}$ in the case of a scalar problem. It declares H_1 for those y values to the right of the threshold η_1 and declares H_0 for those y to the left of η_1. If $\theta_1 > 0$, then the density $f_{\theta_1}(y)$ will always be to the right of $f_0(y)$ as in Figure 4.6, so that the test $\delta_1^{(NP)}$ will be the optimal Neyman–Pearson test of size α.

Before we consider the most powerful tests for $\theta < 0$, we may also write down the detection probability of the detector $\delta_1^{(NP)}$. Indeed, under hypothesis H_1, the

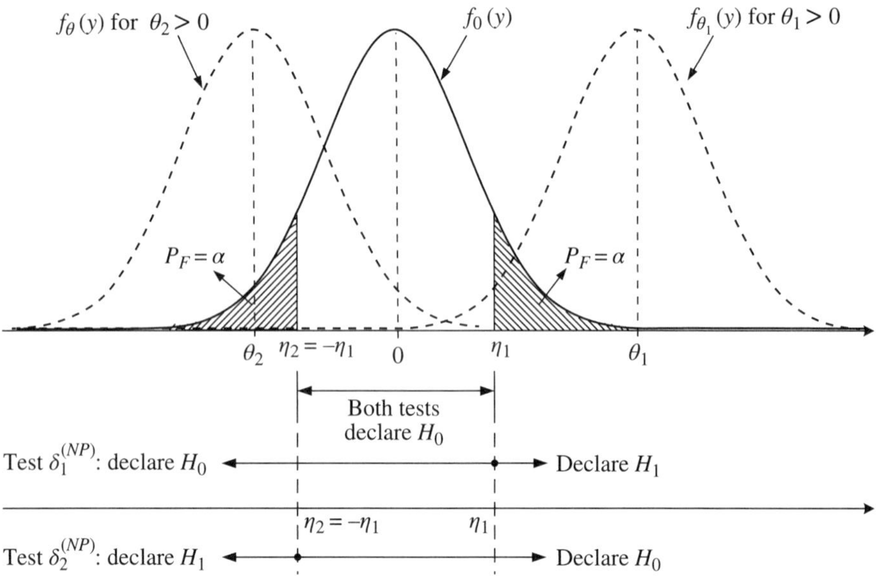

FIGURE 4.6 The test $\delta_1^{(NP)}$ is the uniformly most powerful test against H_0 for all $\theta > 0$. However, test $\delta_2^{(NP)}$ is the uniformly most powerful test against H_0 for all $\theta < 0$. The two tests are different, and thus, none of them is a uniformly most powerful test against H_0 for all $\theta \neq 0$.

decision statistic $T(\mathbf{y}) = \mathbf{s}^T\mathbf{y}$ is still a Gaussian random variable. Indeed, it is easy to see that

$$T(\mathbf{y})|H_1 \sim \mathcal{N}(\theta||\mathbf{s}||^2, \sigma^2||\mathbf{s}||^2). \tag{4.61}$$

Hence, the detection probability of the detector $\delta_1^{(NP)}$ is

$$\begin{aligned} P_{D,1} &= \Pr(T(\mathbf{y}) > \eta_1|H_1) \\ &= Q\left(\frac{\eta_1 - \theta||\mathbf{s}||^2}{\sigma||\mathbf{s}||}\right) \\ &= Q\left(Q^{-1}(\alpha) - \frac{\theta||\mathbf{s}||}{\sigma}\right) \\ &= Q\left(Q^{-1}(\alpha) - \theta\sqrt{SNR}\right). \end{aligned} \tag{4.62}$$

Note that the detection probability of the detector is a function of θ. Indeed, this is the reason why this detector may not be the most powerful detector for some values

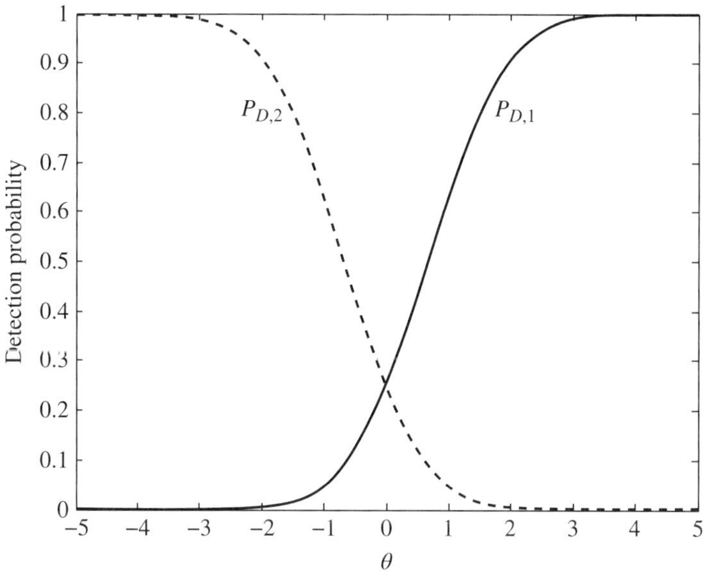

FIGURE 4.7 Detection probabilities $P_{D,1}$ and $P_{D,2}$ of the tests $\delta_1^{(NP)}$ and $\delta_2^{(NP)}$, respectively, as a function of θ for a fixed SNR =0 dB and $\alpha = 0.25$.

of θ. When $\theta > 0$, its effect on the detection probability in (4.62) is to increase the detection probability as can be seen from Figure 4.7. On the other hand, note that when $\theta < 0$, its effect is to decrease the detection probability $P_{D,1}$ as can be seen from (4.62).

Let us revisit (4.55a) to consider the case of $\theta < 0$. For any fixed $\theta < 0$, we can rewrite the optimal test (4.55a) as

$$\delta_2^{(NP)}(\mathbf{y}) = \begin{cases} 1 & < \\ & \text{if } \mathbf{s}^T\mathbf{y} \quad \sigma^2 \dfrac{\log \eta}{\theta} + \theta \dfrac{||\mathbf{s}||^2}{2} \\ 0 & \geq \end{cases}$$

$$= \begin{cases} 1 & < \\ & \text{if } \mathbf{s}^T\mathbf{y} \quad \eta_{2,\theta}, \\ 0 & \geq \end{cases} \tag{4.63}$$

where now $\eta_{2,\theta}$ must be chosen so that the false-alarm probability of the test (4.63) is equal to α. Again, note that the decision statistic in (4.63) is $T(\mathbf{y}) = \mathbf{s}^T\mathbf{y}$, which clearly does not depend on θ under the hypothesis H_0. Hence, the false-alarm probability of test (4.63) for a given threshold does not depend on the particular value of θ, and in turn, the threshold to achieve a required false-alarm probability also does not depend on θ:

$$\begin{aligned}
\alpha &= \Pr(H_1|H_0) \\
&= \Pr\left(\mathbf{s}^T\mathbf{Y} < \eta_{2,\theta}|H_0\right) \\
&= \Pr\left(\mathbf{s}^T\mathbf{W} < \eta_{2,\theta}\right) \\
&= 1 - \Pr\left(\mathbf{s}^T\mathbf{W} > \eta_{2,\theta}\right) \\
&= 1 - Q\left(\frac{\eta_{2,\theta}}{\sigma\,||\mathbf{s}||}\right).
\end{aligned}$$

Hence, as long as $\theta < 0$, the optimal threshold for a size α Neyman–Pearson test is

$$\eta_2 \triangleq \sigma\,||\mathbf{s}||Q^{-1}(1-\alpha), \tag{4.64}$$

and from (4.63), the most powerful size α test for $\theta < 0$ is

$$\delta_2^{(NP)}(\mathbf{y}) = \begin{cases} 1 & \\ & \text{if } \mathbf{s}^T\mathbf{y} \begin{matrix} < \\ \\ \geq \end{matrix} \eta_2 . \\ 0 & \end{cases} \tag{4.65}$$

Since decision statistic used in the test $\delta_2^{(NP)}$ is exactly the same as that in the test $\delta_1^{(NP)}$, the distribution of it under the hypothesis H_1 is still the same Gaussian distribution given in (4.61). Hence, the detection probability of the test $\delta_2^{(NP)}$ is

$$\begin{aligned}
P_{D,2} &= \Pr\left(T(\mathbf{y}) < \eta_2|H_1\right) \\
&= 1 - \Pr\left(T(\mathbf{y}) > \eta_2|H_1\right) \\
&= 1 - Q\left(\frac{\eta_2 - \theta\,||\mathbf{s}||^2}{\sigma\,||\mathbf{s}||}\right) \\
&= 1 - Q\left(Q^{-1}(1-\alpha) - \frac{\theta\,||\mathbf{s}||}{\sigma}\right) \\
&= 1 - Q\left(Q^{-1}(1-\alpha) - \theta\sqrt{SNR}\right),
\end{aligned} \tag{4.66}$$

where we have used (4.64) in obtaining the last step. Note that the dependence of this detection probability on θ is the complete opposite of that of $P_{D,1}$ of the detector $\delta_1^{(NP)}$ in (4.62). Indeed, when $\theta < 0$, the detection probability $P_{D,2}$ increases as can also be seen from Figure 4.7.

Figure 4.6 also shows the decision regions of test $\delta_2^{(NP)}$. It declares H_1 for those y values to the left of the threshold η_2 and declares H_0 for those y to the right of η_2. If $\theta_2 < 0$, then the likelihood $f_{\theta_2}(y)$ will always be to the left of $f_0(y)$ as in Figure 4.6, so that the test $\delta_2^{(NP)}$ will be the optimal Neyman–Pearson test of size α.

Thus, tests $\delta_1^{(NP)}(.)$ and $\delta_2^{(NP)}(.)$ given by (4.59) and (4.65), respectively, are the most powerful tests for the problem (4.51) for $\theta > 0$ and $\theta < 0$. Of course, these two tests are different, and thus, none of them is a UMP test for all $\theta \neq 0$ for the problem (4.51). Furthermore, by the uniqueness of the Neyman–Pearson optimal test, there cannot be any other test that can be the UMP test for all $\theta \neq 0$ and yet be different from $\delta_1^{(NP)}(.)$ and $\delta_2^{(NP)}(.)$, respectively, for $\theta > 0$ and $\theta < 0$ regions. Hence, for the problem (4.51) with $\theta \neq 0$, there is no UMP test. ∎

4.3.2.3 Composite Hypothesis Testing: Locally Most Powerful Neyman–Pearson Detection and the Generalized Neyman–Pearson Lemma The difficulty in finding, or the nonexistence of, UMP Neyman–Pearson tests prompts us to look for other optimality criteria in choosing a suitable detector for the composite hypothesis testing problem (4.5) in the non-Bayesian setting. Within the basic Neyman–Pearson framework, a more general, and very useful, approach is to forego the insistence that the test be UMP.

Indeed, consider the situation in which the null hypothesis H_0 of (4.5) is simple so that H_0 corresponds to $\theta = \theta_0$ with $\Theta_0 = \{\theta_0\}$ and the alternative hypothesis H_1 is composite with $\theta \in \Theta_1$. Since H_0 is simple, the false-alarm probability constraint is simply that $P_F(\delta,\theta_0) \leq \alpha$ where

$$P_F(\delta,\theta_0) = \mathbb{E}\{\delta(\mathbf{Y})|\theta = \theta_0\}. \tag{4.67}$$

Subject to this, our interest is to maximize the detection probability. As we have already pointed out, the difficulty is that the optimal detector can, in general, be a function of the particular value of $\theta \in \Theta_1$.

Consider an arbitrary detector δ of size α. Essentially, the detector δ is a partition of the space of Y values into two disjoint sets: One set of Y values is identified with $\theta = \theta_0$, while the remaining Y values are identified with a $\theta \in \Theta_1$ What $\theta \in \Theta_1$ values are most likely to produce Y values that will more often be mistaken as being one of those Y values corresponding to $\theta = \theta_0$? Under reasonable assumptions, they are those $\theta \in \Theta_1$ values that are in some sense most close to θ_0, that is, those θ values that are in a close neighborhood of θ_0. When $\theta \in \Theta_1$ is really further away from θ_0, the observations Y generated by such a θ may be correspondingly so different from those observations generated by $\theta = \theta_0$, the danger of any reasonable detector mistaking those Y's as being correspond to H_0 could be very small. That is, the detection probability could be very high with any reasonable detector when $\theta \in \Theta_1$ is really further away from θ_0. It is those $\theta \in \Theta_1$ that is most close to θ_0 that might make the detection probability of a detector very sensitive to the exact value of θ.

This observation motivates us to look for a most powerful detector for H_0 versus the composite hypothesis H_1 not uniformly for all $\theta \in \Theta_1$, but only for those $\theta \in \Theta_1$ that are in a close neighborhood of θ_0. Such a test is appropriately named a *locally most powerful* (LMP) or, simply, a *locally optimal* (LO) test. The justification for an LMP test is that the detection probability will be high enough with any reasonable detector for $\theta \in \Theta_1$ that is sufficiently away from θ_0 that it may not be necessary after all to insist on the highest achievable detection probability for those values of θ.

To obtain the structure of an LMP test, we may expand the power function $P_D(\delta,\theta)$ of an arbitrary decision rule δ in a Taylor series expansion around $\theta = \theta_0$ [51]:

$$P_D(\delta,\theta) = P_D(\delta,\theta_0) + (\theta-\theta_0)P'_D(\delta,\theta_0) + \frac{1}{2}(\theta-\theta_0)^2 P''_D(\delta,\theta_0) + \mathcal{O}\left((\theta-\theta_0)^3\right), \tag{4.68}$$

where we have denoted

$$P'_D(\delta,\theta_0) = \frac{\partial}{\partial\theta}\left[P_D(\delta,\theta)\right]\Big|_{\theta=\theta_0} \tag{4.69}$$

and

$$P''_D(\delta,\theta_0) = \frac{\partial^2}{\partial\theta^2}\left[P_D(\delta,\theta)\right]\Big|_{\theta=\theta_0}. \tag{4.70}$$

Observe from (4.67) that the first term in the Taylor series expansion (4.68), $P_D(\delta,\theta_0)$, is nothing but the false-alarm probability $P_F(\delta)$ corresponding to the decision rule δ. Hence, if we are looking to maximize the detection probability subject to the constraint α on false-alarm probability, the first term in the Taylor series expansion will be the same for all decision rules of interest.

To be specific, let us assume that the alternative hypothesis H_1 corresponds to $\theta > \theta_0$. Then as we see from (4.68), for those $\theta \in \Theta_1$ values very close to θ_0, the detection probability of a detector δ is improved by making the slope $P'_D(\delta,\theta_0)$ at $\theta = \theta_0$ large. Hence, if we want the detector that has the best detection probability for θ close to θ_0, then we should select the δ that has the highest slope in its detection probability at $\theta = \theta_0$. Specifically, then the locally optimum detection, or the LMP detection, problem can be written as

$$\max_{\delta} \; P'_D(\delta,\theta_0) \quad \text{subject to} \quad P_F(\delta) = \alpha. \tag{4.71}$$

The solution to the above optimization problem (4.71) will then be a size α test that has the highest detection probability among all size α detection rules for $\theta_0 < \theta < \theta_{\max}$ for some $\theta_{\max} \in \Theta_1$.

As it turned out, there is a simple characterization of the LMP decision rule that solves the constrained optimization problem (4.71). This is a consequence of the following generalization of the Neyman–Pearson lemma [52].

Theorem 4.2 (Generalized Neyman–Pearson Lemma) Let $g(\mathbf{y})$ and $\{h_m(\mathbf{y})\}_{m=1}^{M}$ be all real-valued integrable[1] functions defined on $\mathbb{R}^N$. Then the integrable function $\tilde{\delta}(\mathbf{y})$ defined on $\mathbb{R}^N$ that is the solution to the problem of

[1] A real-valued function $f(x)$ is said to be integrable if $\int_{-\infty}^{\infty} |f(x)|dx < \infty$.

$$\max_{\delta} \int_{\mathbb{R}^N} \delta(\mathbf{y}) g(\mathbf{y}) d\mathbf{y} \tag{4.72}$$

subject to the set of M constraints

$$\int_{\mathbb{R}^N} \delta(\mathbf{y}) h_m(\mathbf{y}) d\mathbf{y} = \alpha_m \qquad \text{for } m = 1, \ldots, M \tag{4.73}$$

is of the form of

$$\tilde{\delta}(\mathbf{y}) = \begin{cases} 1 & \text{if } g(\mathbf{y}) > \sum_{m=1}^{M} \eta_m h_m(\mathbf{y}) \\ \gamma(\mathbf{y}) & \text{if } g(\mathbf{y}) = \sum_{m=1}^{M} \eta_m h_m(\mathbf{y}) \\ 0 & \text{if } g(\mathbf{y}) < \sum_{m=1}^{M} \eta_m h_m(\mathbf{y}) \end{cases} \tag{4.74}$$

where $\eta_m \geq 0$, for $1 \leq m \leq M$, and $0 \leq \gamma(\mathbf{y}) \leq 1$ are chosen to satisfy the M constraints (4.73).

Proof: See [52] for a proof.

To cast the problem (4.71) of finding the LMP decision rule in a form that can be identified with the *generalized Neyman–Pearson lemma*, let us rewrite (4.71) explicitly as

$$\begin{aligned}
&\max_{\delta} P_D'(\delta, \theta_0) \text{ subject to } P_F(\delta) = \alpha \\
&\equiv \max_{\delta} \frac{\partial}{\partial \theta} [P_D(\delta, \theta)] \bigg|_{\theta=\theta_0} \text{ subject to } \int_{\mathbb{R}^N} \delta(y) f_{Y|\theta_0}(y|\theta_0) dy = \alpha \\
&\equiv \max_{\delta} \frac{\partial}{\partial \theta} \left[\int_{\mathbb{R}^N} \delta(y) f_{Y|\theta}(y|\theta) dy \right] \Bigg|_{\theta=\theta_0} \text{ subject to } \int_{\mathbb{R}^N} \delta(y) f_{Y|\theta_0}(y|\theta_0) dy = \alpha \\
&\equiv \max_{\delta} \left[\int_{\mathbb{R}^N} \delta(y) \frac{\partial}{\partial \theta} f_{Y|\theta}(y|\theta) \bigg|_{\theta=\theta_0} dy \right] \text{ subject to } \int_{\mathbb{R}^N} \delta(y) f_{Y|\theta_0}(y|\theta_0) dy = \alpha,
\end{aligned} \tag{4.75}$$

where in the last step we have assumed that the likelihood functions satisfy the required regularity conditions to allow interchange of integration and differentiation. Assuming that this is indeed the case, we can see that (4.75) is a constrained optimization problem of the type (4.72)–(4.73) considered in the generalized Neyman–Pearson lemma with

$$\begin{aligned} M &= 1 \\ \alpha_1 &= \alpha \\ h_1(\mathbf{y}) &= f_{Y|\theta_0}(y|\theta_0) = f_0(y) \\ g(\mathbf{y}) &= \frac{\partial}{\partial\theta} f_{Y|\theta}(y|\theta)\bigg|_{\theta=\theta_0}. \end{aligned}$$

Hence, it follows straightforwardly from the generalized Neyman–Pearson lemma that the LMP size α detector that is the solution to (4.71) must be of the form of

$$\delta^{(LO)}(y) = \begin{cases} 1 & > \\ \gamma \quad \text{if } \frac{\partial}{\partial\theta} f_{Y|\theta}(y|\theta)\Big|_{\theta=\theta_0} & = \eta f_{\theta_0}(y), \\ 0 & < \end{cases} \tag{4.76}$$

where $\eta \geq 0$ and $0 \leq \gamma \leq 1$ are to be chosen so that $P_F(\delta^{(LO)}) = \alpha$.

In many situations, the above LMP design based on maximizing the slope (the first derivative) of the detection probability at $\theta = \theta_0$ leads to useful detectors. But in some situations, it might happen that this slope $P'_D(\delta, \theta_0)$ is exactly zero at $\theta = \theta_0$. The LMP detector formulation in (4.76) then breaks down. However, by revisiting the Taylor series (4.68), we immediately see that in those situations we should maximize the second derivative $P''_D(\delta, \theta_0)$ subject to the false-alarm constraint. A little thought shows that again the detector will have the same form as in (4.76) with $P'_D(\delta, \theta_0)$ replaced by $P''_D(\delta, \theta_0)$. Indeed, we may continue this procedure until we have the first nonzero derivative term in the Taylor series expansion (4.68) and then design an LMP detector of the form of (4.76).

The LMP detector (4.76) for a simple null hypothesis can also equivalently be written as

$$\delta^{(LO)}(y) = \begin{cases} 1 & > \\ \gamma \quad \text{if } \frac{\partial}{\partial\theta} L_\theta(y)\Big|_{\theta=\theta_0} & = \eta, \\ 0 & < \end{cases} \tag{4.77}$$

where

$$L_\theta(y) = \frac{f_{Y|\theta}(y|\theta)}{f_{\theta_0}(y)} = \frac{f_\theta(y)}{f_{\theta_0}(y)}$$

denotes the LR between the simple hypothesis H_0 and the composite hypothesis H_1 for any $\theta \in \Theta_1$. Again, $\eta \geq 0$ and $0 \leq \gamma \leq 1$ are to be chosen so that $P_F(\delta^{(LO)}) = \alpha$.

4.3.3 Another Non-Bayesian Alternative: The Generalized Likelihood Ratio Test

When assumptions under which the generalized Neyman–Pearson lemma can be applied are not satisfied, the LO Neyman–Pearson tests of the form (4.76) and (4.77) may not be easy to find. An alternative approach to obtain a reasonable detector for the composite hypothesis testing problem (4.5) in such situations is to first consider the most likely θ values implied by a given observation y for the two scenarios of $\theta \in \Theta_0$ and $\theta \in \Theta_1$. Then we may treat the problem (4.5) as a simple hypothesis testing problem between the two hypotheses corresponding to these two θ values, whose solution is of course an LR test.

As wc will formally define in the next chapter, the most likely θ value implied by a given observation y is the value of θ that maximizes the likelihood function $f_{Y|\theta}(y|\theta)$ of observation $\mathbf{y}$. This is called the *maximum-likelihood* (ML) estimate of θ. Given a domain of possible values for θ, say, Θ, the ML estimate of parameter θ is thus defined to be

$$\hat{\theta}_{ML}(y) = \max_{\theta \in \Theta} f_{Y|\theta}(y|\theta). \tag{4.78}$$

As we will discuss later in Chapter 5, the ML estimate can be easy to find and may possess many attractive features. For now, let us denote by $\hat{\theta}_{ML,0}(y)$ and $\hat{\theta}_{ML,1}(y)$ the ML estimate of θ under H_0 and H_1 hypotheses, respectively.

Hence, our alternative approach to obtain a reasonable detector for the composite hypothesis testing problem (4.5) is to simply replace the unknown parameter θ in the likelihood functions by its ML estimates under both hypotheses H_0 and H_1. This leads to the following *generalized likelihood ratio* (GLR):

$$\begin{aligned} L_g(y) &= \frac{\max_{\theta \in \Theta_1} f_{Y|\theta}(y|\theta)}{\max_{\theta \in \Theta_0} f_{Y|\theta}(y|\theta)} \\ &= \frac{f_{Y|\theta}(y|\theta = \hat{\theta}_{ML,1}(y))}{f_{Y|\theta}(y|\theta = \hat{\theta}_{ML,0}(y))}. \end{aligned} \tag{4.79}$$

A *generalized likelihood ratio test* (GLRT) for H_0 versus H_1 is then obtained by comparing the above GLR $L_g(y)$ to a suitable threshold.

Example 4.4 (A GLRT for the Hypothesis Testing Problem Defined in Example 4.3) Recall Example 4.3 in which we showed that there is no UMP test for the following problem:

$$\begin{aligned} H_0 &: \mathbf{y} = \mathbf{w} \\ H_1 &: \mathbf{y} = \theta\mathbf{s} + \mathbf{w}, \end{aligned} \tag{4.80}$$

where $\mathbf{y}$, $\mathbf{s}$, and $\mathbf{w} \sim \mathcal{N}(\mathbf{0}, \Sigma)$ are all L-vectors and $\theta \neq 0$. The null hypothesis is simple with $\Theta_0 = \{0\}$, and the alternative hypothesis is composite with $\Theta_1 = \mathbb{R}/\{0\}$. In Example 4.3, we showed that although we can find UMP tests if $\Theta_1 = \mathbb{R}^+$ or $\Theta_1 = \mathbb{R}^-$, for the case of $\Theta_1 = \mathbb{R}/\{0\}$, there is no UMP test. Moreover, for $\Theta_1 = \mathbb{R}/\{0\}$, we cannot apply the LMP test formulation of (4.71) because it is not enough to consider a small neighborhood of θ such that either $\theta > 0$ or $\theta < 0$. Hence, we may instead attempt to derive a GLRT for this problem.

Since the null hypothesis is simple in problem (4.80), we only need to find the ML estimate of θ for the likelihood of $\mathbf{y}$ under H_1:

$$\begin{aligned}\hat{\theta}_{ML,1}(y) &= \max_{\theta\in\Theta_1} f_{\mathbf{Y}|\theta}(\mathbf{y}|\theta)\\ &= \max_{\theta\in\Theta_1} \frac{1}{(2\pi)^{\frac{L}{2}}|\Sigma|^{\frac{1}{2}}} e^{-\frac{1}{2}(\mathbf{y}-\theta\mathbf{s})^T\Sigma^{-1}(\mathbf{y}-\theta\mathbf{s})}\\ &= \min_{\theta\in\Theta_1}(\mathbf{y}-\theta\mathbf{s})^T\Sigma^{-1}(\mathbf{y}-\theta\mathbf{s})\\ &= \frac{\mathbf{s}^T\Sigma^{-1}\mathbf{y}}{\mathbf{s}^T\Sigma^{-1}\mathbf{s}}.\end{aligned} \tag{4.81}$$

Thus, the GLR (4.79) is

$$\begin{aligned}L_g(\mathbf{y}) &= \frac{f_{\mathbf{Y}|\theta}(\mathbf{y}|\theta = \hat{\theta}_{ML,1}(\mathbf{y}))}{f_0(\mathbf{y})}\\ &= \frac{\frac{1}{(2\pi)^{\frac{L}{2}}|\Sigma|^{\frac{1}{2}}} e^{-\frac{1}{2}(\mathbf{y}-\hat{\theta}_{ML,1}\mathbf{s})^T\Sigma^{-1}(\mathbf{y}-\hat{\theta}_{ML,1}\mathbf{s})}}{\frac{1}{(2\pi)^{\frac{L}{2}}|\Sigma|^{\frac{1}{2}}} e^{-\frac{1}{2}\mathbf{y}^T\Sigma^{-1}\mathbf{y}}}\\ &= e^{\hat{\theta}_{ML,1}\mathbf{s}^T\Sigma^{-1}\mathbf{y} - \frac{\hat{\theta}^2_{ML,1}}{2}\mathbf{s}^T\Sigma^{-1}\mathbf{s}}.\\ &= e^{\frac{(\mathbf{s}^T\Sigma^{-1}\mathbf{y})^2}{2\mathbf{s}^T\Sigma^{-1}\mathbf{s}}},\end{aligned} \tag{4.82}$$

where, in the last step, we have substituted from (4.81). Since exp(.) is a monotonic function and $\mathbf{s}^T\Sigma^{-1}\mathbf{s} > 0$ in lieu of the positive definiteness of the covariance matrix Σ, we may then write a GLRT for this problem as

$$\delta^{(g)}(\mathbf{y}) = \begin{cases} 1 & \geq \\ & \text{if } \left(\mathbf{s}^T\Sigma^{-1}\mathbf{y}\right)^2 \quad \log\eta, \\ 0 & < \end{cases} \tag{4.83}$$

where η can be chosen as desired.

Recall that the two Neyman–Pearson optimal tests derived for this same problem in Example 4.3, $\delta_1^{(NP)}(\mathbf{y})$ of (4.59) for $\theta > 0$ region and $\delta_2^{(NP)}(\mathbf{y})$ of (4.65) for $\theta < 0$ region, were both linear detectors: That is, their decision statistics were linear functions of the observation vector $\mathbf{y}$. On the other hand, the GLRT (4.83) is a quadratic detector because it uses a decision statistic that is a quadratic function of the observation vector $\mathbf{y}$. A little thought should convince that this indeed makes sense for a detector that has to be valid for any $\theta \neq 0$.

We may design the above GLRT detector (4.83) to be a size α detector by appropriately choosing the threshold η. Let us denote by T_g the decision statistic in the GLRT detector (4.83):

$$T_g(\mathbf{y}) = \left(\mathbf{s}^T \Sigma^{-1} \mathbf{y}\right)^2 \triangleq Z^2$$

where we have let

$$Z = \mathbf{s}^T \Sigma^{-1} \mathbf{y}.$$

It is easy to see that under hypothesis H_0, Z is a zero-mean Gaussian random variable with variance $\sigma_z^2 = \mathbf{s}^T \Sigma^{-1} \mathbf{s}$:

$$Z|H_0 \sim \mathcal{N}\left(0, \sigma_z^2\right),$$

where

$$\sigma_z^2 = \mathbf{s}^T \Sigma^{-1} \mathbf{s}. \tag{4.84}$$

If we denote the pdf's of Z and T_g under the hypothesis H_0 by $f_{Z,0}(z)$ and $f_{T_g,0}(t)$, respectively, then

$$\begin{aligned} f_{T_g,0}(t) &= \sum_{\{z:z=t^2\}} f_{Z,0}(z) \left| \frac{dz}{dt} \right| \qquad (4.85) \\ &= \begin{cases} \dfrac{1}{\sqrt{2\pi\sigma_z^2 t}} e^{-\frac{t}{2\sigma_z^2}} & \text{if} \quad t \geq 0 \\ 0 & \text{if} \quad t < 0 \end{cases}. \end{aligned}$$

Hence, for $\eta \geq 0$, it follows that

$$\begin{aligned} \Pr\left(T_g(\mathbf{y}) > \eta | H_0\right) &= \int_{\eta}^{\infty} f_{T_g,0}(t) dt \\ &= 2Q\left(\frac{\sqrt{\eta}}{\sigma_z}\right). \end{aligned}$$

Thus, for a size α detector, we should set the threshold η of the detector (4.83) to be

$$\eta = \sigma_z^2 \left(Q^{-1}\left(\frac{\alpha}{2}\right)\right)^2. \tag{4.86}$$

Under hypothesis H_1, Z is still a Gaussian random variable with the same variance σ_z^2 in (4.84). The mean, however, is no longer zero but equals to

$$m_z = \theta \mathbf{s}^T \Sigma^{-1} \mathbf{s}, \tag{4.87}$$

so that

$$Z|H_1 \sim \mathcal{N}\left(m_z, \sigma_z^2\right).$$

Hence, if we denote the pdf's of Z and T_g under the hypothesis H_1 by $f_{Z,1}(z)$ and $f_{T_g,1}(t)$, respectively, similar to (4.85), we obtain

$$\begin{aligned} f_{T_g,1}(t) &= \sum_{\{z:z=t^2\}} f_{Z,1}(z)\left|\frac{dz}{dt}\right| \\ &= \begin{cases} \dfrac{f_{Z,1}(\sqrt{t})}{2\sqrt{t}} + \dfrac{f_{Z,1}(-\sqrt{t})}{2\sqrt{t}} & \text{if } t \geq 0 \\ 0 & \text{if } t < 0 \end{cases} \\ &= \begin{cases} \dfrac{1}{2\sqrt{2\pi\sigma_z^2 t}}\left(e^{-\frac{\left(\sqrt{t}-m_z\right)^2}{2\sigma_z^2}} + e^{-\frac{\left(\sqrt{t}+m_z\right)^2}{2\sigma_z^2}}\right) & \text{if } t \geq 0 \\ 0 & \text{if } t < 0 \end{cases}. \end{aligned} \tag{4.88}$$

Thus, the detection probability of a size α GLRT of the form of (4.83) is

$$\begin{aligned} P_{D,g} &= \Pr\left(T_g(\mathbf{y}) > \eta | H_1\right) \\ &= \int_{\eta}^{\infty} f_{T_g,1}(t)dt \\ &= Q\left(\frac{\sqrt{\eta}-m_z}{\sigma_z}\right) + Q\left(\frac{\sqrt{\eta}+m_z}{\sigma_z}\right) \\ &= Q\left(Q^{-1}\left(\frac{\alpha}{2}\right) - \theta\sigma_z\right) + Q\left(Q^{-1}\left(\frac{\alpha}{2}\right) + \theta\sigma_z\right) \qquad (4.89a) \\ &= Q\left(Q^{-1}\left(\frac{\alpha}{2}\right) - \theta\sqrt{\mathbf{s}^T\Sigma^{-1}\mathbf{s}}\right) + Q\left(Q^{-1}\left(\frac{\alpha}{2}\right) + \theta\sqrt{\mathbf{s}^T\Sigma^{-1}\mathbf{s}}\right), \qquad (4.89b) \end{aligned}$$

where (4.89a) follows by substituting from (4.86) and (4.87) and the last step uses (4.84).

It is interesting to see how the detection performance of this size α GLRT compares with those of $\delta_1^{(NP)}$ and $\delta_2^{(NP)}$ detectors derived in Example 4.3. To that end, let us consider the iid Gaussian noise assumption of $\Sigma = \sigma^2\mathbf{I}_L$, as in Example 4.3. Then the detector (4.83) simplifies to

$$\delta^{(g)}(\mathbf{y}) = \begin{cases} 1 & \geq \\ & \text{if } \left(\mathbf{s}^T\mathbf{y}\right)^2 \quad \sigma^4 \log\eta, \\ 0 & < \end{cases} \tag{4.90}$$

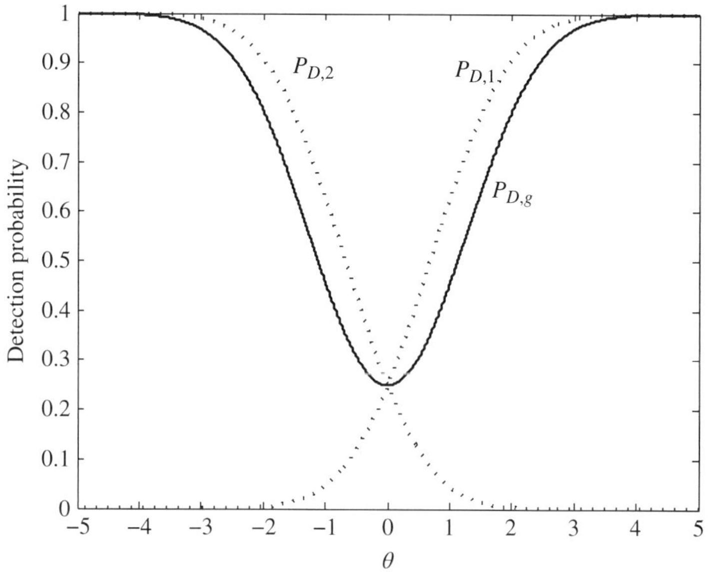

FIGURE 4.8 Comparison of the detection probabilities $P_{D,1}$ and $P_{D,2}$ of the tests $\delta_1^{(NP)}$ and $\delta_2^{(NP)}$, respectively, against that of the generalized likelihood ratio test $\delta^{(g)}$ of (4.90). Dependence is shown as a function of θ for fixed SNR = 0 dB and $\alpha = 0.25$.

showing that the decision statistic in GLRT is exactly the square of those used in the $\delta_1^{(NP)}(\mathbf{y})$ of (4.59) and $\delta_2^{(NP)}(\mathbf{y})$ of (4.65). The detection probability of a size α GLRT in this case can be obtained by simply substituting $\Sigma = \sigma^2 \mathbf{I}_L$ in (4.91):

$$\begin{aligned} P_{D,g} &= Q\left(Q^{-1}\left(\frac{\alpha}{2}\right) - \frac{\theta||\mathbf{s}||}{\sigma}\right) + Q\left(Q^{-1}\left(\frac{\alpha}{2}\right) + \frac{\theta||\mathbf{s}||}{\sigma}\right) \\ &= Q\left(Q^{-1}\left(\frac{\alpha}{2}\right) - \theta\sqrt{SNR}\right) + Q\left(Q^{-1}\left(\frac{\alpha}{2}\right) + \theta\sqrt{SNR}\right). \end{aligned} \tag{4.91}$$

Figure 4.8 shows the above detection probability along with those of the detectors $\delta_1^{(NP)}$ and $\delta_2^{(NP)}$ of Example 4.3. The GLRT strikes a balance between the performance of the two detectors that are UMP in certain regions while performing poorly outside of them. For $\theta > 0$, its detection performance is strictly below that of $\delta_1^{(NP)}$, which is of course the UMP detector for $\theta > 0$. However, for $\theta < 0$, it beats the performance of $\delta_1^{(NP)}$. Clearly, the performance of $\delta_1^{(NP)}$ is unacceptably poor for $\theta < 0$, while the GLRT maintains symmetrically the same performance as for $\theta > 0$. The same is true compared to $\delta_2^{(NP)}$ as well. While its performance is strictly below that of $\delta_2^{(NP)}$ for $\theta < 0$, it beats $\delta_2^{(NP)}$ for $\theta > 0$. ■

4.3.4 Parametric Signal Detection in Additive Noise

In the following, we may apply the concepts developed in the previous section to the specific case of signal detection in additive noise. Of course, the examples we have discussed so far have considered the additive Gaussian noise, but here, we would like to allow both non-Gaussian noise and different assumptions on the signal to be detected.

A common scenario encountered in practice is that of having to detect a signal s_n corrupted by additive noise w_n:

$$y_n = \theta s_n + w_n \quad \text{for } n = 1, \ldots, N, \tag{4.92}$$

where we may assume $\theta > 0$. The additive noise w_n has an associated probability density (or mass) function denoted by $f_{W_n}(w_n)$. Without loss of generality, we may assume that noise w_n has zero mean for all n. When noise is identically distributed, we may let $f_{W_n}(w_n) = f_W(w_n)$ for all $n = 1, \ldots, N$. By defining $\mathbf{y} = (y_1, \ldots, y_N)^T$, $\mathbf{w} = (w_1, \ldots, w_N)^T$, and $\mathbf{s} = (s_1, \ldots, s_N)^T$, we may rewrite (4.92) as a vector observation model:

$$\mathbf{y} = \theta \mathbf{s} + \mathbf{w}. \tag{4.93}$$

The joint density of noise vector $\mathbf{w}$ can then be denoted by $f_{\mathbf{W}}(\mathbf{w})$. If noise samples were to be independent, it is clear that

$$f_{\mathbf{W}}(\mathbf{w}) = \prod_{n=1}^{N} f_{W_n}(w_n). \tag{4.94}$$

The signal detection problem posed in (4.93), or equivalently in (4.92), is to determine when the observation $\mathbf{y}$ indeed contains the signal $\mathbf{s}$. Hence, one may pose a binary hypothesis testing problem as follows:

$$\begin{aligned} H_0 : \mathbf{y} &= \mathbf{w} \\ &\text{versus} \\ H_1 : \mathbf{y} &= \theta \mathbf{s} + \mathbf{w}. \end{aligned} \tag{4.95}$$

In other words, $\theta = 0$ corresponds to the hypothesis H_0, while $\theta > 0$ corresponds to the hypothesis H_1:

$$\begin{aligned} H_0 : \theta &= 0 \\ &\text{versus} \\ H_1 : \theta &> 0. \end{aligned} \tag{4.96}$$

Clearly, the hypothesis H_1 is composite.

Depending on assumptions on what information we have about the signal, noise, and parameter θ, the basic structure of the optimal detector may vary. Perhaps, the most important two cases to be considered on noise are Gaussian and non-Gaussian noise. The detection theory literature in the context of communications usually consider the following three scenarios for the assumptions on the signal $\mathbf{s}$ to be detected [51]:

- Signal $\mathbf{s}$ is known deterministically.
- Signal $\mathbf{s}$ is known deterministically except for a few unknown associated parameters.[2]
- Signal $\mathbf{s}$ is unknown except for its distribution.

In the following, we will consider the problem (4.95) under the first and last assumptions on the signal to be detected since the second case can be handled by appropriate modifications to the methods developed for the other two cases. The reader is referred to [51] for more details and explicit consideration of the second scenario. Since both Bayesian and Neyman–Pearson tests are LR tests, we will present the discussion in terms of one of them, namely, the Neyman–Pearson. The Bayesian tests can be written down by replacing the corresponding thresholds as needed.

4.3.4.1 Detection of a Known Signal in Additive Gaussian Noise Let us start by assuming that signal $\mathbf{s}$ is a known vector (or a sequence of known signal samples) and the additive noise is $\mathbf{W} \sim \mathcal{N}(\mathbf{0}, \Sigma_{\mathbf{W}})$ in problem (4.95). This is the same problem (4.51) considered in Example 4.3 but with $\theta > 0$ under the hypothesis H_1. The LR in this case simplifies as

$$L(\mathbf{y}) = \frac{f_{\mathbf{Y}|\theta}(\mathbf{y}|\theta)}{f_{\mathbf{Y}|\theta=0}(\mathbf{y}|\theta = 0)} \tag{4.97a}$$

$$= \frac{f_{\mathbf{W}}(\mathbf{y} - \mathbf{s}\theta)}{f_{\mathbf{W}}(\mathbf{y})} \tag{4.97b}$$

$$= e^{\mathbf{y}^T \Sigma_{\mathbf{W}}^{-1} \mathbf{s}\theta - \frac{\theta^2}{2} \mathbf{s}^T \Sigma_{\mathbf{W}}^{-1} \mathbf{s}}.$$

The corresponding LLR is

$$\mathcal{L}(\mathbf{y}) = \mathbf{y}^T \Sigma_{\mathbf{W}}^{-1} \mathbf{s}\theta - \frac{\theta^2}{2} \mathbf{s}^T \Sigma_{\mathbf{W}}^{-1} \mathbf{s}, \tag{4.98}$$

[2]A common example of this situation is when $\mathbf{s}$ corresponds to a carrier-modulated signal in which the phase of the carrier is unknown or uncertain. When there does not seem to be a simple way to absorb this phase angle also to the parameter θ (which can in general be a vector θ), it then makes sense to treat $\mathbf{s}$ as being known up to some associated parameters, in this case the phase angle of the carrier. In addition, sometimes, the carrier amplitude may also be not known. If only unknown quantity was the carrier amplitude, however, we can again reduce this to the case of a known signal.

and the optimal detectors can be designed by comparing the LLR to a threshold $\log \eta$, if η is the original threshold for the LR-based detector. After a little rearranging, the optimal detectors can be written as

$$\delta(\mathbf{y}) = \begin{cases} 1 & > \\ \gamma & \text{if } \mathbf{y}^T \Sigma_{\mathbf{W}}^{-1} \mathbf{s} = \dfrac{\log \eta}{\theta} + \dfrac{\theta}{2} \mathbf{s}^T \Sigma_{\mathbf{W}}^{-1} \mathbf{s}, \\ 0 & < \end{cases} \tag{4.99}$$

where $\eta \geq 0$ and $0 \leq \gamma \leq 1$ are to be chosen so that $P_F(\delta) = \alpha$.

The decision statistic used in detector (4.99),

$$T(\mathbf{y}) = \mathbf{y}^T \Sigma_{\mathbf{W}}^{-1} \mathbf{s}, \tag{4.100}$$

is a projection of the observation vector onto the vector

$$\mathbf{x} = \Sigma_{\mathbf{W}}^{-1} \mathbf{s}. \tag{4.101}$$

Such a projection $\mathbf{x}^T \mathbf{y} = \sum_{n=1}^{N} x_n y_n$ is called a *linear correlation* of the two vectors $\mathbf{x}$ and $\mathbf{y}$. Hence, we see that the optimal detector for a known signal in Gaussian noise is a threshold test on the output of a *correlator* in which the observation is correlated with a transformed version $\mathbf{x} = \Sigma_{\mathbf{W}}^{-1} \mathbf{s}$ of the known signal $\mathbf{s}$. The correlator can, however, be implemented as a discrete-time filter with an impulse response function $\mathbf{h}$ where

$$h_n = \begin{cases} x_{N-n} & \text{if } 0 \leq n \leq N-1 \\ 0 & \text{otherwise} \end{cases}. \tag{4.102}$$

Indeed, note that the output at time N of a filter having the impulse response function (4.102) to the input $\mathbf{y}$ is given by

$$\begin{aligned} \sum_{n=-\infty}^{\infty} h_{N-n} y_n &= \sum_{n=1}^{N} h_{N-n} y_n \\ &= \sum_{n'=0}^{N-1} h_{n'} y_{N-n'} \\ &= \sum_{n=0}^{N-1} x_{N-n} y_{N-n} \qquad (4.103\text{a}) \\ &= \sum_{n=1}^{N} x_n y_n = \mathbf{y}^T \mathbf{x} \\ &= T(\mathbf{y}), \qquad (4.103\text{b}) \end{aligned}$$

where (4.103a) follows from the definition of $\mathbf{h}$ given in (4.102). Thus, from (4.103b), we see that indeed the correlation decision statistic $T(\mathbf{y})$ is the output of a filter having the impulse response function (4.102) when we give the observation vector $\mathbf{y}$ as its input. Since the impulse response function of this filter is essentially determined by the transformed (known) signal $\mathbf{x}$, it is called a filter matched to signal $\mathbf{x}$ or the MF for signal $\mathbf{x}$. If noise were to be white so that $\Sigma_{\mathbf{W}} = \sigma_w^2 \mathbf{I}_L$, then the optimal detector for the known signal $\mathbf{s}$ is directly a threshold test on the output of the MF for $\mathbf{s}$.

We have already seen that the decision statistic $T(\mathbf{y})$ used in detector (4.99) is a Gaussian random variable under both hypotheses H_0 and H_1 so that the randomization is unnecessary in (4.99). Indeed, since under H_0

$$T(\mathbf{y})|H_0 \sim \mathcal{N}\left(\mathbf{0}, \mathbf{s}^T \Sigma_{\mathbf{W}}^{-1} \mathbf{s}\right),$$

for any threshold τ,

$$P(T(\mathbf{y}) > \tau|H_0) = Q\left(\frac{\tau}{\sqrt{\mathbf{s}^T \Sigma_{\mathbf{W}}^{-1} \mathbf{s}}}\right). \tag{4.104}$$

Thus, any false-alarm probability $0 \le \alpha \le 1$ can be achieved without the need for randomization by setting $\tau = \sqrt{\mathbf{s}^T \Sigma_{\mathbf{W}}^{-1} \mathbf{s}} Q^{-1}(\alpha)$. Thus, the most powerful α-level test is given by

$$\delta(\mathbf{y}) = \begin{cases} 1 & \quad \geq \\ & \text{if } \mathbf{y}^T \Sigma_{\mathbf{W}}^{-1} \mathbf{s} \quad \sqrt{\mathbf{s}^T \Sigma_{\mathbf{W}}^{-1} \mathbf{s}} Q^{-1}(\alpha). \\ 0 & \quad < \end{cases} \tag{4.105}$$

Since the above test does not depend on the particular value of θ, it follows that this test is indeed the UMP test for any $\theta > 0$. Hence, it is the UMP α-level test for $\theta > 0$ as we saw earlier in Example 4.3.

It is easy to see that

$$T(\mathbf{y})|H_1 \sim \mathcal{N}\left(\theta \mathbf{s}^T \Sigma_{\mathbf{W}}^{-1} \mathbf{s}, \mathbf{s}^T \Sigma_{\mathbf{W}}^{-1} \mathbf{s}\right).$$

Hence, the detection probability of the detector (4.105) is

$$\begin{aligned} P_D &= \Pr\left(T(\mathbf{y}) > \tau|H_1\right) \\ &= Q\left(\frac{\tau - \theta \mathbf{s}^T \Sigma_{\mathbf{W}}^{-1} \mathbf{s}}{\sqrt{\mathbf{s}^T \Sigma_{\mathbf{W}}^{-1} \mathbf{s}}}\right) \\ &= Q\left(Q^{-1}(\alpha) - \theta \sqrt{\mathbf{s}^T \Sigma_{\mathbf{W}}^{-1} \mathbf{s}}\right). \end{aligned} \tag{4.106}$$

4.3.4.2 Detection of a Known Signal in Additive Non-Gaussian Noise Now, let us assume that the additive noise $\mathbf{w}$ in the problem (4.95) has an arbitrary non-Gaussian density $f_{\mathbf{W}}(\mathbf{w})$ although still the signal $\mathbf{s}$ is a known vector as earlier. For an arbitrary noise density function $f_{\mathbf{W}}(\mathbf{w})$, we cannot expect to obtain an LR, or an LLR, that will allow us to easily decouple θ from the decision statistic $T(\mathbf{y})$. Indeed for the composite signal detection problem in (4.95), or (4.96), with a known signal $\mathbf{s}$, (4.97b) is as far as we can go for an arbitrary noise distribution. When this is the case, UMP tests may not exist, or it may be difficult to find or implement them even if they did exist.

To make any progress beyond (4.97b), we need to make some assumptions. Let us first assume that the noise samples are independent for each n, for $1 \leq n \leq N$, so that the corresponding LR can be factorized as

$$L(\mathbf{y}) = \prod_{n=1}^{N} \frac{f_{W_n}(y_n - \theta s_n)}{f_{W_n}(y_n)}, \tag{4.107}$$

to obtain an LLR of the form of

$$\mathcal{L}(\mathbf{y}) = \sum_{n=1}^{N} \log \frac{f_{W_n}(y_n - \theta s_n)}{f_{W_n}(y_n)}. \tag{4.108}$$

The assumption of independent noise samples, thus, essentially allows the decision statistic to be reduced to the form of applying a memoryless (instantaneous) nonlinearity to each sample followed by an accumulator as can be seen from (4.108). If noise samples are nonidentical, however, the memoryless nonlinearity can in general be time varying:

$$\mathcal{L}(\mathbf{y}) = \sum_{n=1}^{N} \mathcal{L}_n(y_n, s_n, \theta), \tag{4.109}$$

where $\mathcal{L}_n(y_n, s_n, \theta) = \log \frac{f_{W_n}(y_n - \theta s_n)}{f_{W_n}(y_n)}$ is the sample LLR that is the memoryless but time-varying nonlinearity.

If the noise samples were to be also identically distributed (iid noise), the nonlinear function to be applied to each observation sample is the same. In other words, the LLR can be written in the form of

$$\mathcal{L}(\mathbf{y}) = \sum_{n=1}^{N} \mathcal{L}(y_n, s_n, \theta), \tag{4.110}$$

where $\mathcal{L}(y_n, s_n, \theta) = \log \frac{f_W(y_n - \theta s_n)}{f_W(y_n)}$ is the time-invariant sample LLR.

Unlike in the case of a known signal in Gaussian noise, the sample LLR in (4.109) or (4.110) may not simplify enough to allow decoupling of θ from y_n easily. Then implementation of an optimal detector may require knowledge of the exact θ. In other words, a detector may only be optimal for the particular θ value assumed in obtaining the optimal detector. For observations that do not correspond to this assumed θ value, the decisions given by the detector may not be optimal. Hence, UMP tests may not exist or may be hard to find.

When noise is non-Gaussian so that UMP tests cannot be found or they do not exist, we may resort to an LO detector or a GLRT detector. These detectors can be particularly attractive in practice since they can lead to simple detector implementations. Moreover, the LO detectors are well suited for the low SNR region operation.

From (4.77), we may readily write down the form of the LO detector for (4.95):

$$\delta^{(LO)}(y) = \begin{cases} 1 & \\ \gamma & \text{if } \left.\frac{\partial}{\partial\theta} L(\mathbf{y},\theta)\right|_{\theta=0} \begin{matrix} > \\ = \\ < \end{matrix} \eta, \\ 0 & \end{cases} \tag{4.111}$$

where

$$\begin{aligned} L(\mathbf{y},\theta) &= \frac{f_{\mathbf{Y}|\theta}(\mathbf{y}|\theta)}{f_0(\mathbf{y})} \\ &= \frac{f_\theta(\mathbf{y})}{f_0(\mathbf{y})} \\ &= \prod_{n=1}^{N} \frac{f_W(y_n - \theta s_n)}{f_W(y_n)} \\ &= \prod_{k-1}^{N} L_n(y_n,\theta) \end{aligned} \tag{4.112}$$

and we have denoted the sample LR as

$$L_n(y_n,\theta) = \frac{f_W(y_n - \theta s_n)}{f_W(y_n)}. \tag{4.113}$$

Note that $L_n(y_n,\theta)|_{\theta=0} = 1$ for all $n = 1,\ldots,N$. Hence,

$$\begin{aligned} \left.\frac{\partial}{\partial\theta} L(\mathbf{y},\theta)\right|_{\theta=0} &= \sum_{j=1}^{N} \left(\prod_{n=1,n\neq j}^{N} L_n(y_n,\theta) \right) \left.\frac{\partial}{\partial\theta} L_j(y_j,\theta)\right|_{\theta=0} \\ &= \sum_{n=1}^{N} \left.\frac{\partial}{\partial\theta} L_n(y_n,\theta)\right|_{\theta=0}. \end{aligned} \tag{4.114}$$

But from (4.113),

$$\left.\frac{\partial}{\partial\theta} L_n(y_n,\theta)\right|_{\theta=0} = -s_n \frac{f'_W(y_n)}{f_W(y_n)} \tag{4.115a}$$

$$= s_n g^{(LO)}(y_n), \tag{4.115b}$$

where we have defined the memoryless LO nonlinearity $g^{(LO)}(x)$ as

$$g^{(LO)}(x) = -\frac{f'_W(x)}{f_W(x)}. \tag{4.116}$$

By substituting (4.115b) and (4.114) in (4.111), we may then write the LO detector as

$$\begin{aligned}
\delta^{(LO)}(y) &= \begin{cases} 1 & \qquad > \\ \gamma & \text{if } \sum_{n=1}^{N} \frac{\partial}{\partial\theta} L_n(y_n,\theta)\Big|_{\theta=0} = \eta \\ 0 & \qquad < \end{cases} \\
&= \begin{cases} 1 & \qquad > \\ \gamma & \text{if } \sum_{n=1}^{N} s_n g^{(LO)}(y_n) = \eta \\ 0 & \qquad < \end{cases} \\
&= \begin{cases} 1 & \qquad > \\ \gamma & \text{if } T_{LO}(\mathbf{y}) = \eta, \\ 0 & \qquad < \end{cases}
\end{aligned} \tag{4.117}$$

where the decision statistic defined as

$$T_{LO}(\mathbf{y}) = \sum_{n=1}^{N} s_n g^{(LO)}(y_n) \tag{4.118}$$

is called the *generalized correlator* with $g^{(LO)}(x)$ given by (4.116).

Note that compared to the MF (linear correlator) that directly correlates the received observation y_n's with the known signal sequence s_n (or a linear transformation of it), the generalized correlator applies a nonlinearity $g^{(LO)}(.)$ to the observation y_n before correlating y_n's with the known signal sequence s_n. Of course, if noise happens to be Gaussian so that $f_W(.)$ is the Gaussian density function, then $g^{(LO)}(y_n) = -\frac{f'_W(y_n)}{f_W(y_n)}$ simplifies to $g^{(LO)}(y_n) = \frac{y_n}{\sigma^2}$ so that we obtain the linear correlation as expected.

It is also worth noting that $L_n(y_n,\theta)|_{\theta=0} = 1$ for any $n = 1,\ldots,N$ implies that $\frac{\partial}{\partial\theta}\log L_n(y_n,\theta)\big|_{\theta=0} = \frac{\partial}{\partial\theta}L_n(y_n,\theta)\big|_{\theta=0}$. Summing this over all n then gives

$$\begin{aligned}
\frac{\partial}{\partial\theta}\log L(\mathbf{y},\theta)\Big|_{\theta=0} &= \sum_{n=1}^{N}\frac{\partial}{\partial\theta}\log L_n(y_n,\theta)\Big|_{\theta=0} = \sum_{n=1}^{N}\frac{\partial}{\partial\theta}L_n(y_n,\theta)\Big|_{\theta=0} \\
&= \frac{\partial}{\partial\theta}L(\mathbf{y},\theta)\Big|_{\theta=0}.
\end{aligned}$$

Hence, one may equivalently write the LO detector (4.117) as

$$\delta^{(LO)}(y) = \begin{cases} 1 & > \\ \gamma & \text{if } \sum_{n=1}^{N} \frac{\partial}{\partial\theta} \log L_n(y_n,\theta)\Big|_{\theta=0} = \eta. \\ 0 & < \end{cases}$$

In summary, the optimal detector for detecting a known signal in Gaussian noise is the linear correlator followed by a thresholding, whereas the optimal detector for detecting a known signal in non-Gaussian noise is the generalized correlator followed by a thresholding.

Example 4.5 (Detection of a Known Signal in Laplacian Noise) Let us consider the composite hypothesis testing problem of (4.95) when noise $\mathbf{w}$ is iid Laplacian (or double-exponential) distributed whose density can be written as

$$f_W(w) = \frac{\lambda}{2} e^{-\lambda|w|} \quad \text{for } -\infty < w < \infty, \tag{4.119}$$

where $\lambda > 0$. It is easy to see that

$$\begin{aligned} L(\mathbf{y},\theta) &= \prod_{n=1}^{N} L_n(y_n,\theta) \\ &= \prod_{n=1}^{N} \frac{f_W(y_n - \theta s_n)}{f_W(y_n)} \\ &= \prod_{n=1}^{N} e^{\lambda(|y_n| - |y_n - \theta s_n|)}, \end{aligned}$$

where $L_n(y_n,\theta) = e^{\lambda(|y_n| - |y_n - \theta s_n|)}$. Hence, the LLR (4.108) is

$$\mathscr{L}(\mathbf{y}) = \lambda \sum_{n=1}^{N} |y_n| - |y_n - \theta s_n|. \tag{4.120}$$

Note that the dependence of sample LLR $\log L_n(y_n,s_n,\theta) = \lambda\,(|y_n| - |y_n - \theta s_n|)$ on θ and y_n cannot be decoupled, meaning that direct implementation of the optimal nonlinearity (4.120) requires the knowledge of θ.

However, clearly, $L_n(y_n,\theta)|_{\theta=0} = 1$ for all $n = 1,\ldots,N$, and

$$\begin{aligned} \frac{\partial}{\partial\theta} L(\mathbf{y},\theta)\Big|_{\theta=0} &= \sum_{n=1}^{N} \frac{\partial}{\partial\theta} L_n(y_n,\theta)\Big|_{\theta=0} \\ &= \lambda \sum_{n=1}^{N} s_n sgn(y_n), \end{aligned}$$

where sgn(.) denotes the sign function, so that the LO detector is of the form of

$$\delta^{(LO)}(y) = \begin{cases} 1 & > \\ \gamma & \text{if } \lambda \sum_{n=1}^{N} s_n sgn(y_n) = \eta. \\ 0 & < \end{cases}$$

The LO nonlinearity in this case is $g^{(LO)}(y_n) = \lambda sgn(y_n)$. The generalized correlator $T_{LO}(\mathbf{y})$ thus correlates the known signal sequence with the sequence $\{\lambda sgn(y_n)\}_{n=1}^{N}$ whose each element only takes the values $\pm\lambda$ depending on the polarity of the signal samples. ■

4.3.4.3 Detection of a Random Signal in Additive Non-Gaussian Noise So far, we have assumed that the signal we try to detect is deterministically known to the detector. There are many situations in which this may not be true. Sometimes, we may know the basic structure of the signal except for perhaps a few signal parameters. An example of this is modulated information signals at a wireless receiver. The receiver may know the basic waveform except, for example, for the signal amplitude and phase. There are also other situations in which the detector may not know anything about the structure of the signal. A good example is the radio astronomical signals that a detector attempts to pick up. Another example is the wideband spectrum sensing problem that we are interested in this book. A wideband cognitive radio that operates over a wide frequency range may not know much about the form of the signals it is attempting to detect, or there may be too many different types of possible signals to account for. In these, and many other, situations, it is common to model the signal of interest as being a random process.

If we model the signal of interest as a random process, we may rewrite the detection problem (4.95) as

$$\begin{aligned} H_0 : \mathbf{Y} &= \mathbf{W} \\ &\text{versus} \\ H_1 : \mathbf{Y} &= \theta\mathbf{S} + \mathbf{W}, \end{aligned} \tag{4.121}$$

where $\mathbf{Y} = (Y_1, \ldots, Y_N)^T$ is the received signal vector, $\mathbf{S} = (S_1, \ldots, S_N)^T$ is the samples of the random signal (process) of interest, and $\mathbf{W} = (W_1, \ldots, W_N)^T$ is taken to be the zero-mean additive noise with, in general, a non-Gaussian multivariate pdf $f_{\mathbf{W}}(\mathbf{w})$. The model (4.121) already implies that the null hypothesis is simple and corresponds to $\theta = 0$. Let us assume that the signal of interest, $\mathbf{S}$, has an associated pdf $f_{\mathbf{S}}(\mathbf{s})$. We will denote by $\bar{\mathbf{s}}$ and $\Sigma_{\mathbf{S}}$, respectively, the mean and covariance matrix of the signal $\mathbf{S}$:

$$\mathbb{E}\{\mathbf{S}\} \triangleq \bar{\mathbf{s}}$$
$$\mathbb{E}\left\{(\mathbf{S} - \bar{\mathbf{s}})(\mathbf{S} - \bar{\mathbf{s}})^T\right\} \triangleq \Sigma_{\mathbf{S}}.$$

As before, we will continue to denote the covariance matrix of noise vector $\mathbf{W}$ as $\Sigma_{\mathbf{W}}$.

The UMP optimal tests, if they exist, must of course be based on the LR:

$$L(\mathbf{y},\theta) = \frac{f_{\mathbf{Y}|\theta}(\mathbf{y}|\theta)}{f_{\mathbf{Y}|\theta=0}(\mathbf{y}|\theta=0)}$$

$$= \frac{\mathbb{E}\{f_{\mathbf{W}}(\mathbf{y}-\theta\mathbf{S})\}}{f_{\mathbf{W}}(\mathbf{y})} \tag{4.122a}$$

$$= \mathbb{E}\left\{\frac{f_{\mathbf{W}}(\mathbf{y}-\theta\mathbf{S})}{f_{\mathbf{W}}(\mathbf{y})}\right\}, \tag{4.122b}$$

where expectation in (4.122a) is taken to be with respect to the distribution of $\mathbf{S}$. Without knowing the densities $f_{\mathbf{W}}(\mathbf{w})$ and $f_{\mathbf{S}}(\mathbf{s})$, however, we may not proceed any further. Even when these densities are known, in general, it may be difficult to decouple the dependence on θ from $\mathbf{y}$ to obtain a UMP test. As we already know, in such situations, we may look for LO detectors.

The LO decision statistic for small θ can be written as

$$\left.\frac{\partial}{\partial\theta}L(\mathbf{y},\theta)\right|_{\theta=0} = \left.\frac{\partial}{\partial\theta}\mathbb{E}\left\{\frac{f_{\mathbf{W}}(\mathbf{y}-\theta\mathbf{S})}{f_{\mathbf{W}}(\mathbf{y})}\right\}\right|_{\theta=0}$$

$$= \left.\frac{\mathbb{E}\left\{\frac{\partial}{\partial\theta}f_{\mathbf{W}}(\mathbf{y}-\theta\mathbf{S})\right\}}{f_{\mathbf{W}}(\mathbf{y})}\right|_{\theta=0}$$

$$= \left.\frac{\mathbb{E}\left\{-\mathbf{S}^T\mathbf{f}'_{\mathbf{W}}(\mathbf{y}-\theta\mathbf{S})\right\}}{f_{\mathbf{W}}(\mathbf{y})}\right|_{\theta=0} \tag{4.123a}$$

$$= -\frac{\mathbb{E}\{\mathbf{S}\}^T\mathbf{f}'_{\mathbf{W}}(\mathbf{y})}{f_{\mathbf{W}}(\mathbf{y})} \tag{4.123b}$$

$$= -\frac{\bar{\mathbf{s}}^T\mathbf{f}'_{\mathbf{W}}(\mathbf{y})}{f_{\mathbf{W}}(\mathbf{y})}, \tag{4.123c}$$

where $\mathbf{f}'_{\mathbf{W}}(\mathbf{w}) = \frac{\partial f_{\mathbf{W}}}{\partial \mathbf{w}} = \left(\frac{\partial f_{\mathbf{W}}}{\partial w_1}, \frac{\partial f_{\mathbf{W}}}{\partial w_2}, \ldots, \frac{\partial f_{\mathbf{W}}}{\partial w_N}\right)^T$ is a column vector and we have used the chain rule for derivatives with respect to vectors.

Note that (4.123c) is simply the vector version of (4.115a) and (4.116), with the mean signal vector $\bar{\mathbf{s}}$ playing the role of deterministically known signal samples. In many cases, however, it might happen that the random signal of interest may have zero mean so that $\bar{\mathbf{s}} = \mathbf{0}$. Then we cannot use (4.123c) to obtain a first-order LO decision statistic. Recall that, in this case, the slope of the detection probability (as a function of θ) at $\theta = 0$ is identically zero.

Of course, we already know how to handle this. We may instead attempt to find the detector that maximizes the second derivative of the detection probability as a function of θ. From the generalized Neyman–Pearson lemma, this amounts to a detector that uses the decision statistic $\left.\frac{\partial^2}{\partial\theta^2}L(\mathbf{y},\theta)\right|_{\theta=0}$. Hence, differentiating (4.123a) once more with respect to θ, we have

$$\left.\frac{\partial^2}{\partial\theta^2}L(\mathbf{y},\theta)\right|_{\theta=0} = \left.\frac{\partial}{\partial\theta}\frac{\mathbb{E}\left\{-\mathbf{S}^T\mathbf{f}'_{\mathbf{W}}(\mathbf{y}-\theta\mathbf{S})\right\}}{f_{\mathbf{W}}(\mathbf{y})}\right|_{\theta=0}$$

$$= \left.\frac{\mathbb{E}\left\{-\frac{\partial}{\partial\theta}\mathbf{S}^T\mathbf{f}'_{\mathbf{W}}(\mathbf{y}-\theta\mathbf{S})\right\}}{f_{\mathbf{W}}(\mathbf{y})}\right|_{\theta=0}$$

$$= \left.\frac{\mathbb{E}\left\{-\left(\frac{\partial}{\partial\theta}\mathbf{f}'_{\mathbf{W}}(\mathbf{y}-\theta\mathbf{S})\right)\mathbf{S}\right\}}{f_{\mathbf{W}}(\mathbf{y})}\right|_{\theta=0}$$

$$= \left.\frac{\mathbb{E}\left\{\left(\mathbf{S}^T\frac{\partial}{\partial(\mathbf{y}-\theta\mathbf{S})}\mathbf{f}'_{\mathbf{W}}(\mathbf{y}-\theta\mathbf{S})\right)\mathbf{S}\right\}}{f_{\mathbf{W}}(\mathbf{y})}\right|_{\theta=0}$$

$$= \left.\frac{\mathbb{E}\left\{\mathbf{S}^T\mathbf{F}''_{\mathbf{W}}(\mathbf{y}-\theta\mathbf{S})\mathbf{S}\right\}}{f_{\mathbf{W}}(\mathbf{y})}\right|_{\theta=0} \tag{4.124a}$$

$$= \frac{\mathbb{E}\left\{\mathbf{S}^T\mathbf{F}''_{\mathbf{W}}(\mathbf{y})\mathbf{S}\right\}}{f_{\mathbf{W}}(\mathbf{y})} \tag{4.124b}$$

$$= \frac{\mathbb{E}\left\{\mathrm{Tr}\left(\mathbf{F}''_{\mathbf{W}}(\mathbf{y})\mathbf{S}\mathbf{S}^T\right)\right\}}{f_{\mathbf{W}}(\mathbf{y})}$$

$$= \frac{\mathrm{Tr}\left(\mathbf{F}''_{\mathbf{W}}(\mathbf{y})\mathbb{E}\left\{\mathbf{S}\mathbf{S}^T\right\}\right)}{f_{\mathbf{W}}(\mathbf{y})} \tag{4.124c}$$

$$= \frac{\mathrm{Tr}\left(\mathbf{F}''_{\mathbf{W}}(\mathbf{y})\left(\Sigma_{\mathbf{S}}+\bar{\mathbf{s}}\bar{\mathbf{s}}^T\right)\right)}{f_{\mathbf{W}}(\mathbf{y})}, \tag{4.124d}$$

where in (4.124a) we have denoted $\mathbf{F}''_{\mathbf{W}}(\mathbf{w}) = \frac{\partial}{\partial\mathbf{w}}\mathbf{f}'_{\mathbf{W}}(\mathbf{w})$ and in (4.124d) we have used the fact that $\mathbb{E}\left\{\mathbf{S}\mathbf{S}^T\right\} = \Sigma_{\mathbf{S}}+\bar{\mathbf{s}}\bar{\mathbf{s}}^T$. Note that both $\mathbf{w}$ and $\mathbf{f}'_{\mathbf{W}}$ are N-vectors so that the second derivative $\mathbf{F}''_{\mathbf{W}}$ is an $N\times N$ matrix given by

$$\mathbf{F}''_{\mathbf{W}}(\mathbf{y}) = \frac{\partial}{\partial\mathbf{y}}\left[\frac{\partial}{\partial\mathbf{y}}f_{\mathbf{W}}(\mathbf{y})\right]$$

$$= \frac{\partial}{\partial\mathbf{y}}\begin{bmatrix}\frac{\partial}{\partial y_1}f_{\mathbf{W}}(\mathbf{y})\\ \frac{\partial}{\partial y_2}f_{\mathbf{W}}(\mathbf{y})\\ \vdots\\ \frac{\partial}{\partial y_N}f_{\mathbf{W}}(\mathbf{y})\end{bmatrix}$$

$$= \begin{bmatrix}\frac{\partial^2}{\partial y_1^2}f_{\mathbf{W}}(\mathbf{y}) & \frac{\partial^2}{\partial y_2 y_1}f_{\mathbf{W}}(\mathbf{y}) & \cdots & \frac{\partial^2}{\partial y_N y_1}f_{\mathbf{W}}(\mathbf{y})\\ \frac{\partial^2}{\partial y_1 y_2}f_{\mathbf{W}}(\mathbf{y}) & \frac{\partial^2}{\partial y_2^2}f_{\mathbf{W}}(\mathbf{y}) & \cdots & \frac{\partial^2}{\partial y_N y_2}f_{\mathbf{W}}(\mathbf{y})\\ \vdots & \vdots & & \vdots\\ \frac{\partial^2}{\partial y_1 y_N}f_{\mathbf{W}}(\mathbf{y}) & \frac{\partial^2}{\partial y_2 y_N}f_{\mathbf{W}}(\mathbf{y}) & \cdots & \frac{\partial^2}{\partial y_N^2}f_{\mathbf{W}}(\mathbf{y})\end{bmatrix}.$$

The LO decision statistic (4.124d) simplifies to a more insightful form if noise $\mathbf{w}$ happens to be iid. It is, however, important to keep in mind that in problems such as wideband spectrum sensing, this assumption might not always be reasonable. Still, if we were to justify independent noise samples, then the joint pdf $f_{\mathbf{W}}(\mathbf{w})$ of noise can be written as the product of the marginal densities as in (4.94). Hence, under the assumption of independent noise, for $n = 1, \ldots, N$,

$$\frac{\partial}{\partial y_n} f_{\mathbf{W}}(\mathbf{y}) = \left(\prod_{i=1, i \neq n}^{N} f_{W_i}(w_i) \right) \frac{\partial}{\partial y_n} f_{W_n}(y_n) \tag{4.125a}$$

$$= f_{\mathbf{W}}(\mathbf{y}) \frac{f'_{W_n}(y_n)}{f_{W_n}(y_n)} \tag{4.125b}$$

and

$$\begin{aligned}
\frac{\partial^2}{\partial y_j y_n} f_{\mathbf{W}}(\mathbf{y}) &= \frac{\partial}{\partial y_j} \left[\left(\prod_{i=1, i \neq n}^{N} f_{W_i}(w_i) \right) f'_{W_n}(y_n) \right] \\
&= \begin{cases} \left(\prod_{i=1, i \neq n}^{N} f_{W_i}(w_l) \right) \frac{\partial}{\partial y_j} f'_{W_n}(y_n) & \text{if } j - n \\ \left(\prod_{i=1, i \neq n, i \neq j}^{N} f_{W_i}(w_i) \right) f'_{W_j}(y_j) f'_{W_n}(y_n) & \text{if } j \neq n \end{cases} \\
&= \begin{cases} f_{\mathbf{W}}(\mathbf{y}) \frac{f''_{W_n}(y_n)}{f_{W_n}(y_n)} & \text{if } j = n \\ f_{\mathbf{W}}(\mathbf{y}) \frac{f'_{W_j}(y_j)\, f'_{W_n}(y_n)}{f_{W_j}(y_j)\, f_{W_n}(y_n)} & \text{if } j \neq n \end{cases} .
\end{aligned} \tag{4.126}$$

From (4.124d) and (4.126), we then have that

$$\begin{aligned}
&\left. \frac{\partial^2}{\partial \theta^2} L(\mathbf{y}, \theta) \right|_{\theta=0} \\
&= \frac{\operatorname{Tr}\left(\mathbf{F}''_{\mathbf{W}}(\mathbf{y}) \left(\Sigma_{\mathbf{S}} + \bar{\mathbf{s}}\bar{\mathbf{s}}^T \right) \right)}{f_{\mathbf{W}}(\mathbf{y})} \\
&= \frac{1}{f_{\mathbf{W}}(\mathbf{y})} \left[\sum_{n=1}^{N} \frac{\partial^2 f_{\mathbf{W}}(\mathbf{y})}{\partial y_n^2} \left(\Sigma_{n,n}^{(s)} + \bar{s}_n^2 \right) + \sum_{n=1}^{N} \sum_{j=1, j \neq n}^{N} \frac{\partial^2 f_{\mathbf{W}}(\mathbf{y})}{\partial y_j y_n} \left(\Sigma_{j,n}^{(s)} + \bar{s}_j \bar{s}_n \right) \right] \\
&= \sum_{n=1}^{N} \frac{f''_{W_n}(y_n)}{f_{W_n}(y_n)} \left(\Sigma_{n,n}^{(s)} + \bar{s}_n^2 \right) + \sum_{n=1}^{N} \sum_{j=1, j \neq n}^{N} \frac{f'_{W_j}(y_j)\, f'_{W_n}(y_n)}{f_{W_j}(y_j)\, f_{W_n}(y_n)} \left(\Sigma_{j,n}^{(s)} + \bar{s}_j \bar{s}_n \right),
\end{aligned} \tag{4.127}$$

where $\Sigma_{j,n}^{(s)}$ denotes the (j,n)th element of the covariance matrix $\Sigma_{\mathbf{S}}$.

Let us define the following instantaneous nonlinearities:

$$g_n^{(LO)}(x) = \frac{f'_{W_n}(x)}{f_{W_n}(x)}, \tag{4.128}$$

and

$$h_n^{(LO)}(x) = \frac{f''_{W_n}(x)}{f_{W_n}(n)}, \tag{4.129}$$

so that we may write the locally optimum decision statistic (4.127) for detecting a random signal in independent, non-Gaussian noise as

$$\left.\frac{\partial^2}{\partial\theta^2}L(\mathbf{y},\theta)\right|_{\theta=0}$$
$$= \sum_{n=1}^{N} h_n^{(LO)}(y_n)\left(\Sigma_{n,n}^{(s)} + \bar{s}_n^2\right) + \sum_{n=1}^{N}\sum_{j=1,j\neq n}^{N} g_j^{(LO)}(y_j)g_n^{(LO)}(y_n)\left(\Sigma_{j,n}^{(s)} + \bar{s}_j\bar{s}_n\right). \tag{4.130}$$

Note that if the random signal were to have a nonzero mean so that $\bar{\mathbf{s}} \neq \mathbf{0}$, then from (4.123c), we can design an LO detector based on the decision statistic

$$\begin{aligned} T_1^{(LO)}(\mathbf{y}) &= \left.\frac{\partial}{\partial\theta}L(\mathbf{y},\theta)\right|_{\theta=0} \\ &= -\sum_{n=1}^{N}\bar{s}_n\frac{f'_{W_n}(y_n)}{f_{W_n}(y_n)} \\ &= -\sum_{n=1}^{N}\bar{s}_n g_n^{(LO)}(y_n). \end{aligned} \tag{4.131}$$

On the other hand, if the random signal were to have a distribution with zero mean, then from (4.130), an LO detector can be designed based on the decision statistic

$$T_2^{(LO)}(\mathbf{y}) = \left.\frac{\partial^2}{\partial\theta^2}L(\mathbf{y},\theta)\right|_{\theta=0}$$
$$= \frac{\mathrm{Tr}\left(\mathbf{F}''_{\mathbf{W}}(\mathbf{y})\,\Sigma_{\mathbf{S}}\right)}{f_{\mathbf{W}}(\mathbf{y})} \tag{4.132a}$$
$$= \sum_{n=1}^{N} h_n^{(LO)}(y_n)\Sigma_{n,n}^{(s)} + \sum_{n=1}^{N}\sum_{j=1,j\neq n}^{N} g_j^{(LO)}(y_j)g_n^{(LO)}(y_n)\Sigma_{j,n}^{(s)}. \tag{4.132b}$$

If signal samples were to be uncorrelated, then the second term in (4.132b) vanishes, simplifying the second-order LO decision statistic to

$$T_2^{(LO)}(\mathbf{y}) = \sum_{n=1}^{N} h_n^{(LO)}(y_n)\Sigma_{n,n}^{(s)}. \tag{4.133}$$

Note the similarity of (4.131) (or (4.123c)) and (4.133) with the case of known signal detection. Both (4.131) and (4.133) are also forms of correlators. However, since we are now dealing with a random signal, there is no deterministically known signal to correlate with the observations. When the signal mean is nonzero, the first-order LO decision statistic (4.131) instead uses the mean vector $\bar{\mathbf{s}}$ of the signal in the correlator. It correlates this mean signal vector with a transformed version of the observations where transformation $g_n^{(LO)}(.)$ is related to the first-order characteristics of the noise density function as can be seen from (4.128). When the signal mean vector $\bar{\mathbf{s}}$ is also zero, the second-order LO decision statistic (4.133) essentially correlates a transformed version of the observations with the variances $\Sigma_{n,n}^{(s)}$ of the signal samples. The transformation function $h_n^{(LO)}(.)$ in this case is determined by the second-order characteristics of the noise density function as can be seen from (4.129). Since variance is related to the signal strength, or energy, a decision statistic of the form of (4.133) is called a *generalized energy detector.*

4.3.4.4 Detection of a Random Signal in Additive Gaussian Noise In the previous section, we considered the general case of detecting a random signal in the presence of possibly non-Gaussian noise. In this section, we may obtain the specific form of the random signal detectors when the noise is Gaussian. Let us assume that the noise pdf $f_{\mathbf{W}}(\mathbf{w})$ in the vector problem of (4.121) is a multivariate Gaussian distribution with zero mean and a covariance matrix $\Sigma_{\mathbf{W}}$. In this case, the LR (4.122a) can explicitly be written as [51]

$$\begin{aligned} L(\mathbf{y}) &= \mathbb{E}\left\{ \frac{\frac{1}{(2\pi)^{\frac{n}{2}}|\Sigma_{\mathbf{W}}|^{\frac{1}{2}}} e^{-\frac{1}{2}(\mathbf{y}-\theta\mathbf{S})^T\Sigma_{\mathbf{W}}^{-1}(\mathbf{y}-\theta\mathbf{S})}}{\frac{1}{(2\pi)^{\frac{n}{2}}|\Sigma_{\mathbf{W}}|^{\frac{1}{2}}} e^{-\frac{1}{2}\mathbf{y}^T\Sigma_{\mathbf{W}}^{-1}\mathbf{y}}} \right\} \\ &= \int_{\mathbb{R}^N} e^{\theta\mathbf{s}^T\Sigma_{\mathbf{W}}^{-1}\mathbf{y}-\frac{\theta^2}{2}\mathbf{s}^T\Sigma_{\mathbf{W}}^{-1}\mathbf{s}} f_{\mathbf{S}}(\mathbf{s})d\mathbf{s}. \end{aligned} \tag{4.134}$$

The optimal tests must be based on this LR statistic. In particular, we may like to obtain a UMP optimal test, if it exists. Although (4.134) is not easy to be evaluated directly, from *mean-value theorem for integrals*[3] [51], there exists an $\hat{\mathbf{s}} \in \mathbb{R}^N$ such that

[3]Let A be a compact and connected set. Let f be continuous on A and g be integrable over A with $g(\mathbf{x}) \geq 0$ for every $\mathbf{x} \in A$. Then, there exists an $\mathbf{x}^* \in A$ such that

$$\int_A f(\mathbf{x})g(\mathbf{x})d\mathbf{x} = f(\mathbf{x}^*)\int_A g(\mathbf{x})d\mathbf{x}.$$

$$L(\mathbf{y}) = e^{\theta \hat{\mathbf{s}}^T \Sigma_{\mathbf{W}}^{-1} \mathbf{y} - \frac{\theta^2}{2} \hat{\mathbf{s}}^T \Sigma_{\mathbf{W}}^{-1} \hat{\mathbf{s}}}. \tag{4.135}$$

In general, $\hat{\mathbf{s}}$ in (4.135) can depend on $\mathbf{y}$ so that $\hat{\mathbf{s}} = \hat{\mathbf{s}}(\mathbf{y})$. Hence, computing the LR decision statistic (4.135) requires first determining $\hat{\mathbf{s}}(\mathbf{y})$, which may not be easy.

The usefulness of (4.135), however, is that it provides a conceptual framework for us to proceed: That is, given the observation $\mathbf{y}$, first obtain the estimate $\hat{\mathbf{s}}(\mathbf{y})$ and then compute the LR according to (4.135). If the hypothesis H_1 were to be also simple, meaning that $\theta = 1$, then the optimal detector will be essentially the same as detecting the known signal $\hat{\mathbf{s}}(\mathbf{y})$. When H_1 is composite, we may consider a new random signal $\tilde{\mathbf{S}} = \theta \mathbf{S}$, and the mean-value theorem will still imply that there is an $\hat{\tilde{\mathbf{s}}}$ for which (4.135) will hold. Thus, provided we may find this estimate of the signal, still, we may compute the LR decision statistic as if we were to detect the known signal $\hat{\tilde{\mathbf{s}}}$. In general, however, it may not be UMP.

When hypothesis H_1 is composite, if we cannot obtain a UMP test, we may look for a test that is either LMP or GLRT. If we can find the estimate $\hat{\mathbf{s}}(\mathbf{y})$ required in the mean-value theorem (4.135), the GLRT based on (4.135) has a particularly simple structure. Indeed, the ML estimate of θ based on the likelihood (ratio) (4.135) is

$$\hat{\theta}_{ML} = \frac{\hat{\mathbf{s}}^T \Sigma_{\mathbf{W}}^{-1} \mathbf{y}}{\hat{\mathbf{s}}^T \Sigma_{\mathbf{W}}^{-1} \hat{\mathbf{s}}}. \tag{4.136}$$

Hence, the GLR can be written as

$$L_G(\mathbf{y}) = e^{\frac{\left(\hat{\mathbf{s}}^T \Sigma_{\mathbf{W}}^{-1} \mathbf{y}\right)^2}{2 \hat{\mathbf{s}}^T \Sigma_{\mathbf{W}}^{-1} \hat{\mathbf{s}}}}. \tag{4.137}$$

There is a striking similarity between (4.137) and the GLR (4.82) for detecting a known (deterministic) signal in Gaussian noise that we derived in Example 4.4. The only difference of (4.137) compared to (4.82) is that instead of the known signal, here, we use the signal estimate $\hat{\mathbf{s}}$ obtained in applying the mean-value theorem.

Of course, the problem with this methods is that, in general, finding the estimate $\hat{\mathbf{s}}(\mathbf{y})$ as required in the mean-value theorem conclusion may not be all that easy. Hence, a more practical approach is to search for an LO detector as in the previous section. Since $f_{\mathbf{W}}(\mathbf{y})$ is the Gaussian pdf corresponding to $\mathbf{W} \sim \mathcal{N}(\mathbf{0}, \Sigma_{\mathbf{W}})$, we have that

$$\begin{aligned} \mathbf{f}'_{\mathbf{W}}(\mathbf{y}) &= \frac{\partial f_{\mathbf{W}}(\mathbf{y})}{\partial \mathbf{y}} \\ &= f_{\mathbf{W}}(\mathbf{y}) \frac{\partial}{\partial \mathbf{y}} \left(-\frac{1}{2} \mathbf{y}^T \Sigma_{\mathbf{W}}^{-1} \mathbf{y} \right) \\ &= -\Sigma_{\mathbf{W}}^{-1} \mathbf{y} f_{\mathbf{W}}(\mathbf{y}). \end{aligned} \tag{4.138}$$

Thus, in the case of Gaussian noise, the LO decision statistic (4.123c) reduces to

$$T_{1,g}^{(LO)}(\mathbf{y}) = \frac{\partial}{\partial\theta} L(\mathbf{y},\theta)\bigg|_{\theta=0},$$
$$= \bar{\mathbf{s}}^T \Sigma_{\mathbf{W}}^{-1} \mathbf{y} \qquad (4.139)$$

and the LO detector can be written as

$$\delta_{1,g}^{(LO)}(y) = \begin{cases} 1 & > \\ \gamma & \text{if } \bar{\mathbf{s}}^T \Sigma_{\mathbf{W}}^{-1} \mathbf{y} = \eta, \\ 0 & < \end{cases}$$

where threshold η is to be chosen to satisfy the false-alarm constraint α. Hence, as we saw in the general noise case of (4.131), the detector treats the mean signal vector as a deterministically known signal. From then on, it is exactly the same detector as for detecting a known signal in Gaussian noise. This surprisingly simple result allows us to conclude that LO detection of a random signal having an arbitrary distribution in Gaussian noise is equivalent to detecting its (known) mean vector in Gaussian noise. The LO detector is nothing but the correlator.

Indeed, if noise samples were to be iid so that $\Sigma_{\mathbf{W}} = \sigma^2\mathbf{I}$, the decision statistic (4.139) becomes

$$T_{1,g}^{(LO)}(\mathbf{y}) = \frac{\bar{\mathbf{s}}^T \mathbf{y}}{\sigma^2}$$
$$= \frac{1}{\sigma^2} \sum_{k=1}^{N} \bar{s}_k y_k, \qquad (4.140)$$

and the LO detector reduces to the MF:

$$\delta_{1,g}^{(LO)}(y) = \begin{cases} 1 & > \\ \gamma & \text{if } \frac{1}{\sigma^2} \sum_{k=1}^{N} \bar{s}_k y_k = \eta. \\ 0 & < \end{cases}$$

If noise was not iid, we can apply a noise whitening transformation so that again the optimal detector is the correlator for the whitened observation sequence with a transformed version of the mean signal vector $\bar{\mathbf{s}}$.

If, on the other hand, the random signal of interest were to have a zero mean so that $\bar{\mathbf{s}}=\mathbf{0}$, then we may obtain a locally optimum detector based on the second-order statistics. Again, noting that $f_{\mathbf{W}}(\mathbf{y})$ is the Gaussian pdf corresponding to

$\mathbf{W} \sim \mathcal{N}(\mathbf{0}, \Sigma_\mathbf{W})$, differentiating (4.138) once more with respect to its argument vector, we obtain

$$\begin{aligned}
\mathbf{F}''_\mathbf{W}(\mathbf{y}) &= \frac{\partial}{\partial \mathbf{y}} \mathbf{f}'_\mathbf{W}(\mathbf{y}) \\
&= -\frac{\partial}{\partial \mathbf{y}} \Sigma_\mathbf{W}^{-1} \mathbf{y} f_\mathbf{W}(\mathbf{y}) \\
&= -\left(\frac{\partial}{\partial \mathbf{y}} \mathbf{y} f_\mathbf{W}(\mathbf{y}) \right) \Sigma_\mathbf{W}^{-T} && (4.141a) \\
&= -\left(\frac{\partial f_\mathbf{W}(\mathbf{y})}{\partial \mathbf{y}} \mathbf{y}^T + f_\mathbf{W}(\mathbf{y}) \frac{\partial \mathbf{y}}{\partial \mathbf{y}} \right) \Sigma_\mathbf{W}^{-1} && (4.141b) \\
&= -\left(\mathbf{f}'_\mathbf{W}(\mathbf{y}) \mathbf{y}^T + f_\mathbf{W}(\mathbf{y}) \mathbf{I} \right) \Sigma_\mathbf{W}^{-1}, && (4.141c)
\end{aligned}$$

where in (4.141a) we have used the fact that if $\mathbf{A}$ does not depend on $\mathbf{x}$, then $\frac{\partial}{\partial \mathbf{x}} \mathbf{A}\mathbf{g} = \frac{\partial \mathbf{g}}{\partial \mathbf{y}} \mathbf{A}^T$; in (4.141b) we have used the product rule for vector differentiation and the fact that $\Sigma_\mathbf{W}$ is symmetric; and in (4.141c), we have used $\frac{\partial \mathbf{y}}{\partial \mathbf{y}} = \mathbf{I}$. Substituting for $\mathbf{f}'_\mathbf{W}(\mathbf{y})$ from (4.138) in (4.141c), we then have that

$$\mathbf{F}''_\mathbf{W}(\mathbf{y}) = \left(\Sigma_\mathbf{W}^{-1} \mathbf{y}\mathbf{y}^T \Sigma_\mathbf{W}^{-1} - \Sigma_\mathbf{W}^{-1} \right) f_\mathbf{W}(\mathbf{y}). \tag{4.142}$$

Combining (4.124d) and (4.142), we may obtain the LO decision statistic for detecting an arbitrarily distributed zero-mean signal in Gaussian noise:

$$\begin{aligned}
T_{2,g}^{(LO)}(\mathbf{y}) &= \left. \frac{\partial^2}{\partial \theta^2} L(\mathbf{y}, \theta) \right|_{\theta=0} \\
&= \mathrm{Tr}\left(\left(\Sigma_\mathbf{W}^{-1} \mathbf{y}\mathbf{y}^T \Sigma_\mathbf{W}^{-1} - \Sigma_\mathbf{W}^{-1} \right) \Sigma_\mathbf{S} \right) \\
&= \mathrm{Tr}\left(\Sigma_\mathbf{W}^{-1} \mathbf{y}\mathbf{y}^T \Sigma_\mathbf{W}^{-1} \Sigma_\mathbf{S} \right) - \mathrm{Tr}\left(\Sigma_\mathbf{W}^{-1} \Sigma_\mathbf{S} \right) \\
&= \mathrm{Tr}\left(\mathbf{y}\mathbf{y}^T \Sigma_\mathbf{W}^{-1} \Sigma_\mathbf{S} \Sigma_\mathbf{W}^{-1} \right) - \mathrm{Tr}\left(\Sigma_\mathbf{S} \Sigma_\mathbf{W}^{-1} \right). && (4.143)
\end{aligned}$$

If noise samples were to be white, so that $\Sigma_\mathbf{W} = \sigma^2 \mathbf{I}$, then the above decision statistic simplifies to

$$\begin{aligned}
T_{2,g}^{(LO)}(\mathbf{y}) &= \frac{1}{\sigma^2} \left(\mathrm{Tr}\left(\frac{\mathbf{y}\mathbf{y}^T \Sigma_\mathbf{S}}{\sigma^2} \right) - \mathrm{Tr}(\Sigma_\mathbf{S}) \right) \\
&= \frac{1}{\sigma^2} \left(\sum_{n=1}^{N} \left(\frac{y_n^2}{\sigma^2} - 1 \right) \Sigma_{n,n}^{(s)} + \sum_{j=1, j \neq n}^{N} \frac{y_n y_j \Sigma_{j,n}^{(s)}}{\sigma^2} \right). && (4.144)
\end{aligned}$$

If, in addition, signal samples were to be also uncorrelated, the second term in (4.144) vanishes and the LO decision statistic reduces to

$$T_{2,g}^{(LO)}(\mathbf{y}) = \frac{1}{\sigma^2}\sum_{n=1}^{N}\left(\frac{y_n^2}{\sigma^2}-1\right)\Sigma_{n,n}^{(s)}. \tag{4.145}$$

When signal samples are white so that $\Sigma_{n,n}^{(s)} = \sigma_s^2$, this becomes an energy detector:

$$T_{2,g}^{(LO)}(\mathbf{y}) = \frac{\sigma_s^2}{\sigma^4}\sum_{n=1}^{N} y_n^2 - N\frac{\sigma_s^2}{\sigma^2}. \tag{4.146}$$

Example 4.6 (Detection of a Gaussian Signal in Gaussian Noise) A common scenario is having to detect a Gaussian-distributed signal in Gaussian noise. Specifically, let us assume that signal $\mathbf{S}$ is a multivariate Gaussian with mean vector $\bar{\mathbf{s}}$ and the covariance matrix $\Sigma_{\mathbf{S}}$: $\mathbf{S} \sim \mathcal{N}(\bar{\mathbf{s}}, \Sigma_{\mathbf{S}})$. We will assume that the signal and noise are independent.

The observation $\mathbf{Y} = \theta\mathbf{S} + \mathbf{W}$ in this case is clearly also Gaussian under both hypotheses. Indeed, under H_1, $\mathbf{Y} \sim \mathcal{N}(\theta\bar{\mathbf{s}}, \theta^2\Sigma_{\mathbf{S}} + \Sigma_{\mathbf{W}})$, whereas under H_0, $\mathbf{Y} \sim \mathcal{N}(\mathbf{0}, \Sigma_{\mathbf{W}})$. Let us define

$$\Sigma_1 = \theta^2\Sigma_{\mathbf{S}} + \Sigma_{\mathbf{W}} \tag{4.147}$$

and

$$\Sigma_0 = \Sigma_{\mathbf{W}}. \tag{4.148}$$

The LR for this problem turns out to be

$$\begin{aligned} L(\mathbf{y}) &= \frac{\dfrac{1}{(2\pi)^{\frac{N}{2}}|\Sigma_1|^{\frac{1}{2}}} e^{-\frac{1}{2}(\mathbf{y}-\theta\bar{\mathbf{s}})' \Sigma_1^{-1}(\mathbf{y}-\theta\bar{\mathbf{s}})}}{\dfrac{1}{(2\pi)^{\frac{N}{2}}|\Sigma_0|^{\frac{1}{2}}} e^{-\frac{1}{2}\mathbf{y}^T\Sigma_0^{-1}\mathbf{y}}} \\ &= \frac{|\Sigma_0|^{\frac{1}{2}}}{|\Sigma_1|^{\frac{1}{2}}} e^{\frac{1}{2}\left(\mathbf{y}^T(\Sigma_0^{-1}-\Sigma_1^{-1})\mathbf{y}+2\theta\bar{\mathbf{s}}^T\Sigma_1^{-1}\mathbf{y}-\theta^2\bar{\mathbf{s}}^T\Sigma_1^{-1}\bar{\mathbf{s}}\right)}. \end{aligned} \tag{4.149}$$

Hence, the LLR is

$$\mathcal{L}(\mathbf{y}) = \frac{1}{2}\log\frac{|\Sigma_0|}{|\Sigma_1|} + \frac{1}{2}\mathbf{y}^T(\Sigma_0^{-1} - \Sigma_1^{-1})\mathbf{y} + \theta\bar{\mathbf{s}}^T\Sigma_1^{-1}\mathbf{y} - \frac{\theta^2}{2}\bar{\mathbf{s}}^T\Sigma_1^{-1}\bar{\mathbf{s}}. \tag{4.150}$$

For an arbitrary θ, there does not seem to be an easy decoupling of it in the above LLR. However, if H_1 hypothesis were to be simple ($\theta = 1$), the LLR (4.150) readily simplifies to

$$\mathcal{L}(\mathbf{y}) = \frac{1}{2}\mathbf{y}^T\mathbf{Q}\mathbf{y} + \bar{\mathbf{s}}^T(\Sigma_{\mathbf{S}} + \Sigma_{\mathbf{W}})^{-1}\mathbf{y} - C,$$

where

$$\begin{aligned}\mathbf{Q} &= \Sigma_0^{-1} - \Sigma_1^{-1} \\ &= \Sigma_{\mathbf{W}}^{-1} - (\Sigma_{\mathbf{S}} + \Sigma_{\mathbf{W}})^{-1} \\ &= \Sigma_{\mathbf{W}}^{-1}\Sigma_{\mathbf{S}}\,(\Sigma_{\mathbf{S}} + \Sigma_{\mathbf{W}})^{-1}\end{aligned}$$

is a positive semidefinite matrix and we have let $C = \frac{1}{2}\bar{\mathbf{s}}^T\Sigma_1^{-1}\bar{\mathbf{s}} - \frac{1}{2}\log\frac{|\Sigma_0|}{|\Sigma_1|}$.

If we further assume that the signal mean is also zero so that $\bar{\mathbf{s}} = \mathbf{0}$, then when both hypotheses are simple, detecting a Gaussian signal in Gaussian noise is through the use of a *quadratic detector* based on the decision statistic $\mathbf{y}^T\mathbf{Q}\mathbf{y}$.

If, on the other hand, hypothesis H_1 is composite, then the above detector is not UMP, and there does not seem to be an easy way to find one. However, we can readily obtain an LO detector by using the methods developed earlier. Indeed, if the mean signal vector $\bar{\mathbf{s}}$ is nonzero, then the LO detector is nothing but the known signal detector (linear correlator) for this mean signal vector $\bar{\mathbf{s}}$. If the mean signal vector $\bar{\mathbf{s}} = \mathbf{0}$, then the LO detector can be designed based on the second derivative statistic in (4.124d). ■

4.4 NONPARAMETRIC SIGNAL DETECTION THEORY

Let us revisit the binary hypothesis testing problem:

$$\begin{aligned} H_0 : Y_n &\sim P_0 \\ &\text{versus} \\ H_1 : Y_n &\sim P_1, \end{aligned}$$

where Y_n, for $n = 1, \ldots, N$, are assumed to be iid. The parametric signal detection theory considered in the previous section assumed that the observation distributions, P_0 and P_1 (or equivalently the corresponding pdf's), under the two hypotheses H_0 and H_1, respectively, are completely known, except for a finite number of possibly unknown parameters. Recall that, even in the case of composite hypothesis testing problems in which under each hypothesis the observations can take a set of possible distributions depending on the value of the parameter θ, the set of distributions under each hypothesis are completely parameterized by θ:

$$\begin{aligned} H_0 :\ Y_n &\sim P_\theta \text{ for } \theta \in \Theta_0 \\ &\text{versus} \\ H_1 :\ Y_n &\sim P_\theta \text{ for } \theta \in \Theta_1. \end{aligned}$$

In other words, the theory we have presented in the previous section applies only if the distributions of data under the two hypotheses can be parameterized by a finite set of parameters. Or, we may say, by a finite-dimensional parameter vector.

However, there are practical situations in which the probability density of the observations may not be parameterized by a finite number of parameters under one, or both, of the hypotheses. Usually, this happens due to unknown noise distributions. For example, in sonar problems, it is common to assume that the noise distribution may have certain symmetry properties but otherwise unknown. In these types of problems, we have only limited and coarse information about the noise distributions that it is not possible to parameterize the noise, or observation, distributions under each hypothesis in terms of a finite-dimensional parameter vector. Hence, the general binary hypothesis testing problem, expressed as

$$\begin{aligned} H_0 : Y_n &\sim P_0 \in \mathcal{P}_0 \\ &\text{versus} \\ H_1 : Y_n &\sim P_1 \in \mathcal{P}_1, \end{aligned} \tag{4.151}$$

where $\mathcal{P}_0$ and $\mathcal{P}_1$ are two disjoint classes of probability distributions, is called a *nonparametric detection* problem when either one or both of the classes $\mathcal{P}_0$ and $\mathcal{P}_1$ cannot be parameterized in terms of a finite-dimensional parameter vector. When they both can be parameterized by a finite-dimensional parameter vector, the problem (4.151) reduces to a parametric problem as assumed in the previous section.

As mentioned earlier, signal detection problems turn out to be nonparametric because of the lack of information on noise density. As an example, consider a binary hypothesis testing problem of the form of

$$\begin{aligned} H_0 : Y_n &= W_n \\ &\text{versus} \\ H_1 : Y_n &= \theta + W_n, \end{aligned} \tag{4.152}$$

where $n = 1, \ldots, N$, and W_n's are iid. This will be a nonparametric detection problem, for example, if all we knew about noise W_n was that the marginal distribution of noise W_n has a zero median or is symmetric about zero.

When we have such limited knowledge about the statistical properties of the observations, then it is not realistic to expect strict performance guarantees from detectors. A very common approach in nonparametric detection is to require that the detector maintains the same constant false-alarm rate (CFAR) over all possible distributions $P_0 \in \mathcal{P}_0$. Such a detector is called a *CFAR* detector. Of course, we may like to have a CFAR detector that also maximizes the detection probability for each $P_1 \in \mathcal{P}_1$. However, when we have limited knowledge about the class $\mathcal{P}_1$ of distributions, this may not be possible. Hence, commonly, the nonparametric detector design criteria is limited to a CFAR or *asymptotically CFAR* criterion.

In the following, we briefly summarize some nonparametric detector formulations from [51] that may possibly be useful in wideband spectrum sensing in cognitive radios.

4.4.1 Signal Detection in Additive Zero-Median Noise: The Sign Test

Let us consider the observation model (4.152) under the assumption that $\theta > 0$ and for $n = 1,\ldots,N$, W_n are iid and have a common distribution with zero median:

$$\Pr(W_n \leq 0) = \Pr(W_n \geq 0) = \frac{1}{2}.$$

Hence, if we define

$$p = \Pr(Y_n \geq 0) \quad \text{for } n = 1,\ldots,N,$$

then the hypothesis testing problem (4.152) with these assumptions is equivalent to

$$\begin{aligned} H_0 &: p = \frac{1}{2} \\ H_1 &: \frac{1}{2} < p < 1. \end{aligned} \tag{4.153}$$

Problem (4.153) is a nonparametric signal detection problem because the class of distributions with zero median cannot be parameterized by a finite set of parameters.

It has been shown that a CFAR detector for the problem (4.153) is given by the following (see [51] for a derivation):

$$\delta^{(s)}(\mathbf{y}) = \begin{cases} 1 & > \\ \gamma & \text{if } T_s(\mathbf{y}) = \eta, \\ 0 & < \end{cases} \tag{4.154}$$

where decision statistic $T_s(.)$ is the number of positive y_n's in the observation sequence $\mathbf{y} = (y_1,\ldots,y_N)^T$, which can be written as

$$T_s(\mathbf{y}) = \sum_{n=1}^{N} q(y_n) \tag{4.155}$$

with $q(.)$ denoting the unit-step function[4] and γ and η being chosen so that $\delta^{(s)}(\mathbf{y})$ have the desired false-alarm probability α. However, note that under the hypothesis H_0, $T_s(.)$ is a binomial $(n, \frac{1}{2})$ random variable. Hence, threshold η must be selected as the smallest integer m that satisfies

$$\frac{1}{2^N} \sum_{n=m}^{N} \frac{N!}{(N-n)!n!} \leq \alpha.$$

[4]The unit-step function $q(x)$ is defined as

$$q(x) = \begin{cases} 1 & \text{if } x > 0 \\ 0 & \text{otherwise} \end{cases}. \tag{4.156}$$

To achieve the exact false-alarm rate of α, the randomization constant γ can then be chosen as

$$\gamma = \frac{\alpha - \frac{1}{2^N}\sum_{n=\eta+1}^{N}\frac{N!}{(N-n)!n!}}{\frac{1}{2^N}\frac{N!}{(N-\eta)!\eta!}}.$$

Note that the test (4.154) only uses very coarse information from the observations y_n. Namely, it only cares whether the observation is positive or negative, and not the exact value of the observations. Making decisions based on such coarse information extracted from the observations makes sense when we only have very limited knowledge regarding the noise that is embedded in those observations. In the problem (4.153), we only know that the noise can be either positive or negative with equal probability $p = \frac{1}{2}$. Hence, it certainly makes sense to make decisions only by considering the sign of the observations. Since the detector (4.154) is based on the signs of the observations, it is called the *sign test*. It is a CFAR detector with the aforementioned threshold and randomization settings for the problem (4.153). Interestingly, it can also be shown that the sign test (4.154) is actually UMP for the problem (4.153) [51].

4.4.2 Signal Detection in Additive Symmetric Noise: The Rank Test

Let us now suppose that we have additional information about noise W_n in hypothesis testing problem (4.152) so that the marginal distributions of W_n can be considered as being symmetric about zero. Thus, W_n, for $n = 1,\ldots,N$, is iid and has a common distribution that is symmetric about zero and $\theta > 0$. Since the class of all symmetric distributions are contained within the class of all zero-median distributions, the sign test (4.154) is clearly a CFAR detector for this case as well. However, it is possible to find other detectors that are specifically CFAR for the symmetric noise. These detectors can perform better than the sign test in the sense of giving better detection probability. Many of these tests are based on the ranking values of the observations and are thus called the *rank tests*.

Indeed, suppose we rank the observations $y_1,\ldots,y_N$ in the increasing order of their absolute value so that $|y_{n_1}| \le |y_{n_2}| \le \ldots \le |y_{n_N}|$: The rank of the n_kth observation, y_{n_k}, within the sequence $y_1,\ldots,y_N$ is k. Then a CFAR test for (4.152) when noise has symmetric marginals is the following rank test called the *Wilcoxon test* [51]:

$$\delta^{(w)}(\mathbf{y}) = \begin{cases} 1 & > \\ \gamma & \text{if } T_w(\mathbf{y}) = \eta, \\ 0 & < \end{cases} \tag{4.157}$$

where the decision statistic $T_w(.)$ is

$$\begin{aligned} T_w(\mathbf{y}) &= \sum_{n=1}^{N} rank(y_n)q(y_n) \\ &= \sum_{k=1}^{N} kq(y_{n_k}). \end{aligned} \tag{4.158}$$

Rank tests such as the Wilcoxon test above can outperform the sign test in many problems of interest. However, this performance gain comes at a price. Indeed, note that computing decision statistic T_w requires waiting for the whole sequence of observations to perform the ranking. Hence, rank tests incur increased computational complexity and delay compared to the sign test.

4.4.3 Signal Detection in Additive Zero Median, Zero Mean, Finite-Variance Noise: The *t*-Test

Let us now suppose that we have additional information about the noise W_n in problem (4.152): In particular, the marginal distributions of W_n are known to have zero mean and finite variance, in addition to being zero median.

Under these conditions, a test that is known to be asymptotically CFAR for the problem (4.152) is the so-called *t*-test:

$$\delta^{(t)}(\mathbf{y}) = \begin{cases} 1 & > \\ \gamma & \text{if } T_t(\mathbf{y}) = \eta, \\ 0 & < \end{cases} \tag{4.159}$$

where decision statistic $T_t(.)$ is

$$T_t(\mathbf{y}) = \frac{\bar{y}}{\sqrt{\bar{y^2}}} \tag{4.160}$$

with $\bar{y}$ and $\bar{y^2}$ being the sample mean and sample variance defined, respectively, as

$$\bar{y} = \frac{1}{N}\sum_{n=1}^{N} y_n \tag{4.161}$$

and

$$\bar{y^2} = \frac{1}{N}\sum_{n=1}^{N} (y_n - \bar{y})^2. \tag{4.162}$$

The *t*-test is in fact UMP if noise W_n were to be zero-mean Gaussian with an unknown variance. It is asymptotically CFAR when we allow the noise distribution to be arbitrary except for being zero mean, zero median, and of finite variance. When noise is non-Gaussian within this class, however, there are many situations in which sign and rank tests may outperform the *t*-test [51].

4.5 SUMMARY

Binary hypothesis testing problem forms the basis for all detection problems. The problem can be reduced to having to choose between two hypotheses H_0 and H_1 based on a given observation. In this case, there are only two possible types of errors a detector can make: false alarms and misses. Since minimizing them are conflicting objectives, any optimality criterion will have to be a trade-off between minimizing these two types of errors. Depending on the optimality criteria used in selecting a detection rule, the form of the optimal detector will be different. Two of the most widely used optimality criteria in signal detection are the Bayesian and Neyman–Pearson criteria. The Bayesian optimality criterion is to minimize a certain cost function of the miss and false-alarm probabilities averaged over the prior probabilities of the hypotheses. The Neyman–Pearson criterion, on the other hand, is to maximize the detection probability subject to a given maximum false-alarm constraint.

When observation probability distribution/density can be parameterized by a finite number of parameters and the form of the parameterized distribution is known, the detection problem is called a parametric detection problem.

- The optimal Bayesian minimum probability of error decision rule can be written as a threshold test on the likelihood ratio:

$$\delta^{(B)}(y) = \begin{cases} 1 & \\ & \text{if } L(y) \begin{matrix} \geq \\ \\ < \end{matrix} \tau_B, \\ 0 & \end{cases}$$

  where $L(y)$ is the likelihood ratio defined as

$$L(y) = \frac{f(y|\theta \in \Theta_1)}{f(y|\theta \in \Theta_0)}$$

  and τ_B is the threshold given by

$$\tau_B = \frac{p_0}{p_1}$$

  with p_0 and p_1 being the prior probabilities of the hypotheses H_0 and H_1, respectively.
- Equivalently, the optimal Bayesian minimum probability of error decision rule can be written as the maximum a posteriori probability detector (MAP detector) as below:

$$\delta^{(B)}(y) = \begin{cases} 1 & \text{if } f(\theta \in \Theta_1|y) \geq f(\theta \in \Theta_0|y) \\ 0 & \text{if } f(\theta \in \Theta_0|y) > f(\theta \in \Theta_1) \end{cases}.$$

- When the two hypotheses are both simple, the optimal Bayesian minimum error probability decision rule simplifies to

$$\delta^{(B)}(y) = \begin{cases} 1 & \geq \\ & \text{if } L(y) \quad \frac{p_0}{p_1}, \\ 0 & < \end{cases}$$

where likelihood ratio is given by

$$L(y) = \frac{f_1(y)}{f_0(y)}$$

with $f_0(y) = f_{Y|\theta=0}(y|\theta = 0)$ and $f_1(y) = f_{Y|\theta=1}(y|\theta = 1)$.

- The Neyman–Pearson optimal decision rule for two hypotheses that are both simple is

$$\delta^{(NP)}(y) = \begin{cases} 1 & > \\ \gamma & \text{if } L(y) = \eta \\ 0 & < \end{cases}$$

where $L(y) = \frac{f_1(y)}{f_0(y)}$ is the likelihood ratio and the optimal threshold $\eta \geq 0$ and the randomization parameter $0 \leq \gamma \leq 1$ are to be chosen such that $P_F(\delta) = \alpha$ for $\alpha > 0$.

- A detector δ is called the uniformly most powerful test for a composite hypothesis testing problem between two hypotheses H_0 and H_1 if it is the solution to the constrained optimization problem

$$\max_{\delta} \; P_D(\delta, \theta) \quad \text{for all } \theta \in \Theta_1 \quad \text{subject to } \sup_{\theta \in \Theta_0} P_F(\delta, \theta) \leq \alpha$$

and δ does not depend on $\theta \in \Theta_1$. The uniformly most powerful tests for a detection problem may only exist under special circumstances.

- A locally most powerful size α detector (to the first order) for a composite hypothesis H_1 against a simple hypotheses H_0 characterized by $\theta = \theta_0$ is a solution to the following constrained optimization problem:

$$\max_{\delta} \; P'_D(\delta, \theta_0) \quad \text{subject to } P_F(\delta) = \alpha,$$

and has the general form of

$$\delta^{(LO)}(y) = \begin{cases} 1 & \quad > \\ \gamma \quad \text{if } \dfrac{\partial}{\partial \theta} f_{Y|\theta}(y|\theta) \Big|_{\theta=\theta_0} & = \eta f_{\theta_0}(y), \\ 0 & \quad < \end{cases}$$

$$= \begin{cases} 1 & \quad > \\ \gamma \quad \text{if } \dfrac{\partial}{\partial \theta} L_\theta(y) \Big|_{\theta-\theta_0} & = \eta, \\ 0 & \quad < \end{cases}$$

where $\eta \geq 0$ and $0 \leq \gamma \leq 1$ are to be chosen so that $P_F(\delta^{(LO)}) = \alpha$ and

$$L_\theta(y) = \frac{f_{Y|\theta}(y|\theta)}{f_{\theta_0}(y)} = \frac{f_\theta(y)}{f_{\theta_0}(y)}.$$

- A generalized likelihood ratio test (GLRT) for a composite hypothesis testing problem between two hypotheses H_0 and H_1 is given by

$$\delta^{(g)}(\mathbf{y}) - \begin{cases} 1 & \quad \geq \\ & \text{if } L_g(y) \quad \log \eta, \\ 0 & \quad < \end{cases}$$

where η can be chosen according to desired criteria and $L_g(y)$ is the generalized likelihood ratio (GLR) defined as

$$L_g(y) = \frac{f_{Y|\theta}(y|\theta = \hat{\theta}_{ML,1}(y))}{f_{Y|\theta}(y|\theta = \hat{\theta}_{ML,0}(y))}.$$

with $\hat{\theta}_{ML,0}(y)$ and $\hat{\theta}_{ML,1}(y)$, respectively, denoting the ML estimates of θ under H_0 and H_1 hypotheses:

$$\hat{\theta}_{ML,0}(y) = \max_{\theta \in \Theta_0} f_{Y|\theta}(y|\theta)$$

$$\hat{\theta}_{ML,1}(y) = \max_{\theta \in \Theta_1} f_{Y|\theta}(y|\theta).$$

When observation distribution/density of the observations under either or both of the hypotheses cannot be parameterized in terms of a finite-dimensional parameter vector, the detection problem is called a nonparametric detection problem. Let us consider a binary hypothesis testing problem of the form of

$$\begin{aligned} H_0 : Y_n &= W_n \\ &\text{versus} \\ H_1 : Y_n &= \theta + W_n, \end{aligned} \tag{4.163}$$

where $n = 1, \ldots, N$, $\theta > 0$, and W_n's are iid. Normally, signal detection problems of the form of (4.163) turn out to be nonparametric due to the lack of information about noise W_n to characterize the noise pdf at least up to a finite-dimensional parameter vector. Since there is very little or coarse information regarding the distributions of observations under different hypotheses in a nonparametric detection problem, it is unrealistic to expect strict performance guarantees from detectors. Hence, in many cases, the only requirement placed on a detector is that it has a constant false-alarm rate (CFAR).

- When noise W_n is known to have a common marginal with zero median, a CFAR detector for the nonparametric problem (4.163) is the sign test

$$\delta^{(s)}(\mathbf{y}) = \begin{cases} 1 & > \\ \gamma \quad \text{if } \sum_{n=1}^{N} u(y_n) = \eta, \\ 0 & < \end{cases}$$

  where, for a false-alarm probability α, η must be chosen so that it is the smallest integer m that satisfies

$$\frac{1}{2^N} \sum_{n=m}^{N} \frac{N!}{(N-n)!n!} \leq \alpha$$

  and randomization constant γ should be

$$\gamma = \frac{\alpha - \frac{1}{2^N} \sum_{n=\eta+1}^{N} \frac{N!}{(N-n)!n!}}{\frac{1}{2^N} \frac{N!}{(N-\eta)!\eta!}}.$$

- When noise W_n is known to have a common marginal that is symmetric about zero, a CFAR detector for the nonparametric problem (4.163) is the following rank test called the *Wilcoxon test*:

$$\delta^{(w)}(\mathbf{y}) = \begin{cases} 1 & > \\ \gamma \quad \text{if } \sum_{n=1}^{N} rank(y_n) q(y_n) = \eta. \cdot \\ 0 & < \end{cases}$$

- When noise W_n is known to have zero mean and a finite variance in addition to the zero median, a CFAR detector for the nonparametric problem (4.163) is the *t-test*:

$$\delta^{(t)}(\mathbf{y}) = \begin{cases} 1 & \qquad > \\ \gamma & \text{if } \dfrac{\bar{y}}{\sqrt{\bar{y^2}}} = \eta, \\ 0 & \qquad < \end{cases}$$

where $\bar{y}$ and $\bar{y}$ are the sample mean and sample variance defined, respectively, as

$$\bar{y} = \frac{1}{N}\sum_{n=1}^{N} y_n$$

and

$$\bar{y^2} = \frac{1}{N}\sum_{n=1}^{N} (y_n - \bar{y})^2.$$

5

INTRODUCTION TO ESTIMATION THEORY

5.1 INTRODUCTION

The binary detection problem of Chapter 4 was concerned with distinguishing between two possible hypotheses, H_0 and H_1. When these were both simple hypotheses, we were concerned with determining whether $\Theta = \theta_0$ or $\Theta = \theta_1$ was true. In this case, the optimal detector essentially determined our best *guess* of true θ corresponding to a given observation y. On the other hand, in the general case of composite hypothesis, testing the optimal detection does not bother about what the exact value of θ could be for a given observation y. Instead, the optimal detectors only attempt to come up with a best guess as to which set Θ_j the θ corresponding to the observation may belong to. Hence, detection problem is concerned only about distinguishing among a set of finite number of choices.

However, in some situations, we may not be satisfied by knowing whether a parameter θ associated with an observation belongs to a certain set of values. We may want to know the exact value itself of the parameter θ corresponding to the given observation. For example, recall the case of θ being the fixed signal amplitude in which hypothesis H_0 corresponds to $\theta = 0$ whereas hypothesis H_1 corresponds to $\theta > 0$. An optimal detector for H_0 versus H_1 only determines whether $\theta = 0$ or $\theta > 0$. When $\theta > 0$, it does not tell us what the exact value of θ is for a given observation. Indeed, for a signal in Gaussian noise, the UMP Neyman–Pearson test does not require the knowledge

Signal Processing for Cognitive Radios, First Edition. Sudharman K. Jayaweera.

of exact θ. Hence, we may derive and implement an optimal detector without caring about what the underlying θ value is. However, as we have seen in the previous chapter, the detection probability of the detector depends on the exact θ value. In particular, for certain values of θ, the detection probability could be unacceptably small. Thus, if we want to guarantee any performance level, we will have to know the exact underlying θ value.

If hypothesis H_1 corresponds to only a finite set of allowed positive values of $\theta > 0$, we could still consider this as a hypothesis testing problem but now with multiple hypotheses. Hence, as long as the parameter of interest θ assumes values on a finite, or a countable, set, determining exact θ can be achieved by using the tools already derived under the theory of detection in Chapter 4. However, if θ is allowed to take a continuum of values, the hypothesis testing formulation can no longer be applied. Now, we have an observation $\mathbf{y}$ that depends on a certain parameter θ that can take a continuum of values in, say, a set Θ. Given a particular observation $\mathbf{y}$, we would like to know the corresponding θ.

Since our model is statistical, θ's effect on the observation $\mathbf{y}$ is only seen through its statistical behavior. Hence, whatever we can come up with as our guess of underlying θ corresponding to a given $\mathbf{y}$ is just a *guess*. In other words, since we do not know a deterministic (and invertible) mapping between $\mathbf{y}$ and θ, we cannot determine the underlying θ corresponding to a given $\mathbf{y}$ deterministically. All we can come up with is a reasonable guess as to what θ might be given that we observed a certain $\mathbf{y}$, knowing how distribution of $\mathbf{Y}$ depends on θ. Thus, we are *estimating* the underlying θ corresponding to an observation based on the observation. Our guess of θ is thus called the *estimate* of θ.

Since the estimate of a parameter θ is computed based on the observation, it is simply a function of observation $\mathbf{y}$. This function, which we denote as $\hat{\theta}(.)$, is called the *estimator*. It takes the input $\mathbf{y}$ and outputs our estimate of θ, which is $\hat{\theta}(\mathbf{y})$. Hence, an estimator is a mapping from the observation space $\mathcal{Y}$ to a parameter space Θ:

$$\hat{\theta} : \mathcal{Y} \longrightarrow \Theta. \tag{5.1}$$

For a given observation $\mathbf{y}$, in general, there can be a discrepancy between our estimate $\hat{\theta}(\mathbf{y})$ of θ and the true θ. This discrepancy, $\theta - \hat{\theta}(\mathbf{y})$, is called the *estimation error*. Since $\mathbf{Y}$ is random, the estimate $\hat{\theta}(\mathbf{Y})$ is also random. Hence, the estimation error $\theta - \hat{\theta}(\mathbf{Y})$ is also a random quantity. Of course, we would like to have our estimate to be as close to the true θ as possible, if not exactly the same, for most of the observations, if not for all. In other words, a good estimator must have a small estimation error. But since it is a random quantity, we can only attempt to achieve this in a certain statistical sense.

How can we design a good estimator and how can we determine whether a particular estimator is better than another, or all other, estimators? Theory of estimation is concerned with answering these questions. Our brief introduction to estimation theory is mostly based on [51] and, to a certain extent, on [56]. An interested reader is referred to [51,56] and references therein for more in-depth coverage.

5.2 RANDOM PARAMETER ESTIMATION: BAYESIAN ESTIMATION

As with detection problems, we may pose the estimation problems also in a Bayesian setting if we can justify the existence and the knowledge of an underlying prior distribution for the parameter θ of interest. Hence, in Bayesian approach, the parameter θ is assumed to be a realization of a random variable Θ that takes values in a set Ξ. We will denote by $f_\Theta(\theta)$ the prior density of the random variable Θ. The observation $\mathbf{Y}$ depends on a particular θ through its distribution. We will assume that associated with its distribution is the density $f_\theta(\mathbf{y}) = f_{\mathbf{Y}|\Theta}(\mathbf{y}|\Theta = \theta)$. Recall that $f_\theta(\mathbf{y})$ is also called the likelihood function of the observation $\mathbf{Y}$.

The Bayesian approach for estimating a parameter is to consider as if there is a certain cost associated with any discrepancy between the true θ and our estimate of it. Naturally, this cost can be taken as a certain function of the estimation error $\theta - \hat{\theta}(\mathbf{y})$. While there are many possible choices for a cost function, the most commonly used choice due to analytical tractability is the squared estimation error $(\theta - \hat{\theta}(\mathbf{y}))^2$. Hence, throughout this book, we will consider the cost to be minimized in designing estimators to be the squared estimation error.

5.2.1 Minimum Mean-Squared Error Estimation

Since we can only expect to minimize the cost in a certain statistical sense, for each fixed θ, we may attempt to minimize the average cost $\mathbb{E}\left\{(\theta - \hat{\theta}(\mathbf{Y}))^2\right\}$ averaged over the conditional distribution of $\mathbf{Y}$ given $\Theta = \theta$. However, the choice of $\hat{\theta}$ that will minimize this average cost will work only for the assumed fixed θ, which we actually do not know. The Bayesian approach recognizes the fact that the estimation error $\theta - \hat{\theta}(\mathbf{y})$ is random not only because the observation is random but also due to the randomness of the parameter Θ itself. Hence, we may also average the cost with respect to the prior distribution of Θ and find the estimator $\hat{\theta}$ that minimizes the average cost, or the *mean-squared error*, $\mathbb{E}\left\{(\Theta - \hat{\theta}(\mathbf{Y}))^2\right\}$. Such an estimator is called a Bayesian *minimum mean-squared error* (MMSE) optimal estimator or, simply, the MMSE estimator:

$$\hat{\theta}_{MMSE} = \min_{\hat{\theta}} \ \mathbb{E}\left\{(\Theta - \hat{\theta}(\mathbf{Y}))^2\right\}. \tag{5.2}$$

By the iterated expectations,

$$\mathbb{E}\left\{(\Theta - \hat{\theta}(\mathbf{Y}))^2\right\} = \mathbb{E}_{\mathbf{Y}}\left\{\mathbb{E}\left\{(\Theta - \hat{\theta}(\mathbf{y}))^2 | \mathbf{Y} = \mathbf{y}\right\}\right\}. \tag{5.3}$$

Hence, for each $\mathbf{y}$, the optimal MMSE estimator can be obtained by minimizing the inner conditional expectation in (5.3) with respect to $\hat{\theta}$:

$$\hat{\theta}_{MMSE}(\mathbf{y}) = \min_{\hat{\theta}} \ \mathbb{E}\left\{(\Theta - \hat{\theta})^2 | \mathbf{Y} = \mathbf{y}\right\}. \tag{5.4}$$

This is a quadratic function of $\hat{\theta}$, and thus, there is a unique global minimum. By differentiating the expanded squared term with respect to $\hat{\theta}$ and setting the derivative to zero, we obtain

$$\mathbb{E}\left\{-2\Theta + 2\hat{\theta}_{MMSE} | \mathbf{Y} = \mathbf{y}\right\} = 0,$$

so that

$$\hat{\theta}_{MMSE}(\mathbf{y}) = \mathbb{E}\{\Theta | \mathbf{Y} = \mathbf{y}\} \tag{5.5}$$

is simply the conditional mean (CM) of Θ given the observation $\mathbf{Y} = \mathbf{y}$. For this reason, the MMSE estimator is also called the *CM estimator*. We may evaluate the CM, and thus the MMSE estimator, as

$$\begin{aligned}\hat{\theta}_{MMSE}(\mathbf{y}) &= \mathbb{E}\{\Theta | \mathbf{Y} = \mathbf{y}\} \\ &= \int \theta f_{\Theta|\mathbf{Y}}(\theta | \mathbf{y}) d\theta,\end{aligned} \tag{5.6}$$

where $f_{\Theta|\mathbf{Y}}(\theta | \mathbf{y})$ denotes the posterior density of parameter Θ given the observation $\mathbf{y}$.

As seen from (5.5), the characterization of the Bayesian MMSE optimal estimator is straightforward. However, its computation requires the ability to obtain the posterior distribution of the parameter Θ given the observation (or the posterior density $f_{\Theta|\mathbf{Y}}(\theta | \mathbf{y})$), which can sometimes be difficult. In particular, when observations come in sequentially (in time) if the posterior distribution and the CM have to be computed each time from scratch taking into account all the observations up to the current time, the computational complexity may become overwhelming. In such situations, it is particularly useful if the posterior distribution, and/or the CM estimator, can be directly obtained from the previous posterior distribution or the CM estimator with a correction term that depends on the new observation. The celebrated *Kalman filter* achieves this when the parameter is allowed to evolve in time according to linear dynamics and the noise is Gaussian [51, 57].

5.2.2 MMSE Estimation of Vector Parameters

The MMSE estimator can easily be derived even if the parameter of interest were to be a vector. In particular, suppose that $\boldsymbol{\theta} \in \mathbb{R}^M$ is an M-dimensional vector and the observation $\mathbf{y} \in \mathbb{R}^N$ is an N-dimensional vector. Suppose that we seek an estimator $\hat{\boldsymbol{\theta}}(.)$ that minimizes the average of the Euclidean, or L_2-, norm of the estimation error:

$$\begin{aligned}
\hat{\boldsymbol{\theta}}_{MMSE} &= \min_{\hat{\theta}} \ \mathbb{E}\left\{\|\hat{\boldsymbol{\theta}}(\mathbf{Y}) - \boldsymbol{\Theta}\|^2\right\} \\
&= \min_{\hat{\theta}} \ \mathbb{E}\left\{\left(\hat{\boldsymbol{\theta}}(\mathbf{Y}) - \boldsymbol{\Theta}\right)^T \left(\hat{\boldsymbol{\theta}}(\mathbf{Y}) - \boldsymbol{\Theta}\right)\right\} \\
&= \min_{\hat{\theta}} \ \mathbb{E}\left\{\sum_{m=1}^{M} (\hat{\theta}_m(\mathbf{Y}) - \Theta_m)^2\right\}.
\end{aligned} \tag{5.7}$$

Hence,

$$\begin{aligned}
\hat{\boldsymbol{\theta}}_{MMSE} &= \sum_{m=1}^{M} \min_{\hat{\theta}} \ \mathbb{E}\left\{(\hat{\theta}_m(\mathbf{Y}) - \Theta_m)^2\right\} \\
&= \sum_{m=1}^{M} \min_{\hat{\theta}} \ \mathbb{E}_{\mathbf{Y}}\left\{\mathbb{E}\left\{(\hat{\theta}_m(\mathbf{y}) - \Theta_m)^2 | \mathbf{Y} = \mathbf{y}\right\}\right\}.
\end{aligned} \tag{5.8}$$

Since the right-hand side of (5.8) is a sum of decoupled mean-squared errors of the individual components of the estimator $\hat{\boldsymbol{\theta}}$, the vector parameter estimation problem reduces to a set of scalar estimator problems. Indeed, from (5.8), it is straightforward to see that

$$\hat{\boldsymbol{\theta}}_{MMSE}(\mathbf{y}) = \mathbb{E}\{\boldsymbol{\Theta} | \mathbf{Y} = \mathbf{y}\}, \tag{5.9}$$

where the mth component of the estimate, for $m = 1, \ldots, M$, is

$$\hat{\theta}_m(y) = \mathbb{E}\{\Theta_m | \mathbf{Y} = \mathbf{y}\}.$$

Example 5.1 (MMSE Estimation of a Gaussian-Distributed Parameter Vector from Jointly Gaussian Observations) Suppose that we have an observation $\mathbf{Y}$ that is jointly Gaussian distributed with a parameter vector $\boldsymbol{\Theta}$. The marginal distribution of $\mathbf{Y}$ is Gaussian with mean vector $\boldsymbol{\mu}_Y$ and covariance matrix Σ_Y: $\mathbf{Y} \sim \mathcal{N}(\boldsymbol{\mu}_Y, \Sigma_Y)$. Similarly, the marginal (prior) distribution of $\boldsymbol{\Theta}$ is Gaussian with mean vector $\boldsymbol{\mu}_\Theta$ and covariance matrix Σ_Θ, so that $\boldsymbol{\Theta} \sim \mathcal{N}\left(\boldsymbol{\mu}_\Theta, \Sigma_\Theta\right)$. The cross-covariance matrix of $\mathbf{Y}$ and $\boldsymbol{\Theta}$ is $\Sigma_{Y\Theta}$ where

$$\Sigma_{Y\Theta} = \mathbb{E}\left\{(\mathbf{Y} - \boldsymbol{\mu}_Y)(\boldsymbol{\Theta} - \boldsymbol{\mu}_\Theta)^T\right\}. \tag{5.10}$$

Since **Y** and Θ are jointly Gaussian distributed, it follows that the augmented vector **X** defined as

$$\mathbf{X} = \begin{bmatrix} \mathbf{Y} \\ \boldsymbol{\Theta} \end{bmatrix} \tag{5.11}$$

is also Gaussian distributed as follows:

$$\mathbf{X} \sim \mathcal{N}\left(\begin{bmatrix} \boldsymbol{\mu}_Y \\ \boldsymbol{\mu}_\Theta \end{bmatrix}, \begin{bmatrix} \Sigma_Y & \Sigma_{Y\Theta} \\ \Sigma_{\Theta Y} & \Sigma_\Theta \end{bmatrix} \right), \tag{5.12}$$

where $\Sigma_{\Theta Y} = \Sigma_{Y\Theta}^T$.

From (5.9), the MMSE estimator of $\boldsymbol{\Theta}$ based on the observation **Y** is simply the CM of $\boldsymbol{\Theta}$ given **Y**. However, from (5.12), it follows that the conditional distribution of $\boldsymbol{\Theta}$ given **Y** is also Gaussian. From the form of the multivariate normal density function (see Appendix B), the mean vector $\mathbb{E}\{\boldsymbol{\Theta}|\mathbf{Y}=\mathbf{y}\}$ and covariance matrix $Cov(\boldsymbol{\Theta}|\mathbf{Y}=\mathbf{y})$ of this conditional distribution can easily be identified as

$$\mathbb{E}\{\boldsymbol{\Theta}|\mathbf{Y}=\mathbf{y}\} = \mathbb{E}\{\boldsymbol{\Theta}\} + \Sigma_{\Theta Y}\Sigma_Y^{-1}(\mathbf{y} - \mathbb{E}\{\mathbf{Y}\}) \tag{5.13}$$

and

$$Cov(\boldsymbol{\Theta}|\mathbf{Y}=\mathbf{y}) = \Sigma_\Theta - \Sigma_{\Theta Y}\Sigma_Y^{-1}\Sigma_{Y\Theta}. \tag{5.14}$$

Hence, from (5.9) and (5.13), it follows that the MMSE estimator of the parameter vector Θ in this case is simply

$$\hat{\theta}_{MMSE}(\mathbf{y}) = \boldsymbol{\mu}_\Theta + \Sigma_{\Theta Y}\Sigma_Y^{-1}(\mathbf{y} - \boldsymbol{\mu}_Y). \tag{5.15}$$

■

Observe from (5.7) that the MMSE estimator of a parameter vector minimizes the sum of the squared errors in estimating each component of the vector parameter $\boldsymbol{\theta}$. In some situations, it might be desired to treat estimation errors in each of the components of the parameter vector differently. In other words, rather than simply summing the estimation errors as in (5.7), we may want to weigh them according to a certain weight function before adding them. Such a weighted squared error can be written as

$$\text{Weighted MSE} = \mathbb{E}\left\{ \left(\hat{\theta} - \boldsymbol{\Theta}\right)^T \mathbf{A} \left(\hat{\theta} - \boldsymbol{\Theta}\right) \right\}, \tag{5.16}$$

where **A** is a symmetric, positive-definite matrix. An estimator that minimizes a weighted mean-squared error as in (5.16) is called a *weighted MMSE estimator* (WMMSE). However, note that

$$\begin{aligned}
&\mathbb{E}\left\{\left(\hat{\theta}(\mathbf{y})-\mathbf{\Theta}\right)^T \mathbf{A}\left(\hat{\theta}(\mathbf{y})-\mathbf{\Theta}\right)\right\} \\
&= \mathbb{E}_{\mathbf{Y}}\left\{\mathbb{E}\left\{\left(\hat{\theta}(\mathbf{y})-\mathbf{\Theta}\right)^T \mathbf{A}\left(\hat{\theta}(\mathbf{y})-\mathbf{\Theta}\right)|\mathbf{Y}=\mathbf{y}\right\}\right\} \\
&= \mathbb{E}_{\mathbf{Y}}\left\{\mathbb{E}\left\{\hat{\theta}(\mathbf{y})^T\mathbf{A}\hat{\theta}(\mathbf{y}) - 2\hat{\theta}(\mathbf{y})^T\mathbf{A}\mathbf{\Theta} + \mathbf{\Theta}^T\mathbf{A}\mathbf{\Theta}|\mathbf{Y}=\mathbf{y}\right\}\right\} \\
&= \mathbb{E}_{\mathbf{Y}}\left\{\hat{\theta}(\mathbf{y})^T\mathbf{A}\hat{\theta}(\mathbf{y}) - 2\hat{\theta}(\mathbf{y})^T\mathbf{A}\mathbb{E}\{\mathbf{\Theta}|\mathbf{Y}=\mathbf{y}\} + \mathbb{E}\left\{\mathbf{\Theta}^T\mathbf{A}\mathbf{\Theta}|\mathbf{Y}=\mathbf{y}\right\}\right\}.
\end{aligned} \tag{5.17}$$

Since the right-hand side of (5.17) is a quadratic function of $\hat{\theta}(\mathbf{y})$, it achieves its minimum at the point at which its gradient with respect to $\hat{\theta}(\mathbf{y})$ vanishes:

$$2\mathbf{A}\hat{\theta}(\mathbf{y}) - 2\mathbf{A}\mathbb{E}\{\mathbf{\Theta}|\mathbf{Y}=\mathbf{y}\} = \mathbf{0},$$

or

$$2\mathbf{A}\hat{\theta}(\mathbf{y}) = 2\mathbf{A}\ \mathbb{E}\{\mathbf{\Theta}|\mathbf{Y}=\mathbf{y}\}. \tag{5.18}$$

Since $\mathbf{A}$ is positive definite, it follows that the estimator that minimizes the weighted mean-squared error of (5.16) is

$$\hat{\theta}_{WMMSE}(\mathbf{y}) = \mathbb{E}\{\mathbf{\Theta}|\mathbf{Y}=\mathbf{y}\}. \tag{5.19}$$

Thus, regardless of the choice of the weight matrix $\mathbf{A}$, the CM vector of $\mathbf{\Theta}$ given $\mathbf{Y}=\mathbf{y}$ is the optimal estimator that minimizes the quadratic cost function of (5.16):

$$\hat{\theta}_{WMMSE}(\mathbf{y}) = \min_{\hat{\theta}}\ \mathbb{E}\left\{\left(\hat{\theta}(\mathbf{Y})-\mathbf{\Theta}\right)^T \mathbf{A}\left(\hat{\theta}(\mathbf{Y})-\mathbf{\Theta}\right)\right\} \tag{5.20a}$$

$$= \mathbb{E}\{\mathbf{\Theta}|\mathbf{Y}=\mathbf{y}\}. \tag{5.20b}$$

5.2.3 Linear Minimum Mean-Squared Error Estimation

Although the MMSE estimator (5.9) is simple to characterize, it might be difficult to compute in closed form. Moreover, even if we were to compute it in closed form, in general, it may turn out to be a complicated function of the observation $\mathbf{y}$. This may lead to unacceptable computational requirements in practice. In many situations, it is particularly convenient if the estimator is a linear function of the observation $\mathbf{y}$. Hence, in the following, we may derive the estimator that minimizes the mean-squared error within the class of estimators that are linear in observation:

$$\hat{\theta}^{(l)}(\mathbf{y}) = A\mathbf{y} + \mathbf{b} \tag{5.21}$$

for some matrix A and vector $\mathbf{b}$ (to be determined). Note that (5.21) is in fact an affine function of the observation rather than strictly being linear.

The linear estimator that minimizes the mean-squared error is called the *linear minimum mean-squared error* (LMMSE) estimator:

$$\hat{\boldsymbol{\theta}}_{LMMSE}(\mathbf{y}) = \min_{\hat{\theta}} \ \mathbb{E}\left\{\left(\hat{\boldsymbol{\theta}}^{(l)}(\mathbf{Y}) - \boldsymbol{\Theta}\right)^T \mathbf{A}\left(\hat{\boldsymbol{\theta}}^{(l)}(\mathbf{Y}) - \boldsymbol{\Theta}\right)\right\}, \tag{5.22}$$

where $\hat{\boldsymbol{\theta}}^{(l)}(.)$ is as defined in (5.21). LMMSE estimators are extremely popular and useful in practice. By substituting the form of (5.21) in (5.22) and taking the gradients with respect to A and $\mathbf{b}$, it can be shown that the optimal A^* and $\mathbf{b}^*$ that achieve the minimum in (5.22) are given by

$$\mathbf{b}^* = \mathbb{E}\{\boldsymbol{\Theta}\} - \mathbf{A}\mathbb{E}\{\mathbf{Y}\} \tag{5.23}$$

and

$$\mathbf{A}^* = \Sigma_{\boldsymbol{\Theta}\mathbf{Y}}\Sigma_{\mathbf{Y}}^{-1}, \tag{5.24}$$

where $\Sigma_{\mathbf{Y}} = Cov(\mathbf{Y})$ is the covariance matrix of observation $\mathbf{Y}$ defined as

$$Cov(\mathbf{Y}) = \mathbb{E}\left\{(\mathbf{Y} - \mathbb{E}\{\mathbf{Y}\})(\mathbf{Y} - \mathbb{E}\{\mathbf{Y}\})^T\right\} \tag{5.25}$$

and $\Sigma_{\boldsymbol{\Theta}\mathbf{Y}} = Cov(\boldsymbol{\Theta}, \mathbf{Y})$ is the cross-covariance of $\boldsymbol{\Theta}$ and $\mathbf{Y}$ defined as

$$Cov(\boldsymbol{\Theta}, \mathbf{Y}) = \mathbb{E}\left\{(\boldsymbol{\Theta} - \mathbb{E}\{\boldsymbol{\Theta}\})(\mathbf{Y} - \mathbb{E}\{\mathbf{Y}\})^T\right\}. \tag{5.26}$$

Substituting (5.25) and (5.26) in (5.21), we may then write the LMMSE estimator that is the solution to (5.22) as

$$\hat{\boldsymbol{\theta}}_{LMMSE}(\mathbf{y}) = \Sigma_{\boldsymbol{\Theta}\mathbf{Y}}\Sigma_{\mathbf{Y}}^{-1}(\mathbf{y} - \mathbb{E}\{\mathbf{Y}\}) + \mathbb{E}\{\boldsymbol{\Theta}\}. \tag{5.27}$$

Example 5.2 (LMMSE Estimation of a Gaussian-Distributed Parameter Vector from Jointly Gaussian Observations) Let us revisit Example 5.1 in which we derived the MMSE estimator of a Gaussian-Distributed parameter vector from jointly Gaussian-Distributed observations $\mathbf{Y}$. Observe from (5.15) that the MMSE estimator is already linear in the observation $\mathbf{Y}$. Hence, in the case of jointly Gaussian parameters and observations, the LMMSE estimator is indeed the MMSE estimator. Restricting the class of allowed estimators to be linear does not cause any loss of performance in this case. ■

5.2.4 Maximum A Posteriori Probability Estimation

Note from (5.6) that, computing the CM of Θ given the observation $\mathbf{y}$ requires evaluating the mean corresponding to the posterior distribution of Θ or the posterior density $f_{\Theta|\mathbf{Y}}(\theta|\mathbf{y})$. Sometimes, computing this integral may be computationally difficult and may not have a closed-form solution, or it may be too complicated to actually implement in practice. In particular, note that this CM can be a nonlinear function of the observation $\mathbf{y}$. In those situations, we may investigate an alternative Bayesian approach that still makes use of the known prior distribution of the parameter Θ.

Such an alternative optimality criterion that still relies on a Bayesian prior can be proposed as to estimate Θ by its most likely value given the observation. This is nothing but the maximum, or the mode, of the posterior density $f_{\Theta|\mathbf{Y}}(\theta|\mathbf{y})$ of Θ and thus is called the *maximum a posteriori probability (MAP) estimator*:

$$\hat{\theta}_{MAP}(\mathbf{y}) = \max_{\theta} f_{\Theta|\mathbf{Y}}(\theta|\mathbf{y}). \tag{5.28}$$

In many situations, MAP estimator might be easier to compute. For this reason, it is widely used in practice.

Since

$$f_{\Theta|\mathbf{Y}}(\theta|\mathbf{y}) = \frac{f_{\mathbf{Y}|\Theta}(\mathbf{y}|\theta) f_{\Theta}(\theta)}{f(\mathbf{y})}, \tag{5.29}$$

we note that the θ that maximizes the posterior density can be found by maximizing the numerator term $f_{\mathbf{Y}|\Theta}(\mathbf{y}|\theta) f_{\Theta}(\theta)$. Since log(.) is a monotonic function of its argument, equivalently, we may maximize the $\log\left(f_{\mathbf{Y}|\Theta}(\mathbf{y}|\theta) f_{\Theta}(\theta)\right) = \log f_{\mathbf{Y}|\Theta}(\mathbf{y}|\theta) + \log f_{\Theta}(\theta)$. Hence, a necessary condition for an estimator to be the MAP estimate is for it to satisfy

$$\frac{\partial}{\partial\theta}\left(\log f_{\mathbf{Y}|\Theta}(\mathbf{y}|\theta) + \log f_{\Theta}(\theta)\right)\bigg|_{\theta=\hat{\theta}} = 0.$$

Thus, the MAP estimator should necessarily satisfy the following *MAP equation*:

$$\frac{\partial}{\partial\theta}\log f_{\mathbf{Y}|\Theta}(\mathbf{y}|\theta)\bigg|_{\theta=\hat{\theta}_{MAP}} = -\frac{\partial}{\partial\theta}\log f_{\Theta}(\theta)\bigg|_{\theta=\hat{\theta}_{MAP}}. \tag{5.30}$$

However, it must be reminded that (5.30) is only a necessary condition and may not be sufficient. Hence, it is possible that an estimator that satisfies the MAP equation may not always be the MAP estimator.

5.3 NONRANDOM PARAMETER ESTIMATION

As we have already discussed in the previous chapter, there are at least two scenarios in which the Bayesian approach may not be justified or workable. For one, it may be the case that the parameter can reasonably be assumed to be a realization of a random variable Θ, but we may not have a known prior distribution f_{Θ} available. More seriously, however, it may be the case that parameter of interest is not a realization of a random variable at all. It may be that the parameter θ is just a fixed parameter taking values in a certain set. All observations $\mathbf{y}$ correspond to this fixed θ value from the domain set Ξ, and we just do not know what this fixed value is. There is no randomness in θ at all, and thus, averaging with respect to the parameter, as in the Bayesian approach, has no meaning.

Conceptually, the case of a nonrandom parameter is perhaps easier than that of the random parameter. However, estimating a nonrandom parameter seems more involved because now the only averaging we may perform in the squared error cost function is the averaging with respect to the observation $\mathbf{Y}$. This gives us a cost function of $\mathbb{E}\left\{(\theta - \hat{\theta}(\mathbf{Y}))^2\right\}$ where θ is the unknown, fixed parameter. But for any fixed $\theta = \theta_1$, we can make this *conditional mean-squared error* $\mathbb{E}\left\{(\theta - \hat{\theta}(\mathbf{Y}))^2\right\}$ exactly zero by setting $\hat{\theta}(\mathbf{y}) = \theta_1$ regardless of the observation. But, since we do not know the actual θ, if the actual θ were to be $\theta_2 \neq \theta_1$, such an estimator will perform poorly.

The cause of the problem is that the conditional mean-squared error $\mathbb{E}\left\{(\theta - \hat{\theta}(\mathbf{Y}))^2\right\}$ is a cost function that cannot be evaluated without knowing the true parameter θ. We may write the conditional mean-squared error as

$$
\begin{aligned}
\mathbb{E}\left\{(\theta - \hat{\theta}(\mathbf{Y}))^2\right\} &= \mathbb{E}\left\{\left(\left(\theta - \mathbb{E}\left\{\hat{\theta}(\mathbf{Y})\right\}\right) - \left(\hat{\theta}(\mathbf{Y}) - \mathbb{E}\left\{\hat{\theta}(\mathbf{Y})\right\}\right)\right)^2\right\} \\
&= \left(\theta - \mathbb{E}\left\{\hat{\theta}(\mathbf{Y})\right\}\right)^2 - 2\mathbb{E}\left\{\left(\theta - \mathbb{E}\left\{\hat{\theta}(\mathbf{Y})\right\}\right)\left(\hat{\theta}(\mathbf{Y}) - \mathbb{E}\left\{\hat{\theta}(\mathbf{Y})\right\}\right)\right\} \\
&\quad + \mathbb{E}\left\{\left(\hat{\theta}(\mathbf{Y}) - \mathbb{E}\left\{\hat{\theta}(\mathbf{Y})\right\}\right)^2\right\} \\
&= \left(b\left(\hat{\theta}(\mathbf{Y})\right)\right)^2 - 2\left(\theta - \mathbb{E}\left\{\hat{\theta}(\mathbf{Y})\right\}\right)\mathbb{E}\left\{\left(\hat{\theta}(\mathbf{Y}) - \mathbb{E}\left\{\hat{\theta}(\mathbf{Y})\right\}\right)\right\} + Var\left(\hat{\theta}(\mathbf{Y})\right) \\
&= \left(b\left(\hat{\theta}(\mathbf{Y})\right)\right)^2 + Var\left(\hat{\theta}(\mathbf{Y})\right),
\end{aligned} \tag{5.31}
$$

where the variance $Var\left(\hat{\theta}(\mathbf{Y})\right)$ of the estimator $\hat{\theta}(\mathbf{Y})$ is defined as

$$
Var\left(\hat{\theta}(\mathbf{Y})\right) = \mathbb{E}\left\{\left(\hat{\theta}(\mathbf{Y}) - \mathbb{E}\left\{\hat{\theta}(\mathbf{Y})\right\}\right)^2\right\} \tag{5.32}
$$

and the *bias* $b\left(\hat{\theta}(\mathbf{Y})\right)$ of the estimator $\hat{\theta}(\mathbf{Y})$ is defined as the

$$
b\left(\hat{\theta}(\mathbf{Y})\right) = \theta - \mathbb{E}\left\{\hat{\theta}(\mathbf{Y})\right\}. \tag{5.33}
$$

From (5.31), it is easily seen that the variance term (second term) can be made the minimum possible value of zero by choosing a constant estimator of the form of $\hat{\theta}(\mathbf{Y}) = \theta_1$ for some fixed θ_1. However, unless $\theta_1 = \theta$, where θ is the true parameter value, this will leave the bias term (first term) of (5.31) fixed at $(\theta - \theta_1)^2$ regardless of the observations. Such a constant estimator clearly is an extreme choice. We may attempt to choose an estimator that minimizes the bias term while constrained to have the variance term zero. But this means that we are searching for an estimator that minimizes the bias term within the class of constant estimators and the solution is clearly to have the constant estimator be equal to the unknown parameter itself. Of course, this is not useful because we do not know the unknown parameter.

The other extreme option is to instead make the first term (bias) of (5.31) always zero and attempt to find the estimator that minimizes the second term (variance) subject to the constraint that the first term be zero. To make the bias term $b\left(\hat{\theta}(\mathbf{Y})\right) = 0$, from (5.33), the estimator should satisfy

$$\mathbb{E}\left\{\hat{\theta}(\mathbf{Y})\right\} = \theta. \tag{5.34}$$

An estimator satisfying (5.34) is called an *unbiased estimator*. Unbiasedness is a desirable property to have in an estimator because it ensures that the average of the estimator over the ensemble of $\mathbf{Y}$ values is equal to the true parameter itself.

Now, subject to the unbiased constraint, we may look for the estimator that minimizes the second term, which is its own variance. Such an estimator is called a *minimum variance unbiased estimator* (MVUE).

Before we discuss the theory of MVUE, it is important to note that so far we have only considered the two extreme ends of (5.31): The constant estimators make the second term zero, and the MVUEs make the first term zero. Of course, it is entirely possible that neither one of them could achieve the real minimum of the averaged conditional squared error in (5.31). In particular, in certain cases, the MVUE may not achieve the minimum possible conditional mean-squared error. It is possible that a biased estimator may still have a small enough variance that will lead to the minimum possible conditional mean-squared error in (5.31). Hence, although we have discussed only the two extremes in which one of the terms in (5.31) is zero, the actual minimum possible value may occur when both terms of (5.31) are nonzero. Once we restrict ourselves to the class of unbiased estimators, we are accepting to look for an estimator that may not be the true minimum conditional mean-squared error estimator.

5.3.1 Theory of *Minimum Variance Unbiased Estimation*

There is an elegant theory based on the concepts of *sufficient statistics* and *complete family of distributions* that characterizes the MVUEs for a wide class of statistical models. In the following, we summarize this theory as presented in [51]. Let us start by defining the necessary concepts.

Definition 5.3.1 (Sufficient Statistics) A function $T(\mathbf{y})$ of $\mathbf{y}$ is a sufficient statistic for the family of distributions $\{P_\theta; \theta \in \Xi\}$ if the distribution of Y conditioned on $T(Y)$ does not depend on θ for $\theta \in \Xi$, when $Y \sim P_\theta$.

Note that the distribution $P_\theta(\mathbf{y})$ of $\mathbf{Y}$ is a function of both $\mathbf{y}$ and the unknown parameter θ. However, conditioned on a sufficient statistic T, the distribution of $\mathbf{Y}$ does not depend on θ any more. Hence, after conditioning on T, it can only be a function of $\mathbf{y}$ and T. Thus, conditioning on a sufficient statistic removes any further explicit dependence on θ of the observations. Then it must be the case that sufficient statistics essentially contain all information about θ contained in the observations, hence the removal of all information of θ from the observations once we know a sufficient statistic. Alternatively, a sufficient statistic $T(\mathbf{y})$ of the observation holds all information regarding θ that was contained in the observation. Thus, it is common to call T simply as a sufficient statistic for θ.

Definition 5.3.2 (Complete Family of Distributions) A family of distributions $\{P_\theta;\theta \in \Xi\}$ is a complete family of distributions if $\mathbb{E}_\theta\{g(\mathbf{Y})\} = 0$ for all $\theta \in \Xi$ implies that $g(\mathbf{Y}) = 0$ with probability 1 whenever $\mathbf{Y} \sim P_\theta$.

The relationship between the two concepts of sufficient statistics and the complete family of distributions is that if $\mathbf{Y} \sim P_\theta$ where $\{P_\theta;\theta \in \Xi\}$ is a complete family of distributions, then the only possible sufficient statistics $T(\mathbf{y})$ for family P_θ are just one-to-one mappings. To see this, suppose that $\mathbf{Y} \sim P_\theta$ and that $\{P_\theta;\theta \in \Xi\}$ is a complete family of distributions. Let us consider the function $g(\mathbf{y}) = \mathbf{y} - \mathbb{E}_\theta\{\mathbf{Y}|T(\mathbf{Y}) = T(\mathbf{y})\}$ where T is a sufficient statistic. By the iterated expectations

$$\begin{aligned}\mathbb{E}_\theta\{g(\mathbf{Y})\} &= \mathbb{E}_\theta\{\mathbf{Y}\} - \mathbb{E}_\theta\{\mathbb{E}_\theta\{\mathbf{Y}|T(\mathbf{Y})\}\}\\ &= \mathbb{E}_\theta\{\mathbf{Y}\} - \mathbb{E}_\theta\{\mathbf{Y}\}\\ &= 0, \end{aligned} \tag{5.35}$$

for any $\theta \in \Theta$. But since $\{P_\theta;\theta \in \Xi\}$ was assumed to be a complete family of distributions, (5.35) implies that $g(\mathbf{y}) = 0$, or equivalently $\mathbf{y} - \mathbb{E}_\theta\{\mathbf{Y}|T(\mathbf{Y}) = T(\mathbf{y})\}$, with probability 1. But by the definition of a sufficient statistic, $\mathbb{E}_\theta\{\mathbf{Y}|T(\mathbf{Y}) = T(\mathbf{y})\}$ does not depend on θ and thus can only be a function of $T(\mathbf{y})$. In other words, $\mathbf{y}$ is a function of $T(\mathbf{y})$. Since obviously $T(\mathbf{y})$ is also a function of $\mathbf{y}$, this then implies that any sufficient statistic T for a complete family of distributions must necessarily be a one-to-one mapping of the observations.

While notion of sufficient statistics attempts to reduce the observations without losing any information it contains about the parameter θ of interest, the notion of completeness of a family of distributions ensures that there is no possible further reduction than the observation itself.

The key to obtaining a characterization of the MVUE for a given problem in terms of these concepts is the following result called the Rao–Blackwell theorem [51].

Theorem 5.1 (Rao–Blackwell theorem) Let T be a sufficient statistic for the family P_θ where $\mathbf{Y} \sim P_\theta$. Suppose that $\hat{g}(\mathbf{y})$ is any unbiased estimate of $g(\theta)$. If

$$\tilde{g}(T(\mathbf{y})) = \mathbb{E}_\theta\left\{\hat{g}(\mathbf{Y})|T(\mathbf{Y}) = T(\mathbf{y})\right\}, \tag{5.36}$$

then

1. $\tilde{g}(T(\mathbf{y}))$ is also an unbiased estimate of $g(\theta)$.
2. $Var_\theta(\tilde{g}(T\mathbf{Y})) \le Var_\theta(\hat{g}(\mathbf{Y}))$, with equality if and only if $\hat{g}(\mathbf{Y}) = \tilde{g}(T(\mathbf{Y}))$ with probability 1 under $\mathbf{Y} \sim P_\theta$.

Proof: See [51] for a proof.

The Rao–Blackwell theorem provides us a way to start from an arbitrary unbiased estimator for a problem and improve that to obtain another still unbiased estimator that has a smaller variance than the original one. To do that, all we need is to have a sufficient statistic. The improved unbiased estimator $\tilde{g}(.)$ is simply the expected value of the original one conditioned on the sufficient statistic, as is given by (5.36). Of course, by the definition of a sufficient statistic, this expectation does not depend on θ, and thus, the improved unbiased estimator $\tilde{g}(.)$ is only a function of the sufficient statistic T, hence the justification for the notation $\tilde{g}(T)$ in (5.36).

In light of this, it then follows that if there is only one function $\tilde{g}(T)$ of a particular sufficient statistic T that is unbiased for $g(\theta)$, then regardless of which unbiased estimator $\hat{g}$ we start with, we should always get the same $\tilde{g}(T)$ after the improvement. Hence, this single function $\tilde{g}(T)$ has a smaller variance compared to all other unbiased estimators $\hat{g}(\mathbf{y})$, and it is also unbiased. Thus, such a $\tilde{g}(T)$ must be an MVUE.

Under what conditions will we have a sufficient statistic $T(\mathbf{y})$ such that there is only one function of T that is unbiased for $g(\theta)$? From the definition of completeness, it follows that all we need is that if $T \sim Q_\theta$, then Q_θ is a complete family of distributions. Indeed, suppose that $\mathbf{Y} \sim P_\theta$, T is a sufficient statistic for P_θ, $T \sim Q_\theta$, and that Q_θ is a complete family of distributions. Let $\tilde{g}_1(T)$ and $\tilde{g}_2(T)$ be two functions of T that are both unbiased for $g(\theta)$. Then for all θ,

$$\begin{aligned}\mathbb{E}_\theta\{\tilde{g}_1(T) - \tilde{g}_2(T)\} &= \mathbb{E}_\theta\{\tilde{g}_1(T(\mathbf{Y}))\} - \mathbb{E}_\theta\{\tilde{g}_2(T(\mathbf{Y}))\} \\ &= g(\theta) - g(\theta) \;=\; 0. \end{aligned} \tag{5.37}$$

Since $T \sim Q_\theta$ and that Q_θ is a complete family of distributions, however, (5.37) implies that $\tilde{g}_1(T) = \tilde{g}_2(T)$ with probability 1. Thus, if T is a sufficient statistic and Q_θ is complete, then there is only one unbiased function of such a sufficient statistic T that can be unbiased for $g(\theta)$. Of course, it will then be the MVUE. Such a sufficient statistic is formally called a *complete sufficient statistic*:

Definition 5.3.3 (Complete Sufficient Statistics) A sufficient statistic $T(\mathbf{y})$ for the family of distributions $\{P_\theta; \theta \in \Xi\}$ is called a complete sufficient statistic if the family of distributions $\{Q_\theta; \theta \in \Xi\}$ is complete where $T \sim Q_\theta$ when $\mathbf{Y} \sim P_\theta$.

Hence, finding an MVUE for $g(\theta)$ can be achieved if we can find a complete sufficient statistic T and any arbitrary unbiased estimate $\hat{g}(\mathbf{y})$ of $g(\theta)$. The MVUE is given by the conditional expectation of this arbitrary $\hat{g}$ conditioned on the complete sufficient statistic T. We may summarize this procedure for finding an MVUE for an arbitrary function $g(\theta)$ of a parameter θ as below:

1. Find any unbiased estimator $\hat{g}(\mathbf{y})$ for $g(\theta)$.
2. Find a complete sufficient statistic $T(\mathbf{y})$ for the family $\{P_\theta; \theta \in \Xi\}$ where $\mathbf{Y} \sim P_\theta(\mathbf{y})$.
3. Find an MVUE by computing $\tilde{g}(T(\mathbf{y})) = \mathbb{E}_\theta\left\{\hat{g}(\mathbf{Y})|T(\mathbf{Y}) = T(\mathbf{y})\right\}$.

Finding a complete sufficient statistic for a given family of observation distributions is facilitated by the following theorem [51]:

Theorem 5.2 (Completeness of exponential families) Suppose that the observations are N-dimensional real-valued vectors $\mathbf{Y} \sim P_\theta(\mathbf{y})$ where $\theta \in \Xi$ is an M-dimensional real-valued parameter vector so that $\Xi \subset \mathbb{R}^M$. If the density f_θ associated with each distribution P_θ is of the form

$$f_\theta(\mathbf{y}) = C(\theta)h(\mathbf{y})e^{\sum_{m=1}^{M} \theta_m T_m(\mathbf{y})}, \tag{5.38}$$

where $C, h, T_1, \ldots, T_M$ are all real-valued functions, then $T(\mathbf{y}) = (T_1(\mathbf{y}), \ldots, T_M(\mathbf{y}))$ is a complete sufficient statistic for the family $\{P_\theta; \theta \in \Xi\}$ if Ξ contains an M-dimensional rectangle.

Proof: See either [51] for an outline of a proof or [58] for a complete proof.

Many commonly encountered distributions such as Gaussian, exponential, and Laplacian distributions are all can be expressed in the exponential family form as in (5.38). Hence, as long as the observation likelihoods can be put in to the form of (5.38), recognizing a complete sufficient statistic and thus finding an MVUE is almost straightforward [51].

Example 5.3 (MVUE with Linear Gaussian Observations) Suppose we are interested in estimating a nonrandom parameter vector $\boldsymbol{\theta} \in \mathbb{R}^M$ based on an observation $\mathbf{y} \in \mathbb{R}^N$ that is known to be related to the parameter according to

$$\mathbf{Y} = H\boldsymbol{\theta} + \mathbf{W}, \tag{5.39}$$

where H is a known $N \times M$ matrix and $\mathbf{W}$ is an N-vector of Gaussian noise such that $\mathbf{W} \sim \mathcal{N}(\mathbf{0}, \Sigma_W)$.

Our first task is to find a complete sufficient statistic for estimating the parameter $\boldsymbol{\theta}$. From (5.39), we observe that

$$\mathbf{Y} \sim \mathcal{N}(H\boldsymbol{\theta}, \Sigma_W). \tag{5.40}$$

Hence, density $f_{\boldsymbol{\theta}}(\mathbf{y})$ is

$$f_{\boldsymbol{\theta}}(\mathbf{y}) = \frac{1}{(2\pi)^{\frac{N}{2}} |\Sigma_W|^{\frac{1}{2}}} e^{-\frac{1}{2}(\mathbf{y}-H\boldsymbol{\theta})^T \Sigma_W^{-1} (\mathbf{y}-H\boldsymbol{\theta})} \tag{5.41a}$$

$$= \frac{e^{-\frac{\boldsymbol{\theta}^T H^T \Sigma_W^{-1} H\boldsymbol{\theta}}{2}}}{(2\pi)^{\frac{N}{2}} |\Sigma_W|^{\frac{1}{2}}} e^{-\frac{\mathbf{y}^T \Sigma_W^{-1} \mathbf{y}}{2}} e^{\boldsymbol{\theta}^T H^T \Sigma_W^{-1} \mathbf{y}}$$

$$= C(\boldsymbol{\theta}) h(\mathbf{y}) e^{\sum_{m=1}^{M} \theta_m T_m(\mathbf{y})}, \tag{5.41b}$$

where in (5.41b) we have let

$$C(\boldsymbol{\theta}) = \frac{e^{-\frac{\boldsymbol{\theta}^T H^T \Sigma_W^{-1} H\boldsymbol{\theta}}{2}}}{(2\pi)^{\frac{N}{2}} |\Sigma_W|^{\frac{1}{2}}}$$

$$h(\mathbf{y}) = e^{-\frac{\mathbf{y}^T \Sigma_W^{-1} \mathbf{y}}{2}}$$

$$T(\mathbf{y}) = \begin{bmatrix} T_1(\mathbf{y}) \\ \vdots \\ T_M(\mathbf{y}) \end{bmatrix} = H^T \Sigma_W^{-1} \mathbf{y}. \tag{5.42}$$

Since $\mathbb{R}^M$ contains an M-dimensional rectangle and (5.41b) is in the form of (5.38), from Theorem 5.2, it follows that $T(\mathbf{y}) = H^T \Sigma_W^{-1} \mathbf{y}$ in (5.42) is a complete sufficient statistic for estimating $\boldsymbol{\theta}$ from $\mathbf{y}$.

From (5.42) and (5.40), we see that

$$\mathbb{E}\{T(\mathbf{y})\} = H^T \Sigma_W^{-1} \mathbb{E}\{\mathbf{Y}\}$$

$$= H^T \Sigma_W^{-1} H\boldsymbol{\theta}. \tag{5.43}$$

Hence, it follows straightforwardly that following function of the complete sufficient statistic $T(\mathbf{y})$ is unbiased for $\boldsymbol{\theta}$:

$$\hat{\boldsymbol{\theta}}(\mathbf{y}) = \left(H^T \Sigma_W^{-1} H\right)^{-1} T(\mathbf{y}). \tag{5.44}$$

Since there can be only one function of a complete sufficient statistic that is unbiased for $\boldsymbol{\theta}$, it follows that (5.44) must be the MVUE of $\boldsymbol{\theta}$:

$$\hat{\boldsymbol{\theta}}_{MVUE}(\mathbf{y}) = \left(H^T \Sigma_W^{-1} H\right)^{-1} T(\mathbf{y}) \tag{5.45a}$$

$$= \left(H^T \Sigma_W^{-1} H\right)^{-1} H^T \Sigma_W^{-1} \mathbf{y}. \tag{5.45b}$$

It is easy to obtain the minimum variance (in fact, the covariance matrix) achieved by this estimator:

$$\begin{aligned}
&\mathrm{Cov}_\theta\left(\hat{\boldsymbol{\theta}}_{MVUE}(\mathbf{Y})\right)\\
&=\mathbb{E}\left\{\left(\hat{\boldsymbol{\theta}}_{MVUE}(\mathbf{Y})-\mathbb{E}\left\{\hat{\boldsymbol{\theta}}_{MVUE}(\mathbf{Y})\right\}\right)\left(\hat{\boldsymbol{\theta}}_{MVUE}(\mathbf{Y})-\mathbb{E}\left\{\hat{\boldsymbol{\theta}}_{MVUE}(\mathbf{Y})\right\}\right)^T\right\}\\
&=\mathbb{E}\left\{\left(\hat{\boldsymbol{\theta}}_{MVUE}(\mathbf{Y})-\theta\right)\left(\hat{\boldsymbol{\theta}}_{MVUE}(\mathbf{Y})-\theta\right)^T\right\} &(5.46a)\\
&=\mathbb{E}\left\{\left(\left(H^T\Sigma_W^{-1}H\right)^{-1}H^T\Sigma_W^{-1}\mathbf{Y}-\theta\right)\left(\left(H^T\Sigma_W^{-1}H\right)^{-1}H^T\Sigma_W^{-1}\mathbf{Y}-\theta\right)^T\right\} &(5.46b)\\
&=\mathbb{E}\left\{\left(\left(H^T\Sigma_W^{-1}H\right)^{-1}H^T\Sigma_W^{-1}\mathbf{W}\right)\left(\left(H^T\Sigma_W^{-1}H\right)^{-1}H^T\Sigma_W^{-1}\mathbf{W}\right)^T\right\} &(5.46c)\\
&=\left(H^T\Sigma_W^{-1}H\right)^{-1}, &(5.46d)
\end{aligned}$$

where (5.46a) follows by the unbiasedness of the estimator $\hat{\boldsymbol{\theta}}_{MVUE}$, (5.46b) uses (5.45b), and (5.46c) follows by using (5.39). ■

5.3.2 Best Linear Unbiased Estimator

Although an MVUE seems a logical choice in estimating a nonrandom parameter, in practice, finding an MVUE may not be easy:

- We may not know the complete distribution function of data Y (or we may not want to assume a model for the data Y).
- Even if we knew the distribution of the data, it may be difficult to find a complete sufficient statistic for estimating the required parameter.

In these situations, we have to resort to alternatives to the theory of minimum variance unbiased estimation. Suppose that observation is an N-vector $\mathbf{Y}=(Y_1,Y_2,\ldots,Y_N)^T$ and the parameter θ to be estimated is related statistically to $\mathbf{Y}$ according to

$$\mathbb{E}\{\mathbf{Y}\}=\mathbf{s}\theta \quad \text{for all } \theta\in\Xi, \tag{5.47}$$

where $\mathbf{s}=(s_1,s_2,\ldots,s_N)$ is an N-vector that is assumed to be known.

Motivated by the LMMSE estimation under the Bayesian framework, we may perhaps look for a linear estimator in this case as well:

$$\hat{\theta}^{(l)}(\mathbf{y})=\sum_{n=1}^{N}a_n y_n=\mathbf{a}^T\mathbf{y}, \tag{5.48}$$

where coefficient vector $\mathbf{a}=(a_1,a_2,\ldots,a_N)^T$ is to be determined.

In particular, we may seek a linear estimate of the form of (5.48) that is unbiased for the parameter θ and has the minimum variance within the class of such unbiased linear estimators. Such an estimator is called the *best linear unbiased estimator* (BLUE).

The requirement of unbiasedness on $\hat{\theta}_{BLUE}(\mathbf{y})$ implies that for all $\theta \in \Xi$,

$$\begin{aligned}\theta &= \mathbb{E}\left\{\hat{\theta}_{BLUE}(\mathbf{Y})\right\} \\ &= \mathbb{E}\left\{\mathbf{a}^T\mathbf{Y}\right\} \\ &= \mathbf{a}^T\mathbb{E}\{\mathbf{Y}\} \\ &= \mathbf{a}^T\mathbf{s}\theta, \end{aligned} \tag{5.49}$$

where we have used (5.47) in the last step. From (5.49), we note that the weight vector $\mathbf{a}$ of the BLUE must satisfy the following normalization condition:

$$\mathbf{a}^T\mathbf{s} = 1. \tag{5.50}$$

We note that the variance of the BLUE (5.48) is

$$\begin{aligned}Var\left(\hat{\theta}_{BLUE}(\mathbf{Y})\right) &= \mathbb{E}\left\{\left(\hat{\theta}_{BLUE}(\mathbf{Y}) - \mathbb{E}\left\{\hat{\theta}_{BLUE}(\mathbf{Y})\right\}\right)^2\right\} \\ &= \mathbb{E}\left\{\left(\mathbf{a}^T\mathbf{Y} - \mathbf{a}^T\mathbb{E}\{\mathbf{Y}\}\right)^2\right\} \\ &= \mathbb{E}\left\{\mathbf{a}^T\left(\mathbf{Y} - \mathbb{E}\{\mathbf{Y}\}\right)\left(\mathbf{Y} - \mathbb{E}\{\mathbf{Y}\}\right)^T\mathbf{a}\right\} \\ &= \mathbf{a}^T Cov(\mathbf{Y})\mathbf{a} \\ &= \mathbf{a}^T\Sigma_{\mathbf{Y}}\mathbf{a}, \end{aligned} \tag{5.51}$$

where $\Sigma_{\mathbf{Y}} = Cov(\mathbf{Y})$ is the covariance matrix of observation $\mathbf{Y}$.

From (5.50) and (5.51), we note that finding the BLUE is equivalent to finding the coefficient vector $\mathbf{a}$ that is the solution to the following constrained optimization problem:

$$\arg\min_{\mathbf{a}} \mathbf{a}^T\Sigma_{\mathbf{Y}}\mathbf{a} \quad \text{subject to } \mathbf{a}^T\mathbf{s} = 1. \tag{5.52}$$

We may easily solve (5.52) by using the Lagrangian method. Indeed, the corresponding Lagrangian L is

$$L = \mathbf{a}^T\Sigma_{\mathbf{Y}}\mathbf{a} + \lambda\left(\mathbf{a}^T\mathbf{s} - 1\right), \tag{5.53}$$

where λ is the Lagrangian multiplier. Setting the gradient of L with respect to $\mathbf{a}$ to be zero gives

$$\frac{\partial L}{\partial \mathbf{a}} = 2\Sigma_{\mathbf{Y}}\mathbf{a} + \lambda\mathbf{s} = \mathbf{0},$$

so that

$$\mathbf{a}^* = -\frac{\lambda}{2}\Sigma_{\mathbf{Y}}^{-1}\mathbf{s}. \tag{5.54}$$

By substituting (5.54) in the constrained equation of (5.52):

$$\mathbf{a}^T\mathbf{s} = -\frac{\lambda}{2}\mathbf{s}^T\Sigma_{\mathbf{Y}}^{-1}\mathbf{s} = 1.$$

This shows that the Lagrangian multiplier is

$$\lambda = -\frac{2}{\mathbf{s}^T\Sigma_{\mathbf{Y}}^{-1}\mathbf{s}}. \tag{5.55}$$

Back substituting (5.55) in (5.54) gives the optimal coefficient vector of the BLUE to be

$$\mathbf{a}^* = \frac{\Sigma_{\mathbf{Y}}^{-1}\mathbf{s}}{\mathbf{s}^T\Sigma_{\mathbf{Y}}^{-1}\mathbf{s}}. \tag{5.56}$$

Note that the fact this indeed is the global minimizer for the problem (5.52) can be verified by using the positive definiteness of the covariance matrix Σ_Y. Using the optimal coefficient vector $\mathbf{a}^*$ of (5.56) in (5.48), we may write down the BLUE as

$$\hat{\theta}_{BLUE}(\mathbf{y}) = \frac{\mathbf{s}^T\Sigma_{\mathbf{Y}}^{-1}\mathbf{y}}{\mathbf{s}^T\Sigma_{\mathbf{Y}}^{-1}\mathbf{s}}. \tag{5.57}$$

It is interesting to observe from (5.57) that computing the BLUE only requires:

- Knowledge of the vector $\mathbf{s}$ (i.e., the mean vector of data $\mathbf{Y}$ up to a scalar amplitude)
- Knowledge of the covariance matrix Σ_Y of data $\mathbf{Y}$

Hence, when it exists, the BLUE can be found with only the knowledge of the first- and second-order statistics of data $\mathbf{Y}$, similar to the LMMSE estimator in the Bayesian framework. (In the Bayesian framework, of course, we also need to know the first- and second-order statistics of the prior of the parameter itself in addition to the first- and second-order statistics of the observations.) This can be a considerable advantage, when we do not know the full distribution function, or the density, of the observations.

However, it must be noted that for some problems we may not be able to find a linear estimator that is unbiased. This is evidenced from (5.47). Clearly, there is no guarantee that a given observation $\mathbf{y}$ may satisfy the relationship (5.47) for a parameter θ that we are interested in estimating. In such cases, we cannot find a BLUE of θ. In fact, it must be clear that if (5.47) is not satisfied for a particular observation model, the linear unbiased estimator criterion itself is not an appropriate criterion for finding an estimator at the first place. However, we may be able to get around this problem by first transforming the data and then finding a BLUE in terms of the transformed data. These discussions can be found in [56].

The minimum variance (among all linear unbiased estimators) achieved by the BLUE is obtained by substituting (5.56) in (5.51):

$$
\begin{aligned}
Var\left(\hat{\theta}_{BLUE}(\mathbf{Y})\right) &= \left(\mathbf{a}^*\right)^T \Sigma_{\mathbf{Y}} \mathbf{a}^* \\
&= \frac{\mathbf{s}^T \Sigma_{\mathbf{Y}}^{-1}}{\mathbf{s}^T \Sigma_{\mathbf{Y}}^{-1} \mathbf{s}} \Sigma_{\mathbf{Y}} \frac{\Sigma_{\mathbf{Y}}^{-1} \mathbf{s}}{\mathbf{s}^T \Sigma_{\mathbf{Y}}^{-1} \mathbf{s}} \\
&= \frac{1}{\mathbf{s}^T \Sigma_{\mathbf{Y}}^{-1} \mathbf{s}}.
\end{aligned} \tag{5.58}
$$

We can also easily generalize the above BLUE to an M-dimensional parameter vector $\boldsymbol{\theta}$ as

$$\hat{\boldsymbol{\theta}}_{BLUE}(\mathbf{y}) = \mathbf{A}\mathbf{y}, \tag{5.59}$$

where $\mathbf{A}$ is an $M \times N$ coefficient matrix to be determined. Note that (5.59) implies that the BLUE of the mth component of the parameter vector $\boldsymbol{\theta}$ is of the form of

$$\hat{\theta}_{m,BLUE}(\mathbf{y}) = \mathbf{a}_m^T \mathbf{y} = \sum_{n=1}^{N} a_{m,n} y_n, \tag{5.60}$$

where $\mathbf{a}_m^T = (a_{m,1}, a_{m,2}, \ldots, a_{m,N})$ is the mth row of the matrix $\mathbf{A}$.

The unbiasedness of $\hat{\boldsymbol{\theta}}_{BLUE}(\mathbf{y})$ requires

$$\mathbb{E}\left\{\hat{\boldsymbol{\theta}}_{BLUE}(\mathbf{Y})\right\} = \mathbf{A}\mathbb{E}\{\mathbf{Y}\} = \boldsymbol{\theta} \quad \text{for all } \boldsymbol{\theta} \in \Xi. \tag{5.61}$$

For (5.61) to be satisfied, the data vector $\mathbf{y}$ should satisfy the condition

$$\mathbb{E}\{\mathbf{Y}\} = \mathbf{S}\boldsymbol{\theta} \quad \text{for all } \boldsymbol{\theta} \in \Xi, \tag{5.62}$$

where $\mathbf{S} = [\mathbf{s}_1, \mathbf{s}_2, \ldots, \mathbf{s}_M]$ is an $N \times M$ matrix that is assumed to be known. Substitution of (5.62) in (5.61) shows that the unbiasedness criterion is equivalent to the following condition on the coefficient matrix $\mathbf{A}$:

$$\mathbf{AS} = \mathbf{I}. \tag{5.63}$$

That is, for $m = 1, \ldots, M$ and $l = 1, \ldots, M$,

$$\mathbf{a}_m^T \mathbf{s}_l = \delta_{ml}, \tag{5.64}$$

where δ_{ml} denotes the Kronecker delta function defined, for integer m and l, as

$$\delta_{ml} = \begin{cases} 1 & \text{if } m = l \\ 0 & \text{if } m \neq l \end{cases}. \tag{5.65}$$

The variance of the BLUE $\hat{\theta}_{m,BLUE}(\mathbf{y})$ of the mth component of the parameter vector $\boldsymbol{\theta}$ is

$$\begin{aligned} Var\left(\hat{\theta}_{m,BLUE}(\mathbf{Y})\right) &= \mathbb{E}\left\{\left(\hat{\theta}_{m,BLUE}(\mathbf{Y}) - \mathbb{E}\left\{\hat{\theta}_{m,BLUE}(\mathbf{Y})\right\}\right)^2\right\} \\ &= \mathbb{E}\left\{\left(\mathbf{a}_m^T\mathbf{Y} - \mathbf{a}_m^T\mathbb{E}\{\mathbf{Y}\}\right)^2\right\} \\ &= \mathbb{E}\left\{\mathbf{a}_m^T(\mathbf{Y} - \mathbb{E}\{\mathbf{Y}\})(\mathbf{Y} - \mathbb{E}\{\mathbf{Y}\})^T\mathbf{a}_m\right\} \\ &= \mathbf{a}_m^T Cov(\mathbf{Y})\mathbf{a}_m \\ &= \mathbf{a}_m^T \Sigma_{\mathbf{Y}}\mathbf{a}_m, \end{aligned} \tag{5.66}$$

where, as before, $\Sigma_Y = Cov(\mathbf{Y})$ is the covariance matrix of data $\mathbf{Y}$.

From (5.64) and (5.66), it is clear that what we have is an M decoupled problems of BLUE design. Hence, following the same procedure as before for a scalar parameter, we can show that the BLUE of a vector parameter $\boldsymbol{\theta}$ is given by

$$\hat{\boldsymbol{\theta}}_{BLUE}(\mathbf{y}) = \left(\mathbf{S}^T\Sigma_{\mathbf{Y}}^{-1}\mathbf{S}\right)^{-1}\mathbf{S}^T\Sigma_{\mathbf{Y}}^{-1}\mathbf{y}. \tag{5.67}$$

Again, determining the BLUE only requires the first- and second-order statistics of data:

- Knowledge of $\mathbf{S}$ (i.e., the mean of data $\mathbf{Y}$)
- Knowledge of the covariance matrix $\Sigma_{\mathbf{Y}}$ of data $\mathbf{Y}$

The covariance matrix of the above BLUE (5.67) can also be shown to be

$$Cov\left(\hat{\boldsymbol{\theta}}_{BLUE}(\mathbf{Y})\right) = \left(\mathbf{S}^T\Sigma_{\mathbf{Y}}^{-1}\mathbf{S}\right)^{-1}. \tag{5.68}$$

Hence, the variance of the mth component of the BLUE is

$$Var\left(\hat{\boldsymbol{\theta}}_{m,BLUE}(\mathbf{Y})\right) = \left[\left(\mathbf{S}^T\Sigma_{\mathbf{Y}}^{-1}\mathbf{S}\right)^{-1}\right]_{m,m}. \tag{5.69}$$

Example 5.4 (BLUE with Linear Observations) Let us revisit the linear observation model of Example 5.3:

$$\mathbf{Y} = H\boldsymbol{\theta} + \mathbf{W}, \tag{5.70}$$

where $\boldsymbol{\theta} \in \mathbb{R}^M$, $\mathbf{Y} \in \mathbb{R}^N$, H is a known $N \times M$ matrix, and $\mathbf{W}$ is an N-vector of noise. However, unlike in Example 5.3, now, suppose that we do not know the exact distribution of the additive noise $\mathbf{W}$ except for its first and second moments. Specifically, suppose that $\mathbf{W}$ is of zero mean and has a covariance matrix of Σ_W.

Since we do not know the exact distribution of data, of course now we cannot apply MVUE approach. However, from (5.70), it follows straightforwardly that

$$\mathbb{E}\{\mathbf{Y}\} = H\boldsymbol{\theta} \tag{5.71}$$

and

$$\begin{aligned} Cov(\mathbf{Y}) &= \mathbb{E}\left\{(\mathbf{Y} - H\boldsymbol{\theta})(\mathbf{Y} - H\boldsymbol{\theta})^T\right\} \\ &= \mathbb{E}\left\{\mathbf{W}\mathbf{W}^T\right\} \\ &= \Sigma_W. \end{aligned} \tag{5.72}$$

Hence, from (5.67), we see that the BLUE $\hat{\boldsymbol{\theta}}_{BLUE}(\mathbf{y})$ of $\boldsymbol{\theta}$ is given by

$$\hat{\boldsymbol{\theta}}_{BLUE}(\mathbf{y}) = \left(\mathbf{H}^T \Sigma_W^{-1} \mathbf{H}\right)^{-1} \mathbf{H}^T \Sigma_W^{-1} \mathbf{y} \tag{5.73}$$

and the minimum variance of the mth component of the estimator is

$$Var\left(\hat{\theta}_{m,BLUE}(\mathbf{Y})\right) = \left[\left(\mathbf{H}^T \Sigma_W^{-1} \mathbf{H}\right)^{-1}\right]_{m,m}. \tag{5.74}$$

Moreover, the covariance matrix of the BLUE $\hat{\boldsymbol{\theta}}_{BLUE}(\mathbf{y})$ of $\boldsymbol{\theta}$ is

$$Cov\left(\hat{\boldsymbol{\theta}}_{BLUE}(\mathbf{Y})\right) = \left(\mathbf{H}^T \Sigma_W^{-1} \mathbf{H}\right)^{-1}. \tag{5.75}$$

Comparison of (5.73) with the MVUE (5.45b) obtained in Example 5.3 shows that the BLUE has the same form as the MVUE. This could have been anticipated a priori in this situation because the MVUE in Example 5.3 happens to be a linear estimator. In general, of course, the BLUE and MVUE will not be the same. ■

5.3.3 Maximum Likelihood Estimation

As we already discussed earlier, although theory of minimum variance unbiased estimation is an elegant framework for nonrandom parameter estimation, in practice, it may be difficult to apply: It might not be easy to find a useful complete sufficient statistic for a problem, and even if we did find a complete sufficient statistic, the rest of the required analysis may not be all that trivial either. One alternative to MVUE is the BLUE that we discussed in the previous section. The BLUE draws a parallel from the LMMSE estimator in the Bayesian framework to the non-Bayesian framework. It is worth recalling that another Bayesian estimator that we saw earlier is the MAP estimator. The maximum likelihood estimator (MLE) can be thought of as drawing a parallel from the Bayesian MAP framework to the non-Bayesian framework.

Recall that the (Bayesian) MAP estimator of (5.28) is that value of the parameter $\theta \in \Xi$ that maximizes the posterior probability of the random parameter Θ given the observation Y. Of course, this MAP computation requires the prior distribution of the random parameter Θ, which we denoted as $f_\Theta(.)$. Obviously, we cannot apply the MAP estimation in the case of a nonrandom parameter because there does not exist such a prior distribution in this case. However, our rationale for the MAP estimator was that it makes sense to estimate θ as that value $\theta \in \Xi$ that has the highest posterior probability given the observation

$$\hat{\theta}_{MAP}(y) = \arg \max_{\theta \in \Xi} f_{\Theta|Y}(\theta | Y = y). \tag{5.76}$$

But then it is also logical to estimate θ as that value $\theta \in \Theta$ that gives the highest posterior probability of actually causing the observation $Y = y$ at the first place:

$$\hat{\theta}_{ML}(y) = \arg \max_{\theta \in \Xi} f_\theta(y), \tag{5.77}$$

where $f_\theta(y)$ is the likelihood function of y given θ. Hence, the estimate (5.77) that maximizes the likelihood function $f_\theta(y)$ is called the *maximum likelihood (ML) estimator* of parameter θ.

Note that since

$$f_{\Theta|Y}(\theta | Y = y) = \frac{f_{Y|\Theta}(y|\theta) f_\Theta(\theta)}{f_Y(y)}, \tag{5.78}$$

one may also think of the ML estimator (5.77) as the result when the parameter θ is treated as though it is uniformly distributed on Ξ. Since uniform distribution over Ξ is the most noninformative prior, it is akin to considering a worst-case prior when the only thing we know is that $\theta \in \Xi$.

Since maximizing $f_\theta(y)$ is equivalent to maximizing $\log f_\theta(y)$, a necessary condition for an estimator to be the ML estimate is that it should satisfy the following *likelihood equation*:

$$\frac{\partial}{\partial \theta} \left[\log \left(f_\theta(y) \right) \right] |_{\theta = \hat{\theta}_{ML}(y)} = 0. \tag{5.79}$$

Of course, satisfying likelihood equation (5.79) is only a necessary condition for an estimate to be an ML estimate. It may not guarantee that the estimate is in fact an ML estimate. For that, we may have to establish that it is also the global maximum of the likelihood function. However, it is known that only a solution to the likelihood equation can be an estimator that achieves the minimum possible estimation error variance among all unbiased estimators. (In the next section, we will derive this bound called the *Cramer–Rao lower bound* (CRLB).) Hence, being a solution to the likelihood equation can be a sign that it can potentially be a good estimator.

It is interesting to observe that ML estimators do not directly have any claim regarding minimizing the conditional mean-squared error $\mathbb{E}\left\{(\hat{\theta}_{ML}(y)-\theta)^2\right\}$, just as MAP estimators do not have any claim to minimizing the MSE. Moreover, unlike in the case of MVUE or the BLUE, we have not even imposed the requirement that the estimator be unbiased. Hence, it is entirely possible that the ML estimators obtained as solutions to the likelihood equation are biased estimators.

In the case of iid observations, however, the solutions to the likelihood equation possess certain properties that make them at least reasonable estimators. Indeed, consider a sequence of i.i.d. observations $Y_1, Y_2, \ldots, Y_N$, each with marginal density $f_\theta(y)$. Let us denote by $\hat{\theta}_N(\mathbf{y})$ the solution to the likelihood equation with sample size N. Interestingly enough, it can be shown that under sufficient regularity conditions, the sequence of solutions $\{\hat{\theta}_N(\mathbf{y})\}$ converges to the true parameter value θ in probability [51]:

$$\lim_{N\to\infty} \hat{\theta}_N(\mathbf{Y}) \underset{=}{\text{i.p.}} \theta. \tag{5.80}$$

That is,

$$\lim_{N\to\infty} P_\theta\left(|\hat{\theta}_N(\mathbf{Y})-\theta| > \epsilon\right) = 0 \;\; \forall \;\; \epsilon > 0. \tag{5.81}$$

The property (5.80) is called the (weak) *consistency*. Hence, ML estimators are consistent under appropriate smoothness and uniqueness conditions. For a discussion of a suitable set of such conditions, the reader is referred to [51].

As mentioned earlier, there is no guarantee that an ML estimator will be an unbiased estimator, let alone minimizing the conditional MSE. However, in the case of iid observations as assumed earlier, it is known that under appropriate regularity conditions, the solutions to the likelihood equation $\{\hat{\theta}_N(\mathbf{y})\}$ are *asymptotically unbiased*:

$$\lim_{N\to\infty} \mathbb{E}\left\{\hat{\theta}_N(\mathbf{Y})\right\} = \theta. \tag{5.82}$$

5.3.4 Performance Bounds: Cramer-Rao Lower Bound

In the previous sections, we have developed several procedures to obtain an estimator of a parameter. Perhaps our ultimate performance criterion is to have an estimator that results in the smallest possible conditional mean-squared error $\mathbb{E}\{(\hat{\theta}(Y)-\theta)^2\}$. However, none of the estimators we have developed directly attempts to minimize this quantity itself. The one that comes closest, at least in terms of the formulation, is the MVUE. However, by imposing the unbiasedness as a prerequisite, we still run the risk of precluding the actual estimator that may minimize the conditional mean-squared error $\mathbb{E}\{(\hat{\theta}(Y)-\theta)^2\}$ from consideration, if it happens to be a biased estimator. On the other hand, ML estimators have nice enough asymptotic properties that do not allow ruling out them being better than MVUEs in certain situations in terms of minimizing the overall MSE.

Given these complications, it is desirable to have ultimate performance bounds that are applicable to all estimators. The following result provides exactly that [51]:

Theorem 5.3 (Information inequality) Consider the problem of estimating a parameter $\theta \in \Xi$, where Ξ is an open interval, based on an observation Y. Suppose that Y has a density function $f_\theta(y)$ for each $\theta \in \Xi$. Let $\hat{\theta}$ be any arbitrary estimate of the parameter θ. Under sufficient regularity conditions on $f_\theta(y)$, the variance of any arbitrary estimator is lower bound as

$$\text{Var}_\theta\left[\hat{\theta}(Y)\right] \geq \frac{\left(\frac{\partial}{\partial\theta}\left[\mathbb{E}_\theta\left\{\hat{\theta}(Y)\right\}\right]\right)^2}{I_\theta}, \tag{5.83}$$

where $Var\left(\hat{\theta}(\mathbf{Y})\right) = \mathbb{E}\left\{\left(\hat{\theta}(\mathbf{Y}) - \mathbb{E}\left\{\hat{\theta}(\mathbf{Y})\right\}\right)^2\right\}$ is the variance of the estimator $\hat{\theta}(\mathbf{Y})$ and

$$I_\theta = \mathbb{E}\left\{\left(\frac{\partial}{\partial\theta}\left[\log\left(f_\theta(Y)\right)\right]\right)^2\right\} = -\mathbb{E}_\theta\left\{\frac{\partial^2}{\partial\theta^2}\left[\log\left(f_\theta(Y)\right)\right]\right\} \tag{5.84}$$

is called the *Fisher information* for estimating θ from Y.

Proof: See [51] for a proof of the information inequality.

The information inequality (5.83) is a lower bound on the variance of any arbitrary estimator for θ. This lower bound is characterized by the quantity defined in (5.84) called the Fisher information. The higher the Fisher information I_θ for the given model, the smaller the lower bound can be.

However, it is important to note that the information inequality is a lower bound on the accuracy of the estimator with respect to itself, rather than the original problem. What we mean by this is that the information inequality lower bound is a lower bound on the variance of an arbitrary estimator, which is a measure of variation of the estimator around its own mean value: $\text{Var}_\theta[\hat{\theta}(Y)] = \mathbb{E}_\theta\left\{\left(\hat{\theta}(Y) - \mathbb{E}\{\hat{\theta}(Y)\}\right)^2\right\}$. From (5.31), recall that this variance term is just one part of the overall conditional MSE $\mathbb{E}\{(\hat{\theta}(Y) - \theta)^2\}$. However, if $\hat{\theta}(\mathbf{Y})$ is an unbiased estimator, then variance itself is the MSE. Hence, in this case, a lower bound on the variance is directly a lower bound on the MSE providing a more informative picture of the situation.

Since

$$\begin{aligned}\frac{\partial}{\partial\theta}\left[\mathbb{E}_\theta\left\{\hat{\theta}(Y)\right\}\right] &= \frac{\partial}{\partial\theta}[\theta] \\ &= 1,\end{aligned} \tag{5.85}$$

for an unbiased estimator, the information inequality (5.83) reduces to

$$\mathrm{Var}_\theta\left[\hat{\theta}(Y)\right] \geq \frac{1}{I_\theta}, \tag{5.86}$$

which is called the CRLB on the variance of an unbiased estimator.

The information inequality and the CRLB tell what we may expect as the ultimate performance of an estimator for a given problem. However, they do not say whether there exists an estimator that achieves the lower bounds given by (5.83) or (5.86). Moreover, even if there is such an estimator, the bounds do not tell us how to find it. Not surprisingly, only under certain special conditions, there will be an estimate that achieves equality in the information inequality (5.83).

Theorem 5.4 (Existence of an Estimator that Achieves the Information Inequality Lower Bound) Within regularity, the information lower bound is achieved by an estimator $\hat{\theta}$ if and only if $\hat{\theta}(y) = T(y)$ with likelihood function having the form of

$$f_\theta(y) = C(\theta)e^{g(\theta)T(y)}\, h(y), \tag{5.87}$$

where C, g, T, and h are all real-valued functions of their arguments and Ξ is an open interval.

Proof: See [51] for an outline.

Note that the class of distributions with density functions of the form of (5.87) is called a *one-parameter exponential family* of distributions. When an estimator achieves the CRLB, it is called an *efficient estimator*. Sometimes, an estimator may only achieve the CRLB asymptotically. In that case, the estimator is called *asymptotically efficient*. Recall that under sufficient regularity conditions, we saw that the ML estimators can be asymptotically unbiased for certain estimation problems. In the same way, ML estimators can be asymptotically efficient for certain problems, under enough regularity conditions.

Observe also from (5.31) that if an estimator is both asymptotically unbiased and the asymptotic variance vanishes, then the asymptotic conditional MSE also vanishes. In this case, we may call that the estimator converges in the sense of mean square to the true parameter. Similar to the weak consistency defined in (5.80), we may call this the *mean-square consistency* or that the estimator is consistent in the sense of mean square:

$$\lim_{N\to\infty} \hat{\theta}_N(\mathbf{Y}) \ \underset{=}{\text{m.s.}} \ \theta. \tag{5.88}$$

That is,

$$\lim_{N\to\infty} \mathbb{E}_\theta\left\{\left|\hat{\theta}_N(\mathbf{Y}) - \theta\right|^2\right\} = 0. \tag{5.89}$$

The CRLB can easily be generalized to the vector case. Indeed, if $\hat{\boldsymbol{\theta}}(\mathbf{y})$ is an unbiased estimator for the parameter vector $\boldsymbol{\theta}$, then the CRLB for the covariance matrix of $\hat{\boldsymbol{\theta}}$ is

$$\mathrm{Cov}\left(\hat{\boldsymbol{\theta}}(\mathbf{Y})\right) \geq \mathbf{J}^{-1}, \tag{5.90}$$

where $\mathrm{Cov}\left(\hat{\boldsymbol{\theta}}(\mathbf{Y})\right) = \mathbb{E}\left\{\left(\hat{\boldsymbol{\theta}}(\mathbf{Y}) - \theta\right)\left(\hat{\boldsymbol{\theta}}(\mathbf{Y}) - \theta\right)^T\right\}$, $\mathbf{J}$ is the *Fisher information matrix* defined as

$$\mathbf{J} = \mathbb{E}\left\{\left[\frac{\partial}{\partial\boldsymbol{\theta}}\log f_\theta(\mathbf{Y})\right]\left[\frac{\partial}{\partial\boldsymbol{\theta}}\log f_\theta(\mathbf{Y})\right]^T\right\} = -\mathbb{E}\left\{\frac{\partial^2}{\partial\boldsymbol{\theta}^2}\log f_\theta(\mathbf{Y})\right\},$$

and the bound (5.90) must be interpreted as $\mathrm{Cov}\left(\hat{\boldsymbol{\theta}}(\mathbf{Y})\right) - J^{-1}$ is a positive semi-definite matrix.

Example 5.5 (CRLB for the linear Gaussian observation model) We may derive the CRLB for estimating the parameter vector $\boldsymbol{\theta}$ in the linear observation model (5.39) of Example 5.3. Since $\mathbf{W} \sim (\mathbf{0}, \Sigma_W)$, from the observation model (5.39), we have that $\mathbf{Y} \sim (H\boldsymbol{\theta}, \Sigma_W)$. Hence,

$$f_\theta(\mathbf{y}) = \frac{1}{(2\pi)^{\frac{N}{2}}|\Sigma_W|^{\frac{1}{2}}} e^{-\frac{1}{2}(\mathbf{y}-H\boldsymbol{\theta})^T\Sigma_W^{-1}(\mathbf{y}-H\boldsymbol{\theta})}$$

and

$$\log f_\theta(\mathbf{y}) = -\frac{N}{2}\log 2\pi - \frac{1}{2}\log|\Sigma_W| - \frac{1}{2}(\mathbf{y} - H\boldsymbol{\theta})^T\Sigma_W^{-1}(\mathbf{y} - H\boldsymbol{\theta}). \tag{5.91}$$

Differentiating (5.91) with respect to the vector $\boldsymbol{\theta}$, we get

$$\frac{\partial}{\partial\boldsymbol{\theta}}\log f_\theta(\mathbf{Y}) = H^T\Sigma_W^{-1}(\mathbf{y} - H\boldsymbol{\theta}). \tag{5.92}$$

Hence,

$$\frac{\partial^2}{\partial\boldsymbol{\theta}^2}\log f_\theta(\mathbf{Y}) = -H^T\Sigma_W^{-1}H. \tag{5.93}$$

Thus, the Fisher information matrix for the linear Gaussian observation model of (5.39) is

$$J = -\mathbb{E}\left\{\frac{\partial^2}{\partial\boldsymbol{\theta}^2}\log f_\theta(\mathbf{Y})\right\} = H^T\Sigma_W^{-1}H, \tag{5.94}$$

so that the CRLB on any unbiased estimator for the parameter vector $\boldsymbol{\theta}$ is

$$CRLB = J^{-1} = \left(H^T \Sigma_W^{-1} H\right)^{-1} \tag{5.95}$$

Comparing (5.95) with the covariance (5.46d) achieved by the MVUE derived in Example 5.3, we note that the MVUE in this case in fact achieves the CRLB and thus is an efficient estimator. ■

5.4 SUMMARY

- When an observation Y depends on a parameter θ that can take a continuum of values, determining θ corresponding to a given observation is called an estimation problem.
- An estimator $\hat{\theta}(.)$ is a mapping from the observation space to a parameter space:

$$\hat{\theta} : \mathcal{Y} \longrightarrow \Theta.$$

  The estimate of θ for a given observation y is the $\hat{\theta}(y)$. The estimation error is given by $\theta - \hat{\theta}(\mathbf{y})$.
- We prefer an estimator that has a small estimation error in some sense. The most common criterion adapted in practice is to look for estimators that have small squared estimation errors $\left(\theta - \hat{\theta}(\mathbf{y})\right)^2$. Of course, since $\hat{\theta}(\mathbf{Y})$ is a random variable, we can only make this squared error small in some statistical sense.
- The Bayesian approach to parameter estimation assumes that the parameter θ corresponding to a given observation is a realization of a random variable Θ with an associated prior distribution. We will assume that this distribution has a corresponding density $f_\Theta(\theta)$. Hence, an estimator is said to be Bayesian minimum mean-squared error (MMSE) optimal if it minimizes the mean-squared error $\mathbb{E}\left\{(\Theta - \hat{\theta}(\mathbf{Y}))^2\right\}$ where expectation is taken with respect to both Θ and $\mathbf{Y}$.

  ◦ The MMSE estimate of a scalar parameter, defined as

$$\hat{\theta}_{MMSE} = \min_{\hat{\theta}} \ \mathbb{E}\left\{(\Theta - \hat{\theta}(\mathbf{Y}))^2\right\},$$

  is given by the conditional mean (CM) of Θ given $\mathbf{Y}$:

$$\hat{\theta}_{MMSE}(\mathbf{y}) = \mathbb{E}\{\Theta | \mathbf{Y} = \mathbf{y}\}.$$

- The MMSE estimate of a vector parameter, defined as

$$\hat{\boldsymbol{\theta}}_{MMSE} = \min_{\hat{\theta}} \ \mathbb{E}\left\{\|\hat{\boldsymbol{\theta}}(\mathbf{Y}) - \boldsymbol{\Theta}\|^2\right\},$$

  is given by the CM of $\boldsymbol{\Theta}$ given $\mathbf{Y}$:

$$\hat{\boldsymbol{\theta}}_{MMSE}(\mathbf{y}) = \mathbb{E}\{\boldsymbol{\Theta}|\mathbf{Y} = \mathbf{y}\},$$

  where the mth component of the estimate, for $m = 1, \ldots, M$, is

$$\hat{\theta}_m(y) = \mathbb{E}\{\Theta_m|\mathbf{Y} - \mathbf{y}\}.$$

- The weighted MMSE estimate of a vector parameter, defined as

$$\hat{\boldsymbol{\theta}}_{WMMSE} = \min_{\hat{\theta}} \ \mathbb{E}\left\{\left(\hat{\boldsymbol{\theta}} - \boldsymbol{\Theta}\right)^T \mathbf{A}\left(\hat{\boldsymbol{\theta}} - \boldsymbol{\Theta}\right)\right\},$$

  where $\mathbf{A}$ is a symmetric, positive-definite matrix, is given by the CM of $\boldsymbol{\Theta}$ given $\mathbf{Y}$:

$$\hat{\boldsymbol{\theta}}_{WMMSE}(\mathbf{y}) = \mathbb{E}\{\boldsymbol{\Theta}|\mathbf{Y} = \mathbf{y}\}.$$

- The linear minimum mean squared error estimator of a vector parameter, defined as

$$\hat{\boldsymbol{\theta}}_{LMMSE}(\mathbf{y}) = \min_{\hat{\theta}^{(l)}} \ \mathbb{E}\left\{\left(\hat{\boldsymbol{\theta}}^{(l)}(\mathbf{Y}) - \boldsymbol{\Theta}\right)^T \mathbf{A}\left(\hat{\boldsymbol{\theta}}^{(l)}(\mathbf{Y}) - \boldsymbol{\Theta}\right)\right\},$$

  where $\hat{\boldsymbol{\theta}}^{(l)}(.)$ is constrained to have the form

$$\hat{\boldsymbol{\theta}}^{(l)}(\mathbf{y}) = A\mathbf{y} + \mathbf{b}$$

  for some yet to be determined matrix A and vector $\mathbf{b}$, is given by

$$\hat{\boldsymbol{\theta}}_{LMMSE}(\mathbf{y}) = \Sigma_{\boldsymbol{\Theta}\mathbf{Y}}\Sigma_{\mathbf{Y}}^{-1}(\mathbf{y} - \mathbb{E}\{\mathbf{Y}\}) + \mathbb{E}\{\boldsymbol{\Theta}\},$$

  where $\Sigma_{\mathbf{Y}} = Cov(\mathbf{Y})$ is the covariance matrix of observation $\mathbf{Y}$ and $\Sigma_{\boldsymbol{\Theta}\mathbf{Y}} = Cov(\boldsymbol{\Theta}, \mathbf{Y})$ is the cross-covariance of $\boldsymbol{\Theta}$ and $\mathbf{Y}$.
- The maximum a posteriori probability (MAP) estimator, defined as

$$\hat{\theta}_{MAP}(\mathbf{y}) = \max_{\theta} f_{\Theta|\mathbf{Y}}(\theta|\mathbf{y}),$$

  is a Bayesian estimator that does not explicitly minimize an MSE cost function as all other Bayesian estimators considered earlier. However, it is a Bayesian estimate since it does still treat the parameter to be estimated as a random quantity with an associated prior distribution. The MAP estimator is a popular choice since it might be easier to compute compared to other Bayesian estimators. A necessary condition for an estimator to be the MAP estimate is for it to satisfy the following MAP equation:

$$\frac{\partial}{\partial\theta}\log f_{\mathbf{Y}|\Theta}(\mathbf{y}|\theta)\bigg|_{\theta=\hat{\theta}_{MAP}} = -\frac{\partial}{\partial\theta}\log f_{\Theta}(\theta)\bigg|_{\theta=\hat{\theta}_{MAP}}.$$

- The non-Bayesian approach only assumes that the parameter corresponding to all observations is some fixed value from a set Ξ, and we just do not know this fixed value. Hence, there is no randomness associated with the parameter, and thus, non-Bayesian approach is also called the nonrandom parameter estimation. Nonrandom parameter estimation can only attempt to minimize the conditional mean-squared error $\mathbb{E}\left\{(\theta-\hat{\theta}(\mathbf{Y}))^2\right\}$ where expectation is taken with respect to $\mathbf{Y}$.
- Since

$$\mathbb{E}\left\{(\theta-\hat{\theta}(\mathbf{Y}))^2\right\} = \left(b\left(\hat{\theta}(\mathbf{Y})\right)\right)^2 + Var\left(\hat{\theta}(\mathbf{Y})\right), \tag{5.96}$$

where $Var\left(\hat{\theta}(\mathbf{Y})\right)$ and $b\left(\hat{\theta}(\mathbf{Y})\right)$ denote, respectively, the variance and bias of an estimator $\hat{\theta}(\mathbf{Y})$ defined as

$$Var\left(\hat{\theta}(\mathbf{Y})\right) = \mathbb{E}\left\{\left(\hat{\theta}(\mathbf{Y}) - \mathbb{E}\left\{\hat{\theta}(\mathbf{Y})\right\}\right)^2\right\}$$

and

$$b\left(\hat{\theta}(\mathbf{Y})\right) = \theta - \mathbb{E}\left\{\hat{\theta}(\mathbf{Y})\right\},$$

nonrandom parameter estimation can be a trade-off of minimizing the contributions from these two components to the conditional MSE.
- A desired property in any estimator is the unbiasedness:

$$\mathbb{E}\left\{\hat{\theta}(\mathbf{Y})\right\} = \theta.$$

- Since bias term in (5.96) vanishes for an unbiased estimator, minimizing the conditional MSE (5.96) is equivalent to minimizing the variance of an unbiased estimator. An unbiased estimator that has the minimum possible variance among all unbiased estimators for a parameter of interest is called a minimum variance unbiased estimator.

  ◦ A function $T(\mathbf{y})$ of $\mathbf{y}$ is called a sufficient statistic for the family of distributions $\{P_\theta; \theta \in \Xi\}$ if the distribution of Y conditioned on $T(Y)$ does not depend on θ

for $\theta \in \Xi$, when $Y \sim P_\theta$. A sufficient statistic can be thought of as containing all information about θ contained in the original observations.

- A family of distributions $\{P_\theta; \theta \in \Xi\}$ is called a complete family of distributions if $\mathbb{E}_\theta\{g(\mathbf{Y})\} = 0$ for all $\theta \in \Xi$ implies that $g(\mathbf{Y}) = 0$ with probability 1 whenever $\mathbf{Y} \sim P_\theta$. If $\mathbf{Y} \sim P_\theta$ where $\{P_\theta; \theta \in \Xi\}$ is a complete family of distributions, then the only sufficient statistics $T(\mathbf{y})$ the family P_θ can have are just one-to-one mappings of the observations. In other words, there is no possible further reduction of the observations.
- A sufficient statistic $T(\mathbf{y})$ for a family of distributions $\{P_\theta; \theta \in \Xi\}$ is called a complete sufficient statistic if the family of distributions $\{Q_\theta; \theta \in \Xi\}$ is complete where $T \sim Q_\theta$ when $\mathbf{Y} \sim P_\theta$.
- There can be only one function of a complete sufficient statistic T that can be unbiased for $g(\theta)$. From the Rao–Blackwell theorem, it then follows that this unique function must be the MVUE of $g(\theta)$.
- Thus, a procedure for finding an MVUE for an arbitrary function $g(\theta)$ of a parameter θ is given by:

  1. Find any unbiased estimator $\hat{g}(\mathbf{y})$ for $g(\theta)$.
  2. Find a complete sufficient statistic $T(\mathbf{y})$ for the family $\{P_\theta; \theta \in \Xi\}$ where $\mathbf{Y} \sim P_\theta(\mathbf{y})$.
  3. Find an MVUE by computing $\tilde{g}(T(\mathbf{y})) = \mathbb{E}_\theta\left\{\hat{g}(\mathbf{Y})|T(\mathbf{Y}) = T(\mathbf{y})\right\}$.

- For many commonly encountered distributions, complete sufficient statistics can be found thanks to the completeness of exponential families stated in Theorem 5.2.

- When information needed to apply theory of minimum variance unbiased estimation is not available, an alternative nonrandom parameter estimation criterion is to constrain the allowed estimators to be linear (or affine, in general) in observation as $\hat{\theta}^{(l)}(\mathbf{y}) = \mathbf{a}^T\mathbf{y}$ for some $\mathbf{a}$ and search for the unbiased estimator with the minimum variance among this class of linear estimators. Such an estimator is called the best linear unbiased estimator (BLUE).

  - Suppose that the N-vector observation $\mathbf{Y}$ and the parameter θ to be estimated are related according to

$$\mathbb{E}\{\mathbf{Y}\} = \mathbf{s}\theta \quad \text{for all } \theta \in \Xi,$$

  where $\mathbf{s}$ is an N-vector that is assumed to be known. The BLUE is then given by

$$\hat{\theta}_{BLUE}(\mathbf{y}) = \frac{\mathbf{s}^T \Sigma_{\mathbf{Y}}^{-1} \mathbf{y}}{\mathbf{s}^T \Sigma_{\mathbf{Y}}^{-1} \mathbf{s}}$$

with the resulting minimum variance

$$Var\left(\hat{\theta}_{BLUE}(\mathbf{Y})\right) = \frac{1}{\mathbf{s}^T \Sigma_{\mathbf{Y}}^{-1} \mathbf{s}}.$$

- If the parameter to be estimated is an M-dimensional vector $\boldsymbol{\theta}$ that is related to the observations according to

$$\mathbb{E}\{\mathbf{Y}\} = \mathbf{S}\boldsymbol{\theta} \quad \text{for all } \boldsymbol{\theta} \in \Xi,$$

where $\mathbf{S}$ is an $N \times M$ matrix that is assumed to be known, the BLUE is given by

$$\hat{\boldsymbol{\theta}}_{BLUE}(\mathbf{y}) = \left(\mathbf{S}^T \Sigma_{\mathbf{Y}}^{-1} \mathbf{S}\right)^{-1} \mathbf{S}^T \Sigma_{\mathbf{Y}}^{-1} \mathbf{y}$$

with the associated estimator covariance matrix

$$Cov\left(\hat{\boldsymbol{\theta}}_{BLUE}(\mathbf{Y})\right) = \left(\mathbf{S}^T \Sigma_{\mathbf{Y}}^{-1} \mathbf{S}\right)^{-1}.$$

- The maximum likelihood (ML) estimate of a parameter θ is defined as the value of the parameter that maximizes the likelihood function $f_\theta(y)$ of the observation:

$$\hat{\theta}_{ML}(y) = \arg \max_{\theta \in \Xi} f_\theta(y).$$

- A necessary condition for an estimator to be the ML estimator is that it should satisfy the likelihood equation:

$$\frac{\partial}{\partial \theta} \left[\log (f_\theta(y))\right] |_{\theta = \hat{\theta}_{ML}(y)} = 0.$$

- Although there is no guarantee that solutions to the likelihood equation, and thus ML estimators, are unbiased, it can be shown that under sufficient regularity conditions, they are consistent estimates in the sense that the sequence of solutions $\{\hat{\theta}_N(\mathbf{y})\}$ to the likelihood equation with N iid samples converges to the true parameter θ in probability:

$$\lim_{N \to \infty} \hat{\theta}_N(\mathbf{Y}) \underset{=}{\text{i.p.}} \theta.$$

- In the case of iid observations with appropriate regularity conditions, the solutions to the likelihood equation $\{\hat{\theta}_N(\mathbf{y})\}$ are asymptotically unbiased:

$$\lim_{N \to \infty} \mathbb{E}\left\{\hat{\theta}_N(\mathbf{Y})\right\} = \theta. \tag{5.97}$$

- Under sufficient regularity conditions, a lower bound on the variance $Var\left(\hat{\theta}(\mathbf{Y})\right) = \mathbb{E}\left\{\left(\hat{\theta}(\mathbf{Y}) - \mathbb{E}\left\{\hat{\theta}(\mathbf{Y})\right\}\right)^2\right\}$ of any arbitrary estimator $\hat{\theta}(\mathbf{Y})$ for a parameter θ, if given by the information inequality

$$\text{Var}_\theta\left[\hat{\theta}(Y)\right] \geq \frac{\left(\frac{\partial}{\partial\theta}\left[\mathbb{E}_\theta\left\{\hat{\theta}(Y)\right\}\right]\right)^2}{I_\theta}, \tag{5.98}$$

  where

$$I_\theta = \mathbb{E}\left\{\left(\frac{\partial}{\partial\theta}\left[\log\left(f_\theta(Y)\right)\right]\right)^2\right\} = -\mathbb{E}_\theta\left\{\frac{\partial^2}{\partial\theta^2}\left[\log\left(f_\theta(Y)\right)\right]\right\}$$

  is called the Fisher information for estimating θ from Y.
- For unbiased estimators, the information inequality (5.98) reduces to the so-called Cramer–Rao lower bound (CRLB):

$$\text{Var}_\theta\left[\hat{\theta}(Y)\right] \geq \frac{1}{I_\theta}.$$

- An unbiased estimator that achieves the CRLB is called an efficient estimator. If an estimator only achieves the CRLB asymptotically, then it is called an asymptotically efficient estimator.
- If $\hat{\boldsymbol{\theta}}(\mathbf{y})$ is an unbiased estimator for a vector parameter $\boldsymbol{\theta}$, then the CRLB for the covariance matrix of $\hat{\boldsymbol{\theta}}$ becomes

$$\text{Cov}\left(\hat{\boldsymbol{\theta}}(\mathbf{Y})\right) \geq \mathbf{J}^{-1},$$

  where $\mathbf{J}$ is the Fisher information matrix defined as

$$\mathbf{J} = \mathbb{E}\left\{\left[\frac{\partial}{\partial\boldsymbol{\theta}}\log f_\theta(\mathbf{Y})\right]\left[\frac{\partial}{\partial\boldsymbol{\theta}}\log f_\theta(\mathbf{Y})\right]^T\right\} = -\mathbb{E}\left\{\frac{\partial^2}{\partial\boldsymbol{\theta}^2}\log f_\theta(\mathbf{Y})\right\}$$

  and the bound must be interpreted as $\text{Cov}\left(\hat{\boldsymbol{\theta}}(\mathbf{Y})\right) - J^{-1}$ is a positive semi-definite matrix.

6

POWER SPECTRUM ESTIMATION

6.1 INTRODUCTION

Power spectral density (PSD), or simply the power spectrum, of a signal is a representation of the spread of signal power as a function of frequency. In this chapter, we review basic concepts and techniques of power spectrum estimation that will be useful in subsequent chapters. We limit ourselves to the so-called *nonparametric spectrum estimation* techniques that do not presume a parameterized form for the signal whose spectrum is to be estimated. The application of interest for us is the wideband spectrum sensing in cognitive radios. In this context, there may be many different signals contained in the signal sensed by a cognitive radio. Hence, assuming a parametric form may not be possible, making nonparametric approaches more suitable.

As we have seen in Chapter 2, cognitive radios are inherently digital devices, and all processing in a cognitive radio is performed on a digital processor called the cognitive engine. Hence, our discussion of power spectrum estimation will also be in discrete time.

Let us start by formally defining the power spectrum of a (power) signal. Suppose that $X[n]$ is a random signal (process) that is *second-order stationary*, or *wide-sense stationary* (wss). This means that

$$\mathbb{E}\{X[n]\} = m_X \quad \text{for all } n \tag{6.1}$$

Signal Processing for Cognitive Radios, First Edition. Sudharman K. Jayaweera.

and

$$\mathbb{E}\{X[n_1]X^*[n_2]\} = R[n_1 - n_2] \quad \text{for all } n_1, n_2 \tag{6.2}$$

where X^* denotes the complex conjugate of X. Note that (6.2) implies that the average (ensemble-averaged) power of the signal $X[n]$ is a constant that is independent of time:

$$\mathbb{E}\{|X[n]|^2\} = P_X \quad \text{for all } n. \tag{6.3}$$

When $X[n]$ is second-order stationary, we may define the following function, called the *autocovariance function*, of $X[n]$:

$$\mathscr{R}_{XX}[m] = \mathbb{E}\left\{(X[n] - \mathbb{E}\{X[n]\})\left(X^*[n-m] - \mathbb{E}\{X^*[n-m]\}\right)\right\}, \tag{6.4}$$

which, due to wss assumption, can also be written as

$$\mathscr{R}_{XX}[m] = \mathbb{E}\left\{X[n]X^*[n-m]\right\} - |m_X|^2. \tag{6.5}$$

Often, $\mathscr{R}_{XX}[m]$ is also referred to as the autocovariance sequence (ACS). To keep notation simple, in the following, we will assume that $m_X = 0$ so that the autocovariance function (6.5) of $X[n]$ becomes

$$\mathscr{R}_{XX}[m] = \mathbb{E}\left\{X[n]X^*[n-m]\right\}. \tag{6.6}$$

When $m_X \neq 0$, (6.6) is in general called the *autocorrelation function* of $X[n]$.

It also follows easily from the wide-sense stationarity of $X[n]$ that

$$\mathbb{E}\left\{X^*[n]X[n+m]\right\} = \mathbb{E}\left\{X[n]X^*[n-m]\right\}, \tag{6.7}$$

so that

$$\mathscr{R}_{X^*X^*}[-m] = \mathscr{R}_{XX}[m]. \tag{6.8}$$

Hence, in general, we may equally well could have defined the autovariance sequence of a discrete-time (or sampled) signal as

$$\mathscr{R}_{XX}[m] = \mathbb{E}\left\{\left(X^*[n] - \mathbb{E}\{X^*[n]\}\right)(X[n+m] - \mathbb{E}\{X[n+m]\})\right\}, \tag{6.9}$$

or when $m_X = 0$,

$$\mathscr{R}_{XX}[m] = \mathbb{E}\left\{X^*[n]X[n+m]\right\}, \tag{6.10}$$

which indeed is the definition used by some authors. However, in this book, we will use (6.4), or (6.6) when $m_X = 0$, as the definition of the ACS.

It is worth observing that

$$\begin{aligned}\mathscr{R}_{XX}[0] &= \mathscr{R}_{XX}[m]\,|_{m=0},\\ &= \mathbb{E}\left\{|X[n]|^2\right\}\\ &= P_X.\end{aligned} \tag{6.11}$$

Hence, the autocorrelation function evaluated at $m = 0$ gives the average power P_X of signal $X[n]$. Moreover, from either (6.6) or (6.8), $\mathscr{R}_{XX}[m]$ satisfies the following symmetry in the lag variable m:

$$\mathscr{R}_{XX}[-m] = \mathscr{R}^*_{XX}[m]. \tag{6.12}$$

The *Wiener–Khinchin theorem* states that under fairly general conditions on the process $X[n]$, the discrete-time Fourier transform (DTFT) of $\mathscr{R}_{XX}[m]$ exists. Let us denote this Fourier transform of the ACS $\mathscr{R}_{XX}[m]$ as $\mathscr{S}_{XX}(F)$:

$$\mathscr{S}_{XX}(F) = \sum_{m=-\infty}^{\infty} \mathscr{R}_{XX}[m]e^{-j2\pi Fm}, \tag{6.13}$$

where $-\frac{1}{2} \leq F \leq \frac{1}{2}$ is called the normalized frequency.

Note that since $\mathscr{S}_{XX}(F)$ is a DTFT of a discrete-time sequence, it is periodic in F with unit period. The Wiener–Khinchin theorem also ensures that under sufficient regularity conditions,[1] the inverse Fourier transform of $\mathscr{S}_{XX}(F)$ is the autocovariance function $\mathscr{R}_{XX}[m]$:

$$\mathscr{R}_{XX}[m] = \int_{-\frac{1}{2}}^{\frac{1}{2}} \mathscr{S}_{XX}(F)e^{j2\pi Fm}dF \quad \text{for } -\infty < m < \infty. \tag{6.14}$$

Observe from (6.14) that

$$\begin{aligned}\int_{-\frac{1}{2}}^{\frac{1}{2}} \mathscr{S}_{XX}(F)dF &= \mathscr{R}_{XX}[0]\\ &= P_X,\end{aligned} \tag{6.15}$$

where in (6.15) we have used (6.11). Hence, the area under the curve $\mathscr{S}_{XX}(F)$ in the frequency domain is the average signal power P_X. Thus, $\mathscr{S}_{XX}(F)$ can be thought

[1] Assuming (6.13) exists, a sufficient condition for (6.14) is that $\mathscr{S}_{XX}(F)$ be square integrable: $\int |\mathscr{S}_{XX}(F)|^2 dF < \infty$ [59–61].

of as giving the spread of average signal power over different frequencies. For this reason, the Fourier transform $\mathscr{S}_{XX}(F)$ of the autocovariance function $\mathscr{R}_{XX}[m]$ of the random signal $X[n]$ is called the power spectral density. By integrating $\mathscr{S}_{XX}(F)$ over all frequencies, we can obtain the average power of the signal $X[n]$. The Wiener–Khinchin theorem states that the power spectrum and the autocovariance are a Fourier transform pair under certain regularity conditions.

Power spectrum estimation is the first step of signal processing a cognitive radio may perform in order to gain spectrum knowledge. By sensing the spectrum, a wideband cognitive radio collects observation samples. Based on this sequence of discrete-time signal samples and assuming that the signal can be modeled as a second-order (wss) stationary random process, the radio may compute the power spectrum corresponding to the sensed signal. However, at this point, it is obvious that if all that the radio has is just a collection of temporal signal samples, computing the PSD function $\mathscr{S}_{XX}(F)$ may not be possible since by definition $\mathscr{S}_{XX}(F)$ is based on the ensemble averages, which requires knowledge of the exact distribution of the signal $X[n]$.

We define the *limit autocorrelation function* of a signal $x[n]$ as

$$R_{xx}[m] = \lim_{N\to\infty} \frac{1}{N} \sum_{n=-\frac{N-1}{2}}^{\frac{N-1}{2}} x[n]x^*[n-m], \tag{6.16}$$

where, for simplicity, we have assumed N to be odd. It is easy to modify (6.16) if N is even. The DTFT of the limit autocorrelation function, denoted by $S_{xx}(F)$, is called the *limit spectrum* [60]:

$$\begin{aligned} S_{xx}(F) &= \mathcal{F}\{R_{xx}[m]\} \\ &= \sum_{m=-\infty}^{\infty} R_{xx}[m]e^{-j2\pi Fm}. \end{aligned} \tag{6.17}$$

If the process $X[n]$ were to be *ergodic*[2], then we know that all time averages of the process are equivalent to corresponding ensemble averages [62]. Hence, if $X[n]$ were to be first- and second-order ergodic, then we will have that

$$R_{xx}[m] = \mathscr{R}_{XX}[m], \tag{6.18}$$

where equality in (6.18) is assumed in the mean-square sense.

[2]A random process is called ergodic if all time and ensemble averages are interchangeable. When first- and second-order time and ensemble averages are interchangeable, we might specifically call the process to be second-order ergodic.

In particular, note from (6.16) to (6.18) that for $m = 0$,

$$P_X = \mathscr{R}_{XX}[m]|_{m=0} = R_{xx}[0]$$
$$= \lim_{N\to\infty} \frac{1}{N} \sum_{n=-\frac{N-1}{2}}^{\frac{N-1}{2}} |x[n]|^2. \tag{6.19}$$

Hence, ensemble-averaged signal power can also be obtained by the temporal average of the squared signal.

When only time samples of data is available, the autocorrelation and the spectrum that we may estimate are the time-averaged autocorrelation and its Fourier transform. Under ergodic assumptions, we may expect them to converge to the actual ensemble-averaged autocorrelation and the true power spectrum. However, this may or may not hold in practice.

6.2 PSD ESTIMATION OF A STATIONARY DISCRETE-TIME SIGNAL

The problem at hand is to estimate the PSD $\mathscr{S}_{XX}(F)$ of the signal $X[n]$. In practice, this is not possible directly since it requires the knowledge of the ensemble-averaged autocovariance function $\mathscr{R}_{XX}[m]$ of the signal. Hence, we may instead attempt to estimate the limit autocorrelation function or directly the limit spectrum and rely on ergodicity to ensure that the limit spectrum is indeed the true PSD $\mathscr{S}_{XX}(F)$ in some sense. In practice, however, we cannot obtain the limit spectrum either because we only have access to a finite number N of samples $\{x[0], x[1], \ldots, x[N-1]\}$ from a single realization of the process $X[n]$. The fact that we have to perform the PSD estimation with a finite-length segment of data from a single realization of the process $X[n]$ means that our estimate have to contend with errors both in the stochastic (ensemble-averaged) sense and statistical (time-averaged) sense.

6.2.1 Correlogram Method

Given the observation data sequence $\{x[0], x[1], \ldots, x[N-1]\}$ of the process $X[n]$, we may define the *time-averaged* ACS of the signal as

$$\hat{R}_{xx}[m;N] = \begin{cases} \frac{1}{N} \sum_{n=m}^{N-1} x[n]x^*[n-m] & \text{if } 0 \le m \le N-1 \\ \hat{R}^*_{xx}[-m;N] & \text{if } -(N-1) \le m \le -1 \end{cases}. \tag{6.20}$$

The estimate $\hat{R}_{xx}[m;N]$ of the autocovariance $\mathscr{R}_{XX}[m]$ is also called the (discrete-time) *correlogram.*

But is this a good estimate of the true ACS $\mathscr{R}_{XX}[m]$? Note that $\hat{R}_{xx}[m;N]$ is a random quantity because it is based on a randomly drawn sample sequence, whereas

$\mathscr{R}_{XX}[m]$ is a deterministic quantity. Hence, we may start by looking at whether $\hat{R}_{xx}[m;N]$ is an unbiased estimate of the true ACS $\mathscr{R}_{XX}[m]$:

$$\begin{aligned}\mathbb{E}\left\{\hat{R}_{xx}[m;N]\right\} &= \frac{1}{N}\sum_{n=m}^{N-1}\mathbb{E}\left\{X(n)X^*(n-m)\right\}\\ &= \frac{N-|m|}{N}\mathscr{R}_{XX}[m] \qquad (6.21a)\\ &= \left(1-\frac{|m|}{N}\right)\mathscr{R}_{XX}[m]. \qquad (6.21b)\end{aligned}$$

Hence, the estimate $\hat{R}_{xx}[m;N]$ is not an unbiased estimate for any finite N. For this reason, correlogram is also referred to as the *standard biased ACS estimate*. Note also that the bias deviation from the true ACS is worse for larger (i.e., closer to N) lag values m. However, it is also clear from (6.21b) that $\hat{R}_{xx}[m;N]$ is asymptotically unbiased for the ACS $\mathscr{R}_{XX}[m]$.

The variance of the estimate $\hat{R}_{xx}(m;N)$ can be shown to be [63]

$$Var\left(\hat{R}_{xx}[m;N]\right) \approx \frac{1}{N}\sum_{n=-\infty}^{\infty}\left(|\mathscr{R}_{XX}[n]|^2 + \mathscr{R}^*_{XX}[n-m]\,\mathscr{R}_{XX}[n+m]\right). \qquad (6.22)$$

If the ACS $\mathscr{R}_{XX}[m]$ is square summable, that is, $\sum_{n=-\infty}^{\infty}|\mathscr{R}_{XX}[n]|^2 < \infty$, then the above variance vanishes as $N \to \infty$. In this case, the correlogram estimate $\hat{R}_{xx}[m;N]$ is also asymptotically efficient (see Section 5.3.4 for details). Moreover, the asymptotic unbiasedness and vanishing variance together then imply that the estimate $\hat{R}_{xx}[m;N]$ is also consistent in the mean-square sense as defined in (5.88) and (5.89).

Thus, correlogram (6.20) at least have satisfactory asymptotic properties as an estimate of the true autocovariance $\mathscr{R}_{XX}[m]$. Thus, we may form a reasonable estimate of the PSD $\mathscr{S}_{XX}(F)$ as the DTFT of the correlogram:

$$\hat{S}_{xx}(F;N) = \mathcal{F}\left\{\hat{R}_{xx}[m;N]\right\} \qquad (6.23a)$$

$$= \sum_{m=-(N-1)}^{N-1}\hat{R}_{xx}[m;N]e^{-j2\pi Fm} \quad \text{for} \quad -\frac{1}{2} \le F \le \frac{1}{2}. \qquad (6.23b)$$

Algorithm 6.1 summarizes the correlogram-based PSD estimation.

Algorithm 6.1

Correlogram Power Spectral Density Estimator

(1) Input Data $\{x[0],x[1],\cdots,x[N-1]\}$.

(2) For $m=-(N-1):N-1$: Compute the correlogram $\hat{R}_{xx}[m;N]$ using (6.20).

(3) Compute the Fourier transform $\hat{S}_{xx}(F;N)$ of the correlogram $\hat{R}_{xx}[m;N]$ using (6.23b).

The natural next question is how good $\hat{S}_{xx}(F;N)$ is as an estimate of the PSD $\mathcal{S}_{XX}(F)$. Let us postpone that discussion until after we introduce another PSD estimator called the *periodogram* in the next section.

6.2.2 Periodogram Method

Let us revisit the DTFT of the correlogram given in (6.23b). We may rewrite this as

$$
\begin{aligned}
\hat{S}_{xx}(F;N) &= \sum_{m=-(N-1)}^{N-1} \hat{R}_{xx}[m;N]e^{-j2\pi Fm} \\
&= \sum_{m=0}^{N-1} \hat{R}_{xx}[m;N]e^{-j2\pi Fm} + \sum_{m=-(N-1)}^{-1} \hat{R}_{xx}[m;N]e^{-j2\pi Fm} \\
&= \sum_{m=0}^{N-1} \hat{R}_{xx}[m;N]e^{-j2\pi Fm} + \sum_{m=1}^{N-1} \hat{R}_{xx}[-m;N]e^{j2\pi Fm} \\
&= \sum_{m=0}^{N-1} \hat{R}_{xx}[m;N]e^{-j2\pi Fm} + \sum_{m=1}^{N-1} \hat{R}^{*}_{xx}[m;N]e^{j2\pi Fm} \qquad (6.24a)
\end{aligned}
$$

$$
= \frac{1}{N}\left[\sum_{m=0}^{N-1}\sum_{n=m}^{N-1} x[n]x^{*}[n-m]e^{-j2\pi Fm} + \sum_{m=1}^{N-1}\sum_{n=m}^{N-1} x^{*}[n]x[n-m]e^{j2\pi Fm}\right] \qquad (6.24b)
$$

$$
\begin{aligned}
&= \frac{1}{N}\left[\left(\sum_{n=0}^{N-1} x[n]x^{*}[n] + \sum_{n=1}^{N-1} x[n]x^{*}[n-1]e^{-j2\pi F}\right.\right. \\
&\quad + \cdots + \sum_{n=N-2}^{N-1} x[n]x^{*}[n-(N-2)]e^{-j2\pi F(N-2)} \\
&\quad \left. + \sum_{n=N-1}^{N-1} x[n]x^{*}[n-(N-1)]e^{-j2\pi F(N-1)}\right) \\
&\quad + \left(\sum_{n=1}^{N-1} x[n-1]x^{*}[n]e^{j2\pi F} + \sum_{n=2}^{N-1} x[n-2]x^{*}[n]e^{j2\pi F(2)}\right. \\
&\quad + \cdots + \sum_{n=N-2}^{N-1} x[n-(N-2)]x^{*}[n]e^{j2\pi F(N-2)} \\
&\quad \left.\left. + \sum_{n=N-1}^{N-1} x[n-(N-1)]x^{*}[n]e^{j2\pi F(N-1)}\right)\right] \qquad (6.24c)
\end{aligned}
$$

$$
\begin{aligned}
&= \frac{1}{N}\left[\left(\sum_{n=0}^{N-1} x[n]x^*[n] + \sum_{n=1}^{N-1} x[n]x^*[n-1]e^{-j2\pi F}\right.\right. \\
&\quad + \cdots + \sum_{n=N-2}^{N-1} x[n]x^*[n-(N-2)]e^{-j2\pi F(N-2)} \\
&\quad \left. + \sum_{n=N-1}^{N-1} x[n]x^*[n-(N-1)]e^{-j2\pi F(N-1)}\right) \\
&\quad + \left(\sum_{n=0}^{N-2} x[n]x^*[n+1]e^{j2\pi F} + \sum_{n=0}^{N-3} x[n]x^*[n+2]e^{j2\pi F(2)}\right. \\
&\quad + \cdots + \sum_{n=0}^{1} x[n]x^*[n+N-2]e^{j2\pi F(N-2)} \\
&\quad \left.\left. + \sum_{n=0}^{0} x[n]x^*[n+N-1]e^{j2\pi F(N-1)}\right)\right],
\end{aligned}
$$

where (6.24a) and (6.24b) follow from (6.20) and (6.24c) is obtained by collecting terms with the same m together. Collecting $x[n]$ terms for each n together, we get

$$
\begin{aligned}
\hat{S}_{xx}(F;N) &= \frac{1}{N}\left[x[0]\left(x^*[0] + x^*[1]e^{j2\pi F} + x^*[2]e^{j2\pi F(2)} + \cdots + x^*[N-1]e^{j2\pi F(N-1)}\right)\right. \\
&\quad + x[1]\left(x^*[1] + x^*[0]e^{-j2\pi F} + x^*[2]e^{j2\pi F} + x^*[3]e^{j2\pi F(2)}\right. \\
&\quad \left. + \cdots + x^*[N-1]e^{j2\pi F(N-2)}\right) + \cdots + x[N-1] \\
&\quad \times \left(x^*[N-1] + x^*[N-2]e^{-j2\pi F} + x^*[N-3]e^{-j2\pi F(2)}\right. \\
&\quad \left.\left. + \cdots + x^*[0]e^{-j2\pi F(N-1)}\right)\right] \\
&= \frac{1}{N}\left[x[0]\left(\sum_{k=0}^{N-1} x^*[k]e^{j2\pi Fk}\right) + x[1]\left(\sum_{k=0}^{N-1} x^*[k]e^{-j2\pi F(1-k)}\right)\right. \\
&\quad \left. + \cdots + x[N-1]\left(\sum_{k=0}^{N-1} x^*[k]e^{-j2\pi F(N-1-k)}\right)\right] \\
&= \frac{1}{N}\sum_{n=0}^{N-1}\sum_{k=0}^{N-1} x[n]x^*[k]e^{-j2\pi F(n-k)} \\
&= \frac{1}{N}\left(\sum_{n=0}^{N-1} x[n]e^{-j2\pi Fn}\right)\left(\sum_{k=0}^{N-1} x[k]e^{j2\pi Fk}\right)^* \\
&= \frac{1}{N}\left|\sum_{n=0}^{N-1} x[n]e^{-j2\pi Fn}\right|^2 \qquad (6.25a) \\
&= \frac{1}{N}\left|X(F;N)\right|^2, \qquad (6.25b)
\end{aligned}
$$

where

$$X(F;N) = \sum_{n=0}^{N-1} x[n]e^{-j2\pi Fn} \tag{6.26}$$

is the DTFT of the sequence $\{x[n]\}_{n=0}^{N-1}$.

From (6.25b), we see that the correlogram method for estimating the PSD can be streamlined: Rather than first computing the correlogram (time-averaged ACS estimate) $\hat{R}_{xx}[m;N]$ and then taking its DTFT as in Algorithm 6.1, we can directly compute the DTFT of the data sequence $\{x[0],\ldots,x[N-1]\}$ itself and use (6.25b) to obtain the same PSD estimate. The estimate (6.25b) is called the periodogram estimate of the PSD of the data sequence $\{x[n]\}_{n=0}^{N-1}$ and will be denoted by $\hat{S}_p(F;N)$:

$$\hat{S}_p(F;N) = \frac{1}{N}|X(F;N)|^2, \tag{6.27}$$

where $X(F;N)$ is the DTFT in (6.26). Algorithm 6.2 summarizes the periodogram spectrum estimation. From (6.23a) and (6.25b), it should be clear that the correlogram and the periodogram estimates of the power spectrum are the same.

Algorithm 6.2

Periodogram Power Spectral Density Estimator
(1) Input Data $\{x[0],x[1],\ldots,x[N-1]\}$.
(2) Compute the DTFT $X(F;N)$ of data using (6.26).
(3) Compute the periodogram estimate $\hat{S}_p(F;N)$ using (6.27).

6.2.3 Performance of the Periodogram PSD Estimate

The quantity we are interested in estimating, that is, $\mathscr{S}_{XX}(F)$, is a nonrandom (deterministic) function of a variable F. Since periodogram $\hat{S}_p(F;N)$ is a function based on the finite N-length random data, it is a random quantity and is also a function of the data length N.

We saw earlier that under fairly regular conditions, the correlogram is an asymptotically unbiased and efficient estimate for the ACS $\mathscr{R}_{XX}[m]$. Moreover, it can be consistent in the sense of mean square. But how does its DTFT, that is, the periodogram, perform as an estimator for the PSD $\mathscr{S}_{XX}(F)$? Let us first consider the stochastic performance.

From (6.23b), we see that

$$\begin{aligned}\mathbb{E}\left\{\hat{S}_p(F;N)\right\} &= \sum_{m=-(N-1)}^{N-1} \mathbb{E}\left\{\hat{R}_{xx}[m;N]\right\} e^{-j2\pi Fm} \\ &= \sum_{m=-(N-1)}^{N-1} \left(1-\frac{|m|}{N}\right) \mathscr{R}_{XX}[m]e^{-j2\pi Fm}.\end{aligned} \tag{6.28}$$

Hence, for any finite N, the periodogram is a biased estimator for $\mathscr{S}_{XX}(F)$. However, (6.28) also shows that periodogram is an asymptotically unbiased estimate for the PSD $\mathscr{S}_{XX}(F)$, because

$$\lim_{N\to\infty} \mathbb{E}\left\{\hat{S}_p(F;N)\right\} = \mathscr{S}_{XX}(F). \tag{6.29}$$

As we already know from Chapter 5, however, unbiasedness alone may not be good enough. In particular, two estimators may be unbiased, while one has a larger variance around the mean and the other having a smaller variance. Of course, we would then prefer the estimator that is not only unbiased, but also have a smaller variance around the mean.

Although the exact computation of the variance of the periodogram estimate is difficult in general, if the sequence $X[n]$ is a Gaussian random process, the variance of the periodogram estimate can be shown to be [63]

$$Var\left(\hat{S}_p(F;N)\right) = \mathscr{S}^2_{XX}(F)\left[1+\left(\frac{\sin 2\pi FN}{N\sin 2\pi F}\right)^2\right]. \tag{6.30}$$

The asymptotic variance of the periodogram estimator, in the Gaussian case, is thus

$$\lim_{N\to\infty} Var\left(\hat{S}_p(F;N)\right) = \mathscr{S}^2_{XX}(F). \tag{6.31}$$

Recall that the correlogram was both an asymptotically efficient and consistent (in the sense of mean square) estimator for the ACS $\mathscr{R}_{XX}[m]$ since it was asymptotically unbiased and its asymptotic variance vanished, at least under the condition of a square summable ACS sequence. On the other hand, (6.29) and (6.31) show that its DTFT, which is the periodogram, is asymptotically unbiased, but its asymptotic variance is equal to $\mathscr{S}^2_{XX}(F)$.

From (5.31), we may write the conditional MSE of the periodogram $\hat{S}_p(F;N)$ as

$$\begin{aligned}
&\mathbb{E}\left\{\left|\hat{S}_p(F;N) - \mathscr{S}_{XX}(F)\right|^2\right\} \\
&= \mathbb{E}\left\{\left|\left(\hat{S}_p(F;N) - \mathbb{E}\left\{\hat{S}_p(F;N)\right\}\right) + \left(\mathbb{E}\left\{\hat{S}_p(F;N)\right\} - \mathscr{S}_{XX}(F)\right)\right|^2\right\} \\
&= Var\left(\hat{S}_p(F;N)\right) + \left(\mathbb{E}\left\{\hat{S}_p(F;N)\right\} - \mathscr{S}_{XX}(F)\right)^2.
\end{aligned}$$

Since the asymptotic variance does not vanish when $X[n]$ is Gaussian, the conditional mean-squared error is also asymptotically nonzero. Hence, the periodogram is not a mean-square consistent estimator for $\mathscr{S}_{XX}(F)$ when $X[n]$ is Gaussian. This, however, is known to be true in general, even if the sequence is non-Gaussian.

Thus, periodogram estimate can have a high variance around the true PSD even asymptotically. Although periodogram is a poor PSD estimator due to its constant asymptotic variance, it must be mentioned that it is a convenient estimator to compute in practice.

In practice, however, we only have a finite number of samples N. Hence, in addition to the stochastic and asymptotic performance as $N \to \infty$, we should also look at its statistical and finite N performance. In particular, as we saw in (6.28), the expected value $\mathbb{E}\left\{\hat{S}_p(F;N)\right\}$ of the periodogram estimate for any finite N is not equal to the true PSD $\mathscr{S}_{XX}(F)$. Instead, from (6.28), we see that the expected value of the periodogram estimate can be written as

$$\mathbb{E}\left\{\hat{S}_p(F;N)\right\} = \mathcal{F}\left\{R_B[m]\right\}, \tag{6.32}$$

where the *windowed autocorrelation sequence* $R_B[m]$ is defined as

$$R_B[m] = \left(1 - \frac{|m|}{N}\right) \mathscr{R}_{XX}[m].$$

Let us define the triangular window (also called the *Bartlett window*) function $w_B[m;N]$ as

$$w_B[m;N] = \begin{cases} 1 - \frac{|m|}{N} & \text{if } -(N-1) \le m \le N-1 \\ 0 & \text{otherwise} \end{cases}. \tag{6.33}$$

Note that $w_B[m;N]$ has a nonzero length of $2N-1$ and its Fourier transform is given by

$$\begin{aligned} W_B(F;N) &= \mathcal{F}\{w_B[m;N]\} \\ &= \frac{1}{N}\left(\frac{\sin \pi FN}{\sin \pi F}\right)^2. \end{aligned} \tag{6.34}$$

We may express the windowed ACS $R_B[m]$ in terms of the Bartlett window as

$$R_B[m] = w_B[m;N]\, \mathscr{R}_{XX}[m]. \tag{6.35}$$

From the properties of the Fourier transform, multiplication (6.35) in time domain leads to convolution in the frequency domain of the corresponding Fourier transforms of $w_B[m;N]$ and $\mathscr{R}_{XX}[m]$. Since DTFT of $\mathscr{R}_{XX}[m]$ is the true PSD $\mathscr{S}_{XX}(F)$, from (6.32) and (6.35), we have that

$$\mathbb{E}\left\{\hat{S}_p(F;N)\right\} = \mathscr{S}_{XX}(F) \star W_B(F;N) \tag{6.36a}$$

$$= \int_{-\frac{1}{2}}^{\frac{1}{2}} W_B(\alpha;N)\,\mathscr{S}_{XX}(F-\alpha)d\alpha \tag{6.36b}$$

$$= \int_{-\frac{1}{2}}^{\frac{1}{2}} W_B(F-\alpha;N)\,\mathscr{S}_{XX}(\alpha)d\alpha. \tag{6.36c}$$

From (6.36c), we see the effect of biasedness of the periodogram estimate for any finite N: The convolution of the true PSD $\mathscr{S}_{XX}(F)$ with $W_B(F;N)$ makes the expected value of the periodogram to be a smoothen version of the true PSD $\mathscr{S}_{XX}(F)$. As long as window $W_B(F;N)$ has a nonzero main-lobe spectral width, convolution of $\mathscr{S}_{XX}(F)$ with it makes the result to be a smeared version of $\mathscr{S}_{XX}(F)$. This smearing causes the spectral estimator to have a broadened spectral width than the actual spectral width of $\mathscr{S}_{XX}(F)$. This is called *spectral broadening*. The smaller the N, the broader the spectral width of $W_B(F;N)$ and the worse the smearing effect on the expected value of the periodogram as can be seen from Figure 6.1. When $N \to \infty$, $W_B(F;N)$ resembles a delta (impulse) function (in frequency domain) so that the convolution (6.36c) will result in $\mathscr{S}_{XX}(F)$ itself, establishing the asymptotic unbiasedness (6.29) of the periodogram.

But this is not all either. In addition to the nonzero main-lobe width, $W_B(F;N)$ also has side lobes (see Figure 6.1). Convolution of $\mathscr{S}_{XX}(F)$ with these side lobes causes the expected value of the periodogram to show nonzero spectral contents even in regions where PSD $\mathscr{S}_{XX}(F)$ is actually zero. This effect is called the *spectral leakage*. From time–frequency duality of Fourier transforms, it follows that the high side lobe level is an indication of abrupt changes in the time-domain function $w_B[m]$. For example, let us denote by $w_R[n]$ the rectangular window function with the same support defined as

$$w_R[n] = \begin{cases} 1 & \text{if } -(N-1) \leq n \leq N-1 \\ 0 & \text{otherwise} \end{cases}.$$

The rectangular window with the same support has higher side-lobe levels because the rectangular function has a much sharper change in time domain than the triangular window. In summary, both main-lobe width (broadening) and the side-lobe level (leakage) of the Bartlett window contribute to the biasedness of the periodogram estimator.

Both spectral broadening and leakage effects limits the *spectral resolution* capability of the periodogram. Spectral resolution is the ability of a spectrum estimator

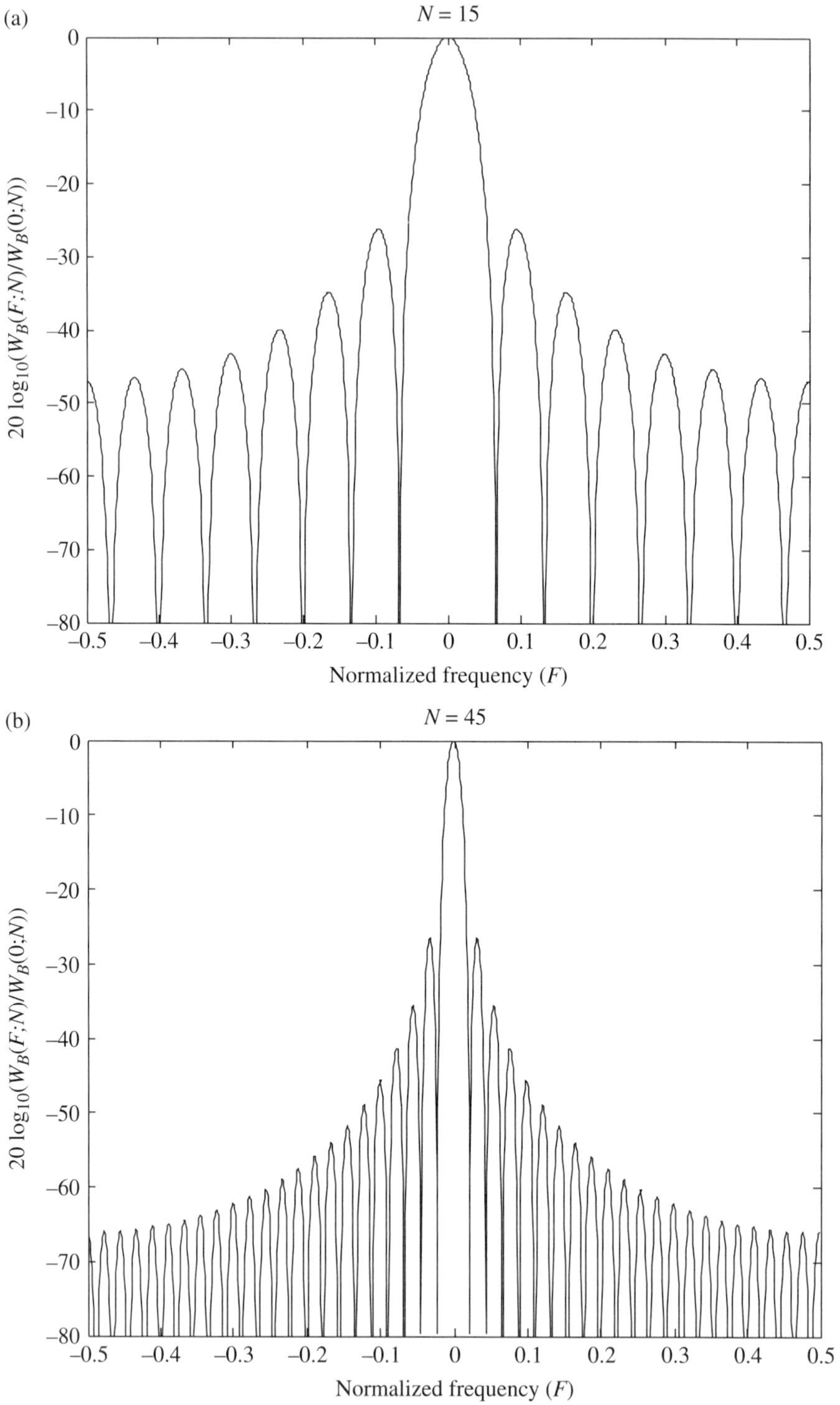

FIGURE 6.1 Bartlett window function. (a) $\frac{W_B(F;N)}{W_B(0;N)}$ in dB for $N = 15$. (b) $\frac{W_B(F;N)}{W_B(0;N)}$ in dB for $N = 45$.

to show distinctly two closely separated spectral components in the true PSD. For instance, suppose there are two closely separated spectral components in a given PSD $\mathscr{S}_{XX}(F)$. Then there is the possibility that the broadening of the periodogram spectrum $\hat{S}_p(F;N)$ due to smoothing and leakage effects will make it difficult to distinguish these two components as distinct. Spectral broadening may lead to them appearing as a single broadened spectral line. The minimum spectral separation that a spectrum estimator can show distinctly is called the spectral resolution of the estimator. In the case of the periodogram, this minimum resolution is limited by the main-lobe width of the Bartlett (triangular) window $W_B(F;N)$, which is approximately $\frac{1}{N}$ if we consider the half-power (3-dB) width of the main lobe. For all its shortcomings, however, this spectral resolution of the periodogram is as small as we can expect from any nonparametric PSD estimator for a given N.

It appears that by making N large, we may reduce the bias of the periodogram estimator and improve the spectral resolution. However, the more serious drawback of the periodogram is its nonzero asymptotic variance given in (6.31). This causes the periodogram estimate to have large fluctuations around the true PSD even when N is very large. Various classical nonparametric PSD estimators found in the literature are attempts to address this limitation of the periodogram. All these methods try to obtain an estimate of the PSD that has a smaller variance than that of the periodogram. For a given data length of N, this variance reduction is almost always achieved by averaging (or smoothing) the periodogram estimate whose inevitable effect is to reduce the spectral resolution capability. This is because any averaging in time domain will necessarily smoothen the final PSD estimator, making it smear even more than the original periodogram. Hence, these techniques are essentially a trade-off between the minimum spectral resolution limit (bias) and the variance of the estimator.

6.3 BLACKMAN–TUKEY ESTIMATOR OF THE POWER SPECTRUM

Recall that the periodogram estimator is the DTFT of the standard autocovariance estimate $\hat{R}_{XX}[m;N]$. Observe from (6.20) that the standard autocovariance estimate must particularly be a poor estimate for large m values because the time-averaging only includes a smaller number of time samples. At the extreme case, we see that for $m = N - 1$, the autocovariance estimate (6.20) is actually based on just one term and there is no time averaging at all. Hence, for large values of m, these autocovariance estimates must have a very high variance around their mean, which is the true autocovariance $\mathscr{R}_{XX}[m]$ (assuming large N). In light of this, we should not be surprised that the periodogram PSD estimator based on such unreliable autocovariance terms performs poorly in terms of having a high variance. Indeed, this is the root cause of the nonzero (and high) asymptotic variance of the periodogram.

The *Blackman–Tukey PSD estimator* tries to overcome this negative effect of poor autocovariance estimates of the correlogram for large m values on the periodogram estimate by giving them a smaller weight. Thus, each standard autocovariance estimate value $\hat{R}_{XX}[m;N]$ is multiplied by a corresponding weight value $w[m]$ before they are used in the DTFT computation:

$$\hat{S}_{BT}(F;N) = \sum_{m=-\infty}^{\infty} w[m]\hat{R}_{XX}[m;N]e^{-j2\pi Fm} \tag{6.37a}$$

$$= \sum_{m=-(M-1)}^{M-1} w[m]\hat{R}_{XX}[m;N]e^{-j2\pi Fm}, \tag{6.37b}$$

where we have assumed that the weight values are nonzero only for $-(M-1) \leq m \leq M-1$ for some $M \leq N$.

Since the effect of giving weights $w[m]$ is equivalent to multiplying $\hat{R}_{XX}[m;N]$ by a window $w[m]$, in the classical literature, this is called *windowing*.[3] Thus, the Blackman–Tukey PSD estimation method is essentially a smoothed periodogram obtained by windowing the correlogram (autocovariance estimate).

Let us denote the Fourier transform of the window/weight function $w[m]$ as $W(F)$. Since Fourier transform of the standard ACS $\hat{R}_{XX}[m;N]$ is the periodogram estimator $\hat{S}_p(F;N)$, from (6.37a), we see that the Blackman–Tukey PSD estimator $\hat{S}_{BT}(F;N)$ is related to the periodogram via

$$\hat{S}_{BT}(F;N) = \mathcal{F}\left\{w[m]\hat{R}_{XX}[m;N]\right\}$$

$$= \int_{-\frac{1}{2}}^{\frac{1}{2}} \hat{S}_p(\alpha;N)W(F-\alpha)d\alpha. \tag{6.38}$$

Clearly, $\hat{S}_{BT}(F;N)$ is a smoothen version of the periodogram $\hat{S}_p(F;N)$, which in turn was a smoothen version of the original PSD $\mathscr{S}_{XX}(F)$ (on average). Earlier, we saw that the smoothing introduced in computing the periodogram led to the minimum spectral resolution limit of $\frac{1}{N}$ in the periodogram. Hence, as long as $W(F)$ has a nonzero main-lobe spectral width, we may expect $\hat{S}_{BT}(F;N)$ to be even more smeared than the periodogram. Hence, the spectral resolution of $\hat{S}_{BT}(F;N)$ must be worse off than that of the periodogram. Of course, we pay this price to achieve some reduction in the variance compared to that of the periodogram. The exact trade-off will depend on the exact form of the window function $w[n]$ used.

It is worth noticing that since PSD $\mathscr{S}_{XX}(F)$ is a real nonnegative quantity, to ensure that the DTFT (6.37a) is also real valued, we need to choose $w[n]$ to be symmetric in n. If $W(F)$ is chosen to be nonnegative, then from (6.38), we see that the Blackman–Tukey PSD estimate will also be nonnegative. However, this is only a

[3]The term windowing is used when the autocorrelation estimate is multiplied by a window of finite nonzero length. When the data sequence itself is multiplied by a finite nonzero length weight sequence, it is referred to as (data) tapering and the weight sequence is called a (data) taper (see [61] for a discussion).

sufficient condition. Indeed, there are some classical windows that do not satisfy this nonnegativity property yet are chosen due to their other desirable properties although they lead to an estimator that may sometimes give negative (and thus unrealistic) PSD estimates.

From (6.38), note that

$$\mathbb{E}\left\{\hat{S}_{BT}(F;N)\right\} = \int_{-\frac{1}{2}}^{\frac{1}{2}} \mathbb{E}\left\{\hat{S}_p(\alpha;N)\right\} W(F-\alpha)d\alpha \tag{6.39a}$$

$$= \int_{-\frac{1}{2}}^{\frac{1}{2}} \left(\int_{-\frac{1}{2}}^{\frac{1}{2}} \mathscr{S}_{XX}(\beta) W_B(\alpha-\beta)d\beta \right) W(F-\alpha)d\alpha \tag{6.39b}$$

$$\approx \int_{-\frac{1}{2}}^{\frac{1}{2}} \mathscr{S}_{XX}(\alpha) W(F-\alpha)d\alpha, \tag{6.39c}$$

where (6.39b) follows from (6.36c) and (6.39c) is obtained by assuming that the length M of the window $w[n]$ is much narrower than the length N of the Bartlett window $w_B[n]$ (the actual length of these windows are $2M-1$ and $2N-1$, respectively). For any reasonable window $w[n]$, we may assume that as $N \to \infty$, $W(F)$ will resemble a Dirac-delta function so that from (6.39c) the Blackman–Tukey PSD estimator will still be asymptotically unbiased.

On the other hand, as before, if we presume that the data sequence is drawn from a Gaussian random process, then it can be shown that the variance of the Blackman–Tukey estimator will be approximately equal to [63]

$$Var\left(\hat{S}_{BT}(F;N)\right) \approx \frac{1}{N} \int_{-\frac{1}{2}}^{\frac{1}{2}} \mathscr{S}_{XX}^2(\alpha) W^2(F-\alpha)d\alpha \tag{6.40a}$$

$$\approx \mathscr{S}_{XX}^2(F) \left(\frac{1}{N} \int_{-\frac{1}{2}}^{\frac{1}{2}} W^2(\alpha)d\alpha \right) \tag{6.40b}$$

$$= \mathscr{S}_{XX}^2(F) \left(\frac{1}{N} \sum_{m=-(M-1)}^{M-1} w^2[m] \right), \tag{6.40c}$$

where (6.40b) follows by further assuming that $W(F)$ is much narrower compared to the true PSD $\mathscr{S}_{XX}(F)$.

Consider a rectangular window function $w[n]$ so that $w[n] = \Pi\left(\frac{n}{2M-1}\right)$, where $\Pi\left[\frac{n}{N}\right]$ is the rectangular function defined as

$$\Pi\left[\frac{n}{N}\right] = \begin{cases} 1 & \text{if } -\frac{N-1}{2} \leq n \leq \frac{N-1}{2} \\ 0 & \text{otherwise} \end{cases}. \tag{6.41}$$

Then the variance (6.40c) turns out to be

$$Var\left(\hat{S}_{BT}(F;N)\right) \approx \frac{2M}{N} \mathscr{S}_{XX}^2(F).$$

If, on the other hand, $w[n]$ was chosen to be a triangular (Bartlett) window of the same nonzero length $2M - 1$ [63],

$$Var\left(\hat{S}_{BT}(F;N)\right) \approx \frac{2M}{3N} \mathscr{S}_{XX}^2(F).$$

From these examples, we see that the general trend is that the variance is proportional to $\frac{M}{N}$. Hence, by choosing a smaller window length M, we may obtain a smaller variance in the Blackman–Tukey PSD estimator. However, since the effective window function that determines the amount of main-lobe smearing is now this $M < N$ length window, the spectral resolution of the Blackman–Tukey estimator will then be limited to roughly $\frac{1}{M}$. Hence, smaller M will improve the variance of the Blackman–Tukey estimator at the expense of degrading the spectral resolution capability. This is the basic trade-off in the window length choice for smoothed periodogram estimates based on windowed autocovariance estimates: The window length is a trade-off between the variance of the PSD estimate and the spectral resolution (bias) of the estimate.[4]

As it turns out, it is not possible to simultaneously minimize the main-lobe width and the side-lobe level of a window. Hence, once the length M of the window is decided based on an acceptable variance–spectral resolution trade-off, one may attempt to further optimize the shape of the window $w[n]$ to obtain a trade-off between spectral broadening (main-lobe smearing) and side-lobe leakage. Table 6.1 shows some commonly used window functions with their main-lobe widths and side-lobe levels [61].

[4]The phenomenon that bias or variance of an estimate can be made smaller only at the expense of making the other larger is called the bias–variance trade-off. The bias–variance trade-off arises in various forms in many statistical problems. In Chapter 16, we will see another instance of a bias–variance trade-off in designing a classifier, or a decision rule, based on a given set of training data. The trade-off that we will encounter there between minimizing empirical risk and maximizing the generalization performance can be interpreted as a bias–variance trade-off.

TABLE 6.1 Commonly used window functions

Window name	Window equation (for $0 \leq n \leq M-1$)	Approx. main-lobe width	Side-lobe level (dB)
Rectangular	$w[n] = 1$	$1/M$	−13
Bartlett	$w[n] = \dfrac{M-n}{M}$	$2/M$	−25
Hanning	$w[n] = 0.5 + 0.5\cos\left(\frac{\pi n}{M}\right)$	$2/M$	−31
Hamming	$w[n] = 0.54 + 0.46\cos\left(\dfrac{\pi n}{M-1}\right)$	$2/M$	−41
Blackman	$w[n] = 0.42 + 0.5\cos\left(\dfrac{\pi n}{M-1}\right) + 0.08\cos\left(\dfrac{\pi n}{M-1}\right)$	$3/M$	−57

It is assumed that window function $w[n] \neq 0$ only for $|n| \leq M-1$ and is symmetric so that $w[-n] = w[n]$ [61].

6.4 OTHER PSD ESTIMATORS BASED ON MODIFIED PERIODOGRAMS

We have already seen that the basic problem in the periodogram as a PSD estimator is its asymptotically constant high variance. Its bias, on the other hand, vanishes asymptotically as $N \longrightarrow \infty$. Since spectral resolution is related to the bias, the spectral resolution of the periodogram can be improved by increasing N. Hence, all other nonparametric approaches to PSD estimation essentially trade off the spectral resolution to reduce the variance. As we saw in the previous section, the Blackman–Tukey PSD estimate is one way to achieve this. In that, the correlogram was windowed by a window of length $M \ll N$ before taking the DTFT. The effect of windowing the correlogram was to average the periodogram in the frequency domain leading to a smoothed periodogram. This reduced the variance while degrading the spectral resolution.

In the following, we discuss two other classical PSD estimators that achieve lower variance than the periodogram by trading off the spectral resolution.

6.4.1 Bartlett PSD Estimator

Given the N-length data sequence $\{x[n]\}_{n=0}^{N-1}$, the Barlett method first segments the data into K nonoverlapping subsequences $x_k[n]$'s for $k = 0, \ldots, K-1$. Each subsequence length is thus $M = \frac{N}{K}$. The kth subsequence is

$$x_k[n] = x[n + kM] \quad \text{for } n = 0, \ldots, M-1. \tag{6.42}$$

Next, K periodograms are computed from each of the M-length subsequences. If we denote by $\hat{S}_{p,k}(F;M)$ the kth periodogram computed from the subsequence $x_k[.]$, then

$$\hat{S}_{p,k}(F;M) = \frac{1}{M}\left|\sum_{n=0}^{M-1} x_k[n]e^{-j2\pi Fn}\right|^2 \quad \text{for } k = 0, \ldots, K-1. \tag{6.43}$$

The Bartlett PSD estimate, denoted as $\hat{S}_B(F;N)$, is obtained by averaging the above K periodograms:

$$\hat{S}_B(F;N) = \frac{1}{K}\sum_{k=0}^{K-1} \hat{S}_{p,k}(F;M). \tag{6.44}$$

It is worth noticing that the overall effect of averaging the periodograms computed from the subsequences as in (6.44) is equivalent to computing a periodogram from a windowed autocovariance estimate, where the window is actually a rectangular window, as in the Blackman–Tukey method. The difference is that the autocovariance estimate in effect here is not the correlogram of the data but a different estimate. Indeed, since periodogram is the Fourier transform of the correlogram, the kth periodogram in (6.43) should also be equal to the Fourier transform of the corresponding correlogram computed from the kth subsequence. If we denote this correlogram by $\hat{R}_{xx,k}[m;M]$, then for $k = 0,\ldots,K-1$,

$$\hat{S}_{p,k}(F;M) = \mathcal{F}\left\{\hat{R}_{xx,k}[m;M]\right\}. \tag{6.45}$$

Note that each $\hat{R}_{xx,k}[m;M]$ is nonzero only for $-(M-1) \leq m \leq M-1$. Thus, (6.44) can be written as

$$\begin{aligned}\hat{S}_B(F;N) &= \frac{1}{K}\sum_{k=0}^{K-1} \mathcal{F}\left\{\hat{R}_{xx,k}[m;M]\right\}\\ &= \mathcal{F}\left\{\frac{1}{K}\sum_{k=0}^{K-1}\hat{R}_{xx,k}[m;M]\right\} &(6.46a)\\ &= \sum_{m=-(M-1)}^{M-1}\left(\frac{1}{K}\sum_{k=0}^{K-1}\hat{R}_{xx,k}[m;M]\right)e^{-j2\pi Fm}. &(6.46b)\end{aligned}$$

Hence, the Bartlett PSD estimator is similar to the Blackman–Tukey estimator with a rectangular window $w[n]$, but in place of the correlogram sequence, a different ACS estimate, namely, $\frac{1}{K}\sum_{k=0}^{K-1}\hat{R}_{xx,k}[m;M]$, is used.

From (6.44) and (6.36c), it follows that the bias of the Bartlett PSD estimate is

$$\begin{aligned}\mathbb{E}\left\{\hat{S}_B(F;N)\right\} &= \frac{1}{K}\sum_{k=0}^{K-1}\mathbb{E}\left\{\hat{S}_{p,k}(F;M)\right\}\\ &= \mathbb{E}\left\{\hat{S}_p(F;M)\right\} &(6.47a)\\ &= \mathscr{S}_{XX}(F)\star W_B(F;M) &(6.47b)\\ &= \int_{-\frac{1}{2}}^{\frac{1}{2}} W_B(F-\alpha;N)\,\mathscr{S}_{XX}(\alpha)d\alpha, &(6.47c)\end{aligned}$$

where (6.47b) uses (6.36a). Observe from (6.47a) that the expected value of the Bartlett PSD is the same as the expected value of a periodogram but with only M-length data. As a result, this mean is equal to the frequency-domain convolution of the true PSD $\mathscr{S}_{XX}(F)$ with a Bartlett (triangular) window of length $2M-1$ instead of length $2N-1$ in the original periodogram. Since $M \ll N$, the effect of using the shorter length Bartlett window in (6.47c) is to have a $W_B(F)$ with a wider main-lobe. Thus, the spectral broadening effect is greater, and the bias of the Bartlett PSD estimator $\hat{S}_B(F;N)$ is larger than that of the periodogram $\hat{S}_p(F;N)$. Equivalently, the spectral resolution of the Bartlett PSD estimator $\hat{S}_B(F;N)$ is on the order of $\frac{1}{M}$ which is worse off compared to that of the periodogram $\hat{S}_p(F;N)$, which is on the order of $\frac{1}{N}$.

From (6.44), we see that the variance of the Bartlett PSD estimate, on the other hand, is

$$\begin{aligned} Var\left(\hat{S}_B(F;N)\right) &= \frac{1}{K^2}\sum_{k=0}^{K-1} Var\left(\hat{S}_{p,k}(F;M)\right) \\ &= \frac{1}{K} Var\left(\hat{S}_p(F;M)\right). \end{aligned} \tag{6.48}$$

Hence, the Bartlett PSD estimator reduces the variance of the periodogram by a factor of $K = \frac{N}{M}$ in return for the degradation of the spectral resolution by the same factor $\frac{N}{M}$ (spectral resolution goes from $\frac{1}{N}$ to $\frac{1}{M}$).

6.4.2 Welch PSD Estimator

The Welch PSD estimator is obtained by introducing two modifications to the Bartlett method [63]: First, the subsequences $x_k[n]$'s of length M are allowed to overlap. Hence, the kth subsequence is

$$x_k[n] = x[n+kL] \quad \text{for} \quad n = 0,\ldots,M-1, \tag{6.49}$$

where $x[kL]$ is the starting data point of the kth sequence and $L \leq M$. Note that if $L = M$, then there is no overlap in successive subsequences and the segmentation is similar to that in the Bartlett method. On the other hand, if $L = \frac{M}{2}$, then there is a 50% overlap between successive subsequences. As a result, now there are more M-length subsequences. Indeed, with the 50% overlap, the number of sequences is now doubled to about $K = \frac{2N}{M}$ compared to that in the Bartlett approach. Recall that the variance reduction in the Bartlett method was by the factor of K, hence the motivation to increase K by allowing for overlap.

Second, the subsequences $x_k[.]$'s are weighted by a temporal window, called a data *taper*, before using them to compute the corresponding periodograms. If we denote by $\hat{S}_{w,k}(F;M)$ the kth periodogram computed from the temporally windowed data subsequence $x_k[.]$, for $k = 0,\ldots,K-1$, then

$$\hat{S}_{w,k}(F;M) = \frac{1}{MP_w}\left|\sum_{n=0}^{M-1} x_k[n]w[n]e^{-j2\pi Fn}\right|^2 \quad \text{for} \quad k = 0,\ldots,K-1, \tag{6.50}$$

where $w[n]$ is the M-length data tapering window and P_w is its average power defined as

$$P_w = \frac{1}{M}\sum_{n=0}^{M-1} |w[n]|^2. \tag{6.51}$$

The Welch PSD estimate is then obtained by averaging the above K periodograms computed in (6.50):

$$\hat{S}_W(F;N) = \frac{1}{K}\sum_{k=0}^{K-1} \hat{S}_{w,k}(F;M). \tag{6.52}$$

From (6.52), the bias of the Welch PSD estimate is

$$\begin{aligned}
\mathbb{E}\left\{\hat{S}_W(F;N)\right\} &= \frac{1}{K}\sum_{k=0}^{K-1}\mathbb{E}\left\{\hat{S}_{w,k}(F;M)\right\} \\
&= \mathbb{E}\left\{\hat{S}_{w,k}(F;M)\right\} \\
&= \frac{1}{MP_w}\mathbb{E}\left\{\left|\sum_{n=0}^{M-1} X_k[n]w[n]e^{-j2\pi Fn}\right|^2\right\} && \text{(6.53a)} \\
&= \frac{1}{MP_w}\sum_{n=0}^{M-1}\sum_{n'=0}^{M-1}\mathbb{E}\left\{X_k[n]X_k^*[n']\right\} w[n]w^*[n']e^{-j2\pi Fn} \\
&= \frac{1}{MP_w}\sum_{n=0}^{M-1}\sum_{n'=0}^{M-1}\mathscr{R}_{XX}[n-n']w[n]w^*[n']e^{-j2\pi F(n-n')}, && \text{(6.53b)}
\end{aligned}$$

where (6.53a) and (6.53b) follow, respectively, from (6.50) and (6.6). Using (6.14) in (6.53b), we have that

$$\mathbb{E}\left\{\hat{S}_W(F;N)\right\} = \frac{1}{MP_w}\sum_{n=0}^{M-1}\sum_{n'=0}^{M-1}\left(\int_{-\frac{1}{2}}^{\frac{1}{2}}\mathscr{S}_{XX}(\alpha)e^{j2\pi\alpha(n-n')}d\alpha\right)w[n]w^*[n']e^{-j2\pi F(n-n')} \tag{6.54a}$$

$$\begin{aligned}
&= \int_{-\frac{1}{2}}^{\frac{1}{2}}\mathscr{S}_{XX}(\alpha)\left(\frac{1}{MP_w}\sum_{n=0}^{M-1}\sum_{n'=0}^{M-1}w[n]w^*[n']e^{-j2\pi(n-n')(F-\alpha)}\right)d\alpha \\
&= \int_{-\frac{1}{2}}^{\frac{1}{2}}\mathscr{S}_{XX}(\alpha)\left(\frac{1}{MP_w}\left(\sum_{n=0}^{M-1}w[n]e^{-j2\pi n(F-\alpha)}\right)\right. \\
&\quad\times\left.\left(\sum_{n'=0}^{M-1}w[n']e^{-j2\pi n'(F-\alpha)}\right)^*\right)d\alpha
\end{aligned}$$

$$= \int_{-\frac{1}{2}}^{\frac{1}{2}} \mathscr{S}_{XX}(\alpha) \left(\frac{1}{MP_w} \left| \sum_{n=0}^{M-1} w[n] e^{-j2\pi n(F-\alpha)} \right|^2 \right) d\alpha$$

$$= \int_{-\frac{1}{2}}^{\frac{1}{2}} \mathscr{S}_{XX}(\alpha) \left(\frac{1}{MP_w} |W(F-\alpha)|^2 \right) d\alpha, \tag{6.54b}$$

where in (6.54b) we have denoted the Fourier transform of the window function $w[n]$ by $W(F) = \mathcal{F}\{w[n]\}$. Let us define a normalized frequency-domain window function $\tilde{W}(F)$ as

$$\tilde{W}(F) = \frac{1}{MP_w} |W(F)|^2. \tag{6.55}$$

From (6.51), it follows that $\tilde{W}(F)$ has unit energy:

$$\int_{-\frac{1}{2}}^{\frac{1}{2}} \tilde{W}(F) dF = \frac{1}{MP_w} \int_{-\frac{1}{2}}^{\frac{1}{2}} |W(F)|^2 \, dF$$

$$= \frac{1}{MP_w} \sum_{n=0}^{M-1} |w[n]|^2$$

$$= 1 \tag{6.56}$$

Substituting (6.56) in (6.54b),

$$\mathbb{E}\left\{ \hat{S}_W(F;N) \right\} = \int_{-\frac{1}{2}}^{\frac{1}{2}} \mathscr{S}_{XX}(\alpha) \tilde{W}(F-\alpha) d\alpha \tag{6.57a}$$

$$= \mathscr{S}_{XX}(F) \star \left(\frac{1}{MP_w} |\mathcal{F}\{w[n]\}|^2 \right). \tag{6.57b}$$

Hence, from (6.57b), we see that the mean of the Welch PSD estimate is the convolution of the true PSD $\mathscr{S}_{XX}(F)$ with the frequency-domain window function $\tilde{W}(F)$ determined by the data taper $w[n]$. Since the window length is M, $W(F)$ will have a main-lobe width of the order $\frac{1}{M}$. Hence, the bias and spectral resolution of the Welch PSD estimator are of the order of $\frac{1}{M}$.

The variance of the Welch estimate when there is 50% overlap in the data subsequences and a triangular data taper is used so that $w[n] = w_B[n]$ is shown to be [63, 64]

$$Var\left(\hat{S}_W(F;N)\right) \approx \frac{9}{8K}\mathscr{S}_{XX}(F) \tag{6.58a}$$

$$= \frac{9M}{16N}\mathscr{S}_{XX}(F). \tag{6.58b}$$

Hence, the variance is reduced compared to both the original periodogram and the Bartlett PSD estimate. By changing the amount of overlap and the choice of the window function, it is possible to achieve different variance values.

6.5 PSD ESTIMATION OF NONSTATIONARY DISCRETE-TIME SIGNALS

So far, we have considered power spectrum estimation of wss processes. For such processes, the first- and second-order statistics in (6.1) and (6.2), respectively, did not depend on the exact time instant $n = n$, and the autocorrelation function in (6.6), for example, only depended on the time lag $m = n_1 - n_2$. But in practice, such stationary assumptions might not hold. This may particularly be true for signals encountered in wireless communications channels.

Suppose we are interested in estimating the spectral content of a nonstationary signal $X[n]$ at time $n = n_0$. As before, we will assume that only a finite-length sample realization of $x[n]$, for $n_0 - \frac{N-1}{2} \le n \le n_0 + \frac{N-1}{2}$, is available to us. Note that we have assumed that N is an odd integer. This is equivalent to assuming that we are given a signal

$$x_{N;n_0}[n] = x[n]\Pi\left[\frac{n-n_0}{N}\right], \tag{6.59}$$

where $\Pi\left[\frac{n}{N}\right]$ is the rectangular function defined in (6.41). When $n_0 = 0$, we have

$$x_{N;0}[n] = x[n]\Pi\left[\frac{n}{N}\right]. \tag{6.60}$$

In this case, we may drop n_0 from the above notation and write

$$x_{N;0}[n] = x_N[n]. \tag{6.61}$$

Unlike in the case of a wss signal, now we assume that second-order statistics computed from $x[n]$ for $n_0 - \frac{N-1}{2} < n < n_0 + \frac{N-1}{2}$ do depend on the particular time instant $n = n_0$. When the time instant n_0 of interest is changed, the statistics will also change, implying that the spectral components must also change.

By generalizing the notion of Fourier transform of a wss process, we may define a *time-variant finite-time discrete-time Fourier transform* (TVFT-DTFT) of the sequence $x[n]$ as [60]

$$X(F, n_0; N) \triangleq \sum_{n=n_0-\frac{N-1}{2}}^{n_0+\frac{N-1}{2}} x[n]e^{-j2\pi Fn}, \tag{6.62}$$

with the corresponding inverse transform formula

$$x[n] = \int_{-\frac{1}{2}}^{\frac{1}{2}} X(F, n_0; N)e^{j2\pi Fn}dF \quad \text{for} \quad n_0 - \frac{N-1}{2} \le n \le n_0 + \frac{N-1}{2}. \tag{6.63}$$

Observe from (6.63) that $X(F, n_0; N)$ is the complex amplitude of the complex exponential having frequency F in the signal $x[n]$ for $n_0 - \frac{N-1}{2} < n < n_0 + \frac{N-1}{2}$. This complex amplitude of the frequency F spectral component depends on the time interval around which time (in this case $n = n_0$) the spectral content of the signal is observed. If the observation interval is changed from the one centered around $n = n_0$ to another one centered around $n = n_1$, the spectral components will also change from $X(F, n_0; N)$ to $X(F, n_1; N)$.

Since we have taken observation interval length to be N, we are assuming that spectral components are indeed the same (and do not change) over a time interval of length of N. Hence, if the signal were to have spectral components that do change within the time duration N, then $X(F, n_0; N)$ will not be able to show that (just like Fourier transform that does not show any spectral components that change at all because $N \to \infty$). Hence, we call N the *temporal resolution* (width) Δn of the TVFT-DTFT

$$\Delta n = N.$$

Any spectral changes within N or less time will not be captured by $X(F, n; N)$.

As we have already seen, when the spectral estimation is based on data over a finite-time interval N, the minimum *spectral resolution*, denoted by ΔF, is limited to

$$\Delta F = \frac{1}{N}.$$

When $X(F, n_0; N)$ in (6.62) depends on time $n = n_0$, we may define a *time-variant periodogram* that is computed from $x[n]$ for $n_0 - \frac{N-1}{2} \le n \le n_0 + \frac{N-1}{2}$, as

$$\hat{S}_p(F, n_0; N) = \frac{1}{N}|X(F, n_0; N)|^2. \tag{6.64}$$

It can be shown that there is a temporal function, called the *time-variant autocorrelation sequence estimate* or the *time-variant correlogram*, that can be computed from $x[n]$ for $n_0 - \frac{N-1}{2} \le n \le n_0 + \frac{N-1}{2}$, such that [60]

$$\hat{S}_p(F, n_0; N) = \mathcal{F}\left\{\hat{R}_{xx}[m, n_0; N]\right\}. \tag{6.65}$$

Indeed,

$$\hat{R}_{xx}[m, n_0; N] = \begin{cases} \frac{1}{N} \sum_{n=n_0-\frac{N-1}{2}+m}^{n_0+\frac{N-1}{2}} x[n]x^*[n-m] & \text{if } 0 \le m \le N-1 \\ \hat{R}^*_{xx}[-m, n_0; N] & \text{if } -(N-1) \le m \le -1 \end{cases}. \tag{6.66}$$

Hence, the time-variant periodogram and the time-variant correlogram form a Fourier transform pair as in (6.65).

6.5.1 Temporally Windowed Observations

With the notation in (6.59), we see that TVFT-DTFT (6.62) is equivalent to the DTFT of the observation signal $x_{N;n_0}[n]$:

$$X(F, n_0; N) = \mathcal{F}\left\{x_{N;n_0}[n]\right\}$$

$$= \sum_{n=-\infty}^{\infty} x[n]e^{-j2\pi Fn}\Pi\left[\frac{n-n_0}{N}\right] \tag{6.67a}$$

$$= x[n_0]e^{-j2\pi Fn_0} \star \Pi\left[\frac{-n_0}{N}\right] \tag{6.67b}$$

$$= x[n_0]e^{-j2\pi Fn_0} \star \Pi\left[\frac{n_0}{N}\right].$$

Hence, $X(F, n; N)$ is the convolution of the spectral component $x[n]e^{-j2\pi Fn}$ in $x[n]$ with the rectangular function centered at time $n = 0$ with a finite width N. The final result is to smoothen, or broaden, the spectral component by an order of $\frac{1}{N}$.

Then it is also possible to replace the rectangular window in (6.59) by a general data tapering function $w_N[n]$ that is nonzero only for $-\frac{N-1}{2} \le n \le \frac{N-1}{2}$. Let us denote by $x_{w;n_0}[n]$ the tapered data obtained by using a data taper $w_N[n]$:

$$x_{w;n_0}[n] = x[n]w_N[n-n_0]. \tag{6.68}$$

The resulting DTFT of the tapered data $x[n]w_N[n-n_0]$ is

$$\begin{aligned} X_w(F,n_0;N) &= \mathcal{F}\{x[n]w_N[n-n_0]\} \\ &= \sum_{n=-\infty}^{\infty} x[n]w_N[n-n_0]e^{-j2\pi Fn} \\ &= \sum_{n=n_0-\frac{N-1}{2}}^{n_0+\frac{N-1}{2}} x[n]w_N[n-n_0]e^{-j2\pi Fn}, \end{aligned} \tag{6.69}$$

where last step follows by noting that $w_N[n]$ is an N-length taper centered at $n=0$. The corresponding time-variant periodogram estimate of the PSD based on the tapered data is

$$\hat{S}_{p,w}(F,n_0;N) = \frac{1}{N}|X_w(F,n_0;N)|^2. \tag{6.70}$$

As before, it is possible to verify that there is a temporal function $\hat{R}_{xx,w}[m,n_0;N]$ satisfying the following Fourier transform relation

$$\hat{S}_{p,w}(F,n;N) = \mathcal{F}\left\{\hat{R}_{xx,w}[m,n_0;N]\right\},$$

where $\hat{R}_{xx,w}[m,n_0;N]$ is obtained by replacing $x[n]$ in (6.66) by $x_{w;n_0}[n]$ [60].

6.5.2 Temporal and Spectral Smoothing of PSD Estimates of Nonstationary Discrete-Time Signals

Recall the time-varying periodogram estimate (6.64) at time $n = n_0$ based on the TVFT-DTFT (6.62) computed from the given data for $n_0 - \frac{N-1}{2} \le n \le n_0 + \frac{N-1}{2}$. As we saw, even in the case of stationary signals, the periodogram estimate does not have a guarantee of consistency. In particular, it does not converge in the mean-square sense to the true spectrum as $N \longrightarrow \infty$. Instead, it has a constant asymptotic variance when $N \longrightarrow \infty$, which, for example, was shown to be equal to $\mathscr{S}_{XX}^2(f)$ in the Gaussian case. Hence, it is to be expected that $\hat{S}_p(F,n_0;N)$ computed as in the preceding text will also be highly erratic around its mean even when $N \longrightarrow \infty$. Moreover, letting $N \longrightarrow \infty$ may not help when data is nonstationary anyway. In that case, taking N large implies trying to obtain a fixed set of spectral components to represent data that might have actually time-varying spectral components.

When the periodogram $\hat{S}_p(F,n_0;N)$ may be too erratic as an estimator, a reasonable approach is to time average (or time smoothen) it. This can be achieved by convolving the time-variant periodogram $\hat{S}_p(F,n_0;N)$ with a rectangular function $\frac{1}{L}\Pi\left[\frac{n}{L}\right]$ of unit area [60].

First, consider the untapered data in (6.59). Let us denote averaging with respect to a variable using the following subscript notation:

$$\hat{S}_p\left[F, n_{[L]}; N\right] = \hat{S}_p[F, n; N] \star \left(\frac{1}{L}\Pi\left[\frac{n}{L}\right]\right) \tag{6.71a}$$

$$= \frac{1}{L}\sum_{l=-\infty}^{\infty} \hat{S}_p[F, l; N]\Pi\left[\frac{n-l}{L}\right]$$

$$= \frac{1}{L}\sum_{l=n-\frac{L-1}{2}}^{n+\frac{L-1}{2}} \hat{S}_p(F, l; N). \tag{6.71b}$$

It has been shown that if the limit autocorrelation function $R_{xx}[m]$ of (6.16) exists, then [60]

$$\lim_{N\to\infty}\lim_{L\to\infty} \hat{S}_p(F, n_{[L]}; N) = S_{xx}(F),$$

where $S_{xx}(F) = \mathcal{F}\{R_{xx}[m]\}$ is the limit spectrum of (6.17). Under the same conditions, it can be shown that if the smoothing window length $L \gg N$, then [60]

$$\hat{S}_p(F, n; N) \star \left(\frac{1}{L}\Pi\left[\frac{n}{L}\right]\right) \approx \hat{S}_p(F, n; L) \star \left(\frac{1}{N}\left(\mathcal{F}\left\{\prod\left(\frac{n}{N}\right)\right\}\right)^2\right). \tag{6.72}$$

Since

$$\mathcal{F}\left\{\Pi\left[\frac{n}{N}\right]\right\} = \frac{\sin \pi F N}{\sin \pi F}, \tag{6.73}$$

we can rewrite (6.72) as

$$\hat{S}_p(F, n; N) \star \left(\frac{1}{L}\Pi\left[\frac{n}{L}\right]\right) \approx \hat{S}_p(F, n; L) \star \left(\frac{\sin^2 \pi F N}{N \sin^2 \pi F}\right). \tag{6.74}$$

Hence, we note that the smoothing in the time domain is equivalent to a certain smoothing (convolution) in the frequency domain. Recall that the data taper in the time domain was $w_N[n] = \Pi\left[\frac{n}{N}\right]$. Essentially, it is this rectangular function used in the data tapering that determines the function to be convolved in the frequency domain and not the rectangular function $\Pi\left[\frac{n}{L}\right]$ used to achieve time averaging.

Now, suppose that a general data tapering window $w_N[n]$ is used to first obtain the tapered data-based periodogram estimate $\hat{S}_{p,w}(F, n; N)$ from the given data for

$n_0 - \frac{N-1}{2} \leq n \leq n_0 + \frac{N-1}{2}$. Suppose that we time average this estimate over a period of odd-length L duration as follows:

$$\hat{S}_{p,w}(F, n_{[L]}; T) = \hat{S}_{p,w}(F, n; N) \star \left(\frac{1}{L} \Pi\left(\frac{n}{L}\right) \right) \tag{6.75a}$$

$$= \frac{1}{L} \sum_{l=-\infty}^{\infty} \hat{S}_{p,w}(F, l; N) \Pi\left(\frac{n-l}{L}\right)$$

$$= \frac{1}{L} \sum_{l=n-\frac{L-1}{2}}^{n+\frac{L-1}{2}} \hat{S}_{p,w}(F, l; N) \tag{6.75b}$$

$$= \frac{1}{LN} \sum_{l=n-\frac{L-1}{2}}^{n+\frac{L-1}{2}} |X_w(F, l; N)|^2, \tag{6.75c}$$

where we have used (6.70) in the last step.

Then under the assumption that limit correlation $R_{xx}[m]$ of (6.16) exists and $L \gg N$, it has been shown that [60]

$$\hat{S}_{p,w}(F, n; N) \star \left(\frac{1}{L} \Pi\left(\frac{n}{L}\right) \right) \approx \hat{S}_p(F, n; L) \star \left(\frac{1}{N} |W_N(F)|^2 \right), \tag{6.76}$$

where $W_N(F)$ is the DTFT of the data taper used:

$$W_N(F) = \sum_{n=-\infty}^{\infty} w_N(n) e^{-j2\pi Fn}$$

$$= \sum_{n=-\frac{N-1}{2}}^{\frac{N-1}{2}} w_N(n) e^{-j2\pi Fn}.$$

Hence, if $L \gg N$, smoothing N-length data-based time-variant periodogram over an L-length time window is equivalent to smoothing (in the sense of convolving with the appropriate frequency-domain function as specified in (6.76)) the L-length data-based time-variant periodogram over about $\frac{1}{N}$-length frequency window in the frequency domain.

6.5.3 DFT-Based PSD Computation

In practice, it is common to compute the DTFT of the given sequence of data $x[n]$ using the discrete Fourier transform (DFT):

$$X(k; N) = \frac{1}{N} \sum_{n=0}^{N-1} x[n] e^{-j2\pi \frac{k}{N} n} \quad \text{for } k = 0, \ldots, N-1.$$

The attractiveness of DFT is that it can be computed efficiently using fast Fourier transform (FFT) algorithms.

Let the DFT computed at time n be $X(k,n;N)$ for $k = 0,\ldots,N-1$ where $F = \frac{k}{N}$. Then, if $L \gg N$, from (6.76),

$$\begin{aligned}
\hat{S}_{p,w}(k,n_{[L]};N) &= \hat{S}_{p,w}(k,n;N) \star \left(\frac{1}{L}\Pi\left(\frac{n}{L}\right)\right) \\
&\approx \hat{S}_p(k,n;L) \star \left(\frac{1}{N}\left|W_N(k)\right|^2\right) \\
&= \frac{1}{LN}\sum_{l=-\frac{M-1}{2}}^{\frac{M-1}{2}} \left|X(k+l,n;N)\right|^2 \left|W_N(l)\right|^2,
\end{aligned}$$

where we have assumed that $W_N(k)$ is nonzero only for frequencies up to an order of $\frac{1}{N} = \frac{M}{L}$ for some integer $M \ll L$. Hence, time-domain smoothing of the periodogram can equivalently be achieved by averaging the periodogram computed from the DFT in the frequency domain. Note that if the time-domain data taper was a *sinc*(.) function[5] so that the $W_N(k)$ is a rectangular function, then the above spectrally smoothed periodogram estimate reduces to

$$\hat{S}_{p,w}(k,n_{[L]};N) = \frac{1}{LN}\sum_{l=-\frac{M-1}{2}}^{\frac{M-1}{2}} \left|X(k+l,n;N)\right|^2.$$

6.6 SPECTRAL CORRELATION OF CYCLOSTATIONARY SIGNALS

So far, we considered wss processes (in which first- and second-order statistics did not depend on the exact time instant $n = n$ and the autocorrelation function only depended on the time lag $m = n_1 - n_2$) and nonstationary processes in which the autocorrelation may in general depend on the exact time. However, many man-made signals, though not wss, at least exhibit certain periodicities in their first- and second-order statistics. This is due to underlying periodic properties inherent in the way these signals are generated. As we will see, such periodicities in the statistics can be unique to a particular type of signal and thus can act as a kind of a spectral fingerprint for the class of signals. The following brief overview of the effect of such underlying periodicities of signals on their spectral content is mostly based on [60,65,66] and the review [67].

Let us denote by $m_X[n]$ and $R_{XX}[n_1,n_2]$ the first- and second-order moments (mean and autocorrelation) of a discrete-time random signal $X[n]$:

$$\mathbb{E}\{X[n]\} = m_X[n] \tag{6.77}$$

[5] $sinc(x) = \frac{\sin \pi x}{\pi x}$.

and

$$\mathbb{E}\{X[n]X^*[n-m]\} = R_{XX}[n,m]. \tag{6.78}$$

Then $X[n]$ is called *wide-sense cyclostationary* (wscs), or second-order cyclostationary, if there exists an $N_0 > 0$ such that [67]

$$m_X[n+N_0] = m_X[n] \quad \text{for all } n \tag{6.79}$$

and

$$\mathscr{R}_{XX}[n+N_0,m] = \mathscr{R}_{XX}[n,m] \quad \text{for all } n,m. \tag{6.80}$$

When this is true for some $N_0 > 0$, then N_0 is called the period of the cyclostationary process and $\mathscr{R}_{XX}[n,m]$ is called a *periodic autocorrelation function* [67]. Note that the mean of a cyclostationary process is not constant, but it is also not an arbitrary function of time n for all n either. It is a periodic function of time n. In the following, we will, however, assume that the mean is zero for all n so that the autocorrelation can be considered as the autocovariance function. Similarly, the autocorrelation function of a cyclostationary signal is not simply a function of the time difference between the two time instants as in the case of a wss signal. Instead, the autocorrelation is both a function of the absolute time as well as the relative time difference between the two time instants. However, it is not completely arbitrary as a function of the absolute time origin but is periodic with period N_0 as a function of absolute time n.

Since $\mathscr{R}_{XX}[n,m]$ is periodic in time n with a period N_0, it can be expanded in a discrete-time Fourier series (DTFS) as follows (assuming the series indeed converges to $\mathscr{R}_{XX}[n,m]$):

$$\mathscr{R}_{XX}[n,m] = \sum_{k=-\frac{N_0-1}{2}}^{\frac{N_0-1}{2}} \mathscr{R}_{XX}^{\alpha_k}[m]e^{j2\pi\alpha_k n}, \tag{6.81}$$

where we have denoted the harmonic frequencies by $\alpha_k = \frac{k}{N_0}$ and the Fourier series coefficients are given by

$$\mathscr{R}_{XX}^{\alpha_k}[m] = \frac{1}{N_0}\sum_{n=-\frac{N_0-1}{2}}^{\frac{N_0-1}{2}} \mathscr{R}_{XX}[n,m]e^{-j2\pi\alpha_k n}. \tag{6.82}$$

In the above, we have assumed that N_0 is odd, although it can be modified easily if N_0 is even as

$$\mathscr{R}_{XX}[n,m] = \sum_{k=-\frac{N_0}{2}+1}^{\frac{N_0}{2}} \mathscr{R}_{XX}^{\alpha_k}[m]e^{j2\pi\alpha_k n}. \tag{6.83}$$

In the following, we will assume N_0 to be odd.

The Fourier coefficients $\mathscr{R}_{XX}^{\alpha_k}[m]$ of (6.82) are called *cyclic autocorrelation functions*, or simply the cyclic autocorrelation, and the harmonic frequencies $\alpha_k = \frac{k}{N_0}$, for $k = 0, \pm 1, \ldots, \pm\frac{N_0-1}{2}$ are called the *cyclic frequencies*. Note that since DTFS coefficients are periodic in the discrete frequencies α_k, we could have selected the set of harmonic frequencies to be $\alpha_k = \frac{k}{N_0}$, for $k = 0, 1, \ldots, N_0 - 1$. Clearly, wss processes are a special case of the cyclostationary processes in which $\mathscr{R}_{XX}^{\alpha_k}[m] \neq 0$ only for $k = 0$. In other words, the only cyclic frequency present in a wss signal is $\alpha_k = 0$. Equivalently, $N_0 = 1$.

There are situations in which $\mathscr{R}_{XX}[n,m]$ is not exactly periodic, yet it can be expressed as [67]

$$\mathscr{R}_{XX}[n,m] = \sum_{\alpha\in\mathcal{A}} \mathscr{R}_{XX}^{\alpha}[m]e^{j2\pi\alpha n}, \tag{6.84}$$

for some countable set $\mathcal{A}$ with corresponding coefficients given by (for odd N)

$$\mathscr{R}_{XX}^{\alpha}[m] = \lim_{N\longrightarrow\infty} \frac{1}{N} \sum_{n=-\frac{N-1}{2}}^{\frac{N-1}{2}} \mathscr{R}_{XX}[n,m]e^{-j2\pi\alpha n}. \tag{6.85}$$

Such an autocorrelation function is called an *almost periodic autocorrelation function* in n. If the autocorrelation function of a process $X[n]$ is almost periodic in this sense, then the process is called *wide-sense almost cyclostationary* or second-order cyclostationary [67].

Again, since the cyclic autocorrelation $\mathscr{R}_{XX}^{\alpha}[m]$ is periodic in the cyclic frequency α with period 1, for convenience, we may choose $\alpha \in [0,1]$ so that $\mathcal{A} = \{\alpha \in [0,1] : R_{XX}^{\alpha}[m] \neq 0\}$.

Cyclostationarity leads to spectral correlation. To see this, denote the DTFT of a *wide-sense almost cyclostationary* process $X[n]$ as

$$X(F) = \sum_{n=-\infty}^{\infty} X[n]e^{-j2\pi Fn}.$$

Let us consider the spectral autocorrelation $\mathbb{E}\{X(F_1)X^*(F_2)\}$ of $X(F)$, for $-\frac{1}{2} \leq F_1, F_2, \leq \frac{1}{2}$:

$$\begin{aligned}
\mathbb{E}\left\{X(F_1)X^*(F_2)\right\} &= \mathbb{E}\left\{\sum_{n=-\infty}^{\infty} X[n]e^{-j2\pi F_1 n}\left(\sum_{n'=-\infty}^{\infty} X[n']e^{-j2\pi F_2 n'}\right)^*\right\} \\
&= \sum_{n=-\infty}^{\infty}\sum_{n'=-\infty}^{\infty} \mathbb{E}\left\{X[n]X^*[n']\right\} e^{-j2\pi F_1 n}e^{j2\pi F_2 n'} \\
&= \sum_{n=-\infty}^{\infty}\sum_{n'=-\infty}^{\infty} \mathscr{R}_{XX}[n,n-n']e^{-j2\pi F_1 n}e^{j2\pi F_2 n'} \qquad (6.86a) \\
&= \sum_{n=-\infty}^{\infty}\sum_{n'=-\infty}^{\infty}\sum_{\alpha\in\mathcal{A}} \mathscr{R}^{\alpha}_{XX}[n-n']e^{j2\pi\alpha n}e^{-j2\pi F_1 n}e^{j2\pi F_2 n'} \qquad (6.86b) \\
&= \sum_{\alpha\in\mathcal{A}}\sum_{n=-\infty}^{\infty}\left(\sum_{m=-\infty}^{\infty} \mathscr{R}^{\alpha}_{XX}[m]e^{-j2\pi F_2 m}\right)e^{j2\pi(\alpha-F_1+F_2)n}, \qquad (6.86c)
\end{aligned}$$

where (6.86a) follows from (6.78) and (6.86b) follows from (6.84). Let us denote the DTFT of the cyclic autocorrelation function $\mathscr{R}^{\alpha}_{XX}[m]$ by $\mathscr{S}^{\alpha}_{XX}(F)$ called the *cyclic spectrum* of $X[n]$:

$$\mathscr{S}^{\alpha}_{XX}(F) = \sum_{m=-\infty}^{\infty} \mathscr{R}^{\alpha}_{XX}[m]e^{-j2\pi Fm}. \qquad (6.87)$$

Substituting (6.87) in (6.86c), we see that

$$\begin{aligned}
\mathbb{E}\left\{X(F_1)X^*(F_2)\right\} &= \sum_{\alpha\in\mathcal{A}} \mathscr{S}^{\alpha}_{XX}(F_2)\sum_{n=-\infty}^{\infty} e^{-j2\pi((F_1-F_2)-\alpha)n} \qquad (6.88a) \\
&= \sum_{\alpha\in\mathcal{A}} \mathscr{S}^{\alpha}_{XX}(F_2)\delta_D\left((F_1-F_2)-\alpha\right). \qquad (6.88b)
\end{aligned}$$

From (6.88b), note that if $\mathscr{R}^{\alpha}_{XX}[m] \neq 0$ only for $\alpha = 0$, then

$$\mathbb{E}\left\{X(F_1)X^*(F_2)\right\} = \mathscr{S}^{0}_{XX}(F_2)\delta_D(F_1-F_2) \qquad (6.89)$$

is nonzero only for $F_1 = F_2$. In other words,

$$\mathbb{E}\left\{X(F_1)X^*(F_2)\right\} = 0 \quad \text{for all } F_1 \neq F_2, \qquad (6.90)$$

implying that $X(F_1)$ and $X(F_2)$ are uncorrelated whenever $F_1 \neq F_2$.

Of course, a process with $\mathscr{R}^{\alpha}_{XX}[m] \neq 0$ only for $\alpha = 0$ is a wss process in which autocorrelation function $\mathscr{R}_{XX}[n,n-m]$ does not depend on absolute time

n: $\mathscr{R}_{XX}[n,n-m] = \mathscr{R}_{XX}[m]$. Furthermore, for $\alpha = 0$, the cyclic autocorrelation function $\mathscr{R}^0_{XX}[m]$ reduces to the autocorrelation function $\mathscr{R}_{XX}[m]$. Thus, in this case, its Fourier transform $\mathscr{S}^0_{XX}(F)$ is the PSD $\mathscr{S}_{XX}(F)$.

On the other hand, if $\mathscr{R}^\alpha_{XX}[m] \neq 0$ for some $\alpha \neq 0$, then (6.88b) implies that

$$\mathbb{E}\left\{X(F_1)X^*(F_2)\right\} \neq 0 \quad \text{whenever} \quad F_1 - F_2 = \alpha. \tag{6.91}$$

Hence, $X(F_1)$ and $X(F_2)$ are correlated whenever $F_1 = F_2 + \alpha$.

Since for an $\alpha \neq 0$ for which $\mathscr{R}^\alpha_{XX}[m] \neq 0$, from (6.88b),

$$\mathbb{E}\left\{X(F_2+\alpha)X^*(F_2)\right\} = \mathscr{S}^\alpha_{XX}(F_2)\delta_D(0), \tag{6.92}$$

the cyclic spectrum $\mathscr{S}^\alpha_{XX}(F)$ is also called the *spectral correlation density function* by analogy with the PSD function [60]. Often, it is simply referred to as the *spectral correlation function* (SCF).

If we consider $\mathscr{S}^\alpha_{XX}(F)$ as a function of the two variables (F,α), then for a fixed $F = F_0$, $\mathscr{S}^\alpha_{XX}(F_0)$ may have nonzero values along the α-axis at most for a countable set of α points. At those places, $\mathscr{S}^\alpha_{XX}(F_0)$ will be equal to the correlation between the two spectral components $X(F_0)$ and $X(F_0+\alpha)$ of $X[n]$.

6.6.1 Spectral Correlation and Spectral Autocoherence

As we just saw earlier, $\mathscr{S}^\alpha_{XX}(F)$ shows the spectral correlation density of $X[n]$. Specifically, at frequency F, it shows the correlation of $X(F)$ and $X(F+\alpha)$, which is a frequency-shifted version of $X(F)$. Similar to time-domain correlation coefficient, we may then characterize the amount of this spectral correlation by defining the following *spectral correlation coefficient*:

$$\begin{aligned} C^\alpha_{XX}(F) &= \frac{\mathscr{S}^\alpha_{XX}(F)}{\sqrt{\mathscr{S}_{XX}(F+\alpha)\,\mathscr{S}_{XX}(F)}} \\ &= \frac{\mathbb{E}\{X(F+\alpha)X^*(F)\}}{\sqrt{\mathscr{S}_{XX}(F+\alpha)\,\mathscr{S}_{XX}(F)}}. \end{aligned} \tag{6.93}$$

Note that $C^\alpha_{XX}(F)$ is the correlation of $X(F+\alpha)$ and $X(F)$ normalized by the (power) spectral density at those two frequencies and can, in general, be complex valued. For this, reason, it is also called the *complex autocoherence*, and its magnitude $|C^\alpha_{XX}(F)|$ is called the *autocoherence*. It can be shown that similar to the time-domain correlation coefficient, the spectral correlation coefficient also satisfies [60]

$$|C^\alpha_{XX}(F)| \leq 1. \tag{6.94}$$

When $C^\alpha_{XX}(F) = 0$, $X[n]$ is said to be completely incoherent at frequency F and cycle frequency α. On the other hand, if $|C^\alpha_{XX}(F)| = 1$, then $X[n]$ is said to be completely coherent at frequency F and cycle frequency α. The strength of spectral correlation, $|C_{XX}^{\alpha}(F)|$, is invariant to linear time-invariant (LTI) transformations [60]. In other words, if $Y[n] = h[n] \star X[n]$, then $|C_{YY}^{\alpha}(F)| = |C_{XX}^{\alpha}(F)|$. The only change is possibly in the phase.

6.6.2 Time-Averaged Spectral Correlation

An intuitive interpretation of the cyclic autocorrelation $\mathscr{R}_{XX}^{\alpha}[m]$ can be made by expressing it as the usual *cross-correlation* between two distinct processes. To that end, let us define the following:

$$U[n] = X[n]e^{\,j2\pi\alpha n}$$
$$V[n] = X[n].$$

In other words, $V[n]$ is a frequency-shifted version of $U[n]$. Then we may write

$$\begin{aligned}
\mathscr{R}_{XX}^{\alpha}[m] &= \lim_{N\longrightarrow\infty}\frac{1}{N}\sum_{n=-\frac{N-1}{2}}^{\frac{N-1}{2}} \mathscr{R}_{XX}[n,m]e^{-j2\pi\alpha n} \\
&= \lim_{N\longrightarrow\infty}\frac{1}{N}\sum_{n=-\frac{N-1}{2}}^{\frac{N-1}{2}} \mathbb{E}\{X[n]X^{*}[n-m]\}e^{-j2\pi\alpha n} \\
&= \lim_{N\longrightarrow\infty}\frac{1}{N}\sum_{n=-\frac{N-1}{2}}^{\frac{N-1}{2}} \mathbb{E}\{U[n]V^{*}[n-m]\} \\
&= \lim_{N\longrightarrow\infty}\frac{1}{N}\sum_{n=-\frac{N-1}{2}}^{\frac{N-1}{2}} \mathscr{R}_{UV}[n,m],
\end{aligned}$$

where $\mathscr{R}_{UV}[n,m]$ is the time-varying cross-correlation between $X[n]$ and its frequency-shifted version in which the frequency shift is equal to α. Hence, the cyclic autocorrelation function $\mathscr{R}_{XX}^{\alpha}[m]$ is nothing but a time-smoothed (time-averaged) measure of this cross-correlation between $X[n]$ and its frequency-shifted version $X[n]e^{-j2\pi\alpha n}$. Of course, the process is said to be cyclostationary if and only if this correlation is nonzero for some $\alpha \neq 0$.

Similar to (6.16), then we may define a temporal (time-averaged) version of cyclic autocorrelation function (6.80) as

$$R_{xx}^{\alpha}[m] = \lim_{N\longrightarrow\infty}\frac{1}{N}\sum_{n=-\frac{N-1}{2}}^{\frac{N-1}{2}} x[n]x^{*}[n-m]e^{-j2\pi\alpha n},$$

which is called the *limit cyclic autocorrelation* function [60].

The DTFT of the limit cyclic autocorrelation is called the *limit cyclic spectrum*:

$$S_{xx}^{\alpha}(F) = \mathcal{F}\left\{R_{xx}^{\alpha}[m]\right\}.$$

6.6.3 Estimation of Spectral Correlation

Now, suppose that we only have a finite data length at time $n = n_0$ for $n_0 - \frac{N-1}{2} < n < n_0 + \frac{N-1}{2}$. Then we may define the TVFT-DTFT $X(F, n_0; N)$ of $X[n]$ using (6.62):

$$X(F, n_0; N) = \sum_{n=n_0-\frac{N-1}{2}}^{n_0+\frac{N-1}{2}} x[n]e^{-j2\pi Fn}. \tag{6.95}$$

From this, we may now define a new version of periodogram, called the *time-variant finite-time cyclic periodogram*, as:

$$\hat{S}_p^{\alpha}(F, n_0; N) = \frac{1}{N} X(F+\alpha, n_0; N)X^*(F, n_0; N). \tag{6.96}$$

As with wss signals, the above time-variant finite-time cyclic periodogram $\hat{S}_p^{\alpha}(F, n_0; N)$ can be shown to be the Fourier transform of a corresponding *cyclic correlogram* [67]:

$$\begin{aligned} \hat{S}_p^{\alpha}(F, n_0; N) &= \mathcal{F}\left\{\hat{R}_{xx}^{\alpha}[m, n_0; N]\right\} \\ &= \sum_{m=-(N-1)}^{N-1} \hat{R}_{XX}^{\alpha}[m, n_0; N]e^{-j2\pi Fm}, \end{aligned} \tag{6.97}$$

where cyclic correlogram $\hat{R}_{xx}^{\alpha}[m, n_0; N]$ is defined as

$$\hat{R}_{xx}^{\alpha}[m, n_0; N] = \frac{1}{N} \sum_{n=n_0-\frac{N-1}{2}+m}^{n_0+\frac{N-1}{2}} x[n]x^*[n-m]e^{-j2\pi\alpha n} \quad \text{for} \quad m = 0, 1, \ldots, N-1$$

and

$$\hat{R}_{xx}^{\alpha}[-m, n_0; N] = \left(\hat{R}_{xx}^{\alpha}[m, n_0; N]\right)^* \quad \text{for} \quad m = 1, \ldots, N-1.$$

The usefulness of the time-variant finite-time cyclic periodogram is due to the fact that, under certain conditions, its temporal and frequency averages converge to the limit cyclic spectrum $S_{xx}^{\alpha}(F)$:

$$S_{xx}^{\alpha}(F) = \lim_{N\longrightarrow\infty} \lim_{L\longrightarrow\infty} \frac{1}{L} \sum_{l=n_0-\frac{L-1}{2}}^{n_0+\frac{L-1}{2}} \hat{S}_p^{\alpha}[F, l; N]$$

and

$$S_{xx}^{\alpha}(F) = \lim_{\Delta F \longrightarrow 0} \lim_{N \longrightarrow \infty} \frac{1}{\Delta F} \int_{F-\frac{\Delta F}{2}}^{F+\frac{\Delta F}{2}} \hat{S}_p^{\alpha}(\nu, n_0; N) d\nu.$$

Hence, given a finite-length data segment, a measure of spectral correlation can be computed by time smoothing the time variant finite-time cyclic periodogram $\hat{S}_p^{\alpha}(F, n_0; N)$:

$$\hat{S}_{XX}^{\alpha}[F, n_{[L]}; N] = \frac{1}{L} \sum_{l=n-\frac{L-1}{2}}^{n+\frac{L-1}{2}} \hat{S}_p^{\alpha}[F, l; N]. \tag{6.98}$$

We have already seen that the temporal and spectral resolution of such a smoothed spectrum is L and $\frac{1}{N}$, respectively. As before, here, the spectral resolution means that the ability to distinguish between two spectral components F_1 and F_2 will only be possible if $|F_1 - F_2| > \frac{1}{N}$. In addition, it has been noted that the *cyclic frequency resolution* $\Delta\alpha$ is also limited to $\frac{1}{L}$ [65].

On the other hand, instead of temporal smoothing, one may obtain a measure of spectral correlation by performing smoothing in the frequency domain:

$$\hat{S}_{XX}^{\alpha}(F_{[\Delta F]}, n; N) = \frac{1}{\Delta F} \int_{F-\frac{\Delta F}{2}}^{F+\frac{\Delta F}{2}} \hat{S}_p^{\alpha}(\nu, n; N) d\nu. \tag{6.99}$$

It can be expected that smoothing in the time domain is equivalent to a certain smoothing in the frequency domain. Indeed, noting that $\hat{S}_{XX}^{\alpha}[F, n_{[L]}; N] = \hat{S}_{XX}^{\alpha}(F, n; N) \star \left(\frac{1}{L}\Pi\left(\frac{n}{L}\right)\right)$, it has been shown that if $L \gg N$, [60]

$$\begin{aligned}\hat{S}_{XX}^{\alpha}[F, n; N] \star \left(\frac{1}{L}\Pi\left(\frac{n}{L}\right)\right) &\approx \hat{S}_{XX}^{\alpha}(F, n; L) \star \left(\frac{1}{N}\left(\mathcal{F}\left\{\prod\left(\frac{n}{N}\right)\right\}\right)^2\right) \\ &= \hat{S}_{XX}^{\alpha}(F, n; L) \star \left(\frac{\sin^2 \pi NF}{N \sin^2 \pi F}\right).\end{aligned} \tag{6.100}$$

6.7 SUMMARY

- The autocovariance function of a second-order stationary (or wide-sense stationary (wss)) signal $X[n]$ is defined as

$$\mathscr{R}_{XX}[m] = \mathbb{E}\left\{(X[n] - \mathbb{E}\{X[n]\})\left(X^*[n-m] - \mathbb{E}\{X^*[n-m]\}\right)\right\},$$
$$= \mathbb{E}\left\{X[n]X^*[n-m]\right\} - |m_X|^2.$$

When signal mean $m_X = 0$, the autocovariance function of $X[n]$ simplifies to

$$\mathscr{R}_{XX}[m] = \mathbb{E}\left\{X[n]X^*[n-m]\right\}, \tag{6.101}$$

which, in general, is called the autocorrelation function of $X[n]$ when $m_X \neq 0$.

- The autocovariance function has the following properties:

$$\mathscr{R}_{XX}[-m] = \mathscr{R}^*_{XX}[m] \quad \text{for all } m \tag{6.102}$$

and

$$\mathscr{R}_{XX}[0] = P_X. \tag{6.103}$$

- The Fourier transform of the autocovariance function is called the power spectral density (PSD) of signal $X[n]$:

$$\mathscr{S}_{XX}(F) = \mathcal{F}\left\{\mathscr{R}_{XX}[m]\right\} = \sum_{m=-\infty}^{\infty} \mathscr{R}_{XX}[m]e^{-j2\pi Fm}.$$

The Wiener–Khinchin theorem states that under sufficient regularity conditions on the process $X[n]$, the PSD and the autocovariance form a Fourier transform pair:

$$\mathscr{R}_{XX}[m] = \int_{-\frac{1}{2}}^{\frac{1}{2}} \mathscr{S}_{XX}(F)e^{j2\pi Fm}dF \quad \text{for } -\infty < m < \infty.$$

- The limit autocorrelation function of a signal $x[n]$ is defined as

$$R_{xx}[m] = \lim_{N\to\infty} \frac{1}{N} \sum_{n=-\frac{N-1}{2}}^{\frac{N-1}{2}} x[n]x^*[n-m].$$

- The limit spectrum is the Fourier transform of the limit autocorrelation function:

$$S_{xx}(F) = \sum_{m=-\infty}^{\infty} R_{xx}[m]e^{-j2\pi Fm}.$$

- The time-averaged autocovariance sequence estimate based on $\{x[0], x[1], \ldots, x[N-1]\}$ is called the correlogram:

$$\hat{R}_{xx}[m;N] = \begin{cases} \frac{1}{N}\sum_{n=m}^{N-1} x[n]x^*[n-m] & \text{if } 0 \le m \le N-1 \\ \hat{R}^*_{xx}[-m;N] & \text{if } -(N-1) \le m \le -1 \end{cases}.$$

- For any finite N, correlogram is a biased estimate of the true autocovariance sequence $\mathscr{R}_{XX}[m]$. However, it is asymptotically unbiased for $\mathscr{R}_{XX}[m]$. If $\mathscr{R}_{XX}[m]$ is square summable, then the asymptotic variance of the correlogram vanishes, making it both mean-square consistent and asymptotically efficient.
- Fourier transform of correlogram is the periodogram $\hat{S}_p(F;N)$, which can also be written as

$$\hat{S}_p(F;N) = \frac{1}{N}\left|X(F;N)\right|^2,$$

where $X(F;N)$ is the DTFT of the sequence $\{x[n]\}_{n=0}^{N-1}$.
- Periodogram is an asymptotically unbiased estimate. However, for any finite N, it is biased:

$$\mathbb{E}\left\{\hat{S}_p(F;N)\right\} = \int_{-\frac{1}{2}}^{\frac{1}{2}} W_B(F-\alpha;N)\,\mathscr{S}_{XX}(\alpha)d\alpha.$$

where $W_B(F;N)$ is the Fourier transform of the triangular window function of length $2N-1$.
- For any finite N, the expected value of the periodogram shows spectral broadening due to nonzero main-lobe width (smearing/smoothing) and side lobes (leakage) of $W_B(F;N)$.
- The minimum spectral separation that a spectrum estimator can show distinctly is called the spectral resolution of the estimator. The spectral resolution of the periodogram is limited to about $\frac{1}{N}$.
- A serious drawback of the periodogram is that, in general, it has a high variance and its asymptotic variance can be a fixed nonzero value. For example, in the case of a Gaussian signal $X[n]$, the asymptotic variance of the periodogram estimator is given by

$$\lim_{N\to\infty} Var\left(\hat{S}_p(F;N)\right) = \mathscr{S}^2_{XX}(F).$$

- The Blackman–Tukey PSD estimator weights each standard autocovariance estimate value $\hat{R}_{XX}[m;N]$ by a weight value $w[m]$ before it is used in computing the DTFT:

$$\hat{S}_{BT}(F;N) = \sum_{m=-(M-1)}^{M-1} w[m]\hat{R}_{XX}[m;N]e^{-j2\pi Fm},$$

where weight values are nonzero only for $-(M-1) \le m \le M-1$ for some $M \le N$.

- The Blackman–Tukey PSD estimator $\hat{S}_{BT}(F;N)$ is a smoothen version of the periodogram:

$$\hat{S}_{BT}(F;N) = \int_{-\frac{1}{2}}^{\frac{1}{2}} \hat{S}_p(\alpha;N)W(F-\alpha)d\alpha,$$

where $W(F)$ is the Fourier transform of the window $w[n]$.

- For a window length on the order of $M < N$, the Blackman–Tukey estimator reduces the variance of the periodogram by a factor of approximately $\frac{M}{N}$. However, this variance improvement comes at the price of degrading its spectral resolution capability by the same factor compared to that of the periodogram. Hence, window length M has to be chosen as a trade-off between the variance improvement and spectral resolution degradation.
- Given an N-length data sequence $\{x[n]\}_{n=0}^{N-1}$, the Barlett method for PSD estimation first segments the data into K nonoverlapping $M = \frac{N}{K}$-length subsequences $x_k[n]$'s for $k = 0, \ldots, K-1$:

$$x_k[n] = x[n+kM] \quad \text{for } n = 0, \ldots, M-1.$$

The Bartlett PSD estimate $\hat{S}_B(F;N)$ is obtained by averaging the K periodograms computed from these K subsequences

$$\hat{S}_B(F;N) = \frac{1}{K}\sum_{k=0}^{K-1} \hat{S}_{p,k}(F;M),$$

where $\hat{S}_{p,k}(F;M)$ is the kth periodogram computed from the subsequence $x_k[.]$:

$$\hat{S}_{p,k}(F;M) = \frac{1}{M}\left|\sum_{n=0}^{M-1} x_k[n]e^{-j2\pi Fn}\right|^2 \qquad \text{for } k = 0, \ldots, K-1.$$

- The variance of the Bartlett PSD estimator $\hat{S}_B(F;N)$ is smaller by a factor of $K = \frac{N}{M}$ compared to that of the periodogram. However, its spectral resolution is on the order of $\frac{1}{M} = \frac{K}{N}$, which is worse off compared to that of the periodogram by the same factor.

- Given an N-length data sequence $\{x[n]\}_{n=0}^{N-1}$, the Welch method for PSD estimation first segments the data into K possibly overlapping subsequences $x_k[n]$'s of length M:

$$x_k[n] = x[n+kL] \quad \text{for } n = 0, \ldots, M-1,$$

where $x[kL]$ is the starting data point of the kth sequence and $L \leq M$ so that subsequences overlap.

The Welch PSD estimate is obtained by averaging the K periodograms computed from weighted subsequences $x_k[n]$'s:

$$\hat{S}_W(F;N) = \frac{1}{K}\sum_{k=0}^{K-1} \hat{S}_{w,k}(F;M),$$

where $\hat{S}_{w,k}(F;M)$ is the kth periodogram computed from the tapered data subsequence $x_k[.]$ for $k = 0, \ldots, K-1$:

$$\hat{S}_{w,k}(F;M) = \frac{1}{MP_w}\left|\sum_{n=0}^{M-1} x_k[n]w[n]e^{-j2\pi Fn}\right|^2 \qquad \text{for } k = 0, \ldots, K-1,$$

where $w[n]$ is the M-length data tapering window and P_w is its average power defined as

$$P_w = \frac{1}{M}\sum_{n=0}^{M-1} |w[n]|^2.$$

- The spectral resolution of the Welch PSD estimator is of the order of $\frac{1}{M}$. The variances of the Welch estimator can be smaller than both the periodogram and the Bartlett PSD estimate depending on the amount of allowed overlapping and the tapering function used.
- The time-variant periodogram at time $n = n_0$ computed from $x[n]$ given for $n_0 - \frac{N-1}{2} \leq n \leq n_0 + \frac{N-1}{2}$ is

$$\hat{S}_p(F, n_0; N) = \frac{1}{N}|X(F, n_0; N)|^2,$$

where $X(F, n_0; N)$ is the time-variant finite-time discrete-time Fourier transform (TVFT-DTFT) of the sequence $x[n]$ defined as

$$X(F, n_0; N) \triangleq \sum_{n=n_0-\frac{N-1}{2}}^{n_0+\frac{N-1}{2}} x[n]e^{-j2\pi Fn}.$$

- The time-variant periodogram is the Fourier transform of the time-variant correlogram defined as

$$\hat{R}_{xx}[m,n_0;N] = \begin{cases} \frac{1}{N} \sum\limits_{n=n_0-\frac{N-1}{2}+m}^{n_0+\frac{N-1}{2}} x[n]x^*[n-m] & \text{if } 0 \leq m \leq N-1 \\ \hat{R}^*_{xx}[-m,n_0;N] & \text{if } -(N-1) \leq m \leq -1 \end{cases}.$$

- The temporal and spectral resolutions of time-variant periodogram, respectively, are given by

$$\Delta n = N,$$

and

$$\Delta F = \frac{1}{N}.$$

- Variance of time-variant periodogram can be reduced by smoothing or averaging either in the time or frequency domain, which is essentially equivalent (with proper smoothing window choices).
- A random signal $X[n]$ is called wide-sense cyclostationary (wscs) if there exists an $N_0 > 0$ such that

$$m_X[n+N_0] = m_X[n] \quad \text{for all } n$$

and

$$\mathscr{R}_{XX}[n+N_0,m] = \mathscr{R}_{XX}[n,m] \quad \text{for all } n,m.$$

- When this is true for some $N_0 > 0$, N_0 is called the period of the cyclostationary process, and $\mathscr{R}_{XX}[n,m]$ is called a periodic autocorrelation function.
- In many such situations, $\mathscr{R}_{XX}[n,m]$ can be expressed as

$$\mathscr{R}_{XX}[n,m] = \sum_{\alpha \in \mathcal{A}} \mathscr{R}^{\alpha}_{XX}[m]e^{j2\pi\alpha n}, \tag{6.104}$$

for some countable set $\mathcal{A}$ with corresponding coefficients given by (for odd N)

$$\mathscr{R}^{\alpha}_{XX}[m] = \lim_{N\longrightarrow\infty} \frac{1}{N} \sum_{n=-\frac{N-1}{2}}^{\frac{N-1}{2}} \mathscr{R}_{XX}[n,m]e^{-j2\pi\alpha n}.$$

- When $\mathscr{R}_{XX}[n,m]$ can be expressed as in (6.104) while it is not exactly a periodic autocorrelation function, it is called an almost periodic autocorrelation function and the process $X[n]$ called wide-sense almost cyclostationary.

- The coefficients $\mathscr{R}_{XX}^{\alpha}[m]$ are called cyclic autocorrelation functions and the frequencies $\alpha \in [0,1]$ are called the cyclic frequencies. The wss processes are a special case of the cyclostationary processes in which $\mathscr{R}_{XX}^{\alpha}[m] \neq 0$ only for $\alpha = 0$. Thus, the only cyclic frequency present in a wss signal is $\alpha = 0$, and for $\alpha = 0$, $\mathscr{R}_{XX}^{0}[m]$ reduces to the autocorrelation function $\mathscr{R}_{XX}[m]$.
- The Fourier transform of the cyclic autocorrelation function $\mathscr{R}_{XX}^{\alpha}(m)$ is called the cyclic spectrum of $X[n]$:

$$\mathscr{S}_{XX}^{\alpha}(F) = \sum_{m=-\infty}^{\infty} \mathscr{R}_{XX}^{\alpha}[m]e^{-j2\pi Fm}.$$

- Cyclic spectrum $\mathscr{S}_{XX}^{\alpha}(F)$ is a measure of correlation of $X(F)$ and $X(F+\alpha)$. For this reason, it is also called the spectral correlation function (SCF). When $\alpha = 0$, cyclic spectrum reduces to the PSD $\mathscr{S}_{XX}(F)$.
- The spectral correlation coefficient, or the complex autocoherence, is defined as

$$C_{XX}^{\alpha}(F) = \frac{\mathscr{S}_{XX}^{\alpha}(F)}{\sqrt{\mathscr{S}_{XX}(F+\alpha)\, \mathscr{S}_{XX}(F)}}.$$

  The magnitude $|C_{XX}^{\alpha}(F)|$ of the spectral correlation coefficient is called the autocoherence and satisfies

$$|C_{XX}^{\alpha}(F)| \preceq 1.$$

- The limit cyclic autocorrelation function and the limit cyclic spectrum are defined, respectively, as

$$R_{xx}^{\alpha}[m] = \lim_{N\longrightarrow\infty} \frac{1}{N} \sum_{n=-\frac{N-1}{2}}^{\frac{N-1}{2}} x[n]x^{*}[n-m]e^{-j2\pi\alpha n}$$

  and

$$S_{xx}^{\alpha}(F) = \mathcal{F}\left\{R_{xx}^{\alpha}[m]\right\}$$

- The time-variant finite-time cyclic periodogram is defined as

$$\hat{S}_{p}^{\alpha}(F,n_0;N) = \frac{1}{N}X(F+\alpha,n_0;N)X^{*}(F,n_0;N),$$

  where $X(F,n_0;N)$ is the TVFT-DTFT computed at time $n = n_0$ based on data given for $n_0 - \frac{N-1}{2} < n < n_0 + \frac{N-1}{2}$.

- Time-variant finite-time cyclic periodogram $\hat{S}_p^{\alpha}(F, n_0; N)$ is the Fourier transform of the cyclic correlogram computed as

$$\hat{R}_{xx}^{\alpha}[m, n_0; N] = \frac{1}{N} \sum_{n=n_0-\frac{N-1}{2}+m}^{n_0+\frac{N-1}{2}} x[n]x^*[n-m]e^{-j2\pi\alpha n} \quad \text{for } m = 0, 1, \ldots, N-1$$

and

$$\hat{R}_{xx}^{\alpha}[-m, n_0; N] = \left(\hat{R}_{xx}^{\alpha}[m, n_0; N]\right)^* \quad \text{for } m = 1, \ldots, N-1.$$

- Temporal and spectral domain averages of the time-variant finite-time cyclic periodogram are known to converge to the limit cyclic spectrum $S_{xx}^{\alpha}(F)$, thereby justifying their use as estimates of the SCF.

7

MARKOV DECISION PROCESSES

7.1 INTRODUCTION

In Chapter 2, we defined cognitive radios as software-defined radios with built-in cognition and intelligence. A distinctive feature of intelligence is the ability to apply and reason with knowledge. We saw in Chapter 2 that a cognitive radio acquires spectrum knowledge through a process called spectrum sensing. Simply possessing knowledge, however, is not intelligence. An intelligent cognitive radio must be able to use the accumulated knowledge to make better decisions. This is called reasoning.

It is not hard to see how the need for reasoning arises in the operation of a cognitive radio. Indeed, the spectrum sensing problem itself will require the cognitive radio to decide which spectrum band or channel to sense at any given time. While a fixed protocol or random guessing can be alternative ways of making these selections, they certainly do not sound very intelligent. Instead, we expect that a cognitive radio uses the spectrum knowledge it had gained in selecting a particular spectrum band or a channel for sensing. Hence, the cognitive radio must be able to reason based on knowledge it has about the spectrum environment which band may be worth sensing in order to meet its performance objectives. Similarly, cognitive radios need to make decisions on how to perform actual communications based on its knowledge of spectrum environment and the user performance objectives. Knowledge is not perfect. Hence, in its reasoning, a cognitive radio needs to take into account possible uncertainties introduced as a result of imperfect knowledge. The process of reasoning from

Signal Processing for Cognitive Radios, First Edition. Sudharman K. Jayaweera.

knowledge and evidence to arrive at conclusions is inference, which leads to reasoned decisions.

Detection and estimation methods we saw in Chapters 4 and 5 can be viewed as making reasoned decisions. The knowledge in these cases is simply the observations. The cognitive radio makes decisions about a particular hypothesis or the value of a parameter based on these observations. However, the reasoning becomes more challenging, and interesting, if the radio has to predict what may be the true hypothesis in the next time instant $n+1$ before actually having the observation at time $n+1$. The radio has to now reason based on observations up to time n and make a reasoned decision regarding the true hypothesis at time $n+1$. This is called *prediction*, or an educated guess. A cognitive radio may need to make such decisions regarding the future state of nature (in this case, the state of its RF environment and user needs), in order for it to select one action over another. For example, in the case of spectrum sensing, it may need to select which channel to sense. If the objective of the cognitive radio is to find a channel that may be idle, or free of interference from other transmissions, then the radio needs to predict which channel may be idle during the next time period. This is a reasoned decision. The action to start transmitting on a particular channel during the next time period will be based on this decision. The selection scheme of a particular action based on reasoned decisions is called a *decision rule* and denoted by $\pi(.)$.

A decision rule as discussed so far only addresses an action selection for a single future time instant: For example, the action to be selected at next time instant $n+1$ when the current time is n. We may call this one-step decision-making. In many situations of practical interest, however, it is natural that we have to look beyond just a single future time instant. For example, a chess player will map out a sequence of possible moves in his/her head to follow up with depending on what the opponent will do in response to his/her immediate next move. This sequence of moves is a contingency plan. For each time instant, the chess player thinks of what he/she will do depending on possible moves the opposing player may make. The chess player does not know exactly what the opposing player will do at the future time instant in question. But he/she plans out a response for each of the possible moves (or at least for few of them). Each other's decisions and actions drives the chess game along a certain path. Hence, the chess game can be viewed as a process driven by the player decisions, or a *decision process*. Similarly, the spectrum environment of a cognitive radio evolves according to certain dynamics. Radio's actions can influence these dynamics as well as its own observations of the environment. The radio will select a certain action at a given time based on what it knows about the state of its RF environment and user needs at current time. This action may effect its observation as well as the resulting new environment state. The radio will choose the next action by incorporating this new observation into its knowledge.

Making a sequence of contingency plans in a decision process is called *planning*. Planning results in a *policy* which specifies a response (action) for each of the possible states of nature (or opponent actions) the planner might encounter at each of the future time instants of interest. Hence, a policy is a sequence of decision rules, one for each future time instant of interest, and can be denoted as $\underline{\pi} = (\pi_1(.), \pi_2(.), \ldots)$.

Many of the inference problems encountered by a cognitive radio can be viewed as planning problems. In spectrum sensing, for example, the radio would like to find a spectrum band that might stay free of other interferers for a certain duration. Hence, it is important to look ahead more than just one time instant. In case an interferer appears in the current spectrum band, it might want to have the option to switch to a new spectrum band with minimum disruption to its communications session and minimum hardware reconfigurability. Hence, this requires planning and a policy. Similarly, if the cognitive radio is attempting to avoid a jammer that is pursuing it, the radio may want to plan out its actions for several time instants into future to have the best chance of evading the jammer.

The literature on inference and planning is vast. Similar to detection theory, estimation theory, and the power spectrum estimation in the previous chapters, this topic itself can take up a whole book. In this chapter, thus, we limit ourselves to a particular type of decision processes in which the state dynamics are *Markov*. Such a process, called a *Markov decision process* (MDP), makes sense in many situations as a reasonable model and have in fact found applications in a wide range of practical problems. As we will see later in the book, many of the decision-making problems encountered in a cognitive radio can also be cast as MDPs, making their study justifiable. The presentation in this chapter is based mostly on books [68] and [69] and several other materials such as [70].

7.2 MARKOV DECISION PROCESSES

Detection and estimation theories discussed in Chapters 4 and 5, respectively, can be adapted for making one-step decisions. For example, many of the one-step problems can be reduced to either the MAP or ML criteria and solve by computing the posterior distributions of quantities of interest given the available observations. When we need to make a sequence of decisions into the future or come up with a policy, we need to formulate the problem in a tractable form. In general, these can be hard problems. A particularly interesting class of problems is the MDPs.

An MDP is a decision process in which the next state $S[n+1]$ of the environment, or the system, is completely determined by the current state of the system denoted by $S[n]$ and the action (or the decision) taken at current time a_n. More specifically, we may define a *finite-state MDP* in which both state and action can take only a finite set of possible choices by giving the following [68]:

- A finite set $\mathcal{S}$ of system states.
- A finite set $\mathcal{A}$ of available actions. In each state $s \in \mathcal{S}$, a subset $\mathcal{A}_s \subseteq \mathcal{A}$ of actions might be available.
- A nonnegative function $p_n(s'|s,a)$ denoting the conditional probability that the system is in state s' at time epoch $n+1$, when the action $a \in \mathcal{A}$ is chosen in state $s \in \mathcal{S}$ at time n.

- A real-valued function $r_n(s,a)$ defined for state $s \in \mathcal{S}$ and action $a \in \mathcal{A}$ to denote the reward received at time n if action a is taken while in state s.

There are several ways to both simplify and generalize the earlier description of an MDP. One of the important simplifications is to assume that both transition probabilities and the reward values, $p_n(s'|s,a)$ and $r_n(s,a)$, respectively, can be taken to be time invariant so that they may be denoted as $p(s'|s,a)$ and $r(s,a)$. Then the MDP is called a *stationary MDPs*. Another simplification is to assume that the reward $r_n(s,a)$ does not depend on the particular action taken but only on the state s. In this case, we may write it as $r_n(s)$. If, in addition, the MDP is stationary, we can write the rewards as $r(s)$ independent of time.

On the other hand, in some situations, the reward may not only depend on the current state s and the currently selected action a but also on which state s' the system ends up. In this case, we may need to use the more elaborate notation $r_n(s,s',a)$. Moreover, although we have defined an MDP earlier by explicitly assuming that both the state and action sets are finite, most of the theory that we discuss in the following can be generalized to either countable or even compact action sets with certain necessary conditions on the reward functions. In many cases, the set of states can also be generalized to be countable. Further, these sets of states and actions can also be allowed to be time dependent themselves. However, we may not delve into such generalizations in order to keep this discussion concise.

In fact, in this chapter, we will exclusively consider MDPs in which the transition probabilities and rewards are stationary:

$$\Pr(S[n+1]=s^{'}|S[n]=s, a_n=a)=p(s'|s,a)$$

and

$$r_n(S[n]=s, a_n=s)=r(s,a),$$

where $s^{'}, s \in \mathcal{S}$ and $a \in \mathcal{A}$. Hence, the rewards do not depend on the specific time instant but only on the particular (state, action) pair.

Let us consider an MDP that starts at time $n=0$ in an initial state of $S[0]=s_0$. For example, suppose $s_0=1$ as in Figure 7.1. At time $n=0$, there is no other information available; hence, we may take the action $a_{\max} \in \mathcal{A}$ that gives the largest reward $r(s_0,a_{\max})$ for state $s_0 \in \mathcal{S}$. In the case of Figure 7.1, $a_{\max}=a_1$ since $r(1,a_1)=10$ is the largest reward from state 1. But once we make the decision π_0 to take the action $a_{\max}$, that is, $\pi_0=a_{\max}$, we transition into next state $S[1]=s'$ according to the transition probabilities $p(s'|s_0,a_{\max})$. After we choose the action $\pi_0=a_{\max}$, we do not have much control over the next state $S[1]$. It will be dictated by the conditional probability distribution (or the conditional pmf, to be exact) $p(s'|s_0,a_{\max})$. Hence, if we are to make a decision plan, a policy, at time $n=0$ that will specify what to do at time $n=1$, then it needs to specify what action to be taken at time $n=1$ for any of the possible $S[1]=s' \in \mathcal{S}$ states that we may end up at time $n=1$. Hence, as mentioned earlier, decision rule for time $n=1$ is a contingency plan. Depending on which state

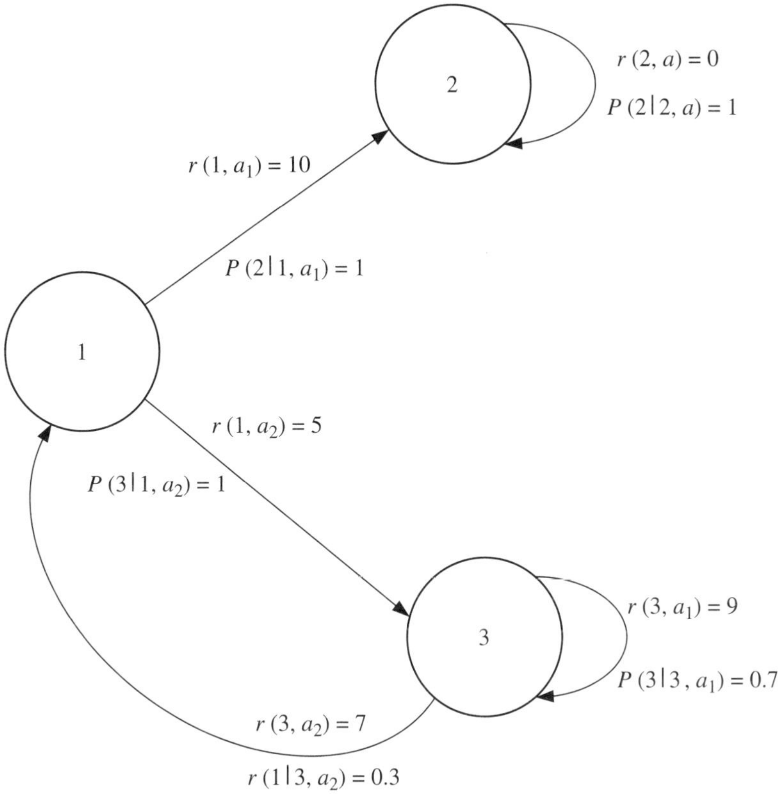

FIGURE 7.1 States and rewards in a Markov decision process.

s' we will end up at time $n = 1$, we will pick an action as specified by the contingency plan, which is the decision rule for time $n = 1$. Hence, the decision rule for each time instant n in a policy, as pre-computed at time $n = 0$, is not a single action a_n. Rather, it is a set of actions of the form $\pi_n(s)$ for $s \in \mathcal{S}$ with each $\pi_n(s) \in \mathcal{A}$. We call such a decision a *strategy* or a decision strategy, and thus a policy is a sequence of strategies for each time instant n.

Consider again the MDP that starts in state $S[0] = s_0 = 1$ at the initial time instant $n = 0$ and our inclination to select the action $a_0 = a_{\max}$ that gives the largest reward $r(s_0, a_{\max})$. Suppose there is a particular state $\tilde{s} \in \mathcal{S}$ in which the reward $r(\tilde{s}, a)$ is particularly low regardless of the action $a \in \mathcal{A}$ taken. Indeed, state 2 in Fig. 7.1 is such a state since reward $r(2, a) = 0$ for all actions a. Now, if the action $a_{\max}$ has a high likelihood of taking us to that state $\tilde{s}$ from current state s_0, that is, $p(\tilde{s}|s_0, a_{\max})$ is relatively high, then it may not be a good idea to take $a_{\max}$ action at current time $n = 0$. In Fig. 7.1, actually, this probability is exactly unity for action a_1. It may be even worse if the probability of escaping the state $\tilde{s}$ once you are in it is also very low: That is, $p(\tilde{s}|\tilde{s}, a)$ is also close to unity for all $a \in \mathcal{A}$. In the MDP in Fig. 7.1, this probability is also exactly equal to unity. Since $r(\tilde{s},a) = 0$ for all $a \in \mathcal{A}$, then once we reach the state $\tilde{s}$, the chance of us receiving any future rewards can be very small. In this case, if there is another action available at $n = 0$ that gives slightly less reward than that of action $a_{\max}$, but has a high chance of taking us to a next state $S[1]$ from which we will only be transitioning among states with really high rewards, then selecting that action at time $n = 0$ in favor of $a_{\max}$ is intuitively more appealing. The choice of action a_2 from state 1 in Fig. 7.1 is such a scenario.

The preceding discussion shows that in forming a decision policy, we need to think about the whole time horizon of interest in order to make any single decision strategy for a particular time instant. This is what makes the decision processes interesting as well as challenging. First, we will formulate and solve the MDP problem assuming a finite decision-making horizon. Later, we may extend the formulation to an infinite decision-making horizon.

7.3 FINITE-HORIZON MDPs

Let us consider making decisions over a finite time horizon T (T is a positive integer). At time $n = 0$, we need to specify a decision strategy $\pi_n(.)$ that specifies an action to be taken depending on the possible state $S[n]$ for each time instant $n = 0, \cdots, T-1$. All these action strategies from $n = 0$ to $n = T-1$ taken together, we have a decision policy $\underline{\pi} = (\pi_0(.), \cdots, \pi_{T-1}(.))$ where $\pi_n(s) \in \mathcal{A}$ for all $s \in \mathcal{S}$ and $n = 0, \cdots, T-1$. The decision process is assumed to terminate at time $n = T$.

7.3.1 Definitions

7.3.1.1 Decision Strategies So far, for simplicity, we have assumed that the actions are deterministic in the sense that each $\pi_n(s) \in \mathcal{A}$. Hence, decision rules are of the form of

$$\pi_n : \mathcal{S} \to \mathcal{A}.$$

Since actions only depend on current state and they are deterministically characterized, such decision rules are called *deterministic Markov decision rules*. The policies based on deterministic Markov rules are called the *deterministic Markov decision policies*.

Similar to randomized decisions we encountered in Chapter 4, we may in general allow these actions to be also randomized. We may denote randomized decisions as $\pi_n(s,a)$ where π_n is now defined as

$$\pi_n : \mathcal{S} \times \mathcal{A} \to [0,1]. \tag{7.1}$$

In (7.1), $\pi_n(s,a)$ is interpreted as the probability that action a is to be executed if in state s at time n:

$$\pi_n(s,a) = \Pr(a_n = a | S[n] = s). \tag{7.2}$$

Note that, although we use the same notation π_n for both deterministic and randomized decision rules, there should not be a confusion due to the nature of the arguments. Decision rules of the form of (7.1) or (7.2) are called *randomized Markov decision rules*, and policies based on them are called *randomized Markov decision policies*. We will denote by Π^{MR} and Π^{MD}, respectively, the sets of all randomized and deterministic Markov decision policies.

A decision process is Markov if next state is completely determined by the current state and the selected action. However, in general, actions can be selected based on the whole history of the process up to that time. At initial time $n = 0$, the history is simply the current state $\underline{h}_0 = (s_0)$. For $n > 0$, history at time $n = n$ is of the form of

$$\underline{h}_n = (\underline{h}_{n-1}, a_{n-1}, s[n]). \tag{7.3}$$

Let $\mathcal{H}_n$ denote the set of all possible histories at time n. In general, a decision rule π_n for time n can be a mapping from the set of histories and actions to the probabilities:

$$\pi_n : \mathcal{H}_n \times \mathcal{A} \to [0,1], \tag{7.4}$$

with the interpretation that

$$\pi_n(\underline{h}, a) = \Pr(a_n = a | \underline{h}_n = \underline{h}).$$

Hence, if $\underline{\pi} = (\pi_0, \ldots, \pi_{T-1})$ denotes a policy, then each $\pi_n(.,.)$ is a randomized decision strategy that can depend on whole history up to time n. Each $\pi_n(.,.)$ specifies the probability with which different actions must be chosen at time n given the history $\underline{h}_n$. We will call such decisions (7.4) as *history-dependent randomized decision rules* and denote by Π^{HR} the set of all history-dependent randomized decision rules.

On the other hand, a deterministic decision rule π_n, for $n = 0, \ldots, T-1$, can be considered as a mapping

$$\pi_n : \mathcal{H}_n \to \mathcal{A}$$

with the interpretation that $\pi_n(\underline{h}_n) \in \mathcal{A}$ denotes the action to be taken at time n if the history at time n is $\underline{h}_n$. We will call such decisions as *history-dependent deterministic decision rules* and denote the set of all history-dependent deterministic decision rules by Π^{HD}.

7.3.1.2 Reward, Expected Total Return, and Expected Total Discounted Return

In the case of a stationary finite-horizon MDP, specifying a reward function amounts to specifying a function $r_n(s,a) = r(s,a)$ for $0 \leq n \leq T-1$ and a terminal reward $r_T(s)$ at $n = T$. Different policies $\underline{\pi} = (\pi_0, \ldots, \pi_{T-1})$ will obviously result in different sequences of rewards $\{r(s[n], \pi_n)\}_{n=0}^{T}$. The objective of a decision policy is to gain better rewards. Then there needs to be a way for us to differentiate our preference between two given policies, say, $\underline{\pi}^{(1)} = (\pi_0^{(1)}, \ldots, \pi_{T-1}^{(1)})$ and $\underline{\pi}^{(2)} = (\pi_0^{(2)}, \ldots, \pi_{T-1}^{(2)})$.

While there are many possible criteria one can think of, one reasonable criterion is to consider the *total return* (also called the *total reward*) associated with a policy $\underline{\pi}$, defined as

$$R_T^{\underline{\pi}} = \sum_{n=0}^{T} r_n(S[n], \pi_n(.)), \tag{7.5}$$

where $\underline{\pi} = (\pi_0, \ldots, \pi_{T-1})$ and it is understood that $r_n(S[n], \pi_n(.)) = r(S[n], \pi_n(.))$ for $0 \leq T - 1$ with $r_T(S[T], \pi_T(.)) = r_T(S[T])$ for all $S[T] \in \mathcal{S}$ and for all $\pi_T(.)$. The total return $R_T^{\underline{\pi}}$ in (7.5) is a random quantity for two reasons: First, the initial state $S[0]$ is random, and second, even with a fixed initial state, the subsequent states are all randomly determined by the action π_n and the state $S[n]$ sequence. Hence, we may consider the *expected total return* (ETR) associated with a particular decision policy $\underline{\pi}$:

$$\overline{R_T^{\underline{\pi}}} = \mathbb{E}_{\underline{\pi}}\left\{\sum_{n=0}^{T} r_n(S[n], \pi_n(.))\right\} \tag{7.6a}$$

$$= \mathbb{E}_{\underline{\pi}}\left\{\sum_{n=0}^{T-1} r_n(S[n], \pi_n(.)) + r_T(S_T)\right\}, \tag{7.6b}$$

where $\mathbb{E}_{\underline{\pi}}\{.\}$ denotes the expectation taken with respect to the state transitions corresponding to the policy $\underline{\pi}$.

A slightly modified criterion called the *expected total discounted return* (ETDR) is defined as

$$\overline{R_{T,\gamma}^{\underline{\pi}}} = \mathbb{E}_{\underline{\pi}}\left\{\sum_{n=0}^{T} \gamma^n r_n(S[n], \pi_n(.))\right\} \tag{7.7a}$$

$$= \mathbb{E}_{\underline{\pi}}\left\{\sum_{n=0}^{T-1} \gamma^n r_n(S[n], \pi_n(.)) + \gamma^T r_T(S[T])\right\}, \tag{7.7b}$$

where $0 < \gamma \leq 1$ is called the discount factor. Note that when $\gamma = 1$ the ETDR reduces to ETR $\overline{R_T^{\underline{\pi}}}$.

In the case of a finite-horizon MDP, both ETR and ETDR can be used as the criterion for designing decision rules. However, in the case of infinite-horizon MDPs that we will discuss later, the ETR may run into convergence issues. Hence, for ease of generalization to infinite-horizon situation, we may always assume expected total discounted reward as the basis for choosing one policy over another. Of course, we would prefer a policy that gives the largest possible expected total discounted reward.

7.3.1.3 Value Function and Optimal Policies We may express the ETDR $\overline{R^{\underline{\pi}}_{T,\gamma}}$ in (7.7a) as

$$\overline{R^{\underline{\pi}}_{T,\gamma}} = \mathbb{E}_{S_0}\left\{\mathbb{E}_{\underline{\pi}}\left\{\sum_{n=0}^{T}\gamma^n r_n(S[n],\pi_n(.))\middle| S_0\right\}\right\}$$
$$= \sum_{s\in\mathcal{S}} V^{\underline{\pi}}_T(s)\Pr(S[0]=s), \tag{7.8}$$

where we have defined the *state-value function of policy $\underline{\pi}$* as

$$V^{\underline{\pi}}_T(s) = \mathbb{E}_{\underline{\pi}}\left\{\sum_{n=0}^{T}\gamma^n r_n(S[n],\pi_n(.))\middle| S[0]=s\right\} \tag{7.9a}$$
$$= \mathbb{E}_{\underline{\pi}}\left\{\sum_{n=0}^{T-1}\gamma^n r_n(S[n],\pi_n(.)) + \gamma^T r_T(S_T)\middle| S[0]=s\right\}. \tag{7.9b}$$

In designing a policy for an MDP, there are two possible scenarios to be considered at initial time $n = 0$: In one scenario, we may not know the initial state $S[0] = s \in \mathcal{S}$ when we design a policy. In that case, we need to select a policy to maximize the ETDR (7.8), which is the average of the state-value function $V^{\underline{\pi}}_T(S[0])$ over the distribution $\Pr(S[0]=s)$ of the initial state. Hence, choosing a policy $\underline{\pi}$ that maximizes $\overline{R^{\underline{\pi}}_{T,\gamma}}$ in this case is the same as choosing a policy $\underline{\pi}$ that maximizes $V^{\underline{\pi}}_T(s)$ for each possible initial state s having a non-zero probability $\Pr(S[0]=s)$. The second scenario is that we may know the initial state $S[0] = s$. In that case, the ETDR (7.8) that we need to consider reduces simply to the state-value function $V^{\underline{\pi}}_T(s)$. Hence, we just need to select a policy that maximizes $V^{\underline{\pi}}_T(s)$ for the given initial state s.

Thus, under either of the scenarios, the selection of policies needs to be to maximize the state-value function $V^{\underline{\pi}}_T(s)$. Thus, in the following, we may choose to assume that the initial state $S[0] = s$ is given and base our policies on the state-value function $V^{\underline{\pi}}_T(s)$ in (7.9a).

A policy $\underline{\pi}^*$ is called an *optimal policy* for an MDP, if, given $S[0] = s$,

$$V^{\underline{\pi}^*}_T(s) \geq V^{\underline{\pi}}_T(s) \quad \text{for all } \underline{\pi} \in \Pi^{HR}, \tag{7.10}$$

where Π^{HR} denotes the set of all history-dependent randomized decision policies.

The *value of the MDP* is defined to be the value of the state-value function with an optimal policy:

$$V^*_T(s) = \sup_{\underline{\pi}\in\Pi^{HR}} V^{\underline{\pi}}_T(s) \quad \text{for } s \in \mathcal{S}. \tag{7.11}$$

Often, $V^*_T(s)$ is called the *optimal value function*.

From (7.10) and (7.11), the state-value function emerges as the key quantity involved in determining optimal policies. Our objective is to find an optimal policy for a given MDP at time $n = 0$. To do this, first of all, we need to be able to evaluate the state-value function corresponding to any given policy $\underline{\pi}$. This is called *policy evaluation*. Next, we need a way to compute the optimal state-value function for a given MDP and find an optimal policy that achieves this optimal value, if it exists.

In (7.10), we have defined an optimal policy as a policy that is as good as any other possible policy from the class of randomized history-dependent policies Π^{HR}. Hence, to verify whether a given policy is an optimal policy, we need to compare its state-value function against all policies in Π^{HR}. However, recall that we have assumed that both rewards and state transition probabilities only depend on the current state and not the current history:

$$r_n(\underline{h}_n, a_n) = r_n(s[n], a_n) \tag{7.12a}$$

$$\Pr(S[n+1] = s^{'} | \underline{h}_n, a_n) = \Pr(S[n+1] = s^{'} | s[n], a_n), \tag{7.12b}$$

where $\underline{h}_n = (\underline{h}_{n-1}, \underline{a}_{n-1}, s[n])$. When rewards and transition probabilities only depend on the current and next states and the current action as in (7.12a) and (7.12b) but not on the full history, they are called *Markov rewards* and *Markov probabilities*, respectively.

An important simplification that arises in (7.10) when the rewards and transition probabilities are Markov is that we do not have to search over all history-dependent randomized decision policies Π^{HR} to find the optimal value function. Instead, we may limit the search to only over the set of all randomized Markov decision policies Π^{MR}, which is a smaller subset of Π^{HR} [68].

7.3.2 Optimal Policies for MDPs

7.3.2.1 Bellman Equation For an arbitrary policy $\underline{\pi}$ and for $0 \leq m \leq T$, let us define the following function:

$$v_m^{\underline{\pi}}(\underline{h}_{T-m}) = \mathbb{E}_{\underline{\pi}} \left\{ \sum_{i=T-m}^{T} \gamma^{i-(T-m)} r_i(S[i], \pi_i(.)) \middle| \underline{h}_{T-m} \right\} \tag{7.13a}$$

$$= \mathbb{E}_{\underline{\pi}} \left\{ \sum_{i=T-m}^{T-1} \gamma^{i-T+m} r_i(S[i], \pi_i) + \gamma^m r_T(\underline{h}_T) \middle| \underline{h}_{T-m} \right\} \tag{7.13b}$$

with the boundary condition for $m = 0$ as

$$v_0^{\underline{\pi}}(\underline{h}_T) = r_T(\underline{h}_T) = r_T(s[T]). \tag{7.14}$$

Comparing (7.13a) with (7.9a), it is clear that $v_m^{\underline{\pi}}(\underline{h}_{T-m})$ is the expected total discounted reward of the policy $\underline{\pi} = (\pi_1, \pi_2, \ldots, \pi_T)$ if we started the decision process at time $n = T - m$ with the knowledge of the initial state $\underline{h}_{T-m}$ instead of starting at time $n = 0$. In evaluating $v_m^{\underline{\pi}}(\underline{h}_{T-m})$, we only need the last m decision rules of the policy $\underline{\pi}$. Stated differently, $v_m^{\underline{\pi}}(.)$ is the value achieved by the policy $\underline{\pi}$ when there are m steps to go before termination of the decision process. For this reason, $v_m^{\underline{\pi}}(.)$ is called the *value-to-go function* or, sometimes, simply the value function.

The state-value function $V_T^{\underline{\pi}}(s)$, on the other hand, is the value of the policy $\underline{\pi}$ when there are T steps to go before termination of the decision process. Hence, if $m = T$ and if we assume $\underline{h}_0 = s$, then

$$v_T^{\underline{\pi}}(s) = V_T^{\underline{\pi}}(s). \tag{7.15}$$

Hence, we have a straightforward method to obtain the state-value function for any given policy $\underline{\pi}$, if we can evaluate the value-to-go function $v_m^{\underline{\pi}}(.)$ for time $m = T$. As it turned out, there is a simple (backward) recursive procedure to compute the sequence of value-to-go functions $\{v_m^{\underline{\pi}}(\cdot)\}$ starting with the boundary condition (7.14) for $m = 0$. Indeed, with $n = T - m$ and $\underline{h}_n = (\underline{h}_{n-1}, a_{n-1}, s[n])$, we have that

$$\begin{aligned}
v_m^{\underline{\pi}}(\underline{h}_n) &= \mathbb{E}_{\underline{\pi}}\left\{\sum_{i=n}^{T} \gamma^{i-n} r_i\left(S[i], \pi_i(.)\right) \middle| \underline{h}_n\right\}, \quad \text{where } n = T - m \\
&= \mathbb{E}_{\underline{\pi}}\left\{ r_n\left(s[n], \pi_n(\underline{h}_n)\right) + \sum_{i=n+1}^{T} \gamma^{i-n} r_i(S[i], \pi_i(.)) \middle| \underline{h}_n\right\} \\
&= r_n\left(s[n], \pi_n(\underline{h}_n)\right) + \mathbb{E}_{\underline{\pi}}\left\{\gamma \sum_{i=n+1}^{T} \gamma^{i-n-1} r_i(S[i], \pi_i(.)) \middle| \underline{h}_n\right\} \\
&= r_n\left(s[n], \pi_n(\underline{h}_n)\right) + \gamma \mathbb{E}_{\underline{\pi}}\left\{\sum_{i=n+1}^{T} \gamma^{i-(n+1)} r_i(S[i], \pi_i(.)) \middle| \underline{h}_n, \pi_n(\underline{h}_n)\right\} \\
&= r_n\left(s[n], \pi_n(\underline{h}_n)\right) \\
&\quad + \gamma \sum_{s'} \mathbb{E}_{\underline{\pi}}\left\{\sum_{i=n+1}^{T} \gamma^{i-(n+1)} r_i(S[i], \pi_i(.)) \middle| \underline{h}_n, \pi_n(\underline{h}_n), S[n+1] = s'\right\} \\
&\quad \times \Pr(S[n+1] = s' \mid \underline{h}_n, \pi_n(\underline{h}_n)) \\
&= r_n\left(s[n], \pi_n(\underline{h}_n)\right) + \gamma \sum_{s'} \mathbb{E}_{\underline{\pi}}\left\{\sum_{i=n+1}^{T} \gamma^{i-(n+1)} r_i(S[i], \pi_i(.)) | \underline{h}_{n+1}\right\} \\
&\quad \times \Pr(S[n+1] = s' \mid S[n] = s[n], \pi_n(\underline{h}_n)) &(7.16a) \\
&= r_n\left(s[n], \pi_n(\underline{h}_n)\right) + \gamma \sum_{s'} v_{m-1}^{\underline{\pi}}\left(\underline{h}_{n+1}\right) \Pr(s' \mid s[n], \pi_n(\underline{h}_n)), &(7.16b)
\end{aligned}$$

where we have let $\underline{h}_{n+1} = \left(\underline{h}_n, \pi_n(\underline{h}_n), S[n+1] = s'\right)$ in (7.16a). From (7.16b), we have the following recursion, called the *Bellman equation*, to evaluate the value-to-go function for $m = 0, \cdots, T$, where $n = T - m$:

$$\begin{aligned} v_m^{\pi}(\underline{h}_n) &= r_n\left(s[n], \pi_n(\underline{h}_n)\right) + \gamma \sum_{s' \in \mathcal{S}} \Pr(S[n+1] = s' \,|\, s[n], \pi_n(\underline{h}_n)) \\ &\quad \times v_{m-1}^{\pi}\left((\underline{h}_n, \pi_n(\underline{h}_n), s')\right) \\ &= r_n\left(s[n], \pi_n(\underline{h}_n)\right) + \gamma \sum_{s' \in \mathcal{S}} p(s' | s[n], \pi_n(\underline{h}_n)) v_{m-1}^{\pi}\left((\underline{h}_n, \pi_n(\underline{h}_n), s')\right) \end{aligned} \tag{7.17}$$

with the initialization

$$v_0^{\pi}(\underline{h}_T) = r_T(s[T]) \quad \text{for all} \quad \underline{h}_T \in \mathcal{H}_T. \tag{7.18}$$

The Bellman equation (7.17) gives a method for evaluating the state-value function of a given policy (policy evaluation) since (7.17) shows how to recursively compute the value-to-go $v_m^{\pi}(\cdot)$ till we reach $v_T^{\pi}(\cdot)$ as needed in (7.15).

7.3.2.2 ***Bellman-Optimality Equations*** Of course, we are interested in finding an optimal policy. Thus, computing the optimal value function $V_T^*(s)$ in (7.11) is of particular interest. From (7.15), it is clear that a policy that maximizes $v_T^{\pi}(s)$ also maximizes $V_T^{\pi}(s)$. Hence, we may find an optimal policy by finding a policy $\underline{\pi}$ that maximizes the value-to-go function $v_T^{\pi}(s)$. Hence, let us consider maximizing $v_m^{\pi}(\underline{h}_{T-m})$ for each $m = 1, \cdots, T$ over all allowed policies.

However, from the definition of the value-to-go function (7.13a), $v_m^{\pi}(\underline{h}_{T-m})$ only involves the portion of the policy $\underline{\pi} = (\pi_0, \cdots, \pi_{T-1})$ from time $n = T - m$ to $n = T - 1$. , that is, those decision rules $\pi_{T-m}, \cdots, \pi_{T-1}$. Let us denote by $v_m^*(\underline{h}_{T-m})$ the supremum of $v_m^{\pi}(\underline{h}_{T-m})$ over all decision rule sequences $\pi_{T-m}, \cdots, \pi_{T-1}$:

$$v_m^*(\underline{h}_{T-m}) = \sup_{\underline{\pi} \in \Pi_m^{HR}} v_m^{\pi}(\underline{h}_{T-m}), \tag{7.19}$$

where Π_m^{HR} denotes the set of all randomized history-dependent decision policies of length m.

Similarly, $v_{m-1}^{\pi}(\underline{h}_{T-m+1})$ only involves the portion of the policy $\underline{\pi} = (\pi_0, \cdots, \pi_{T-1})$ from time $n = T - m + 1$ to $n = T - 1$. Hence, a policy that maximizes $v_{m-1}^{\pi}(\underline{h}_{T-m+1})$ for any $m = 0, \cdots, T - 1$, is a sequence of decision rules $\pi_{T-m+1}^*, \cdots, \pi_{T-1}^*$ that maximizes the value-to-go when there are $m - 1$ steps before the decision process terminates. Let us denote by $v_{m-1}^*(\underline{h}_{T-m+1})$ this supremum.

Now, if we consider the value-to-go function $v_m^{\pi}(.)$, which is for the case when there are m steps to go before the process terminates, all we need to do is to add to the beginning of the sequence of decision rules $\pi_{T-m+1}^*, \cdots, \pi_{T-1}^*$ from $n = T - m + 1$ to $n = T - 1$, a decision rule π_{T-m}^* for $n = T - m$. This is because what happens from time $n = T - m$ to time $n = T - m + 1$ does not change the fact that $\pi_{T-m+1}^*, \cdots, \pi_{T-1}^*$ is the sequence of decision rules that maximizes the value-to-go function from $n = T - m + 1$

to $n = T - 1$. The Bellman equation (7.17) shows exactly how $v_m^*(\underline{h}_{T-m})$ should be related to π_{T-m}^* and $v_{m-1}^*(\underline{h}_{T-m+1})$, which is summarized in the following theorem (Theorem 4.3.2 of [68]).

Theorem 7.1 (Optimality/Bellman-Optimality equations) For $m = 0, \ldots, T-1$, $v_m^*(h_n)$ with $n = T - m$ satisfies the following equations:

$$v_m^*(\underline{h}_n) = \sup_{a \in \mathcal{A}} \left[r_n(s[n], a) + \gamma \sum_{s' \in \mathcal{S}} \Pr(S[n+1] = s' | s[n], a_n = a) v_{m-1}^*(\underline{h}_n, a, s') \right], \tag{7.20}$$

where $\underline{h}_n = (\underline{h}_{n-1}, a_{n-1}, s[n])$ and

$$v_0^*(\underline{h}_T) = r_T(s[T]), \tag{7.21}$$

for $\underline{h}_T = (\underline{h}_{T-1}, a_{T-1}, s[T])$.

Proof: Proof is immediate from the proceeding discussion. For a proof by induction, see the proof of Theorem 4.3.2 in [68].

The equation (7.20), called the *optimality equation* or the *Bellman-optimality equation*, is the key to solving MDP problems efficiently.

Theorem 7.1 and the optimality equation (7.20) can be simplified by taking into account our assumption of Markov rewards and state transition probabilities as stated in (7.12a) and (7.12b), respectively, which means that rewards and state transition probabilities depend on $\underline{h}_n = (\underline{h}_{n-1}, a_{n-1}, s[n])$ only through the current state $s[n]$. When this is the case, it is not difficult to see that the value-to-go function $v_m^*(\underline{h}_{T-m})$ will also depend on $\underline{h}_n$ only through $s[n]$[1] so that $v_m^*(\underline{h}_{T-m}) = v_m^*(s[T-m])$. Thus, when rewards and state transition probabilities are Markov, we may rewrite the optimality equation (7.20) as, for $m = 0, \ldots, T-1$ and $n = T - m$,

$$\begin{aligned} v_m^*(s[n]) &= \sup_{a \in \mathcal{A}} \left[r_n(s[n], a) + \gamma \sum_{s' \in \mathcal{S}} \Pr(S[n+1] = s' | s[n], a_n = a) v_{m-1}^*(s') \right] \\ &= \sup_{a \in \mathcal{A}} \left[r_n(s[n], a) + \gamma \sum_{s' \in \mathcal{S}} p(s' | s[n], a_n = a) v_{m-1}^*(s') \right] \end{aligned} \tag{7.22}$$

with the boundary condition

$$v_0^*(s[T]) = r_T(s[T]). \tag{7.23}$$

[1] This is indeed shown in the part (a) of Theorem 4.4.2 of [68].

From (7.15) and (7.19), we note that by solving the Bellman-optimality equations, we may obtain the optimal value function $V_T^*(s)$, because

$$\begin{aligned} V_T^*(s) &= \sup_{\pi \in \Pi^{HR}} V_T^{\pi}(s) \\ &= \sup_{\pi \in \Pi^{HR}} v_T^{\pi}(s) \\ &= v_T^*(s). \end{aligned} \tag{7.24}$$

Recall that when rewards and transition probabilities are Markov, we can reduce the search for an optimal policy to policies over Π^{MR}, the set of randomized Markov policies. In addition, when the supremum in (7.22) is achieved for each state, it is further known that there is actually an optimal policy within the class of deterministic decision rules. By combining these, we have the following theorem (Theorem 4.4.2 of [68]).

Theorem 7.2 (Existence of an optimal deterministic Markov policy) When rewards and transition probabilities are Markov as in (7.12a) and (7.12b), the optimal value function can be found by limiting the search to only over deterministic Markov policies:

$$V_T^*(s) = \sup_{\pi \in \Pi^{HR}} V_T^{\pi}(s) = \sup_{\pi \in \Pi^{MD}} V_T^{\pi}(s), \tag{7.25}$$

for any $s \in \mathcal{S}$, provided that for each $s[n] \in \mathcal{S}$, for $m = 0, \cdots, T$, there exists $a^* \in \mathcal{A}$ such that

$$\begin{aligned} & r_n(s[n], a^*) + \gamma \sum_{s' \in \mathcal{S}} \mathrm{P}(S[n+1] = s' | s[n], a^*) v_{m-1}^*(\underline{h}_n, a^*, s') \\ &= \sup_{a \in \mathcal{A}} \left[r_n(s[n], a) + \gamma \sum_{s' \in \mathcal{S}} \mathrm{P}(S[n+1] = s' | s[n], a) v_{m-1}^*(\underline{h}_n, a, s') \right], \end{aligned} \tag{7.26}$$

where $n = T - m$.

Proof: See the proof of Theorem 4.4.2 in [68].

Indeed, it is known that (7.26) holds when $\mathcal{S}$ and $\mathcal{A}$ are both finite, as we have assumed above. When $\mathcal{A}$ is not finite but compact, then a set of sufficient conditions to guarantee (7.26) is given in Proposition 4.4.3 of [68]. In the following, however, we will assume that $\mathcal{A}$ and $\mathcal{S}$ are both finite sets so that the suprema are achieved and thus can be replaced by maxima.

7.3.2.3 Solving Optimality Equations: Backward Induction (Dynamic Programming) From the previous section, we know that under the Markov rewards and state transition probabilities, there is an optimal policy within the deterministic Markov policies as long as $\mathcal{S}$ and $\mathcal{A}$ are both finite. Then, we may compute the optimal value function $V_T^*(s)$ easily by solving the optimality equation (7.22) for $m = 0, \cdots, T$ and then using (7.24). This *backward induction* algorithm, also called the *dynamic programming*, is given in Algorithm 7.1.

Algorithm 7.1

Backward Induction/Dynamic Programming

(1) For $m=0$, set

$$v_T^*(s) = r_T(s) \quad \forall s \in \mathcal{S}$$

(2) If $m=T$, stop. Otherwise go to step (3).

(3) Let $m=m+1$.

For each $s \in \mathcal{S}$, compute

$$v_m^*(s) = \max_{a\in\mathcal{A}} \left[r_{T-m}(s,a) + \gamma \sum_{s' \in \mathcal{S}} p(s'|s,a) v_{m-1}^*(s') \right] \tag{7.27}$$

and

$$a_{T-m}^*(s) = \arg\max_{a\in\mathcal{A}} \left[r_{T-m}(s,a) + \gamma \sum_{s' \in \mathcal{S}} p(s'|s,a) v_{m-1}^*(s') \right] \tag{7.28}$$

and go to step (2).

From (7.24), it is clear that the solution of the backward induction algorithm 7.1 provides the optimal value function $V_T^*(s) = v_T^*(s)$. In addition, we see that we may obtain an optimal policy $\underline{\pi}^* = (\pi_0^*, \ldots, \pi_{T-1}^*)$ by setting

$$\pi_n^*(s) = a_n^*(s) \quad \text{for all } s \in \mathcal{S}, \tag{7.29}$$

where a_n^* is given by (7.28).

To appreciate the efficiency of backward induction procedure, consider the computational complexity of evaluating the state-value function $V_T^{\underline{\pi}}(s)$ over all possible policies $\underline{\pi} \in \Pi^{MD}$ by brute force. If the number of states and actions are $|\mathcal{S}|$ and $|\mathcal{A}|$, respectively, then at each time instant n, for $n = 1, \ldots, T-1$, there are $|\mathcal{A}|^{|\mathcal{S}|}$ number of decisions we need to consider (for $n = 0$, there are only $|\mathcal{A}|$ number of decisions to be considered since we assume that the initial state $S[0] = s$ is given). Hence, the number of Π^{MD} policies to consider is about $|\mathcal{A}|^{(T-1)|\mathcal{S}|}$, which is exponential in both T and $|\mathcal{S}|$. The brute-force approach requires computing the state-value function $V_T^{\underline{\pi}}(s)$ for this many number of different policies $\underline{\pi}$ in order to determine the optimal one.

On the other hand, for each time instant $m = 1, \cdots, T$, the backward induction algorithm only involves computing the term inside the square brackets of (7.27) for $|\mathcal{A}|$ number of times. The term inside the square brackets requires about $|\mathcal{S}|^2$ multiplications. Hence, for example, the total multiplications required to obtain the optimal value function from the backward induction algorithm are about $(T-1)|\mathcal{A}||\mathcal{S}|^2$, which is linear in T and quadratic in $|\mathcal{S}|$. This is an enormous amount of savings since, especially, the length of time horizon can be large. It is also not uncommon that the number of states $|\mathcal{S}|$ is at least moderately large in many scenarios.

It is important to realize though that the earlier backward induction algorithm only works if we know the MDP model (i.e., transition probabilities and the rewards) completely. On the other hand, the assumptions of finite state and action sets can easily be dropped. Indeed, the earlier results hold in general, for either countable or compact action and state sets. We refer interested readers to [68] for more detailed discussions on these issues.

7.4 INFINITE-HORIZON MDPs

There are situations in which it may be more reasonable to assume an infinite time horizon $T \to \infty$ because decisions have to be made for future time instants $n = 0, 1, \ldots,$ without knowing a clear termination time. This may happen due to various reasons. For instance, it may be the case that the process $S[n]$ itself is indeed non-terminating: Once we move from $n \Leftarrow n$ to $n \Leftarrow n+1$, the structure of the decision problem stays the same. However, this can also happen not because the process is non-terminating but simply because we do not know the terminating time. Another alternative scenario that may force us to treat as $T \longrightarrow \infty$ is if the terminal time T is modeled as a random variable. When the terminating time $T \longrightarrow \infty$, we call the MDP an *infinite-horizon MDP*.

A policy $\underline{\pi}$ for an infinite-horizon MDP is a sequence of decision rules of the form $\underline{\pi} = (\pi_0, \pi_1, \ldots,)$. In the following, however, we will limit our discussion to stationary and finite MDP, that is,both $\mathcal{S}$ and $\mathcal{A}$ are finite (or at least countable), and for any given policy $\underline{\pi} = (\pi_0, \pi_1, \ldots,)$, both state transition probabilities and instant rewards are time invariant.

We may think of using the ETR $\overline{R_T^{\underline{\pi}}}$ and ETDR $\overline{R_{T,\gamma}^{\underline{\pi}}}$ defined in (7.6a) and (7.7a), respectively, as the basis for preferring one policy over another. As $T \to \infty$, however, there is the possibility that the ETR (7.6a) could diverge even when all immediate rewards are finite. On the other hand, it is straightforward to verify that whenever the immediate reward values are all bounded, the ETDR (7.7a) is always finite. Indeed, if there is an $M < \infty$ such that $|r_n(s[n], a_n)| \leq M$ for all n, then from (7.7a),

$$\overline{R_{\infty,\gamma}^{\underline{\pi}}} = \mathbb{E}_{\underline{\pi}} \left\{ \sum_{n=0}^{\infty} \gamma^n r_n(S[n], \pi_n(.)) \right\} \tag{7.30a}$$

$$\leq \frac{M}{1-\gamma} \tag{7.30b}$$

for all $\gamma \in [0, 1)$. Hence, thc ETDR is always a well-defined objective function in the case of infinite-length time horizons. Thus, in this book, we will exclusively take the

ETDR defined as

$$\overline{R_{\gamma}^{\underline{\pi}}} = \lim_{T \longrightarrow \infty} \mathbb{E}_{\underline{\pi}} \left\{ \sum_{n=0}^{T} \gamma^n r_n(S[n], \pi_n(.)) \right\}, \tag{7.31}$$

where $\gamma \in [0, 1)$ as the optimality criterion for infinite-horizon MDPs.

It is worth noting that the ETDR does indeed have a sound justification as the objective function to be used in an infinite-horizon MDP. To see this, let us write (7.31) as

$$\begin{aligned} \overline{R_{\gamma}^{\underline{\pi}}} &= \mathbb{E}_{\underline{\pi}} \{r_0(s[0], \pi_0(.)\} + \gamma \mathbb{E}_{\underline{\pi}} \{r_1(s[1], \pi_1(.)\} \\ &\quad + \gamma^2 \mathbb{E}_{\underline{\pi}} \left\{ \sum_{n=0}^{\infty} \gamma^i r_{2+n}(s[2+n], \pi_{2+n}(.)) \right\}. \end{aligned} \tag{7.32}$$

Observe from (7.32) that when policy $\underline{\pi}$ is used at time $n = 0$, it results in an immediate reward of $\mathbb{E}_{\underline{\pi}} \{r_0(s[0], \pi_0(s[0])\}$ at time $n = 0$ while it promises a reward of $\mathbb{E}_{\underline{\pi}} \{r_1(s[1], \pi_1(s[1])\}$ at time $n = 1$ and so on.

It is reasonable to ask whether a reward amount of r at time $n \Leftarrow n+1$ has the value same as a reward of r at time $n \Leftarrow n$. For example, suppose that the reward r was money. Then an r amount of reward available at time $n \Leftarrow n$ can be invested at a return rate of $\alpha \geq 0$, resulting in a reward amount of $(1+\alpha)r$ at time $n \Leftarrow n+1$. Hence, having a reward r at time $n \Leftarrow n$ actually has more value than having the same reward r at time $n \Leftarrow n+1$. Alternatively, a reward r at time $n \Leftarrow n+1$ is really equivalent to having a reward of $\frac{r}{1+\alpha}$ at time $n \Leftarrow n$. If we let $\gamma = \frac{1}{1+\alpha}$, that is a reward of γr at time $n \Leftarrow n$, where $\gamma \in [0, 1)$.

For an infinite-horizon MDP with an initial state $S[n] = s$, we may define its state-value function as the ETDR associated with a policy $\underline{\pi} \in \Pi^{HR}$ similar to (7.9a):

$$V^{\underline{\pi}}(s) = \lim_{T \to \infty} \mathbb{E}_{\underline{\pi}} \left\{ \sum_{n=0}^{T} \gamma^n r(s[n], \pi_n(\underline{h}_n)) | S[0] = s \right\}, \tag{7.33}$$

where $0 \leq \gamma < 1$ and $\underline{h}_n = (\underline{h}_{n-1}, a_{n-1}, s[n])$.

If there is an M such that

$$\sup_{s \in \mathcal{S}} \sup_{a \in \mathcal{A}} |r(s, a)| < M < \infty,$$

then the limit (7.33) exists and we can write

$$V^{\underline{\pi}}(s) = \mathbb{E}_{\underline{\pi}} \left\{ \sum_{n=0}^{\infty} \gamma^n r(s[n], \pi_n(\underline{h}_n)) | S[0] = s \right\}. \tag{7.34}$$

As in (7.10) for finite-horizon MDPs, we may now define a policy $\underline{\pi}^*$ as optimal (for a given $0 \leq \gamma < 1$), if

$$V^{\underline{\pi}^*}(s) \geq V^{\underline{\pi}}(s) \quad \text{for all } \ s \in \mathcal{S} \text{ and for all } \underline{\pi} \in \Pi^{HR}. \tag{7.35}$$

The value of an MDP that starts in initial state $S[0] = s \in \mathcal{S}$ is given by the optimal value function defined as

$$V^*(s) = \sup_{\underline{\pi} \in \Pi^{HR}} V^{\underline{\pi}}(s). \tag{7.36}$$

We call an optimal policy exists for the infinite-horizon MDP, if there is a policy $\underline{\pi}^* \in \Pi^{HR}$ such that

$$V^{\underline{\pi}^*}(s) = V^*(s) \quad \text{for all } s \in \mathcal{S}. \tag{7.37}$$

7.4.1 Stationary Optimal Policies for Infinite-Horizon MDPs

Searching for an optimal policy over all randomized history-dependent policies Π^{HR} as in (7.35) is especially challenging in the case of an infinite-horizon MDPs. However, it is known that if $\underline{\pi} \in \Pi^{HR}$, an arbitrary history-dependent policy for a stationary MDP, then for each starting state $s \in \mathcal{S}$, there exists a randomized Markov policy $\underline{\pi}^s \in \Pi^{MR}$ with the same ETDR (see Theorem 5.5.3 of [68]):

$$V^{\underline{\pi}^s}(s) = V^{\underline{\pi}}(s), \quad \text{for all } 0 \leq \gamma < 1. \tag{7.38}$$

Thanks to this result, we may limit our search for optimal polices in (7.35) to randomized Markov policies of the form of

$$\underline{\pi} = (\pi_0(.), \pi_1(.), \ldots,), \tag{7.39}$$

where each π_n, for $n = 0, 1, \cdots$, is

$$\pi_n : \mathcal{S} \times \mathcal{A} \longrightarrow [0, 1].$$

In addition to being stationary (i.e., stationary rewards and transition probabilities), if the MDP is of finite state (or at least countable) and has bounded rewards, then it can be shown that for any $0 \leq \gamma \leq 1$, if there is an optimal randomized Markov policy, then there is always an optimal deterministic Markov policy as well. Hence, we may further restrict the search for optimal policies in (7.35) to deterministic Markov Π^{MD} policies so that each π_n in (7.39) is such that $\pi_n(s[n]) \in \mathcal{A}$. Hence, for a finite-state, stationary MDP with bounded rewards, the optimal value (7.36) of the MDP can be written as

$$V^*(s) = \sup_{\underline{\pi} \in \Pi^{MD}} V^{\underline{\pi}}(s). \tag{7.40}$$

Consider again a policy (7.39) for an infinite-horizon MDP. A policy of the form of (7.39) prescribes a decision rule $\pi_n(s[n])$ for each time instant n. Considering the fact that the time horizon extends up to infinity, both computing and storing such a sequence of decision rules $\pi_n(.)$ for each n are difficult. What will really be convenient is if we may limit our attention to policies of the form of

$$\underline{\pi} = (\pi, \pi, \cdots,), \tag{7.41}$$

where $\pi(.)$ is a deterministic Markov decision rule such that

$$\pi : \mathcal{S} \longrightarrow \mathcal{A}. \tag{7.42}$$

A decision policy of the form of (7.41) is called a *stationary policy*, and if π is a deterministic Markov rule as in (7.42), then it is called a *deterministic stationary Markov decision policy*. Let us denote by Π_s^{MD} the set of all deterministic stationary Markov decision policies of the form of $\underline{\pi} = (\pi, \pi, \cdots)$ with $\pi : \mathcal{S} \to \mathcal{A}$.

We may denote the state-value function $V^{\underline{\pi}}(s)$ of an MDP with a stationary policy $\underline{\pi} = (\pi, \pi, \cdots,)$ as $V^{\underline{\pi}}(s) = V^{\pi}(s)$, for all $s \in \mathcal{S}$.

The following theorem from [68] states conditions under which we may limit the search for optimal policies in (7.35) to deterministic stationary Markov policies

Theorem 7.3 (Existence of a Stationary Optimal Policy) If a stationary MDP has bounded rewards and finite state and action sets $\mathcal{A}$ and $\mathcal{S}$, then it has a deterministic stationary Markov policy that is optimal.

Proof: See the proof of Theorem 6.2.10 in [68].

Hence, when conditions needed in Theorem 7.3 are held, we may write the optimal value of a stationary MDP as

$$V^*(s) = \sup_{\underline{\pi} \in \Pi_s^{MD}} V^{\pi}(s), \tag{7.43}$$

where $\underline{\pi} = (\pi, \pi, \cdots)$.

7.4.1.1 *Policy Evaluation: Bellman Equation for an Infinite-Horizon MDP*

Let $\underline{\pi} = (\pi, \pi, \cdots)$ be an arbitrary stationary policy. From the definition of the state-value function in (7.34) and (7.43),

$$\begin{aligned}
V^{\pi}(s) &= V^{\underline{\pi}}(s) \\
&= \mathbb{E}_{\underline{\pi}} \left\{ \sum_{n=0}^{\infty} \gamma^n r(s[n], \pi(s[n]) | S[0] = s \right\} \\
&= r(S[0] = s, \pi(s)) + \mathbb{E}_{\underline{\pi}} \left\{ \sum_{n=1}^{\infty} \gamma^n r(s[n], \pi(s[n])) | S[0] = s \right\}
\end{aligned}$$

$$= r(s,\pi(s)) + \gamma \mathbb{E}_{\underline{\pi}} \left\{ \sum_{n=0}^{\infty} \gamma^n r(s[n+1], \pi(s[n+1])) | S[0] = s \right\}$$

$$= r(s,\pi(s)) + \gamma \sum_{s' \in \mathcal{S}} \mathbb{E}_{\underline{\pi}} \left\{ \sum_{n=0}^{\infty} \gamma^n r(s[n+1], \pi(s[n+1])) \middle| S[1] = s' \right\}$$

$$\times \Pr\left(S[1] = s' \middle| S[0] = s, \pi(s) \right)$$

$$= r(s,\pi(s)) + \gamma \sum_{s' \in \mathcal{S}} p(s' | s, \pi(s)) \mathbb{E}_{\underline{\pi}} \left\{ \sum_{n=0}^{\infty} \gamma^n r(s[n+1], \pi(s[n+1])) \middle| S[1] = s' \right\}$$

$$= r(s,\pi(s)) + \gamma \sum_{s' \in \mathcal{S}} p(s' | s, \pi(s)) V^{\underline{\pi}}(s') \tag{7.44a}$$

$$= r(s,\pi(s)) + \gamma \sum_{s' \in \mathcal{S}} p(s' | s, \pi(s)) V^{\pi}(s'), \tag{7.44b}$$

which is called the *Bellman equation for the value function* $V^{\pi}(\cdot)$ *of a stationary policy* π, which is the counterpart to (7.17) we obtained for a finite-horizon MDP. By solving the Bellman equation (7.44b), we may obtain the value of a given stationary policy π.

Note that evaluation of (7.44b) is straightforward since it is just a set of $|\mathcal{S}|$ linear equations in the $|\mathcal{S}|$ variables $V^{\pi}(s)$ for each $s \in \mathcal{S}$. Indeed, suppose that the number of possible states is M so that

$$M = |\mathcal{S}|.$$

For simplicity, we may label these M states by $\mathcal{S} = \{1, \cdots, M\}$. For a given stationary policy π, let us define the $M \times M$ matrix of transition probabilities P_{π} such that

$$[P_{\pi}]_{i,j} = \Pr(S[n+1] = j | S[n] = i, \pi(i)) \quad \text{for } 1 \leq i,j \leq M, \tag{7.45}$$

where $[P_{\pi}]_{i,j}$ denotes the (i,j)th element of the matrix P_{π}. If we define

$$\mathbf{r}_{\pi} = (r(1,\pi(1)), r(2,\pi(2)), \ldots, r(M,\pi(M)))^T$$

and

$$\mathbf{v}^{\pi} = \left(V^{\pi}(1), V^{\pi}(2), \ldots, V^{\pi}(M) \right)^T,$$

then we may write the M equations in (7.44b) in matrix notation as [68]

$$\mathbf{v}^{\pi} = \mathbf{r}_{\pi} + \gamma P_{\pi} \mathbf{v}^{\pi}. \tag{7.46}$$

This is the Bellman equation for the stationary policy π in matrix form. Hence, the closed-form solution to the Bellman equation is immediate:

$$\mathbf{v}^{\pi} = (I - \gamma P_{\pi})^{-1}\, \mathbf{r}^{\pi}. \tag{7.47}$$

Solution (7.47) is called the *policy evaluation* (or the prediction problem). If the model is completely known (i.e., P_{π} and $\mathbf{r}_{\pi}$), then $\mathbf{v}^{\pi}$ can be computed directly from (7.47). Sometimes, however, it is convenient to solve for $\mathbf{v}^{\pi}$ iteratively. From (7.44b), we may indeed write an iterative policy evaluation method, for $k = 1, 2, \cdots$, as

$$V_{k+1}^{\pi}(s) = r(s, \pi(s)) + \gamma \sum_{s' \in \mathcal{S}} p(s' | s, \pi(s)) V_k^{\pi}(s'). \tag{7.48}$$

Iteration (7.48) indeed converges to $V^{\pi}(s)$ whenever $V^{\pi}(s)$ exists:

$$V_k^{\pi}(s) \to V^{\pi}(s). \tag{7.49}$$

7.4.2 Bellman-Optimality Equations

As in the case of finite-horizon MDPs, our interest is in finding an optimal policy for the infinite-horizon MDP. In particular, we search for a stationary optimal policy. From (7.43) and (7.44b), we see that the optimal value function satisfies

$$\begin{aligned}
V^*(s) &= \sup_{\pi \in \Pi_S^{MD}} V^{\pi}(s) \\
&= \sup_{\pi \in \Pi_S^{MD}} \left\{ r(s, \pi(s)) + \gamma \sum_{s' \in \mathcal{S}} p(s' | s, \pi(s)) V^{\pi}(s') \right\} \\
&= \sup_{\pi \in \Pi_S^{MD}} \left\{ r(s, \pi(s)) + \gamma \sum_{s' \in \mathcal{S}} p(s' | s, \pi(s)) V^*(s') \right\} \\
&= \sup_{a \in \mathcal{A}} \left\{ r(s, a) + \gamma \sum_{s' \in \mathcal{S}} p(s' | s, a) V^*(s') \right\},
\end{aligned} \tag{7.50}$$

which is called the *Bellman-optimality equations* or, simply, the *optimality equations*. The optimal value function $V^*(\cdot)$ is thus a fixed point of the optimality equation (7.50). Unlike the Bellman equation (7.44b) for a given policy π, the optimality equation (7.50) is not necessarily linear due to the max (or sup) operation appearing on the right-hand side. Hence, unlike policy evaluation with (7.44b), solving the optimality equation to find the optimal value function may not be straightforward. Indeed, there is no closed-form solution for (7.50). Due to the nonlinear nature of the optimality equation (7.50), finding its solution $V^*(s)$, and subsequently an optimal policy, requires more work than simple policy evaluation to compute $V^{\pi}(s)$.

It can, however, be shown that if $\mathcal{S}$ is finite (or countable) and $\{r(s,a)\}$ is bounded, then for any $0 \leq \gamma < 1$, there exists a unique V^* that is the solution to the fixed-point equation (7.50)[2] [68].

Since, under the assumed conditions, there exists a stationary deterministic Markov policy $\boldsymbol{\pi}^* \in \Pi_s^{MD}$ that is optimal, where $\boldsymbol{\pi}^* = (\pi^*, \pi^*, \ldots)$, we note that

$$V^*(s) = V^{\pi^*}(s) \tag{7.51}$$

and $V^{\pi^*}(s)$ must also satisfy the optimality equation (7.50). Indeed, clearly, any π^* policy is optimal if and only if V^{π^*} is a solution to the optimality equation (7.50) (see Theorem 6.2.6 [68]).

In the following, we discuss several commonly used approaches for solving the Bellman-optimality equations (7.50).

7.4.2.1 Policy Iteration *Policy iteration* is a simple, yet effective, way to compute the optimal value function and an optimal policy. It combines the policy evaluation (7.47) that we already discussed with an additional step called the *policy improvement*.

Let us define the following function $Q^\pi(s,a)$ that is called the *action-value function* of a policy π:

$$Q^\pi(s,a) = \mathbb{E}_\pi \left\{ \sum_{n=0}^{\infty} \gamma^n r(s[n], \pi(.)) \middle| S[0] = s, \pi_0 = a \right\}, \tag{7.52}$$

where action at time $n = 0$ is a but for all other $n > 0$ is determined by the stationary decision rule π. Hence, $Q^\pi(s,a)$ is the expected total discounted reward starting from the initial state $S[0] = s$, but taking the action $\pi_0 = a$ at time $n = 0$, and then following the stationary policy π from there on. Note that with a stationary policy π, clearly,

$$Q^\pi(s, \pi(s)) = V^\pi(s). \tag{7.53}$$

From the definition (7.52) of $Q^\pi(s,a)$, we have that

$$\begin{aligned} Q^\pi(s,a) &= r(s,a) + \gamma \mathbb{E}_\pi \left\{ \sum_{n=0}^{\infty} \gamma^n r(s[n+1], \pi_{n+1}) \middle| S[0] = s, \pi_0 = a \right\} \\ &= r(s,a) + \gamma \mathbb{E}_\pi \left\{ \sum_{n=0}^{\infty} \gamma^n r(s[n+1], \pi) \middle| S[0] = s, \pi_0 = a \right\} \end{aligned} \tag{7.54a}$$

[2]This is Theorem 6.2.5 of [68]. The proof is established by using the *Banach fixed-point theorem*, which states that if f is a contraction mapping $f : \mathbb{A} \longrightarrow \mathbb{A}$ where $\mathbb{A}$ is a closed subset of a Banach space (a complete normed space), then there exists a unique $x^* \in \mathbb{A}$ such that $f(x^*) = x^*$. The result may be generalized to broader classes of state spaces such as $\mathbb{R}^n$ under additional regularity conditions.

$$= r(s,a) + \gamma \sum_{s' \in \mathcal{S}} \mathbb{E}_\pi \left\{ \sum_{n=0}^{\infty} \gamma^n r(s[n+1], \pi) \middle| S[1] = s' \right\} p(s'|s,a)$$

$$= r(s,a) + \gamma \sum_{s' \in \mathcal{S}} V^\pi(s') p(s'|s,a), \tag{7.54b}$$

where (7.54a) is due to the stationarity of the policy π and (7.54b) follows by noting that the expectation in the preceding step is indeed the value function of the policy π with the initial state s'.

Now, suppose we define a new policy π' according to

$$\pi'(s) = \arg\max_a Q^\pi(s,a) \quad \text{for all } s \in \mathcal{S} \tag{7.55a}$$

$$= \arg\max_a \left[r(s,a) + \gamma \sum_{s' \in \mathcal{S}} p(s'|s,a) V^\pi(s') \right]. \tag{7.55b}$$

Let the value functions corresponding to the two policies, π' and π, be $V^{\pi'}$ and V^π, respectively. Then (7.55b) implies that

$$r(s,\pi'(s)) + \gamma \sum_{s' \in \mathcal{S}} p(s'|s,\pi'(s)) V^\pi(s') \geq r(s,\pi(s)) + \gamma \sum_{s' \in \mathcal{S}} p(s'|s,\pi(s)) V^\pi(s')$$

$$= V^\pi(s), \tag{7.56}$$

where (7.56) follows by the Bellman equation (7.44b). In vector notation, (7.56) can be written as

$$\mathbf{r}_{\pi'} + \gamma P_{\pi'} \mathbf{v}^\pi \geq \mathbf{v}^\pi,$$

so that

$$(\mathbf{I} - \gamma P_{\pi'})^{-1} \mathbf{r}_{\pi'} \geq \mathbf{v}^\pi, \tag{7.57}$$

where $\mathbf{x} \geq \mathbf{y}$ means componentwise $x_i \geq y_i$. But note from (7.47) that solution to the Bellman equation for the new policy π' defined as in (7.55a) is

$$\mathbf{v}^{\pi'} = (\mathbf{I} - \gamma P_{\pi'})^{-1} \mathbf{r}_{\pi'}. \tag{7.58}$$

From (7.57) and (7.58), we see that

$$V^{\pi'}(s) \geq V^\pi(s) \quad \text{for all } s \in \mathcal{S}. \tag{7.59}$$

Hence, the value of the new policy π' defined in (7.55a) is better or as good as the old policy π for all states $s \in \mathcal{S}$. This gives us a way to obtain an improved policy from a given policy. Indeed, given a policy π, we may define a new policy π' by defining for each state s the action

$$\pi'(s) = \arg\max_a Q^{\pi}(s,a). \tag{7.60}$$

This π' is our new policy and, from (7.59), it is at least as good as π or better for each state $s \in \mathcal{S}$. Hence, π' is an improved policy over π and the process (7.60) is called the *policy improvement*.

But now, we can repeat the same steps with this new policy π' to obtain an even better policy. In each iteration, either we should have $V^{\pi'}(s) > V^{\pi}(s)$ or $V^{\pi'}(s) = V^{\pi}(s)$. When $V^{\pi'}(s) = V^{\pi}(s)$, we have from (7.53) that

$$\begin{aligned} V^{\pi'}(s) &= Q^{\pi'}(s,\pi'(s)) \\ &= \max_a Q^{\pi}(s,a) \\ &= \max_a \left[r(s,a) + \gamma \sum_{s' \in \mathcal{S}} p(s'|s,a) V^{\pi}(s') \right] \\ &= \max_a \left[r(s,a) + \gamma \sum_{s' \in \mathcal{S}} p(s'|s,a) V^{\pi'}(s') \right]. \end{aligned} \tag{7.61}$$

But (7.61) is the same as (7.50), which is the optimality equation for policy π'. Hence, when $V^{\pi'}(s) = V^{\pi}(s)$, $V^{\pi'}$ is a solution to the optimality equation. Thus, $V^{\pi'}$ must be the optimal value function V^* and so π' and π must both be optimal policies.

In summary, starting with an arbitrary policy π, the *policy iteration algorithm* (see Algorithm 7.2) can be stated as:

1. First, do policy evaluation for π to obtain $V^{\pi}(s)$.
2. Obtain a new policy via policy improvement:

$$\begin{aligned} \pi'(s) &= \arg\max_a Q^{\pi}(s,a) \\ &= \arg\max_a \left[r(s,a) + \gamma \sum_{s' \in \mathcal{S}} p(s'|s,a) V^{\pi}(s') \right]. \end{aligned}$$

3. Repeat the above two steps till convergence.

Since there are only a finite number of deterministic policies when both $|\mathcal{S}|$ and $|\mathcal{A}|$ are finite and in each iteration we obtain a policy that is strictly better than the previous policy, this algorithm must converge (or stop) by giving $\pi' = \pi$, at which point we know that $\pi' = \pi$ is optimal.

Algorithm 7.2

```
Policy Iteration Algorithm
 (1) Initialize π(s) ∈ 𝒜 arbitrarily for all s ∈ 𝒮.
 (2) Policy evaluation:

     V^π(s) = r(s,a) + γ Σ_{s'∈𝒮} p(s'|s,a) V^π(s').

 (3) Policy improvement: for each s ∈ 𝒮, let

     π'(s) = argmax_a [r(s,a) + γ Σ_{s'∈𝒮} p(s'|s,a) V^π(s')].

 (4) If π' = π, stop. Otherwise set π = π' and go to step (2).
```

7.4.2.2 Value Iteration During each iteration of the policy iteration algorithm 7.2, there is a policy evaluation step (step (2)). In step (3) of the policy iteration algorithm, we update to a new policy π', and then in step (2) of the next iteration, we need to compute the new value function $V^{\pi'}$. Although simple enough, computing $V^{\pi'}$ itself may take time. Alternatively, as noted earlier, we may use iterative policy evaluation (7.48) to evaluate $V^{\pi'}$, which may still take time for convergence.

A simple modification to avoid this need for policy evaluation in the policy iteration algorithm 7.2 is to instead interweave iterative policy evaluation (7.48) with the policy improvement step (step (3)) in the policy iteration algorithm. In other words, in step (2) of each iteration of the policy iteration algorithm, we may implement an iterative policy evaluation step (7.48) rather than solving for $V^{\pi'}$ exactly. Following this, in step (3), we may do policy improvement as we already have. But then, we actually don't need step (2) at all, and we have Algorithm 7.3 called the *value iteration algorithm.*

Algorithm 7.3

```
Value Iteration Algorithm
 (1) Set k = 0, and choose ε > 0 and V_k(s) ∈ ℝ for s ∈ 𝒮 arbitrarily.
 (2) For each s ∈ 𝒮, let
     V_{k+1}(s) = max_a [r(s,a) + γ Σ_{s'∈𝒮} p(s'|s,a) V_k(s')].
 (3) If ||V_{k+1} − V_k|| < (1−γ)/(2γ) ε, go to step (4). Otherwise set k ← k+1 and go
to step (2).
 (4) For each s ∈ 𝒮, set
     π(s) = argmax_a [r(s,a) + γ Σ_{s'∈𝒮} p(s'|s,a) V_{k+1}(s')]
     and stop.
```

The value iteration algorithm 7.3 is simple to implement, yet indeed it can be shown that $V_k(s) \to V^*(s)$, for each $s \in \mathcal{S}$, as long as $V^*(s)$ exists, which is true under bounded rewards and finite state and action sets. The stopping criterion in step (3) of the value iteration algorithm ensures that the policy π computed in step (4) when algorithm terminates is an ϵ-optimal policy [68].

Definition 7.4.1 (ϵ-Optimal Policies) A policy π_ϵ^* is called an ϵ-optimal policy if

$$V^{\pi_\epsilon^*}(s) \geq V^*(s) - \epsilon \quad \text{for all } s \in \mathcal{S}.$$

The value of an ϵ-optimal policy is thus at most within ϵ of the optimal value function for any state $s \in \mathcal{S}$. The stopping criterion in step (3) of the value iteration algorithm ensures that the value of the policy found when the algorithm terminates is within ϵ of the optimal value function for any state $s \in \mathcal{S}$.

7.4.2.3 Generalized Policy Iteration Clearly, the value iteration algorithm 7.3 is a special case of the policy iteration algorithm 7.2: The policy iteration algorithm consists of two steps, policy evaluation and policy improvement. The value iteration algorithm simplifies the policy evaluation step of policy iteration algorithm by performing just one iterative policy evaluation step.

But then, we may change these two steps in other ways as well. For example, in step (3) of the policy iteration algorithm, rather than updating the policy π for all states, we may just update it for a single state and obtain a new policy π'. That is the new policy $\pi'(s) = \pi(s)$ for $\forall s \in \mathcal{S}$ except for a single state $s = \tilde{s}$. After that, we may return to step (2) and do the policy evaluation. Here also, we could either do full policy evaluation (as in the policy iteration algorithm) or just one or more iterative policy evaluation steps. As we see, there are many modifications one can think of, and all these are, in general, called *generalized policy iteration (GPI) algorithms* [69]. Thus, value iteration is just one possible GPI algorithm.

7.5 PARTIALLY OBSERVABLE MARKOV DECISION PROCESSES

So far, we have been concerned with obtaining optimal decision rules of the form of $\pi_n^*(s[n])$ for an MDP. Hence, our assumption has been that at each time instant n, we can observe state $S[n] = s$ of the Markov process and then execute the decision rule $\pi_n^*(s)$. Such MDPs in which the state is exactly known to the decision-maker are aptly called the *completely observable Markov decision processes* (COMDPs). Hence, optimal policy approaches of the previous sections are only applicable to COMDPs.

But it is exceedingly common in practice that the exact state of the process is not directly observable by the decision-maker. Instead, one normally gets to observe a quantity, say, $Y[n]$, at each time instant $n = n$ that is (statistically) related to the true state $S[n]$ of the system at time n. At time $n = n$, the decision-maker has knowledge of the observation $Y[n]$, but not $S[n]$. At time $n = 0$, then there is no point specifying a decision policy for the time horizon of interest as a function of true state $s[n]$ (as we have done so far), because we will not know the true state at each time instant to make use of such a policy.

Of course, motivated by estimation theory ideas from Chapter 5, a natural thought may be to first estimate $S[n]$ from $Y[n]$ and then pick the decision at time n based on the estimated state $\hat{s}[n] \in \mathcal{S}$. While this makes sense, it is an ad hoc scheme that does not have any claims of optimality and it may perform poorly in many situations. The deficiency lies in the fact that the unobservability of the state $S[n]$ is not taken into account in designing a policy at the first place. Hence, our goal in this section is to investigate optimal policies for this situation in a systematic way.

An MDP in which the state is not directly observable is called a *partially observable markov decision process* (POMDP). POMDP models are suitable in situations in which there is uncertainty about the true state of the system due to either full state not directly being accessible or the full state not exactly being observable. In Chapter 9, we will see that the wideband spectrum sensing problem in a cognitive radio is a POMDP due to both these reasons: The full state is not directly accessible due to hardware limitations on how wide a spectrum range the radio can sense at any time. Moreover, sensing observations even in a limited portion of the spectrum are also never exact because of the inherent noise in communications receivers. In Chapter 13, we will see that the problem of primary channel sensing in dynamic spectrum sharing cognitive radios can also be modeled as a POMDP due to the latter reason.

In the following, we will limit our discussion to stationary POMDPs with finite state and action spaces, which can formally be defined by specifying:

- A finite set $\mathcal{S} = \{s_1, \cdots, s_M\}$ of states with the associated a priori distribution $\mathbf{p} = (p_1, \cdots, p_M)$ on the initial state $S[0]$ where $p_m = \Pr(S[0] = s_m)$ for $s_m \in \mathcal{S}$ and $m = 1, \cdots, M$.
- A finite set $\mathcal{A}$ of available actions. In each state $s \in \mathcal{S}$, a subset $\mathcal{A}_s \subseteq \mathcal{A}$ might be available.
- A nonnegative function $p(s'|s,a)$ denoting the probability that the system is in state s' at the next time instant, when the action $a \in \mathcal{A}$ is executed in state $s \in \mathcal{S}$ at the current time. That is, for all n, $\Pr(S[n+1] = s'|S[n] = s, a_n = a) = p(s'|s,a)$. We may denote these transition probabilities by a set of $M \times M$ matrices $\{P(a)\}_{a \in \mathcal{A}}$ indexed by each action $a \in \mathcal{A}$. The (i,j)th entry of matrix $P(a)$ is $[P(a)]_{i,j} = p(s_j|s_i,a)$ for $s_i \in \mathcal{S}$ and $s_j \in \mathcal{S}$.
- A real-valued function $r(s,a)$ defined on state $s \in \mathcal{S}$ and action $a \in \mathcal{A}$ denoting the immediate reward received if action a is taken while in state s.
- A random variable $Y[n]$, called the observation, taking values in a finite set $\mathcal{Y} = \{y_1, \cdots, y_L\}$.
- A nonnegative function $q(y|s,a)$ denoting the probability that $Y[n] = y$ given that action $a_{n-1} = a$ is taken at time $n-1$ resulting in the underlying state $S[n] = s$ at time n. That is, for all n, $\Pr(Y[n] = y|S[n] = s, a_{n-1} = a) \triangleq q(y|s,a)$. Note that the observation $Y[n]$ at time n is not simply determined by the underlying state $s[n]$ alone of the Markov process. It is allowed to be effected by the action taken at the previous time instant $n-1$.

Thus, in a POMDP, we have two stochastic processes: the process $S[n]$ of states and the process $Y[n]$ of observations. The process $\{S[n]\}$ is called the *core process*, while $\{Y[n]\}$ is called the *observation process*.

7.5.1 Definitions

Unlike in the case of COMDPs, the information history that is available to the decision-maker in a POMDP is not the sequence of states, actions, and rewards.

Instead, it is the sequence of observations, actions, and the associated rewards. Hence, if we again denote by $\underline{h}_n$ the complete history available for making a decision at time n in a POMDP, then, for $n = 0, 1, 2, \cdots$,

$$\underline{h}_n = \left(\underline{h}_{n-1}, a_{n-1}, y[n]\right)$$

with the initialization $\underline{h}_0 = (\mathbf{p}, y[0])$, where $\mathbf{p}$ denotes the distribution of the initial state. Hence, we may write a decision rule for a POMDP for time n given the history $\underline{h}_n$ to be of the form of $\pi_n\left(\underline{h}_n\right)$.

At time $n = 0$, the planning problem for a POMDP is to design a sequence of decision rules of the form of $\pi_n\left(\underline{h}_n\right)$ for each n to obtain a policy $\underline{\pi} = \left(\pi_0\left(\underline{h}_0\right), \pi_1\left(\underline{h}_1\right), \cdots, \right)$. Of course, we would like to have an optimal policy in some sense.

7.5.1.1 A Sufficient Statistic for Optimal Decision-Making: The Belief State

Earlier, we saw that when rewards and state transition probabilities are Markov, the optimal policies for a COMDP can be found within the class of Markov decision policies. These were policies in which each decision rule $\pi_n(.)$ is only a function of the current state $s[n]$ rather than the whole history $\underline{h}_n$ up to time n. We have already assumed that our POMDP has Markov rewards and state transition probabilities. It is of interest to know then whether we can search for optimal policies without having to consider the whole history $\underline{h}_n$ up to time n.

Let us denote the a posteriori probability of state $S[n] = s_j$ at time n, for each $s_j \in \mathcal{S}$ and $j = 1, \cdots, M$, given the complete history $\underline{h}_n$ up to time n as

$$\begin{aligned} b_j[n] &= \Pr\left(S[n] = s_j \mid \underline{h}_n\right) \\ &= \Pr\left(S[n] = s_j \mid \left(\underline{h}_{n-1}, a_{n-1}, y[n]\right)\right). \end{aligned} \quad (7.62)$$

We may collect these a posteriori probabilities into a vector

$$\mathbf{b}[n] = \left(b_1[n], \cdots, b_M[n]\right)^T.$$

Hence, $\mathbf{b}[n]$ is the posterior distribution, or the posterior pmf, of the state $S[n]$ given all knowledge up to time n.

Note that $b_j[n]$ represents our posterior belief that the current state $S[n]$ is $s_j \in \mathcal{S}$ given all available evidence up to time n. For this reason, the M-vector $\mathbf{b}[n]$ is called the *belief state vector*, or the *belief state*, at time n. Each element b_j of the belief vector $\mathbf{b}$ is a probability: $b_j \in [0, 1]$ for $j = 1, \cdots, M$. Thus, for all n, the belief state vector $\mathbf{b}[n]$ is a probability vector (or a probability mass function) that takes values in the $(M - 1)$-dimensional probability simplex $\mathcal{P}^{M-1}$ defined as

$$\mathcal{P}^{M-1} = \left\{ \mathbf{x} \in \mathbb{R}^M; \sum_{i=1}^{M} x_i = 1, x_i \geq 0 \text{ for } i = 1, \cdots, M \right\}. \quad (7.63)$$

From estimation theory of Chapter 5, we can infer that the posterior distribution of state $S[n]$ given history is all we need in forming optimal state estimators. Thus, not surprisingly, the belief state vector $\mathbf{b}[n]$ is indeed all we need to make optimal decisions in a POMDP.

Theorem 7.4 (Sufficient Statistics for Optimal Decision-Making in a POMDP) The belief state vector $\mathbf{b}[n] \in \mathcal{P}^{M-1}$ is a sufficient statistic for the POMDP optimal decision-making problem at time n, for $n = 0, 1, 2, \cdots$.

Proof: A proof of this can be found in Appendix A of [71] or in [72]. The fact that the belief vector is a sufficient statistic for the POMDP problem was first established in [73].

7.5.1.2 An Equivalent COMDP for a POMDP Since $\mathbf{b}[n]$ is a sufficient statistic for decision-making in a POMDP, instead of taking into account the whole history $\underline{h}_n$, at each time instant n, we may base our decisions on $\mathbf{b}[n]$. Hence, rather than history-dependent decision rules $\pi_n(\underline{h}_n)$, we may limit ourselves to decision rules of the form of $\pi_n(\mathbf{b}[n])$. All we need to know at time n is the belief state vector $\mathbf{b}[n]$ rather than the complete history $\underline{h}_n$.

Note that for each $s_j \in \mathcal{S}$, for $j = 1, \cdots, M$, and $\underline{h}_{n+1} = \left(\underline{h}_n, a_n = a, Y[n+1] = y\right)$ for $a \in \mathcal{A}$ and $y \in \mathcal{Y}$,

$$\begin{aligned} b_j[n+1] &= \Pr\left(S[n+1] = s_j \mid \underline{h}_{n+1}\right) \\ &= \frac{\Pr\left(S[n+1] = s_j, \underline{h}_{n+1}\right)}{\Pr\left(\underline{h}_{n+1}\right)} \\ &= \frac{\Pr\left(S[n+1] = s_j, \underline{h}_{n+1}\right)}{\sum_{j'-1}^{M} \Pr\left(S[n+1] = s_{j'}, \underline{h}_{n+1}\right)} \\ &= \frac{\Pr\left(Y[n+1] = y \mid S[n+1] = s_j, \underline{h}_n, a_n = a\right) \Pr\left(S[n+1] = s_j, \underline{h}_n, a_n = a\right)}{\sum_{j'=1}^{M} \Pr\left(Y[n+1] = y \mid S[n+1] = s_{j'}, \underline{h}_n, a_n = a\right) \Pr\left(S[n+1] = s_{j'}, \underline{h}_n, a_n = a\right)}. \end{aligned} \tag{7.64}$$

But by definition,

$$\begin{aligned} \Pr\left(Y[n+1] = y \mid S[n+1] = s_j, \underline{h}_n, a_n = a\right) &= \Pr\left(Y[n+1] = y \mid S[n+1] = s_j, a_n = a\right) \\ &= q\left(y \mid s_j, a\right) \end{aligned} \tag{7.65}$$

and

$$\begin{aligned} \Pr\left(S[n+1] = s_j, \underline{h}_n, a_n = a\right) &= \sum_{i=1}^{M} \Pr\left(S[n+1] = s_j, S[n] = s_i, \underline{h}_n, a_n = a\right) \\ &= \sum_{i=1}^{M} \Pr\left(S[n+1] = s_j \mid S[n] = i, \underline{h}_n, a_n = a\right) \\ &\quad \times \Pr\left(S[n] = s_i, \underline{h}_n, a_n = a\right) \end{aligned}$$

$$= \sum_{i=1}^{M} \Pr\left(S[n+1] = s_j \,|S[n] = s_i, a_n = a\right)$$
$$\times \Pr\left(S[n] = s_i | \underline{h}_n, a_n = a\right) \Pr\left(\underline{h}_n, a_n = a\right)$$
$$= \sum_{i=1}^{M} p(j|i,a) \Pr\left(S[n] = s_i | \underline{h}_n\right) \Pr\left(\underline{h}_n, a_n = a\right) \qquad (7.66a)$$
$$= \sum_{i=1}^{M} [P(a)]_{i,j} b_i[n] \Pr\left(\underline{h}_n, a_n = a\right), \qquad (7.66b)$$

where (7.66a) follows from the definition of the state transition probabilities of the core process and (7.66b) follows from the definition of the belief state (7.62). Substituting (7.65) and (7.66b) in (7.64), we have, for $j = 1, \cdots, M$,

$$b_j[n+1] = \frac{q\left(y|s_j, a\right) \sum_{i=1}^{M} [P(a)]_{i,j} b_i[n]}{\sum_{j'=1}^{M} q\left(y|s_{j'}, a\right) \sum_{i=1}^{M} [P(a)]_{i,j'} b_i[n]}. \qquad (7.67)$$

Hence, for each $j = 1, \cdots, M$, the belief $b_j[n+1]$ is completely determined by the belief state vector $\mathbf{b}[n]$ at time n along with the action $a_n = a$ taken at time n and the observation $Y[n+1] = y$ at time $n+1$:

$$b_j[n+1] = \Pr\left(S[n+1] \ = \ s_j \mid \underline{h}_{n+1}\right)$$
$$= \Pr\left(S[n+1] = s_j \mid \left(\underline{h}_n, a_n, y_{n+1}\right)\right)$$
$$= \Pr\left(S[n+1] = s_j \mid \mathbf{b}[n], a_n, y_{n+1}\right) \qquad (7.68)$$
$$\triangleq T_j(\mathbf{b}[n], a_n, y_{n+1}). \qquad (7.69)$$

If we let $\mathbf{T}(\mathbf{b}[n], a_n, y_{n+1}) \triangleq (T_1(\mathbf{b}[n], a_n, y_{n+1}), \cdots, T_M(\mathbf{b}[n], a_n, y_{n+1}))^T$, then we may denote the mapping from $\mathbf{b}[n]$ to $\mathbf{b}[n+1]$ under the action $a_n = a$ and observation $Y_{n+1} = y$ by $\mathbf{T}$ so that

$$\mathbf{b}[n+1] = \mathbf{T}(\mathbf{b}[n], a_n, y_{n+1}). \qquad (7.70)$$

Sometimes, we may also denote this new belief state by $\mathbf{b}^{(\mathbf{b}[n], a_n, y_{n+1})}$:

$$\mathbf{b}[n+1] = \mathbf{b}^{(\mathbf{b}[n], a_n, y_{n+1})} \triangleq \mathbf{T}(\mathbf{b}[n], a_n, y_{n+1}). \qquad (7.71)$$

For later use, from (7.68), we may rewrite the belief state update in a slightly different form as below:

$$b_j[n+1] = \Pr\left(S[n+1] = s_j | \mathbf{b}[n], Y[n+1] = y, a_n = a\right)$$
$$= \frac{\Pr\left(S[n+1] = s_j, b[n], Y[n+1] = y, a_n = a\right)}{\Pr\left(\mathbf{b}[n], Y[n+1] = y, a_n = a\right)}$$

$$
\begin{aligned}
&= \frac{\Pr\left(Y[n+1]=y|S[n+1]=s_j, b[n], a_n=a\right)\Pr\left(S[n+1]=s_j, b[n], a_n=a\right)}{\Pr\left(\mathbf{b}[n], Y[n+1]=y, a_n=a\right)} \\
&= \frac{\Pr\left(Y[n+1]=y|S[n+1]=s_j, a_n=a\right)\Pr\left(S[n+1]=s_j|b[n], a_n=a\right)}{\Pr\left(Y[n+1]=y|\mathbf{b}[n], a_n=a\right)} \\
&= \frac{q\left(y|s_j, a\right)\Pr\left(S[n+1]=s_j|b[n], a_n=a\right)}{\Pr\left(Y[n+1]=y|\mathbf{b}[n], a_n=a\right)}. \qquad (7.72)
\end{aligned}
$$

Recall that $r(s,a)$ denotes the immediate reward from taking action a when in state s. However, in a POMDP, we cannot evaluate the immediate reward $r(s,a)$ since we do not have access to the actual state. We may instead make decisions based on the *expected immediate reward* given that we know the belief state $\mathbf{b} = (b_1, \cdots, b_M)^T$, where expected immediate reward $r(\mathbf{b}, a)$ is defined as

$$
\begin{aligned}
r(\mathbf{b}, a) &= \mathbb{E}\{r(S,a)|\mathbf{b}\} = \sum_{j=1}^{M} b_j r(s_j, a) \\
&= \mathbf{b}^T \mathbf{r}(a) \qquad (7.73)
\end{aligned}
$$

with

$$\mathbf{r}(a) = (r_1(a), \cdots, r_M(a))^T$$

and, for $\mathcal{S} = \{s_1, \cdots, s_M\}$,

$$r_j(a) = r(s_j, a).$$

Hence, as far as optimal decision-making is concerned, we can completely forego keeping track of history $\underline{h}_n$. Instead, we may just keep track of the belief state vector $\mathbf{b}[n]$ and make our decisions based on it.

From (7.70), we see that the belief vector $\mathbf{b}[n]$ can be considered as the *continuous-valued* state of a new Markov process, where state vector $\mathbf{b}[n]$ takes values in the $(M-1)$-dimensional probability simplex $\mathcal{P}^{M-1}$ defined in (7.63). Since this state $\mathbf{b}[n]$ of the new Markov process is computable at each time instant (from the actions, observations, and the previous state $\mathbf{b}[n-1]$), it is a COMDP with the associated reward process of expected immediate rewards $r(\mathbf{b}, a)$ given in (7.73).

Hence, we have an optimal decision-making problem for a COMDP where the fully observable state is the belief state $\mathbf{b}[n]$: We have converted the original POMDP with the partially observable state $S[n]$ into a COMDP problem with the new state $\mathbf{b}[n]$. However, note that unlike the original POMDP, the new MDP is not of finite state. The state vector $\mathbf{b}[n]$ takes values in the $(M-1)$-dimensional probability simplex $\mathcal{P}^{M-1}$ which is an infinite set. This is the price we have paid for converting the original finite-state POMDP into an *equivalent COMDP*. As we will see shortly, this is indeed a significant price.

7.5.1.3 Optimal Policies and Value Functions Since we know that $\mathbf{b}[n]$ is a sufficient statistic for decision-making in a POMDP, in the following, we may denote allowed decision rules as $\pi_n(\mathbf{b}[n])$ rather than $\pi_n\left(\underline{h}_n\right)$.

Let us consider a finite-horizon POMDP starting at time $n = 0$. Then, a non-stationary policy $\underline{\pi} = (\pi_0(.), \cdots, \pi_{T-1}(.)$ for such a POMDP is a sequence of action functions such that $\pi_n(\mathbf{b}[n]) \in \mathcal{A}$ denoting the action to be executed at time n if the belief state at time $n \Leftarrow n$ is $\mathbf{b}[n]$.

Suppose that we are given the initial belief state $\mathbf{b}[0] = \mathbf{b} \in \mathcal{P}^M$ at time $n = 0$. We may define the value of a finite-horizon POMDP under a policy $\underline{\pi}$ as the expected discounted total reward under the policy $\underline{\pi}$:

$$V_T^{\underline{\pi}}(\mathbf{b}) = \mathbb{E}_{\underline{\pi}}\left\{\sum_{n=0}^{T} \gamma^n r(\mathbf{b}[n], \pi_n(\mathbf{b}[n])) \mid \mathbf{b}[0] = \mathbf{b}\right\}.$$

As before, when $T \rightarrow \infty$, we have an infinite-horizon POMDP in which case we may denote the value function as $V^{\underline{\pi}}(\mathbf{b})$, where

$$V^{\underline{\pi}}(\mathbf{b}) = \lim_{T \longrightarrow \infty} \mathbb{E}_{\underline{\pi}}\left\{\sum_{n=0}^{T} \gamma^n r(\mathbf{b}[n], \pi_n(\mathbf{b}[n])) \mid \mathbf{b}[0] = \mathbf{b}\right\}.$$

The optimal value function, given the initial state $\mathbf{b}[0] = \mathbf{b}$ at time $n = 0$, of a POMDP is defined as

$$V_T^*(\mathbf{b}) = \sup_{\underline{\pi}} V_T^{\underline{\pi}}(\mathbf{b})$$

and a policy $\underline{\pi}^*$ is called an optimal policy if

$$V_T^{\underline{\pi}^*}(\mathbf{b}) = V_T^*(\mathbf{b}) \quad \text{for all } \mathbf{b} \in \mathcal{P}^{M-1}.$$

7.5.2 Policy Evaluation for a Finite-Horizon POMDP

As in the case of completely observable COMDPs, the value function of a finite-horizon POMDP can be expressed as a backward induction (dynamic programming) by defining a value-to-go function $v_m(.)$, for $m = 0, \cdots, T$, as

$$v_m^{\underline{\pi}}(\mathbf{b}) = \mathbb{E}_{\pi}\left\{\sum_{n=T-m}^{T} \gamma^{n-(T-m)} r(\mathbf{b}[n], \pi_n(\mathbf{b}[n])) \mid \mathbf{b}[T-m] = \mathbf{b}\right\}. \tag{7.74}$$

As before, $v_m^{\underline{\pi}}(\mathbf{b})$ is the value of the policy $\underline{\pi}$ when there are only m steps to go before decision process terminates, where $m = 0, \cdots, T$. For $m = T$, clearly,

$$V_T^{\underline{\pi}}(\mathbf{b}) = v_T^{\underline{\pi}}(\mathbf{b}), \tag{7.75}$$

for any policy $\underline{\pi}$ and belief state $\mathbf{b}$.

In theory, $v_m^{\underline{\pi}}(.)$ for any given policy $\underline{\pi}$ can be computed using a backward induction similar to that in the case of a COMDP. First, note that for a finite-horizon POMDP, at time $n = T$, there is no action to be taken. Hence, the immediate reward $r_T(.)$ is only a function of the state $s[n]$ of the core process at time $n = T$. Let us thus define $r_T(s_j, a) = r_T(s_j)$. As a result, the expected immediate reward at the terminating stage $n = T$ is also only a function of the final belief state $\mathbf{b}[T]$:

$$\begin{aligned}\mathbb{E}\{r_T(S[T], a)|\mathbf{b}[T]\} &= \mathbb{E}\{r_T(S[T])|\mathbf{b}[T]\}\\ &= \sum_{j=1}^{M} r_T(s_j) b_j[T]\\ &= \mathbf{r}_T^T \mathbf{b}[T],\end{aligned} \tag{7.76}$$

where we have defined the vector of terminating rewards $\mathbf{r}_T$ as

$$\mathbf{r}_T = (r_T(s_1), r_T(s_2), \cdots, r_T(s_M))^T. \tag{7.77}$$

Hence, for $m = 0$, for any policy $\underline{\pi}$, we have

$$\begin{aligned}v_0^{\underline{\pi}}(\mathbf{b}) &= \mathbb{E}_{\pi}\{r(\mathbf{b}[T], \pi_T(\mathbf{b}[T])) \mid \mathbf{b}[T] = \mathbf{b}\}\\ &= \mathbb{E}\{r_T(S[T]) \mid \mathbf{b}[T] = \mathbf{b}\}\\ &= \mathbf{r}_T^T \mathbf{b}[T].\end{aligned} \tag{7.78}$$

For $m = 1, \cdots, T$, for any policy $\underline{\pi}$, with $n = T - m$,

$$\begin{aligned}v_m^{\underline{\pi}}(\mathbf{b}) &= \mathbb{E}_{\underline{\pi}}\left\{\sum_{i=n}^{T} \gamma^{i-n} r_i(\mathbf{b}[i], \pi_i(\mathbf{b}[i]))\,\middle|\, \mathbf{b}[n] = \mathbf{b}\right\} \quad \text{where } n = T - m\\ &= \mathbb{E}_{\underline{\pi}}\left\{r_n(\mathbf{b}, \pi_n(\mathbf{b})) + \sum_{i=n+1}^{T} \gamma^{i-n} r_i(\mathbf{b}_i, \pi_i(\mathbf{b}[i])) | \mathbf{b}[n] = \mathbf{b}\right\}\\ &= r_n(\mathbf{b}, \pi_n(\mathbf{b})) + \gamma \mathbb{E}_{\underline{\pi}}\left\{\sum_{i=n+1}^{T} \gamma^{i-(n+1)} r_i(\mathbf{b}[i], \pi_i(\mathbf{b}[i]))\,\middle|\, \mathbf{b}[n] = \mathbf{b}, \pi_n(\mathbf{b}[n])\right\}\\ &= r_n(\mathbf{b}, \pi_n(\mathbf{b})) + \gamma \sum_{y \in \mathcal{Y}} \Pr(Y_{n+1} = y | \mathbf{b}[n] = \mathbf{b}, \pi_n(\mathbf{b}))\\ &\quad \times \mathbb{E}_{\underline{\pi}}\left\{\sum_{i=n+1}^{T} \gamma^{i-(n+1)} r_i(\mathbf{b}[i], \pi_i(\mathbf{b}[i]))\,\middle|\, \mathbf{b}[n] = \mathbf{b}, Y_{n+1} = y, \pi_n(\mathbf{b})\right\},\end{aligned} \tag{7.79}$$

where, the last step follows by conditioning over the next observation $Y[n+1]$ where $n = T - m$. Note that the conditional expectation in the second term of (7.79) depends on the conditioned variables $\mathbf{b}[n] = \mathbf{b}$, $Y_{n+1} = y$ and $\pi_n(\mathbf{b})$ only through their effect on determining the initial belief state $\mathbf{b}[n+1]$ in the sum. Since, from (7.70), $\mathbf{b}[n+1] = \mathbf{T}(\mathbf{b}[n], \pi_n(\mathbf{b}[n]), Y[n+1] = y)$, we have that

$$\begin{aligned}
v_m^{\underline{\pi}}(\mathbf{b}) &= r_n(\mathbf{b}, \pi_n(\mathbf{b})) + \gamma \sum_{y \in \mathcal{Y}} \Pr\left(Y[n+1] = y | \mathbf{b}[n], \pi_n(\mathbf{b}[n])\right) \\
&\quad \times \mathbb{E}_{\underline{\pi}} \left\{ \sum_{i=n+1}^{T} \gamma^{i-(n+1)} r_i(\mathbf{b}[i], \pi_i(\mathbf{b}[i])) \middle| \mathbf{b}[n+1] = \mathbf{T}(\mathbf{b}, \pi_n(\mathbf{b}), y) \right\} \\
&= r_n(\mathbf{b}, \pi_n(\mathbf{b})) + \gamma \sum_{y \in \mathcal{Y}} v_{m-1}^{\underline{\pi}}(\mathbf{T}(\mathbf{b}, \pi_n(\mathbf{b}), y)) \Pr(Y[n+1] = y | \mathbf{b}[n] = \mathbf{b}, \pi_n(\mathbf{b})) \\
&= r_n(\mathbf{b}, \pi_n(\mathbf{b})) + \gamma \sum_{y \in \mathcal{Y}} v_{m-1}^{\underline{\pi}}(\mathbf{T}(\mathbf{b}, \pi_n(\mathbf{b}), y)) \Pr(Y[n+1] = y | \mathbf{b}[n] = \mathbf{b}, \pi_n(\mathbf{b})) \\
&= r_n(\mathbf{b}, \pi_n(\mathbf{b})) + \gamma \sum_{y \in \mathcal{Y}} v_{m-1}^{\underline{\pi}}\left(\mathbf{b}^{(\mathbf{b}, \pi_n(\mathbf{b}), y)}\right) \sigma\left(y | \mathbf{b}, \pi_n(\mathbf{b})\right), \qquad (7.80)
\end{aligned}$$

where we have used the notation (7.73) and introduced the new notation

$$\Pr(Y[n+1] = y | \mathbf{b}[n] = \mathbf{b}, a_n = a) \triangleq \sigma\left(y | \mathbf{b}, a\right).$$

However, note that

$$\begin{aligned}
\sigma\left(y | \mathbf{b}, a\right) &= \Pr(Y[n+1] = y | \mathbf{b}[n], a_n = a) \\
&= \sum_{i'} \Pr(Y[n+1] = y | S[n+1] = s_{i'}, \mathbf{b}[n], a_n = a) \\
&\quad \times \Pr(S[n+1] = s_{i'} | \mathbf{b}[n], a_n = a) \\
&= \sum_{i'} \Pr(Y[n+1] = y | S[n+1] = s_{i'}, a_n = a) \Pr(S[n+1] = s_{i'} | \mathbf{b}[n], a_n = a) \\
&= \sum_{i'} q\left(y | s_{i'}, a\right) \sum_{i} \Pr(S[n+1] = s_{i'} | S[n] = s_i, \mathbf{b}[n], a_n = a) \\
&\quad \times \Pr(S[n] = s_i | \mathbf{b}[n], a_n = a) \qquad (7.81a) \\
&= \sum_{i'} q\left(y | s_{i'}, a\right) \sum_{i} \Pr(S[n+1] = s_{i'} | S[n] = s_i, a_n = a) \Pr(S[n] = s_i | \mathbf{b}[n]) \\
&= \sum_{i'} q\left(y | s_{i'}, a\right) \sum_{i} [P(a)]_{i,i'} b_i[n], \qquad (7.81b)
\end{aligned}$$

where (7.81a) is due to the definition that $\Pr(Y[n] = y | S[n] = s, a_n = a) = q\left(y | s, a\right)$.

Hence, we may evaluate the value-to-go function $v_m^{\underline{\pi}}(.)$ for a policy $\underline{\pi}$ as a backward induction by using (7.81b) in (7.80) with the initialization (7.78). Once this is done, we may obtain the value function $V_T^{\underline{\pi}}(\mathbf{b})$ of the policy $\underline{\pi}$ by using (7.75).

7.5.3 Optimality Equations for a Finite-Horizon POMDP

It is clear that (7.75) holds regardless of the policy in question. Hence, if we denote the optimal value-to-go function associated with an optimal policy that maximizes the expected discounted reward from time $n = T - m$ to $n = T$ by $v_m^*(\mathbf{b})$, then we have

$$V_T^*(\mathbf{b}) = v_T^*(\mathbf{b}). \tag{7.82}$$

The following theorem (Theorem 2.2 of [74]) gives the optimality equations for a finite-horizon POMDP based on its equivalent COMDP representation in terms of the belief vector state $\mathbf{b}$.

Theorem 7.5 Let the rewards vector for any action $a \in \mathcal{A}$ be $\mathbf{r}(a) = (r(s_1, a), r(s_2, a), \cdots, r(s_M, a))^T$ with the terminating rewards of the form of $r(s[T], a) = r(s[T])$. If we denote the terminating rewards vector by

$$\mathbf{r}_T = (r(s_1), \cdots, r(s_M))^T,$$

then

$$v_0^*(\mathbf{b}) = \mathbf{b}^T \mathbf{r}_T \quad \text{for all} \quad \mathbf{b} \in \mathcal{P}^{M-1}$$

and, for $m = 1, \cdots, T$,

$$v_m^*(\mathbf{b}) = \max_{a \in \mathcal{A}} \left\{ \mathbf{b}^T \mathbf{r}(a) + \gamma \sum_{y \in \mathcal{Y}} v_{m-1}^* \left(\mathbf{b}^{(\mathbf{b},a,y)} \right) \sigma\left(y | \mathbf{b}, a\right) \right\}, \tag{7.83}$$

where $\mathbf{b}^{(\mathbf{b},a,y)} = \mathbf{T}(\mathbf{b}, a, y) = \left(b_1^{(\mathbf{b},a,y)}, b_2^{(\mathbf{b},a,y)}, \cdots, b_M^{(\mathbf{b},a,y)} \right)$ with, for $1 \le j \le M$,

$$\begin{aligned} b_j^{(\mathbf{b},a,y)} &= T_j(\mathbf{b}, a, y) \\ &= \Pr\left(S[n+1] = s_j \mid \mathbf{b}[n] = \mathbf{b}, a_n = a, Y[n+1] = y \right) \end{aligned}$$

and

$$\sigma\left(y | \mathbf{b}, a\right) = \Pr\left(Y[n+1] = y \mid \mathbf{b}[n] = \mathbf{b}, a_n = a\right).$$

The optimal value function $V_T^*(\mathbf{b})$ is given by

$$V_T^*(\mathbf{b}) = v_T^*(\mathbf{b}). \tag{7.84}$$

Proof: The proof follows straightforwardly from the definition of the value-to-go function. At the terminating stage, that is, at time $n = T$, there is no action to be taken and the reward is $r_T(s,a) = r_T(s)$. Hence, for any belief state $\mathbf{b}[T]$ at the terminal stage, the expected reward is $\mathbf{r}_T^T\mathbf{b}[T]$ and thus

$$v_0^*(\mathbf{b}) = \mathbf{b}^T\mathbf{r}_T \quad \text{for all } \mathbf{b} \in \mathcal{P}^{M-1}.$$

Suppose now that the belief state is $\mathbf{b}$ when there are m stages to go before the termination so that $n = T - m$. Then, the selection of action $a_n = a$ at time $n = T - m$ will result in a value-to-go of

$$\begin{aligned} v_m(\mathbf{b}) &= \mathbf{b}^T\mathbf{r}(a) + \gamma \\ &\qquad \times \text{(expected future reward with } m-1 \text{ stages to go before termination)} \\ &= \mathbf{b}^T\mathbf{r}(a) + \gamma \sum_{y\in\mathcal{Y}} v_{m-1}^{\pi}\left(\mathbf{b}^{(\mathbf{b},\pi_n(\mathbf{b}),y)}\right)\sigma\,(y|\mathbf{b},a), \end{aligned} \tag{7.85}$$

where (7.85) follows from (7.80) with $\mathbf{b}^{(\mathbf{b},a,y)} = \mathbf{T}(\mathbf{b},a,y)$.

To achieve the maximum value-to-go at this stage, then we should choose the action a_n to be that $a \in \mathcal{A}$ that will maximize the sum of the two terms on the right-hand side of (7.85). Suppose we have already determined the optimal value-to-go function $v_{m-1}^*(.)$ when there are $(m-1)$ stages to go. Then from (7.85), $v_m^*(.)$ is given by

$$v_m^*(\mathbf{b}) = \max_{a\in\mathcal{A}} \left\{ \mathbf{b}^T\mathbf{r}(a) + \gamma \sum_{y\in\mathcal{Y}} v_{m-1}^*(\mathbf{b}^{(\mathbf{b},a,y)})\sigma\,(y|\mathbf{b},a) \right\}. \tag{7.86}$$

The assertion (7.84) is already established in (7.82), completing the proof.

7.5.4 Optimal Policy Computation for a Finite-Horizon POMDP

With $n = T - m$, we may explicitly write the backward induction formula (7.83) of Theorem (7.5) as

$$v_m^*(\mathbf{b}[n]) = \max_{a\in\mathcal{A}} \left\{ \mathbf{b}[n]^T\mathbf{r}(a) + \gamma \sum_{y\in\mathcal{Y}} \Pr(y[n+1] = y \mid \mathbf{b}[n], a) v_{m-1}^*(\mathbf{T}(\mathbf{b}[n], a, y)) \right\} \tag{7.87}$$

with the initialization

$$v_0^*(\mathbf{b}) = \mathbf{b}^T\mathbf{r}_T \quad \text{for all} \quad \mathbf{b} \in \mathcal{P}^{M-1}, \tag{7.88}$$

where $\mathbf{T}(\mathbf{b}[n], a, y) = \mathbf{b}[n+1]$.

Essentially, (7.87) states that if the belief state at time n is $\mathbf{b}[n]$, then the highest value-to-go is achieved by choosing the action a that maximizes the immediate expected reward $\mathbf{b}[n]^T\mathbf{r}(a)$ plus the expected value of the optimal value-to-go function at the next time instant with the new belief vector $\mathbf{b}[n+1] = \mathbf{T}(\mathbf{b}[n], a_n, y[n+1])$ at time $n+1$ resulting from taking action a at time n and observing $y[n+1]$. Hence, the optimal value-to-go at time n is given by the maximum of, where the maximum is over the choice of the actions, immediate expected reward $\mathbf{b}[n]^T\mathbf{r}(a)$ plus the discounted value-to-go at time $n+1$ with the new belief vector $\mathbf{b}[n+1] = \mathbf{T}(\mathbf{b}[n], a_n, y[n+1])$ at time $n+1$. Thus, we can evaluate $v_m^*(.)$ if we already knew $v_{m-1}^*(.)$.

It may (falsely) appear that this backward induction (7.87) can be used to obtain the optimal value function of a POMDP as easily as (7.22) in the case of a COMDP. However, it is not. Recall that in the case of a COMDP, the backward induction algorithm (7.22) works because at time n we observe the state $\mathbf{s}[n]$ that can only take values in a finite set. Similarly, the next state $\mathbf{s}[n+1]$ can also only take the values in that same finite set. Hence, to evaluate $v_m^*(\mathbf{s}[n])$ for a given $\mathbf{s}[n]$, all we need is to compute $v_{m-1}^*(\mathbf{s}[n+1])$ when there are $m-1$ steps to go before termination for the finite set of values the sate $\mathbf{s}[n+1]$ can take.

In the case of a POMDP, on the other hand, $v_{m-1}^*(\mathbf{b}[n+1])$ is a function of the belief vector $\mathbf{b}[n+1]$ that takes values on the $(M-1)$-dimensional probability simplex. Hence, we need to compute $v_{m-1}^*(\mathbf{b}[n+1])$ over an argument that takes values on a continuum since $\mathbf{b}[n+1] = \mathbf{b}^{(\mathbf{b},a,y)}$ can be any value in $\mathcal{P}^{M-1}$ for a given $\mathbf{b}[n]$. For example, even in the case of a 2-state Markov process, $\mathbf{b}$ takes values on the $[0, 1]$ line segment (which is the 1-dimensional probability simplex). Hence, the backward induction (7.87) cannot possibly be evaluated since it is not possible to compute $v_{m-1}^*(\mathbf{b}[n+1])$ for all $\mathbf{b}[n+1] \in \mathcal{P}^{M-1}$.

That is unless there is some structure in the optimal value function that might facilitate characterizing it for all $\mathbf{b}[n+1] \in \mathcal{P}^{M-1}$ in a concise way. For example, suppose that for any $m = 0, \cdots, T$, with $n = T-m$, although $\mathbf{b}[n]$ can take a continuum of values in a probability simplex, we can partition this simplex in to a finite collection of subsets $\{\mathcal{B}_m\}_i$ so that

$$v_m^*(\mathbf{b}[n]) = \mathbf{b}[n]^T\boldsymbol{\beta}^{(m,i)} \quad \text{if } \mathbf{b}[n] \in \mathcal{B}_{m,i}, \tag{7.89}$$

where $\left\{\boldsymbol{\beta}^{(m,i)}\right\}_i$ is a finite collection of vectors for each m and $\bigcup_i \mathcal{B}_{m,i} = \mathcal{P}^{M-1}$. Essentially, (7.89) states that $v_m^*(\mathbf{b}[n])$ is piecewise linear on each of the $\mathcal{B}_{m,i}$'s with possibly different slopes $\boldsymbol{\beta}^{(m,i)}$'s. If (7.89) is true, then we can indeed characterize the optimal value-to-go function for each m by simply giving the finite collections of $\left\{\left(\mathcal{B}_{m,i}, \boldsymbol{\beta}^{(m,i)}\right)\right\}_i$ for each $m = 0, \cdots, T$.

The following important result regarding the POMDPs due to [71], in fact, establishes that the optimal value function of any POMDP can indeed be represented as in (7.89).

Theorem 7.6 (Piecewise Linearity and Convexity of the Optimal Value Function of a POMDP) The optimal value-to-go function, and consequently the optimal value function, of a POMDP is piecewise linear and convex (PWLC) as a function of belief state $\mathbf{b} \in \mathcal{P}^{M-1}$.

Proof: To see (7.89) inductively, first consider $m = 0$. From (7.88),

$$v_0^*(\mathbf{b}) = \mathbf{r}_T^T\mathbf{b} \quad \text{for any } \mathbf{b} \in \mathcal{P}^{M-1}. \tag{7.90}$$

Hence, clearly, we may choose $\mathcal{B}_{0,1} = \mathcal{P}^{M-1}$ so that $\{\mathcal{B}_0\}_i = \{\mathcal{B}_{0,1}\}$ and $\boldsymbol{\beta}^{(0,1)} = \mathbf{r}$.

Now suppose that when there are $m-1$ stages to go before termination, we have a collection $\left\{\left(\mathcal{B}_{m-1,i}, \boldsymbol{\beta}^{(m,i)}\right)\right\}_i$ so that $\bigcup_i \mathcal{B}_{m-1,i} = \mathcal{P}^{M-1}$ and

$$v_{m-1}^*(\mathbf{b}[n+1]) = \mathbf{b}[n+1]^T\boldsymbol{\beta}^{(m-1,i)} \quad \text{if } \mathbf{b}[n+1] \in \mathcal{B}_{m-1,i}, \tag{7.91}$$

where $n = T - m$.

Then, for any $\mathbf{b}$, we may express the value-to-go function $v_m^*(\mathbf{b})$ (7.87) when there are m steps to go before termination as

$$v_m^*(\mathbf{b}) = \max_{a\in\mathcal{A}}\left\{\mathbf{r}^T(a)\mathbf{b} + \gamma\sum_{y\in\mathcal{Y}}\Pr(Y[n+1] = y \mid \mathbf{b}[n] = \mathbf{b}, a_n = a)v_{m-1}^*(\mathbf{T}(\mathbf{b}, a, y))\right\}, \tag{7.92}$$

where $\mathbf{b}[n+1] = \mathbf{T}(\mathbf{b}, a, y) = \mathbf{b}^{(\mathbf{b},a,y)}$. But from (7.91), then there is an index $i = i_{(\mathbf{b},y,a)}$ such that

$$\begin{aligned}
v_{m-1}^*(\mathbf{T}(\mathbf{b}, y, a)) &= \left(\boldsymbol{\beta}^{(m-1,i_{(\mathbf{b},y,a)})}\right)^T\mathbf{b}^{(\mathbf{b},y,a)} \quad \text{where } \mathbf{b}^{(\mathbf{b},y,a)} \in \mathcal{B}_{m-1,i_{(\mathbf{b},y,a)}}\\
&= \sum_{j=1}^{M}\beta_j^{(m-1,i_{\mathbf{b},y,a})}b_j^{(\mathbf{b},y,a)} \quad \text{where } \mathbf{b}^{(\mathbf{b},y,a)} \in \mathcal{B}_{m-1,i_{\mathbf{b},y,a}}.
\end{aligned} \tag{7.93}$$

However, recall from (7.72) that

$$\begin{aligned}
b_j^{(\mathbf{b},y,a)} &= T_j(\mathbf{b}, y, a)\\
&= \frac{q(y|s_j, a)\sum_i[P(a)]_{i,j}b_i}{\Pr\left(Y[n+1] = y|\mathbf{b}[n] = \mathbf{b}, a_n = a\right)}.
\end{aligned} \tag{7.94}$$

Substituting (7.94) and (7.93) in (7.92), we have that

$$\begin{aligned}
v_m^*(\mathbf{b}) &= \max_{a\in\mathcal{A}}\left\{\mathbf{b}^T\mathbf{r}(a) + \gamma\sum_{y\in\mathcal{Y}}\sum_{j=1}^{M}\beta_j^{(m-1,i_{(\mathbf{b},y,a)})}q(y|s_j, a)\sum_i[P(a)]_{i,j}b_i\right\}\\
&= \max_{a\in\mathcal{A}}\left\{\mathbf{b}^T\mathbf{r}(a) + \gamma\sum_{y\in\mathcal{Y}}\sum_{j=1}^{M}\beta_j^{(m-1,i_{(\mathbf{b},y,a)})}q(y|s_j, a)\left([P(a)]_{:,j}\right)^T\mathbf{b}\right\}\\
&= \max_{a\in\mathcal{A}}\left\{\mathbf{b}^T\mathbf{r}(a) + \gamma\sum_{y\in\mathcal{Y}}\sum_{j=1}^{M}\mathbf{b}^T[P(a)]_{:,j}\,q(y|s_j, a)\beta_j^{(m-1,i_{(\mathbf{b},y,a)})}\right\},
\end{aligned} \tag{7.95}$$

where we have denoted the jth column of the matrix $P(a)$ by $[P(a)]_{:,j}$. Let us define a set of $M \times M$ diagonal matrices $\left\{Q^{(y)}(a)\right\}_{y\in\mathcal{Y}}$, one for each $y \in \mathcal{Y}$, such that the jth diagonal entry of the matrix $Q^{(y)}(a)$ is

$$\left[Q^{(y)}(a)\right]_{j,j} = q(y|s_j, a).$$

With this, (7.95) can be written as

$$\begin{aligned}
v_m^*(\mathbf{b}) &= \max_{a\in\mathcal{A}} \left\{ \mathbf{b}[n]^T \mathbf{r}(a) + \gamma \sum_{y\in\mathcal{Y}} \mathbf{b}^T \sum_j [P(a)]_{:,j} \left[Q^{(y)}(a)\right]_{j,j} \beta_j^{(m-1, i_{(\mathbf{b},y,a)})} \right\} \\
&= \max_{a\in\mathcal{A}} \left\{ \mathbf{b}^T \mathbf{r}(a) + \gamma \sum_{y\in\mathcal{Y}} \mathbf{b}^T P(a) Q^{(y)}(a) \boldsymbol{\beta}^{(m-1, i_{(\mathbf{b},y,a)})} \right\} \\
&= \max_{a\in\mathcal{A}} \left\{ \mathbf{b}^T \left(\mathbf{r}(a) + \gamma \sum_{y\in\mathcal{Y}} P(a) Q^{(y)}(a) \boldsymbol{\beta}^{(m-1, i_{(\mathbf{b},y,a)})} \right) \right\} \\
&= \max_{a\in\mathcal{A}} \left\{ \mathbf{b}[n]^T \boldsymbol{\beta}^{(m,\mathbf{b},a)} \right\} \\
&= \mathbf{b}[n]^T \boldsymbol{\beta}^{(m,\mathbf{b},a^*(\mathbf{b}))},
\end{aligned} \tag{7.96}$$

where we have let

$$\boldsymbol{\beta}^{(m,\mathbf{b},a)} \triangleq \mathbf{r}(a) + \gamma \sum_{y\in\mathcal{Y}} P(a) Q^{(y)}(a) \boldsymbol{\beta}^{(m-1, i_{(\mathbf{b},y,a)})}, \tag{7.97}$$

$$\boldsymbol{\beta}^{(m,\mathbf{b},a^*(\mathbf{b}))} \triangleq \mathbf{r}\left(a^*(\mathbf{b})\right) + \gamma \sum_{y\in\mathcal{Y}} P\left(a^*(\mathbf{b})\right) Q^{(y)}\left(a^*(\mathbf{b})\right) \boldsymbol{\beta}^{(m-1, i_{(\mathbf{b},y,a^*(\mathbf{b}))})} \tag{7.98}$$

and

$$a^*(\mathbf{b}) \triangleq \arg\max_{a\in\mathcal{A}} \mathbf{b}^T \left(\mathbf{r}(a) + \gamma \sum_{y\in\mathcal{Y}} P(a) Q^{(y)}(a) \boldsymbol{\beta}^{(m-1, i_{(\mathbf{b},y,a)})} \right) \tag{7.99}$$

is the optimal action to be taken if the belief state is $\mathbf{b}$ when there are m stages to go before termination. The β-vector that characterizes value of the optimal value function resulting from this action choice is $\boldsymbol{\beta}^{(m,\mathbf{b},a^*(\mathbf{b}))}$.

However, observe from (7.98) that the number of possible distinct $\boldsymbol{\beta}^{(m,\mathbf{b},a^*(\mathbf{b}))}$ vectors is at most

$$|\mathcal{A}| \times (\text{number of } \boldsymbol{\beta}^{(m-1,i)} \text{ vectors})^L, \tag{7.100}$$

where $L = |\mathcal{Y}|$. Hence, if the set of $\left\{\boldsymbol{\beta}^{(m-1,i)}\right\}_i$ vectors is finite at stage $m-1$, then so is the set of $\left\{\boldsymbol{\beta}^{(m,i')}\right\}_{i'}$ vectors at stage m.

Hence, (7.96) shows that if there is a finite collection of sets and vectors $\left\{\left(\mathcal{B}_{m-1,i}, \boldsymbol{\beta}^{(m-1,i)}\right)\right\}_i$ such that for all $\mathbf{b} \in \mathcal{B}_{m-1,i}$ we can write $v^*_{m-1}(\mathbf{b}) = \mathbf{b}^T \boldsymbol{\beta}^{(m-1,i)}$, then we can always find a finite collection of sets and vectors $\left\{\left(\mathcal{B}_{m,i'}, \boldsymbol{\beta}^{(m,i')}\right)\right\}_{i'}$ such that for all $\mathbf{b} \in \mathcal{B}_{m,i'}$ we can write $v^*_m(\mathbf{b}) = \mathbf{b}^T \boldsymbol{\beta}^{(m,i')}$. Since we have already shown that $v^*_0(\mathbf{b}) = \mathbf{b}^T \mathbf{r}_T$, it follows from mathematical induction that this is true for all $m = 0, \cdots, T$. Hence, for each m, $v^*_m(\mathbf{b})$ is piecewise linear.

We omit the proof of convexity here, which can be found in [71].

Note that at each m, for $m = 0, \cdots, T$, we have a partition $\left\{\mathcal{B}_{m,i}\right\}_i$ of the probability simplex $\mathcal{P}^{M-1}$. Associated with each set $\mathcal{B}_{m,i}$ of belief vectors is a β-vector $\boldsymbol{\beta}^{(m,i)}$ such that for all $\mathbf{b} \in \mathcal{B}_{m,i}$ we can write

$$v^*_m(\mathbf{b}) = \mathbf{b}^T \boldsymbol{\beta}^{(m,i)}. \tag{7.101}$$

Then, instead of having to compute $v^*_m(\mathbf{b})$ for a continuum of values $\mathbf{b}$, we may simply find the finite collection $\left\{\mathcal{B}_{m,i}\right\}_i$ of subsets of $\mathcal{P}^{M-1}$ and the corresponding vectors $\boldsymbol{\beta}^{(m,i)}$ associated with each of these sets for each step m.

Let us denote by Λ_m the finite set of $\boldsymbol{\beta}^{(m,i)}$ vectors associated with the partition $\left\{\mathcal{B}_{m,i}\right\}_i$ at step m so that what we need to compute at each step m is $\left(\left\{\mathcal{B}_{m,i}\right\}_i, \Lambda_m\right)$.

7.5.4.1 Exhaustive Search From Theorem 7.6, we see that due to PWLC property of the optimal value function, for any $\mathbf{b}$, we may write

$$v^*_m(\mathbf{b}) = \mathbf{b}^T \boldsymbol{\beta}, \tag{7.102}$$

for some finite set of β-vectors. Since

$$v^*_m(\mathbf{b}) = \max_{a \in \mathcal{A}} \left[\mathbf{b}^T \mathbf{r}(a) + \gamma \sum_y \sigma(y|\mathbf{b}, a) v^*_{m-1}(\mathbf{T}(\mathbf{b}, y, a)) \right], \tag{7.103}$$

if we knew a set Λ_{m-1} of β-vectors that characterizes the optimal $m-1$ step-to-go value function $v^*_{m-1}(.)$, then we can easily obtain a set Λ_m of β-vectors for characterizing $v^*_m(.)$ when there are m-teps to go. Indeed, following (7.98), a straightforward way to generate such a set is [74]

$$\Lambda_m = \left\{ \mathbf{r}(a) + \gamma \sum_{y \in \mathcal{Y}} P(a) Q^{(y)}(a) \boldsymbol{\beta}_y \ : a \in \mathcal{A},\ \boldsymbol{\beta}_y \in \Lambda_{m-1} \right\}. \tag{7.104}$$

An exhaustive search algorithm can be designed to essentially implement the above idea. From (7.90), we may start at $m = 0$ with the initialization

$$\Lambda_0 = \{\mathbf{r}_T\} \tag{7.105}$$

and then recursively compute the sets Λ_m of β-vectors for each m, for $m = 1, \cdots, T$, from (7.104). Since Λ_m is a finite set for each m, in theory, these sets can be pre-computed and stored. Then, we can express (7.96) as

$$v_m^*(\mathbf{b}) = \max_{\boldsymbol{\beta} \in \Lambda_m} \mathbf{b}^T \boldsymbol{\beta},$$

which defines the optimal action choice a_n^* for any $\mathbf{b}$ and for each time step $n = T - m$ through (7.98) and (7.99).

Although in theory this exhaustive search scheme gives the optimal value function and the optimal finite-horizon policy for a POMDP, clearly, the computational complexity can be prohibitive. Indeed, although finite, the size of the set Λ_m grows exponentially with the time horizon T and the number of possible observations $L = |\mathcal{Y}|$ as can be seen from (7.100) and (7.104). Hence, in practice, this computational complexity may render most problems intractable unless the time horizon T is very small.

7.5.4.2 *Modified Exhaustive Search with Pruning*

Observe that definition of Λ_m in (7.104) is the worst case in the sense that it includes all possible $\boldsymbol{\beta}^{(m,i')}$ vectors generated by combinations of $\boldsymbol{\beta}^{(m-1,i)}$ vectors in Λ_{m-1} and $a \in \mathcal{A}$. In the worst case, of course, it is possible that there may be some $\mathbf{b}[n]$ for which each one of these $\boldsymbol{\beta}^{(m,i')}$ vectors may characterize the optimal value-to-go function $v_m^*(.)$ at step m. However, in many situations, most of these vectors may be redundant: Some of these $\boldsymbol{\beta}$-vectors in Λ_m defined as in (7.104) may not characterize the optimal value-to-go function $v_m^*(\mathbf{b})$ for any $\mathbf{b}$. Then there is no need to compute these redundant vectors as in (7.104). Moreover, keeping them in Λ_m when we compute Λ_{m+1} will also make the computational load at $m \Leftarrow m+1$ to unnecessarily grow. Indeed, it is keeping those redundant vectors that makes the set Λ_m to grow with time step m. We may reduce the computational burden by keeping at each step m the only β-vectors that characterize the optimal policy $v_m^*(\mathbf{b})$ for some belief state $\mathbf{b}$.

Let us call that a vector $\boldsymbol{\beta}$ is *dominated by* $\boldsymbol{\beta}'$, if, for all $\mathbf{b}$, $\mathbf{b}^T\boldsymbol{\beta} \le \mathbf{b}^T\boldsymbol{\beta}'$. Then, for some arbitrary collection of β-vectors Λ, we may define the *dominant region* of a $\boldsymbol{\beta} \in \Lambda$ as

$$R(\boldsymbol{\beta}, \Lambda) = \left\{\mathbf{b} :\ \mathbf{b}^T\boldsymbol{\beta} > \mathbf{b}^T\boldsymbol{\beta}', \mathbf{b} \in \mathcal{P}^{M-1}, \boldsymbol{\beta}' \in \Lambda - \{\boldsymbol{\beta}\}\right\}. \tag{7.106}$$

A vector that is dominated by other vectors for all $\mathbf{b} \in \mathcal{P}^{M-1}$ will have an empty dominant region. Clearly, such a vector $\boldsymbol{\beta}$ does not define an optimal policy for any belief state $\mathbf{b}$. Hence, we may improve the efficiency of exhaustive search by removing such dominated vectors from Λ_m at each step m. We define the *parsimonious set* Λ_m^* to be the minimal set of $\boldsymbol{\beta}$-vectors at each step m.

Definition 7.5.1 A set Λ_m of β-vectors is called a parsimonious representation of the PWLC optimal value function $v_m^*(.)$ if $R(\boldsymbol{\beta}, \Lambda_m)$ is nonempty for all $\boldsymbol{\beta} \in \Lambda_m$ and the collection $\{R(\boldsymbol{\beta}, \Lambda_m)\}_{\boldsymbol{\beta}}$ is a partition of $\mathcal{P}^{M-1}$.

Now suppose that we have a *parsimonious representation* Λ_{m-1}^* of the optimal value function v_{m-1}^*. The exhaustive search approach of (7.104) can first be used to generate the set of all possible β-vectors that will give a representation of v_m^*:

$$\Lambda_m = \left\{ \mathbf{r}(a) + \gamma \sum_y P(a) Q^y(a) \boldsymbol{\beta}_y \;\; a \in \mathcal{A}, \boldsymbol{\beta}_y \in \Lambda_{m-1}^* \right\}. \tag{7.107}$$

Of course, this Λ_m may not be a parsimonious set for v_m^*. Hence, all efficient algorithms for solving POMDPs essentially focus on how to efficiently compute a parsimonious set Λ_m^* from a given parsimonious set Λ_{m-1}^*. Some of the algorithms found in the literature, however, fail to construct an exact parsimonious set.

One such approach is to first generate Λ_m from a parsimonious Λ_{m-1}^* and then prune it by removing those $\boldsymbol{\beta}$-vectors in Λ_m that are dominated by others to obtain a parsimonious Λ_m^* [74]. This approach, called the *modified exhaustive search with pruning* [74], first generate the set Λ_m as in (7.107) for a given set Λ_{m-1}^*. Then, it prunes the dominated vectors from this set by checking whether for all belief states the value-to-go with a particular $\boldsymbol{\beta}_1 \in \Lambda_m$ is lower than that with $\boldsymbol{\beta}_2$ for some $\boldsymbol{\beta}_2 \in \Lambda_m$. If so, then the algorithm removes $\boldsymbol{\beta}_1$ from Λ_m. This test to find dominated vectors is implemented as the following linear program:

For each $\boldsymbol{\beta} \in \Lambda_m$, solve [74]

$$\begin{aligned} \min_{\mathbf{b} \in \mathcal{P}^{M-1}} \quad & x - \mathbf{b}^T \boldsymbol{\beta} \\ & \text{s. t. } \mathbf{b}^T \boldsymbol{\beta}' \le x \text{ for all } \boldsymbol{\beta}' \in \Lambda_m - \{\boldsymbol{\beta}\} \\ & \sum_{j=1}^{M} b_j = 1 \\ & b_j \ge 0 \text{ for } j = 1, \cdots, M. \end{aligned}$$

If $x \neq 0$, then remove $\boldsymbol{\beta}$ from Λ_m.

The modified exhaustive search with pruning algorithm can reduce the computational effort by reducing the size of the set of $\boldsymbol{\beta}$-vectors in each step. Essentially, rather than increasing the size of the set by a factor $|\mathcal{A}|$, which becomes the base of the exponential term, pruning may decrease it to something smaller. However, in the worst case, the modified algorithm is still exponential in the time horizon and thus can be intractable. Moreover, observe that at each m, the algorithm needs to compute the full set Λ_m of β-vectors given in (7.107) before pruning it to obtain the parsimonious representation Λ_m^*.

7.5.4.3 Sondik's One-Pass Algorithm As mentioned earlier, the modified exhaustive search with pruning approach can be seriously wasteful: It first computes the redundant set Λ_m in (7.107) that may contain many dominated vectors that will be removed by the dominance test in the second step of the algorithm. An alternative approach for efficiently computing optimal POMDP solutions thus is to avoid computing the set Λ_m all together and try to directly obtain the parsimonious set Λ_m^* from Λ_{m-1}^*. However, so far, there is no algorithm that in general achieves this exactly. Many of the algorithms that follow this alternative approach construct an intermediate set that does not explicitly seem to be constructing Λ_m. Still, usually, this intermediate set can be bigger than the desired Λ_m^*. Hence, again, a pruning step is still included to remove the dominated vectors from the intermediate set so as to construct Λ_m^*. Sondik's *one-pass algorithm* is one such algorithm [70, 71, 75].

We will need the following new notation:

$$\Lambda_m^a = \left\{ \mathbf{r}(a) + \gamma \sum_{y \in \mathcal{Y}} P(a) Q^{(y)}(a) \boldsymbol{\beta}_y \ : \ \boldsymbol{\beta}_y \in \Lambda_{m-1} \right\}. \tag{7.108}$$

Comparing with (7.98)) and (7.104), it is clear that Λ_m^a is the set of β-vectors at step m generated from a given set Λ_{m-1} if we restrict the action to be taken at step m to be $a \in \mathcal{A}$.

Suppose we have the parsimonious set Λ_{m-1}^* of β-vectors that characterizes the optimal $m-1$ step value-to-go function $v_{m-1}^*(\mathbf{b})$ for all $\mathbf{b} \in \mathcal{P}^{M-1}$. For step m, the one-pass algorithm attempts to construct a parsimonious Λ_m^* from Λ_{m-1}^* without having to construct the set Λ_m of (7.107), by following these steps [70]:

1. Start with an arbitrary $\mathbf{b} \in \mathcal{P}^{M-1}$ and find the optimal β-vector denoted by $\boldsymbol{\beta}^{(m,\mathbf{b},a^*(\mathbf{b}))}$ corresponding to the optimal action $a^*(\mathbf{b})$: This can be achieved by using (7.99) and (7.98) that give $a^*(\mathbf{b})$ and $\boldsymbol{\beta}^{(m,\mathbf{b},a^*(\mathbf{b}))}$, respectively, in terms of Λ_{m-1}^*. The corresponding optimal $v_m^*(\mathbf{b})$ is then computed as in (7.96).
2. Find the set of all belief vectors for which $\boldsymbol{\beta}^{(m,\mathbf{b},a^*(\mathbf{b}))}$ vector is the optimal $\boldsymbol{\beta}$-vector: Although we computed $\boldsymbol{\beta}^{(m,\mathbf{b},a^*(\mathbf{b}))}$ as the optimal β-vector for belief state vector $\mathbf{b}$, it is possible that $\boldsymbol{\beta}^{(m,\mathbf{b},a^*(\mathbf{b}))}$ can be the optimal β-vector for other belief state vectors as well. Indeed, the set of belief vectors for which $\boldsymbol{\beta}^{(m,\mathbf{b},a^*(\mathbf{b}))}$ is the optimal $\boldsymbol{\beta}$-vector is the dominant region $R\left(\boldsymbol{\beta}^{(m,\mathbf{b},a^*(\mathbf{b}))}, \Lambda_m^*\right)$ of the vector $\boldsymbol{\beta}^{(m,\mathbf{b},a^*(\mathbf{b}))}$. Computing this set directly, however, is difficult. Hence, instead, we compute the set of all belief vectors $\mathbf{b}$ for which $\boldsymbol{\beta}^{(m,\mathbf{b},a^*(\mathbf{b}))}$ is the optimal $\boldsymbol{\beta}$-vector if we constrained the action to be taken also to be $a^*(\mathbf{b})$. Let us denote this set by $R\left(\boldsymbol{\beta}^{(m,\mathbf{b},a^*(\mathbf{b}))}, \Lambda_m^{a^*(\mathbf{b})}\right)$ where $\Lambda_m^{a^*(\mathbf{b})}$ is as defined in (7.108).
3. Remove the extra belief vectors in the set $R\left(\boldsymbol{\beta}^{(m,\mathbf{b},a^*(\mathbf{b}))}, \Lambda_m^{a^*(\mathbf{b})}\right)$ till it reduces to the dominant region $R\left(\boldsymbol{\beta}^{(m,\mathbf{b},a^*(\mathbf{b}))}, \Lambda_m^*\right)$ of the vector $\boldsymbol{\beta}^{(m,\mathbf{b},a^*(\mathbf{b}))}$: It is clear that $R\left(\boldsymbol{\beta}^{(m,\mathbf{b},a^*(\mathbf{b}))}, \Lambda_m^*\right) \subseteq R\left(\boldsymbol{\beta}^{(m,\mathbf{b},a^*(\mathbf{b}))}, \Lambda_m^{a^*(\mathbf{b})}\right)$. The goal can be achieved

by imposing constraints on the belief vectors that should be considered in constructing the set $R\left(\boldsymbol{\beta}^{(m,\mathbf{b},a^*(\mathbf{b}))}, \Lambda_m^{a^*(\mathbf{b})}\right)$.

4. Repeat the above steps till the union of the constructed dominant regions forms a partition of $\mathcal{P}^{M-1}$.

Finding $a^*(\mathbf{b})$ and $\boldsymbol{\beta}^{(m,\mathbf{b},a^*(\mathbf{b}))}$ needed in step 1 is already clear: We may use (7.99) and (7.98), respectively. The set $R\left(\boldsymbol{\beta}^{(m,\mathbf{b},a^*(\mathbf{b}))}, \Lambda_m^{a^*(\mathbf{b})}\right)$ of belief vectors $\mathbf{b}$ for which $\boldsymbol{\beta}^{(m,\mathbf{b},a^*(\mathbf{b}))}$ is the optimal $\boldsymbol{\beta}$-vector if we constrained the action to be taken also to be $a^*(\mathbf{b})$ can also be computed easily after we have computed $\Lambda_m^{a^*(\mathbf{b})}$ using (7.108).

However, $R\left(\boldsymbol{\beta}^{(m,\mathbf{b},a^*(\mathbf{b}))}, \Lambda_m^{a^*(\mathbf{b})}\right)$ is not the same as the dominant region $R\left(\boldsymbol{\beta}^{(m,\mathbf{b},a^*(\mathbf{b}))}, \Lambda_m^*\right)$ of $\boldsymbol{\beta}^{(m,\mathbf{b},a^*(\mathbf{b}))}$ unless $\boldsymbol{\beta}^{(m,\mathbf{b},a^*(\mathbf{b}))}$ is the optimal $\boldsymbol{\beta}$-vector for all $\mathbf{b}' \in R\left(\boldsymbol{\beta}^{(m,\mathbf{b},a^*(\mathbf{b}))}, \Lambda_m^{a^*(\mathbf{b})}\right)$. It is certainly possible that this may not hold for some of the belief vectors $\mathbf{b}' \in R\left(\boldsymbol{\beta}^{(m,\mathbf{b},a^*(\mathbf{b}))}, \Lambda_m^{a^*(\mathbf{b})}\right)$ for several reasons. For example, there could be $\mathbf{b}' \in R\left(\boldsymbol{\beta}^{(m,\mathbf{b},a^*(\mathbf{b}))}, \Lambda_m^{a^*(\mathbf{b})}\right)$ such that the optimal action at $\mathbf{b}'$ is not $a^*(\mathbf{b})$ but a different action $a^*(\mathbf{b}') \neq a^*(\mathbf{b})$. Then the optimal $\boldsymbol{\beta}$-vector at $\mathbf{b}'$ is $\boldsymbol{\beta}^{(m,\mathbf{b}',a^*(\mathbf{b}'))}$, which can certainly be different from $\boldsymbol{\beta}^{(m,\mathbf{b},a^*(\mathbf{b}))}$. Note that it is possible that even with two different optimal actions $a^*(\mathbf{b}) \neq a^*(\mathbf{b}')$, we may have $\boldsymbol{\beta}^{(m,\mathbf{b}',a^*(\mathbf{b}'))} = \boldsymbol{\beta}^{(m,\mathbf{b},a^*(\mathbf{b}))}$. However, in general, this may or may not happen, and thus $\mathbf{b}'$ may not belong in the dominant region $R\left(\boldsymbol{\beta}^{(m,\mathbf{b},a^*(\mathbf{b}))}, \Lambda_m^*\right)$ of $\boldsymbol{\beta}^{(m,\mathbf{b},a^*(\mathbf{b}))}$. This shows that, in general,

$$R\left(\boldsymbol{\beta}^{(m,\mathbf{b},a^*(\mathbf{b}))}, \Lambda_m^*\right) \subseteq R\left(\boldsymbol{\beta}^{(m,\mathbf{b},a^*(\mathbf{b}))}, \Lambda_m^{a^*(\mathbf{b})}\right). \tag{7.109}$$

To remove those additional belief vectors in the set $R\left(\boldsymbol{\beta}^{(m,\mathbf{b},a^*(\mathbf{b}))}, \Lambda_m^{a^*(\mathbf{b})}\right)$ to reduce it to $R\left(\boldsymbol{\beta}^{(m,\mathbf{b},a^*(\mathbf{b}))}, \Lambda_m^*\right)$, the one-pass algorithm imposes constraints on the belief vectors that should be considered in constructing the set $R\left(\boldsymbol{\beta}^{(m,\mathbf{b},a^*(\mathbf{b}))}, \Lambda_m^{a^*(\mathbf{b})}\right)$.

The first constraint is to only consider belief vectors $\mathbf{b}'$ in the region in which $\boldsymbol{\beta}^{(m,\mathbf{b},a^*(\mathbf{b}))}$ dominates all $\boldsymbol{\beta}^{(m,\mathbf{b},a)}$ vectors for $a \in \mathcal{A}$:

$$R(\boldsymbol{\beta}^{(m,\mathbf{b},a^*(\mathbf{b}))}, \left\{\boldsymbol{\beta}^{(m,\mathbf{b},a)} : a \in \mathcal{A} - \{a^*(\mathbf{b})\}\right\}). \tag{7.110}$$

This, of course, makes sense because if $\boldsymbol{\beta}^{(m,\mathbf{b},a^*(\mathbf{b}))}$ is not even better than $\boldsymbol{\beta}^{(m,\mathbf{b},a)}$ vectors, for $a \neq a^*(\mathbf{b})$, at a particular belief state, then clearly this belief state does not belong in $R\left(\boldsymbol{\beta}^{(m,\mathbf{b},a^*(\mathbf{b}))}, \Lambda_m^*\right)$. This is shown in Fig. 7.2, assuming that $M = 2$, where it can be seen that over a portion of the dominant region $R\left(\boldsymbol{\beta}^{(m,\mathbf{b},a^*(\mathbf{b}))}, \Lambda_m^{a^*(\mathbf{b})}\right)$ a vector $\boldsymbol{\beta}^{(m,\mathbf{b},a)}$ with $a \neq a^*(\mathbf{b})$ is better than $\boldsymbol{\beta}^{(m,\mathbf{b},a^*(\mathbf{b}))}$: That is, the portion to the right of the intersection point of the two lines corresponding to $\boldsymbol{\beta}^{(m,\mathbf{b},a)}$ and $\boldsymbol{\beta}^{(m,\mathbf{b},a^*(\mathbf{b}))}$.

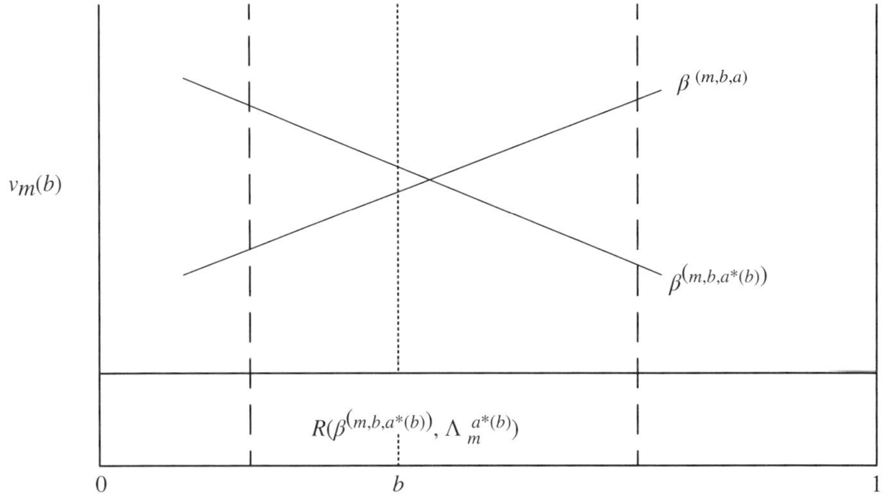

FIGURE 7.2 A scenario in which $\boldsymbol{\beta}^{(m,\mathbf{b},a^*(\mathbf{b}))}$ cannot be the optimal β-vector over a portion of the set $R\left(\boldsymbol{\beta}^{(m,\mathbf{b},a^*(\mathbf{b}))}, \Lambda_m^{a^*(\mathbf{b})}\right)$ of $\mathbf{b}$ vectors. This is evidenced by the fact that in the region to the right of the intersection point of the two lines corresponding to vectors $\boldsymbol{\beta}^{(m,\mathbf{b},a)}$ and $\boldsymbol{\beta}^{(m,\mathbf{b},a^*(\mathbf{b}))}$, the β-vector $\boldsymbol{\beta}^{(m,\mathbf{b},a)}$ is already better than $\boldsymbol{\beta}^{(m,\mathbf{b},a^*(\mathbf{b}))}$.

Constraint (7.110), however, is still not sufficient.[3] There could be some belief vectors $\mathbf{b}' \in R(\boldsymbol{\beta}^{(m,\mathbf{b},a^*(\mathbf{b}))}, \left\{\boldsymbol{\beta}^{(m,\mathbf{b},a)} : a \in \mathcal{A} - \{a^*(\mathbf{b})\}\right\}$ for which the optimal action $a^*(\mathbf{b}')$ at the belief $\mathbf{b}'$ is different from $a^*(\mathbf{b})$. Hence, the optimal $\boldsymbol{\beta}$-vectors could also be different at these two belief states: $\boldsymbol{\beta}^{(m,\mathbf{b}',a^*(\mathbf{b}'))} \neq \boldsymbol{\beta}^{(m,\mathbf{b},a^*(\mathbf{b}))}$. Then, clearly, it is possible that

$$(\mathbf{b}')^T \boldsymbol{\beta}^{(m,\mathbf{b},a^*(\mathbf{b}))} \geq (\mathbf{b}')^T \boldsymbol{\beta}^{(m,\mathbf{b},a)} \quad \text{for all} \quad a \in \mathcal{A} - a^*(\mathbf{b}) \tag{7.111}$$

while

$$v_m^*(\mathbf{b}') = (\mathbf{b}')^T \boldsymbol{\beta}^{(m,\mathbf{b}',a^*(\mathbf{b}'))}, \tag{7.112}$$

where $\boldsymbol{\beta}^{(m,\mathbf{b}',a^*(\mathbf{b}'))} \neq \boldsymbol{\beta}^{(m,\mathbf{b},a)}$ for any $a \in \mathcal{A}$. An example of such a scenario is shown in Fig. 7.3 in the case of $M = 2$. First, at $\mathbf{b}'$, the value achieved with the vector $\boldsymbol{\beta}^{(m,\mathbf{b},a^*(\mathbf{b}))}$ is better than that with $\boldsymbol{\beta}^{(m,\mathbf{b},a)}$ for all $a \in \mathcal{A}$ (Fig. 7.3 shows one representative line). Moreover, the value achieved with the vector $\boldsymbol{\beta}^{(m,\mathbf{b},a^*(\mathbf{b}))}$ is also better than that with the vector $\boldsymbol{\beta}^{(m,\mathbf{b}',a^*(\mathbf{b}))}$. That is if we constrained the action to be chosen at $\mathbf{b}'$ at step m to be $a^*(\mathbf{b})$ and used the β-vector that characterizes the value function at step m if the

[3]However, as noted in [70], constraint (7.110) is the only constraint suggested in the one-pass algorithm description given in [71]. The second set of constraints were suggested later by others once it was realized that this is not enough to ensure that the resulting region is indeed the optimal region [70].

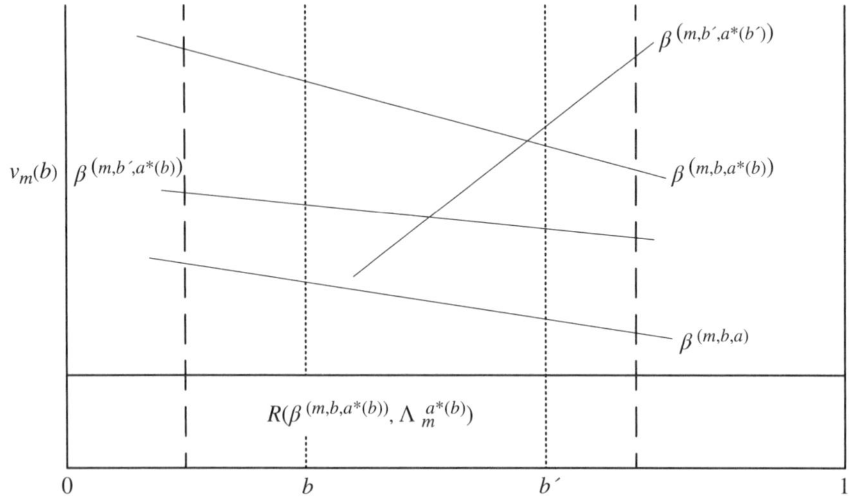

FIGURE 7.3 A scenario in which $\boldsymbol{\beta}^{(m,\mathbf{b},a^*(\mathbf{b}))}$ is not the optimal β-vector at some point $\mathbf{b}' \in R\left(\boldsymbol{\beta}^{(m,\mathbf{b},a^*(\mathbf{b}))}, \Lambda_m^{a^*(\mathbf{b})}\right)$ even if $\boldsymbol{\beta}^{(m,\mathbf{b},a^*(\mathbf{b}))}$ is better than all $\boldsymbol{\beta}^{(m,\mathbf{b},a)}$ vectors for $a \in \mathcal{A}$ as well as $\boldsymbol{\beta}^{(m,\mathbf{b}',a^*(\mathbf{b}))}$ at a point $\mathbf{b}'$. The optimal action at $\mathbf{b}'$ happened to be different from that at $\mathbf{b}$, that is, $a^*(\mathbf{b}') \neq a^*(\mathbf{b}')$. As a result, there is an optimal β-vector $\boldsymbol{\beta}^{(m,\mathbf{b}',a^*(\mathbf{b}'))}$ at $\mathbf{b}'$ that is different from $\boldsymbol{\beta}^{(m,\mathbf{b}',a^*(\mathbf{b}))}$ and giving rise to a strictly higher value-to-go at $\mathbf{b}'$.

optimal actions were used from $m-1$ step till termination (see (7.97)). However, in this example, the optimal action $a^*(\mathbf{b}')$ at $\mathbf{b}'$ is happened to be different from $a^*(\mathbf{b})$ so that $a^*(\mathbf{b}') \neq a^*(\mathbf{b}')$ and thus the optimal β-vector $\boldsymbol{\beta}^{(m,\mathbf{b}',a^*(\mathbf{b}'))}$ at $\mathbf{b}'$ is also different from $\boldsymbol{\beta}^{(m,\mathbf{b}',a^*(\mathbf{b}))}$.

To avoid this, it has been suggested to further impose the restriction that we only consider those $\mathbf{b}' \in R\left(\boldsymbol{\beta}^{(m,\mathbf{b},a^*(\mathbf{b}))}, \Lambda_m^{a^*(\mathbf{b})}\right)$ that are also in each of the dominant regions $R\left(\boldsymbol{\beta}^{(m,\mathbf{b},a)}, \Lambda_m^a\right)$ for all $a \in \mathcal{A}$ [70]:

$$\bigcap_a R\left(\boldsymbol{\beta}^{(m,\mathbf{b},a)}, \Lambda_m^a\right). \tag{7.113}$$

However, constraint (7.113) is too restrictive. For example, (7.113) will remove any belief state $\mathbf{b}' \in R\left(\boldsymbol{\beta}^{(m,\mathbf{b},a^*(\mathbf{b}))}, \Lambda_m^*\right)$ that has a β-vector $\boldsymbol{\beta}^{(m,\mathbf{b}',a)} \neq \boldsymbol{\beta}^{(m,\mathbf{b},a)}$ for some $a \in \mathcal{A}$ such that $(\mathbf{b}')^T\boldsymbol{\beta}^{(m,\mathbf{b}',a)} > (\mathbf{b}')^T\boldsymbol{\beta}^{(m,\mathbf{b},a)}$. However, it is possible that $(\mathbf{b}')^T\boldsymbol{\beta}^{(m,\mathbf{b},a^*(\mathbf{b}))} > (\mathbf{b}')^T\boldsymbol{\beta}^{(m,\mathbf{b},a)}$ for all such $a \in \mathcal{A}$ so that still $\boldsymbol{\beta}^{(m,\mathbf{b},a^*(\mathbf{b}))}$ is the optimal β-vector at this $\mathbf{b}'$ as shown in Fig. 7.4.

As a result, it is possible that the region obtained by imposing constraints (7.110) and (7.113) on $R\left(\boldsymbol{\beta}^{(m,\mathbf{b},a^*(\mathbf{b}))}, \Lambda_m^{a^*(\mathbf{b})}\right)$ could be smaller than the dominant region $R\left(\boldsymbol{\beta}^{(m,\mathbf{b},a^*(\mathbf{b}))}, \Lambda_m^*\right)$ of $\boldsymbol{\beta}^{(m,\mathbf{b},a^*(\mathbf{b}))}$. However, there does not appear to be any better alternative constraint set.

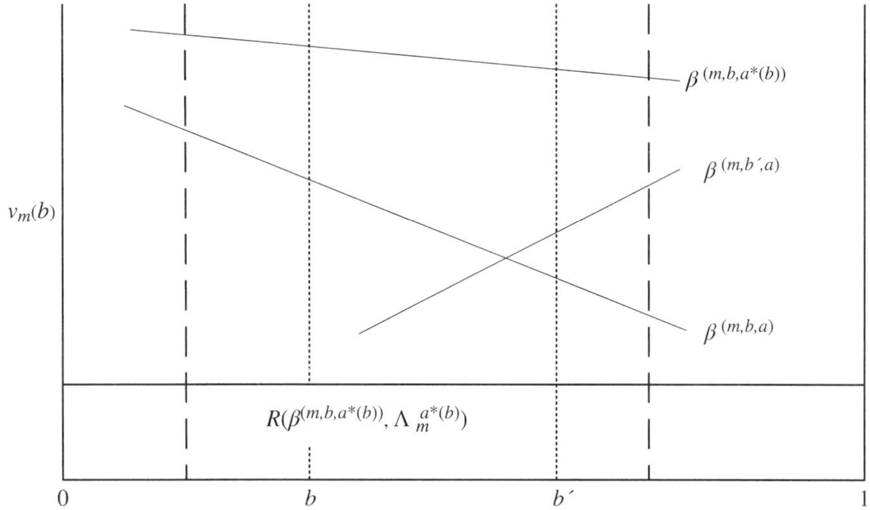

FIGURE 7.4 A scenario in which constraint (7.113) becomes too restrictive. Although $\boldsymbol{\beta}^{(m,\mathbf{b}',a)} \neq \boldsymbol{\beta}^{(m,\mathbf{b},a)}$ for some $a \in \mathcal{A}$ and $\boldsymbol{\beta}^{(m,\mathbf{b}',a)}$ is better than $\boldsymbol{\beta}^{(m,\mathbf{b},a)}$ at $\mathbf{b}'$, still, the vector $\boldsymbol{\beta}^{(m,\mathbf{b},a^*(\mathbf{b}))}$ is the optimal β-vector at this $\mathbf{b}'$. However, constraint (7.113) will remove $\mathbf{b}'$ from consideration.

7.5.4.4 The Witness Algorithm The witness algorithm is another POMDP solution approach that attempts to directly compute a parsimonious Λ_m^* from Λ_{m-1}^*. Instead of the value-to-go functions, however, the witness algorithm is formulated in terms of the Q-functions.

Definition 7.5.2 The function $Q_m(s,a)$ is the value of taking action a at step m and following an optimal policy thereafter for the remaining $m-1$ steps:

$$Q_m(s,a) = r(s,a) + \gamma \sum_{s'} P(s'|s,a) v_{m-1}^*(s'). \tag{7.114}$$

Note that these Q-functions are much similar to the action-value functions $Q^{\pi}(s,a)$ of a stationary policy π for an infinite-horizon MDP defined in (7.52). This is clear by comparing (7.114) with (7.54b). Their difference emerges because $Q^{\pi}(s,a)$ is valid for an arbitrary stationary decision rule π and that the infinite-horizon value functions at step m and $m-1$ are essentially the same.

The Q-function, $Q_m(\mathbf{b},a)$, is then given by

$$Q_m(\mathbf{b},a) = \mathbf{b}^T \mathbf{r}(a) + \gamma \sum_{y \in \mathcal{Y}} \Pr(y|\mathbf{b},a) v_{m-1}^*(\mathbf{T}(\mathbf{b},y,a)).$$

Clearly,

$$v_m^*(\mathbf{b}) = \max_{a\in\mathcal{A}} Q_m(\mathbf{b}, a). \tag{7.115}$$

It can be shown that $Q_m(\mathbf{b}, a)$ is also PWLC. Indeed, recall from (7.108) that Λ_m^a denotes the set of all possible β-vectors generated at step m from a given parsimonious Λ_{m-1}^* when the action at step m is restricted to be a. Hence, for all $\mathbf{b}$, and any $a \in \mathcal{A}$,

$$Q_m(\mathbf{b}, a) = \mathbf{b}^T\boldsymbol{\beta} \quad \text{for some } \boldsymbol{\beta} \in \Lambda_m^a. \tag{7.116}$$

We will denote the β-vectors that are in Λ_m^a by $\boldsymbol{\beta}^{(m,a)}$.

Thus, Λ_m^a is a representation of the $Q_m(\mathbf{b}, a)$, just as Λ_m is a representation of the value function $v_m(\mathbf{b})$. Similarly, Λ_m^a is not a parsimonious representation of $Q_m(\mathbf{b}, a)$.

The approach taken by the witness algorithm is to construct a parsimonious set, denoted by $\Lambda_m^{*,a}$, of β-vectors that characterizes the Q-function $Q_m(\mathbf{b}, a)$. Hence, with a parsimonious $\Lambda_m^{*,a}$, for all $\mathbf{b}$, and any $a \in \mathcal{A}$,

$$Q_m(\mathbf{b}, a) = \mathbf{b}^T\boldsymbol{\beta} \quad \text{for some } \boldsymbol{\beta} \in \Lambda_m^{*,a}, \tag{7.117}$$

and the β-vector in (7.117) is called the maximal β-vector for belief vector $\mathbf{b}$ at time step m if action is constrained to be a. It is easy to see that this maximal vector is indeed $\boldsymbol{\beta}^{(m,\mathbf{b},a)}$ given by (7.97).

From (7.115), it is clear that every optimal β-vector must be included in $\bigcup_{a\in\mathcal{A}} \Lambda_m^{*,a}$: That is, if $\boldsymbol{\beta} \in \Lambda_m$, then $\boldsymbol{\beta} \in \bigcup_{a\in\mathcal{A}} \Lambda_m^{*,a}$. Hence, $\Lambda_m \subseteq \bigcup_{a\in\mathcal{A}} \Lambda_m^{*,a}$. Thus, witness algorithm constructs a parsimonious Λ_m^* by applying pruning to $\bigcup_{a\in\mathcal{A}} \Lambda_m^{*,a}$ to remove the extraneous β-vectors in $\Lambda_m^{*,a}$.

Before we go on to describe the algorithm, let us define a *neighbor* of a β-vector $\boldsymbol{\beta} \in \Lambda_m^a$.

Definition 7.5.3 (Neighbor of a β-Vector) Consider a vector $\boldsymbol{\beta} \in \Lambda_m^a$:

$$\boldsymbol{\beta} = \mathbf{r}(a) + \gamma \sum_{y\in\mathcal{Y}} P(a)Q^{(y)}(a)\boldsymbol{\beta}_y,$$

where $\boldsymbol{\beta}_y \in \Lambda_{m-1}$ for all $y \in \mathcal{Y}$. Then, a vector $\mathbf{v}$ is called a neighbor of $\boldsymbol{\beta}$ if

$$\mathbf{v} = \mathbf{r}(a) + \gamma \sum_{\substack{y\in\mathcal{Y}\\ y\neq y'}} P(a)Q^{(y)}(a)\boldsymbol{\beta}_y + P(a)Q^{(y')}(a)\tilde{\boldsymbol{\beta}}_{y'},$$

for $\tilde{\boldsymbol{\beta}}_{y'} \neq \boldsymbol{\beta}_{y'}$ and $\tilde{\boldsymbol{\beta}}_{y'} \in \Lambda_{m-1}$.

Note that all neighbors of a β-vector $\boldsymbol{\beta} \in \Lambda_m^a$ are also in Λ_m^a. Moreover, it is easy to see that any β-vector $\boldsymbol{\beta} \in \Lambda_m^a$ has only $(|\Lambda_{m-1}| - 1)L$ number of neighbors, where

$L = |\mathcal{Y}|$ [70]. Let us denote by $\Omega(\boldsymbol{\beta})$ the set of neighbors of $\boldsymbol{\beta}$. The following theorem from [70] highlights the usefulness of the concept of neighbors.

Theorem 7.7 (Neighbor Theorem) For any $\boldsymbol{\beta}^{(m,a)} \in \Lambda_m^a$, denote by $\Omega\left(\boldsymbol{\beta}^{(m,a)}\right)$ its set of neighbors. For any $\boldsymbol{\beta}^{(m,a)} \in \Lambda_m^a$, there exists a belief $\mathbf{b}$ and a $\tilde{\boldsymbol{\beta}}^{(m,a)} \in \Lambda_m^a$ such that

$$\mathbf{b}^T \tilde{\boldsymbol{\beta}}^{(m,a)} \geq \mathbf{b}^T \boldsymbol{\beta}^{(m,a)} \tag{7.118}$$

if and only if there exists a $\mathbf{v} \in \Omega\left(\boldsymbol{\beta}^{(m,a)}\right)$ such that

$$\mathbf{b}^T \mathbf{v} \geq \mathbf{b}^T \boldsymbol{\beta}^{(m,a)}. \tag{7.119}$$

Proof: A proof can be found in [70].

The witness algorithm is organized in terms of an outer loop and an inner loop. The outer loop of the algorithm is essentially a value iteration: At each time step m, given a parsimonious Λ_{m-1}^* from step $m-1$, first compute the sets $\Lambda_m^{*,a}$ for each action a. Then, construct the parsimonious set $\Lambda_m^* = prune\left(\bigcup_{a\in\mathcal{A}} \Lambda_m^{*,a}\right)$ where function $prune(.)$ is a scheme to remove the dominated β-vectors.

The key idea of the witness algorithm lies in its inner loop that efficiently constructs the sets $\Lambda_m^{*,a}$, for each action a, directly from the parsimonious set Λ_{m-1}^*. Consider action a in step m. The inner loop initializes a holding set of vectors $\mathcal{Q}$ with a single vector that is the maximal β-vector for some arbitrary belief vector $\mathbf{b}$ at time step m if action is constrained to be a. Recall that this can easily be done by picking an arbitrary $\mathbf{b}$ and then using (7.97). It also initializes a set $\hat{\Lambda}_m^{*,a}$ to be the empty set. The algorithm will keep adding vectors to this set $\hat{\Lambda}_m^{*,a}$ till it becomes the minimal representation $\Lambda_m^{*,a}$ of Λ_m^a.

After these initializations, the inner loop runs through the following iteration till the holding set of vectors $\mathcal{Q}$ becomes empty: In each iteration, a vector $\mathbf{v}$ from the holding set $\mathcal{Q}$ is picked and checked whether there is any belief state $\mathbf{b}$ for which this $\mathbf{v}$ is a better vector than any of the vectors currently in $\hat{\Lambda}_m^{*,a}$. If there is no such $\mathbf{b}$, then that vector $\mathbf{v}$ is a redundant vector, so it is removed from the holding set $\mathcal{Q}$ and we start a new iteration by picking a different vector from $\mathcal{Q}$. On the other hand, if we found a belief vector $\mathbf{b}$ such that this $\mathbf{v}$ is better than any of the vectors already in $\hat{\Lambda}_m^{*,a}$, this shows that none of the vectors currently in $\hat{\Lambda}_m^{*,a}$ is maximal for this belief state with action a. The belief vector $\mathbf{b}$ found is thus a *witness* to the fact that $\hat{\Lambda}_m^{*,a}$ in current form is not yet the $\Lambda_m^{*,a}$, hence the name of the algorithm. The algorithm then finds the maximal vector $\boldsymbol{\beta}^{(m,\mathbf{b},a)}$ for this particular witness $\mathbf{b}$ with action a, and add $\boldsymbol{\beta}^{(m,\mathbf{b},a)}$ to $\hat{\Lambda}_m^{*,a}$.

At this point, the witness algorithm performs two important things before going into the next iteration: It puts $\mathbf{v}$ back into the holding set $\mathcal{Q}$. Not only that, it also adds all of the neighbors of the newly added vector $\boldsymbol{\beta}^{(m,\mathbf{b},a)}$ to the holding set $\mathcal{Q}$. Only after these, it will go back to a new iteration by picking a new $\mathbf{v}$ from the holding set $\mathcal{Q}$. The algorithm terminates if the holding set $\mathcal{Q}$ is empty.

First, according to the way the algorithm operates, it only cares to find one belief vector $\mathbf{b}$, say, $\mathbf{b}_1$, such that none of the current vectors in $\hat{\Lambda}_m^{*,a}$ is maximal at this belief vector $\mathbf{b}$. In other words, it only finds just one witness to the fact that $\hat{\Lambda}_m^{*,a}$ is not yet the optimal set. Consequently, we only add the maximal β-vector for this one witness belief vector $\mathbf{b}_1$. There is the possibility that there may be other belief vectors at which also none of the current vectors in $\hat{\Lambda}_m^{*,a}$ is maximal. Not only that, even the new maximal vector for $\mathbf{b}_1$ we just added to $\hat{\Lambda}_m^{*,a}$ might not also be optimal for some of these other belief vectors. Thus, we need to recheck (in a new iteration) whether still there are some belief vectors $\mathbf{b}$ for which this $\mathbf{v}$ is better than what is already in this $\hat{\Lambda}_m^{*,a}$, hence the reason for putting $\mathbf{v}$ back in the holding set at the end of the iteration.

Second, note that in each iteration of the witness inner loop, we add a β-vector that is maximal at a newfound witness belief state. Of course, we do not know whether this new β-vector will be maximal for all remaining belief states for which also we do not have maximal β-vectors in the current $\hat{\Lambda}_m^{*,a}$. The only way to be sure is to check this for all $\mathbf{b}$ vectors, which requires checking for dominance against all β-vectors in Λ_m^a. Of course, witness algorithm's objective is to avoid having to construct Λ_m^a. However, Theorem 7.7 ensures that it is actually possible to verify whether there exists any belief state at which a given β-vector is not the maximal vector by testing against another set of β-vectors. These are the neighbors of the given vector. Indeed, the neighbor theorem states that if there is a belief state at which there is a better β-vector than a β-vector we currently have, then one of the neighbors of the current β-vector will also be a better β-vector at this belief state. Hence, in order to verify whether a given vector $\boldsymbol{\beta}$ is maximal at a particular belief state, we do not have to check for all vectors in Λ_m^a. Instead, it is enough to check for all the neighbor vectors of the β-vectors we currently have in $\hat{\Lambda}_m^{*,a}$. Since there are only a finite number of, exactly $(|\Lambda_{m-1}^*| - 1)L$, neighbor vectors for a given $\boldsymbol{\beta}^{(m,a)} \in \Lambda_m^a$, it is possible to verify whether there can be any belief state at which the newly added β-vector might not be maximal by adding all its neighbors to the holding set $\mathcal{Q}$, hence, the reason for adding all the neighbors of a newly added maximal β-vector of a witness to the holding set at the end of an iteration.

The following theorem from [70], whose proof we omit, establishes the fact that indeed witness algorithm correctly constructs the parsimonious representation $\Lambda_m^{*,a}$ of the Q-function Q_m^a.

Theorem 7.8 The witness algorithm correctly constructs the parsimonious representation $\Lambda_m^{*,a}$ of the Q-function Q_m^a in a finite number of iterations.

Proof: A complete proof can be found in [70].

The core idea of witness algorithm lies in the efficient way it constructs the sets $\Lambda_m^{*,a}$ without having to construct Λ_m^a sets first. Indeed, it is known that its time complexity is polynomial in problem parameters and $|\Lambda_m^{*,a}|$. Hence, if $|\Lambda_m^{*,a}|$ is polynomial, then the witness algorithm will have polynomial time complexity. However, there is still the possibility that $|\Lambda_m^{*,a}|$ could be exponentially larger than $|\Lambda_m^*|$, making the worst-case complexity of the witness algorithm to be exponential. The justification of the approach lies in the fact that in many applications of interest, this may not happen.

7.5.5 Infinite-Horizon POMDPs

When $T \to \infty$, we have an infinite-horizon POMDP. In most cases, we are interested in obtaining a stationary policy $\boldsymbol{\pi} = (\pi, \pi, \cdots)$ so that $\pi_n = \pi$ for all n. The optimal infinite-horizon POMDP value function with a stationary policy π is defined as

$$V^*(\mathbf{b}) = \sup_{\pi} V^{\pi}(\mathbf{b}), \tag{7.120}$$

where $\boldsymbol{\pi} = (\pi, \pi, \cdots)$,

If the optimal value function exists, the following theorem (Theorem 2.2 of [74]) states that the optimal value function $V^*(\mathbf{b})$ satisfies a fixed-point equation similar to (7.50) in the case of an infinite-horizon MDP.

Theorem 7.9 The optimal value function $V^*(\mathbf{b})$ of an infinite-horizon POMDP satisfies

$$V^*(\mathbf{b}) = \max_{a \in \mathcal{A}} \left[\mathbf{b}^T \mathbf{r}(a) + \gamma \sum_{y \in \mathcal{Y}} \sigma(y|\mathbf{b}, a)\, V^*\left(\mathbf{b}^{(\mathbf{b},a,y)}\right) \right],$$

where $\mathbf{b}^{(\mathbf{b},a,y)} = \mathbf{T}(\mathbf{b}, a, y) = \left(b_1^{(\mathbf{b},a,y)}, b_2^{(\mathbf{b},a,y)}, \cdots, b_M^{(\mathbf{b},a,y)}\right)$ with, for $1 \leq j \leq M$,

$$\begin{aligned} b_j^{(\mathbf{b},a,y)} &= T_j(\mathbf{b}, a, y) \\ &= \Pr\left(S[n+1] = s_j \mid \mathbf{b}[n] = \mathbf{b}, a_n = a, Y[n+1] = y\right) \\ &= \frac{q\left(y|s_j, a\right) \sum_{i=1}^{M} [P(a)]_{i,j}\, b_i[n]}{\sum_{j'=1}^{M} q\left(y|s_{j'}, a\right) \sum_{i=1}^{M} [P(a)]_{i,j'}\, b_i[n]} \end{aligned} \tag{7.121}$$

and

$$\begin{aligned} \sigma(y|\mathbf{b}, a) &= \Pr(Y[n+1] = y \mid \mathbf{b}[n] = \mathbf{b}, a_n = a) \\ &= \sum_{i'} q(y|s_{i'}, a) \sum_{i} [P(a)]_{i,i'}\, b_i[n], \end{aligned} \tag{7.122}$$

where $\mathbf{r}(a) = (r(s_1, a), r(s_2, a), \cdots, r(s_M, a))^T$ for all $a \in \mathcal{A}$.

Note that (7.121) and (7.122), respectively, are simply restatements of (7.67) and (7.81b).

Solving the infinite-horizon POMDP problem poses additional difficulty since the optimal value function loses one of its key properties that all finite-horizon POMDP solvers of the previous section relied on. Indeed, although the optimal infinite-horizon POMDP value function is still convex, it may no longer be linear [76]. In general, however, one can approximate an infinite-horizon nonlinear optimal value function by a properly chosen piecewise linear function [74]. Hence, the infinite-horizon optimal value function can be approximated by a finite-horizon value function computed for a sufficiently long horizon T. Indeed, by choosing T large enough, one may achieve arbitrarily close approximation to the optimal infinite-horizon value function [76–78].

Sometimes, however, it may be the case that the infinite-horizon optimal value function is indeed piecewise linear, although this may be difficult to verify a priori. It is known that the value functions associated with the so-called *finitely transient* policies are always piecewise linear [74, 78].

Definition 7.5.4 (Finitely Transient Policies) A stationary policy π is called a finitely transient policy if after some finite period of time the belief state $\mathbf{b}$ will not lie at a point in $\mathcal{P}^{M-1}$ at which $\pi(\mathbf{b})$ is discontinuous.

In other words, finitely transient policies are policies that ensure belief states at which the policy is discontinuous will not recur after some finite number of steps [78].

Hence, a situation in which the optimal value function of an infinite-horizon POMDP will be piecewise linear is if the POMDP is known to have an optimal policy that is finitely transient. Of course, this is only a sufficient condition and there are piecewise linear value functions that are not the result of finitely transient policies. Moreover, as we stated earlier, some infinite-horizon optimal value functions are not piecewise linear at all.

If indeed the POMDP is known to have an optimal policy that is finitely transient, then it has been shown that policy iteration can be used to obtain an optimal policy [78]. When the finitely transient condition is not satisfied, these can still provide an approximate value function that is piecewise linear and thus an approximately optimal policy. In these algorithms, the policy evaluation step can be achieved by using algorithms such as one pass and witness of the previous section.

Of course, one may then also use a value iteration algorithm, as in the case of COMDP's, also to obtain an infinite-horizon optimal value function as in Algorithm 7.4.

Algorithm 7.4

POMDP Value Iteration Algorithm

(1) Initialize $n=1$ and $V_0(\mathbf{b})=0$ for all $\mathbf{b}$.

(2) Let $n=n+1$, and update:

$$V_n(\mathbf{b}) = \max_{a\in\mathcal{A}} \left[\mathbf{b}^T \mathbf{r}(a) + \gamma \sum_{y\in\mathcal{Y}} \sigma(y|\mathbf{b},a)\, V_{n-1}\left(\mathbf{b}^{(\mathbf{b},a,y)}\right) \right].$$

(3) Go to step (2), if $\max_{\mathbf{b}} |V_n(\mathbf{b}) - V_{n-1}(\mathbf{b})| > \epsilon$

Otherwise, set $V^*(\mathbf{b}) = V_n(\mathbf{b})$ and extract an optimal infinite-horizon policy as

$$\pi^*(\mathbf{b}) = \arg\max_{a \in \mathcal{A}} \left[\mathbf{b}^T \mathbf{r}(a) + \gamma \sum_{y \in \mathcal{Y}} \sigma(y|\mathbf{b},a) V_{n-1}\left(\mathbf{b}^{(\mathbf{b},a,y)}\right) \right].$$

Here, the value iteration is used to compute a sequence of optimal value functions for finite-horizon POMDPs with increasing horizon length until the difference between two successive value functions is small enough. Note that the value function computation in step 2 can be achieved by using an algorithm for a finite-horizon POMDP discussed in the previous section. It can be verified that when Algorithm 7.4 terminates with $\max_{s \in \mathcal{S}} |v_n(s) - v_{n-1}(s)| < \epsilon$, the value function computed by the value iteration algorithm is within $\frac{2\epsilon\gamma}{1-\gamma}$ of the optimal value function [76]. From this approximate value function, an ϵ-optimal policy can be extracted.

7.6 SUMMARY

- A Markov decision process (MDP) is a decision process in which the next state $S[n+1]$ of the environment is completely determined by the current state $S[n]$ and the action taken at current time a_n. When the number of possible states is finite, it is called a finite-state MDP.
- An MDP with both transition probabilities $p_n(s'|s,a)$ and rewards $r_n(s,a)$ time invariant is called a stationary MDP.
- A decision rule, or a strategy, for an MDP at a given time is a function that prescribes an action, or a set of actions, for each possible state $s \in \mathcal{S}$. A sequence of strategies, one for each time instant, forms a decision policy.
- A decision rule of the form of

$$\pi_n : \mathcal{S} \rightarrow \mathcal{A}$$

  that deterministically prescribes an action based only on current state is called a deterministic Markov rule. The policies based on deterministic Markov rules are called the deterministic Markov policies. The set of all deterministic Markov decision rules is denoted by Π^{MD}.
- A decision rule of the form of

$$\pi_n : \mathcal{S} \times \mathcal{A} \rightarrow [0,1]$$

  that specifies the probability with which a particular action is to be selected based only on current state is called a randomized Markov decision rule. The set of all randomized Markov decision rules is denoted by Π^{MR}. The policies based on randomized Markov decision rules are called the randomized Markov policies.
- The history at initial time $n = 0$ is simply the current state $\underline{h}_0 = (s_0)$. For $n > 0$, history at time $n = n$ is defined as

$$\underline{h}_n = (\underline{h}_{n-1}, a_{n-1}, s[n]).$$

- A decision rule of the form of

$$\pi_n : \mathcal{H}_n \rightarrow \mathcal{A},$$

where $\mathcal{H}_n$ denotes the set of all possible histories at time n, is called a deterministic history-dependent decision strategy. The set of all deterministic history-dependent decision rules is denoted by Π^{HD}.

- A decision rule of the form of

$$\pi_n : \mathcal{H}_n \times \mathcal{A} \rightarrow [0, 1],$$

where $\mathcal{H}_n$ denotes the set of all possible histories at time n, is called a randomized history-dependent decision rule. The set of all randomized history-dependent decision rules is denoted by Π^{HR}.

- The expected total return (ETR) associated with a decision policy $\underline{\pi} = (\pi_0, \ldots, \pi_{T-1})$ is

$$\overline{R_T^{\underline{\pi}}} = \mathbb{E}_{\underline{\pi}} \left\{ \sum_{n=0}^{T} r_n(S[n], \pi_n(.)) \right\},$$

where $\mathbb{E}_{\underline{\pi}}\{.\}$ denotes the expectation taken with respect to the state transitions corresponding to the policy $\underline{\pi}$, $r_n(S[n], \pi_n(.)) = r(S[n], \pi_n(.))$ for $0 \leq T - 1$ with $r_T(S[T], \pi_T(.)) = r_T(S[T])$ for all $S[T] \in \mathcal{S}$ and for all $\pi_T(.)$.

- The expected total discounted return (ETDR) associated with a decision policy $\underline{\pi} = (\pi_0, \ldots, \pi_{T-1})$ is

$$\overline{R_{T,\gamma}^{\underline{\pi}}} = \mathbb{E}_{\underline{\pi}} \left\{ \sum_{n=0}^{T} \gamma^n r_n(S[n], \pi_n(.)) \right\},$$

where $0 < \gamma \leq 1$ is called the discount factor.

- The state-value function of policy $\underline{\pi}$ is defined as

$$V_T^{\underline{\pi}}(s) = \mathbb{E}_{\underline{\pi}} \left\{ \sum_{n=0}^{T} \gamma^n r_n(S[n], \pi_n(.)) \middle| S[0] = s \right\}.$$

- A policy $\underline{\pi}^*$ is called an optimal policy for an MDP, if, given $S[0] = s$,

$$V_T^{\underline{\pi}^*}(s) \geq V_T^{\underline{\pi}}(s) \quad \text{for all } \underline{\pi} \in \Pi^{HR},$$

and the value of the MDP is defined to be the value of the state-value function with an optimal policy:

$$V_T^*(s) = \sup_{\underline{\pi} \in \Pi^{HR}} V_T^{\underline{\pi}}(s) \quad \text{for } s \in \mathcal{S}.$$

- When rewards and transition probabilities are time invariant, an optimal policy can be found by limiting the search to the set of all randomized Markov decision policies Π^{MR}.
- For a finite-horizon MDP, the value-to-go function $v_m^{\pi}(.)$, for $0 \leq m \leq T$, associated with a policy $\underline{\pi}$ is defined as

$$v_m^{\pi}(\underline{h}_{T-m}) = \mathbb{E}_{\underline{\pi}}\left\{ \sum_{i=T-m}^{T} \gamma^{i-(T-m)} r_i(S[i], \pi_i(.)) \middle| \underline{h}_{T-m} \right\}$$

with the boundary condition for $m = 0$

$$v_0^{\pi}(\underline{h}_T) = r_T(\underline{h}_T) \;=\; r_T(s_T).$$

- The state-value function $V_T^{\pi}(s)$ of a policy $\underline{\pi}$ can be obtained by the value-to-go function:

$$V_T^{\pi}(s) = v_T^{\pi}(s).$$

- The Bellman equation gives a recursion to evaluate the value-to-go function of a policy $\underline{\pi}$: For $m = 0, \cdots, T$, with $n = T - m$,

$$\begin{aligned} v_m^{\pi}(\underline{h}_n) &= r_n\left(s[n], \pi_n(\underline{h}_n)\right) \\ &\quad + \gamma \sum_{s' \in \mathcal{S}} \Pr(\, S[n+1] = s' \,|\, s[n], \pi_n(\underline{h}_n)) v_{m-1}^{\pi}\left(\left(\underline{h}_n, \pi_n(\underline{h}_n), s'\right)\right) \end{aligned}$$

with the initialization

$$v_0^{\pi}(\underline{h}_T) = r_T(s_T) \quad \text{for all} \quad \underline{h}_T \in \mathcal{H}_T.$$

- When rewards and state transition probabilities are Markov, for $m = 0, \ldots, T-1$, $v_m^*(s)$ with $n = T - m$ satisfies the following called the Bellman-optimality equations:

$$v_m^*(s[n]) = \sup_{a \in \mathcal{A}} \left[r_n(s[n], a) + \gamma \sum_{s' \in \mathcal{S}} p(s' | s[n], a) v_{m-1}^*(s') \right]$$

with the boundary condition

$$v_0^*(s_T) = r_T(s_T).$$

- The backward induction algorithm solution to the optimality equations is called dynamic programming.

- By solving the Bellman-optimality equations, we may obtain the optimal value function $V_T^*(s)$:

$$V_T^*(s) = v_T^*(s).$$

- An MDP with $T \rightarrow \infty$ is called an infinite-horizon MDP.
- Due to possible issues with convergence, in the case of infinite-horizon MDPs it is convenient to use the ETDR defined as

$$\overline{R_\gamma^{\underline{\pi}}} = \lim_{T \longrightarrow \infty} \mathbb{E}_{\underline{\pi}} \left\{ \sum_{n=0}^{T} \gamma^n r_n(S[n], \pi_n(.)) \right\},$$

where $\gamma \in [0, 1)$ as the optimality criterion.
- If the rewards are bounded, then the state-value function associated with a policy $\underline{\pi} \in \Pi^{HR}$ can be written as

$$V^{\underline{\pi}}(s) = \mathbb{E}_{\underline{\pi}} \left\{ \sum_{n=0}^{\infty} \gamma^n r(s[n], \pi_n(\underline{h}_n)) | S[0] = s \right\}.$$

- The value of an MDP that starts in initial state $S[0] = s \in \mathcal{S}$ is defined as

$$V^*(s) = \sup_{\underline{\pi} \in \Pi^{HR}} V^{\underline{\pi}}(s).$$

- A policy $\underline{\pi}^*$ is optimal for an infinite-horizon MDP if

$$V^{\underline{\pi}^*}(s) \geq V^{\underline{\pi}}(s) \quad \text{for all } s \in \mathcal{S} \quad \text{and for all } \underline{\pi} \in \Pi^{HR}$$

and an optimal policy exists for the infinite-horizon MDP if there is a policy $\underline{\pi}^* \in \Pi^{HR}$ such that

$$V^{\underline{\pi}^*}(s) = V^*(s) \quad \text{for all } s \in \mathcal{S}.$$

- A decision policy of the form of $\underline{\pi} = (\pi, \pi, \cdots,)$ is called a stationary policy. If a stationary MDP has bounded rewards and finite state and action sets $\mathcal{A}$ and $\mathcal{S}$, then it has an optimal policy within the class of deterministic stationary Markov policies. We may denote a stationary policy by π.
- The value $V^\pi(s)$ of a stationary policy π can be evaluated by solving the following Bellman equation:

$$V^\pi(s) = r(s, \pi(s)) + \gamma \sum_{s' \in \mathcal{S}} p(s'|s, \pi(s)) V^\pi(s'). \tag{7.123}$$

- Writing the Bellman equation (7.123) in matrix notation as

$$\mathbf{v}^{\pi} = \mathbf{r}_{\pi} + \gamma P_{\pi} \mathbf{v}^{\pi},$$

the closed-form solution to the Bellman equation can be obtained as follows:

$$\mathbf{v}^{\pi} = (I - \gamma P_{\pi})^{-1} \mathbf{r}^{\pi}.$$

- The optimal value function of an infinite-horizon policy satisfies the following fixed-point equation called the Bellman-optimality equation:

$$V^*(s) = \sup_{a \in \mathcal{A}} \left\{ r(s,a) + \gamma \sum_{s' \in \mathcal{S}} p(s'|s,a) V^*(s') \right\}. \quad (7.124)$$

If $\mathcal{S}$ is finite and $\{r(s,a)\}$ is bounded, then the above Bellman–optimality equation has a unique solution V^* for any $0 \leq \gamma < 1$.

- Policy iteration is an algorithm that iterates between policy evaluation and policy improvement to compute the optimal value function and an optimal policy for an infinite-horizon MDP:

1. Initialize $\pi(s) \in \mathcal{A}$ arbitrarily for all $s \in \mathcal{S}$.
2. Policy evaluation: For each $s \in \mathcal{S}$,

$$V^{\pi}(s) = r(s,a) + \gamma \sum_{s' \in \mathcal{S}} p(s'|s,a) V^{\pi}(s').$$

3. Policy improvement: For each $s \in \mathcal{S}$,

$$\begin{aligned} \pi'(s) &= \arg\max_a Q^{\pi}(s,a) \\ &= \arg\max_a \left[r(s,a) + \gamma \sum_{s' \in \mathcal{S}} p(s'|s,a) V^{\pi}(s') \right]. \end{aligned}$$

4. If $\pi' = \pi$, stop. Otherwise, set $\pi = \pi'$ and go to step (2).

- Value iteration is a simpler algorithm to evaluate the optimal value function and obtain an optimal policy for an infinite-horizon MDP. Instead of evaluating the value function of the current policy, it replaces the step (2) of the policy iteration algorithm by just one iterative policy evaluation step:

1. Set $k = 0$, and choose $\epsilon > 0$ and $V_k(s) \in \mathbb{R}$ for $s \in \mathcal{S}$ arbitrarily.
2. For each $s \in \mathcal{S}$, let

$$V_{k+1}(s) = \max_a \left[r(s,a) + \gamma \sum_{s' \in \mathcal{S}} p(s'|s,a) V_k(s') \right].$$

3. If $||\mathbf{V}_{k+1} - \mathbf{V}_k|| < \frac{1-\gamma}{2\gamma}\epsilon$, go to step (4). Otherwise, set $k \leftarrow k+1$ and go to step (2).
4. For each $s \in \mathcal{S}$, set

$$\pi(s) = \arg\max_a \left[r(s,a) + \gamma \sum_{s' \in \mathcal{S}} p(s'|s,a) V_{k+1}(s') \right]$$

and stop.

- An MDP in which the state is not directly observable is called a *partially observable Markov decision process (POMDP).*
- The information history available for decision-making in a POMDP at time $n = 0, 1, 2, \cdots$, is

$$\underline{h}_n = \left(\underline{h}_{n-1}, a_{n-1}, y[n]\right)$$

with the initialization $\underline{h}_0 = (\mathbf{p}, y[0])$, where $\mathbf{p}$ denotes the distribution of the initial state.
- The belief state vector $\mathbf{b}[n] \in \mathcal{P}^M$ is a sufficient statistic for the POMDP optimal decision-making problem at time n, where

$$\mathbf{b}[n] = (b_1[n], \cdots, b_M[n]))^T$$

with

$$b_j[n] = \Pr\left(S[n] = s_j \mid \underline{h}_n\right).$$

- A given POMDP can be converted into an equivalent belief state COMDP defined in terms of the state update equation

$$\mathbf{b}[n+1] = \mathbf{T}(\mathbf{b}[n], a_n, y_{n+1}),$$

where $\mathbf{T}(\mathbf{b}[n], a_n, y_{n+1}) \triangleq (T_1(\mathbf{b}[n], a_n, y_{n+1}), \cdots, T_M(\mathbf{b}[n], a_n, y_{n+1}))^T$ with

$$T_j(\mathbf{b}[n], a_n, y_{n+1}) = \frac{q\left(y|s_j, a\right) \sum_{i=1}^M [P(a)]_{i,j} b_i[n]}{\sum_{j'=1}^M q\left(y|s_{j'}, a\right) \sum_{i=1}^M [P(a)]_{i,j'} b_i[n]}.$$

and corresponding rewards

$$r(\mathbf{b},a)=\mathbf{b}^T\mathbf{r}(a),$$

where $\mathbf{r}(a)=(r_1(a),\cdots,r_M(a))^T$ with $r_j(a)=r(s_j,a)$.

- The value of a finite-horizon POMDP under a policy $\underline{\pi}$ is defined as

$$V_T^{\underline{\pi}}(\mathbf{b})=\mathbb{E}_{\underline{\pi}}\left\{\sum_{n=0}^{T}\gamma^n r(\mathbf{b}[n],\pi_n(\mathbf{b}[n]))\mid \mathbf{b}[0]=\mathbf{b}\right\}.$$

- Given the initial state $\mathbf{b}[0]=\mathbf{b}$, the optimal value function of a POMDP is defined as

$$V_T^*(\mathbf{b})=\sup_{\underline{\pi}} V_T^{\underline{\pi}}(\mathbf{b}).$$

- A policy $\underline{\pi}^*$ is called an optimal policy for a POMDP if

$$V_T^{\underline{\pi}^*}(\mathbf{b})=V^*(\mathbf{b})\qquad \text{for all } \mathbf{b}\in\mathcal{P}^{(M-1)}.$$

- If we let $v_m^{\underline{\pi}}(\mathbf{b})$ to be the value-to-go with policy $\underline{\pi}$ when there are m steps before termination, defined as

$$v_m^{\underline{\pi}}(\mathbf{b})=\mathbb{E}_{\pi}\left\{\sum_{n=T-m}^{T}\gamma^{n-(T-m)}r(\mathbf{b}[n],\pi_n(\mathbf{b}[n]))\mid \mathbf{b}[T-m]=\mathbf{b}\right\},$$

then

$$V_T^{\underline{\pi}}(\mathbf{b})=v_T^{\underline{\pi}}(\mathbf{b}), \tag{7.125}$$

for any policy $\underline{\pi}$ and belief state $\mathbf{b}$.

- The value-to-go function $v_m^{\underline{\pi}}(\mathbf{b})$ can be evaluated according to the following backward induction:

$$v_m^{\underline{\pi}}(\mathbf{b})=r_n(\mathbf{b},\pi_n(\mathbf{b}))+\gamma\sum_{y\in\mathcal{Y}}v_{m-1}^{\underline{\pi}}\left(\mathbf{b}^{(\mathbf{b},\pi_n(\mathbf{b}),y)}\right)\sigma\left(y|\mathbf{b},\pi_n(\mathbf{b})\right),$$

where $\mathbf{b}^{(\mathbf{b}[n],a_n,y_{n+1})}=\mathbf{T}(\mathbf{b}[n],a_n,y_{n+1})$ and

$$\sigma\left(y|\mathbf{b},a\right)=\sum_{i'}q\left(y|s_{i'},a\right)\sum_{i}[P(a)]_{i,i'}\,b_i[n],$$

with the initialization

$$v_0^{\underline{\pi}}(\mathbf{b})=\mathbf{r}_T^T\mathbf{b}[T].$$

- The optimal value function can be obtained by the optimal value-to-go function:

$$V_T^*(\mathbf{b}) = v_T^*(\mathbf{b}).$$

- The optimal value-to-go function can be evaluated by the following backward induction optimality equations: For $m = 1, \cdots, T$,

$$v_m^*(\mathbf{b}) = \max_{a \in \mathcal{A}} \left\{ \mathbf{b}^T \mathbf{r}(a) + \gamma \sum_{y \in \mathcal{Y}} v_{m-1}^* \left(\mathbf{b}^{(\mathbf{b},a,y)} \right) \sigma\left(y | \mathbf{b}, a \right) \right\}, \tag{7.126}$$

where $\mathbf{r}(a) = (r(s_1, a), r(s_2, a), \cdots, r(s_M, a))^T$, $\mathbf{b}^{(\mathbf{b},a,y)} = \mathbf{T}(\mathbf{b}, a, y) = \Big(b_1^{(\mathbf{b},a,y)},$ $b_2^{(\mathbf{b},a,y)}, \cdots, b_M^{(\mathbf{b},a,y)} \Big)$ with, for $1 \le j \le M$,

$$\begin{aligned} b_j^{(\mathbf{b},a,y)} &= T_j(\mathbf{b}, a, y) \\ &= \Pr\left(S[n+1] = s_j \mid \mathbf{b}[n] = \mathbf{b}, a_n = a, Y[n+1] = y \right) \end{aligned}$$

and

$$\sigma\left(y | \mathbf{b}, a \right) = \Pr\left(Y[n+1] = y \mid \mathbf{b}[n] = \mathbf{b}, a_n = a \right),$$

with the initialization

$$v_0^*(\mathbf{b}) = \mathbf{b}^T \mathbf{r}_T \quad \text{for all } \mathbf{b} \in \mathcal{P}^M,$$

where $\mathbf{r}_T = (r(s_1), \cdots, r(s_M))^T$.

- Both optimal value-to-go function $v_m^*(\mathbf{b})$, for each $m = 1, \cdots, T$, and the optimal value function $V_T^*(\mathbf{b})$ of a POMDP are piecewise linear and convex (PWLC) as a function of belief state $\mathbf{b} \in \mathcal{P}^{M-1}$. Hence, for each $m = 1, \cdots, T$, the value-to-go function $v_m^*(\mathbf{b})$ can be characterized by giving a finite partition $\left\{ \mathcal{B}_{m,i} \right\}_i$ of the probability simplex $\mathcal{P}^{M-1}$ and a collection of vectors $\boldsymbol{\beta}^{(m,i)}$ such that for all $\mathbf{b} \in \mathcal{B}_{m,i}$,

$$v_m^*(\mathbf{b}) = \mathbf{b}^T \boldsymbol{\beta}^{(m,i)}.$$

  If we let Λ_m denote the finite set of $\boldsymbol{\beta}^{(m,i)}$ vectors associated with the partition $\left\{ \mathcal{B}_{m,i} \right\}_i$, then the optimal value-to-go function $v_m^*(\mathbf{b})$ at step m is characterized by computing $\left(\left\{ \mathcal{B}_{m,i} \right\}_i, \Lambda_m \right)$.

- A set Λ_m^* of β-vectors is called a parsimonious representation of the PWLC optimal value function $v_m^*(.)$ if the dominant region $R(\boldsymbol{\beta}, \Lambda_m^*)$ is non-empty for all $\boldsymbol{\beta} \in \Lambda_m^*$ and the collection of $\left\{ R(\boldsymbol{\beta}, \Lambda_m^*) \right\}_{\boldsymbol{\beta}}$ is a partition of $\mathcal{P}^{M-1}$, where

$$R(\boldsymbol{\beta}, \Lambda) = \left\{ \mathbf{b} : \; \mathbf{b}^T \boldsymbol{\beta} > \mathbf{b}^T \boldsymbol{\beta}', \mathbf{b} \in \mathcal{P}^{M-1}, \boldsymbol{\beta}' \in \Lambda - \{\boldsymbol{\beta}\} \right\}. \tag{7.127}$$

- There are several algorithms that attempt to compute $\left(\left\{\mathcal{B}_{m,i}\right\}_i, \Lambda_m\right)$ in a backward induction based on the optimality equation (7.126). One class of these algorithms first computes all the β-vectors that can be generated by the collection of β-vectors characterizing the value function at the previous stage $m-1$, and then tries to extract a parsimonious Λ_m^* from this:

  - Exhaustive search computes all possible β-vectors Λ_m that can be generated from a given Λ_{m-1}:

$$\Lambda_m = \left\{ \mathbf{r}(a) + \gamma \sum_{y \in \mathcal{Y}} P(a) Q^{(y)}(a) \boldsymbol{\beta}_y \ : a \in \mathcal{A},\ \boldsymbol{\beta}_y \in \Lambda_{m-1} \right\},$$

    with the initialization $\Lambda_0 = \{\mathbf{r}_T\}$. Then

$$v_m^*(\mathbf{b}) = \max_{\boldsymbol{\beta} \in \Lambda_m} \mathbf{b}^T \boldsymbol{\beta}.$$

    Exhaustive search complexity is proportional to the size of the set Λ_m, which grows exponentially with the time horizon T and the number of possible observations $L = |\mathcal{Y}|$.
  - Modified exhaustive search with pruning first generates Λ_m, from a given parsimonious representation Λ_{m-1}^* and then uses a linear program to identify and remove the dominated vectors from Λ_m to obtain a parsimonious representation Λ_m^*. Although it can considerably reduce the computational complexity of the exhaustive search approach, in the worst case, the modified algorithm is still exponential in the time horizon T.

- Another class of algorithms attempts to directly generate a parsimonious Λ_m^* from a given parsimonious Λ_{m-1}^* without computing the set Λ_m of all β-vectors that can be generated by Λ_{m-1}^*. Sondik's one-pass algorithm and the witness algorithm are examples of algorithms that belong to this category.
- The optimal value function $V^*(\mathbf{b})$ of an infinite-horizon POMDP satisfies the following fixed-point equation:

$$V^*(\mathbf{b}) = \max_{a \in \mathcal{A}} \left[\mathbf{b}^T \mathbf{r}(a) + \gamma \sum_{y \in \mathcal{Y}} \sigma\left(y|\mathbf{b}, a\right) V^*\left(\mathbf{b}^{(\mathbf{b},a,y)}\right) \right],$$

  where $\mathbf{b}^{(\mathbf{b},a,y)} = \mathbf{T}(\mathbf{b}, a, y) = \left(b_1^{(\mathbf{b},a,y)}, b_2^{(\mathbf{b},a,y)}, \cdots, b_M^{(\mathbf{b},a,y)}\right)$ with, for $1 \le j \le M$,

$$b_j^{(\mathbf{b},a,y)} = \frac{q\left(y|s_j, a\right) \sum_{i=1}^M [P(a)]_{i,j} b_i[n]}{\sum_{j'=1}^M q\left(y|s_{j'}, a\right) \sum_{i=1}^M [P(a)]_{i,j'} b_i[n]}$$

and

$$\sigma(y|\mathbf{b},a) = \sum_{i'} q(y|s_{i'},a) \sum_{i} [P(a)]_{i,i'}\, b_i[n],$$

where $\mathbf{r}(a) = (r(s_1,a), r(s_2,a), \cdots, r(s_M,a))^T$ for all $a \in \mathcal{A}$.

- A stationary policy π for an infinite-horizon POMDP is called a finitely transient policy if after some finite period of time the resulting belief state $\mathbf{b}$ will not be at a point in $\mathcal{P}^{M-1}$ at which $\pi(\mathbf{b})$ is discontinuous.
- The optimal infinite-horizon value functions are always convex. However, they may not be piecewise linear. If, however, an infinite-horizon POMDP is known to have an optimal policy within the class of finitely transient policies, then its value function will be piecewise linear.
- In general, an infinite-horizon optimal value function can arbitrarily closely be approximated by a piecewise linear finite-horizon value function computed for a sufficiently long horizon T. Both policy iteration and value iteration can thus be used to compute a sequence of value functions for a finite-horizon POMDP with increasing horizon length until the change is negligible as an approximation to the infinite-horizon optimal value function.

8

BAYESIAN NONPARAMETRIC CLASSIFICATION

8.1 INTRODUCTION

Spectral activity identification problem can be either a *classification* or a *clustering* problem. Once a wideband cognitive radio has detected a set of spectral activities located at certain frequencies, determining the type of each signal requires a classification stage as shown in Figure 2.4. Suppose that we are given a set of observations $\{y_i\}_{i=1}^N$. For simplicity, let us assume that each $y_i \in \mathbb{R}$. Each y_i corresponds to a specific class θ_i. The general classification problem can then be stated as

$$\text{Determine } \{\theta_i\}_{i=1}^N \quad \text{given data } \{y_i\}_{i=1}^N.$$

Classification approaches can be broadly divided into two classes: supervised and unsupervised. In *supervised classification*, there are initial training samples that can be used as a guide in classifying the given data $\{y_i\}_{i=1}^N$. For example, suppose that the training data $\left\{\left(y_i^{(t)}, \theta_i^{(t)}\right)\right\}_{i=1}^{N_t}$ were given where $\theta_i^{(t)}$ denotes the class of the training data sample $y_i^{(t)}$. Note that the distinct values present in the set of $\left\{\theta_i^{(t)}\right\}_{i=1}^{N_t}$ represent the distinct classes. Let $\phi_k^{(t)}$, for $k \in \{1, \ldots, K^{(t)}\}$, denote the $K^{(t)}$ distinct values taken by the $\theta_i^{(t)}$'s of the training data set. In other words, according to the training data, there are $K^{(t)}$ possible classes. The supervised classification approach classifies the

Signal Processing for Cognitive Radios, First Edition. Sudharman K. Jayaweera.

data $\{y_i\}_{i=1}^N$ into these same known $K^{(t)}$ classes with associated labels $\phi_k^{(t)}$. For obvious reasons, supervised classification is also called the *training-based classification*.

On the other hand, *unsupervised classification* approaches do not assume training data. The only input available is the data set $\{y_i\}_{i=1}^N$ itself, which must be divided into a set of classes, or clusters, without any further information. Note that in general, even how many classes are there may not be specified, although many algorithms assume this is known. Since the problem is to group similar data points into the same cluster, unsupervised classification problem is one of *clustering problem*.

The spectral activity identification problem in a wideband cognitive radio can be either a supervised or an unsupervised classification (or clustering) problem. In practice, it is likely that a cognitive radio will have to support both these approaches. In some RF environments, the cognitive radio may already have a priori knowledge of which types of systems are in operation and their characteristics. Then the inputs $\{y_i\}_{i=1}^N$ corresponding to the detected spectral activities can be identified by simply classifying them into those already known systems. In other situations, the cognitive radio may detect a set of spectral activities within the sensed spectrum but not know what systems are really active in that spectrum range at that time. Hence, the problem may be that of a clustering problem. It can be expected that the clustering algorithm will be followed by an *association algorithm* that will associate the detected clusters with some known systems. This stage can again be posed as a supervised classification problem. However, while some of the clusters may correspond to known systems, others may not. There may be signals belonging to systems that a wideband cognitive radio has not encountered previously. Hence, we may need a framework that allows for such situations as well.

When dependence of the observations on the clusters is statistical, and that the cluster labels are assumed to be randomly drawn so that $\boldsymbol{\theta} = (\theta_1, \ldots, \theta_N)^T$ is a random vector, then we can pose the classification problem in a Bayesian framework. By presuming a prior distribution for $\boldsymbol{\theta}$, then a solution for the supervised classification problem, for example, can be based on the maximum a posteriori probability (MAP) criterion of (5.28):

$$\boldsymbol{\theta}^*_{MAP} = \underset{\boldsymbol{\theta} \in \left\{\phi_1^{(t)}, \ldots, \phi_{K^{(t)}}^{(t)}\right\}^N}{\arg\max} \Pr(\boldsymbol{\theta} | \{y_i\}_{i=1}^N), \tag{8.1}$$

where $\Pr(\boldsymbol{\theta} | \{y_i\}_{i=1}^N)$ is the a posteriori probability of the classes given the data. Hence, we may call the classification solution (8.1) the *MAP classifier*.

From Bayes, rule, it follows that

$$\boldsymbol{\theta}^* = \underset{\boldsymbol{\theta}}{\arg\max} \frac{P(\{y_i\}_{i=1}^N | \boldsymbol{\theta}) G(\boldsymbol{\theta})}{P(\{\mathbf{y}_i\}_{i=1}^N)}, \tag{8.2}$$

where $G(.)$ denotes the prior distribution of the cluster labels and $P(\{y_i\}_{i=1}^N | \boldsymbol{\theta})$ is called the *class-conditional probability*. As we saw in Chapters 4 and 5, when we do not have a prior distribution on $\boldsymbol{\theta}$, we may (for the moment) assume that it is uniform so that the MAP classifier reduces to the following *maximum likelihood (ML) classifier*:

$$\boldsymbol{\theta}^*_{ML} = \underset{\boldsymbol{\theta}\in\left\{\phi_1^{(t)},\ldots,\phi_{K(t)}^{(t)}\right\}^N}{\arg\max} \quad P(\{y_i\}_{i=1}^N|\boldsymbol{\theta}). \tag{8.3}$$

As we discussed in Chapter 5, of course, the ML classifier can also be interpreted as a classifier that treats the clusters $\boldsymbol{\theta}$ as nonrandom but unknown.

Depending on what information we have on the required distribution (or density) functions, classification approaches can be divided broadly into either being parametric or nonparametric. In particular, classification approaches that assume the knowledge of exact distributions are called *parametric classification* approaches. The most common situation is to assume the knowledge of class-conditional densities (likelihoods) and treat the cluster parameters as nonrandom unknowns. An example parametric classifier in this case is the ML classifier (8.3).

On the other hand, classification approaches that do not assume the knowledge of distributions/densities are in general called the *nonparametric classification* approaches. These approaches may use observations $\{y_i\}_{i=1}^N$ themselves to first estimate the required distributions or densities, without assuming any parametric form for them, and use these estimated densities to either obtain an ML or MAP estimate as in (8.3) or (8.2), respectively. In this case, nonparametric classification essentially reduces to a problem of *density estimation*.

One approach for density estimation focuses on the class-conditional densities or the likelihoods. A very general class of approaches that falls in this category is the *Parzen window-based density estimation* [79–81]. Here, class-conditional densities are estimated first and then are used to perform either the ML or the MAP classification. An alternative approach completely foregoes the unknown class-conditional densities and, instead, directly estimates the posterior probabilities required in the MAP classifier. The *K-nearest neighbor classification* is one of the most commonly used approaches of this type [79, 82–84]. Both these approaches to nonparametric density estimation, however, require the knowledge of the number of classes or clusters. Not only that, in most density estimation approaches, the class priors are assumed to be uniform, or class parameters θ_i's are treated as just unknown deterministic quantities.

As a result, both class-conditional density (likelihood) estimation and direct posterior density estimation approaches to nonparametric estimation do not address one important issue involved in the case of spectral activity identification in a wideband cognitive radio. It is likely that information about the type of systems active in a given subband at a given time, number of such active systems, and number of signals active in each system may all be unknowns to a cognitive radio a priori. Hence, in many situations, a wideband cognitive radio may detect a set of spectral activities $\{y_i\}_{i=1}^N$ but not have prior information to assist with how many classes are there in these observations and how many signals belong to each class. In those situations, the problem faced by the cognitive radio is that of a clustering problem. However, since the number of active clusters present in the data set $\{y_i\}_{i=1}^N$ is unknown, none of the above-discussed approaches can directly be used to solve the problem. In particular, the ML classifiers that assume the clusters to be a priori uniformly distributed may not be justified. Certainly, since each cluster in wideband spectrum sensing represents a different system, or a class of signals, these cluster probabilities may not be uniform. For example, consider signals in the 2.4 GHz ISM band. It is unreasonable to assume that microwave oven radiation and WiFi signals in this band will have the same probability of occurring. The microwave oven radiation may only be present when the oven is actually used, which can be only a few times of the day, whereas WiFi signals can likely be on most of the time. In addition, the types of clusters that are active can be different from one time to another. Sometimes, the data set $\{y_i\}_{i=1}^N$ may correspond to signals from just WiFi and microwave oven radiation, while at other times there may be Bluetooth signals included in the mix as well. The Bayesian solution to this problem is to treat the θ_i's as a realization of a random process.

Of course, the MAP estimator (8.1) and (8.2) already treats θ_i's as being random with a prior distribution $G(\theta)$. However, in case of spectral activity classification problem in a wideband cognitive radio, we still need further generalization. Recall that the types of active systems and the number of active systems both can vary from time to time. A fixed distribution $G(\theta)$ defined over a finite discrete support set (denoting a finite set of cluster labels) cannot model such a scenario since the distribution of clusters has to be defined by the data $\{y_i\}_{i=1}^N$ itself. In other words, the distinct set of clusters $\{\phi_k\}_{k=1}^K$ assumed by the set of cluster labels $\{\theta_i\}_{i=1}^N$ corresponding to each data set $\{y_i\}_{i=1}^N$ may be different. Hence, we would like to have the ability to specify a prior $G(\theta)$ that is specifically supported on the set $\{\phi_k\}_{k=1}^K$ corresponding to the data. A natural model for this is the *Bayesian nonparametric* approach in which the prior $G(\theta)$ can be allowed to be a realization of a random distribution. In contrast, when either a fixed distribution or a distribution characterized by a finite set of parameters is used as the prior $G(\theta)$, we may call it as a Bayesian parametric classifier.

As mentioned above, it is common to refer to approaches that do not assume parameterized knowledge of class-conditional, or likelihood, distributions as nonparametric classifiers. When likelihoods are known at least within a finite-dimensional parameter vector, the resulting classifier is called a parametric classifier. As we will see, in the Bayesian nonparametric approach, it is common to assume that the likelihoods are known within a finite-dimensional parameter vector conditioned on the cluster knowledge. Hence, one may see them as parametric classifiers in one sense. However, instead of the likelihoods, the Bayesian nonparametric approach considers the cluster prior $G(\theta)$ as an unknown distribution. Moreover, being a Bayesian approach, the unknown cluster prior itself is treated as a random quantity (i.e., a random distribution). Hence, although the class-conditional densities are assumed to be known within a finite-dimensional parameter vector conditioned on the cluster, the clusters themselves are allowed to be unbounded. In this sense, Bayesian nonparametric approaches can also indeed be viewed as nonparametric classifiers.

Any wideband cognitive radio will have to support a combination of several types of signal classification algorithms to handle different situations the radio may encounter. For example, as mentioned above, a cognitive radio will have to include supervised classification algorithms to best handle classifying detected signals into known systems. In other situations, it may have to perform unsupervised clustering of detected spectral activities. Many of the well-known classification algorithms can easily be adapted as signal classifiers when the problem is posed as either a parametric or a nonparametric classification problem. The case that perhaps, needs more attention is the above situation of unknown, distinct, and variable number of classes that does not directly fall under either the parametric or nonparametric classifiers treated in many of the standard references and textbooks. Hence, this chapter specifically is an introduction to the Bayesian nonparametric classification rather than an in-depth introduction to the general classification approaches. For the latter, we refer the reader to many good textbooks devoted to the topic of classification and clustering such as [79, 85–87].

Consider again the set of observations $\{\mathbf{y}_i\}_{i=1}^N$ that needs to be classified and let their classification be $\{\theta_i\}_{i=1}^N$. Some of these θ_i's may be distinct, while others may be the same implying the same system. Let ϕ_k, for $k \in \{1, \ldots, K\}$, denote the K distinct values taken by the θ_i's, for $i = 1, \ldots, N$. Neither the ϕ_k's nor K is known a priori. Thus, in order to allow a cognitive radio to operate in RF environments in which it does not know the number of active systems or the number of active signals belonging to each system, or both, we need clustering approaches that do not presume the number of classes a priori. In the Bayesian nonparametric approach, θ_i's will be treated as random and being drawn from a certain unknown distribution $G(\theta)$, which itself is random. Under these assumptions, we look for an estimate of $\boldsymbol{\theta} = (\theta_1, \cdots, \theta_N)^T$ of the form of

$$\boldsymbol{\theta}^* = \arg\max_{\boldsymbol{\theta}} P(\boldsymbol{\theta} | \{\mathbf{y}_i\}_{i=1}^N).$$

If, however, K is known, there are well-known clustering algorithms one may use. One of the most commonly used is the *K-means algorithm* [88–90]. In this algorithm, the observations $\{y_i\}_{i=1}^N$ are classified, or divided, into a set of K classes, and the corresponding θ_i's are determined. The value of K is prespecified. The obvious limitation is that the algorithm needs to know the number of classes K a priori. There are variations of K-means algorithm that attempt to estimate K from data itself. These algorithms can be viewed as an alternative to the Bayesian nonparametric approach albeit with many limitations in comparison. One such variation is the so-called X-means algorithm [91]. However, X-means still needs to specify a finite range of values for the number of classes. Moreover, another drawback of both K-means and X-means is that they assume underlying (spherical) Gaussian likelihoods [91].

In the case of classification of detected spectral activity in wideband spectrum sensing, both the above assumptions may be too restrictive. Bayesian nonparametric classifiers that allow the number of clusters to be a random variable that can grow with the complexity of the data set itself seem well suited for this scenario [92–95]. A very general Bayesian nonparametric clustering framework can be developed based on the *Dirichlet process mixture model* (DPMM) [96]. A DPMM-based classifier assumes no prior knowledge of the number of clusters. Neither does it require the distributions to be spherical Gaussians. It can infer the number of clusters (called the mixture components) from the data itself. Such DPMM-based Bayesian nonparametric approaches have been previously applied for galaxy clustering [97], speaker diarization [98], speaker adaptation [99], image segmentation [100], and compressive sensing [101].

After briefly discussing the K-means and X-means clustering algorithms, in this chapter, we will focus on the DPMM-based Bayesian nonparametric classification of detected spectral activities since it allows for arbitrary noise distributions and can estimate the unknown parameters of corresponding probability distribution from data itself in order to best fit the observed model. Our discussion of the DPMM classifier will follow [97]. However, we will further explicitly generalize it to the multidimensional case as in [102]. In Chapter 11, we will apply the DPMM classifier to spectral activity identification in a wideband cognitive radio.

8.2 *K*-MEANS CLASSIFICATION ALGORITHM

The K-means algorithm is essentially a vector-generalized version of the *Lloyd–Max algorithm* for scalar quantization [89, 90, 103]. The vector-generalized Lloyd–Max algorithm can be stated as follows: Suppose that random vector $\mathbf{y}$ has a probability density function (pdf) $f_{\mathbf{Y}}(\mathbf{y})$ having support $\mathcal{S} \subseteq \mathbb{R}^d$. Let $\mathcal{S}_1, \ldots, \mathcal{S}_K$ be a partition of $\mathcal{S}$ into K disjoint sets. Denote by $\boldsymbol{\mu}_1, \ldots, \boldsymbol{\mu}_K$, where $\boldsymbol{\mu}_k \in \mathcal{S}_k$, for $k = 1, \ldots, K$, a set of *representative points* from these disjoint sets. Let us define the classification function $\theta(\mathbf{y}) = \boldsymbol{\mu}_k$ if $\mathbf{y} \in \mathcal{S}_k$. The vector-generalized version of the Lloyd–Max algorithm is an iterative algorithm to solve the problem of minimizing the mean squared error (MSE) defined as

$$\min \mathbb{E}\left\{ |\mathbf{Y} - \theta(\mathbf{Y})|^2 \right\}. \tag{8.4}$$

The minimization is to be carried out jointly over the sets $\mathcal{S}_1, \ldots, \mathcal{S}_K$ and the representation points $\boldsymbol{\mu}_1, \ldots, \boldsymbol{\mu}_K$.

Note that minimizing the MSE (8.4) is equivalent to maximizing the likelihood of $\mathbf{y}$ if $\mathbf{Y}$ is assumed to be Gaussian distributed with unknown mean $\boldsymbol{\mu}_k$ and the same (known) covariance matrix for all k. In other words, the data is spherical Gaussian distributed with means $\boldsymbol{\mu}_1, \ldots, \boldsymbol{\mu}_K$ and a common covariance matrix. The solution to (8.4) thus leads to the ML classifier, and it is equivalent to the MAP classifier (8.1) if all clusters were to have uniform probability.

The solution to (8.4) can be specified in terms of two steps:

1. Suppose first that we are given the K representation points $\boldsymbol{\mu}_1, \ldots, \boldsymbol{\mu}_K$. Then, the problem reduces to specifying the optimal sets $\mathcal{S}_1, \ldots, \mathcal{S}_K$ such that the MSE (8.4) is minimized. It is easily seen that optimal way to specify the sets in this case is to assign $\mathbf{y}$ to set $\mathcal{S}_{\hat{k}}$ if

$$\hat{k} = \arg\min_k |\mathbf{y} - \boldsymbol{\mu}_k|^2.$$

In other words, the optimal sets are the minimum distance, or Voronoi, regions with respect to the given representation points $\boldsymbol{\mu}_1, \ldots, \boldsymbol{\mu}_K$.

2. On the other hand, if we were given the sets $\mathcal{S}_1, \ldots, \mathcal{S}_K$, it can be shown that the optimal representation points that will minimize the MSE (8.4) are given by the conditional mean of the given sets so that, for $k = 1, \ldots, K$,

$$\boldsymbol{\mu}_k = \mathbb{E}\{\mathbf{Y}|\mathcal{S}_k\}. \tag{8.5}$$

The Lloyd–Max algorithm is an alternate iteration of the above two steps.

Algorithm 8.1

```
Vector-generalized Lloyd-Max Algorithm
  (1) Initialize μ_1,...,μ_K so that each μ_k ∈ S for k=1,...,K.
  (2) FOR y ∈ S:
     Assign y to S_k if k̂ = argmin_k |y − μ_k|.
  (3)  FOR k=1,...,K:
     Set μ_k = E{Y|S_k}.
  (4) IF MSE > ε, go to Step (2). Otherwise Stop.
```

The K-means algorithm, shown in Algorithm 8.2, is essentially an implementation of the above Lloyd–Max algorithm [88–90]. Given the observation data $\{\mathbf{y}_i\}_{i=1}^N$, where each $\mathbf{y}_i \in \mathbb{R}^d$, and the number of classes K (an integer), the K-means algorithm partitions the data into K disjoint sets $\mathcal{S}_1, \ldots, \mathcal{S}_K$ with centroids $\mu_1, \ldots, \mu_K$ such that the *total squared error (TSE) distortion* defined as

$$TSE = \sum_{i=1}^{N} |\mathbf{y}_i - \theta(\mathbf{y}_i)|^2$$

is minimized. Note that as before, $\theta(\mathbf{y}) = \boldsymbol{\mu}_k$ if $\mathbf{y} \in \mathcal{S}_k$. The centroid of cluster k is defined as

$$\boldsymbol{\mu}_k = \frac{1}{|\mathcal{S}_k|} \sum_{\mathbf{y} \in \mathcal{S}_k} \mathbf{y}_i,$$

which is a sample approximation to the conditional expectation (8.5) in the Lloyd–Max algorithm.

Algorithm 8.2

```
K-means Algorithm
  (1) Initialize μ_1,...,μ_K to random values (e.g. For k=1,...,K set
μ_k = y_i where i = U({1,2,...,K})).
  (2) FOR i=1:N:
     Assign y_i to S_k if k̂ = argmin_k |y − μ_k|.
  (3)  FOR k=1,...,K:
     Set μ_k = (1/|S_k|) Σ_{y∈S_k} y_i.
  (4) IF convergence condition is not satisfied, go to Step (2).
Otherwise Stop.
```

The K-means algorithm stops when the new cluster centroids have not changed from the old ones. At this point, the algorithm has converged to a local optimal of the TSE cost function. It is to be noted that the algorithm is indeed unsupervised in the sense that it does not require any training data. Through iteration, the algorithm forms a set of K clusters of the given data so that the TSE is minimized with respect to the centroids of the clusters.

8.3 X-MEANS CLASSIFICATION ALGORITHM

The X-means algorithm treats the problem of unknown number of clusters as a *model selection problem*. Note that if the given set of observations $\{\mathbf{y}_i\}_{i=1}^N$ belongs to K clusters, then these observations can be viewed as generated by a model with K parameters. Let the kth model be the model with k parameters. When the number of clusters K is unknown, then the problem is to determine which model k is the true model. Hence, this is a model selection problem. In general, the model selection problem is to select the model that best fits the given data. However, models with more parameters may better fit the data at the risk of overfitting and model complexity. Hence, it is desirable to select the model that gives the best trade-off between explaining data and overfitting. Of course, the best trade-off will depend on the criterion used to quantify these two aspects. Widely used criteria include the *Akaike information criterion* (AIC) and *Bayesian information criterion* (BIC) and their variations.

In order to define these concepts precisely, let us consider the Bayesian approach to model selection. Suppose we are given observations $\{\mathbf{y}_i\}_{i=1}^N$ that may possibly belong to one of $K_{\max}$ possible models. Let us denote the kth model by M_k. We may assume that model M_k is characterized by k number of parameters, denoted by the k-length vector $\boldsymbol{\theta}_k$ taking values on a parameter space Θ_k, for $k = 1, \ldots, K_{\max}$. Since θ_k is assumed to characterize the observation dependence on model M_k completely, we may denote the likelihood function $f\left(\{\mathbf{y}_i\}_{i=1}^N | \boldsymbol{\theta}_k, M_k\right)$ given the model M_k and the parameter vector θ_k simply by $f\left(\{\mathbf{y}_i\}_{i=1}^N | \boldsymbol{\theta}_k\right)$.

Then the Bayesian MAP approach to model selection will be to pick the model M_k that has the largest posterior probability $\Pr(M_k | \{\mathbf{y}_i\}_{i=1}^N)$ given the observations. However, we may write

$$\Pr(M_k | \{\mathbf{y}_i\}_{i=1}^N) = \frac{\Pr(M_k)}{f(\{\mathbf{y}_i\}_{i=1}^N)} \int_{\Theta_k} f\left(\{\mathbf{y}_i\}_{i=1}^N | \boldsymbol{\theta}_k\right) g_k(\boldsymbol{\theta}_k) d\boldsymbol{\theta}_k,$$

where we have defined $g_k(\boldsymbol{\theta}_k) = f(\boldsymbol{\theta}_k | M_k)$, $f\left(\{\mathbf{y}_i\}_{i=1}^N | \boldsymbol{\theta}_k\right) = f\left(\{\mathbf{y}_i\}_{i=1}^N | \boldsymbol{\theta}_k, M_k\right)$ and $\Pr(M_k)$ is the prior probability of the model M_k. Observe that maximizing the posterior probability $\Pr(M_k | \{\mathbf{y}_i\}_{i=1}^N)$ is equivalent to minimizing $-2 \log \Pr(M_k | \{\mathbf{y}_i\}_{i=1}^N)$. The BIC is indeed to select the model that minimizes this quantity $-2 \log \Pr(M_k | \{\mathbf{y}_i\}_{i=1}^N)$ asymptotically when N is large [104]. Under the assumption that conditional prior $g_k(\boldsymbol{\theta}_k)$ of θ_k given M_k has sufficient regularity, it has been shown that when N is large enough, $-2 \log \Pr(M_k | \{\mathbf{y}_i\}_{i=1}^N)$ converges to the following (dropping the terms that are bounded as N tends to infinity) [105]:

$$\text{BIC}_k = -2 \log f\left(\{\mathbf{y}_i\}_{i=1}^N | \hat{\boldsymbol{\theta}}_k\right) + k \log N, \tag{8.6}$$

where $\hat{\boldsymbol{\theta}}_k$ is the ML estimate of parameter vector $\boldsymbol{\theta}_k$ given by

$$\hat{\boldsymbol{\theta}}_k = \arg\max_{\boldsymbol{\theta}_k \in \Theta_k} f\left(\{\mathbf{y}_i\}_{i=1}^N | \boldsymbol{\theta}_k\right).$$

Hence, the Bayesian approach to model selection is to select the model M_k with k parameters that minimizes the BIC in (8.6).

The BIC was proposed in [105] as a Bayesian alternative to the so-called AIC defined as [106]

$$\text{AIC}_k = -2 \log f\left(\{\mathbf{y}_i\}_{i=1}^N | \hat{\boldsymbol{\theta}}_k\right) + 2k. \tag{8.7}$$

The AIC (8.7) approximately minimizes the Kullback–Leibler distance between the true distribution $f\left(\{\mathbf{y}_i\}_{i=1}^N\right)$ of observations and the distribution (likelihood) given θ_k denoted by $f\left(\{\mathbf{y}_i\}_{i=1}^N | \hat{\boldsymbol{\theta}}_k\right)$ [104, 107]. The Kullback–Leibler distance, also called the *relative entropy*, between two density functions $f(x)$ and $g(x)$ with respect to $f(x)$ is defined as [26]

$$D(f||g) = \mathbb{E}_f\left\{\frac{f(X)}{g(X)}\right\} = \int \frac{f(x)}{g(x)} f(x) dx.$$

Note that although Kullback–Leibler distance is a measure of the separation (dissimilarity) between two distributions, it is not a true distance measure due to its lack of symmetry with respect to the two distributions [26].

The X-means algorithm modifies the K-means algorithm so that rather than the actual K value, the algorithm can be initialized by specifying a possible range of values for K, say, $K \in [K_{\min}, \ldots, K_{\max}]$ [91]. Given the observation data $\{\mathbf{y}_i\}_{i=1}^N$, first, the K-means algorithm is run with $K = K_{\min}$ to determine the optimal cluster centroids. Next, the algorithm determines whether it is better to split any of the clusters into two clusters. For this, X-means algorithm splits each of the clusters into two clusters by replacing the original centroid with two new centroids and then running a K-means algorithm within each cluster with $K = 2$. The two new centroids for the initialization of each of the K-means algorithms are obtained by perturbing an original centroid in two opposite directions along a randomly chosen vector by an amount proportional to the size of the respective cluster.

Next, the algorithm performs a model selection test to determine whether the two new clusters are a better model than the original single cluster in each of the cases. The original algorithm [91] makes use of the BIC given in (8.6) for this although one may conceivably use any other suitable criterion as well. Depending on the outcome, either an original single cluster is kept or it is replaced by two of its children, completing one iteration of the algorithm. Now, the number of clusters K is updated to the current value of clusters. If $K > K_{max}$ or the cluster assignment did not change from the last iteration, the algorithm will stop. Otherwise, it will start a new iteration by running a new K-means algorithm iteration using the current centroids as initial clusters and repeating the aforementioned procedure.

Algorithm 8.3

```
X-means Algorithm
 (1) Initialize K = K_min.
 (2) Run K-means algorithm.
 (3)  FOR k = 1,...,K:
        Replace each centroid μ_k by two centroids μ_k^(1) and μ_k^(2).
        Run K-means algorithm with K = 2 over the cluster S_k.
        Replace or retain each centroid based on the model selection
criteria.
 (4) IF convergence condition is not satisfied, go to Step (2).
Otherwise Stop.
```

The X-means algorithm, similar to its parent algorithm K-means, presumes underlying data to be spherical Gaussian. However, it is known that X-means can be significantly faster than running K-means over several different K values.

8.4 DIRICHLET PROCESS MIXTURE MODEL

In the case of spectral activity classification, the Gaussian assumption of data may be questionable. Moreover, K-means and its variants are ML classifiers that assume the clusters are a priori uniformly distributed. Certainly, as argued earlier, this does not have to be the case in practice.

A Bayesian nonparametric alternative may help overcome both these drawbacks. It would be particularly interesting if we could model each observation y_i being correspond to a cluster specified by an associated parameter θ where the distribution $G(\theta)$ of θ is itself random. This not only allows a nonuniform prior on clusters but also allows for the distribution of clusters to be determined by the data itself. As we detail in this section, this is exactly what the so-called DPMM does. In addition, DPMM also allows the likelihoods to be non-Gaussian, thereby removing one of the limiting underlying assumptions of K-means and X-means algorithms.

8.4.1 Dirichlet Process

A vector $(X_1,\ldots,X_n)$ is said to be distributed according to a *Dirichlet distribution* with parameters $(a_1,\ldots,a_n)$, which we may denote as $(X_1,\ldots,X_n) \sim Dir(a_1,\ldots,a_n)$, if

$$f(x_1,\ldots,x_n|a_1,\ldots,a_n) = \frac{\Gamma\left(\sum_{i=1}^{n} a_i\right)}{\prod_{i=1}^{n}\Gamma(a_i)} \prod_{i=1}^{n} x_i^{a_i-1}, \tag{8.8}$$

subject to $\sum_{i=1}^{n} x_i = 1$, with $x_i > 0$, $a_i > 0$, for all $i = 1,\ldots,n$ [108].

For $\alpha_0 > 0$ and G_0, a probability measure, a *Dirichlet process* $DP(\alpha_0, G_0)$ is defined to be the distribution of a random probability measure G that is defined over a measurable space $(\Theta, \mathcal{B})$ such that for any finite measurable partition $(A_1, \ldots, A_r)$ of Θ, the random vector $(G(A_1), \ldots, G(A_r))$ is distributed as a finite-dimensional Dirichlet distribution with parameters $(\alpha_0 G_0(A_1), \ldots, \alpha_0 G_0(A_r))$ [92, 109, 110]:

$$(G(A_1), \ldots, G(A_r)) \sim Dir(\alpha_0 G_0(A_1), \ldots, \alpha_0 G_0(A_r)). \tag{8.9}$$

We use the notation $G \sim DP(\alpha_0, G_0)$ to denote that the probability measure G is drawn from the Dirichlet process $DP(\alpha_0, G_0)$. Then G is called a *random probability measure* whose distribution is given by the Dirichlet process $DP(\alpha_0, G_0)$ [92]. The outcomes G of a Dirichlet process are random discrete probability distributions, in contrast with simple random variables or random processes that are usually encountered in probabilistic models.

The parameter G_0, called the *base measure*, is the mean of the Dirichlet process $DP(\alpha_0, G_0)$. Hence, a priori, G_0 is the best guess for the random probability measure G. The parameter α_0, on the other hand, is a measure of the strength of this a priori belief and is called the *scale parameter*. Large values of α_0 imply that the random measure G is more likely to be closer to G_0 [96].

There are several properties of the Dirichlet process that make it a suitable candidate for the Bayesian nonparametric clustering problems. First, G is a discrete probability measure with probability 1. Note that this is true even if the base measure G_0 is nondiscrete. Second, if $(\theta_1, \ldots, \theta_n)$ is drawn from a Dirichlet process $G \sim DP(\alpha_0, G_0)$, then the posterior distribution of G given $(\theta_1, \ldots, \theta_N)$ is also a Dirichlet distribution of the form

$$G|(\theta_1, \ldots, \theta_N) \sim DP(\alpha_0, G_0 + \sum_{n=1}^{N} \delta_D(\theta_n)).$$

Proofs of these properties can be found in [108, 111].

8.4.2 Construction of the Dirichlet Process

The direct construction approach to Dirichlet process was developed in [111]. This method constructs the random probability distribution $G \sim DP(\alpha_0, G_0)$ based on the so-called *stick-breaking process*, shown in Figure 8.1, that can be summarized as follows [92, 111]:

1. Generate independent i.i.d. sequences $\{\pi'_k\}_{k'=1}^{\infty}$ and $\{\phi_k\}_{k=1}^{\infty}$ such that

$$\begin{cases} \pi'_k|\alpha_0, G_0 \sim beta(1, \alpha_0) \\ \phi_k|\alpha_0, G_0 \sim G_0 \end{cases}, \tag{8.10}$$

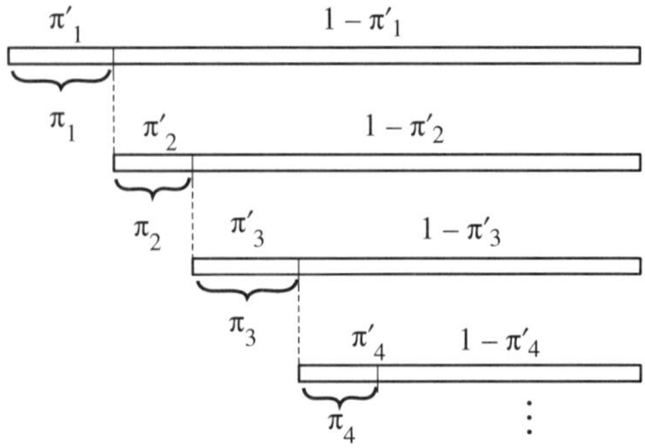

FIGURE 8.1 The stick-breaking process. Note that the values above the stick are the ratios (not the actual lengths).

where beta(a,b) is the beta distribution whose pdf is given by (see Appendix B for details)

$$f(x,a,b) = \frac{x^{a-1}(1-x)^{b-1}}{\int_0^1 u^{a-1}(1-u)^{b-1}du}.$$

2. Define

$$\pi_k = \pi'_k \prod_{l=1}^{k-1}(1-\pi'_l). \tag{8.11}$$

   It can be shown that $\sum_{k=1}^{\infty} \pi_k = 1$ with probability 1.

3. Define

$$G = \sum_{k=1}^{\infty} \pi_k \delta_{\phi_k}, \tag{8.12}$$

   where δ_{ϕ_k} denotes a probability mass concentrated at ϕ.

The method above is called the stick-breaking construction, or the stick-breaking process, because the π_k's can be interpreted as the fraction of a remaining length of an originally unit-length stick that breaks off at each time k, as shown in Figure 8.1. In each step, the remaining portion of the stick is divided according to ratio $\pi'_k : 1-\pi'_k$, and the portion with π'_k proportion is broken off as the length π_k. The randomness in G stems from the random nature of both the weights π_k and the weight positions ϕ_k in (8.12).

A sample distribution G of a Dirichlet process is illustrated in Figure 8.2, using the steps described earlier in the stick-breaking method. The fact that G constructed as in the preceding text is indeed a random probability measure distributed according

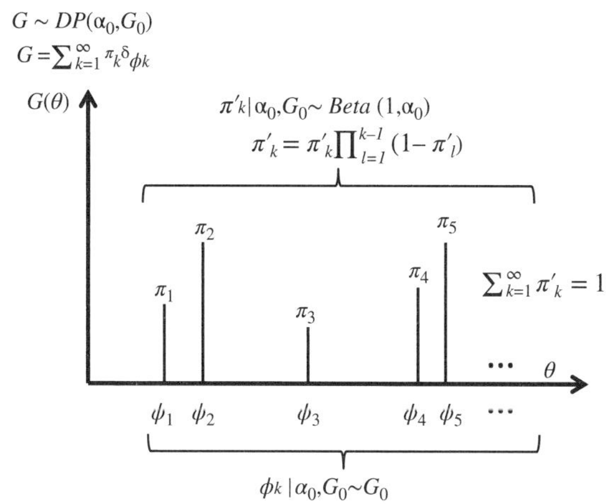

FIGURE 8.2 One realization of the Dirichlet process.

to $DP(\alpha_0, G_0)$ was established in [111]. Since G has an infinite discrete support (i.e., $\{\phi_k\}_{k=1}^{\infty}$), this makes it a suitable candidate for Bayesian nonparametric classification priors in which the number of clusters is unknown and, possibly, unbounded a priori.

However, due to the infinite sum in G in (8.12), in many applications, it may not be practical to construct G directly by using the stick-breaking process as in the preceding text. There are alternative approaches to constructing a Dirichlet process that do not define G explicitly as in (8.12), but instead characterizes the predictive distribution of the drawings θ from a $DP(\alpha_0, G_0)$. To do this, let $\theta_1, \theta_2, \ldots$ to be i.i.d. random variables distributed according to G where $G \sim DP(\alpha_0, G_0)$. Since, given G, θ_i's are distributed according to G, we may write

$$\theta_i | G \sim G. \tag{8.13}$$

Note that random variables $\theta_1, \theta_2, \ldots$ are conditionally independent only if G is given. If G is integrated out, $\theta_1, \theta_2, \ldots$ are no longer conditionally independent but can be characterized as

$$\theta_i | \theta_1, \ldots, \theta_{i-1}, \alpha_0, G_0 \sim \sum_{k=1}^{K_{i-1}} \frac{m_k}{i-1+\alpha_0} \delta_{\phi_k} + \frac{\alpha_0}{i-1+\alpha_0} G_0 \tag{8.14}$$

with

$$\theta_1 \sim G_0, \tag{8.15}$$

where $\{\phi_k\}_{k=1}^{K_{i-1}}$ denote the K_{i-1} distinct values assumed by $\theta_1, \ldots, \theta_{i-1}$ and m_k is the number of θ_j's, for $1 \leq j \leq i-1$, that are equal to ϕ_k for $1 \leq k \leq K_{i-1}$.

Note that (8.14) can also be written explicitly as

$$\theta_i|\theta_1,\ldots,\theta_{i-1},\alpha_0,G_0 = \begin{cases} \phi_k & \text{with probability} \quad \dfrac{m_k}{i-1+\alpha_0} \\ \sim G_0 & \text{with probability} \quad \dfrac{\alpha_0}{i-1+\alpha_0}G_0 \end{cases} \tag{8.16}$$

with $\theta_1 \sim G_0$. The above representation (8.14) (or (8.16)) is the so-called *Polya urn model* [112].

When the exact values ϕ_k's assumed by θ_i's are not of interest but how many of them are equal to each ϕ_k, the above representation is called the *Chinese restaurant process* (CRP) [92, 113]. The CRP is best explained by considering a (Chinese) restaurant[1] with an unbounded number of tables. A customer entering the restaurant is denoted by θ_i, whereas the distinct tables at which the customers sit are denoted by ϕ_k. The ith customer sits at the table indexed by ϕ_k, with a probability proportional to the number of customers m_k already seated there, and sits at a new table with a probability proportional to α_0 [92].

Note that the exact values of ϕ_k's are not really relevant to the CRP unlike in the case of the Polya urn model. They can be any label. Either way, these constructions provide practical approaches to sample the values θ from a distribution G that is drawn from a certain Dirichlet process. The fact that such sequentially generated random variables θ_i's are indeed equivalent to an iid drawings from a random distribution G relies on the concept of *infinitely exchangeable sequences* and *de Finetti's theorem*. A meaningful discussion of these will take us much further from the objective of developing a Bayesian nonparametric classification approach for wideband cognitive radios. An interested reader is referred to [114–117].

Observe that the conditional distribution (8.14) is not necessarily discrete since G_0 might be a continuous distribution (in contrast with G, which is discrete with probability 1). Moreover, (8.14) and (8.16) highlight the fact that θ_i's drawn from a Dirichlet process G indeed exhibit a clustering behavior since a certain value assumed by a previous θ_i, say, ϕ_k, is likely to recur with a nonnegative probability (due to the point mass functions in the conditional distribution (8.14)). In fact, the CRP representation illustrates that those ϕ_k values that have repeated frequently are more likely to be taken by the next θ. Still, the possible number of distinct θ_i values can be infinite, in general, since there is a nonnegative probability that the new θ_i value can be distinct from all the previous values assumed by $\theta_1,\ldots,\theta_{i-1}$. This conforms with the definition of G as a probability mass function over an infinite discrete support set.

8.4.3 DPMM

Let us consider a set of observations, say, $\{\mathbf{y}_i\}_{i=1}^N$, where each $\mathbf{y}_i \in \mathbb{R}^d$. According to the DPMM, these random observations $\mathbf{y}_i$ are considered to be drawn from a mixture

[1] This perhaps is better understood as a food court with open seating and a large number of tables rather than a restaurant. Customers can walk in and sit anywhere they want. Of course, usually, many are regulars, so they would tend to walk up to tables where their friends are already sitting and join them with a certain probability.

model in which the mixture components correspond to random variables θ_i's drawn from a distribution G, which, in turn, is drawn from a certain Dirichlet process. We may formally state this as [92]

$$\begin{aligned} G &\sim DP(\alpha_0, G_0) \\ \theta_i | G &\sim G \\ \mathbf{y}_i | \theta_i &\sim f_{\theta_i}(\mathbf{y}_i) \end{aligned} \quad . \tag{8.17}$$

Hence, θ_i's act as parameters that determine the likelihood of data. To be precise, the actual parameters are the ϕ_k values assumed by θ_i's. As before, we may denote by ϕ_k, for $1 \leq k \leq K_N \leq N$, these distinct values assumed by θ_i's where K_N denotes the number of mixture components in the N-sample data. Each different ϕ_k introduces a different mixture component that results in a different likelihood function $f_{\theta_i}(\mathbf{y}_i)$ for data $\mathbf{y}_i$. Since G is drawn from a Dirichlet process, it is a discrete probability distribution with an infinite support set. Hence, the model (8.17) allows for the mixture to consist of an infinitely many mixture components and endows the parameters of the mixture model with a nonparametric prior.

Our interest in the Dirichlet process arises from the suitability of the DPMM for developing a Bayesian nonparametric classification framework for spectral activity identification in wideband spectrum sensing. The classification problem of interest to us is that given a set of observations $\{\mathbf{y}_i\}_{i=1}^N$, for each $\mathbf{y}_i \in \mathbb{R}^d$, we need to separate them into a set of different groups (or classes or clusters) without a priori knowing the number of classes. Moreover, we would like to allow likelihoods of data from different clusters to be of different types of distributions rather than just multivariate Gaussians with different means and covariances.

From the stick-breaking construction (8.10–8.12), it is clear that ϕ_k's that are the parameters of the likelihood functions are drawn from the basc distribution G_0. Hence, we may construct a suitable Dirichlet process so that ϕ_k's assumed by θ_i's represent parameters of a likelihood function corresponding to a distinct cluster. This gives a Bayesian nonparametric classification approach that allows the number of clusters to be treated as a random variable whose support set can grow with the complexity of the data set itself as opposed to parametric classifiers, which may assume that the number of clusters is fixed [93–96].

8.5 BAYESIAN NONPARAMETRIC CLASSIFICATION BASED ON THE DPMM AND THE GIBBS SAMPLING

Consider a sequence of observations $\mathbf{y}_{1:N} \triangleq \{\mathbf{y}_i\}_{i=1}^N$ where each $\mathbf{y}_i \triangleq [y_{i,1}, \ldots, y_{i,d}]^T \in \mathbb{R}^d$. Suppose that these observations are drawn from a mixture model and we do not know the number of mixture components. Further, we are interested in allowing the number of mixture components to grow with the complexity of the data itself. Then, it is natural for us to consider that the data $\{\mathbf{y}_i\}_{i=1}^N$ are generated from a DPMM of the form of (8.17). This means that the mixture components θ_i are drawn from a (discrete) distribution G, which itself is a random distribution drawn from a Dirichlet process: $G \sim DP(\alpha_0, G_0)$, for $G = \sum_{k=1}^{\infty} \pi_k \delta_{\phi_k}$. Here, ϕ_k's are the distinct values assumed by θ_i's so that they characterize the distinct mixture components represented in data $\{\mathbf{y}_i\}_{i=1}^N$ with π_k's being the corresponding probabilities of each of the distinct mixture components (or the distinct clusters). Recall also from (8.16) that the ϕ_k's are generated by the base prior $G_0(.)$. Hence, this framework gives us freedom to select an appropriate base prior so as to represent desired dependence of observations on the cluster properties. In the following, we will assume that G_0 has an associated density g_0.

The problem at hand is to estimate the mixture component $\hat{\theta}_i$ for each observation $\mathbf{y}_i$, for all $i \in \{1, \ldots, N\}$. We are particularly interested in finding the MAP estimates of θ_i $(i = 1, \ldots, N)$, given the observations $\mathbf{y}_{1:N}$. However, usually, it is hard to find analytical MAP estimates of θ_i's since the joint distribution of $(\theta_1, \ldots, \theta_N)$, given $\mathbf{y}_{1:N}$, is, in general, unknown. As an alternative, we may try to obtain the individual MAP estimates if we could compute the distribution of θ_i given $\mathbf{y}_{1:N}$. When this is also not possible, we may use Monte Carlo methods. A particularly simple Markov chain Monte Carlo (MCMC) approach to obtain the joint conditional distribution of $(\theta_1, \ldots, \theta_N)$ given $(\mathbf{y}_1, \ldots, \mathbf{y}_N)$ is the *Gibbs sampling* that draws samples from the marginal conditional distribution of each parameter θ_i, given $\mathbf{y}_{1:N}$ and all the other parameters $\{\theta_j\}_{j \neq i}$, where $\{\theta_j\}_{j \neq i}$ are kept fixed at the values of their most recent estimates [118]. The assumption of the DPMM data allows us to write down this posterior distribution of $\theta_i | \{\theta_j\}_{j \neq i}, \mathbf{y}_{1:N}$ as [96]:

$$\theta_i | \{\theta_j\}_{j \neq i}, \mathbf{y}_{1:N} \begin{cases} = \theta_j & \text{with probability} \quad q_j \\ \sim f(\theta_i | \mathbf{y}_i) & \text{with probability} \quad q_0 \end{cases}, \tag{8.18}$$

where q_j, for $1 \leq j < i$, and q_0 are given, respectively, by

$$q_j = \frac{f_{\theta_j}(\mathbf{y}_i)}{\alpha_0 f(\mathbf{y}_i) + \sum_{j=1, j \neq i}^{N} f_{\theta_j}(\mathbf{y}_i)} \tag{8.19}$$

and

$$q_0 = \frac{\alpha_0 f(\mathbf{y}_i)}{\alpha_0 f(\mathbf{y}_i) + \sum_{j=1, j \neq i}^{N} f_{\theta_j}(\mathbf{y}_i)}, \tag{8.20}$$

with

$$f(\mathbf{y}_i) = \int_{\theta} f_{\theta}(\mathbf{y}_i) g_0(\theta) d\theta \tag{8.21}$$

and

$$f_{\theta}(\mathbf{y}_i) \triangleq f(\mathbf{y}_i | \theta_i = \theta) \tag{8.22}$$

denoting, respectively, the marginal distribution of $\mathbf{y}_i$ and the likelihood of observations for any θ, where θ_i is the parameter corresponding to observation $\mathbf{y}_i$.

From (8.18), note that the assumption of an underlying DPMM for the cluster parameters θ_i's implies that the parameter θ_i corresponding to data $\mathbf{y}_i$ is equal to one of the previous θ_j's with probability q_j, for $1 \leq j \leq i-1$, or it is a new value drawn according to the conditional distribution $f(\theta_i|\mathbf{y}_i)$ with probability q_0.

Since (8.18) is exactly the conditional distribution that is needed to implement a Gibbs sampling procedure, it has been shown that indeed (8.18) can be used to implement a MCMC approach based on Gibbs sampling to obtain the joint conditional distribution of $(\theta_1,\ldots,\theta_N)$ given $(\mathbf{y}_1,\ldots,\mathbf{y}_N)$ [96]. Since it is normally assumed that the likelihood $f_\theta(\mathbf{y}_i)$ is known, then the evaluation of (8.18) only requires the ability to obtain the posterior distribution $f(\theta_i|\mathbf{y}_i)$ in a closed form. This, can be achieved in a particularly convenient way if θ_i has a *conjugate prior* for the likelihood $f_{\theta_i}(\mathbf{y}_i)$. Note that the prior and posterior are called conjugate distributions if the posterior distribution $P(\theta|y)$ is in the same family of distributions as the prior distribution $P(\theta)$. When this is true, the prior is called a conjugate prior for the given likelihood. If, moreover, the conjugate prior for a given likelihood is the same distribution as that of the likelihood, then such a family of distributions is called *self-conjugate*.[2]

Conjugate prior for a given likelihood is not necessarily unique since it depends on the family of distributions of interest. For many useful likelihood distributions, there are known conjugate priors. See Appendix C for a collection of conjugate priors for commonly encountered univariate likelihood functions. An interested reader is referred to [119] for a more detailed list of likelihoods and their corresponding conjugate priors and posterior update formulas.

If we assume a conjugate prior in a DPMM model, then the base prior $g_0(\theta_i)$ and the posterior $f(\theta_i|\mathbf{y}_i)$ will belong to the same family of distributions. For instance, if both base prior $g_0(\theta_i)$ and the likelihood function $f_{\theta_i}(\mathbf{y}_i)$ were to be Gaussian, then the posterior distribution $f(\theta_i|\mathbf{y}_i)$ will also be Gaussian. Hence, Gaussian prior is conjugate to Gaussian likelihood. Of course, Gaussian family is then self-conjugate as well.

In many applications of DPMM, it is common to assume conjugate priors due to the computational advantage afforded by them [92,96,100]. When nonconjugate priors are assumed, simple Gibbs sampling-based approach below requires numerical integration that can lead to unacceptably high computational complexity. Although special approaches to handle DPMM clustering (estimation) with nonconjugate priors have been developed in [120–124], in the following, we will always assume conjugate priors so that a clustering approach with reasonable time complexity can be achieved.

The Gibbs sampling algorithm to obtain the joint conditional distribution of $(\theta_1,\ldots,\theta_N)$ given $(\mathbf{y}_1,\ldots,\mathbf{y}_N)$ can be initialized with arbitrary estimates of θ_i's:

$$(\theta_1,\ldots,\theta_N) = (\theta_1^{(0)},\ldots,\theta_N^{(0)}) \;=\; \theta_{1:N}^{(0)},$$

for some arbitrary initial values.

[2]http://en.wikipedia.org/wiki/Conjugate_prior

Then it draws samples from the conditional distribution of each parameter θ_i, given $\mathbf{y}_{1:N}$, and the other parameters $\{\theta_j\}_{j\neq i}$, where $\{\theta_j\}_{j\neq i}$ are fixed at the values of their most recent estimates [118]. One Gibbs iteration contains the following N steps:

$$\begin{aligned}
&\text{Draw a sample } \theta_1^{(1)} \text{ from } \theta_1|\theta_2 = \theta_2^{(0)}, \theta_3 = \theta_3^{(0)}, \ldots, \theta_N = \theta_N^{(0)}, \mathbf{y}_{1:N}.\\
&\text{Draw a sample } \theta_2^{(1)} \text{ from } \theta_2|\theta_1 = \theta_1^{(1)}, \theta_3 = \theta_3^{(0)}, \ldots, \theta_N = \theta_N^{(0)}, \mathbf{y}_{1:N}.\\
&\vdots\\
&\text{Draw a sample } \theta_N^{(1)} \text{ from } \theta_N|\theta_1 = \theta_1^{(1)}, \theta_2 = \theta_2^{(1)}, \ldots, \theta_{N-1} = \theta_{N-1}^{(1)}, \mathbf{y}_{1:N}.
\end{aligned} \tag{8.23}$$

Note that each step in the above sampling algorithm only requires a conditional distribution of the form of (8.18). Hence, when conditional marginal distributions (8.18) of θ_i's can be characterized easily, the above Gibbs sampling scheme is particularly convenient to implement [96].

Simple as it may, it is known that these samples indeed converge in probability to the actual posterior distribution $P(\theta_{1:N}|\mathbf{y}_{1:N})$ of $(\theta_1, \ldots, \theta_N)$ given $\mathbf{y}_{1:N}$ [97]. Let us denote by $Q_I(\theta_{1:N}^{(0)}, A)$ the probability that after one iteration the Algorithm (8.64) produces a sample value that is contained in the measurable set A, with an initial value $\theta_{1:N}^{(0)}$:

$$Q_I(\theta_{1:N}^{(0)}, A) = P\left\{\theta_{1:N}^{(1)} \in A|\theta_{1:N}^{(0)}\right\}.$$

Hence, $Q_I(.,.)$ is called the transition kernel of the Markov chain. Similarly, for $s > 0$, we may define

$$Q_I^s(\theta_{1:N}^{(0)}, A) = P\left\{\theta_{1:N}^{(s)} \in A|\theta_{1:N}^{(0)}, s\right\}.$$

Theorem 1 of [97] establishes that for almost all starting values of $\theta_{1:N}^{(0)}$, the probability measure Q_I^s (defined over a measurable space $\Omega \supset A$) converges in total variation norm to the posterior distribution $P(\theta_{1:N}|\mathbf{y}_{1:N})$:

$$\lim_{s\to\infty} \|Q_I^s(\theta_{1:N}^{(0)}, .) - P(\theta_{1:N}|\mathbf{y}_{1:N})\| = 0,$$

ensuring that for large s Gibbs sampling outcomes $\theta_{1:N}^{(s)}$ will indeed be distributed according to the actual posterior distribution $P(\theta_{1:N}|\mathbf{y}_{1:N})$ of $\theta_{1:N}$ given $\mathbf{y}_{1:N}$. This justifies the use of the Gibbs sampling outcomes (8.64) to construct the posterior distribution of $\theta_{1:N}$ given $\mathbf{y}_{1:N}$ and use it for finding an estimation of $\theta_{1:N}$.

8.5.1 DPMM-Based Classification of Scalar Observations

Let us first consider the simple case of scalar observations in which each $y_i \in \mathbb{R}$. In this case, it can be verified that the conditional distribution (8.18) of θ_i reduces to the following [96]:

$$\theta_i | \{\theta_j\}_{j \neq i}, \mathbf{y} = \begin{cases} \theta_j & \text{with prob.} \quad q_j \\ \sim f(\theta_i | y_i) & \text{with prob.} \quad q_0 \end{cases}, \tag{8.24}$$

where

$$q_j = \frac{f_{\theta_j}(y_i)}{\alpha_0 f(y_i) + \sum_{n=1, n \neq i}^{N} f_{\theta_n}(y_i)}, \tag{8.25a}$$

$$q_0 = \frac{\alpha_0 f(y_i)}{\alpha_0 f(y_i) + \sum_{n=1, n \neq i}^{N} f_{\theta_n}(y_i)} \tag{8.25b}$$

$$f(\theta_i | y_i) = \frac{f_{\theta_i}(y_i) g_0(\theta_i)}{f(y_i)}, \tag{8.25c}$$

and

$$f(y_i) = \int f_\theta(y_i) g_0(\theta) d\theta. \tag{8.26}$$

8.5.1.1 Scalar Gaussian Observations with a Uniform Prior on the Means and Fixed Variance A particularly interesting example is that of a Gaussian mixture. When the observation is thought of as correspond to a mix of several different Gaussian distributions, it is called Gaussian mixture model (GMM). Indeed, suppose that $\{y_i\}_{i=1}^{N}$ is drawn from a Gaussian mixture formed by a set of mixture components where each mixture component is characterized by the value of a parameter $\theta \in \mathbb{R}$. In particular, let us assume that given θ_i each y_i is independent and has a Gaussian likelihood with mean θ_i and a known variance σ_i^2: $y_i | \theta_i \sim \mathcal{N}(\theta_i, \sigma_i^2)$ so that

$$f_{\theta_i}(y_i) = \frac{1}{\sqrt{2\pi\sigma_i^2}} e^{-\frac{(y_i - \theta_i)^2}{2\sigma_i^2}}. \tag{8.27}$$

The GMM is quite commonly used in clustering and classification approaches. However, normally, it is assumed that the number of mixture components is known. In contrast, the DPMM framework allows us to treat the number of mixture components to be any (random) value. The actual value will be determined by the data itself. For example, suppose that the only information we have is that the parameter θ can assume values in a certain range, say, from $\theta_{\min}$ to $\theta_{\max}$. Thus, θ can take any value in this range, and correspondingly, we may have any number of distinct mixture components in the observed data. We may model this scenario by assuming a base prior $g_0(\theta)$ on the mean of the observation likelihood that is uniform over the range $[\theta_{\min}, \theta_{\max}]$:

$$g_0(\theta_i) \sim \mathcal{U}[\theta_{\min}, \theta_{\max}]. \tag{8.28}$$

Using (8.27) and (8.28) in (8.26), the marginal distribution $f(y_i)$ can be evaluated to be

$$f(y_i) = \frac{1}{\theta_{\max} - \theta_{\min}} \left[Q\left(\frac{\theta_{\min} - y_i}{\sigma_i}\right) - Q\left(\frac{\theta_{\max} - y_i}{\sigma_i}\right) \right]. \tag{8.29}$$

From (8.25c) and (8.29),

$$f(\theta_i|y_i) = \begin{cases} \dfrac{1}{\sqrt{2\pi\sigma_i^2}} \dfrac{1}{f(y_i)} e^{-\frac{(y_i-\theta_i)^2}{2\sigma_i^2}} & \text{if } \theta_{\min} \leq \theta_i \leq \theta_{\max} \\ 0 & \text{otherwise} \end{cases}. \tag{8.30}$$

Note that the uniform prior is not conjugate for the Gaussian likelihood. Hence, $f(\theta_i|y_i)$ of (8.30) is not very convenient to use in practice. However, if the support of the prior G_0 is sufficiently large, then we may be able to approximate $f(\theta_i|y_i)$ as being normal so that $f(\theta_i|y_i) \sim \mathcal{N}(y_i, \sigma_i^2)$.

8.5.1.2 Scalar Gaussian Observations with a Gaussian Prior on the Means and Fixed Variance As can be seen from Appendix C, the most convenient conjugate prior in the case of Gaussian likelihood is the Gaussian prior. Hence, let us consider the same case as in the preceding section in which $y_i \in \mathbb{R}$ are independent, $\theta_i \in \mathbb{R}$ with $y_i|\theta_i \sim \mathcal{N}(\theta_i, \sigma_i^2)$ for σ_i^2 known.

However, now suppose that we assume a base prior on the mean θ_i's that is normal as below:

$$g_0(\theta_i) \sim \mathcal{N}(m, \tau\sigma_i^2),$$

where m and $\tau > 0$ are, respectively, the mean and (variance) scale parameters of the prior distribution. It is straightforward to show that the posterior distribution of θ_i given y_i is

$$f(\theta_i|y_i) = \frac{1}{\sqrt{2\pi X\sigma_i^2}} e^{-\frac{(\theta_i - x_i)^2}{2X\sigma_i^2}},$$

where

$$x_i = \frac{m + \tau y_i}{1 + \tau}$$

and

$$X = \frac{\tau}{1 + \tau}.$$

Hence, when the cluster base prior and observation likelihood are both Gaussian, the posterior distribution of θ_i given y_i is also a Gaussian of the form of

$$\theta_i | y_i \sim \mathcal{N}\left(x_i, X\sigma_i^2\right).$$

The marginal distribution $f(y_i)$ of y_i required for evaluating (8.24) is also Gaussian in this case. Indeed,

$$\begin{aligned} f(y_i) &= \int_{-\infty}^{\infty} \frac{1}{\sqrt{2\pi\sigma_i^2}} e^{-\frac{(y_i-\theta)^2}{2\sigma_i^2}} \frac{1}{\sqrt{2\pi\tau\sigma_i^2}} e^{-\frac{(\theta-m)^2}{2\tau\sigma_i^2}} d\theta \\ &= \frac{1}{\sqrt{2\pi\sigma_i^2(1+\tau)}} e^{-\frac{(y_i-m)^2}{2\sigma_i^2(1+\tau)}}, \end{aligned}$$

so that

$$f(y_i) \sim \mathcal{N}\left(m, \sigma_i^2(1+\tau)\right).$$

Example 8.1 (Scalar GMM Observations with $K=2$ and a Known Variance) Let us consider an $N = 100$ observations drawn from a GMM with $K = 2$ mixture components. The two Gaussians are centered at $\theta_1 = 0$ and $\theta_2 = 1$. Figure 8.3 shows the classification performance with a known fixed observation variance of $\sigma_i = 0.1$ for the DPMM with uniform prior, DPMM with Gaussian prior, and X-means algorithms. In the case of DPMM-based approaches, we have taken $\theta_i \sim U(-2,2)$ for the uniform prior, whereas we have assumed $\theta_i \sim \mathcal{N}(0.5,1)$ for the Gaussian prior. In this case, all the methods lead to a classification accuracy of 100%.

In Figure 8.4, we have performed the same procedure as in the preceding text but now with the fixed variance increased to $\sigma_i = 0.25$. Here, we have changed the assumed Gaussian prior to be $\theta_i \sim \mathcal{N}(0.5, 6.25)$. Figure 8.4 shows that the DPMM with uniform prior has a classification accuracy of 96%, while both the DPMM with Gaussian prior and the X-means algorithm achieve an accuracy of 98%. ■

The two cluster situation of the above example is of course one of the simplest situations one may assume. In the following, we consider a little more difficult problem.

Example 8.2 (Scalar GMM Observations with $K=4$ and a Known Variance) Here, we consider a situation similar to the scalar GMM data classification problem of Example 8.1 but with $K = 4$ mixture components. The four Gaussians are characterized by their means $\theta_1 = 0$, $\theta_2 = 1$, $\theta_3 = 2$, and $\theta_4 = 3$. Figure 8.5 corresponds to the case of their common standard deviation being $\sigma_i = 0.1$. Again, we apply DPMM with uniform prior, DPMM with Gaussian prior, and X-means classification algorithms. We have assumed $\theta_i \sim U(-2,4)$ for the uniform prior whereas $\theta_i \sim \mathcal{N}(1.5,1)$

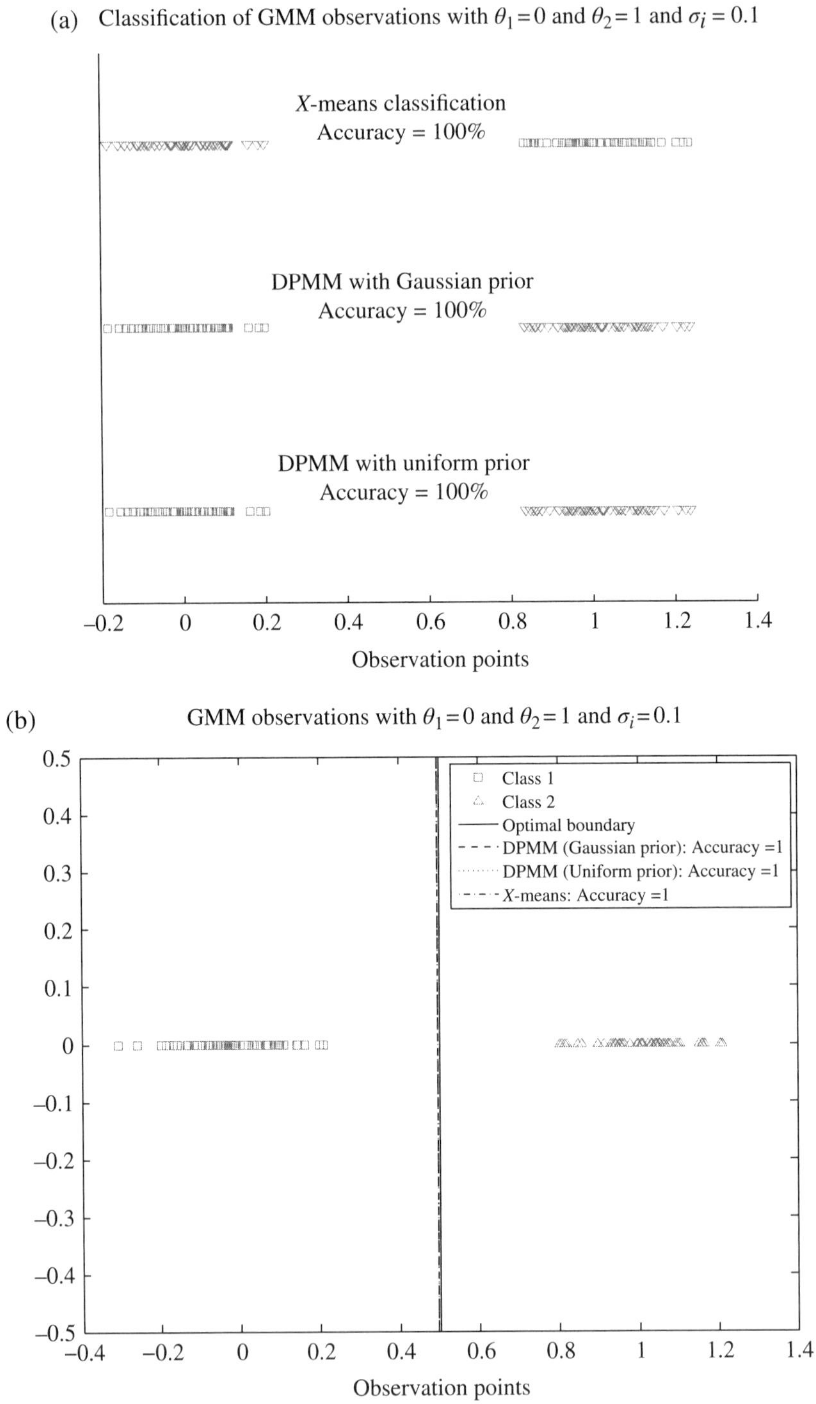

FIGURE 8.3 Classification of $K = 2$ GMM observations with $\theta_1 = 0$, and $\theta_2 = 1$ and known $\sigma_i = 0.1$. (a) Classification results with X-means and DPMM classifiers. (b) Classification boundaries with X-means and DPMM classifiers.

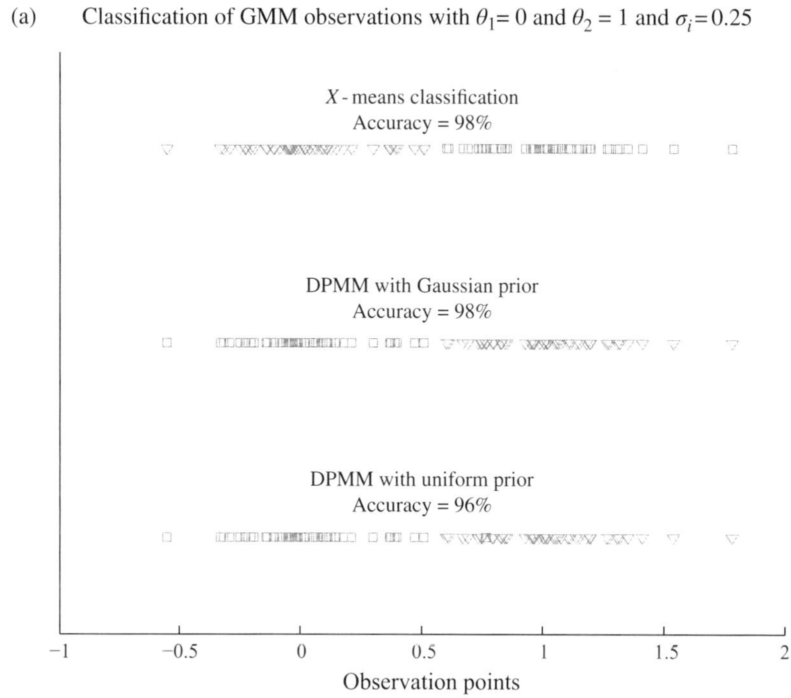

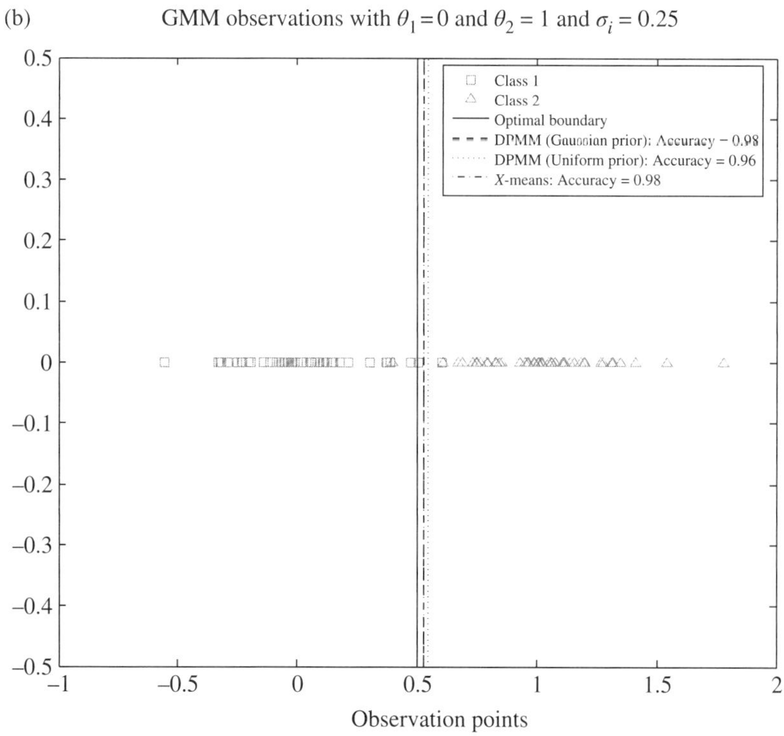

FIGURE 8.4 Classification of $K = 2$ GMM observations with $\theta_1 = 0$ and, $\theta_2 = 1$ and known $\sigma_i = 0.25$. (a) Classification results with *X*-means and DPMM classifiers. (b) Classification boundaries with *X*-means and DPMM classifiers.

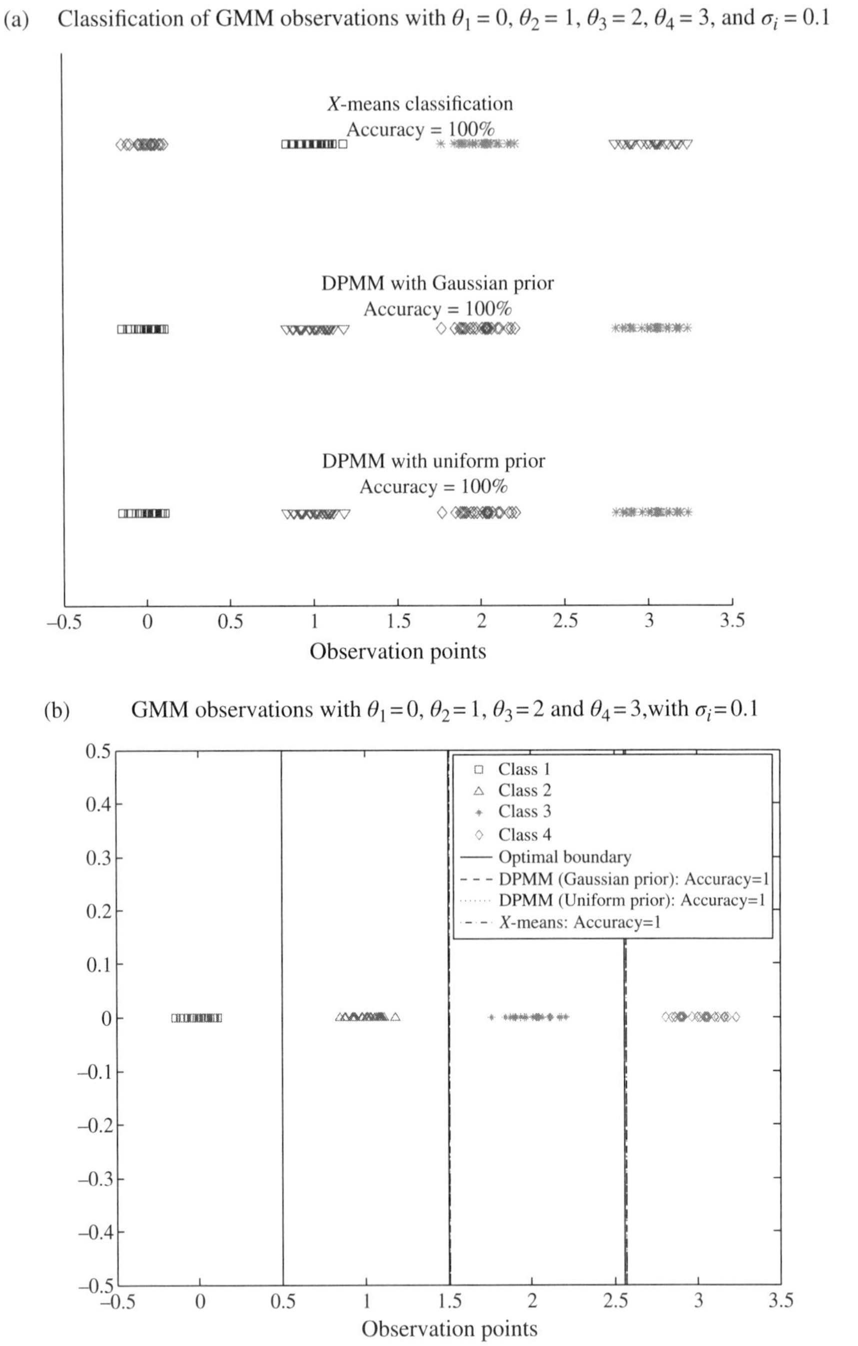

FIGURE 8.5 Classification of $K = 4$ GMM observations with $\theta_1 = 0$, $\theta_2 = 1$, $\theta_3 = 2$, and $\theta_4 = 3$ and known $\sigma_i = 0.1$. (a) Classification results with X-means and DPMM classifiers. (b) Classification boundaries with X-means and DPMM classifiers.

for the Gaussian prior. Figure 8.5 shows that in this case all three methods lead to a classification accuracy of 100%.

Figure 8.6 shows the performance with the noise variance increased to $\sigma_i = 0.25$. Again we have changed the Gaussian prior to $\theta_i \sim \mathcal{N}(1.5, 6.25)$ in the DPMM with Gaussian prior algorithm. It is seen that the DPMM with uniform prior has a classification accuracy of 85% and the DPMM with Gaussian prior has a classification accuracy of 91%. However, the X-means achieves an accuracy of only 73% in this case. The reason is that with $K = 4$, X-means algorithm was not able to correctly estimate the number of clusters. Indeed, it is seen that X-means algorithm estimates the number of clusters to be $\hat{K} = 3$. ■

From the above two examples, we see that essentially all methods have similar performance for $K = 2$ GMM clusters, while the performance of all algorithms degrades as the variance of the data increases. When the number of clusters is $K = 4$, however, the DPMM-based classification significantly outperforms the X-means algorithm. As we observed from Figure 8.6, the reason is mainly due to erroneous estimation of the number of clusters K by the X-means algorithm.

8.5.1.3 Scalar Gaussian Observations with a Gaussian Prior on the Means with an Unknown Variance In the case of GMM, a commonly encountered scenario is that not only the means of the likelihoods but also the variances are distinct and unknown. Such a model can provide more flexibility in practice to model observations corresponding to different classes compared to having just different means.

Thus, consider a sequence of data $y_{1:N}$, where each $y_i \in \mathbb{R}$ as before, that are drawn from a DPMM and are normally distributed given their mixture component parameters $\theta_{1:N} \triangleq \{\theta_i\}_{i=1}^N$. However, now let us assume that each θ_i represents a mixture component with a distinct mean μ_i as well as a distinct variance σ_i^2. Under the DPMM formulation, then both mean and the variance of the conditional distribution of y_i (i.e., the likelihood) have to be treated as random and unknown.

We may model this scenario by letting $\theta_i = (\mu_i, \sigma_i^2)$, for $i = 1, \ldots, N$. Note that given θ_i, the observation y_i is still Gaussian:

$$y_i|\theta_i \sim \mathcal{N}(\mu_i, \sigma_i^2). \tag{8.31}$$

Since $\theta_i|G \sim G$, now the distribution G of the parameters $\theta_{1:N}$ is a probability measure on $\mathbb{R} \times \mathbb{R}^+$. For $G \sim DP(\alpha_0, G_0)$ to model this scenario, the base prior G_0 needs to be a bivariate distribution over $\mathbb{R} \times \mathbb{R}^+$. Moreover, we would like this to be a conjugate prior to the Gaussian likelihood assumed in (8.31). It can be verified (see Appendix C) that a conjugate prior G_0 for a Gaussian likelihood with both random mean and variance is the normal/inverse-gamma prior according to which the unknown mean has a normal prior, while the reciprocal of the unknown variance has a gamma prior (see Appendix B for the definition of the gamma distribution $G(\alpha, \beta)$ that is characterized by a scale parameter α and a shape parameter β). Thus, with

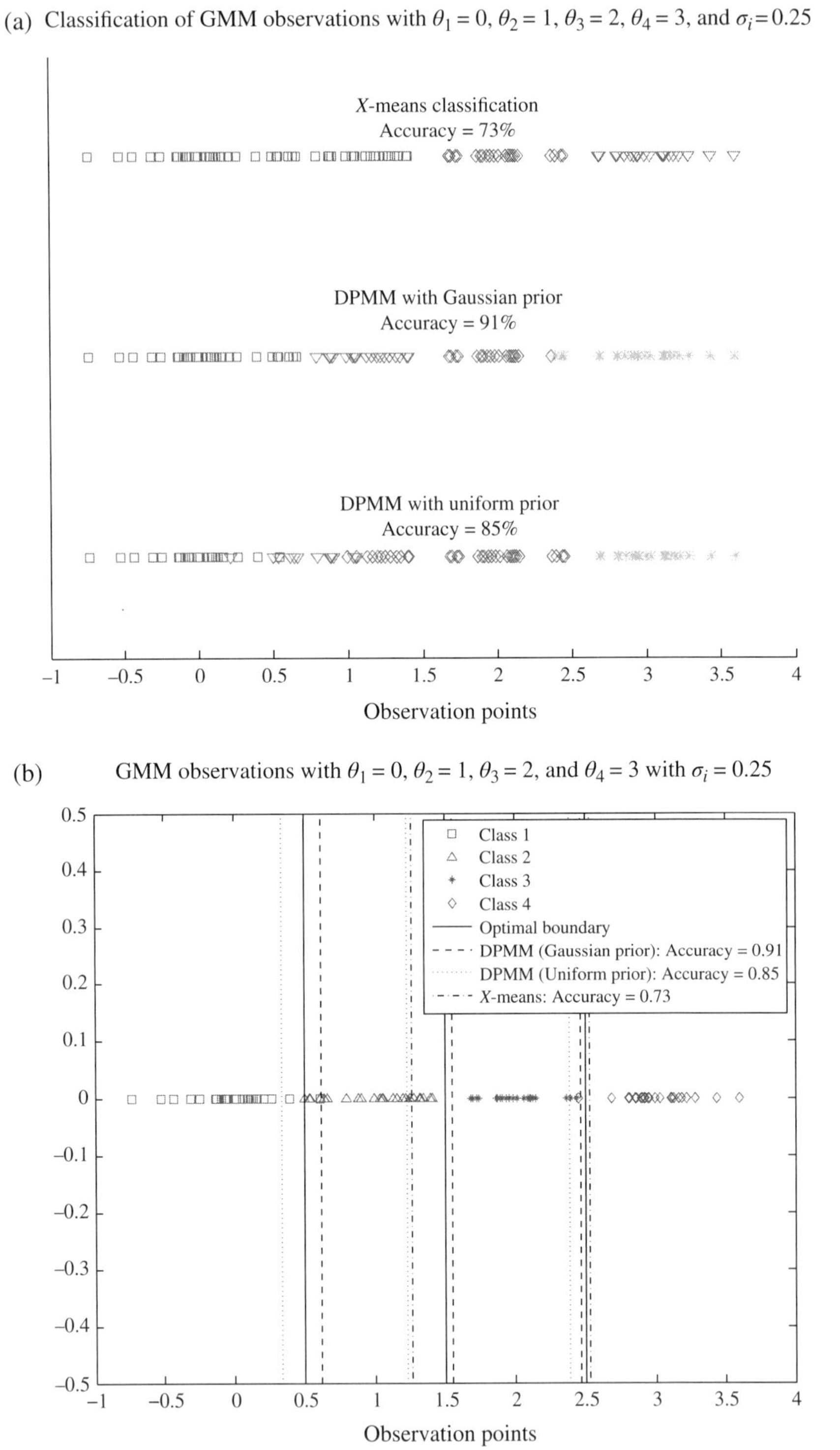

FIGURE 8.6 Classification of $K = 4$ GMM observations with $\theta_1 = 0$, $\theta_2 = 1$, $\theta_3 = 2$, and $\theta_4 = 3$ and known $\sigma_i = 0.25$. (a) Classification results with X-means and DPMM classifiers. (b) Classification boundaries with X-means and DPMM classifiers.

conjugate normal/inverse-gamma prior for θ_i, new mean and variance values in the DPMM (8.16) or (8.18) are drawn according to

$$\frac{1}{\sigma_i^2} \sim G(S/2, s/2) \tag{8.32}$$

and

$$\mu_i|\sigma_i^2 \sim \mathcal{N}(m, \tau\sigma_i^2), \tag{8.33}$$

for some scale and shape parameters $S/2$ and $s/2$, respectively, and mean and variance scale factor m and τ, respectively. The shape parameter s of the gamma distribution is related to the uncertainty about the observation errors such that s reflects the confidence in the estimated observation error. A large value of s corresponds to the case in which the observation error can be estimated accurately (e.g., within a certain range), while a small value of s corresponds to the case in which there is little knowledge available about the magnitude of observation error. Parameter S is the prior estimate of the measurement errors [97]. A large value of τ implies a large dispersion among the cluster means, whereas parameter m is a prior estimate of these means [97].

With a slight abuse of notation, we may write the posterior marginal distribution (8.24) of θ_i, for each $i = 1, \ldots, N$, given the observations $y_{1:N}$ and $\{\theta_j\}_{j\neq i}$ in the following compact form [97]:

$$\theta_i|\{\theta_j\}_{j\neq i}, y_{1:N} \sim q_0 f(\theta_i|y_i) + \sum_{j=1, j\neq i}^{N} q_j \delta_{\theta_j}, \tag{8.34}$$

where $f(\theta_i|y_i)$ denotes the posterior bivariate normal/inverse-gamma distribution whose components can be shown to be

$$\frac{1}{\sigma_i^2}|y_i \sim \Gamma\left(\frac{S_i}{2}, \frac{1+s}{2}\right)$$

and

$$\mu_i|\sigma_i^2, y_i \sim \mathcal{N}\left(x_i, X\sigma_i^2\right),$$

where

$$\begin{aligned} S_i &= S + \frac{(y_i - m)^2}{1+\tau} \\ X &= \frac{\tau}{1+\tau} \\ x_i &= \frac{m + \tau y_i}{1+\tau}. \end{aligned} \tag{8.35}$$

Using (8.25a) and (8.25b), respectively, it can be shown that the required probabilities q_0 and q_j in (8.34) are given by

$$q_0 \propto \frac{\alpha_0 c(s)}{\sqrt{M}} \left(1 + \frac{(y_i - m)^2}{sM}\right)^{-(1+s)/2}$$

and, for $j \in \{1, \ldots, N\}$ with $j \neq i$,

$$q_j \propto \frac{1}{\sqrt{2\sigma_j^2}} e^{\frac{-(y_j-\mu_j)^2}{2\sigma_j^2}},$$

subject to

$$\sum_{j=1, j\neq i}^{N} q_j = 1$$

with

$$M = \frac{1+\tau}{s} S$$

and

$$c(s) = s^{-1/2} \Gamma\left(\frac{1+s}{2}\right) \Gamma\left(\frac{s}{2}\right). \tag{8.36}$$

Example 8.3 (Scalar GMM Observations with $K = 2$ and Both Unknown Means and an Unknown Variance) As an example of both unknown cluster means and unknown covariance scenario, let us consider $N = 100$ observations drawn from a GMM with $K = 2$ clusters. Two unknown cluster means of the GMM are taken to be $\mu_1 = 0$ and $\mu_2 = 1$. We assume an observation model with an unknown variance σ_i^2. Figure 8.7(a) and (b) show the classification results with the X-means algorithm as well as with the DPMM-based algorithm for $\sigma_i^2 = 0.01$ and $\sigma_i^2 = 0.05$.

In applying the DPMM-based classification algorithm, we assume that the observations are independent Gaussian distributed given the cluster parameters θ_i's, for $i = 1, \ldots, N$, so that $y_i|\theta_i \sim \mathcal{N}(\mu_i, \sigma_i^2)$. We arbitrarily set the hyper-parameters of the model to be $S = 0.1$, $s = 4$, $m = 1.5$, and $\tau = 1$. The rate of creating new clusters in the Dirichlet process prior (scale parameter of the Dirichlet process) is set to $\alpha_0 = 1$.

Figure 8.7(a) shows that when $\sigma_i^2 = 0.01$, both DPMM and X-means achieve perfect classification. Similarly, Figure 8.7(b) shows that with $\sigma_i^2 = 0.05$, the DPMM achieves an accuracy of 99%, while the X-means achieves 98%. Note that the Gibbs sampling will require more iterations for convergence when the variance of the data is unknown, compared to the known variance case. ■

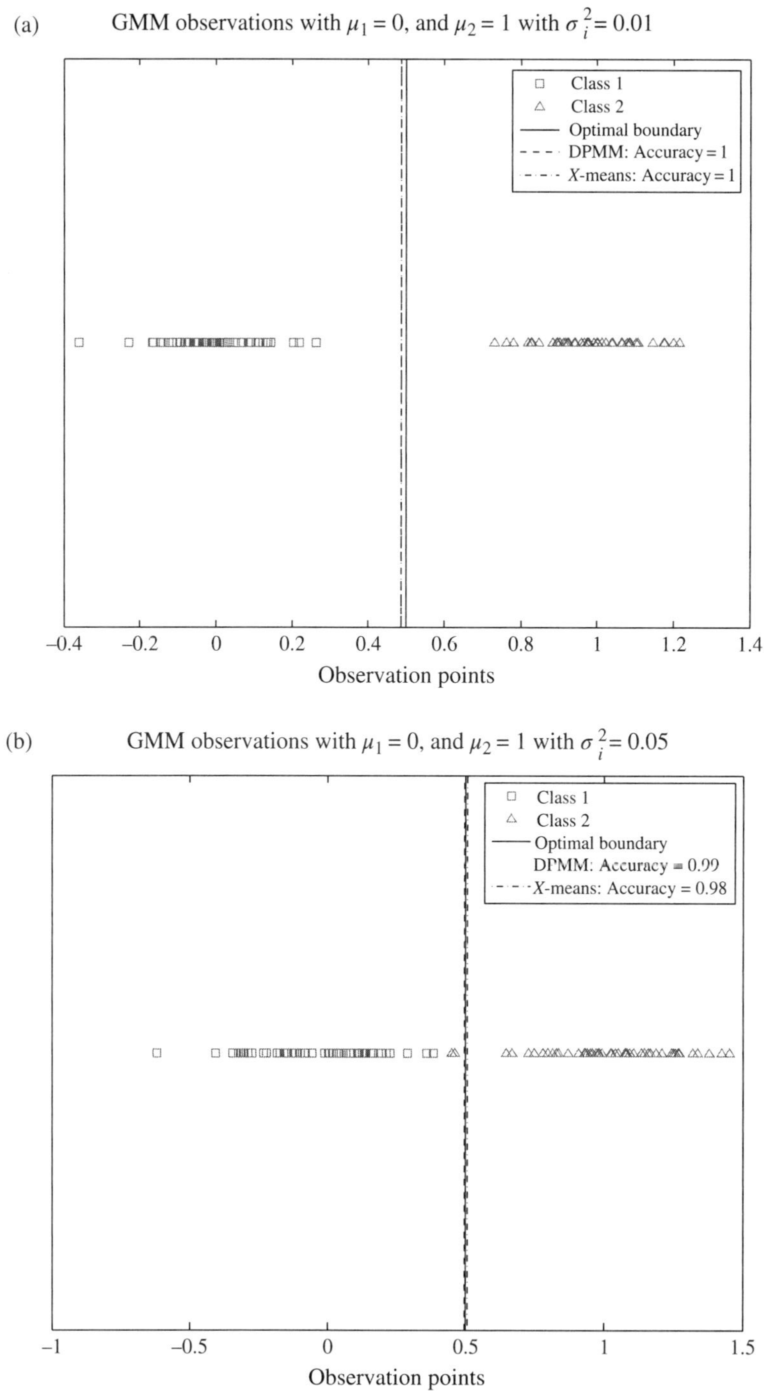

FIGURE 8.7 Classification of $K = 2$ GMM observations with unknown means $\mu_1 = 0$ and $\mu_2 = 1$ and an unknown common variance. (a) Classification results with X-means and DPMM classifiers when the unknown common variance $\sigma_i^2 = 0.01$. (b) Classification results with X-means and DPMM classifiers when the unknown common variance $\sigma_i^2 = 0.05$.

As before, in the next example, let us assume a somewhat more challenging classification scenario to highlight the performance difference of different algorithms.

Example 8.4 (Scalar GMM Observations with K=4 and Both Unknown Means and an Unknown Variance) In Figure 8.8, we show the classification algorithms for a scalar Gaussian case but now with $K = 4$ clusters with both unknown means and variance. Unknown cluster means of the GMM model from which the observations are generated are $\mu_1 = 0$, $\mu_2 = 1$, $\mu_3 = 2$, and $\mu_4 = 3$.

Figure 8.8(a) and (b), respectively, show the performance of the X-means and the DPMM-based classification algorithms when the unknown variance $\sigma_i^2 = 0.01$ and $\sigma_i^2 = 0.05$. The prior distributions for the means and the variance in DPMM model are chosen to be the same as in the aforementioned $K = 2$ case. As can be seen from Figure 8.8(a), when $\sigma_i^2 = 0.01$, both DPMM and X-means achieve perfect classification. However, when $\sigma_i^2 = 0.05$, Figure 8.8(b) shows that the DPMM achieves an accuracy of 87%, while the X-means algorithm achieves only about 74%. ■

8.5.2 DPMM-Based Classification of Multidimensional Gaussian Observations

Even when observations are assumed to be vector valued, the multivariate normal likelihoods with conjugate priors lead to closed-form analytical expressions for the conditional marginal distributions of θ_i's characterized by (8.18). Thus, we may easily generalize the aforementioned two cases of unknown mean with known variance and unknown mean with unknown variance to the vector observations.

8.5.2.1 Multivariate Gaussian Observations with a Gaussian Prior on the Means with a Known Covariance Let us assume a sequence of observations $\mathbf{y}_{1:N}$, where each $\mathbf{y}_i \in \mathbb{R}^d$, for some $d > 1$, that are drawn from a DPMM. Suppose further that given its mixture component parameter θ_i, each $\mathbf{y}_i$ is independent and is multi-variate normal distributed:

$$\mathbf{y}_i|\theta_i \sim \mathcal{N}\left(\boldsymbol{\mu}_i, \mathbf{V}_i\right), \tag{8.37}$$

where $\theta_i \triangleq \boldsymbol{\mu}_i$ for $i \in \{1, \ldots, N\}$.

Hence, this models a situation in which each observation is drawn from a multivariate Gaussian distribution, but there are several possible multivariate Gaussian distributions specified by distinct mean vectors and covariance matrices. While it is assumed that the covariance matrices $\mathbf{V}_i$ of each of the multivariate distributions are known, the mean vectors $\boldsymbol{\mu}_i$'s are, however, unknown and assumed to be random.

As in the scalar case, the most convenient conjugate prior for the multivariate normal likelihood (8.37) is the Gaussian prior [119]. Thus, let us assume that each θ_i has a Gaussian prior of the form of $\theta_i \sim \mathcal{N}(\mathbf{m}, \tau\mathbf{V}_i)$ for $\tau > 0$. It is easy to verify that the posterior distribution of the mean θ_i given $\mathbf{y}_i$ is also a Gaussian of the form of

$$\theta_i|\mathbf{y}_i \sim \mathcal{N}\left(\mathbf{x}_i, X\mathbf{V}_i\right), \tag{8.38}$$

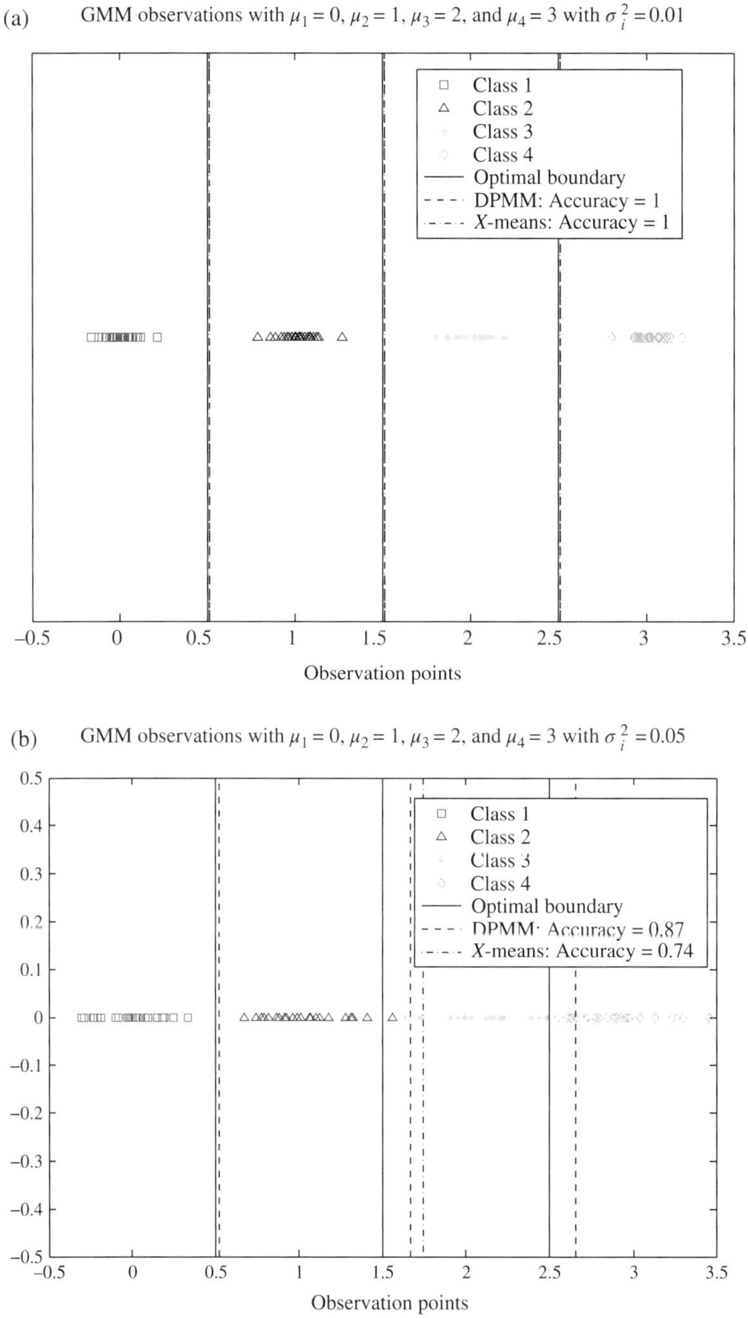

FIGURE 8.8 Classification of $K = 4$ GMM observations with unknown means $\theta_1 = 0$, $\theta_2 = 1$, $\theta_3 = 2$, and $\theta_4 = 3$ and an unknown common variance. (a) Classification results with *X*-means and DPMM classifiers when the unknown common variance $\sigma_i^2 = 0.01$. (b) Classification results with *X*-means and DPMM classifiers when the unknown common variance $\sigma_i^2 = 0.05$.

where

$$\mathbf{x}_i = \frac{\mathbf{m} + \tau \mathbf{y}_i}{1 + \tau} \tag{8.39}$$

and

$$X = \frac{\tau}{1 + \tau}, \tag{8.40}$$

with $\mathbf{V}_i$ denoting the assumed to be known covariance matrix.

Example 8.5 (Multivariate GMM Observations with $K = 2$ and a Known Covariance Matrix) Let us consider an example with two-dimensional (2-D) Gaussian vector observations ($d = 2$). Suppose that each 2-D vector $\mathbf{y}_i$ is drawn from a GMM with $K = 2$ mixture components. The cluster centers are $\theta_1 = [0, 0]^T$ and $\theta_2 = [1, 1]^T$, and we assume a known common circular covariance matrix:

$$\mathbf{V}_i = \Sigma = \begin{bmatrix} 0.1 & 0 \\ 0 & 0.1 \end{bmatrix}. \tag{8.41}$$

To apply the Bayesian nonparametric classification, we model parameters θ_i's as drawn from a Gaussian prior distribution $\theta_i \sim \mathcal{N}(\mathbf{m}, \tau \mathbf{V}_i)$ where $\mathbf{m} = [0.5, 0.5]^T$ and $\tau = 100$.

Figure 8.9(a) shows the classification performance of the X-means and DPMM with the Gaussian prior algorithms. We see that the DPMM algorithm achieves a classification accuracy of 98%, while the X-means achieves an equally well accuracy of 97%. Good performance of the X-means algorithm is perhaps not surprising since the observations are drawn from a circular Gaussian observation model and the X-means algorithm itself is constructed based on the K-means algorithm that assumes an underlying circular Gaussian observation model of the data. The X-means algorithm will show good performance as long as it is able to accurately estimate the number of clusters. However, its performance can deteriorate drastically if the number of clusters is not estimated correctly.

In Figure 8.9(b), we show the performance of these algorithms when the observations have the following noncircular covariance matrix:

$$\mathbf{V}_i = \mathbf{V} = \begin{bmatrix} 0.2 & 0 \\ 0 & 0.05 \end{bmatrix}. \tag{8.42}$$

Everything else has been kept the same as in the previous example. Figure 8.9(b) shows that the DPMM still achieves a classification accuracy of 98%, while the X-means performance drops to 92%. Although the X-means still has a satisfactory performance in this case, it does not perfectly adapt to noncircular Gaussian data, in contrast with the DPMM. ■

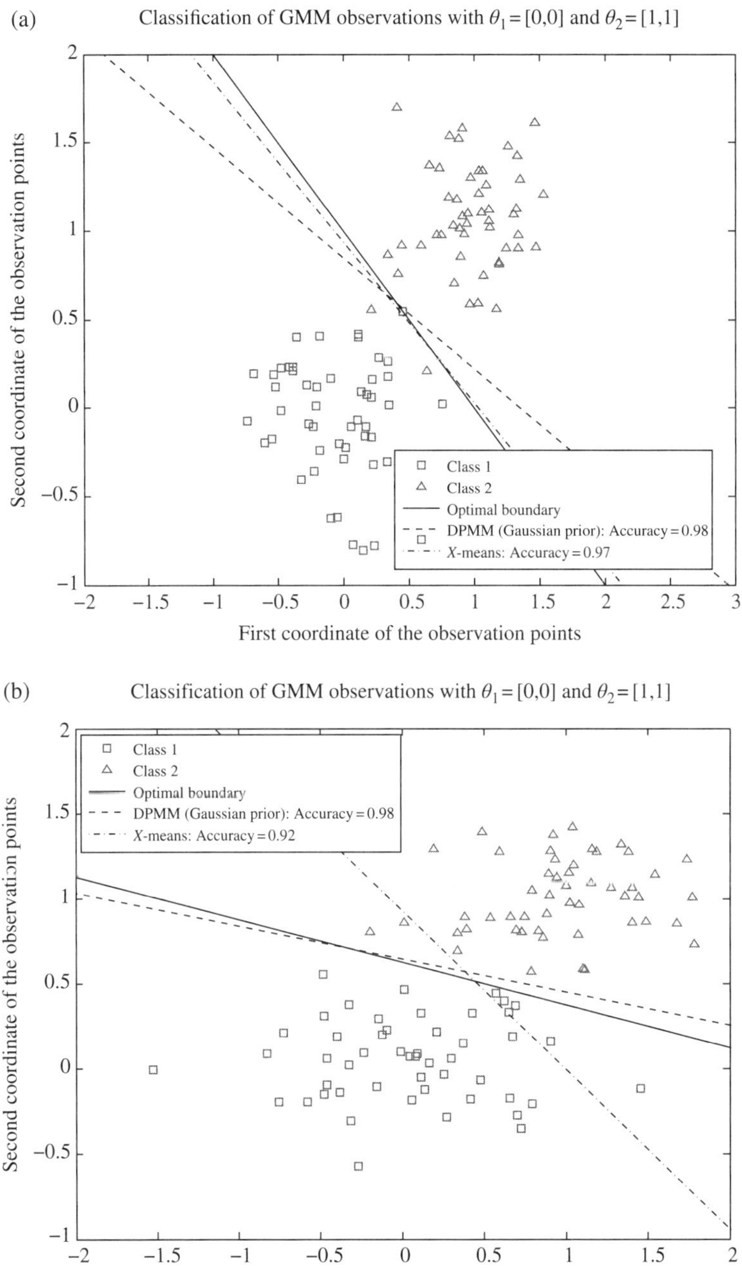

FIGURE 8.9 Classification of circular and noncircular 2-dimensional GMM observations with $K = 2$ mixture components where $\theta_1 = (0, 0)$, and $\theta_2 = (1, 1)$ and the covariance matrix $\mathbf{V}_i = \mathbf{V}$ is known: performance of X-means algorithm and DPMM-based algorithm with a Gaussian prior. (a) Classification results with a known circular covariance matrix $\mathbf{V}_i$. (b) Classification with a known noncircular covariance matrix $\mathbf{V}_i$.

In the next example, again, we will make the classification problem more interesting by allowing for $K = 4$ distinct clusters.

Example 8.6 (Multivariate GMM Observations with $K = 4$ and a Known Covariance Matrix) In this example, let us consider a little more challenging classification problem with 2-D observations. The observations are now allowed to be drawn from a GMM with $K = 4$ mixture components. The cluster centers are $\theta_1 = [0,0]^T$, $\theta_2 = [1,1]^T$, $\theta_3 = [2,2]^T$, and $\theta_4 = [3,3]^T$. In Figure 8.10, we show classification performance with both circular (with the covariance matrix (8.41)) and noncircular (with the covariance matrix (8.42)) covariance matrices.

We will also assume that the parameters θ_i's have a Gaussian prior distribution such that $\theta_i \sim \mathcal{N}(\mathbf{m}, \tau\mathbf{V}_i)$ where $\mathbf{m} = [1.5, 1.5]^T$ and $\tau = 100$.

Performance of the two algorithms with circular Gaussian observations is shown in Figure 8.10(a) where both DPMM and X-means achieve a classification accuracy of 97%. On the other hand, Figure 8.10(b) shows that for noncircular Gaussian observations, the DPMM algorithm achieves a classification accuracy of 99%, while the performance of the X-means algorithm drops to 92%. The reason here is mainly due to the fact that the X-means algorithm is not able to accurately match the underlying noncircular Gaussian distribution of the data compared to the DPMM algorithm. This is not surprising since X-means was not designed to do that, while DPMM allows taking into account an arbitrary covariance matrix.

Indeed, if noise becomes stronger, the performance of X-means algorithm can degrade even more dramatically compared to the DPMM-based classification. For the same example of $K = 4$ earlier, let us replace the circular Gaussian noise covariance (8.41) with the following:

$$\mathbf{V}_i = \mathbf{V} = \begin{bmatrix} 0.15 & 0 \\ 0 & 0.15 \end{bmatrix}. \tag{8.43}$$

For the noncircular Gaussian noise, let us use the following in place of the (8.42):

$$\mathbf{V}_i = \begin{bmatrix} 0.4 & 0 \\ 0 & 0.025 \end{bmatrix}. \tag{8.44}$$

Figure 8.11 shows that while DPMM performance is almost insensitive to this noise increase (although it will if noise is much stronger), the performance of X-means has already degraded significantly. Indeed, from Figure 8.11(a), we note that with stronger circular Gaussian noise, the DPMM achieves a classification accuracy of 91%, while the X-means performance has dropped to about 73%. Things become even worse when stronger noncircular Gaussian noise is considered as in Figure 8.11(b). The DPMM performance is almost identical to the earlier case with the covariance matrix (8.42). However, the X-means algorithm achieves only 56% accuracy. This shows the superiority of the DPMM-based classification algorithm in classifying observations drawn from noncircular observation models as well as its better noise performance compared to the X-means algorithm. ■

(a) Classification of GMM observations with $\theta_1=[0,0]$, $\theta_2=[1,1]$, $\theta_3=[2,2]$, and $\theta_4=[3,3]$

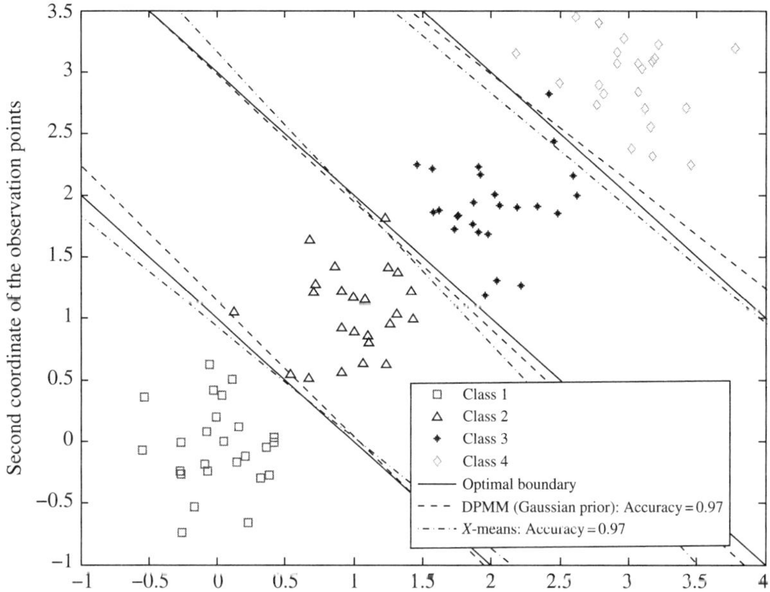

(b) Classification of GMM observations with $\theta_1=[0,0]$, $\theta_2=[1,1]$, $\theta_3=[2,2]$, and $\theta_4=[3,3]$

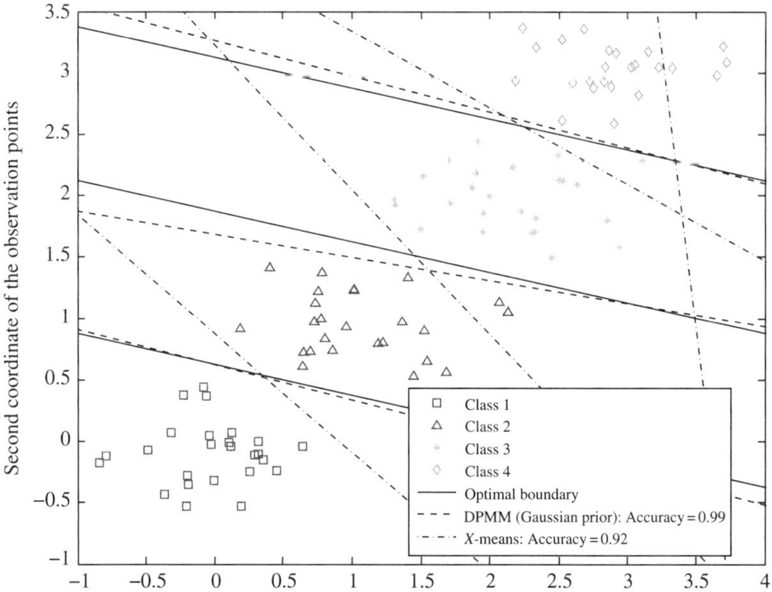

FIGURE 8.10 Classification of circular and noncircular 2-dimensional GMM observations with $K=4$ mixture components where $\theta_1=(0,0)$ and, $\theta_2=(1,1)$, $\theta_3=(2,2)$, and $\theta_4=(3,3)$ and the covariance matrix $\mathbf{V}_i=\mathbf{V}$ is known: performance of X-means algorithm and DPMM-based algorithm with a Gaussian prior. (a) Classification results with a known circular covariance matrix $\mathbf{V}_i$. (b) Classification with a known noncircular covariance matrix $\mathbf{V}_i$.

(a) Classification of GMM observations with $\theta_1=[0,0]$, $\theta_2=[1,1]$, $\theta_3=[2,2]$, and $\theta_4=[3,3]$

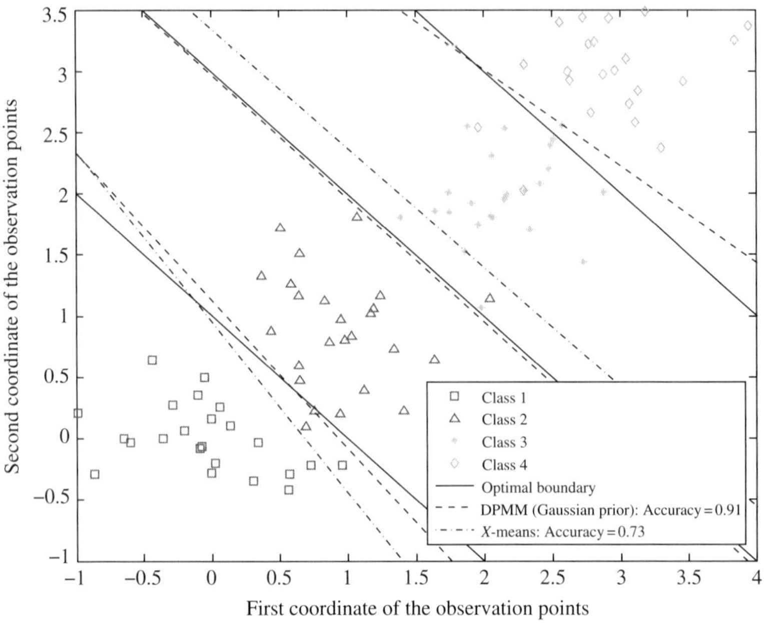

(b) Classification of GMM observations with $\theta_1=[0,0]$, $\theta_2=[1,1]$, $\theta_3=[2,2]$, and $\theta_4=[3,3]$

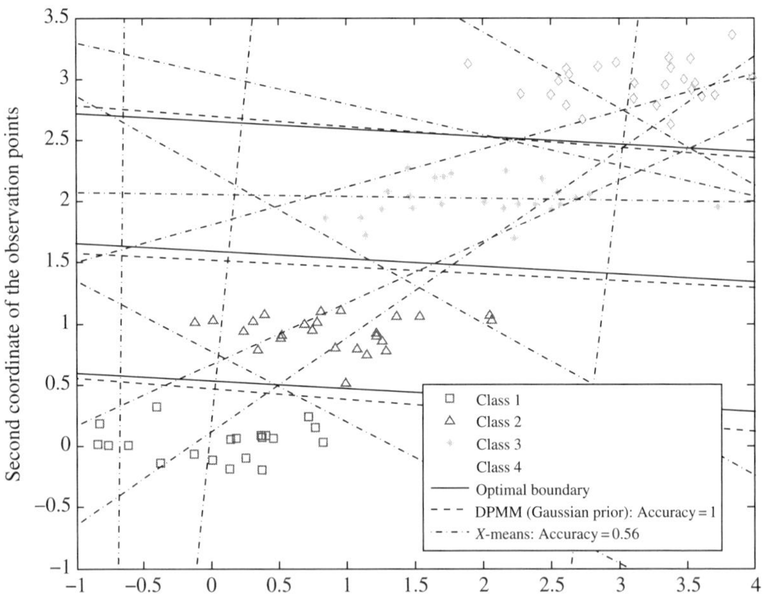

FIGURE 8.11 Classification of circular and noncircular 2-dimensional GMM observations with $K = 4$ mixture components where $\theta_1 = (0,0)$, $\theta_2 = (1,1)$, $\theta_3 = (2,2)$, and $\theta_4 = (3,3)$ and the covariance matrix $\mathbf{V}_i = \mathbf{V}$ is known: performance of X-means algorithm and DPMM-based algorithm with a Gaussian prior with increased noise strength. (a) Classification results with a known circular covariance matrix $\mathbf{V}_i$. (b) Classification with a known noncircular covariance matrix $\mathbf{V}_i$.

8.5.2.2 Multivariate Gaussian Observations with a Gaussian Prior on the Means with an Unknown Covariance Although there may be some situations in which the aforementioned model of a known (or fixed) covariance matrix can be justified, perhaps the more natural assumption is to allow the mixture covariances to be also different from one mixture component to another. Hence, let us now assume a sequence of d-dimensional observations $\mathbf{y}_{1:N}$ that are drawn from a DPMM and that are independent and normally distributed, given their mixture component parameters $\theta_{1:N} \triangleq \{\theta_i\}_{i=1}^N$. As before, we let

$$\mathbf{y}_i|\theta_i \sim \mathcal{N}\left(\boldsymbol{\mu}_i, \mathbf{V}_i\right), \tag{8.45}$$

where we now assume that the mixture parameter θ_i, for $i = 1, \ldots, N$, to be

$$\theta_i = (\boldsymbol{\mu}_i, \mathbf{V}_i). \tag{8.46}$$

This model allows each observation to be treated as drawn from a mixture of multivariate Gaussian distributions, where each multivariate Gaussian is characterized by a distinct mean vector and a covariance matrix.

From [119], we note that the natural conjugate base prior G_0 for θ when θ represents both mean vector and the covariance matrix of a multivariate Gaussian observation as in (8.46) is the Gaussian/inverse-Wishart conjugate prior $G_0(\theta_i)$ for θ_i where

$$\mathbf{V}_i^{-1} \sim W_d\left(\mathbf{S}, s\right) \tag{8.47}$$

and

$$\boldsymbol{\mu}_i|\mathbf{V}_i \sim \mathcal{N}(\mathbf{m}, \tau\mathbf{V}_i) \tag{8.48}$$

for some mean $\mathbf{m}$ and a scale factor $\tau > 0$ and $W_d(\mathbf{V}, n)$ denotes a Wishart distribution characterized by a positive definite scale matrix V and a number of degrees of freedom n. See Appendix B for the form of the Wishart density and how it is related to the distribution of the sample covariance matrix of multivariate Gaussian vectors.

Note that a large value of τ implies a large dispersion among the cluster means, whereas $\mathbf{m}$ represents a prior estimate of these means [97]. One may question whether these parameters may always be known. However, as shown in [97], for example, it is possible to infer the values of τ and $\mathbf{m}$ also from the Gibbs sampling itself. The parameter s reflects the confidence in the estimated observation error, or the value of the covariance matrix $\mathbf{V}_i$. A large value of s implies high confidence in the value of $\mathbf{V}_i$ being closer to its prior estimate $\mathbf{S}$, whereas a small value of s corresponds to the case in which little knowledge is available about $\mathbf{V}_i$ [97].

With these assumptions, the posterior distribution $f(\theta_i|\mathbf{y}_i)$ of θ_i can be shown to be given by

$$\mathbf{V}_i^{-1} \sim W\left(\mathbf{S}_i, 1+s\right) \tag{8.49}$$

$$\boldsymbol{\mu}_i|\mathbf{V}_i \sim \mathcal{N}\left(\mathbf{x}_i, X\mathbf{V}_i\right), \tag{8.50}$$

where

$$\mathbf{S}_i = \mathbf{S} + \frac{(\mathbf{y}_i - \mathbf{m})(\mathbf{y}_i - \mathbf{m})^T}{1+\tau},$$
$$\mathbf{x}_i = \frac{\mathbf{m} + \tau \mathbf{y}_i}{1+\tau}$$

and X is as given in (8.35).

Using this characterization of the posterior marginal distribution of θ_i, we can show that probabilities q_0 and q_j required in evaluating (8.18) are given by

$$q_0 \propto \alpha_0 c(s) \left(\frac{1}{1+\tau}\right)^{\frac{d}{2}} |\mathbf{S}|^{1/2} |\mathbf{S}_i|^{-(1+s)/2} \tag{8.51}$$

and, for $j \in \{1, \ldots, N\}$ with $j \neq i$,

$$q_j \propto \frac{1}{\sqrt{2|\mathbf{V}_j|}} e^{\frac{-(\mathbf{y}_j - \boldsymbol{\mu}_j)^T \mathbf{V}_j^{-1} (\mathbf{y}_j - \boldsymbol{\mu}_j)}{2}}, \tag{8.52}$$

subject to

$$\sum_{j=1, j\neq i}^{N} q_j = 1,$$

where

$$c(s) = \pi^{-\frac{d}{2}} \frac{\Gamma_d\left(\frac{1+s}{2}\right)}{\Gamma_d\left(\frac{s}{2}\right)}$$

with multivariate gamma function $\Gamma_d(s)$ is as defined by (B.2) in Appendix B.

In the next example, we use the above conditional marginal distribution to implement a Gibbs sampling procedure.

Example 8.7 (Multivariate GMM Observations with K=4 and an Unknown Covariance Matrix) As an example, let us consider $N = 100$ observations that are drawn from a 2-D GMM with $K = 4$ clusters centered at $[0,0]^T$, $[1,1]^T$, $[2,2]^T$, and $[3,3]^T$. The observations are independent Gaussian distributed, given the cluster parameters θ_i's. We let $\theta_i = (\boldsymbol{\mu}_i, \mathbf{V}_i)$, for $i = 1, \ldots, N$ and denote $\mathbf{y}_i|\theta_i \sim \mathcal{N}(\boldsymbol{\mu}_i, \mathbf{V}_i)$. We assume that both cluster means and covariance matrix are unknown and random.

Figure 8.12(a) shows the performance of DPMM and X-means classification algorithms when the cluster covariance matrix is taken to be the circular covariance matrix $\mathbf{V}_i = 0.1\mathbf{I}$. In applying the DPMM-based algorithm, we have set the hyperparameters to be $s = 4$, $\mathbf{m} = [1.5, 1.5]^T$, $\tau = 1$, and $\mathbf{S} = 0.2\mathbf{I}$. The rate of creating new clusters is

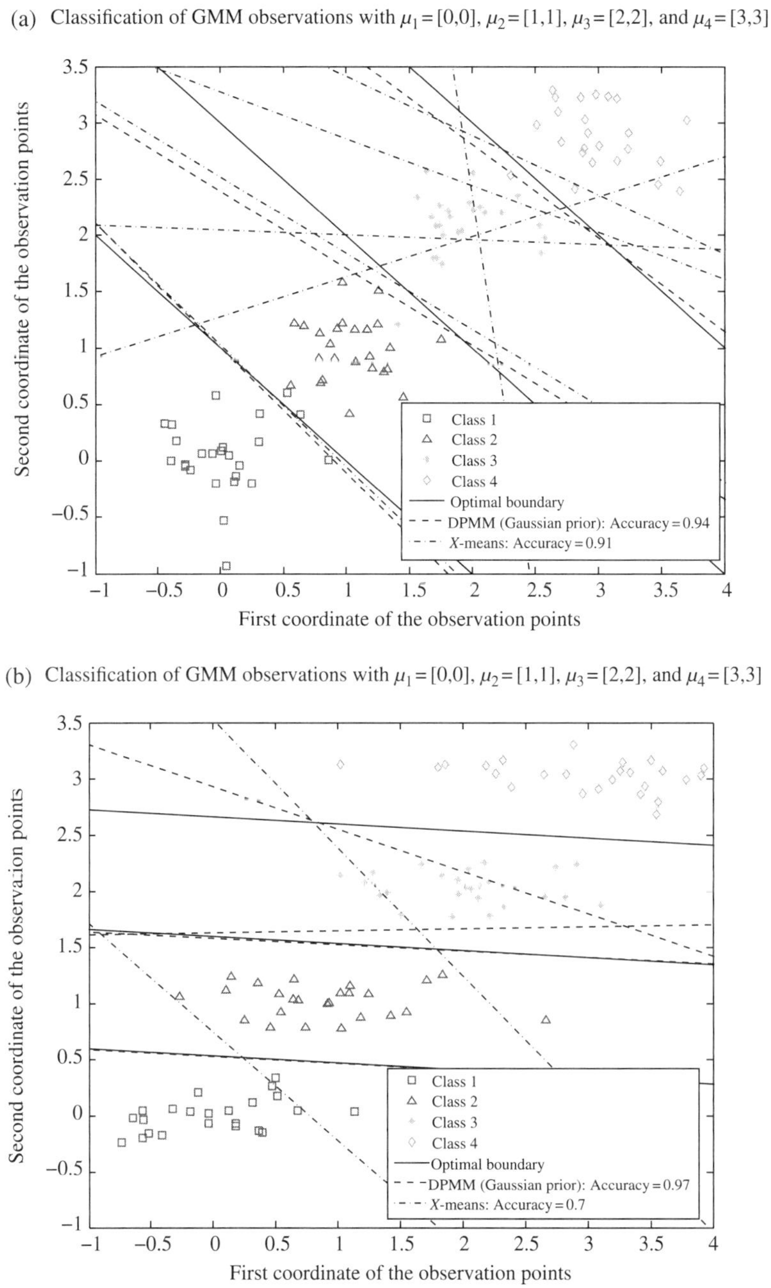

FIGURE 8.12 Classification of circular and noncircular 2-dimensional $K = 4$ GMM observations with unknown means $\theta_1 = (0,0)$, $\theta_2 = (1,1)$, $\theta_3 = (2,2)$, and $\theta_4 = (3,3)$ and an unknown covariance matrix. (a) Classification performance of X-means and DPMM-based algorithms when the unknown covariance matrix is circular and equal to $\mathbf{V}_i = 0.1\mathbf{I}$. (b) Classification performance of X-means and DPMM-based algorithms when the unknown covariance matrix is circular and equal to $\mathbf{V}_i = diag(0.4, 0.025)$.

taken to be $\alpha_0 = 2$ in the Dirichlet process prior. Figure 8.12(a) shows that DPMM and X-means algorithms achieve close enough, 94% and 91%, respectively, classification accuracy in this case.

In Figure 8.12(b), however, we have shown the classification performance when noise is taken to be noncircular Gaussian distributed with the covariance matrix[3] (8.44) in Example 8.6:

$$\mathbf{V}_i = \begin{bmatrix} 0.4 & 0 \\ 0 & 0.025 \end{bmatrix}. \tag{8.53}$$

We have kept all hyperparameters in the DPMM as before except now we have taken $\mathbf{S}$ to be

$$\mathbf{S} = \begin{bmatrix} 2 & 0 \\ 0 & 0.625 \end{bmatrix}. \tag{8.54}$$

Figure 8.12(b) shows that in this case the classification accuracy of DPMM approach is 97%, while that of the X-means algorithm has dropped to about 70%. This again shows the limitations of the X-means performance in the presence of noncircular observations. ■

8.5.3 DPMM-Based Classification of Possibly Non-Gaussian Multidimensional Observations

One of the limitations of the aforementioned approach is that it presumes that the observations $\mathbf{y}_{1:N}$ are multivariate normal distributed, given the cluster parameters θ_i's. In practical applications, it is likely that observations belonging to different clusters may have different likelihoods that are not necessarily limited to multivariate Gaussians. For example, suppose we have two mixture components as identified by two parameters $\theta_1 = \phi_1$ and $\theta_2 = \phi_2$. In the following, we discuss a framework to allow relaxing the Gaussian likelihood constraint so that for example, the observations corresponding to the cluster marked by θ_2 are allowed to have a non-Gaussian likelihood.

To keep analytical tractability, however, we need to modify the base priors accordingly if we were to modify the likelihood $f_{\theta_i}(\mathbf{y}_i)$. This is to ensure that the prior still remains a conjugate prior for the assumed likelihood so that a closed-form expression for the posterior distribution of θ_i, as in (8.18), may be derived. For example, a Gaussian prior on the mean is conjugate for the Gaussian likelihood. However, if we were to assume a gamma likelihood function, then the Gaussian prior is no longer a conjugate prior for the shape parameter of gamma likelihood. Thus, modifying the likelihood $f_{\theta_i}(\mathbf{y}_i)$ needs to be coordinated with modifying the base prior.

[3]Note that $diag(\mathbf{v})$ denotes a diagonal matrix whose diagonal entries are given by the elements of the vector $\mathbf{v}$.

As before, let parameter θ_i denote the distribution parameter (or a parameter vector) of the likelihood of observation $\mathbf{y}_i$. However, suppose that each likelihood $f_{\theta_i}(\mathbf{y}_i)$ may belong to one of L different possible types of distributions. The type of the parameter θ can be very different from one class of likelihoods to another. Within each class, the parameter is assumed to be random as we have done in the previous section. However, we now introduce an index $Z_i \in \{1,\dots,L\}$ to specify the type of the likelihood $f_{\theta_i}(\mathbf{y}_i)$. Thus, we may treat θ_i (or the base prior) itself as belonging to a *mixture model* of L components in which each component stands for a random parameter drawn from a certain set $\mathcal{S}_l$, for $l = 1,\dots,L$. The set $\mathcal{S}_l$ is the all possible values the parameter θ can take when the likelihood is of the lth type.

For example, we may allow the likelihoods to be either a Gaussian or a gamma distribution by letting $L = 2$. If we assume that different classes with Gaussian likelihood are due to both different means and covariances, then $\mathcal{S}_1$ can be taken as the set of possible means and covariances. Similarly, if we assume that different classes with gamma likelihood for observation are caused by different parameter values assumed by the gamma $\Gamma\left(\frac{S}{2},\frac{s}{2}\right)$ distribution, then $\mathcal{S}_2$ may contain the set of possible values for S and s (Figure 8.13).

We may introduce a Bayesian prior on the type of likelihood itself by assuming a discrete prior distribution for the type index Z_i such that for $l = 1,\dots,L$,

$$\Pr(Z_i = l) \triangleq \kappa_l.$$

Given a certain type $Z_i = l$ for the likelihood of $\mathbf{y}_i$, we may denote the conditional base prior for θ_i as

$$\theta_i|\{Z_i = l\} \sim G_0^{(l)}(\theta_i) \quad \text{for } \theta_i \in \mathcal{S}_l.$$

Stated differently, θ_i is a mixture of types where each type is represented in the mixture with a probability κ_l, for $l = 1,\dots,L$.

Let us denote the likelihood function of the observation $\mathbf{y}_i$, given that $Z_i = l$ as

$$f_\theta^{(l)}(\mathbf{y}_i) \triangleq f(\mathbf{y}_i|\theta_i = \theta, Z_i = l) \quad \text{for } \theta_i \in \mathcal{S}_l.$$

Thus, we can write

$$\mathbf{y}_i|\{\theta_i, Z_i\} \sim \sum_{l=1}^{L} f_{\theta_i}^{(l)}(\mathbf{y}_i)\mathcal{I}_{\{Z_i=l\}}, \tag{8.55}$$

where the indicator function $\mathcal{I}_A$ is defined as $\mathcal{I}_A = 1$ if the event A is true and 0 otherwise. Note that the distribution of $\mathbf{y}_i|\{\theta_i, Z_i\}$ is defined for $\theta_i \in \mathcal{S}_{Z_i}$ such that θ_i is a valid parameter for the Z_ith distribution model.

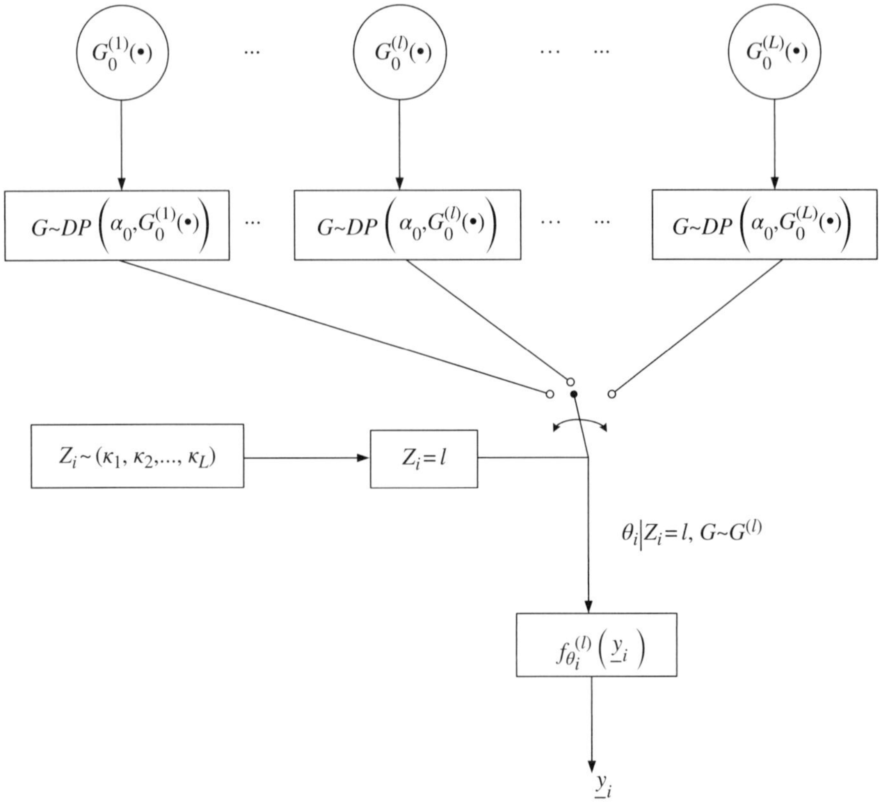

FIGURE 8.13 Non-Gaussian likelihoods can be allowed while still preserving the conjugate prior property by introducing an auxiliary mixing parameter to the DPMM.

The marginal distribution of observation $\mathbf{y}_i$ required for evaluating (8.18) can then be written as

$$f(\mathbf{y}_i) = \sum_{l=1}^{L} \kappa_l \int_{\theta \in \mathcal{S}_l} f_{\theta}^{(l)}(\mathbf{y}_i) G_0^{(l)}(\theta) d\theta. \tag{8.56}$$

Then the posterior distribution of θ_i, given the observation $\mathbf{y}_i$, can be defined over the set $\mathcal{S} \triangleq \bigcup_{l=1}^{L} \mathcal{S}_l$ according to

$$\begin{aligned} f(\theta_i|\mathbf{y}_i) &= \sum_{l=1}^{L} f(\theta_i, Z_i = l|\mathbf{y}_i) \\ &= \sum_{l=1}^{L} f(\theta_i|\mathbf{y}_i, Z_i = l) \Pr(Z_i = l|\mathbf{y}_i) \\ &= \sum_{l=1}^{L} \hat{\kappa}_{l,i} f(\theta_i|\mathbf{y}_i, Z_i = l), \end{aligned} \tag{8.57}$$

where $\hat{\kappa}_{l,i}$ is the posterior probability of the lth mixture component of types given the observation $\mathbf{y}_i$:

$$\begin{aligned}\hat{\kappa}_{l,i} &\triangleq \Pr(Z_i = l|\mathbf{y}_i)\\ &= \frac{\Pr(Z_i = l)f(\mathbf{y}_i|Z_i = l)}{\sum_{l'=1}^{L}\Pr(Z_i = l')f(\mathbf{y}_i|Z_i = l')}\\ &= \frac{\kappa_l f(\mathbf{y}_i|Z_i = l)}{\sum_{l'=1}^{L}\kappa_{l'} f(\mathbf{y}_i|Z_i = l')}\\ &= \frac{\kappa_l \int_{\theta\in\mathcal{S}_l} f_\theta^{(l)}(\mathbf{y}_i)g_0^{(l)}(\theta)d\theta}{\sum_{l'=1}^{L}\kappa_{l'}\int_{\theta\in\mathcal{S}_{l'}} f_\theta^{(l')}(\mathbf{y}_i)g_0^{(l')}(\theta)d\theta}.\end{aligned} \tag{8.58}$$

It is to be understood that $f(\theta_i|\mathbf{y}_i, Z_i = l) = 0$ if $\theta_i \notin \mathcal{S}_l$.

By substituting (8.56) and (8.57) in (8.18), we can again obtain the posterior distribution of $\theta_i|\{\theta_j\}_{j\neq i}, \mathbf{y}_{1:N}$ required to implement a Gibbs sampling approach to estimate θ_i's. If $\hat{\kappa}_{l,i}$ cannot be evaluated in closed form as in (8.58), they can also be estimated during the Gibbs sampling itself. From (8.57), the posterior $f(\theta_i|\mathbf{y}_i)$ can be implemented by first sampling Z_i from the discrete set $\{1,\ldots,L\}$ according to the posterior type probabilities $\{\hat{\kappa}_{l,i}\}_{l=1}^{L}$ and then sampling θ from $\mathcal{S}_{Z_i}$ using the posterior distribution $f(\theta_i|\mathbf{y}_i, Z_i)$. Of course, if $G_0^{(l)}(.)$ is conjugate for the likelihood $f_{\theta_i}^{(l)}(\mathbf{y}_i)$, for all $l \in \{1,\ldots,L\}$, then the posterior in (8.57) can be expressed in closed form to facilitate this.

8.5.3.1 A Mixture of Lognormal and Gaussian-Distributed Observations with Likelihoods Characterized by Unknown Random Parameters Let us consider the case of $L = 2$ in order to develop a model that can accommodate two different types of likelihoods for observations. In particular, suppose that $\mathbf{y}_i|\theta_i$ is a mixture of Gaussian and lognormal distributions where $\mathbf{y}_i = [y_{i,1},\ldots,y_{i,d}]^T \in \mathbb{R}^d$. Let the distribution type index $Z_i \in \{1,2\}$ with $Z_i = 1$ and $Z_i = 2$ corresponding, respectively, to Gaussian and lognormal likelihoods. Note that within each likelihood, the observations may still belong to an arbitrary number of distinct clusters depending on the exact value assumed by θ_i corresponding to $\mathbf{y}_i$. For analytical tractability, we will select base priors for each type to be conjugate to the corresponding likelihood. We also assume that the elements of $\mathbf{y}_i$'s are independent in the case of non-Gaussian observation models.

For $l = 1$, as in Section 8.5.2.2, we may define $\mathcal{S}_1 \triangleq \mathbb{R}^d \times \mathbb{R}^{d\times d}$ to be the set of possible parameters of the Gaussian likelihood function corresponding to

$$\theta_i|\{Z_i = 1\} \triangleq (\boldsymbol{\mu}_i, \mathbf{V}_i).$$

For obvious reasons, we select the base prior $G_0^{(1)}(.)$ with this θ to be the normal/inverse-Wishart conjugate prior for the Gaussian likelihood $f_{\theta_i}^{(1)}(\mathbf{y}_i)$ so that the posterior $f(\theta_i|\mathbf{y}_i, Z_i = 1)$ and the marginal $\int_{\theta\in\mathcal{S}_1} f_\theta^{(1)}(\mathbf{y}_i)G_0^{(1)}(\theta)d\theta$ can be computed as described in Section 8.5.2.2.

Next, let the second type of likelihoods be lognormal. In particular, we will assume that each $y_{i,k}$ for $1 \leq k \leq d$ is independent lognormal distributed according to $y_{i,k}|\{\theta_i = \theta, Z_i = 2\} \sim \ln\mathcal{N}(\rho_k, \xi_k^2)$ where $\boldsymbol{\xi} = (\xi_1, \ldots, \xi_d)^T$ are the shape parameters and $\boldsymbol{\rho} = (\rho_1, \ldots, \rho_d)^T$ are the mean parameters of the lognormal likelihood function. In the following, we may assume that the different classes with the lognormal likelihoods are caused by distinct mean parameters ρ_k. Thus, we will assume fixed shape parameters $\{\xi_k\}_{k=1}^d$, and for $l = 2$, we define $\mathcal{S}_2 \triangleq \mathbb{R}^d$ such that

$$\theta_i|\{Z_i = 2\} \triangleq \boldsymbol{\rho},$$

where $\boldsymbol{\rho} = [\rho_1, \ldots, \rho_d]^T$ are the mean parameters of a lognormal likelihood function of the form of

$$f_\theta^{(2)}(\mathbf{y}_i) = \prod_{k=1}^{d} \frac{1}{y_{i,k}\sqrt{2\pi\xi_k^2}} e^{-\frac{(\ln y_{i,k} - \rho_k)^2}{2\xi_k^2}}, \tag{8.59}$$

where we let $\theta = \boldsymbol{\rho}$. From Appendix C, a convenient conjugate prior for $\boldsymbol{\rho}$ is an independent normal distribution of the form

$$\rho_k \sim \mathcal{N}\left(\rho_{0,k}, \xi_{0,k}^2\right),$$

so that

$$g_0^{(2)}(\theta) = g_0^{(2)}(\boldsymbol{\rho}) = \prod_{k=1}^{d} \frac{1}{\sqrt{2\pi\xi_{0,k}^2}} e^{-\frac{(\rho_k - \rho_{0,k})^2}{2\xi_{0,k}^2}}, \tag{8.60}$$

where $\rho_{0,k}$ and $\xi_{0,k}$, for $k = 1, \ldots, d$, are the parameters of the base prior.

With this conjugate prior, using Appendix C, the posterior distribution of θ_i can be shown to be

$$\rho_k | y_{i,k} \sim \mathcal{N}(\nu_k, \psi_k),$$

so that

$$f(\theta_i|\mathbf{y}_i, Z_i = 2) = f(\boldsymbol{\rho}|\mathbf{y}_i) = \prod_{k=1}^{d} \frac{1}{\sqrt{2\pi\psi_k}} e^{-\frac{(\rho_k - \nu_k)^2}{2\psi_k}}, \tag{8.61}$$

where

$$\nu_k = \frac{\xi_k^2 \rho_{0,k} + \xi_{0,k}^2 \log y_{i,k}}{\xi_{0,k}^2 + \xi_k^2}$$

and

$$\psi_k = \xi_{0,k}^2 + \xi_k^2.$$

The marginal distribution of $\mathbf{y}$ required for evaluating (8.18) can also be evaluated as

$$f(\mathbf{y}_i|Z_i = 2) = \prod_{k=1}^{d} \frac{1}{2\pi y_{i,k}\sqrt{\xi_k^2 \xi_{0,k}^2}} \int_{-\infty}^{\infty} e^{-\frac{(\ln y_{i,k}-\rho)^2}{2\xi_k^2}} e^{-\frac{(\rho-\rho_{0,k})^2}{2\xi_{0,k}^2}} d\rho, \tag{8.62}$$

which may need numerical computation.

Example 8.8 (A K=4 Mixture of Gaussian and Lognormal Observations with Unknown Parameters on Likelihoods) As an example, let us consider 2-D observations $\mathbf{y}_i \in \mathbb{R}^2$ drawn from the DPMM with Gaussian and lognormal observations discussed in Section 8.5.3.1, where $\Pr(Z_i = 1) = \Pr(Z_i = 2) = 0.5$ for all $i = 1, \ldots, N$. Since $d = 2$, $\mathcal{S}_1 \triangleq \mathbb{R}^2 \times \mathbb{R}^{2\times 2}$ and $\mathcal{S}_2 \triangleq \mathbb{R}^2$. Consider a situation in which the observations $\{\mathbf{y}_i\}_{i=1}^N$ are drawn from a mixture model of $K = 4$ components.

Two of the mixture components generate Gaussian likelihoods with respective means $\boldsymbol{\mu}_1 = [10, 10]^T$ and $\boldsymbol{\mu}_2 = [11, 11]^T$ but with a common covariance matrix $\mathbf{V}_i = diag(0.4, 0.025)$ of (8.44) as used in the Examples 8.6 and 8.7. Note that this covariance implies that the two components of the Gaussian observation vector are also independent. The other two mixture components generate lognormal distributed observations with respective mean parameters $\boldsymbol{\rho}_3 = [\log(12), \log(12)]^T$, and $\boldsymbol{\rho}_4 = [\log(13), \log(13)]^T$. Both lognormal models are assumed to have fixed shape parameters $\boldsymbol{\xi}_3 \in \mathbb{R}^2$ and $\boldsymbol{\xi}_4 \in \mathbb{R}^2$. We may denote $\boldsymbol{\xi}_l = (\xi_{l,1}, \xi_{l,2})^T$, for $l = 3, 4$, where $\xi_{l,d}$ denotes the shape parameter of the dth coordinate of observation $\mathbf{y}_i$ that is drawn from the lth mixture model.

In the DPMM-based classification approach, we apply Gibbs sampling to estimate both the distribution index Z_i and the corresponding parameter θ_i of each observation $\mathbf{y}_i$. Moreover, in applying the DPMM-based algorithm, for possibly Gaussian observations, the parameter $\theta_i = (\boldsymbol{\mu}_i, \mathbf{V}_i)$ is assumed to admit a normal/inverse-Wishart conjugate prior with parameters $\mathbf{m} = [10, 10]^T$, $\tau = 1$, $s = 4$, and $\mathbf{S} = diag(0.8, 0.05)$. For possibly lognormal observations, on the other hand, we assume a Gaussian conjugate prior for the parameter $\theta_i = \boldsymbol{\rho}_i$ with $\rho_{0,k} = \log(10)$. In the following, we will evaluate classification performance for two possible values for the variance, $\xi_{0,k}$, for $k = 1, 2$, of this Gaussian conjugate base prior for the mean of the lognormal likelihoods.

Figure 8.14 shows the classification performance of the X-means and DPMM-based algorithms in the aforementioned assumed scenario. As expected, in this case, the DPMM indeed does outperform the X-means algorithm. For example, Figure 8.14(a) shows that the DPMM-based algorithm is able to perfectly classify the observation points by correctly identifying the observation model of each of the data

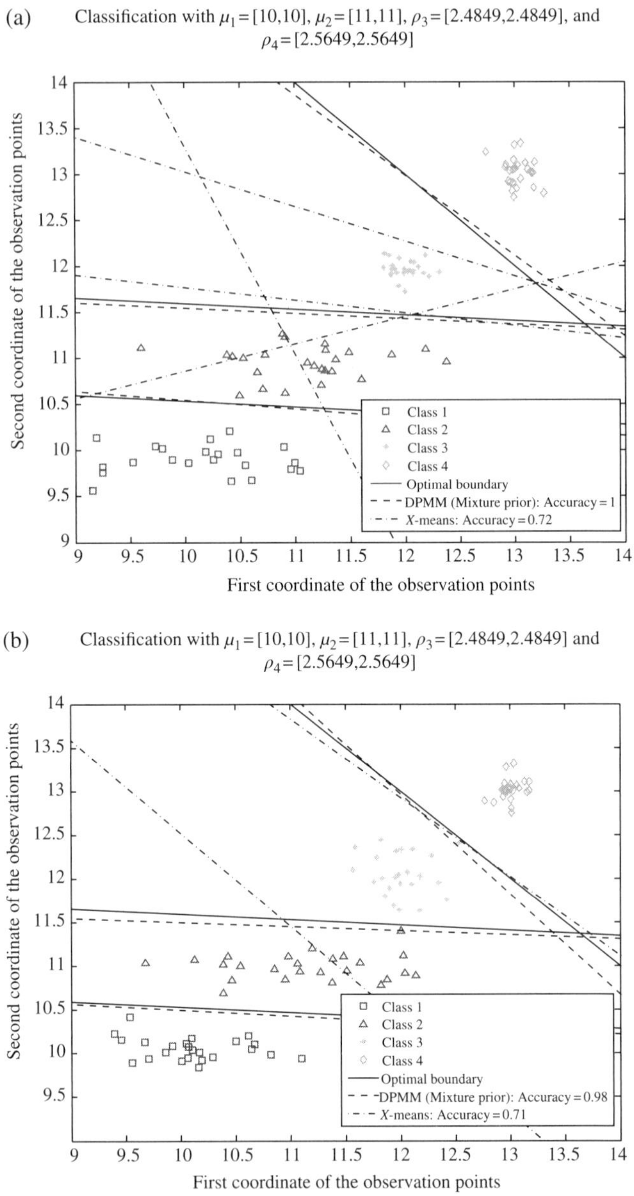

FIGURE 8.14 Classification performance of the X-means and DPMM-based algorithms for 2-dimensional data from a $K = 4$ mixture of Gaussian and lognormal observation models. (a) Classification performance of X-means and DPMM-based algorithms when variances of the two lognormal data clusters are $\boldsymbol{\xi}_3 = (0.01, 0.01)$ and $\boldsymbol{\xi}_4 = (0.01, 0.01)$. (b) Classification performance of X-means and DPMM-based algorithms when variances of the two lognormal data clusters are $\boldsymbol{\xi}_3 = (0.02, 0.02)$ and $\boldsymbol{\xi}_4 = (0.01, 0.01)$. In both (a) and (b), the Gaussian data is drawn from two clusters with unknown means $\boldsymbol{\mu}_1 = (10, 10)$ and $\boldsymbol{\mu}_2 = (11, 11)$ and an unknown covariance matrix $\mathbf{V}_i = diag(0.4, 0.025)$. Log-normal data in both (a) and (b) are drawn from two clusters having mean parameters of $\boldsymbol{\rho}_3 = (\log(12), \log(12))$, and $\boldsymbol{\rho}_4 = (\log(13), \log(13))$.

points $\mathbf{y}_i$ under the assumed parameters. However, the X-means algorithm achieves only a classification accuracy of about 72%, in this case. Note that this is not surprising, and of course, it is not a fair comparison in the sense that the X-means algorithm was not defined to handle non-Gaussian observations. The point is that when observations are indeed non-Gaussian, an algorithm that explicitly accounts for that fact can provide better performance. The DPMM-based approach does just that.

In Figure 8.14(b), we have increased the noisiness of the third mixture component in the observations by letting $\xi_3 = 0.02$. The performance degradation of both algorithms is almost negligible. ■

Though DPMM-based approach seems a powerful tool for Bayesian nonparametric classification, arguably, its drawback is the complexity involved and the need to carefully select certain hyperparameters. In particular, there is the need for specifying the base prior and the parameter α_0. As we already discussed, it is computationally convenient to select the base prior to be conjugate for the assumed observation likelihood. However, in practice, this has to be properly justified. The choice of α_0 seems to be somewhat arbitrary. However, without enough insight into the application scenario, it may be difficult to specify a suitable value for α_0. For this reason, sometimes, it is also modeled as having a certain prior distribution. Then, during the Gibbs sampling procedure, the posterior distribution of α_0 is used to sample also α_0. Of course, it is then convenient to select a prior for α_0 that may indeed facilitate easy computation of its posterior. Such issues have been investigated in [102, 125, 126].

8.6 SUMMARY

- Given a set of observations $\{y_i\}_{i=1}^N$ with each y_i corresponding to a specific class θ_i, the general classification problem is to

$$\text{Determine } \{\theta_i\}_{i=1}^N \text{ given data } \{y_i\}_{i=1}^N.$$

- In supervised classification, a set of initial training data is provided to identify the distinct classes. Later data $\{y_i\}_{i=1}^N$ are then classified into these same known classes. Thus, supervised classification is also called the training-based classification. Unsupervised classification approach does not assume training data. Since the problem is to group similar data points into the same cluster, unsupervised classification problem is one of clustering problem.
- The MAP classifier is defined as

$$\boldsymbol{\theta}^*_{MAP} = \underset{\boldsymbol{\theta} \in \left\{\phi_1^{(t)},\ldots,\phi_{K^{(t)}}^{(t)}\right\}^N}{\arg\max} \Pr(\boldsymbol{\theta} | \{y_i\}_{i=1}^N),$$

where $\Pr(\boldsymbol{\theta} | \{y_i\}_{i=1}^N)$ is the a posteriori probability of the classes given the data and $\phi_k^{(t)}$, for $k \in \{1, \ldots, K^{(t)}\}$, are the $K^{(t)}$ distinct values taken by the $\theta_i^{(t)}$'s of a given training data set.

- The ML classifier is defined as

$$\boldsymbol{\theta}^*_{ML} = \underset{\boldsymbol{\theta} \in \left\{\phi_1^{(t)},\ldots,\phi_{K(t)}^{(t)}\right\}^N}{\arg\max} P(\{y_i\}_{i=1}^N | \boldsymbol{\theta}).$$

- Classification approaches that assume knowledge of exact distributions are called parametric classifier approaches, while those that do not are called nonparametric classifier approaches. Nonparametric classification methods may use observations $\{y_i\}_{i=1}^N$ themselves to first estimate the required distributions or densities, without assuming any parametric form for them, and use these estimated densities to either obtain an ML or MAP estimate. In this case, nonparametric classification essentially reduces to a problem of density estimation.
- There are two approaches to density estimation in classification schemes: either to estimate the class-conditional densities (i.e., likelihoods) or directly estimate the posterior probabilities required in the MAP classifier. Parzen window-based density estimation is an example of the former, while K-nearest neighbor classification algorithm is an example of the latter.
- Most nonparametric classification approaches assume that the number of clusters is known and that the class priors are either uniformly distributed or unknown deterministic quantities.
- Bayesian nonparametric approaches allow the number of classes to be unknown and random. Moreover, the class prior $G(\theta)$ is modeled as an unknown random distribution.
- The K-means algorithm is an approach for clustering a given set of observations into a given number of classes so that the total squared error (TSE) distortion defined as

$$TSE = \sum_{i=1}^{N} |\mathbf{y}_i - \theta(\mathbf{y}_i)|^2$$

is minimized.
- Given the observations $\{\mathbf{y}_i\}_{i=1}^N$ and the number of classes K, the K-means algorithm partitions the data into K disjoint sets $\mathcal{S}_1,\ldots,\mathcal{S}_K$ with centroids $\mu_1,\ldots,\mu_K$ as follows:

1. Initialize $\boldsymbol{\mu}_1,\ldots,\boldsymbol{\mu}_K$ to random values (e.g., for $k = 1,\ldots,K$ set $\boldsymbol{\mu}_k = \mathbf{y}_i$ where $i = \mathcal{U}(\{1,2,\ldots,K\})$).
2. For $i = 1,\ldots,N$, assign $\mathbf{y}_i$ to $\mathcal{S}_{\hat{k}}$ if $\hat{k} = \arg\min_k |\mathbf{y} - \boldsymbol{\mu}_k|$.
3. For $k = 1,\ldots,K$, set $\boldsymbol{\mu}_k = \frac{1}{|\mathcal{S}_k|} \sum_{\mathbf{y} \in \mathcal{S}_k} \mathbf{y}_i$.
4. If the new cluster centroids $\boldsymbol{\mu}_k$'s have not changed from the old ones, stop. Otherwise, go to step 2.

- The Bayesian information criterion (BIC) for model selection is to select the model M_k that minimizes

$$\mathrm{BIC}_k = -2\log f\left(\{\mathbf{y}_i\}_{i=1}^N|\hat{\boldsymbol{\theta}}_k\right) + k\log N, \tag{8.63}$$

where $\hat{\boldsymbol{\theta}}_k$ is the ML estimate of parameter vector $\boldsymbol{\theta}_k$ given by

$$\hat{\boldsymbol{\theta}}_k = \arg\max_{\boldsymbol{\theta}_k \in \Theta_k} f\left(\{\mathbf{y}_i\}_{i=1}^N|\boldsymbol{\theta}_k\right).$$

Hence, the Bayesian approach to model selection is to select the model M_k with k parameters that minimizes the BIC in (8.63).

- The Akaike information criterion for model selection is to select the model M_k that minimizes

$$\mathrm{AIC}_k = -2\log f\left(\{\mathbf{y}_i\}_{i=1}^N|\hat{\boldsymbol{\theta}}_k\right) + 2k.$$

- The X-means algorithm is a variation of the K-means algorithm that does not require the number of classes K to be specified a priori. Instead, the algorithm attempts to estimate the number of clusters K by treating it as a model selection problem. Algorithm is initialized by setting $K = K_{\min}$. In each iteration of the algorithm, the K-means algorithm is run with the current value of K. Next, each cluster is split into two clusters, and K-means algorithm is run within each cluster with $K = 2$. A model selection criterion such as BIC is then used to determine whether the two new clusters are a better model than the original single cluster in each of the cases. Algorithm stops if either $K > K_{\max}$ or the cluster assignment did not change from the last iteration.
- A vector $(X_1, \ldots, X_n)$ is said to be distributed according to a Dirichlet distribution with parameters $(a_1, \ldots, a_n)$ if

$$f(x_1, \ldots, x_n|a_1, \ldots, a_n) = \frac{\Gamma\left(\sum_{i=1}^n a_i\right)}{\prod_{i=1}^n \Gamma(a_i)} \prod_{i=1}^n x_i^{a_i - 1},$$

subject to $\sum_{i=1}^n x_i = 1$, with $x_i > 0$, $a_i > 0$, for all $i = 1, \ldots, n$.

- For $\alpha_0 > 0$ and G_0, a probability measure, a Dirichlet process $DP(\alpha_0, G_0)$ is defined to be the distribution of a random probability measure G that is defined over a measurable space $(\Theta, \mathcal{B})$ such that for any finite measurable partition $(A_1, \ldots, A_r)$ of Θ, the random vector $(G(A_1), \ldots, G(A_r))$ is distributed as a finite-dimensional Dirichlet distribution with parameters $(\alpha_0 G_0(A_1), \ldots, \alpha_0 G_0(A_r))$.
- Drawings G from a Dirichlet process, denoted as $G \sim DP(\alpha_0, G_0)$, are random discrete probability distributions. Parameters α_0 and G_0 are called, respectively, the scale parameter and the base measure/prior. G_0 is the mean of the Dirichlet process $DP(\alpha_0, G_0)$, and thus, a priori, it is the best guess for the random

probability measure G. Parameter α_0 is a measure of the strength of this a priori belief.

- If G is a drawing from a Dirichlet process, then G is a discrete probability measure with probability 1.
- If $(\theta_1,\ldots,\theta_n)$ is drawn from a Dirichlet process $G \sim DP(\alpha_0, G_0)$, then the posterior distribution of G given $(\theta_1,\ldots,\theta_N)$ is also a Dirichlet distribution of the form

$$G|(\theta_1,\ldots,\theta_N) \sim DP(\alpha_0, G_0 + \sum_{n=1}^{N} \delta_{\theta_n}).$$

- The stick-breaking process allows a direct construction of the Dirichlet process so that a drawing from a Dirichlet process can be written as

$$G = \sum_{k=1}^{\infty} \pi_k \delta_{\phi_k},$$

where

$$\pi_k = \pi'_k \prod_{l=1}^{k-1} (1 - \pi'_l)$$

with $\{\pi'_k\}_{k'=1}^{\infty}$ are generated iid according to $\pi'_k|\alpha_0, G_0 \sim Beta(1, \alpha_0)$ and $\{\phi_k\}_{k=1}^{\infty}$ are generated iid according to $\phi_k|\alpha_0, G_0 \sim G_0$.

- Alternative approaches to constructing a Dirichlet process characterize the predictive distribution of the drawings θ from a $DP(\alpha_0, G_0)$ rather than $G \sim DP(\alpha_0, G_0)$.
- The Polya urn model specifies the conditional distribution of successive drawings θ from a $DP(\alpha_0, G_0)$:

$$\theta_i|\theta_1,\ldots,\theta_{i-1},\alpha_0,G_0 \sim \sum_{k=1}^{K_{i-1}} \frac{m_k}{i-1+\alpha_0} \delta_{\phi_k} + \frac{\alpha_0}{i-1+\alpha_0} G_0$$

with $\theta_1 \sim G_0$, where $\{\phi_k\}_{k=1}^{K_{i-1}}$ denote the K_{i-1} distinct values assumed by $\theta_1,\ldots,\theta_{i-1}$ and m_k is the number of θ_j's, for $1 \leq j \leq i-1$, that are equal to ϕ_k for $1 \leq k \leq K_{i-1}$.

- When the exact values ϕ_k's assumed by θ_i's are not of interest but how many of them are equal to each ϕ_k, the above representation is called the Chinese restaurant process (CRP).

- A Dirichlet process mixture model (DPMM) can be defined as follows:

$$\begin{aligned} G &\sim DP(\alpha_0, G_0) \\ \theta_i|G &\sim G \\ \mathbf{y}_i|\theta_i &\sim f_{\theta_i}(\mathbf{y}_i) \end{aligned}.$$

- The Bayesian nonparametric approach to classification models observations $\{\mathbf{y}_i\}_{i=1}^N$ as generated from a DPMM. Mixture components θ_i's represent the cluster of the observation $\mathbf{y}_i$. Then ϕ_k values assumed by θ_i's characterize the distinct clusters contained in data $\{\mathbf{y}_i\}_{i=1}^N$. The classification problem is to estimate the mixture component $\hat{\theta}_i$ for each observation $\mathbf{y}_i$, for all $i \in \{1,\ldots,N\}$. Computing a joint MAP estimate $(\hat{\theta}_1,\ldots,\hat{\theta}_N)$ will require the joint conditional density $P(\theta_1,\ldots,\theta_N|\mathbf{y}_1,\ldots,\mathbf{y}_N)$.
- An approach to obtain the joint conditional distribution of $(\theta_1,\ldots,\theta_N)$ given $(\mathbf{y}_1,\ldots,\mathbf{y}_N)$ is the Gibbs sampling that draws samples from the marginal conditional distribution of each parameter θ_i, given $\mathbf{y}_{1:N}$ and all the other parameters $\{\theta_j\}_{j\neq i}$, where $\{\theta_j\}_{j\neq i}$ are kept fixed at the values of their most recent estimates. This posterior distribution of $\theta_i|\{\theta_j\}_{j\neq i},\mathbf{y}_{1:N}$ is readily available due to the assumption of the DPMM:

$$\theta_i|\{\theta_j\}_{j\neq i},\mathbf{y}_{1:N} \begin{cases} =\theta_j & \text{with probability} \quad q_j \\ \sim f(\theta_i|\mathbf{y}_i) & \text{with probability} \quad q_0 \end{cases},$$

where q_j, for $1 \leq j < i$, and q_0 are given, respectively, by

$$q_j = \frac{f_{\theta_i}(\mathbf{y}_i)}{\alpha_0 f(\mathbf{y}_i) + \sum_{j=1,j\neq i}^N f_{\theta_j}(\mathbf{y}_i)}$$

and

$$q_0 = \frac{\alpha_0 f(\mathbf{y}_i)}{\alpha_0 f(\mathbf{y}_i) + \sum_{j=1,j\neq i}^N f_{\theta_j}(\mathbf{y}_i)},$$

with

$$f(\mathbf{y}_i) = \int_\theta f_\theta(\mathbf{y}_i) g_0(\theta) d\theta$$

and

$$f_\theta(\mathbf{y}_i) \triangleq f(\mathbf{y}_i|\theta_i = \theta)$$

denoting, respectively, the marginal distribution of $\mathbf{y}_i$ and the likelihood of observations for any θ, where θ_i is the parameter corresponding to observation $\mathbf{y}_i$.

- Gibbs sampling is initialized with arbitrary estimates of θ_i's:

$$(\theta_1, \ldots, \theta_N) = (\theta_1^{(0)}, \ldots, \theta_N^{(0)}) = \theta_{1:N}^{(0)}.$$

  One Gibbs iteration is formed by the following N steps:

$$\begin{aligned}
&\text{Draw a sample } \theta_1^{(1)} \text{ from } \theta_1 | \theta_2 = \theta_2^{(0)}, \theta_3 = \theta_3^{(0)}, \ldots, \theta_N = \theta_N^{(0)}, \mathbf{y}_{1:N}.\\
&\text{Draw a sample } \theta_2^{(1)} \text{ from } \theta_2 | \theta_1 = \theta_1^{(1)}, \theta_3 = \theta_3^{(0)}, \ldots, \theta_N = \theta_N^{(0)}, \mathbf{y}_{1:N}.\\
&\vdots\\
&\text{Draw a sample } \theta_N^{(1)} \text{ from } \theta_N | \theta_1 = \theta_1^{(1)}, \theta_2 = \theta_2^{(1)}, \ldots, \theta_{N-1} = \theta_{N-1}^{(1)}, \mathbf{y}_{1:N}.
\end{aligned} \tag{8.64}$$

- Samples generated by the Gibbs sampling algorithm indeed converge in probability to the actual posterior distribution $P(\theta_{1:N}|\mathbf{y}_{1:N})$ of $(\theta_1, \ldots, \theta_N)$ given $\mathbf{y}_{1:N}$.
- It is convenient in practice, for computational reasons, to ensure that the base prior $G(\theta)$ of parameters is a conjugate prior for the likelihood of the observations. This ensures that the posterior distribution $f(\theta_i|\mathbf{y}_i)$ of the parameters is of the same family of distributions as $G(\theta)$.
- By defining a mixture model for the base prior, it is possible to allow observations to have different types of likelihoods while still preserving the conjugate prior condition.

PART III

SIGNAL PROCESSING IN COGNITIVE RADIOS

9

WIDEBAND SPECTRUM SENSING

9.1 INTRODUCTION

As noted in Chapter 3, the early focus of cognitive radios was mostly directed at achieving DSS and improving the spectrum utilization [15, 127], as opposed to more general definition of the concept envisioned in [1, 20]. However, in recent years, the focus of research on cognitive radios has returned to its original emphasis: *radios that are, simply, cognitive* or, more accurately, radios that are intelligent and cognitive [15, 16, 128]. While various forms of DSS for improving wireless spectrum utilization still remain an important application area, there is a growing recognition of much broader, and more ambitious, potential of cognitive radios.

For example, in recent years, cognitive radios have been proposed for e-health, telemedicine, and Wireless Medical Telemetry Services (WMTS) [38, 39]; for emergency and first-responder communications [41–43]; and for jamming and anti-jamming systems [45, 47, 50]. These applications are not merely motivated by improving spectrum utilization. Instead, cognitive communication is used to achieve some of the key performance objectives and to overcome certain inherent challenges. These include system and regulatory constraints, reliability, security, robustness, and energy and spectral efficiency. For instance, according to FCC regulations, WMTS communications are allowed in three noncontiguous spectrum bands: TV channel 37 (608–614 MHz), lower L band (1395–1400 MHz), and upper L band (1427–1432 MHz). Cognitive radios can provide a way to operate over these

Signal Processing for Cognitive Radios, First Edition. Sudharman K. Jayaweera.

multiple noncontiguous frequency bands using possibly multiple radio interfaces (multimode) as per user requirements. In military communications applications, on the other hand, cognitive radio capability can be leveraged to effectively jam enemy communications or to defeat a sophisticated jammer. In applications such as emergency, first-responder, and WMTS communications, intelligent cognitive radios offer great promise in ensuring reliability and robustness. Better spectrum utilization, though important, is not the major concern in these and some of the other emerging applications of cognitive radios.

Cognitive radios add something new to existing radios: the spectrum awareness. As we argued in Chapter 2, it is the ability to be aware of the RF environment and user needs that makes the cognitive radios cognitive. Add intelligence to that and we have a radio that can adapt itself in response to these spectrum conditions and user needs. However, it is the spectrum awareness that enables this intelligent behavior at the first place. Indeed, it is this ability to be aware of its spectrum that have made cognitive radios an ideal approach to improve spectrum utilization via DSS. The ability to be aware of spectrum and to intelligently adapt the radio operation in response to the perceived state of the spectrum are the defining features that have made cognitive radios an attractive technology for many of the emerging applications mentioned above [16].

It is then natural to start any discussion of a cognitive radio with the spectrum awareness. The process of gaining spectrum awareness is called the spectrum knowledge acquisition. In DSA cognitive radio literature, this is commonly referred to as spectrum sensing. From the early days of cognitive radios, there was a clear recognition of the importance of accurate spectrum knowledge to the success of DSA. Although in recent years the FCC has shown the inclination not to rely completely on spectrum sensing by the radios in DSA applications but to make use of spectrum databases, still, this does not take anything away from the fundamental importance of spectrum sensing to the general cognitive radios that go beyond DSS applications.

Indeed, wideband cognitive radios introduced in Chapter 2 will share the spectrum with other users over a wide range of possibly noncontiguous frequencies. These cognitive radios have the ability to communicate simultaneously over multiple radio interfaces spanning these noncontiguous spectrum bands by reconfiguring their operating mode in real time, in response to their RF environment and the user needs. Hence, real time spectrum awareness is critical to the operation of wideband cognitive radios. Effective and efficient real time spectrum sensing, or the spectrum knowledge acquisition, is a prerequisite for the radio's cognition. The radio must be able to scan all its allowed spectrum, detect ongoing spectral activities, and identify for which systems and services do they belong to. This instantaneous spectrum knowledge over the complete range of frequencies on which the radio can communicate is the basis for self-orienting the radio's mode of operation. In a WMTS system, for example, such spectrum knowledge will allow a cognitive radio to identify what signals are priority signals that must not be interfered with as well as spectrum opportunities with appropriate characteristics to possibly meet its own transmission quality requirements. On the other hand, in a military application, knowledge obtained by detection and classification of signals may help the radio discern which signals are possible jammers and enemy transmissions so that certain frequency bands can be avoided.

In this chapter, we start our discussion of spectrum knowledge acquisition problem in the context of a wideband cognitive radio. Hence, this can also be termed wideband spectrum sensing. As will be clear shortly, wideband spectrum sensing is the type of spectrum sensing that will be performed by future cognitive radios that have the capability to operate over multiple noncontiguous spectrum bands spread over a large range of frequencies. As we saw in Chapter 2, the spectrum knowledge acquisition problem in a wideband cognitive radio consists of several steps, and in this chapter, we will go in depth into one of these subproblems. The remaining steps will be covered in the next two chapters, Chapters 10 and 11. The specific spectrum sensing problem encountered in DSS cognitive radios will be discussed in Chapters 12 and 13.

9.2 WIDEBAND SPECTRUM SENSING PROBLEM

The spectrum sensing problem posed by a wideband cognitive radio is different from that of the spectrum sensing problem that has been normally considered in the context of DSS cognitive radios. First, most DSS systems presume a spectrum band licensed to a primary system. Hence, spectrum sensing is limited to the spectrum occupied by this particular primary system or the primary service. A common example is the cognitive radios aimed at utilizing the idle spectrum opportunities over the spectrum allocated for broadcast TV systems: IEEE 802.22 wireless regional area networks (WRAN) that are allowed to operate in the 54–862 MHz spectrum range. However, in the case of a wideband cognitive radio, the spectrum of interest is wideband, possibly on the order of a several hundred MHz to even a few GHz. Second, this spectrum of interest will likely be noncontiguous. Indeed, when we allow the operating spectrum to be wide as in the aforementioned, then it is unreasonable that such a wide frequency range will, or can, be allocated to radio operation contiguously. Even with new regulations and laws, there will still be many legacy systems that simply cannot be moved or changed. Operation over noncontiguous frequency bands is indeed not completely alien to today's wireless systems. For instance, the spectrum that has been made available for cognitive WMTS radio operation is made up of TV channel 37 (608–614 MHz), lower L band (1395–1400 MHz), and upper L band (1427–1432 MHz). There are also many IEEE802.11 WLAN standards that are designed to span frequencies in either 2.4–2.4875 GHz industrial, scientific, and medical (ISM) band or the 5.47–5.725 GHz range. Hence, future wideband cognitive radios will have to operate over noncontiguous spectrum bands spread over a wide frequency range. Thus, their wideband spectrum sensing will have to be over such noncontiguous frequency bands.

Once we allow the spectrum of interest to be made of noncontiguous bands spread over a wide frequency range, then we also need to anticipate more than just one type of signals. This is a key difference compared to spectrum sensing in DSS cognitive radios. In those systems, the spectrum sensing is aimed at detecting the primary users. However, in the case of wideband spectrum sensing, there can be different types of

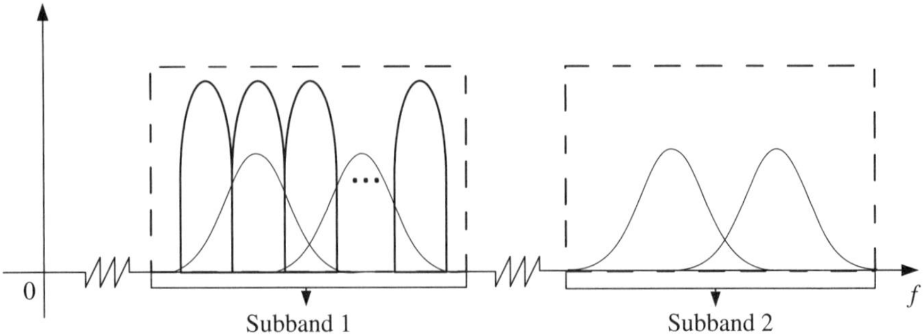

FIGURE 9.1 There can be different types of signals belonging to different systems within different subbands and sometimes even within the same subband.

signals belonging to possibly different systems in each of the noncontiguous bands and even within the same noncontiguous band (see Figure 9.1). For instance, the upper and lower L bands that are available for WMTS communications may also include utility telemetry, government (FAA and Air Force) radar, as well as radio astronomy signals [38,39]. Hence, a cognitive radio used for a WMTS application will have to be able to detect and identify all these different types of signals. Finally, it is important to note that the objective of wideband spectrum sensing is also different from that of the spectrum sensing in DSS context. In particular, it is not enough simply to find a spectrum white space. The objective is to gain the instantaneous spectrum knowledge as completely as possible. Hence, not only signal/spectral activity detection but also signal classification is of importance.

In Chapter 2, we divided the wideband spectrum sensing problem into the following three subproblems:

1. *Wideband spectrum scanning problem*: How to efficiently and effectively scan a very wide spectrum range
2. *Spectral activity detection problem*: How to detect signals from a wideband received signals
3. *Signal classification problem*: How to identify the detected signals

In this chapter, we explore the first of these subproblems, namely, the wideband spectrum scanning problem. In the next two chapters, we will discuss the other two subproblems.

9.3 WIDEBAND SPECTRUM SCANNING PROBLEM

Spectrum sensing for DSS is one of the most mature topics in all of cognitive radio literature. As we saw in Chapter 3, DSS, and DSA in particular, assumes that the cognitive radios operate in frequency bands owned by primary systems. The objective of cognitive radios is to access unused spectrum opportunities without causing harmful interference to these primary users. Hence, spectrum sensing problem in the context of DSS is confined to sensing a particular primary channel or a set of channels that belongs to a single primary system. Thus, the spectrum of interest can span a few tens of, or at most a few hundreds of, MHz wide spectrum. Moreover, these cognitive radios are mostly assumed to be operating in a single mode of operation.

In contrast, wideband spectrum sensing problem is concerned with scanning a much wider, and likely noncontiguous, spectrum. Consider, for example, two noncontiguous frequency bands of interest where one is in the 2.4–2.4875 GHz ISM band and the other is in the 5.47–5.725 GHz band. In most cases, a single wideband antenna spanning 2.4–5.725 GHz range may not be realistic. A possible solution is to use a reconfigurable antenna. In one configuration, the antenna may cover 2.4–2.4875 GHz band, and in another configuration, it may cover 5.47–5.725 GHz band. In general, hence, the wide noncontiguous spectrum of interest will need to be divided into a set of bands that can be sensed by the particular mode of a real time reconfigurable antenna. At any given time, the radio can only sense the spectrum range covered by the particular configuration mode of the antenna.

In each configuration setting, however, the antenna is still a wideband antenna since the response range is in the order of hundred MHz or more. It is unlikely that a cognitive radio will be able to process such a wideband signal in real time. It may still be too wide a range to convert into digital form in real time. This is because the sampling rate is required to be at least as large as the Nyquist frequency in order to avoid aliasing while also providing for sufficient resolution to avoid excessive quantization noise (see the discussion of analog-to-digital (A/D) conversion in Chapter 1 and refer to Appendix A for the sampling theorem and its derivation). Thus, we will call the maximum spectrum width that the cognitive radio can sense and process in real time a *subband*. Hence, in general, we will assume that the spectrum range covered by each configuration of the RF antenna is further divided into a set of subbands. This poses a problem of optimally dividing the wide spectrum of interest into a set of subbands that can be sensed by physically feasible reconfigurable antennas with enough gain and fast response and be processed by realistic A/D converters. This is called the *spectrum segmentation and subbanding*.

At any given time, by definition, the radio is only able to observe, sample, and process one subband. Continuing with the previous example of a radio operating over the two bands of 2.4–2.4875 GHz and 5.47–5.725 GHz, suppose that the radio is currently sensing a subband in the 2.4–2.4875 GHz range. The radio needs to make a decision regarding which subband it should sense during the next sensing period. This decision will depend in part on the user needs and a priori knowledge the radio might have on the characteristics of the subbands. There are also other issues that the radio needs to take into account. For example, sensing either the same or a different subband in the same frequency range of 2.4–2.4875 GHz may require less energy and time delay costs as compared to changing to a subband in the range of 5.47–5.725 GHz. This is because the latter requires reconfiguring the RF antenna and front-end mode that can be much slower, while the former may be achieved by baseband adaptation within the software-defined radio platform. Hence, subband selection problem has to take into account these issues in its formulation.

9.4 SPECTRUM SEGMENTATION AND SUBBANDING

One solution to handling wideband spectrum is to segment the spectrum of interest into a set of subbands. Let us start by assuming a cognitive radio with a reconfigurable RF front-end that can operate in L possible configurations. In each of these configurations, the radio can sense different frequency bands $\mathcal{W}_l$, for $l = 1,\ldots,L$, where bandwidth of each band is $W_l = |\mathcal{W}_l|$. Note that frequency ranges $\mathcal{W}_l$ and their widths will primarily be determined by the constraints and limitations in reconfigurable antenna and front-end design and the spectrum of interest to the radio. Hence, spectrum bands $\mathcal{W}_l$, for $l = 1,\ldots,L$, can be noncontiguous. The total operating bandwidth of the cognitive radio is thus $B_T = \sum_{l=1}^{L} W_l$ where, for simplicity of discussion, we have assumed that the spectrum ranges covered by each configuration are nonoverlapping.

Usually, bandwidths W_l, for $l = 1,\ldots,L$, can be still too wide to be directly converted into digital form given the constraints on the A/D converter. Suppose that the maximum possible output bit rate of the A/D converter is R_b. Given the total bandwidth B_T of spectrum of interest to the radio (which can be made up of noncontiguous segments), then the spectrum segmentation problem is to choose the optimal width of a subband and the total number of subbands N_b subject to the given maximum possible bit rate R_b of the A/D converter.

The signal detection and all subsequent processing performance is affected by the quantization noise of the A/D converter. Hence, to maintain a tolerable level of quantization noise, there will have to be a minimum number of quantization bits per sample. Let us denote this as $q_{\min}$ bits per sample. Then the maximum sampling rate $f_{\max}$ possible with the assumed A/D converter is

$$f_{\max} \leq \frac{R_b}{q_{\min}}. \tag{9.1}$$

According to the sampling theorem (see Appendix A), this maximum sampling rate $f_{\max}$ constraints the maximum possible bandwidth of a subband to be $\frac{f_{\max}}{2}$, hence the need for further segmenting each of the frequency bands W_l, for $l = 1,\ldots,L$, into smaller width subbands. As mentioned earlier, a subband is the basic unit of spectrum that the cognitive radio can sense and process at a time. Assuming that the subband bandwidth $B_j = B$, for $j = 1,\ldots,N_b$, is the same for all subbands, then we have that

$$B \leq \frac{f_{\max}}{2}, \tag{9.2}$$

and consequently, the number of subbands will be

$$N_b = \frac{B_T}{B}$$
$$\geq \frac{2B_T}{f_{\max}} \tag{9.3}$$
$$\geq \frac{2B_T q_{\min}}{R_b}, \tag{9.4}$$

where (9.3) uses (9.2) and (9.4) follows from (9.1).

Segmentation of each frequency band $\mathcal{W}_l$ into a set of subbands can be achieved by using a tunable RF band-pass filter (BPF) or a bank of BPF. Let us denote by N_l the number of subbands within the spectrum covered by the lth configuration of the reconfigurable antenna. If we denote by $\mathcal{N}_l$ this set of subbands, then $|\mathcal{N}_l| = N_l$. We will assume that these subbands are nonoverlapping and they all have the equal bandwidth B.

We have a lower bound for the number of subbands from (9.4). However, it is not desirable to arbitrarily increase the number of subbands and make the bandwidth of a subband very small. How small we may make the bandwidth of a subband, and consequently how large the number of subband can be, will be determined by the signal dynamics (ON–OFF behavior) and the required spectral resolution. In order to be able to detect signals that may exhibit ON–OFF behavior, a cognitive radio may need to revisit each subband at least once every T_0 duration for some T_0. Let us assume uniform sensing of subbands so that each subband will then be revisited with a minimum period of T_0. If the total number of subbands is N_b, then the maximum possible dwell time in each subband is

$$T_D \leq \frac{T_0}{N_b}. \tag{9.5}$$

However, recall from Chapter 6 that the minimum possible spectral resolution Δf is inversely proportional to the dwell time T_D (since T_D is proportional to the number of samples N for a given sampling rate f_s). Hence, from (9.5),

$$\Delta f \approx \frac{1}{T_D}$$
$$\geq \frac{N_b}{T_0},$$

so that

$$N_b \leq \Delta f T_0. \tag{9.6}$$

If the cognitive radio needs to ensure a minimum spectral resolution of $\Delta f \leq \Delta f_0$ for some Δf_0, then from (9.6), the maximum possible number of subbands is upper bounded by

$$N_b \leq T_0 \Delta f_0. \tag{9.7}$$

From (9.4) and (9.7), we have a constraint on the range of possible number of subbands to be

$$\frac{2B_T q_{\min}}{R_b} \leq N_b \leq T_0 \Delta f_0. \tag{9.8}$$

Note that the upper bound is characterized by the maximum tolerable revisit time T_0 and the required spectral resolution, whereas the lower bound is determined by the A/D converter characteristics. Hence, it is certainly possible that for certain combinations of these parameters, there may not be a feasible value for N_b. This means that either we need to redesign our hardware (i.e., the A/D converter) or we need to lower our performance expectations (i.e., we may have to miss certain signals that come and go rapidly or not be able to distinguish between closely spaced signals).

Example 9.1 (Spectrum segmentation) As an example, consider a cognitive radio having an A/D converter with maximum output bit rate of $R_b = 160$ Mbps. Suppose the minimum required bits per sample is $q_{\min} = 8$ bits and we are interested in sensing a bandwidth of $B_T = 100$ MHz. If each subband must be revisited at least once every $T_0 = 20$ ms, then we have that

$$10 \leq N_b \leq 20 \times 10^{-3} \times \Delta f_0.$$

Hence, we need to divide the total bandwidth into at least 10 subbands. On the other hand, this minimum number of subbands corresponds to the minimum spectral resolution achievable:

$$\Delta f_0 \geq \frac{10}{20 \times 10^{-3}} = 500\,\text{Hz} = 0.5\,\text{kHz}.$$

If the A/D converter is to stay fixed, and if we need to achieve a better spectral resolution than that of 0.5 kHz, then we need to allow the revisit time T_0 to be larger. But this will run the risk of missing certain spectral activities in a subband because they may change before the cognitive radio gets to sense them. ■

9.5 WIDEBAND SPECTRUM SENSING RECEIVER

Let us suppose that the spectrum of interest is segmented into an N_b number of disjoint, still wide, subbands, as shown in Figure 9.2. We allow these subbands to be centered at arbitrary frequencies $f_{c_1}, \ldots, f_{c_{N_b}}$, and have bandwidths of $B_1, \ldots, B_{N_b}$.

The wideband spectrum sensing receiver of a cognitive radio is assumed to sense a single subband at any given time. Which subband it will sense at a given time is to be the topic of the next section. For the moment, we may assume that the cognitive radio sequentially picks one of the N_b subbands to sense. Thus, let us assume that at time t, the cognitive radio senses the subband $m \in \{1, \ldots, N_b\}$. At any given time t, the number of signals contained in a subband can be time varying. Hence, let $K_m(t)$

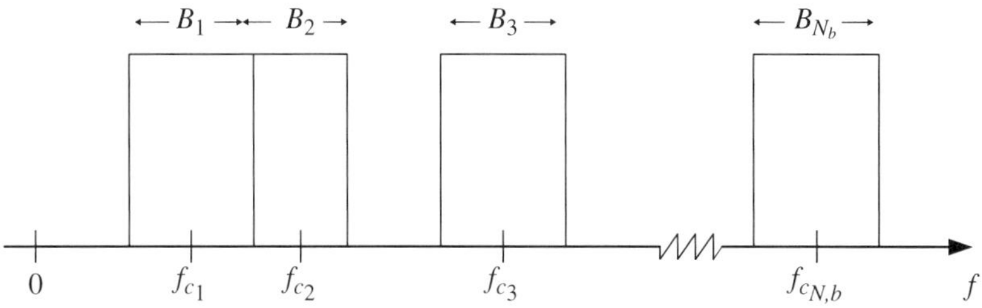

FIGURE 9.2 The wide spectrum of interest is divided into N_b disjoint subbands for the purpose of spectrum sensing.

denote the number of signals in the subband m, for $m = 1, \ldots, N_b$, at time t. We will denote the (complex) baseband representation of the kth of these signals by $x_k(t)$, which is being modulated to the carrier frequency f_k, for $1 \le k \le K_m(t)$. For these signals to be in the subband m, then $f_{c_m} - \frac{B_m}{2} \le f_k \le f_{c_m} + \frac{B_m}{2}$, for $1 \le k \le K_m(t)$.

In the following discussion, we will drop the frequency subband index m since it applies to all subbands. Thus, for example, we will write $K(t)$ in place of $K_m(t)$.

It is a reasonable assumption that the channel between the kth signal source in the subband of interest and the cognitive radio receiver can be modeled as being linear time variant (LTV). Let us denote the baseband equivalent LTV impulse response of this channel by $h_k(\tau, t)$: That is, the response of the channel between the kth signal source and the cognitive radio at time $t = t$ to an impulse that stimulates this channel at time $t - \tau$ is $h_k(\tau, t)$. Thus, the baseband received signal component due to the kth signal source is

$$y_k(t) = \int_{-\infty}^{\infty} x_k(t - \tau) h_k(\tau, t) d\tau. \tag{9.9}$$

Let us denote the in-phase and quadrature components of $y_k(t)$ as $y_{k,R}(t)$ and $y_{k,I}(t)$ so that

$$y_k(t) = y_{k,R}(t) + j y_{k,I}(t). \tag{9.10}$$

The received signal at the cognitive radio is a noise-contaminated version of the superposition of these signals in the RF of the subband of interest:

$$r(t) = \Re\left\{\sum_{k=1}^{K(t)} y_k(t) e^{j2\pi f_k t}\right\} + n(t), \tag{9.11}$$

where $n(t)$ is the zero-mean additive noise.

As discussed in Chapter 1, most, if not all, signal processing in a cognitive radio will be performed in digital baseband. Hence, the received signal $r(t)$ needs to be down-converted and A/D converted before any cognitive signal processing may be performed.

9.5.1 Homodyne Receiver Configuration

Let us first assume a homodyne receiver front-end of the form of Figure 1.8 in which the received signal will be directly down-converted to the baseband and A/D converted. Let us assume that the mth subband of interest is centered at frequency $f = f_{c_m}$, for $1 \leq m \leq N_b$. The received RF signal from the antenna will first be passed through an RF BPF. We may assume that the passband of this BPF is exactly the frequency range of the mth subband in question that is centered at frequency $f = f_{c_m}$, for $1 \leq m \leq N_b$. Note that, with this BPF in place, there is some flexibility in designing the reconfigurable RF antenna front-end for spectrum segmentation. For example, one antenna configuration setting may actually span several subbands, and the selection of the desired subband can be achieved by having a tunable BPF at this RF stage. On the other hand, if the RF antenna spans exactly one subband in each of its configurations, then the BPF might just be also tuned to this subband. In either case, this RF BPF helps remove unwanted signals before passing it through a high-frequency low noise amplifier (LNA) that amplifies the signal to a level that is suitable for the RF mixer. By removing the unwanted frequency signals, the RF BPF at the input of the LNA can help prevent the amplifier saturation that may lead to signal distortion.

The (real-valued) received RF signal $r(t)$ of (9.11) is the output of the RF amplifier before the RF mixer. We may assume that noise $n(t)$ is white Gaussian within the passband with a double-sided PSD of $S_n(f) = \frac{N_0}{2}$. If we denote the (baseband) in-phase and quadrature components of this passband noise as $n_c(t)$ and $n_s(t)$, respectively, then we may write (in the sense of mean square)

$$n(t) = n_c(t)\cos\left(2\pi f_{c_m} t\right) - n_s(t)\sin\left(2\pi f_{c_m} t\right) \tag{9.12a}$$

$$= \Re\left\{(n_c(t) + jn_s(t))\, e^{j2\pi f_{c_m} t}\right\}. \tag{9.12b}$$

If $n(t)$ is zero mean, then both $n_c(t)$ and $n_s(t)$ are also zero mean. Furthermore, if $n(t)$ is Gaussian as we assumed, then $n_c(t)$ and $n_s(t)$ are jointly Gaussian. Suppose that the bandwidth of the RF BPF filter is B_m around the center frequency f_{c_m} of subband m. In general,

$$S_{n_c}(f) = S_{n_s}(f) = \begin{cases} S_n(f - f_{c_m}) + S_n(f + f_{c_m}) & \text{for } |f| \leq \frac{B_m}{2} \\ 0 & \text{otherwise} \end{cases}.$$

Since $S_n(f)$ is white in its passband, however, in this case, it is easy to see that

$$S_{n_c}(f) = S_{n_s}(f) = \begin{cases} N_0 & \text{for } |f| \leq \frac{B_m}{2} \\ 0 & \text{otherwise} \end{cases}.$$

The next stage of a homodyne receiver consists of in-phase and quadrature mixers as in Figure 1.8. For processing on the mth subband, the local oscillator output is the reference signal $\cos\left(2\pi f_{c_m} t\right)$. For convenience, let us assume that the amplifiers before the A/D converters have a gain of two so that the output of the low-pass filter (LPF) that follows the mixer in the in-phase branch can be written as

$$y_I(t) = 2 \times LPF\left\{r(t)\cos\left(2\pi f_{c_m} t\right)\right\},$$

where from (9.11) and (9.12b)

$$r(t) = \left(\sum_{k=1}^{K(t)} y_{k,R}(t)\cos\left(j2\pi f_k t\right) - y_{k,I}(t)\sin\left(j2\pi f_k t\right)\right) + \left(n_c(t)\cos\left(2\pi f_{c_m} t\right) - n_s(t)\sin\left(2\pi f_{c_m} t\right)\right). \tag{9.13}$$

Hence, we may write

$$\begin{aligned} y_I(t) &= \sum_{k=1}^{K(t)} y_{k,R}(t)\cos\left(j2\pi\left(f_k - f_{c_m}\right)t\right) - y_{k,I}(t)\sin\left(j2\pi\left(f_k - f_{c_m}\right)t\right) + n_c(t) \\ &= \sum_{k=1}^{K(t)} \Re\left\{y_k(t)e^{j2\pi\left(f_k - f_{cm}\right)t}\right\} + n_c(t), \end{aligned} \tag{9.14}$$

where (9.14) follows by using (9.10) and we have assumed that $2f_{c_m} \gg f_{c_m}$ and $f_k + f_{c_m} \gg f_{c_m}$. Similarly, the output of the LPF that follows the mixer in the quadrature branch is

$$\begin{aligned} y_Q(t) &= -2 \times LPF\left\{r(t)\sin\left(2\pi f_{c_m} t\right)\right\} \\ &= \sum_{k=1}^{K(t)} y_{k,R}(t)\sin\left(j2\pi\left(f_k - f_{c_m}\right)t\right) + y_{k,I}(t)\cos\left(j2\pi\left(f_k - f_{c_m}\right)t\right) + n_s(t) \\ &= \sum_{k=1}^{K(t)} \Im\left\{y_k(t)e^{j2\pi\left(f_k - f_{cm}\right)t}\right\} + n_s(t). \end{aligned} \tag{9.15}$$

The pair $(y_I(t), y_Q(t))$ in (9.14) and (9.15) are the in-phase and quadrature components, respectively, of the sensed signal at time t in the subband m. After low-pass filtering, they are passed on to the A/D converter that was discussed in Example 1.2.

We may denote by $y(t)$ the complex baseband output of the baseband demodulation process that is the input to the A/D converter:

$$\begin{aligned} y(t) &= y_I(t) + jy_Q(t) \\ &= \sum_{k=1}^{K(t)} \Re\left\{y_k(t)e^{j2\pi(f_k - f_{c,m})t}\right\} + j\sum_{k=1}^{K(t)} \Im\left\{y_k(t)e^{j2\pi(f_k - f_{c,m})t}\right\} + (n_c(t) + jn_s(t)) \\ &= \sum_{k=1}^{K(t)} y_k(t)e^{j2\pi(f_k - f_{c,m})t} + w(t), \end{aligned} \tag{9.16}$$

where $w(t) = n_c(t) + jn_s(t)$ denotes the complex baseband noise. Due to our earlier assumptions on $n(t)$, $n_c(t)$, and $n_s(t)$, this complex baseband noise $w(t)$ is zero-mean Gaussian with independent real and imaginary components.

9.5.2 Super Heterodyne Digital Receiver Configuration

For practical reasons, it is common for modern communications receivers to use heterodyne receivers instead of homodyne receivers. In a heterodyne receiver, the received band-pass-filtered signal $r(t)$ is first down-converted to an intermediate frequency (IF) before further down-converting to the baseband. In fact, as shown in Figure 9.3, in general, there can even be several IF stages. An implementation of a wideband spectrum sensing receiver in a cognitive radio thus may resemble Figure 9.3 with the number of IF stages be determined by the particular hardware choices. However, in the following, without loss of generality, we may assume a single IF stage at an IF f_I.

For instance, to bring the received signal corresponding to subband m to the IF f_I, the first stage local oscillator frequency $f_{LO,1}$ will be chosen to be $f_{LO,1} = f_{c_m} + f_I$. Following this first stage mixer is an IF BPF and an IF amplifier that will restore the signal amplitude to a level that is suitable for the next stage. In general, the next stage may be another IF stage or the (final) baseband mixer. However, as shown in

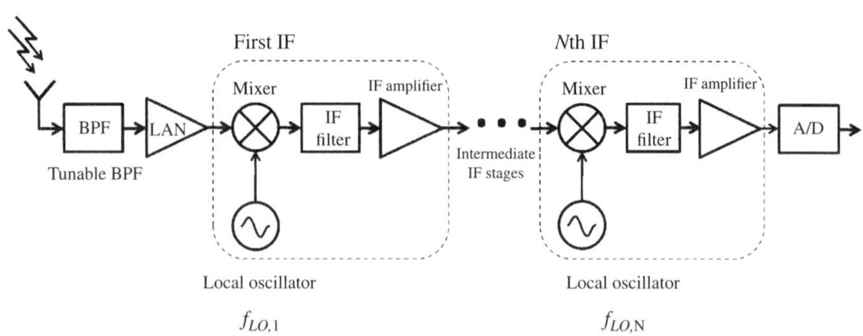

FIGURE 9.3 A multistage superheterodyne receiver for wideband spectrum sensing.

Figure 9.3, the sampling may also be performed in an IF rather than at the baseband. Indeed, in many cases, it can be advantageous to perform IF sampling and A/D followed by digital down-conversion to bring the signal to the digital baseband.

The IF filter is usually required to have high stop-band attenuation to get rid of any unwanted frequencies at its input. Designing a variable frequency filter with such strict constraints can be difficult. Thus, it is convenient to down-convert all subbands to the same IF f_I so that a common fixed IF filter and an IF amplifier can be used. Reconfiguring the cognitive radio to sense different subbands can be achieved by either changing the antenna configuration and the RF BPF or the RF BPF alone, as discussed earlier. In any case, the RF BPF thus necessarily must be of variable frequency. The first stage local oscillator must also be of variable frequency to down-convert different subbands all into the same IF.

The advantage of the superheterodyne receiver is that only the RF BPF, the RF amplifier, and the local oscillator of the first stage of the receiver need to be tunable in order to sense different subbands. Hardware components beyond this point are shared by all subband signals. The only exception is the LPF that follows the baseband mixer, if the bandwidths of subbands were not identical. In this case, we will need these LPFs and the baseband amplifiers to be of variable bandwidths. In either event, the output from the IF mixer is a signal centered at IF f_I (assuming a single IF stage), which is IF filtered and amplified before passing on to the baseband mixer or to the A/D converter as in Figure 9.3. In the former case, the baseband mixer is an I/Q mixer that uses a local oscillator frequency f_I. The in-phase and quadrature branches of the baseband demodulator are then low-pass filtered, amplified, and passed on to the A/D converter.

9.5.3 A/D Conversion and the Discrete-Time Received Signal Model

A/D conversion was discussed in Example 1.2. Let us assume that the A/D conversion is performed in the baseband and focus on the particular subband m for $1 \leq m \leq N_b$. Ideally, sampling at the first step of A/D conversion must be performed in such a way as not to lose any information in the original continuous-time (complex) baseband signal $y(t)$ of (9.16). Since the maximum frequency in the baseband signal is $\frac{B_m}{2}$, sampling rate must be at least B_m on each of the I and Q branches. Combining both I and Q components, thus, the output rate of the sampler will be at least $2B_m$ samples/second.

Let us denote the equivalent complex-valued discrete-time output from the sampling stage of the A/D converter as

$$y[n] = y(nT_s) = y(t)|_{t=nT_s}, \tag{9.17}$$

where T_s is the sampling period that is assumed to be chosen so that there is no aliasing. Moreover, for simplicity of notation, we will assume that the subband bandwidths are all equal to $B_m = B$, for $1 \leq m \leq N_b$, so that the sampling rates $f_{s,m} = f_s$ and sampling periods $T_{s,m} = T_s$ for all subbands $1 \leq m \leq N_b$. Of course, these assumptions can easily be dropped. Under ideal sampling, the discrete-time output signal can then be written as

$$y[n] = \sum_{k=1}^{K[n]} y_k[n] e^{j2\pi F_{k,m} n} + w[n], \tag{9.18}$$

where we have let $K[n] = K(nT_s)$, $w[n] = w(nT_s)$, $y_k[n] = y_k(nT_s)$ and $F_{k,m}$ is the normalized frequency of the kth signal in the mth subband:

$$F_{k,m} = \frac{f_k - f_{c_m}}{f_s} = \left(f_k - f_{c_m}\right) T_s.$$

It can be verified that the discretized noise $w[n] \sim \mathcal{N}_c\left(0, N_0\right)$.

9.6 SUBBAND SELECTION PROBLEM IN WIDEBAND SPECTRUM SENSING

Let us assume a cognitive radio that is designed to sense a spectrum of interest divided into N_b number of subbands. Each subband may generally contain multiple communications channels possibly belonging to different systems. Hence, each subband can itself be wideband in the traditional sense. Moreover, sets of these subbands may be in frequency ranges spanned by different modes of a reconfigurable RF front-end where we assume there are L such modes. An example is shown in Figure 9.4 where $L = 4$.

A subband represents the unit of a spectrum that can be sensed and processed by the cognitive radio at any time. Thus, there is the question of which subband the cognitive radio should scan at any particular time. This is called the *subband selection problem*. For best sensing performance over the spectrum of interest, an efficient decision policy is needed to decide which subband should be scanned at any given time based on the knowledge of signal dynamics of the spectrum bands and the performance objectives of the user. Moreover, compared to switching from one subband to another within the same configuration mode of the RF front-end, switching to one that requires reconfiguring the RF front-end may be much slower.

There may be signals belonging to more than one system in each of these subbands. As a result, each subband may consist of channels belonging to different systems that have different properties. Let us assume that there are $M_1, \ldots, M_{N_b}$ number of possible channels in each of the N_b subbands, respectively. The total number of channels is then $M = \sum_{i=1}^{N_b} M_i$. We may denote by $\mathcal{M}_i$ the set of channel indices belonging to the subband i for $i = 1, \ldots, N_b$. In the following discussion, we may denote the jth channel in the ith subband as the (i,j)th channel. If we enumerate the channels in all the N_b subbands from 1 to M, then the (i,j)th channel is mapped to index k, for $1 \leq k \leq M$, such that

$$k = j + \sum_{i'=0}^{i-1} M_{i'}, \tag{9.19}$$

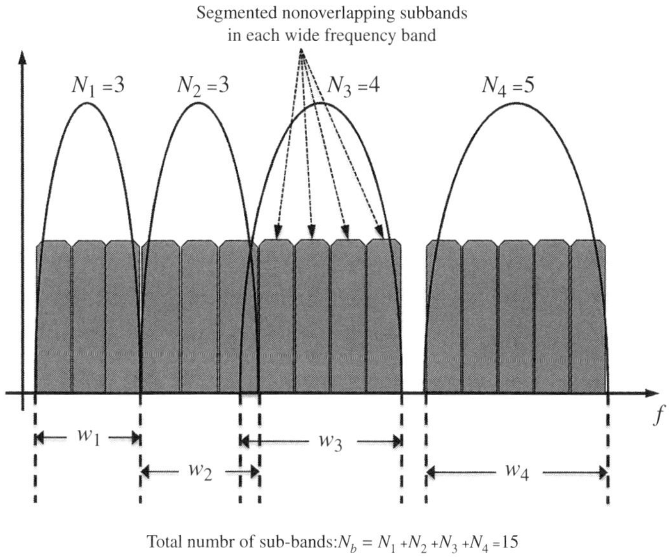

FIGURE 9.4 An illustration of wide frequency bands and further segmented subbands in each wide frequency band.

where, we define, $M_0 = 0$. Alternatively, the kth channel, for $1 \leq k \leq M$, is the jth channel in the ith subband where

$$i = 1 + \max_{i \in \{0,1,\ldots,N_b-1\}} i \quad \text{s.t.} \quad \sum_{i'=0}^{i-1} M_{i'} \leq k - 1 \tag{9.20}$$

and

$$j = k - \sum_{i'=0}^{i-1} M_{i'}. \tag{9.21}$$

Subband selection in wideband spectrum sensing is a decision-making problem. Essentially, the cognitive radios need to predict the subband that will most likely have the desired qualities to meet the user performance objectives during the next time instant. This requires the radio to have a predictive model for the dynamics of the subband qualities. We have not yet defined what we really meant by the quality of a subband. The subband quality, or qualities, of interest will depend on the particular application in question. In the following, we may develop a subband selection policy framework assuming a somewhat simple quality metric in order to demonstrate the basic approach. For a particular application, this can be refined or replaced.

9.6.1 Subband Dynamics

Since subbands can be seen as a collection of channels and channels are what support user traffic, the building block of a subband model can be taken as a channel dynamic model.

9.6.1.1 Channel Markov Models For simplicity, let us assume that the aspect of the channel that is of importance to the cognitive radio is whether the channel is used by another user or not. Hence, we may define each channel to be in one of the two possible states at any given time: As shown in Figure 9.5, the state *busy* (state 0) indicates the channel is occupied by other radios, and the state *idle* (state 1) indicates no radio activities over that channel and it is available for the cognitive radio to access. It is important to observe that the state 0 in this model does not correspond to channels being occupied by signals belonging to a particular primary system, as is the case in DSS cognitive radio networks of Chapter 3. In wideband spectrum sensing, there is no such thing as a single primary system. Channels may belong to an arbitrary set of systems.

We may assume that the state of a channel does not change within a single time slot. This allows the CR to determine the corresponding state of a channel by spending only a short period of time at the beginning of each time slot. Let us denote by $S_{i,j}[n] \in \{0, 1\}$ the true state of the (i,j)th channel (the jth channel in the ith subband) at time n, for $j \in \{1,\ldots,M_i\}$ and $i \in \{1,\ldots,N_b\}$.

A fairly reasonable approach to model the idle/busy state dynamics of a channel is to assume that these changes are (first-order) Markov in that the next state only depends on the current state of the channel. In many cases of interest, this model can be justified [129–133]. If a higher-order Markov model may be better suited to model the idle/busy dynamics of a communications channel, the following approach can easily be modified. Hence, we will assume that the state dynamics of each channel is a two-state Markov chain as in Figure 9.5, where states 0 and 1 correspond to busy and idle, respectively.

As mentioned earlier, a subband may contain several such channels. It is likely that the states of individual channels within a given subband may not be independent nor have identical statistics. For example, the independence assumption can be questionable when there is some level of channel-access control applied to channels

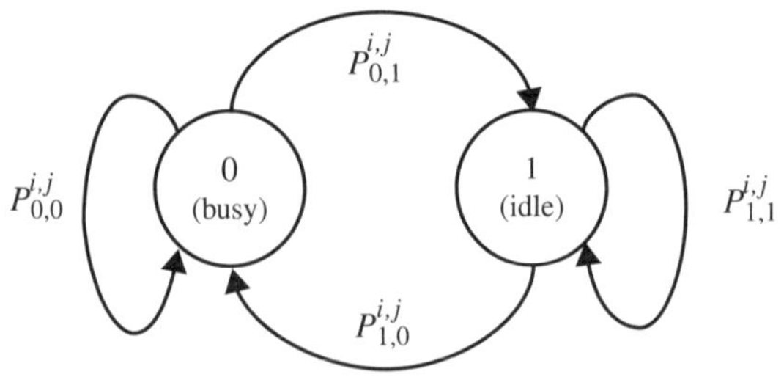

FIGURE 9.5 The Markov chain model for a single communications channel.

belonging to the same communications system: Under a channel-access control protocol, traffic allocated to each of the channels may be centrally determined, causing the idle/busy states of the channels to be correlated. On the other hand, channels that belong to distinct systems, or used by users with different types of payloads, may exhibit nonidentical channel dynamics [134]: That is, the state transition probabilities and the stationary distributions can be different. Hence, in general, we may have to allow the channel states to be nonidentical and dependent.

The Markov chain model of a channel state can be characterized by defining the state transition probabilities. For simplicity, we will assume these transition probabilities to be time invariant. Thus, the transition probability of the (i,j)th channel state from state s to state s' is defined as

$$p^{i,j}_{s,s'} = \Pr\{S_{i,j}[n+1] = s' \mid S_{i,j}[n] = s\}, \quad \forall s, s' \in \{0,1\}. \tag{9.22}$$

The 2×2 *transition probability matrix* of the (i,j)th channel Markov model is denoted by

$$P^{i,j} = \begin{pmatrix} p^{i,j}_{0,0} & p^{i,j}_{0,1} \\ p^{i,j}_{1,0} & p^{i,j}_{1,1} \end{pmatrix}. \tag{9.23}$$

Let $p^{i,j}_0$ and $p^{i,j}_1$ denote the stationary probabilities of busy and idle states of the (i,j)th channel, respectively. Then, the stationary distribution vector of the (i,j)th channel, denoted by $\boldsymbol{p}^{i,j} = \left[p^{i,j}_0, p^{i,j}_1\right]$, is given by the solution to

$$\boldsymbol{p}^{i,j} = \boldsymbol{p}^{i,j} P^{i,j}. \tag{9.24}$$

9.6.1.2 Subband Markov Models Let us consider the ith subband with M_i number of channels. The state of this subband is defined to be the M_i-length vector $\mathbf{S}_i = (S_{i,1}, \ldots, S_{i,M_i})$ where each $S_{i,j} \in \{0,1\}$ denotes the state of the jth channel, for $1 \leq j \leq M_i$, of the ith subband. Let us denote by $P^{(i)}$ the state transition probability matrix of the ith subband. The Markov state transition probabilities of the subband state $\mathbf{S}_i$ can, in theory, be computed from the given channel transition probabilities of the individual channels $p^{i,j}_{s,s'}$ for $1 \leq j \leq M_i$ and $s, s' \in \{0,1\}$. Since state $\mathbf{S}_i$ can take 2^{M_i} possible values, the matrix $P^{(i)}$ is $2^{M_i} \times 2^{M_i}$.

Example 9.2 (Independent Channels) When the channels have independent state dynamics, the channel transition matrix $P^{(i)}$ is easily computed as

$$P^{(i)} = P^{i,1} \otimes P^{i,2} \otimes \ldots \otimes P^{i,M_i}, \tag{9.25}$$

where $\otimes$ denotes the Kronecker product. ■

The overall spectrum of interest to the radio is made of N_b subbands. Hence, the state of the overall spectrum can be denoted by $\mathbf{S} = (\mathbf{S}_1, \ldots, \mathbf{S}_{N_b})$, which is a length

$M = \sum_{i=1}^{N_b} M_i$ vector of binary elements. Hence, the state $\mathbf{S}[n]$ at time n can take 2^M possible vector values. Let us enumerate these states as $\mathbf{s}_1, \ldots, \mathbf{s}_{2^M}$ and denote the set of these possible states as $\mathcal{S}$. The corresponding $2^M \times 2^M$ state transition probability matrix P corresponding to the overall spectrum state $\mathbf{S}$ can be obtained from the subband state transition probability matrices $P^{(i)}$'s. The (m, m')th element of matrix P, denoted as $p_{m,m'}$, is the transition probability of the overall system state from $\mathbf{s}_m$ to $\mathbf{s}_{m'}$, where both $\mathbf{s}_m, \mathbf{s}_{m'} \in \mathcal{S}$ and $1 \leq m, m' \leq 2^M$:

$$\Pr\left(S[n+1] = \mathbf{s}_{m'} | S[n] = \mathbf{s}_m\right) = p_{m,m'}. \tag{9.26}$$

Example 9.3 (A 2-Subband System) Let us consider a spectrum of interest segmented into two subbands: $N_b = 2$. Suppose one of the subbands, say, $i = 1$, contains $M_1 = 3$ channels, while the other subband contains $M_2 = 2$ channels.

Then, the first subband state $\mathbf{S}_1$ can take $2^3 = 8$ different possible states. The state of second subband denoted by $\mathbf{S}_2$ can similarly be in one of $2^2 = 4$ different possible states each time. The overall system state $\mathbf{S}$ can take $32 = 2^5 = 2^{M_1+M_2}$ possible values. ■

9.6.2 A POMDP Model for Subband Selection

9.6.2.1 Actions and Observations When the spectrum state is $\mathbf{S}[n]$ at time n, the radio can select one of the N_b subbands for sensing. Hence, the action space $\mathcal{A}$ is the set of subband indices:

$$\mathcal{A} = \{1, 2, \ldots, N_b\}. \tag{9.27}$$

As mentioned in Section 7.5, in many applications, it is normally the case that the decision-maker cannot directly observe the system state (called the state of the core process). Instead, the decision-maker only has access to an observable that is a function of the system state. The subband selection for sensing is a situation similar to this, because a cognitive radio can observe only a single subband at a time, out of the total N_b subbands that make up the complete state vector $\mathbf{S}[n]$. Hence, the observation is inherently partial. Even the partial state $\mathbf{S}_a$, for $a \in \mathcal{A}$, corresponding to the action a is usually observed in the presence of noise and other anomalies.

If the cognitive radio selects the action $a \in \mathcal{A}$ when the spectrum state is $\mathbf{S}[n] = (\mathbf{S}_1[n], \ldots, \mathbf{S}_{N_b}[n])$ at time $n = n$, it will sense the subband a, for $1 \leq a \leq N_b$, during the nth sensing period. Hence, the observation $Y[n]$ resulting from taking action a is an observation of the state $\mathbf{S}_a[n]$ of the ath subband at time n. Recall that the state $\mathbf{S}_a[n]$ is the vector of the states of M_a channels that make up the ath subband. Let us assume that the cognitive radio obtains an M_a length measurement vector of the form of $\mathbf{r}_a = (r_{a,1}, \ldots, r_{a,M_a})^T$ following its selection of action a, where $r_{a,j}$ is statistically related to the state $S_{i,j}[n]$ of the (i,j)th channel. Based on each $r_{a,j}$, the cognitive radio is assumed to make a binary decision on the true state of the (i,j)th channel. We denote this decision by $Y_{a,j}$ for $1 \leq j \leq M_a$. This is a detection of the channel state based on

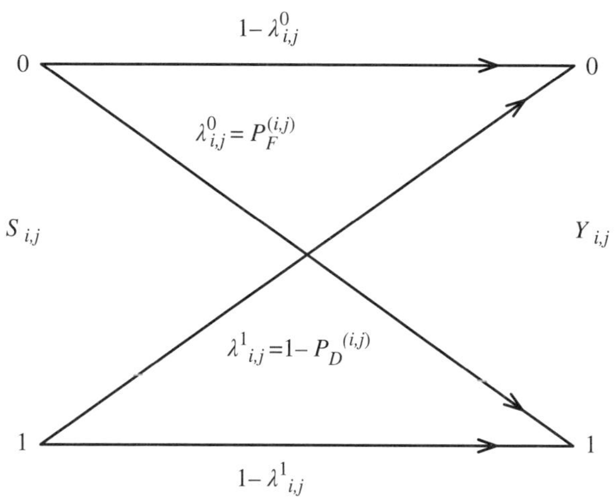

FIGURE 9.6 The state $S_{i,j}$ of the (i,j)th channel and the cognitive radio's observation $Y_{i,j}$ if it were to sense that channel can be viewed as the input and output of an equivalent binary asymmetric channel.

the observation $r_{a,j}$. Hence, $Y_{a,j}$ is to be considered as the output of a detector $\delta(.)$[1] that declares the state of the (a,j)th channel as either 0 or 1 based on the input $r_{a,j}$ so that $y_{a,j} = \delta(r_{a,j})$.

Instead of the original measurement vector $\mathbf{r}_a$, we will take the vector of detections $\mathbf{y}_a = (y_{a,1}, \ldots, y_{a,M_a})^T$ to be the observation $Y[n]$ resulting from selecting action a at time n. Hence, while elements of $\mathbf{r}_a$ can, in general, be continuous valued, $\mathbf{y}_a$ is made of components that only take binary values. Hence, the observation vector $\mathbf{y}_a$ takes values on the finite set $\mathcal{Y}_a = \{0, 1\}^{M_a}$, allowing us to apply the theory of POMDPs as discussed in Section 7.5.

Suppose that the false-alarm and detection probabilities of the (i,j)th channel are denoted by $P_F^{(i,j)}$ and $P_D^{(i,j)}$, respectively. These essentially define the observation process probabilities as an equivalent *binary asymmetric channel* (BAC) as shown in Figure 9.6. The input of this equivalent BAC is the state of the channel and the output is the detector output $y_{i,j}$. The crossover probabilities of the BAC for the (i,j)th channel are simply

$$\begin{aligned}\lambda^0_{i,j} &= \Pr\left(Y_{i,j}[n] = 1 | S_{i,j}[n] = 0\right)\\ &= P_F^{(i,j)}\end{aligned}$$

[1] We could have let $\delta(.)$ to be different for each channel (i,j), but this will make the notation unnecessarily more complicated. The discussion in this chapter can easily be generalized to handle such modifications.

and

$$\begin{aligned}\lambda_{i,j}^{1} &= \Pr\left(Y_{i,j}[n] = 0 | S_{i,j}[n] = 1\right) \\ &= 1 - P_D^{(i,j)}.\end{aligned}$$

For $i \in \mathcal{A}$, $s \in \{0, 1\}$ and $y \in \{0, 1\}$, then we may define the observation process probabilities as

$$\begin{aligned}q_{i,j}(y \mid s, i) &\triangleq \Pr\left(Y_{i,j}[n] = y \mid S_{i,j}[n] = s, a_n = i\right) \\ &= \begin{cases}\Pr\left(Y_{i,j} = 0 \mid S_{i,j}[n] = s\right) & \text{if} \quad y = 0 \\ \Pr\left(Y_{i,j} = 1 \mid S_{i,j}[n] = s\right) & \text{if} \quad y = 1\end{cases},\end{aligned}$$

where

$$\begin{aligned}q_{i,j}(0 \mid s, i) &= \Pr\left(Y_{i,j}[n] = 0 \mid S_{i,j}[n] = s, a_n = i\right) \\ &= \begin{cases}1 - P_F^{(i,j)} & \text{if} \quad s = 0 \\ 1 - P_D^{(i,j)} & \text{if} \quad s = 1\end{cases} \\ &= \begin{cases}1 - \lambda_{i,j}^{0} & \text{if} \quad s = 0 \\ \lambda_{i,j}^{1} & \text{if} \quad s = 1\end{cases}\end{aligned}$$

and

$$\begin{aligned}q_{i,j}(1 \mid s, i) &= \Pr\left(Y_{i,j}[n] = 1 \mid S_{i,j}[n] = s, a_n = i\right) \\ &= \begin{cases}P_F^{(i,j)} & \text{if} \quad s = 0 \\ P_D^{(i,j)} & \text{if} \quad s = 1\end{cases} \\ &= \begin{cases}\lambda_{i,j}^{0} & \text{if} \quad s = 0 \\ 1 - \lambda_{i,j}^{1} & \text{if} \quad s = 1\end{cases}.\end{aligned}$$

We assume that the measurement $r_{i,j}$ only depends on the state $S_{i,j}$ of the (i,j)th channel and that the detector that declares the state of the channel (i,j) is solely based on $r_{i,j}$. Hence, $r_{i,j}$ is conditionally independent of all other channel states given the state of the (i,j)th channel. Consequently, $y_{i,j}$ is also conditionally independent of all other channel states given the state of the (i,j)th channel. Hence, we have that

$$
\begin{aligned}
\Pr(Y[n] &= \mathbf{y} \mid S[n] = \mathbf{s}_m, a_n = a) \\
&= \begin{cases} \Pr(\mathbf{Y}_a[n] = \mathbf{y} \mid S[n] = \mathbf{s}_m, a_n = a) & \text{if } \mathbf{y} \in \mathcal{Y}_a \\ 0 & \text{otherwise} \end{cases} \\
&= \begin{cases} \Pr\left((Y_{a,1}[n], \ldots, Y_{a,M_a})^T[n] = \mathbf{y} \mid S[n] = \mathbf{s}_m, a_n = a\right) & \text{if } \mathbf{y} \in \mathcal{Y}_a \\ 0 & \text{otherwise} \end{cases} \\
&= \begin{cases} \prod_{j=1}^{M_a} \Pr\left(Y_{a,j}[n] = y_j \mid S_{a,j}[n] = s_{a,j}^{(m)}, a_n = a\right) & \text{if } \mathbf{y} \in \mathcal{Y}_a \\ 0 & \text{otherwise} \end{cases} \qquad (9.28) \\
&= \begin{cases} \prod_{j=1}^{M_a} q_{a,j}\left(y_j \mid s_{a,j}^{(m)}, a\right) & \text{if } \mathbf{y} \in \mathcal{Y}_a \\ 0 & \text{otherwise} \end{cases}, \qquad (9.29)
\end{aligned}
$$

where $\mathcal{Y}_a = \{0,1\}^{M_a}$, $s_{a,j}^{(m)} \in \{0,1\}$ is the state of the (a,j)th channel when the state vector is $\mathbf{S} = \mathbf{s}_m \in \mathcal{S}$ and (9.28) follows by invoking the assumption that observations on individual channels are independent conditioned on their respective states.

9.6.2.2 State Dynamics Recall from our definition of a Markov decision process in Chapter 7 that, in general, the actions influence the state dynamics of the Markov system: $\Pr(S[n+1] = s' | S[n] = s, a_n = a) = p(s'|s,a)$. If this was not the case, we really do not have any control over the future rewards. All we can do is at each time instant pick the decision that maximizes the immediate reward $r(s,a)$. In this case, the decision processes are trivial in the sense that optimal finite- or infinite-horizon decision policies can be designed as strategies that are optimal for each time instant.

We have defined the state of the spectrum in terms of idle/busy characteristics of channels. Hence, the state changes can only be affected by transmissions. From (9.27), the action of a cognitive radio is to select a subband for sensing. But sensing is a passive operation in the sense that it does not involve any transmissions. Note that, here, we are making the simplifying assumption that the cognitive radio in question either does not transmit itself while sensing or the transmission decision it is currently executing is independent of the sensing decision to be made regarding the next sensing interval. Hence, under this assumption, the subband selection is a trivial decision process: The subband selection decisions of a cognitive radio do not change the actual spectrum state:

$$
\begin{aligned}
\Pr(S[n+1] = \mathbf{s}_{m'} | S[n] = \mathbf{s}_m, a_n = a) &= \Pr(S[n+1] = \mathbf{s}_{m'} | S[n] = \mathbf{s}_m) \\
&= p_{m,m'}, \qquad (9.30)
\end{aligned}
$$

for all $a \in$ and $\mathbf{s}_{m'}, \mathbf{s}_m \in \mathcal{S}$. Hence, if the subband selection problem were to be a completely observable MDP, then decision-making could be much simpler.

However, from Section 7.5, we know that when only an observation related to the actual state is observable in an MDP, the problem is one of POMDP. As we saw earlier, this is indeed the case in a subband selection problem since we can only observe $Y_i[n]$ corresponding to the state $\mathbf{S}_i[n]$ of the ith subband, but not the overall state $\mathbf{S}[n]$ of the spectrum at time n. Actions in a POMDP can influence not only the actual state dynamics but also the observation process: The radio can only observe the subband it selects to sense.

Hence, the subband selection problem is a special type of a POMDP. It is somewhat simpler than the general POMDP problem considered in Section 7.5 with a finite-state core process because actions only effect what is observed but not the state dynamics of the actual core Markov process.

9.6.2.3 Belief-State MDP As mentioned earlier, the observation $Y[n]$ in the wideband spectrum sensing is inherently partial because at any given time the cognitive radio can only observe one subband out of the total of N_b subbands. Even the partial observation corresponding to the selected subband may not be exactly the state of the channels in that subband due to noise, interference, and other anomalies. Hence, decisions on subband selection for sensing have to be based on the POMDP framework of Section 7.5 in which decisions can be made based on the history at time n, denoted by $\underline{h}_n$.

However, as shown in Figure 9.7, subband selection decisions at time n have to to be made before observing $Y[n]$. Hence, the observation $Y[n]$ corresponding to state $S[n]$ is not available when making action selection for time period n. The information history $\underline{h}_n$ available to the decision-maker at time n is thus

$$\underline{h}_n = (\underline{h}_{n-1}, a_{n-1}, y[n-1]), \tag{9.31}$$

for $n = 0, 1, \ldots$, with $\underline{h}_0 = \phi$ for ϕ denoting the empty set.

In Chapter 7, we saw that a sufficient statistic for decision-making at time $n = n$ is the belief state of the system at time n based on the history at time n. Let us denote the belief of state being $\mathbf{s}_m$, for $1 \leq m \leq 2^M$ and $\mathbf{s}_m \in \mathcal{S}$, given the history $\underline{h}_n$ at time n as

$$b_m[n] = \Pr\left(\mathbf{S}[n] = \mathbf{s}_m | \underline{h}_n\right). \tag{9.32}$$

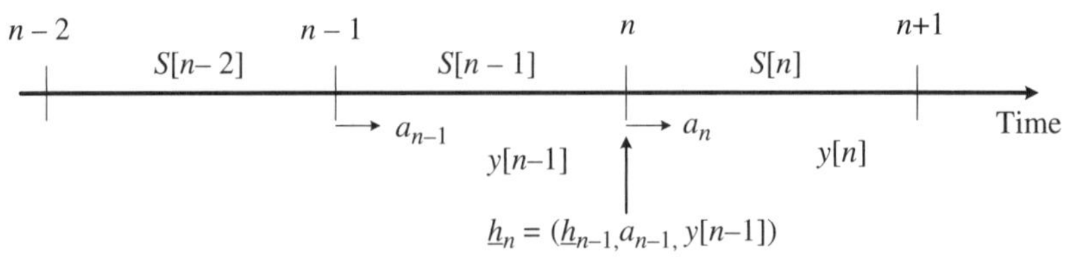

FIGURE 9.7 History available for decision-making at time n does not include the observation during the nth sensing period.

We define the belief-state vector at time n as

$$\mathbf{b}[n] = (b_1[n], \ldots, b_{2^M}[n])^T \in \mathcal{P}^{2^M-1},$$

where $\mathcal{P}^{2^M-1}$ denotes the $(2^M - 1)$-dimensional probability simplex (see (7.63) for definition).

Similar to (7.67) and (7.70) in Chapter 7, we can easily show that $\mathbf{b}[n+1]$ is completely determined by $\mathbf{b}[n]$, the action a_n chosen at time n, and the resulting observation y_n. Indeed, for $m = 1, \ldots, 2^M$ and $\mathbf{s}_m \in \mathcal{S}$,

$$\begin{aligned}
b_m[n+1] &= \Pr\left(S[n+1] = \mathbf{s}_m \mid \underline{h}_{n+1}\right) \\
&= \frac{\sum_{m'=1}^{2^M} \Pr\left(S[n+1] = \mathbf{s}_m, \underline{h}_{n+1} | S[n] = \mathbf{s}_{m'}\right) \Pr\left(S[n] = \mathbf{s}_{m'}\right)}{\Pr\left(\underline{h}_{n+1}\right)} \\
&= \frac{\dfrac{\sum_{m'=1}^{2^M} \Pr\left(S[n+1] = \mathbf{s}_m | S[n] = \mathbf{s}_{m'}, \underline{h}_{n+1}\right) \Pr\left(\underline{h}_{n+1} | S[n] = \mathbf{s}_{m'}\right)}{\Pr\left(S[n] = \mathbf{s}_{m'}\right)}}{\Pr\left(\underline{h}_{n+1}\right)} \\
&= \sum_{m'=1}^{2^M} \Pr\left(S[n+1] = \mathbf{s}_m | S[n] = \mathbf{s}_{m'}, \underline{h}_{n+1}\right) \Pr\left(S[n] = \mathbf{s}_{m'} | \underline{h}_{n+1}\right) \\
&= \sum_{m'=1}^{2^M} \Pr\left(S[n+1] = \mathbf{s}_m | S[n] = \mathbf{s}_{m'}, \underline{h}_n, a_n, y[n]\right) \Pr\left(S[n] = \mathbf{s}_{m'} \left| \underline{h}_{n+1}\right.\right).
\end{aligned} \tag{9.33}$$

However, as noted earlier, since the state transitions of the underlying channel dynamics are not affected by the actions chosen by the secondary cognitive radios or their observations,[2]

$$\begin{aligned}
\Pr\left(S[n+1] = \mathbf{s}_m | S[n] = \mathbf{s}_{m'}, \underline{h}_n, a_n, y[n]\right) &= \Pr\left(S[n+1] = \mathbf{s}_m | S[n] = \mathbf{s}_{m'}\right) \\
&= p_{m',m},
\end{aligned} \tag{9.34}$$

where the last step follows from the definition (9.26) of the overall system transition probability $p_{m',m}$ for $\mathbf{s}_m, \mathbf{s}_{m'} \in \mathcal{S}$. Hence,

$$b_m[n+1] = \sum_{m'=1}^{2^M} p_{m',m} \Pr\left(S[n] = \mathbf{s}_{m'} \left| \underline{h}_{n+1}\right.\right). \tag{9.35}$$

From (9.35), it is clear that the belief state at time $n+1$ is purely a prediction from the *filtered state estimate* $\Pr\left(S[n] = \mathbf{s}' \left| \underline{h}_{n+1}\right.\right)$ available at time n. The reason is that the

[2]In general, the actions selected in an MDP do effect the state transitions so that $\Pr\left(S[n+1] = s | S[n] = s', \underline{h}_n, a_n = a, y[n]\right) = \Pr\left(S[n+1] = s | S[n] = s', a_n = a\right) = p(s|s', a)$.

belief about state $S[n+1]$ at time $n+1$ is formed before having the sensing observation at time $n+1$. Instead, the latest observation available is $Y[n]$, which relates to $S[n]$. The required filtered state estimate is

$$\begin{aligned}
\Pr\left(S[n]=\mathbf{s}_{m'} \mid \underline{h}_{n+1}\right) &= \Pr\left(S[n]=\mathbf{s}_{m'} \;\middle|\; \underline{h}_n, a_n, y[n]\right) \\
&= \frac{\Pr\left(y[n] \mid S[n]=\mathbf{s}_{m'}, \underline{h}_n, a_n\right)\Pr\left(S[n]=\mathbf{s}_{m'}, \underline{h}_n, a_n\right)}{\Pr\left(\underline{h}_n, a_n, y[n]\right)} \\
&= \frac{\Pr\left(y[n] \mid S[n]=\mathbf{s}_{m'}, a_n\right)\Pr\left(S[n]=\mathbf{s}_{m'} \mid \underline{h}_n, a_n\right)}{\Pr\left(y[n] \mid \underline{h}_n, a_n\right)} \\
&= \frac{\Pr\left(y[n] \mid S[n]=\mathbf{s}_{m'}, a_n\right)\Pr\left(S[n]=\mathbf{s}_{m'} \mid \underline{h}_n, a_n\right)}{\sum_{m''=1}^{2M}\Pr\left(y[n] \mid S[n]=\mathbf{s}_{m''}, \underline{h}_n, a_n\right)\Pr\left(S[n]=\mathbf{s}_{m''} \mid \underline{h}_n, a_n\right)} \\
&= \frac{\Pr\left(y[n] \mid S[n]=\mathbf{s}_{m'}, a_n\right)\Pr\left(S[n]=\mathbf{s}_{m'} \mid \underline{h}_n, a_n\right)}{\sum_{m''=1}^{2M}\Pr\left(y[n] \mid S[n]=\mathbf{s}_{m''}, a_n\right)\Pr\left(S[n]=\mathbf{s}_{m''} \mid \underline{h}_n, a_n\right)} \\
&= \frac{\Pr\left(y[n] \mid S[n]=\mathbf{s}_{m'}, a_n\right)\Pr\left(S[n]=\mathbf{s}_{m'} \mid \underline{h}_n\right)}{\sum_{m''=1}^{2M}\Pr\left(y[n] | S[n]=\mathbf{s}_{m''}, a_n\right)\Pr\left(S[n]=\mathbf{s}_{m''} \mid \underline{h}_n\right)} \\
&= \frac{\Pr\left(y[n] \mid S[n]=\mathbf{s}_{m'}, a_n\right) b_{m'}[n]}{\sum_{m''=1}^{2M}\Pr\left(y[n] | S[n]=\mathbf{s}_{m''}, a_n\right) b_{m''}[n]}
\end{aligned} \tag{9.36}$$

since, by definition, $b_{m'}[n] = \Pr\left(S[n]=\mathbf{s}_{m'} \mid \underline{h}_n\right)$ and $\Pr\left(y[n] \mid S[n]=\mathbf{s}_{m'}, a_n\right)$ is given by (9.29).

From (9.35) and (9.36), for $m = 1, \ldots, 2^M$,

$$\begin{aligned}
b_m[n+1] &= \sum_{m'=1}^{2^M} p_{m',m} \frac{\Pr\left(y[n] \mid S[n]=\mathbf{s}_{m'}, a_n\right) b_{m'}[n]}{\sum_{m''=1}^{2M}\Pr\left(y[n] \mid S[n]=\mathbf{s}_{m'}, a_n\right) b_{m''}[n]} \\
&= \frac{\sum_{m'=1}^{2M} p_{m',m}\Pr\left(y[n] \mid S[n]=\mathbf{s}_{m'}, a_n\right) b_{m'}[n]}{\sum_{m''=1}^{2M}\Pr\left(y[n] \mid S[n]=\mathbf{s}_{m'}, a_n\right) b_{m''}[n]} \\
&\triangleq T_m\left(\mathbf{b}[n], a_n, y[n]\right),
\end{aligned} \tag{9.37}$$

where we have denoted the mapping from $\mathbf{b}[n]$ to $b_m[n+1]$ by $T_m(., a, y)$.

We may denote the transformation from $\mathbf{b}[n]$ to $\mathbf{b}[n+1]$ by $\mathbf{T}(., a, y)$ so that $\mathbf{b}[n+1] = \mathbf{T}(\mathbf{b}[n], a_n, y[n])$ with $\mathbf{T} = (T_1, \ldots, T_{2^M})$. Hence, the belief-state vector $\mathbf{b}[n]$ is itself a Markov process: $\mathbf{b}[n+1] = \mathbf{T}(\mathbf{b}[n], a_n, y[n])$. From Section 7.5, we know that $\mathbf{b}[n]$ is a sufficient statistic for the decision-making problem in the original POMDP. Hence, the cognitive radio may form its subband selection decisions based on the Markov state $\mathbf{b}[n]$. However, note that this belief state is a probability vector that can take a continuum of values in the $(2^M - 1)$-dimensional probability simplex.

9.6.3 An Optimal Subband Selection Policy for Spectrum Sensing

Given the history $\underline{h}_n$ at time $n = n$, the cognitive radio can select one of the N_b subbands for sensing. Let us denote the immediate reward associated with selecting action $a \in \mathcal{A}$ when in state $\mathbf{S}$ by $R(\mathbf{S}, \mathbf{S}', a, Y)$ where we have let it to be a function of also the next state of the system $\mathbf{S}'$ and the observation Y resulting by taking action a. Observe that we have also assumed that the immediate reward function $R(.)$ is time invariant. Recall that the observation Y corresponding to taking the action $a \in \{1, \ldots, N_b\}$ when the Markov system is in state $\mathbf{S}$ is the M_a-length binary vector $\mathbf{Y} = (Y_{a,1}, \ldots, Y_{a,M_a})^T$, which is the output of the detectors for each channel in the subband a.

Of course, the immediate reward $R(\mathbf{S}, \mathbf{S}', a, Y)$ cannot be determined by a radio that does not have access to the exact state $\mathbf{S}$ of the system. As noted earlier, however, the cognitive radio can compute its belief of the system state being in each of the possible 2^M states based on the history $\underline{h}_n$. Given this belief vector $\mathbf{b}[n] = (b_1[n], \ldots, b_{2^M}[n])^T \in \mathcal{P}^{2^M-1}$ at time n, similar to (7.73), the cognitive radio instead evaluates the expected immediate reward defined as

$$\begin{aligned} r(\mathbf{b}, a) &= \mathbb{E}\left\{R(\mathbf{S}, \mathbf{S}', a, Y) | \mathbf{b}\right\} \\ &= \sum_m \mathbb{E}\left\{R(\mathbf{S} = \mathbf{s}_m, \mathbf{S}', a, Y)\right\} \Pr\left(\mathbf{S} = \mathbf{s}_m | \mathbf{b}\right) \\ &= \sum_m \mathbb{E}\left\{R(\mathbf{S} = \mathbf{s}_m, \mathbf{S}', a, Y)\right\} b_m \\ &= \sum_m r_m(a) b_m \\ &= \mathbf{r}(a)^T \mathbf{b}, \end{aligned}$$

where

$$\mathbf{r}(a) = \left(r_1(a), \ldots, r_{2^M}(a)\right)^T$$

is a 2^M-length vector and, for $1 \leq m \leq 2^M$,

$$r_m(a) = \mathbb{E}\left\{R(\mathbf{S} = \mathbf{s}_m, \mathbf{S}', a, Y)\right\} \tag{9.38}$$

is the expected reward for taking action a when the system state is $\mathbf{S} = \mathbf{s}_m$. The expectation in (9.38) is with respect to the core and observation processes.

Example 9.4 (A Reward Function) As an example, we may consider the following immediate reward function that is suitable for a cognitive radio whose objective is to sense the subband that contains the most transmission opportunities:

$$R(\mathbf{S}, \mathbf{S}', a, Y) = \sum_{j=1}^{M_a} B_{a,j} S_{a,j} Y_{a,j},$$

where $B_{a,j}$ is the bandwidth of the jth channel in the ath subband. The corresponding expected reward function is

$$\begin{aligned} r_m(a) &= \mathbb{E}\left\{R(\mathbf{S}=\mathbf{s}_m,\mathbf{S}',a,Y)\right\} \\ &= \mathbb{E}\left\{\sum_{j=1}^{M_a} B_{a,j}S_{a,j}Y_{a,j}|\mathbf{S}=\mathbf{s}_m\right\} \\ &= \sum_{j=1}^{M_a} B_{a,j}s_{a,j}^{(m)}\mathbb{E}\left\{Y_{a,j}|S_{a,j}=s_{a,j}^{(m)}\right\} \\ &= \sum_{j=1}^{M_a} B_{a,j}s_{a,j}^{(m)}P_D^{(a,j)}, \end{aligned} \tag{9.39}$$

where, as before, $s_{a,j}^{(m)} \in \{0,1\}$ is the state of the (a,j)th channel when the state vector is $\mathbf{S}=\mathbf{s}_m$ and we have used the fact that

$$\mathbb{E}\left\{Y_{a,j}|S_{a,j}=s_{a,j}^{(m)}\right\} = \begin{cases} P_D^{(a,j)} & \text{if} \quad s_{a,j}^{(m)}=1 \\ P_F^{(a,j)} & \text{if} \quad s_{a,j}^{(m)}=0 \end{cases}$$

with $P_D^{(a,j)}$ and $P_F^{(a,j)}$ denoting, respectively, the detection and false-alarm probabilities of the (a,j)th channel. ■

Let us denote the subband selection decision rule at time n by $\pi_n(\mathbf{b}[n]) \in \mathcal{A}$. Hence, π_n is a mapping $\pi_n : \mathcal{P}^{2^M-1} \longrightarrow \{1,\ldots,N_b\}$. A subband selection (decision) policy $\underline{\pi}$ is a sequence of such decision functions, one for each time instant: $\underline{\pi} = (\pi_0(.),\ldots,\pi_{T-1})$. When $T \longrightarrow \infty$, we have an infinite-horizon problem.

Under the above assumptions, we seek an optimal subband selection policy π for sensing in the sense of maximizing the expected discounted total reward:

$$V_T^{\underline{\pi}}(\mathbf{b}) = \mathbb{E}_{\underline{\pi}}\left\{\sum_{n=0}^{T}\gamma^n r(\mathbf{b}[n],\pi_n) \mid \mathbf{b}[0]=\mathbf{b}\right\}, \tag{9.40}$$

where $\gamma \in (0,1]$ is a discount factor. Note that when no other information is available, the initial belief state $\mathbf{b}[0]=\mathbf{b}$ at time $n=0$ can be taken as the stationary distribution $\mathbf{p}$ of the overall system.

For infinite-horizon decision-making, the approach can be modified by using the theory developed in Chapter 7. Indeed, recall from Section 7.5 that policy $\underline{\pi}$ is said to be stationary if $\pi_n(.)$ does not depend on time instant n. In the following, we will assume that the channel dynamics, rewards, and action space are time-invariant, so that we may expect the existence of a stationary optimal policy $\underline{\pi} = (\pi,\pi,\ldots,)$ such that $\pi_n = \pi$ for all n when $T \longrightarrow \infty$. As usual, the infinite-horizon value function will be denoted by $V^{\pi}(\mathbf{b})$:

$$V^{\pi}(\mathbf{b}) = \mathbb{E}_{\pi}\left\{\sum_{n=0}^{\infty} \gamma^n r(\mathbf{b}[n], \pi) \mid \mathbf{b}[0] = \mathbf{b}\right\}. \tag{9.41}$$

The cognitive radio thus seeks an optimal (stationary) policy π^* such that

$$\pi^* = \arg\max_{\pi} V^{\pi}(\mathbf{b}). \tag{9.42}$$

Let us denote by $V^*(\mathbf{b})$ the optimal value function:

$$V^*(\mathbf{b}) = V^{\pi^*}(\mathbf{b}). \tag{9.43}$$

From Theorem 7.9, the optimal value function $V^*(\mathbf{b})$ satisfies the following optimality equation:

$$V^*(\mathbf{b}) = \max_{a \in \mathcal{A}}\left\{\mathbf{r}(a)^T\mathbf{b} + \gamma \sum_{y \in \mathcal{Y}_a} \Pr(y \mid \mathbf{b}, a) V^*(\mathbf{b}^{(new)})\right\}, \tag{9.44}$$

where $\mathbf{b}^{(new)} = \mathbf{T}(\mathbf{b}, a, \mathbf{y}) = (b_1^{(new)}, b_2^{(new)}, \ldots, b_{2M}^{(new)})$ is given by the transformation (9.37), for $m = 1, \ldots, 2^M$, as

$$b_m^{(new)} = T_m(\mathbf{b}, a, y) \tag{9.45}$$

and

$$\begin{aligned}\Pr(y \mid \mathbf{b}, a) &= \mathbb{E}\{\Pr(y \mid \mathbf{s}, a) \mid \mathbf{b}\} \\ &= \sum_{m=1}^{2^M} \Pr(Y[n] = \mathbf{y} \mid S[n] = \mathbf{s}_m, a_n = a)\, b_m,\end{aligned} \tag{9.46}$$

with $\Pr(Y[n] = \mathbf{y} \mid S[n] = \mathbf{s}_m, a_n = a)$ given by (9.29).

As we discussed in Section 7.5.5, solving (9.44) may not be possible without further assumptions since $\mathbf{b}$ takes a continuum of values on the $(2^M - 1)$-dimensional probability simplex. As we mentioned in Chapter 7, one such situation is when the optimal value function (9.44) can be piecewise linear and convex. While convexity is always assured, the linearity of the infinite-horizon value function may or may not be true. However, as we discussed in Section 7.5.5, it will be linear if the POMDP is known to have an optimal policy that is finitely transient (see Definition 7.5.4). When this is true, we may use the POMDP value iteration given in Algorithm 7.4 to compute the optimal infinite-horizon value function and an optimal policy. If, on the other hand, the POMDP does not have an optimal policy that is finitely transient, the optimal value function $V^*(\mathbf{b})$ will not be piecewise linear. However, in that case, still Algorithm 7.4 may be used to obtain an ϵ-optimal policy.

Whether the decision-making problem for spectrum sensing is best modeled as an infinite-horizon or a finite-horizon POMDP may depend on the application scenario. In general, one may consider that the cognitive radio is unaware of when it may be turned off. Hence, decision-making horizon may be think of as being infinite. However, this may not be the case always. If, indeed, the decision-making horizon T were to be finite, instead of Theorem 7.9, we may use Theorem 7.5 that specifies a backward induction procedure for evaluating the optimal value function. However, as we discussed in Section 7.5, computing the optimal value function using Theorem 7.5 will still require efficient computational procedures due to exponential complexity in time horizon length and number of observations. In Chapter 7, we discussed several of these algorithms. Indeed, note that even the POMDP value iteration algorithm suggested earlier for computing the infinite-horizon value function requires an efficient finite-horizon value evaluation algorithm since essentially it solves a sequence of finite-horizon problems with increasing horizon length T. We refer the reader to Chapter 7 for these details.

9.6.4 A Reduced-Complexity Optimal Sensing Decision-Making Algorithm with Independent Channels

Although it may not always be justifiable, an assumption that may hold sometimes in practice is that of independent channel dynamics. In the extreme case of all channels having independent idle/busy state dynamics, it can be shown that one may derive a new sufficient-statistic vector for the subband selection decision problem that has only a dimension of $M = \sum_i^{N_b} M_i$, instead of 2^M as in the general case assumed in the previous section. Indeed, when channels are independent, a sufficient statistic for decision-making is simply the collection of M beliefs of each channel's state given the history, where belief is the posterior probability of state given the history. Since each channel state is binary, we only need to provide one belief value for each channel (either the belief on state being idle or busy). Thus, an M-length vector of each channel's belief state is sufficient. This dramatically reduces the computational complexity from being exponential in the total number of channels M to only being linear in M. This independent channel model was used in [129] in the context of DSS cognitive radios as we will discuss later in Chapter 13.

Let us keep track of the belief of each channel state k, for $k = 1, \ldots, M$, being idle (i.e., state 1):

$$\omega_k[n] = \Pr\left(S_{i,j}[n] = 1 \mid \underline{h}_n\right),$$

where (i,j) pair for a given k is given by (9.20) and (9.21). Determining k for a given channel (i,j) can be achieved using (9.19).

Note that $\omega_k[n]$ is simply the probability of channel k being idle (i.e., in state 1) at time n, given the history $\underline{h}_n$. We could have equally well defined $\omega_k[n]$ to be the probability of channel k being busy (i.e., in state 0) given the history $\underline{h}_n$. Since the sum of these two probabilities equals unity, they are equivalent. Then, due to the independence of the channel states, a sufficient statistic for the subband selection problem is the M-length vector $\boldsymbol{\omega}$ defined as

$$\boldsymbol{\omega} = (\omega_1, \ldots, \omega_M). \tag{9.47}$$

For channel k corresponding to the (i,j)th channel as in (9.19), we have that

$$\begin{aligned}
\omega_k[n+1] &= \Pr\left(S_{i,j}[n+1] = 1 \,|\underline{h}_{n+1}\right) \\
&= \Pr\left(S_{i,j}[n+1] = 1|S_{i,j}[n] = 1, \underline{h}_{n+1}\right)\Pr\left(S_{i,j}[n] = 1|\underline{h}_{n+1}\right) \\
&\quad + \Pr\left(S_{i,j}[n+1] = 1|S_{i,j}[n] = 0, \underline{h}_{n+1}\right)\Pr\left(S_{i,j}[n] = 0|\underline{h}_{n+1}\right) \\
&= p_{11}^{(i,j)} \Pr\left(S_{i,j}[n] = 1 \left|\underline{h}_{n+1}\right.\right) + p_{01}^{(i,j)} \Pr\left(S_{i,j}[n] = 0 \left|\underline{h}_{n+1}\right.\right),
\end{aligned} \tag{9.48}$$

where (9.48) is due to the special property observed in (9.34) because channel Markov dynamics are independent of actions. Since state and observation sequences of individual channels are independent from those of the other channels and if subband i is not the one selected for sensing by the action a_n then the belief state of channels in subband i can only be updated based on the Markov evolution model, we then have from (9.48) that

$$\begin{aligned}
\omega_k[n+1] &= \Pr\left(S_{i,j}[n+1] = 1 \,|\underline{h}_{n+1}\right) \\
&= p_{11}^{(i,j)} \Pr\left(S_{i,j}[n] = 1 \left|\underline{h}_n, a_n, \mathbf{y}[n]\right.\right) + p_{01}^{(i,j)} \Pr\left(S_{i,j}[n] = 0 \left|\underline{h}_n, a_n, \mathbf{y}[n]\right.\right) \\
&= \begin{cases}
\begin{array}{l} p_{11}^{(i,j)} \Pr\left(S_{i,j}[n] = 1 \left|\underline{h}_n, a_n, y_{i,j}[n]\right.\right) \\ \quad + p_{01}^{(i,j)} \Pr\left(S_{i,j}[n] = 0 \left|\underline{h}_n, a_n, y_{i,j}[n]\right.\right) \end{array} & \text{if} \quad a_n = i \\
p_{11}^{(i,j)} \Pr\left(S_{i,j}[n] = 1 \left|\underline{h}_n\right.\right) + p_{01}^{(i,j)} \Pr\left(S_{i,j}[n] = 0 \left|\underline{h}_n\right.\right) & \text{if} \quad a_n \neq i
\end{cases} \qquad (9.49\text{a}) \\
&= \begin{cases}
\begin{array}{l} p_{11}^{(i,j)} \dfrac{\Pr\left(Y_{i,j}[n] = y_{i,j}[n] \mid S_{i,j}[n] = 1, a_n = i\right)\omega_k[n]}{\Pr(y_{i,j}[n]|\underline{h}_n, a_n = i)} \\ + p_{01}^{(i,j)} \dfrac{\Pr\left(Y_{i,j}[n] = y_{i,j}[n] \mid S_{i,j}[n] = 0, a_n = i\right)(1-\omega_k[n])}{\Pr(y_{i,j}[n]|\underline{h}_n, a_n = i)} \end{array} & \text{if} \quad a_n = i \\
p_{11}^{(i,j)} \omega_k[n] + p_{01}^{(i,j)} (1-\omega_k[n]) & \text{if} \quad a_n \neq i
\end{cases} \\
&= \begin{cases}
\dfrac{\begin{array}{l} p_{11}^{(i,j)} \Pr\left(Y_{i,j}[n] = y_{i,j}[n] \mid S_{i,j}[n] = 1, a_n = i\right)\omega_k[n] \\ + p_{01}^{(i,j)} \Pr\left(Y_{i,j}[n] = y_{i,j}[n] \mid S_{i,j}[n] = 0, a_n = i\right)(1-\omega_k[n]) \end{array}}{\Pr(y_{i,j}[n]|\underline{h}_n, a_n = i)} & \text{if} \quad a_n = i \\
p_{11}^{(i,j)} \omega_k[n] + p_{01}^{(i,j)} (1-\omega_k[n]) & \text{if} \quad a_n \neq i
\end{cases} \\
&\triangleq T_k^{(\omega)}(\omega_k[n], y_{i,j}[n], a_n),
\end{aligned} \tag{9.49b}$$

where we have used $T_k^{(\omega)}(\omega_k[n], y, a)$ to denote the transformation from $\omega_k[n]$ to $\omega_k[n+1]$ under the influence of action a and observation y:

$$T_k^{(\omega)}(\omega_k[n], y, a)$$
$$\triangleq \begin{cases} \dfrac{p_{11}^{(i,j)} \Pr\left(Y_{i,j}[n] = y \mid S_{i,j}[n] = 1, a_n = a\right) \omega_k[n] + p_{01}^{(i,j)} \Pr\left(Y_{i,j}[n] = y \mid S_{i,j}[n] = 0, a_n = a\right) (1 - \omega_k[n])}{\Pr(y|\underline{h}_n, a_n = a)} & \text{if} \quad a = i\,, \\ p_{11}^{(i,j)} \omega_k[n] + p_{01}^{(i,j)} (1 - \omega_k[n]) & \text{if} \quad a \neq i \end{cases} \tag{9.50}$$

where (i,j) are related to k through (9.20) and (9.21).

From (9.49b), it follows that the updating of each individual channel belief state is decoupled from those of the others. We may denote the updated belief vector as

$$\boldsymbol{\omega}[n+1] \triangleq \mathbf{T}^{(\omega)}(\boldsymbol{\omega}[n], y, a_n),$$

where the kth element of the transformation vector $\mathbf{T}^{(\omega)}(.,.,.)$ is given by $T_k^{(\omega)}(\omega_k[n], y, a)$ in (9.50). The expected rewards can also be evaluated in terms of the ω_k's as shown in the following example.

Example 9.5 (Expected Rewards with Independent Channel Dynamics) Let us consider the same immediate reward function $R(\mathbf{S}, \mathbf{S}', a, Y)$ of Example 9.4. The expected immediate reward $r(\boldsymbol{\omega}, a)$ can now be written as

$$\begin{aligned} r(\boldsymbol{\omega}, a) &= \mathbb{E}\left\{\sum_{j=1}^{M_a} S_{a,j} Y_{a,j} B_{a,j} | \boldsymbol{\omega}\right\} \\ &= \sum_{j=1}^{M_a} B_{a,j} \mathbb{E}\left\{S_{a,j} Y_{a,j} | \boldsymbol{\omega}\right\} \\ &= \sum_{j=1}^{M_a} B_{a,j} \mathbb{E}\left\{Y_{a,j} | S_{a,j} = 1, \boldsymbol{\omega}\right\} \Pr(S_{a,j} = 1 | \boldsymbol{\omega}) \\ &= \sum_{j=1}^{M_a} B_{a,j} P_D^{(a,j)} \Pr(S_{a,j} = 1 | \boldsymbol{\omega}) \\ &= \sum_{j=1}^{M_a} B_{a,j} P_D^{(a,j)} \omega_{j + \sum_{i=1}^{a-1} M_i} &&(9.51) \\ &= \sum_{k=1}^{M} r_k(a) \omega_k &&(9.52) \\ &= \mathbf{r}^T(a) \boldsymbol{\omega}, \end{aligned}$$

where (9.51) follows by noting that $\Pr(S_{a,j} = 1|\boldsymbol{\omega}) = w_k$ with $k = j + \sum_{i=0}^{a-1} M_i$ and (9.52) follows by defining the vector of rewards $\mathbf{r} = (r_1, \ldots, r_M)$ such that

$$r_k(a) = \begin{cases} B_{a,j} P_D^{(a,j)} & \text{if} \quad 1 \le j = k - \sum_{i=1}^{a-1} M_i \le M_a \\ 0 & \text{otherwise} \end{cases}.$$

■

Since $\boldsymbol{\omega}$ is a sufficient statistic for the decision-making, at each time instant n, we seek a (stationary) decision rule of the form of $\pi(\boldsymbol{\omega}) \in \mathcal{A}$ that maximizes the expected discounted total return (in the case of an infinite-horizon)

$$\pi^* = \arg\max_{\pi} V^{\pi}(\boldsymbol{\omega}), \tag{9.53}$$

where

$$\begin{aligned} V^{\pi}(\boldsymbol{\omega}_0) &= \mathbb{E}_{\boldsymbol{\pi}} \left\{ \sum_{n=0}^{\infty} \gamma^n r(\boldsymbol{\omega}[n], \pi_n) \mid \boldsymbol{\omega}[0] = \boldsymbol{\omega}_0 \right\} \\ &= \mathbb{E}_{\pi} \left\{ \sum_{n=0}^{\infty} \gamma^n r(\boldsymbol{\omega}[n], \pi) \mid \boldsymbol{\omega}[0] = \boldsymbol{\omega}_0 \right\}. \end{aligned}$$

The initial belief state $\boldsymbol{\omega}_0$ may now be constructed from the stationary distributions $\mathbf{p}^{i,j}$ of the individual channels.

Theorem 7.9 again shows that we may solve for an optimal stationary policy by noting that the optimal value function $V^*(\boldsymbol{\omega})$ satisfies the optimality equation:

$$V^*(\boldsymbol{\omega}) = \max_{a \in \mathcal{A}} \left\{ \mathbf{r}(a)^T \boldsymbol{\omega} + \gamma \sum_{y \in \mathcal{Y}_a} \Pr(y \mid \boldsymbol{\omega}, a) V^*(\boldsymbol{\omega}^{(new)}) \right\}, \tag{9.54}$$

where $\boldsymbol{\omega}^{(new)} = (\omega_1^{(new)}, \ldots, \omega_M^{(new)}) = \mathbf{T}^{(\omega)}(\boldsymbol{\omega}, a, \mathbf{y})$ with $\omega_k^{(new)} = T_k^{(\omega)}(\omega_k, a, y)$, for $k = 1, \ldots, M$, given by (9.50) and

$$\begin{aligned} \Pr(y \mid \boldsymbol{\omega}, a) &= \mathbb{E}\{\Pr(\mathbf{y}_a \mid \mathbf{s}) \mid \boldsymbol{\omega}\} \\ &= \mathbb{E}\left\{ \prod_{j=1}^{M_a} \Pr(y_{a,j} \mid s_{a,j}) \mid \boldsymbol{\omega} \right\} \\ &= \prod_{j=1}^{M_a} \mathbb{E}\left\{ \Pr(y_{a,j} \mid s_{a,j}) \mid \omega_{j+\sum_{i=0}^{a-1}} \right\} \\ &= \prod_{j=1}^{M_a} \left[\Pr(y_{a,j} \mid s_{a,j} = 1) \omega_{j+\sum_{i=0}^{a-1}} + \Pr(y_{a,j} \mid s_{a,j} = 0) \left(1 - \omega_{j+\sum_{i=0}^{a-1}}\right) \right]. \end{aligned}$$

As in the previous section, the optimal value function (9.54) can be evaluated using POMDP value iteration algorithm given in Algorithm 7.4. If, however, the decision-making problem is better modeled as having a finite horizon T, then we may use the backward induction equation given in Theorem 7.5 as the basis for obtaining the optimal value function and an optimal policy. Again, we will have to resort to an efficient computation algorithm such as witness algorithm or modified exhaustive search with pruning discussed in Chapter 7 to perform this computation.

9.6.5 A Reduced Complexity Optimal Sensing Decision-Making Algorithm with Independent Subbands

Perhaps a more reasonable model that is in between the two assumptions of all dependent channels across subbands and all independent channels is the model that will assume channel dynamics are only dependent within each subband. In this case, the state transition dynamics of channels within subband i, say, may depend on each other. However, these dynamics will be independent of those channels in other subbands.

It is not difficult to see that in this case a sufficient statistic can be represented by a length $\sum_{i=1}^{N_b} 2^{M_i}$ vector as opposed to the $2^{\sum_{i=1}^{N_b} M_i}$ length vector required when all channel dynamics are possibly dependent across all subbands. Since optimal policy problem formulation and computation can be developed by appropriately modifying the steps in the previous two sections, we do not elaborate these any further here.

9.6.6 Optimal Myopic Sensing Decision Policies

For reasons discussed in Chapter 7, optimal sequential policy searches as in the previous sections can be computationally demanding in practice. A popular approach, especially in many telecommunications applications, is to obtain a computationally feasible policy by focusing on the instantaneous performance rather than the performance over a whole time horizon of interest. These are called either *myopic* or greedy policies.

The subband selection problem can easily be formulated as a myopic optimization problem. In the general channel and subband dynamic case, an optimal myopic policy can be defined by

$$a_n^* = \arg\max_{a \in \{1,\dots,N_b\}} r(\mathbf{b}[n], a), \tag{9.55}$$

where $\mathbf{b}[n]$ is the belief of state vector at time instant n given the history $\underline{h}_n$ up to time n and $r(\mathbf{b}[n], a) = \mathbf{r}^T(a)\mathbf{b}[n]$ is the expected immediate reward. Myopic decision-making only requires the radio to keep track of the evolution of the belief vector $\mathbf{b}[n]$.

In the case of independent channel dynamics, we may express the optimal myopic policies in terms of the reduced-dimension sufficient-statistic vector $\boldsymbol{\omega}$ as

$$a_n^* = \arg\max_{a \in \{1,\dots,N_b\}} r(\boldsymbol{\omega}[n], a).$$

In this case, we only need to keep track of the evolution of $\boldsymbol{\omega}[n]$ in time, which can be done by keeping track of individual $\omega_k[n]$'s.

Example 9.6 (Optimal Myopic Policies for a Specific Reward Function) For the immediate reward function $R(\mathbf{S},\mathbf{S}',a,Y) = \sum_{j=1}^{M_a} B_{a,j}S_{a,j}Y_{a,j}$ of Example 9.4, the optimal myopic decisions can explicitly be written as

$$a_n^* = \underset{a\in\{1,\ldots,N_b\}}{\arg\max}\ \mathbf{r}^T(a)\mathbf{b}[n]$$

$$= \underset{a\in\{1,\ldots,N_b\}}{\arg\max} \sum_{m=1}^{2^M} r_m(a)b_m[n],$$

where $\mathbf{r}(a) = (r_1(a),\ldots,r_{2^M}(a))^T$. Substituting for $r_m(a)$ from (9.39), we have that

$$a_n^* = \underset{a\in\{1,\ldots,N_b\}}{\arg\max} \sum_{j=1}^{M_a} B_{a,j}P_D^{(a,j)} \sum_{m=1}^{2^M} s_{a,j}^{(m)} b_m[n] \tag{9.56}$$

$$= \underset{a\in\{1,\ldots,N_b\}}{\arg\max} \sum_{j=1}^{M_a} B_{a,j}P_D^{(a,j)} \Pr\left(S_{a,j} = 1|\mathbf{b}[n]\right), \tag{9.57}$$

where the last step follows by noting that the quantity $\sum_{m=1}^{2^M} s_{a,j}^{(m)} b_m[n]$ is the posterior probability of the (a,j)th channel being idle at time n given the belief vector $\mathbf{b}[n]$.

In the case of independent channel dynamics, the posterior probability of the (a,j)-th channel being idle at time n given the belief vector $\mathbf{b}[n]$ is exactly the belief $\omega_k[n]$ where $k = j + \sum_{i=0}^{a-1} M_i$. Hence, the optimal myopic policy (9.57) reduces to

$$a_n^* = \underset{a\in\{1,\ldots,N_b\}}{\arg\max} \sum_{j=1}^{M_a} B_{a,j}P_D^{(a,j)} \omega_{j+\sum_{i=0}^{a-1} M_i}.$$

■

9.7 A REDUCED COMPLEXITY OPTIMAL SUBBAND SELECTION FRAMEWORK WITH AN ALTERNATIVE REWARD FUNCTION

As noted earlier, the complexity of the optimal formulation in the previous sections is exponential in the number of channels in a subband. Consequently, it is exponential in the total number of channels in the spectrum range of interest to the radio. This can be too overwhelming a computation even for a cognitive radio.

The exponential complexity, however, is a result of the way we defined the state vector of the system: that is, as the vector of states of all channels in the system. In turn, this particular state vector was needed because of the reward function we used as the basis for selecting subbands. The rationale was that the quality of a subband is the sum of the qualities of the individual channels making up the subband. This forces us to have a state vector formed of the individual states of all channels in a subband and then in all the subbands. While such a state vector is useful in formulating a reward sequence that is a function of the individual properties of the channels, this may not be necessary if the subband selection can be based on a direct property of a subband as a whole, rather than as an aggregate property of individual channels. In such situations, we may be able to define a state that is a direct characteristic of the subband itself rather than made up of components corresponding to the individual channels. This can result in a much simpler framework for subband selection with reduced complexity, as we discuss in this section.

To motivate a reward function that is a direct property of a subband, let us revisit the immediate reward function defined in Example 9.4:

$$R(\mathbf{S},\mathbf{S}',a,Y) = \sum_{j=1}^{M_a} B_{a,j} S_{a,j} Y_{a,j}.$$

The resulting expected immediate reward when the overall system state is in state $\mathbf{S} = \mathbf{s}_m$ is

$$\begin{aligned} r_m(a) &= \sum_{j=1}^{M_a} \mathbb{E}\left\{B_{a,j} S_{a,j} Y_{a,j} | \mathbf{S} = \mathbf{s}_m\right\} \\ &= \sum_{j=1}^{M_a} s_{a,j}^{(m)} P_D^{(a,j)} B_{a,j}. \end{aligned} \tag{9.58}$$

Note that (9.58) depends on individual channel properties $P_D^{(a,j)}$ and $B_{a,j}$ within the subband $a \in \mathcal{A}$. If, however, these were the same for all channels within the subband a so that $P_D^{(a,j)} = P_D^{(a)}$ and $B_{a,j} = B^{(a)}$ independent of the channel indices, then (9.58) reduces to

$$r_m(a) = P_D^{(a)} B^{(a)} \sum_{j=1}^{M_a} s_{a,j}^{(m)}. \tag{9.59}$$

Now, (9.59) is simply the number of idle channels in the subband a if the overall system state is $\mathbf{S} = \mathbf{s}_m$, except for a scale factor $P_D^{(a)} B^{(a)}$ that is a function of the subband index a. To evaluate this reward, in turn, we do not need the full state $\mathbf{S}$ as defined earlier. Instead, it is enough to have access to the *number of idle channels* in each subband. Motivated by this observation, we may define a new state and a state dynamic model for a subband.

9.7.1 A New Model for Subband Dynamics

Let us define a new state for the subband i as N_i, where $N_i[n]$ denotes the number of idle channels in the subband i at time n:

$$N_i[n] \triangleq \text{Number of idle channels in the subband } i \text{ at time } n.$$

If we assume that the objective of the cognitive radio is to achieve communications, then how many channels in a subband are idle is a reasonable quantity to focus on in making sensing decisions: A radio would like to sense a subband that has a high likelihood of having a large number of idle channels because such a subband may present it with more options in how best to achieve its communications objectives.

This new subband state variable N_i can take possible values in $\{0, 1, \ldots, M_i\}$ where M_i is the number of channels in the subband i. Hence, state N_i can only take $M_i + 1$ possible values as opposed to 2^{M_i} possible values in the previous formulation. This is a substantial reduction in complexity.

It is easily seen that given the state $\mathbf{S}_i$ of the previous section is Markov, N_i is also Markov with $M_i + 1$ number of possible states as shown in Figure 9.8. Let us denote by $\tilde{\mathbf{P}}^{(i)}$ the $(M_i + 1) \times (M_i + 1)$ state transition probability matrix of the ith subband where now we take the state of the subband i at time n to be $N_i[n]$:

$$\tilde{\mathbf{P}}^{(i)} = \begin{pmatrix} \tilde{p}^{(i)}_{0,0} & \tilde{p}^{(i)}_{0,1} & \cdots & \tilde{p}^{(i)}_{0,M_i} \\ \tilde{p}^{(i)}_{1,0} & \tilde{p}^{(i)}_{1,1} & \cdots & \tilde{p}^{(i)}_{1,M_i} \\ \vdots & \vdots & \ddots & \vdots \\ \tilde{p}^{(i)}_{M_i,0} & \tilde{p}^{(i)}_{M_i,1} & \cdots & \tilde{p}^{(i)}_{M_i,M_i} \end{pmatrix}, \tag{9.60}$$

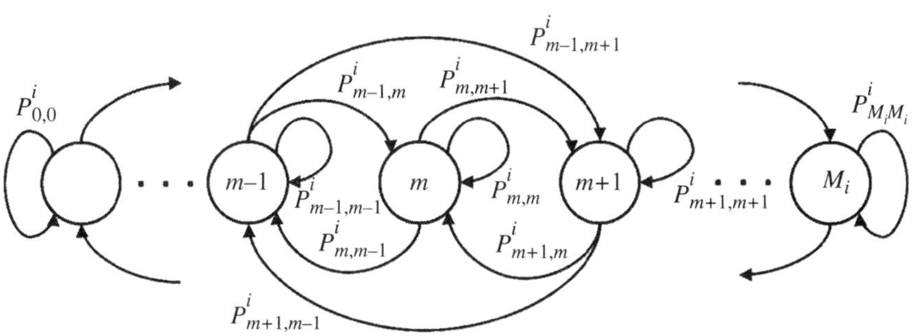

FIGURE 9.8 The Markov chain model of subband dynamics in terms of single channel transition probability parameters when state is defined to be the number of idle channels in the subband.

with transition probability of the Markov state $N_i[n]$ from state k to state k' defined as

$$\tilde{p}_{k,k'}^{(i)} = \Pr\{N_i[n+1] = k' \mid N_i[n] = k\},\ \forall k,k' \in \{0,1,\ldots,M_i\}. \tag{9.61}$$

In theory, we may compute these transition probability parameters from the knowledge of the individual channel Markov models in the subband. Indeed, for all $k,k' \in \{0,\ldots,M_i\}$,

$$\begin{aligned}
\tilde{p}_{k,k'}^{(i)} &= \Pr\left\{ N_i[n+1] = k' \middle| N_i[n] = k \right\} \\
&= \Pr\left\{ \sum_{j=1}^{M_i} S_{i,j}[n+1] = k' \middle| \sum_{j=1}^{M_i} S_{i,j}[n] = k \right\} \\
&= \sum_{\mathcal{A}_k^{(i)}} \sum_{\mathbf{s}_i : \sum_j^{M_i} s_{i,j} = k'} \left[\left(\prod_{j \in \mathcal{A}_k^{(i)}} p_{1,s_{i,j}}^{(i,j)} \right) \left(\prod_{j \in \left(\mathcal{A}_k^{(i)}\right)^c} p_{0,s_{i,j}}^{(i,j)} \right) \right],
\end{aligned} \tag{9.62}$$

where $\mathcal{A}_k^{(i)}$ denotes a subset of channels in the ith subband with cardinality k and $\left(\mathcal{A}_k^{(i)}\right)^c$ is the relative complement of $\mathcal{A}_k^{(i)}$ with respect to the set of all channels in the ith subband. The summation in (9.62) is taken over all possible $\mathcal{A}_k^{(i)}$'s and all possible combinations of states $(s_{i,1},\ldots,s_{i,M_i})$ such that $\sum_j^{M_i} s_{i,j} = k'$ where $s_{i,j} \in \{0,1\}$ for all $j \in \{1,\ldots,M_i\}$.

If we denote by vector $\tilde{\boldsymbol{p}}^{(i)} = \left[\tilde{p}_0^{(i)},\ldots,\tilde{p}_{M_i}^{(i)}\right]$ the stationary distribution of this Markov process N_i, then it is characterized by

$$\tilde{\boldsymbol{p}}^{(i)} = \tilde{\boldsymbol{p}}^{(i)} \tilde{\mathbf{P}}^{(i)},$$

where $\tilde{\boldsymbol{p}}^{(i)} = (\tilde{p}_0^{(i)},\ldots,\tilde{p}_{M_i}^{(i)})$ is the stationary probabilities of the states $0,1,\ldots,M_i$ in Figure 9.8.

Similar to the transition probabilities $\tilde{p}_{k,k'}^{(i)}$, the stationary probabilities $\tilde{p}_k^{(i)}$ for $k \in \{0,\ldots,M_i\}$ can also be expressed in terms of the stationary probabilities of the overall system state of the previous section:

$$\tilde{p}_k^{(i)} = \Pr\left\{ \sum_{j=1}^{M_i} S_{i,j}[n] = k \right\}. \tag{9.63}$$

For example, when channels within the subband i have independent Markov dynamics, (9.63) can be written as

$$\tilde{p}_k^{(i)} = \sum_{\mathcal{A}_k^{(i)}} \left(\prod_{j \in \mathcal{A}_k^{(i)}} p_1^{(i,j)} \right) \left(\prod_{j \in \left(\mathcal{A}_k^{(i)}\right)^c} p_0^{(i,j)} \right), \tag{9.64}$$

where $p_1^{(i,j)}$ denotes the stationary probability of channel (i,j) being in idle state.

Computing the transition probabilities $\tilde{p}_{k,k'}^{(i)}$ and the stationary distribution $\tilde{p}_k^{(i)}$ of subband i from (9.62) and (9.64), respectively, has high complexity and may not be possible altogether if individual channel Markov model parameters are not available. Indeed, evaluating them may be wasteful when we are not interested in the overall system state $\mathbf{S}$ as in the previous section. However, it is not necessary to go through (9.62) and (9.64) to compute $\tilde{p}_{k,k'}^{(i)}$'s and $\tilde{p}_k^i$'s. Instead, in practice, we may directly estimate these subband Markov model parameters themselves.

We define the new overall system state made of all N_b subbands as

$$\mathbf{N} = (N_1, \ldots, N_{N_b}), \tag{9.65}$$

where N_i is the ith subband state. Since each $N_i \in \{0, 1, \ldots, M_i\}$, the new state space consists of $\prod_{i=1}^{N_b}(M_i + 1)$ number of elements.

For the simplicity of the following discussion, let us assume that all subbands have the same $M_i = \bar{M} - 1$ number of channels, for $\bar{M} \geq 1$. Note that the assumption $\bar{M} \geq 1$ is reasonable because the subband must at least contain one channel. Then the number of possible states of the system is $\bar{M}^{N_b}$. Since the total number of channels in the overall system is $M = N_b(\bar{M} - 1)$, this should be compared with the total number of states of $2^{N_b(\bar{M}-1)}$ in the previous formulation. For the purpose of the following discussion, let us denote by $\tilde{\mathcal{S}}$ the set of $\bar{M}^{N_b}$ possible states of the system where we enumerate these states as $\tilde{\mathcal{S}} = \{\tilde{\mathbf{n}}_1, \ldots, \tilde{\mathbf{n}}_{\bar{M}^{N_b}}\}$.

9.7.2 A Simplified Reward Function and a Reduced-Complexity Optimal Policy

Suppose that the cognitive radio is to make subband selection decisions based on this new state vector $\mathbf{N}[n]$ at time n. As before, when in state $\mathbf{N}[n]$ at time n, the radio can select one of the N_b subbands for sensing. Hence, the action space is still $\mathcal{A} = \{1, 2, \ldots, N_b\}$. The observation $Z[n]$ resulting from taking action $a \in \mathcal{A}$ is a noisy measurement of the true state $N_a[n]$ of the number of idle channels in subband a. As before, the observation is inherently partial since the cognitive radio can only observe the (noisy) state of only the subband a instead of all subbands. We may assume that the observation $Z[n]$ is formed from the outputs of the individual channel status detectors.

With these definitions, we may define a new immediate reward function as $R(\mathbf{N}, \mathbf{N}', a, Z)$. It is the reward associated with selecting action $a \in \mathcal{A}$ when in state $\mathbf{N}$ and observing $Z[n]$.

Example 9.7 (A New Reward Function Associated with Subband Qualities) In parallel with the immediate reward function $R(\mathbf{S},\mathbf{S}',a,Y)$ of Example 9.4, we may define a new immediate reward function with the new state vector $\mathbf{N}$ as

$$R(\mathbf{N},\mathbf{N}',a,Z) = N_a. \tag{9.66}$$

In other words, the immediate reward for choosing action $a \in \{1,\ldots,N_b\}$ when in state $\mathbf{N}$ is simply equal to the number of idle channels in the subband a. Note that, as in (9.59), we could have in general introduced a scaling factor to (9.66) that depends on the subband index a. ■

As before, the immediate reward $R(\mathbf{N},\mathbf{N}',a,Z)$ cannot be evaluated by a radio that does not have access to the exact state $\mathbf{N}$ of the system. Instead, the radio only has access to the history of the decision process denoted by $\underline{\tilde{h}}_n$ which we now define as

$$\underline{\tilde{h}}_n = \left(\underline{\tilde{h}}_{n-1}, a_{n-1}, z[n-1]\right)$$

with $\underline{\tilde{h}}_0 = \phi$. Given the history $\underline{\tilde{h}}_n$ at time n, the belief of system being in state $\mathbf{N}[n] = \mathbf{n}_m \in \tilde{\mathcal{S}}$, for $1 \le m \le \bar{M}^{N_b}$, is defined as

$$\tilde{b}_m[n] = \Pr\left(\mathbf{N}[n] = \mathbf{n}_m | \underline{\tilde{h}}_n\right).$$

Then a sufficient statistic for the decision-making problem, given the history $\underline{\tilde{h}}_n$ at time n, is the belief sate vector defined as $\tilde{\mathbf{b}}[n] = (\tilde{b}_1[n],\ldots,\tilde{b}_{\bar{M}^{N_b}}[n])^T$.

We may then compute the expected immediate reward of taking action a when the belief state is $\tilde{\mathbf{b}}$:

$$\begin{aligned} r(\tilde{\mathbf{b}},a) &= \mathbb{E}\left\{R(\mathbf{N},\mathbf{N}',a,Z)|\tilde{\mathbf{b}}\right\} \\ &= \sum_m \mathbb{E}\left\{R(\mathbf{N}=\mathbf{n}_m,\mathbf{N}',a,Z)\right\}\tilde{b}_m \\ &= \mathbf{r}(a)^T\tilde{\mathbf{b}}, \end{aligned} \tag{9.67}$$

where

$$\mathbf{r}(a) = \left(r_1(a),\ldots,r_{\bar{M}^{N_b}}(a)\right) \tag{9.68}$$

with

$$r_m(a) = \mathbb{E}\left\{R(\mathbf{N}=\mathbf{n}_m,\mathbf{N}',a,Z)\right\}, \tag{9.69}$$

for $1 \le m \le r_{\bar{M}^{N_b}}$.

Example 9.8 (New Expected Rewards: Expected Number of Idle Channels in a Subband) Continuing with the immediate reward function (9.66) of Example 9.7, we have from (9.67).

$$r(\tilde{\mathbf{b}}, a) = \mathbb{E}\left\{N_a | \tilde{\mathbf{b}}\right\} \\ = \mathbf{r}(a)^T \tilde{\mathbf{b}}. \tag{9.70}$$

Hence, the expected reward for selecting action a (i.e., subband a) when the belief state is $\tilde{\mathbf{b}}$ is simply the expected number of idle channels in the subband a given the belief state $\tilde{\mathbf{b}}$. Moreover, in this case,

$$\begin{aligned} r_m(a) &= \mathbb{E}\left\{R(\mathbf{N} = \mathbf{n}_m, \mathbf{N}', a, Z)\right\} \\ &= \mathbb{E}\left\{N_a | \mathbf{N} = \mathbf{n}_m\right\} \\ &= n_a^{(m)}, \end{aligned} \tag{9.71}$$

where $n_a^{(m)}$ is the state of the subband a when the overall system state vector is $\mathbf{N} = \mathbf{n}_m \in \tilde{\mathcal{S}}$. By the definition of our state, $n_a^{(m)}$ denotes the number of idle channels in subband a according to the state vector $\mathbf{n}_m$. ■

As before, it can be verified that the above belief-state vector $\tilde{\mathbf{b}}[n]$ is Markov in that $\tilde{\mathbf{b}}[n+1]$ is completely determined by $\tilde{\mathbf{b}}[n]$, a_n, and $z[n]$. Indeed, following the same steps used in deriving (9.37), it can be shown that, for $\mathbf{n}_m \in \tilde{\mathcal{S}}$ for $m = 1, \ldots, \bar{M}^{N_b}$,

$$\begin{aligned} \tilde{b}_m[n+1] &= \Pr\left(N[n+1] = \mathbf{n}_m \middle| \tilde{h}_{n+1}\right) \\ &= \Pr\left(N[n+1] = \mathbf{n}_m \middle| \underline{\tilde{h}}_n, a_n, z[n]\right) \\ &= \sum_{m'=1}^{\bar{M}^{N_b}} p_{m',m} \frac{\Pr\left(z[n] \mid N[n] = \mathbf{n}_{m'}, a_n\right) \tilde{b}_{m'}[n]}{\sum_{m''=1}^{\bar{M}^{N_b}} \Pr\left(z[n] \mid N[n] = \mathbf{n}_{m''}, a_n\right) \tilde{b}_{m''}[n]} \\ &\triangleq \tilde{T}_m\left(\tilde{\mathbf{b}}[n], a_n, z[n]\right), \end{aligned} \tag{9.72}$$

where $\tilde{T}_m(., a, z)$ denotes the mapping from $\tilde{\mathbf{b}}[n]$ to $\tilde{b}_m[n+1]$. Then we may denote the mapping of the belief-state vector $\tilde{\mathbf{b}}[n]$ to $\tilde{\mathbf{b}}[n+1]$ by

$$\tilde{\mathbf{T}}(., a, y) = \left(\tilde{T}_1(., a, y), \ldots, \tilde{T}_{\bar{M}^{N_b}}(., a, y)\right). \tag{9.73}$$

Hence, we can formulate a new POMDP framework for making subband selection decisions for spectrum sensing by defining decision rules as functions of the new belief vector $\tilde{\mathbf{b}}[n]$. Thus, in the case of an infinite decision-making horizon, a stationary policy for the subband selection is a policy $\underline{\pi} = (\pi_0(\tilde{\mathbf{b}}[0]), \pi_1(\tilde{\mathbf{b}}[1]), \ldots,) = (\pi(\tilde{\mathbf{b}}[0]), \pi(\tilde{\mathbf{b}}[1]), \ldots,)$ where $\pi_n(\tilde{\mathbf{b}}[n]) = \pi(\tilde{\mathbf{b}}[n])$ for all n. The new expected discounted total return of such a policy then gives rise to the following value function:

$$V^{\pi}(\tilde{\mathbf{b}}) = \mathbb{E}_{\pi}\left\{\sum_{n=0}^{\infty}\gamma^n r(\tilde{\mathbf{b}}[n],\pi(\tilde{\mathbf{b}}[n]))\,\middle|\,\tilde{\mathbf{b}}[0]=\tilde{\mathbf{b}}\right\}.$$

The policy-making problem now is to seek an optimal (stationary) policy π^* such that

$$\pi^* = \arg\max_{\pi} V^{\pi}(\tilde{\mathbf{b}}). \tag{9.74}$$

As before, from Theorem 7.9, the optimal value function $V^*(\tilde{\mathbf{b}})$ satisfies the following optimality equation:

$$V^*(\tilde{\mathbf{b}}) = \max_{a\in\mathcal{A}}\left\{\mathbf{r}(a)^T\tilde{\mathbf{b}} + \gamma\sum_{z\in\mathcal{Z}_a}\Pr(z\mid\tilde{\mathbf{b}},a)V^*(\tilde{\mathbf{b}}^{(new)})\right\} \tag{9.75}$$

where $\mathcal{Z}_a = \{0,1,\ldots,M_a+1\}$, $\tilde{\mathbf{b}}^{(new)} = \tilde{\mathbf{T}}(\tilde{\mathbf{b}},a,\mathbf{z}) = \left(\tilde{T}_1(\tilde{\mathbf{b}},a,y),\ldots,\tilde{T}_{\bar{M}^{N_b}}(\tilde{\mathbf{b}},a,y)\right)$ with $\tilde{b}_m^{(new)} = \tilde{T}_m(\tilde{\mathbf{b}},a,z)$ as defined in (9.72) and

$$\begin{aligned}\Pr(z\mid\tilde{\mathbf{b}},a) &= \mathbb{E}\left\{\Pr(z\mid\mathbf{n},a)\mid\tilde{\mathbf{b}}\right\}\\ &= \sum_{m=1}^{\bar{M}^{N_b}}\Pr(Z[n]=z|\mathbf{N}[n]=\mathbf{n}_m,a)\tilde{b}_m.\end{aligned}$$

Note that $\Pr(Z[n]=z|\mathbf{N}[n]=\mathbf{n}_m,a)$ can be computed based on a set of given observation probabilities that relates subband states and actions to the observations.

9.7.3 A Reduced Complexity Optimal Policy for Independent Subbands

Let us denote by $\tilde{\mathbf{b}}_i[n]$ the belief vector of the state of the ith subband at time n given the history $\underline{\tilde{h}}_n$. If we assume that the dynamics of the subbands are independent of each other, then in fact we can also limit this history to the history on channel i denoted as $\underline{\tilde{h}}_{i,n}$. Since subband i can assume M_i+1 number of states,

$$\tilde{\mathbf{b}}_i[n] = \left(\tilde{b}_{i,1}[n],\ldots,\tilde{b}_{i,M_i+1}[n]\right),$$

where

$$\tilde{b}_{i,m}[n] = \Pr(N_i[n]=m-1|\underline{\tilde{h}}_{i,n})$$

is the belief that the state of the ith subband at time n is $m-1$ (i.e., the number of idle channels is $m-1$) for $1\le m\le M_i+1$.

As in Sections 9.6.4 and 9.6.5, then it can easily be verified that a sufficient statistic for decision-making when the dynamics of the subbands are independent of each other is the collection of N_b vectors $\left\{\tilde{\mathbf{b}}_i[n]\right\}_{i=1}^{N_b}$, where each vector is of length[3] $(M_i + 1)$, for $1 \le i \le N_b$. We may define a new belief vector for this scenario as

$$\tilde{\boldsymbol{\beta}}[n] = \left(\tilde{\mathbf{b}}_1[n], \ldots, \tilde{\mathbf{b}}_{N_b}[n]\right). \tag{9.76}$$

For simplicity, if we assume all $M_i = \bar{M} - 1$, for $1 \le i \le N_b$, then the dimensionality of this sufficient statistic is simply $\bar{M}N_b$ compared to that of $\bar{M}^{N_b}$ of the full state vector of the previous section. An optimal subband selection policy for this scenario can be obtained following the same methodology discussed in Sections 9.6.4 and 9.6.5 with little modification. We omit these details here to avoid repetition.

9.7.4 Optimal Myopic Policies with Reduced Dimensional Subband State Vectors

Similar to Section 9.6.6, we may also consider myopic policies based on the new state definition $N[n]$ of the system. Such myopic policies can drastically reduce the computational complexity as we have already discussed. Moreover, given the special property of the subband selection problem that renders state transitions independent of actions, it can be argued that performance loss due to myopic policies can be relatively small in many cases.

In the case of general channel and subband dynamics, an optimal myopic policy with the new state definition of a subband can be defined by

$$a_n^* = \underset{a \in \{1,\ldots,N_b\}}{\arg\max}\ r\left(\tilde{\mathbf{b}}[n], a\right),$$

where $\tilde{\mathbf{b}}[n]$ is the belief of state vector at time n given the history $\underline{\tilde{h}}_n$ up to time n and $r\left(\tilde{\mathbf{b}}[n], a\right) = \mathbf{r}^T(a)\tilde{\mathbf{b}}[n]$ is the expected immediate reward with $\mathbf{r}$ as defined in (9.68)–(9.69).

In the case of independent subband dynamics, an optimal myopic policy can be defined in terms of the reduced-dimensional sufficient-statistic vector $\boldsymbol{\beta}$ of (9.76) as

$$a_n^* = \underset{a \in \{1,\ldots,N_b\}}{\arg\max}\ r\left(\tilde{\boldsymbol{\beta}}[n], a\right).$$

Note that the expected immediate rewards $r\left(\tilde{\boldsymbol{\beta}}[n], a\right)$ can easily be characterized following a method similar to that used to obtain $r(\boldsymbol{\omega}[n], a)$ earlier in Example 9.5 of Section 9.6.4.

[3] In fact, only a length M_i is sufficient since the sum of the elements in the $(M_i + 1)$-length vector is equal to 1.

Example 9.9 (Optimal Myopic Policies for a Specific Reward Function) For the immediate reward function (9.66) considered in Examples 9.7 and 9.8, an optimal myopic policy can be written as

$$\begin{aligned} a_n^* &= \arg\max_{a\in\{1,\ldots,N_b\}} \mathbf{r}^T(a)\tilde{\mathbf{b}}[n] \\ &= \arg\max_{a\in\{1,\ldots,N_b\}} \sum_{m=1}^{\bar{M}^{N_b}} r_m(a)\tilde{b}_m[n], \end{aligned}$$

where $\mathbf{r}(a) = (r_1(a), \ldots, r_{\bar{M}^{N_b}}(a))^T$. Substituting for $r_m(a)$ from (9.71), we have that

$$a_n^* = \arg\max_{a\in\{1,\ldots,N_b\}} \sum_{m=1}^{\bar{M}^{N_b}} n_a^{(m)}\tilde{b}_m[n] \tag{9.77}$$

$$= \arg\max_{a\in\{1,\ldots,N_b\}} \mathbb{E}\left\{N_a|\tilde{\mathbf{b}}[n]\right\}. \tag{9.78}$$

Hence, the optimal myopic policy is particularly simple in this case: Pick the subband that has the highest number of expected idle channels given the current belief vector $\tilde{\mathbf{b}}[n]$.

In the case of independent subband dynamics, the optimal myopic policy (9.78) can be expressed in the form of

$$\begin{aligned} a_n^* &= \arg\max_{a\in\{1,\ldots,N_b\}} \mathbb{E}\left\{N_a|\tilde{\boldsymbol{\beta}}[n]\right\} \\ &= \arg\max_{a\in\{1,\ldots,N_b\}} \mathbb{E}\left\{N_a|\tilde{\mathbf{b}}_a[n]\right\}. \end{aligned}$$

■

9.8 MACHINE-LEARNING AIDED SUBBAND SELECTION POLICIES

A common assumption in all subband selection policies derived so far is that the knowledge of the system model is available. This is true regardless of whether the decisions are based on the full state defined in terms of individual channel states or they are based on the reduced dimensional state defined in terms of direct subband states. Even in the case of myopic policies, the transition probabilities of the Markov models are required for the belief state to be updated. In practice, however, it is likely that we do not know the channel Markov models nor the subband Markov models.

Moreover, as we saw in Chapter 7, the optimal policy determination for POMDP is a computationally demanding task due to the continuous state space of the belief state. For example, for a finite-horizon problem, we may have to use an algorithm such as either the Sondik's one-pass algorithm or the witness algorithm. These algorithms still pose a significant computational problem that makes optimal POMDP solutions almost impractical for realistic values of numbers of subbands and channels in spectrum of interest. This is exacerbated by the fact that these policies may need to be computed in real time in the case of a cognitive radio. Even if one were to be able to somehow bear this computational demand, still, these algorithms require the full knowledge of the transition probabilities of the Markov models as mentioned earlier.

The dynamism of wireless medium can make things worse. Not only we may not know these model parameters a priori, but they may likely be time varying. Hence, any use of the solution algorithms of Chapter 7 will have to be accompanied with estimating or updating these parameters.

An alternative to such parameter and model estimation is to bypass the step of subband Markov model estimation and directly attempt to learn a decision policy. In general, these approaches are called machine learning. While there are many machine learning approaches that can be used to learn a policy directly, one method that is particularly suitable for underlying Markov dynamics is the *reinforcement learning* (RL) that we will discuss in Chapter 16.

There are several advantages to learning a policy rather than attempting to compute it. As already mentioned, not requiring the knowledge of transition probabilities of the Markov models is one of them. Moreover, policy learning with RL can also help dealing with time varying or nonstationary RF environments, at least to some extent. Indeed, since RL will directly try to learn a policy based on the actual observations, as data distributions drift, the policies will also drift with them. Of course, for any learning algorithm to successfully converge, we may need sufficient ergodicity in the underlying statistical phenomena. At least as long as the changes are slow enough, learning approach may be able to effectively learn a policy that is in-line with the actual underlying distribution.

Finally, it is worth noting that even when transition probability model is known, learning a policy may be simpler than attempting to compute an optimal policy. In particular, if it is known that at least under certain conditions, the learning algorithm can converge to an optimal policy, this may lead to significant savings in terms of computational complexity.

9.8.1 Q-Learning

In the following, we may discuss a form of RL called Q-learning [135]. The Q-learning algorithm maintains a table of Q-values where Q-value is a quantification of taking a particular action when in a given state. Hence, in a completely observable MDP, the Q-table entries are defined to be $Q(s,a)$: the value of taking action a when in state s. The learning algorithm updates the Q-values based on actions and rewards resulted by taking them without any knowledge of Markov model parameters as follows:

$$\begin{aligned} Q(s[n-1],a_{n-1}) \leftarrow & (1-\alpha)Q(s[n-1],a_{n-1}) \\ & +\alpha\left[r_n\left(s[n-1],a_{n-1}\right)+\gamma \max_a Q(s[n],a)\right], \end{aligned} \tag{9.79}$$

where $Q(s[n-1],a_{n-1})$ and $r_n\left(s[n-1],a_{n-1}\right)$ denote, respectively, the Q-value and the observed reward corresponding to taking action a_{n-1} when in state $s[n-1]$ at time

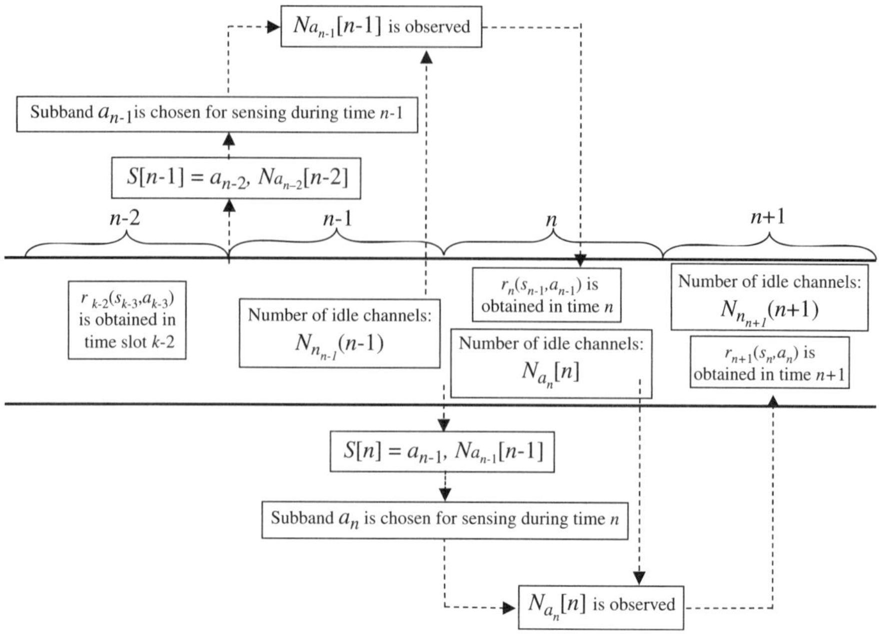

FIGURE 9.9 Q-learning when model parameters are unknown.

$n-1$ as shown in Figure 9.9. In (9.79), $0 < \alpha < 1$ and $0 \leq \gamma < 1$ denote the *learning rate* and the discount factor, respectively. The (action) subband selection when in a given state $s[n]$ at time n is now performed based directly on these Q-values:

$$a_n^* = \arg\max_a Q(s[n], a). \tag{9.80}$$

It is easy to see that Q-learning algorithm (9.79) above attempts to *reinforce* the actions that lead to better immediate and future outcomes from a given state, hence the name RL. This simple algorithm is known to converge to the optimal Q-function and an optimal policy with probability 1 in a single learning agent MDP under fairly common assumptions on stochastic convergence [69, 135–137].

In the subband selection problem, the action a_n refers to the subband index $a \in \{1, \ldots, N_b\}$ chosen to be sensed during sensing epoch n and $s[n]$ is the subband the cognitive radio is currently in. Note that $r_n(s[n-1], a_{n-1})$ is the delayed reward observed at time n as a result of taking the action a_{n-1} in state $s[n-1]$. It is evaluated after sensing the selected subband and finding out what exactly was the reward for the chosen action. In the case of the immediate reward (9.66) of Examples 9.7–9.9 in which the expected reward was the expected number of idle channels in a subband, this delayed reward will be the actual number of idle channels that was found in the selected subband by sensing the subband.

RL approaches such as Q-learning, however, may get trapped in local optima unless all (state, action) pairs are updated consistently. Otherwise, clearly, those actions that have been tried out from a given state have the unfair advantage of getting selected due to their already high Q-values (think of a Q-table initialized with all zeros). Thus, Q-learning-based decision policies will overexploit the already learned $(state, action)$ combinations without letting the agent explore other possible options. Hence, some provision for *exploration* has to be built into the policy learning algorithm. Exploration refers to searching new opportunities, whereas a policy such as (9.80) is an example of *exploitation* that selects the actions giving the immediate best reward.

Indeed, allowing for a certain level of exploration in RL frameworks may help avoid being trapped in local maxima. Hence, often in RL techniques, an *exploration rate* $\epsilon \in (0, 1)$ is defined, such that at each decision epoch the learning agent selects the action characterized by (9.80) with probability $1 - \epsilon$ and uniformly chooses an action out of all the possible actions with probability ϵ:

$$a^* = \begin{cases} \arg\max_{a \in \mathcal{A}} Q(s,a) & \text{with probability } 1-\epsilon \\ \sim \mathcal{U}(\mathcal{A}) & \text{with probability } \epsilon \end{cases}, \tag{9.81}$$

where $\mathcal{U}(\mathcal{A})$ denotes the uniform distribution over the action set $\mathcal{A}$. Choosing a high exploration rate ϵ may help the agent to quickly *understand* the environment. However, it may also reduce the overall performance due to excessive exploring and not exploiting the already learned optimal actions. On the other hand, a low exploration rate may increase the required time for the algorithm to converge to the optimal solution. Hence, success of RL strategies requires a careful balance between the exploration and exploitation [69].

The uniform action selection while exploring as in (9.81), however, may not always be necessary and may not be the best for certain applications. It must be said that the effectiveness of any learning approach is highly dependent on the problem scenario. By utilizing side information available regarding the structure of the problem, learning algorithms can be made more efficient and effective. In particular, there are various other ways to select an action during the exploration step in (9.81). For example, rather than the uniform selection, one may use an entirely different distribution function for the probabilities of selecting actions. One such approach is to select the actions with probabilities proportional to a *Boltzmann distribution* on Q-values [69]:

$$\Pr(a|Q(.,.)) = \frac{e^{\frac{Q(s[n],a)}{\tau}}}{\sum_{a \in \mathcal{A}} e^{\frac{Q(s[n],a)}{\tau}}}, \tag{9.82}$$

where τ is a parameter that can be adjusted as desired. Note that when $\tau \longrightarrow \infty$, these probabilities essentially resemble a uniform distribution as before. On the other hand, smaller τ values may create high variation among action selection probabilities (9.82).

Of course, (9.82) is not the only possible choice. Moreover, exploration may not always have to be according to a distribution that gives nonzero probabilities to all possible actions either. In fact, when certain actions are already known to be really poor choices from a certain state, it may be more effective to exclude them from action selection. One extreme of this would be to select the second best action. However, such fine-tuning has to be performed in the context of the actual learning problem.

9.8.2 *Q*-Learning in a POMDP: A *Q*-Learning Algorithm for Subband Selection

The *Q*-learning algorithm of (9.79) cannot directly be used for learning a subband selection policy because the problem is defined as a POMDP rather than an MDP. In Chapter 16, we will discuss several variations of the *Q*-learning algorithm that have been proposed for learning an optimal, or near-optimal, policy for a POMDP. In the following, however, we may discuss an alternative that will allow using the *Q*-learning directly to learn a subband selection policy under certain simplifying assumptions. It may be possible to modify this framework in practice to relax some of these assumptions.

Obtaining an optimal subband selection policy for a wideband cognitive radio is made difficult by two facts: First, it may be that the state transition probabilities are unknown. The *Q*-learning can help resolve this. Second is the computational complexity due to the fact that the problem is a POMDP. Perhaps this is the more pressing concern. An alternative is to define a *Q*-table based on a state that is observable to the cognitive radio. Moreover, it is important to note that the spectrum sensing environment of a cognitive radio may not always be exactly Markov or single agent. In particular, there may be several cognitive radios who independently try to learn a decision policy violating the single-agent environment. However, it has been shown that *Q*-learning can lead to near-optimal or at least satisfactory policies for some learning problems in multi agent (decentralized) POMDP environments of cognitive radios [138].

In the following, we discuss a *Q*-learning-aided subband selection approach for a wideband cognitive radio when no knowledge of channel or subband transition probabilities is available. For simplicity, let us assume that there are no sensing errors so that when the cognitive radio senses a particular subband m at time n, it is able to observe the exact state of that particular subband, which is $N_m[n]$. Note that, even with this assumption, the problem is still a POMDP because the cognitive radio can only observe the exact state of the sensed subband and not all the other subbands. Since the observation is partial, as mentioned earlier, one approach is to define a quantized belief-state-based *Q*-table. However, we may follow an alternative approach in which we define an observable pseudostate based on which the cognitive radio may make its decisions and use *Q*-learning to learn a policy based on these new state values.

Indeed, let us define the state at time $n = n$ as

$$s[n] = \left(a_{n-1}, N_{a_{n-1}}[n-1]\right). \tag{9.83}$$

Note that since a_{n-1} is the action selected at time $n-1$, it is the subband that was sensed during the $(n-1)$th sensing epoch. As a result, the decisions on which subband to be sensed at time n have to be made based on the history available to the radio up to and including the $(n-1)$th sensing period. In place of the full history, the state defined in (9.83) uses the immediate past information included in the history: the subband that was selected at time $n-1$ and the number of idle channels that was observed to be in that subband during the $(n-1)$th sensing period.

Example 9.10 (A Partial State for Q-Learning-Based Subband Selection) As an example, consider a spectrum environment with $N_b = 2$. Suppose that the two subbands contain $M_1 = 2$ and $M_2 = 3$ number of channels. Then, state N_1 can take possible values in $\{0,1,2\}$ and the state N_2 can take possible values in $\{0,1,2,3\}$. Hence, the state $s[n]$ defined in (9.83) can take the possible values (tuples) of

$$\{(1,0),(1,1),(1,2),(2,0),(2,1),(2,2),(2,3)\}.$$

Hence, $s[n]$ defined in (9.83) can take $N_b + \sum_{i=1}^{N_b} M_i$ possible values. ■

With state (9.83) and the available action set being the set of subband indices $\mathcal{A} = \{1,\ldots,N_b\}$, we may then use Q-learning (9.79) to learn a subband selection policy of the form of (9.81). The delayed reward can be chosen as an appropriate function that can be evaluated from available information. In the following example, we show one such implementation that takes into account the hardware implementation issues in addition to the actual transmission opportunities the cognitive radio may find by selecting a particular subband. It also demonstrates how Q-learning can be implemented in terms of an observable pseudostate definition, as that in (9.83), when the system is a POMDP and the transition probability model is also unknown.

Example 9.11 (Q-Learning-Based Subband Selection with Hardware Constraints) We consider a spectrum environment with N_b subbands where the ith subband, for $1 \leq i \leq N_b$, contains M_i number of channels. Suppose that the cognitive radio can only sense one subband at a time so that the action set is $\mathcal{A}$, but it can accurately observe the number of idle channels in the observed subband. Hence, we take the state of a Q-learning-based policy determination framework to be $s[n]$ defined in (9.83).

The delayed reward $r_n(s[n-1], a_{n-1})$ at time n is the result of taking action $a_{n-1} \in \mathcal{A}$ when in state $s[n-1]$ at time $n-1$. As in previous examples, a part of this reward is taken to be the number of idle channels $N_{a_{n-1}}[n-1]$ that was found to be actually idle in the selected subband a_{n-1} during the $(n-1)$-th sensing period. This makes sense if we assume that the basic interest of the cognitive radio is to achieve high-throughput communication and essentially the value of each channel is the same on average.

However, now we may also take into account the cost of switching among subbands. Recall that the wideband cognitive radio spans the N_b subbands by utilizing an RF front-end that can be reconfigured in any one of the L possible modes. In each

RF configuration mode l, for $1 \leq l \leq L$, the radio spans a spectrum range of $\mathcal{W}_l$ which we divided into a set of N_l number of subbands. Earlier, we denoted the set of indices of the subbands that are accessed by the lth configuration mode of the RF front-end to be $\mathcal{N}_l$. Suppose that at time $n = n$, the cognitive radio front-end is currently configured to be in the l'th mode and it is sensing a subband $i' \in \mathcal{N}_l$. Depending on the subband selection decision, there are three possible scenarios that can happen at time $n+1$:

1. The cognitive radio may decide to keep sensing the same subband i' during time $n+1$. In this case, there is no need for any hardware switching or filter reconfiguration in either IF or baseband. The only energy that is needed to achieve the desired subband selection is the actual energy needed for sensing the subband i'.
2. The cognitive radio may decide to sense a different subband i that is still within the same RF front-end configuration mode l' so that $i \in \mathcal{N}_{l'}$, during time $n+1$. In this case, still there is no need for any hardware switching for RF front-end reconfiguration. The only switching and reconfigurations needed can be filter reconfiguration in IF and baseband. Hence, in addition to the energy needed for sensing the subband i, there may be additional energy needed for such IF and baseband switching to achieve the desired subband selection.
3. The cognitive radio may decide to sense a different subband i that is in the frequency range spanned by a different RF front-end configuration mode l, during time $n+1$: $i \in \mathcal{N}_l$. In this case, not only the IF and baseband switching but also the RF front-end hardware needs to be reconfigured to achieve the desired switching. Hence, in addition to the energy needed for sensing the subband i and IF and baseband switching, there may be additional energy needed to achieve the RF front-end switching.

Hence, in order to take into account the imbalance among different decisions, we may include in the delayed reward function a term that depends on these switching energy costs. Let us indeed denote by $c_s(i', i)$ the switching energy cost from subband i' to subband i, where

$$c_s(i', i) = \begin{cases} c_{3,i}, & \text{if } i' = i \\ c_2 + c_{3,i}, & \text{if } i' \in \mathcal{N}_{l'},\ i \in \mathcal{N}_{l'},\ \text{but } i' \neq i \\ c_1 + c_{3,i}, & \text{if } i' \in \mathcal{N}_{l'},\ i \in \mathcal{N}_l,\ \text{and } l' \neq l \end{cases}, \tag{9.84}$$

where c_1 denotes the energy cost for switching among different RF configuration modes, c_2 denotes the energy cost for switching among different subbands within the same RF configuration mode, and $c_{3,i}$ denotes the energy cost required for spectrum sensing in the ith subband.

Since hardware reconfiguration may require more energy consumption, we may assume that $c_1 > c_2$. Energy needed to achieve sensing in different subbands $c_{3,i}$ may be different due to several possible reasons. One of them is the fact that in different subbands, the received SNR values may be different, causing the radio to spend different amounts of time sensing different subbands to achieve the same desired level of sensing accuracy. The SNR values on different subbands, in turn, may be different due to the nature of signals encountered in different subbands, interference and propagation characteristics, as well as antenna response patterns. In addition, different frequency bands may impose different operating constraints on a cognitive radio. For instance, some licensed frequency bands may impose more stringent collision conditions with licensed users requiring a cognitive radio to spend more time on spectrum sensing in order to achieve the required level of probability of detecting licensed signals (if any), thereby lowering the risk of colliding with them. On the other hand, in some unlicensed frequency bands, such as the ISM bands, for example, a cognitive radio may only be required to spend comparatively less time to detect a transmission opportunity since a higher collision probability might be acceptable. One way to account for these variations is to let the collision probability with incumbent users to be dependent on the subband index. If we insist on maintaining the same detection probability, this then will require that the cognitive radio spends a subband-dependent T_0^i time duration for sensing the ith subband. The longer this duration, the larger $c_{3,i}$ will be. A detailed implementation that takes into consideration these issues has been discussed in [139, 140]. In practice, it is possible that c_1 and c_2 may also not be constants but depend on the subband and configuration mode indices.

Suppose that the state at time $n-1$ is $s[n-1] = \left(a_{n-2}, N_{a_{n-2}}[n-2]\right) = \left(i', N_{i'}[n-2]\right)$, as shown in Figure 9.9. Let us denote by $a_{n-1} = i$ the action selected when in this state $s[n-1]$ at time $n-1$. This implies that the radio senses subband i during time $n-1$ and thus will observe the number of idle channels $N_{a_{n-1}}[n-1] = N_i[n-1]$ in subband i during time $n-1$. Then the delayed reward $r_n\left(s[n-1], a_{n-1}\right)$ resulting from selecting action a_{n-1} when in state $s[n-1]$ at time $n-1$ is the reward achieved by deciding to sense subband i during time period $n-1$ while radio sensing the subband i' during time $n-2$:

$$\begin{aligned} r_n\left(s[n-1], a_{n-1}\right) &\triangleq r_n\left(a_{n-2}, a_{n-1}, N_{a_{n-1}}[n-1]\right) \\ &\triangleq r_n\left(i', i, N_i[n-1]\right), \end{aligned} \tag{9.85}$$

where

$$r_n\left(i', i, N_i\right) = \min\{N_i, G\}\, \bar{r}(i, N_i) - \beta c_s(i', i), \tag{9.86}$$

with $\bar{r}(i, N_i)$ denoting a per-channel-normalized reward of finding N_i number of idle channels in subband i, G is the maximum number of idle channels that the cognitive radio can possibly use simultaneously due to hardware limitations on its transmission circuitry, $c_s(i', i)$ is given by (9.84), and β is a relative weight parameter.

In the following, we may assume that

$$\bar{r}(i, N_i) = K_i, \tag{9.87}$$

for some subband-dependent parameter K_i. For example, K_i can be the average bandwidth of an individual channel in subband i. Perhaps, a little better choice may be the product of average channel bandwidth and the detection probability of channel/signal activities in subband i. In either case, with $\bar{r}(i,N_i)$ as given by (9.88), the delayed reward (9.86) can be written as

$$r_n\left(i',i,N_i[n-1]\right) = K_i \min\{N_i[n-1],G\} - \beta c_s(i',i), \tag{9.88}$$

so that we may write

$$\begin{aligned} r_n(s[n-1],a_{n-1}) &= r_n\left(a_{n-2},a_{n-1},N_{a_{n-1}}[n-1]\right) \\ &= K_{a_{n-1}} \min\left\{N_{a_{n-1}}[n-1],G\right\} - \beta c_s(a_{n-2},a_{n-1}). \end{aligned} \tag{9.89}$$

With this delayed reward function, we may then use the Q-learning algorithm (9.79) with an appropriate initialization to learn a table of Q-values $Q(s,a)$. With our chosen definition of the state in (9.83), this Q-table will have $N_b + \sum_{i=1}^{N_b} M_i$ number of rows and N_b number of columns (corresponding to N_b possible action choices). Subband selection decisions can then be made based on (9.81) in order to allow for sufficient amount of exploration. ■

9.9 SUMMARY

- Perhaps the defining features of a cognitive radio are the ability to be aware of spectrum and to intelligently adapt radio operation in response to the perceived state of the spectrum and user needs. Hence, spectrum knowledge acquisition, or spectrum sensing as it is often called, is of fundamental importance to a cognitive radio.
- Wideband spectrum sensing problem is different from the spectrum sensing problem addressed in the context of DSS cognitive radios due to its additional complexity. We may divide the wideband spectrum sensing problem into three subproblems:

  1. *Wideband spectrum scanning problem*: How to efficiently and effectively scan a very wide spectrum range
  2. *Spectral activity detection problem*: How to detect signals from a wideband received signals
  3. *Signal classification problem*: How to identify the detected signals

- Spectrum of interest to a wideband cognitive radio may not be covered by a single RF antenna. In that case, a real time reconfigurable antenna that may span portions of the spectrum of interest in its different modes of operation may be a practical solution.
- To facilitate real time wideband spectrum sensing, even the spectrum range spanned by the antenna in one of its modes may have to be divided into a set of smaller spectrum bands. The maximum spectrum width that the cognitive radio

can sense and process in real time is called a subband. Hence, by definition, the radio is only able to observe, sample, and process one subband at a time.

- Spectrum segmentation and subbanding refers to the problem of optimally dividing the wide spectrum of interest into a set of subbands that can be sensed by physically feasible reconfigurable antennas and processed by realistic analog-to-digital converters.
- Given a total bandwidth B_T, a maximum output bit rate R_b of A/D converter, a minimum required quantization resolution $q_{\min}$, a minimum required spectral resolution Δf_0, and a maximum subband revisit time T_0, the number of subbands N_b can be shown to be constrained as

$$\frac{2B_T q_{\min}}{R_b} \leq N_b \leq T_0 \Delta f_0.$$

- A wideband spectrum sensing receiver can be designed based on either a homodyne or a heterodyne configuration. In a homodyne receiver, the received RF subband signal will be directly down-converted to the baseband and A/D conversion will be performed in the baseband. A heterodyne receiver can be more flexible in that the received RF subband signal can be down-converted either to one or more intermediate frequencies (IF). A/D conversion can either be performed in baseband or in an IF.
- Subband selection problem is one of the basic decision-making problems a wideband cognitive radio has to address. It refers to the question of which subband the cognitive radio should scan at any particular time so that the radio can best fulfill its performance objectives given the state of the RF spectrum.
- In order to determine which subband it should sense, the cognitive radio needs to have a model for the spectrum dynamics, which can be defined as the collective dynamics of suitably chosen states of subbands. An obvious model for a subband state is the collection of states of individual channels within the subband.
- The state of an individual channel can be defined as being either busy (state 0) or idle (state 1). A reasonable model for these state dynamics is a (first-order) Markov process. Hence, dynamics of the state $S_{i,j}[n] \in \{0, 1\}$ of the (i,j)th channel at time n, for $j \in \{1, \ldots, M_i\}$ and $i \in \{1, \ldots, N_b\}$, can be characterized by giving a 2×2 channel transition matrix $P^{i,j}$.
- The state of the ith subband is defined to be the M_i-length vector $\mathbf{S}_i = (S_{i,1}, \ldots, S_{i,M_i})$ where each $S_{i,j} \in \{0, 1\}$ denotes the state of the jth channel, for $1 \leq j \leq M_i$, of the ith subband. This subband state $\mathbf{S}_i[n]$ is a Markov chain with a $2^{M_i} \times 2^{M_i}$ state transition probability matrix $P^{(i)}$ whose entries can be computed from the given channel transition probability matrices $P^{i,j}$ for $1 \leq j \leq M_i$. The state of the overall spectrum can be defined as length $M = \sum_{i=1}^{N_b} M_i$ vector $\mathbf{S} = (\mathbf{S}_1, \ldots, \mathbf{S}_{N_b})$. The $2^M \times 2^M$ state transition probability matrix P corresponding to the overall spectrum state $\mathbf{S}$ can be obtained from the subband state transition probability matrices $P^{(i)}$'s. State $\mathbf{S}[n]$ at time n takes values in set $\mathcal{S} = \left\{\mathbf{s}_1, \ldots, \mathbf{s}_{2^M}\right\}$.

- Since cognitive radio can only observe one subband at a time, and even the state of this subband is observed in the presence of noise, subband selection problem turns out to be a POMDP problem. The action available to the cognitive radio is the selection of subband indices $a \in \{1, \ldots, N_b\}$. The observations are the possibly erroneous binary states of the channels within the sensed subband. Information available for decision-making at time n is the history $\underline{h}_n = (\underline{h}_{n-1}, a_{n-1}, y[n-1])$. As usual, by defining the belief

$$b_m[n] = \Pr\left(\mathbf{S}[n] = \mathbf{s}_m | \underline{h}_n\right)$$

  and the belief-state vector

$$\mathbf{b}[n] = (b_1[n], \ldots, b_{2^M}[n])^T \in \mathcal{P}^{2^M-1},$$

  we can transform the original POMDP into a belief-state POMDP.
- For a suitably defined immediate reward function, we may then define optimal policies for subband selection as maximizing the expected discounted total return. Depending on whether the decision-making horizon is finite or infinite, optimal policies are given, respectively, by Theorem 7.5 or 7.9. In either case, in theory, the algorithms introduced in Chapter 7 can be used to compute these optimal policies. The complexity of these optimal policies, however, will be exponential in the total number of channels in the whole spectrum of interest. This means that it is exponential in both number of subbands N_b and number of channels in each subband.
- Complexity of optimal subband selection policy computation can be reduced if we were to assume that channel dynamics are independent of one another. In this case, the complexity can be reduced to being linear in the total number of channels. However, this may not always be a valid assumption. A more reasonable assumption can be that the dynamics of channels in different subbands are independent of each other in which case complexity will be exponential in the number of channels in a subband but linear in the number of subbands N_b.
- Myopic, or greedy, policies focus on decision rules that maximize the instantaneous reward rather than that over the whole decision-making horizon. These policies are considerably less demanding computationally compared to optimal policies over the whole decision horizon:
  - An optimal myopic subband selection policy can be defined by

$$a_n^* = \underset{a \in \{1,\ldots,N_b\}}{\arg\max}\; r(\mathbf{b}[n], a) = \underset{a \in \{1,\ldots,N_b\}}{\arg\max}\; \mathbf{r}^T(a)\mathbf{b}[n].$$

  - When channel dynamics are independent of each other, a reduced complexity optimal myopic policy can be defined in terms of the M-length belief state

$\boldsymbol{\omega}[n]$ as below:

$$a_n^* = \underset{a \in \{1,\ldots,N_b\}}{\arg\max}\ r(\boldsymbol{\omega}[n], a)\,.$$

- An alternative approach to obtain reduced complexity subband selection policies is to redefine the state of the spectrum. In particular, it may make sense to define the state of a subband in terms of a direct characteristic of a subband rather than a collective of individual channel properties within a subband. When communications opportunities within a subband are the main aspect of interest to the radio, a new state $N_i[n]$ for the subband i can be defined as the number of idle channels in the subband i at time n:

$$N_i[n] \triangleq \text{Number of idle channels in the subband } i \text{ at time } n.$$

- State N_i of subband i can only take $M_i + 1$ possible values. Hence, the overall spectrum state can take $\prod_{i=1}^{N_b}(M_i + 1)$ number of possible values. If all subbands are assumed to have an equal number $\bar{M} - 1$ of channels, the complexity is thus about $\bar{M}^{N_b}$.
- All different forms of policies discussed earlier, can be formulated in terms of this new subband state. For example, optimal finite- or infinite-horizon policies can be formulated and solved, at least in theory, the same way as before. Of particular interest can be the situation of independent subband dynamics. In this case, the dimensionality of the sufficient statistic for the spectrum state can be reduced to be linear in both number of channels in a subband and the number of subbands. Optimal myopic policies can also be defined in terms of the new state definition.
- When parameters of the Markov model, in particular the state transition probabilities, are unknown, it is possible to use learning algorithms to learn an optimal, or near-optimal, policy. Even when model parameters are known, it can be computationally advantageous to learn an optimal policy since these algorithms can be much simpler to implement. A class of learning algorithms that is known to be particularly suitable for Markov environments is the reinforcement learning.
- A commonly used reinforcement learning algorithm is the Q-learning, in which a table of Q-values, where $Q(s,a)$ is the value of taking action a when in state s, is updated as follows:

$$\begin{aligned} Q(s[n-1], a_{n-1}) \leftarrow\ & (1-\alpha) Q(s[n-1], a_{n-1}) \\ & + \alpha \left[r_n(s[n-1], a_{n-1}) + \gamma \max_a Q(s[n], a) \right], \end{aligned}$$

where $Q(s[n-1], a_{n-1})$ and $r_n(s[n-1], a_{n-1})$ denote, respectively, the Q-value and the observed reward corresponding to taking action a_{n-1} when in state $s[n-1]$ at time $n-1$ and $0 < \alpha < 1$ and $0 \le \gamma < 1$ denote the learning rate and the discount factor, respectively. The action selection when in a given state $s[n]$ at time n is performed as

$$a_n^* = \arg\max_a Q(s[n], a).$$

- To ensure convergence of Q-learning and other similar reinforcement algorithms, it is important that there is a balance between exploration and exploitation. This can be achieved by allowing for a random action choice with a probability of $0 < \epsilon < 1$ for an appropriately chosen ϵ:

$$a^* = \begin{cases} \arg\max_{a \in \mathcal{A}} Q(s, a) & \text{with probability } 1 - \epsilon \\ \sim \mathcal{U}(\mathcal{A}) & \text{with probability } \epsilon \end{cases},$$

where $\mathcal{U}(\mathcal{A})$ denotes the uniform distribution over the action set $\mathcal{A}$. The uniform action selection during exploration is, however, not always necessary and can be replaced by more effective application-specific alternatives.

- When transition probabilities are unknown, Q-learning-type algorithms may be developed to learn at least a reasonable policy for a POMDP by proper choice of an observable state.

10

SPECTRAL ACTIVITY DETECTION IN WIDEBAND COGNITIVE RADIOS

10.1 INTRODUCTION

In Chapter 9, we discussed how wideband spectrum scanning can be achieved in a cognitive radio: The spectrum range of interest to the radio is first segmented into a set of N_b subbands, each of which may still be wide enough to contain several different channels. At a given time instant n, a cognitive radio can sense only one of these subbands. Hence, if the subband i, for $1 \leq i \leq N_b$, is sensed at time n, the radio's spectrum observation is a signal $Y[n]$ that is a noise-corrupted version of the superposition of signals within subband i. The radio needs to detect signals contained in this sensing observation $Y[n]$.

If the radio scans a single channel, then detecting whether there is any signal activity in the channel based on a sensed signal is straightforward: The particular channel is isolated by means of a bandpass filter (BPF), so the sensed signal $Y[n]$ only contains frequencies within the span of this channel. Depending on the type of a priori knowledge available regarding the signal in this particular channel, one of the detection procedures discussed in Chapter 4 can, for instance, be used to make a detection decision. In particular, if the signal was assumed to be simply a Gaussian-distributed random signal with a certain average power and the only contamination is an additive Gaussian noise, then the optimal detection will simply be to compute the received signal power and compare it to a threshold. Indeed, as shown in Figure 10.1(a), this is basically what a cognitive radio designed to achieve dynamic spectrum access over the idle broadcast TV bands may do: The radio scans a (narrowband) primary channel

Signal Processing for Cognitive Radios, First Edition. Sudharman K. Jayaweera.

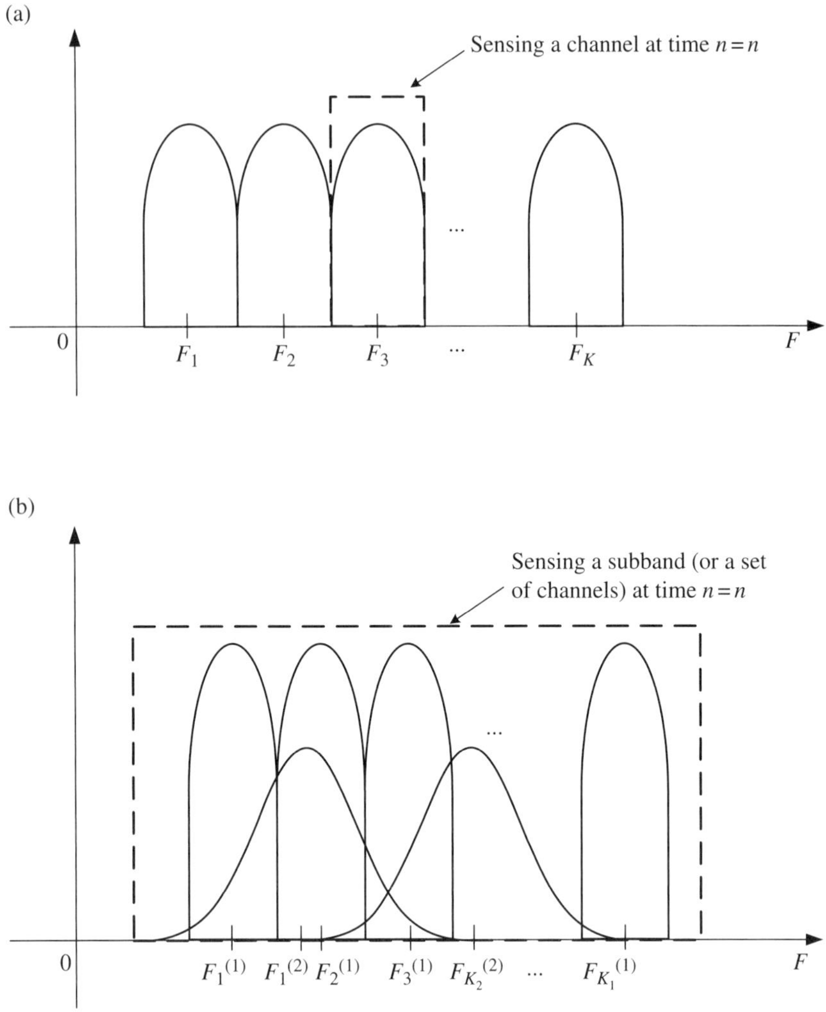

FIGURE 10.1 Wideband spectrum sensing versus spectrum sensing for dynamic spectrum sharing. (a) In spectrum sensing for dynamic spectrum sharing, the cognitive radio scans a single channel or a set of well-defined channels. (b) In wideband spectrum sensing, the cognitive radio scans a subband that may consist of several signals of different types as indicated by the superscripts of center frequencies.

to detect the presence or absence of a primary signal, which is a TV signal. Hence, the spectral characteristics (e.g., carrier frequency and bandwidth, etc.) of signals of interest are known parameters.

However, this approach needs to be modified for a wideband cognitive radio. As we discussed in Chapter 9, at any given time, in wideband spectrum sensing, a cognitive radio scans a wide spectrum subband (not just a selected primary channel). Within this subband, there may be several different active systems as shown in Figure 10.1(b). Hence, the sensed signal $Y[n]$ may contain several signals of possibly different types centered at different frequencies. For example, in the 2.4GHz ISM band, there may be WiFi (IEEE 802.11b), Bluetooth, as well as other unknown signals such as microwave oven radiation. Even if this band is separated into two or more subbands, there will likely still be many channels of different systems within a single subband. Another example is encountered in WMTS systems operating in the lower and upper L bands (1395–1400MHz and 1427–1432MHz, respectively): These bands may contain an unknown mix of utility telemetry signals, government radar signals, as well as radio astronomy signals [39]. Hence, the signals to be sensed are usually not homogeneous and, in general, can be centered at unknown positions within the sensed subband.

The objectives of a wideband autonomous cognitive radio may not be simply dynamic spectrum access either. Hence, determining whether a particular channel is ON or OFF may not be sufficient spectrum knowledge. For example, a future wideband cognitive radio may have the ability to transmit over a combination of different transmission modes (protocols) simultaneously. To fully capitalize on this ability, the radio may need to determine all available spectrum opportunities within a subband so it can formulate an optimal multimode transmission strategy. In a military application, a cognitive radio may need to interpret all signal activities within a spectrum subband to determine whether there are any signals that can possibly be belonging to an enemy communications system or be possibly jamming signals. Hence, in general, a wideband autonomous cognitive radio needs to detect all spectral activities within the sensed spectrum. Once detected, they can then be classified so as to determine their origins.

In this chapter, we focus on the problem of *detecting* unknown signals located at unknown carrier frequencies over a possibly wide spectrum subband. This is called the wideband spectral activity detection problem.

10.2 OPTIMAL WIDEBAND SPECTRAL ACTIVITY DETECTION

Given that the spectrum of interest is divided into N_b subbands, spectrum sensing is performed on one subband at a time, as discussed in the previous chapter. Hence, without loss of any generality, we may focus on spectral activity detection in the ith subband, for $1 \leq i \leq N_b$. Since detection and classification procedures apply to each of the subbands in the same way, we may consider an arbitrary subband i and, for simplicity of notation, drop the subband index i from the following discussion.

During an arbitrary sensing period, suppose that the cognitive radio collects N signal samples denoted by $\mathbf{y} = (y_1, \ldots, y_N)^T$ on the selected subband. Let us denote by K the number of signals active within this subband during this sensing duration. We will assume that the signal environment over the subband of interest contains the same number of signals (and they are the same signals) during the time it takes to collect the N signal samples. For this to be reasonable, of course, the user dynamics in the subband of interest must be at a slower rate compared to the sensing duration. The kth baseband signal is a basic signal pulse $\mathbf{s}_k$ located at a center frequency F_k for $-\frac{1}{2} \leq F_k \leq \frac{1}{2}$, for $k = 1, \ldots, K$. Hence, for $k = 1, \ldots, K$, the kth signal can be denoted as $\mathbf{x}_k = \left(x_{k,1}, \ldots, x_{k,N}\right)^T$ where $x_{k,n} = x_k[n] = s_{k,n} e^{j2\pi F_k n}$ for $1 \leq n \leq N$. Thus, the equivalent discrete-time complex baseband-sensed signal is

$$\begin{aligned} \mathbf{y} &= \sum_{k=1}^{K} \mathbf{x}_k + \mathbf{w} \\ &= \mathbf{x} + \mathbf{w}, \end{aligned} \tag{10.1}$$

where $\mathbf{w} = (w_1, \ldots, w_N)^T \sim \mathcal{N}_c\left(\mathbf{0}, N_0 \mathbf{I}_N\right)$ and $\mathbf{x} = \sum_{k=1}^{K} \mathbf{x}_k$.

Since we do not know anything about the signals $\{x_k\}_{k=1}^{K}$, it makes sense to treat them as stochastic signals. For simplicity of notation, we may set their means to be zero. To motivate a reasonable approach to wideband signal detection, let us further assume that these signals were modeled as being Gaussian and wide-sense stationary (wss) so that their power spectral densities exist. If we knew the covariance matrix Σ_X of the signal $\mathbf{x}$, then detection problem for the signal $\mathbf{x}$ is nothing but a stochastic signal detection problem:

$$\begin{aligned} &H_0 \text{ (signal absent)} : \mathbf{y} \sim \mathcal{N}\left(\mathbf{0}, \Sigma_0\right) \\ &\qquad\qquad\qquad \text{versus} \\ &H_1 \text{ (signal present)} : \mathbf{y} \sim \mathcal{N}\left(\mathbf{0}, \Sigma_1\right). \end{aligned} \tag{10.2}$$

But since $\Sigma_0 = \Sigma_W$ and $\Sigma_1 = \Sigma_X + \Sigma_W$, where $\Sigma_W = N_0 \mathbf{I}_N$, (10.2) is equivalent to the following hypothesis testing problem:

$$\begin{aligned} &H_0 : \mathbf{y} \sim \mathcal{N}\left(\mathbf{0}, \Sigma_W\right) \\ &\qquad\quad \text{versus} \\ &H_1 : \mathbf{y} \sim \mathcal{N}\left(\mathbf{0}, \Sigma_X + \Sigma_W\right). \end{aligned} \tag{10.3}$$

From Chapter 4, we know that optimal tests for (10.3) are based on the LLR $L(\mathbf{y}) = \frac{f_{\mathbf{Y}|H_1}(\mathbf{y}|H_1)}{f_{\mathbf{Y}|H_0}(\mathbf{y}|H_0)}$. In fact, (10.3) is nothing but Example 4.6 with $\theta = 1$. Thus, from (4.150), the log-likelihood ratio (LLR) for this scenario is

$$\begin{aligned} \mathscr{L}(\mathbf{y}) &= \log L(\mathbf{y}) \\ &= \frac{1}{2} \mathbf{y}^\dagger \left(\Sigma_0^{-1} - \Sigma_1^{-1}\right) \mathbf{y} + C \\ &= \frac{1}{2} \mathbf{y}^\dagger Q \mathbf{y} + C, \end{aligned} \tag{10.4}$$

where

$$
\begin{aligned}
Q &= \Sigma_0^{-1} - \Sigma_1^{-1} \\
&= \Sigma_0^{-1} - (\Sigma_X + \Sigma_0)^{-1} \\
&= \Sigma_0^{-1}\Sigma_X(\Sigma_X + \Sigma_0)^{-1} && (10.5a)\\
&= N_0^{-1}\Sigma_X(\Sigma_X + N_0 I_N)^{-1} && (10.5b)
\end{aligned}
$$

and $C = \frac{1}{2}\log\frac{|\Sigma_0|}{|\Sigma_1|}$. Hence, optimal detectors may be designed so that they compare the decision statistic

$$T(\mathbf{y}) = \mathbf{y}^\dagger Q\mathbf{y} \tag{10.6}$$

to a threshold.

Computing the decision statistic $T(\mathbf{y})$, however, requires the knowledge of matrix Q, which in turn requires the knowledge of the signal covariance matrix Σ_X. Since $\mathbf{x} = \sum_{k=1}^{K}\mathbf{x}_k$, assuming that the K signals are mutually independent, we have that $\Sigma_X = \sum_{k=1}^{K}\Sigma_{X_k}$. Thus, to perform an optimal threshold test based on the LLR statistic in (10.4), we need to know the covariance matrices Σ_{X_k} of all the signals X_k for $k = 1,\ldots,K$. In general, we may not have such knowledge.

In order to motivate an approach to get around this issue, recall that we have assumed wss signals. Hence, both Σ_0 and Σ_1 are Hermitian Toeplitz matrices, and so the matrix Q is also Hermitian Toeplitz. Hence, we may denote the (n,n')th element of the $N\times N$ matrix Q, $[Q]_{n,n'}$, as $[Q]_{n,n'} = q_{n-n'}$ with $q_{n-n'} = q^*_{n-n'}$, for $1 \le n,n' \le N$. With this notation, we can express the decision statistic $T(\mathbf{y})$ in (10.6) as (see [51] for details)

$$T(\mathbf{y}) = \sum_{n=1}^{N}\sum_{n'=1}^{N} y_n y^*_{n'} q_{n'-n}. \tag{10.7}$$

Now, suppose that N is large enough to assume that the sequence $\{q_m\}$ is specified for all m. Let us denote the discrete-time Fourier transform (DTFT) of sequence $\{q_m\}$ by $S_Q(F)$, assuming it exists:

$$S_Q(F) = \sum_{m=-\infty}^{\infty} q_m e^{-j2\pi Fm} \quad \text{for} \quad -\frac{1}{2} \le F \le \frac{1}{2}.$$

Since Q is a positive-definite matrix, $S_Q(F)$ represents the power spectral density (PSD) of a signal having $\{q_m\}$ as its covariance sequence (or Q as the covariance matrix). Assuming that the Wiener–Khinchine theorem holds, the inverse Fourier transform of $S_Q(F)$ will exist and be well defined, so that

$$q_m = \int_{-\frac{1}{2}}^{\frac{1}{2}} S_Q(F) e^{j2\pi Fm} dF \quad \text{for} \quad -\infty < m < \infty.$$

Hence, the LLR decision statistic $T(\mathbf{y})$ of (10.7) can equivalently be written as

$$\begin{aligned}
\frac{1}{N} T(\mathbf{y}) &= \frac{1}{N} \sum_{n=1}^{N} \sum_{n'=1}^{N} y_n y_{n'}^* \int_{-\frac{1}{2}}^{\frac{1}{2}} S_Q(F) e^{j2\pi F(n'-n)} dF \\
&= \frac{1}{N} \int_{-\frac{1}{2}}^{\frac{1}{2}} \left(\sum_{n=1}^{N} y_n e^{-j2\pi Fn} \right) \left(\sum_{n'=1}^{N} y_{n'}^* e^{j2\pi Fn'} \right) S_Q(F) dF \\
&= \int_{-\frac{1}{2}}^{\frac{1}{2}} \frac{1}{N} \left| \sum_{n=1}^{N} y_n e^{-j2\pi Fn} \right|^2 S_Q(F) dF \\
&= \int_{-\frac{1}{2}}^{\frac{1}{2}} \hat{S}_y(F) S_Q(F) dF,
\end{aligned} \tag{10.8}$$

where, from (6.27), we observe that $\hat{S}_y(F)$ is the periodogram PSD estimate of the sensed signal $\mathbf{y}$:

$$\hat{S}_y(F) = \frac{1}{N} \left| \sum_{n=1}^{N} y_n e^{-j2\pi Fn} \right|^2. \tag{10.9}$$

From (10.8), we see that the optimal detector is equivalent to first computing the periodogram PSD estimate $\hat{S}_y(F)$ of the received signal and correlating it with the PSD $S_Q(F)$. The resulting statistic is then compared to a threshold to determine whether the signal $\mathbf{x}$ is present or not.

Observe from (10.5a) that if signal power is small, then we may approximate Q to be

$$\begin{aligned}
Q &= \Sigma_0^{-1} \Sigma_X (\Sigma_X + \Sigma_0)^{-1} \\
&\approx \Sigma_0^{-1} \Sigma_{\mathbf{X}} \Sigma_0^{-1},
\end{aligned} \tag{10.10}$$

so that

$$Q \approx \frac{1}{N_0^2} \Sigma_{\mathbf{X}}. \tag{10.11}$$

Hence, in this case the PSD corresponding to matrix Q is approximately

$$S_Q(F) \approx \frac{1}{N_0^2} \mathscr{S}_X(F), \tag{10.12}$$

where $\mathscr{S}_X(F)$ is the PSD of the signal $\mathbf{x}$. Thus, in the low SNR region, the LLR test statistic is approximately equivalent to

$$\frac{1}{N} T(\mathbf{y}) \approx \frac{1}{N_0^2} \int_{-\frac{1}{2}}^{\frac{1}{2}} \hat{S}_y(F) \mathscr{S}_X(F) dF. \tag{10.13}$$

In other words, when SNR is low, the decision static is equivalent to correlating the received signal periodogram power spectrum with the signal power spectrum. Note that the integral (10.13) will essentially be only over the support of the spectrum of the signal of interest $\mathbf{x}$. This conforms with the traditional energy detector used for narrow-band signal detection: First, filter the received signal with a BPF having a passband wide enough to just pass the signal of interest. Then integrate the output energy of the BPF and compare it to a threshold.

However, we have made no real progress because knowing $S_Q(F)$ is the same as having to know the matrix Q. Still, the frequency-domain implementation (10.8) is useful in motivating a meaningful spectral activity detector for wideband spectrum sensing. To that end, suppose in addition to noise $\mathbf{w}$, signal $\mathbf{x}$ is also white and Gaussian with $\Sigma_X = \sigma_s^2 \mathbf{I}_N$, then the matrix Q of (10.5b) is diagonal and is given by

$$Q = \frac{\sigma_s^2}{N_0 \left(N_0 + \sigma_s^2\right)} \mathbf{I}_N, \tag{10.14}$$

so that

$$S_Q(F) = \frac{\sigma_s^2}{N_0 \left(N_0 + \sigma_s^2\right)} \quad \text{for} \quad -\frac{1}{2} \leq F \leq \frac{1}{2}. \tag{10.15}$$

This reduces the decision statistic (10.8) to

$$\frac{1}{N} T(\mathbf{y}) = \frac{\sigma_s^2}{N_0 \left(N_0 + \sigma_s^2\right)} \int_{-\frac{1}{2}}^{\frac{1}{2}} \hat{S}_y(F) dF.$$

Hence, if both noise and signals were white, detecting the presence of signal $\mathbf{x}$ is equivalent to integrating the power under the estimated periodogram PSD $\hat{S}_y(F)$ of $\mathbf{y}$ and comparing it to a threshold. It is true that this is an extreme case in which all signals are superposed in both time and frequency. Signals $\{\mathbf{x}_k\}$'s, for $k = 1, \ldots, K$, all have constant PSD throughout the whole subband. Hence, we can detect the presence or absence of $\mathbf{x}$, but we may not be able to distinguish among individual signals $\mathbf{x}_k$'s.

At the other extreme, consider the situation in which each signal x_k is a sinusoid located at a normalized frequency F_k:

$$x_k[n] = \sqrt{2P_k} \cos 2\pi F_k n. \tag{10.16}$$

Then it follows that $\mathscr{S}_X(F) = \sum_{k=1}^{K} P_k \left(\delta_D(F - F_k) + \delta_D(F + F_k)\right)$, so that from (10.12)

$$S_Q(F) \approx \frac{1}{N_0^2} \sum_{k=1}^{K} P_k \left(\delta_D(F - F_k) + \delta_D(F + F_k)\right),$$

assuming that the SNR is low.

Then from (10.13), the decision statistic for detecting spectral activities in low SNR region reduces to

$$\begin{aligned} \frac{1}{N} T(\mathbf{y}) &\approx \frac{1}{N_0^2} \sum_{k=1}^{K} \int_{-\frac{1}{2}}^{\frac{1}{2}} \hat{S}_y(F) P_k \left(\delta_D(F - F_k) + \delta_D(F + F_k)\right) dF \\ &= \sum_{k=1}^{K} \frac{P_k}{N_0^2} \left(\hat{S}_y(F_k) + \hat{S}_y(-F_k)\right). \end{aligned} \tag{10.17}$$

Consider the case of $K = 1$ in which the decision statistic (10.17) becomes

$$\frac{1}{N} T(\mathbf{y}) = \frac{P_1}{N_0^2} \left(\hat{S}_y(F_1) + \hat{S}_y(-F_1)\right).$$

This still tells us that we need to know the frequency F_1 of the single sinusoidal signal $x_1[n]$ that we are interested in. Indeed, if we knew F_1, we could just evaluate $\hat{S}_y(F)$ at $F = F_1$ and perform the threshold test. But if we do not know Q, or equivalently the frequency F_1, we can compute the periodogram $\hat{S}_y(F)$ for all F and then find at which frequency F it exceeds the required threshold. For example, Figure 10.2(a) shows $\hat{S}_y(F)$ when $K = 1$ with $F_1 = 0.2$ where $N = 100$ and $SNR = -5B$, where SNR is defined as $SNR = \frac{P_1}{N_0}$. At least under ideal conditions, the frequency at which $\hat{S}_y(F)$ will exceed the desired threshold will be $F = F_1$, if indeed the signal $x_1[n]$ is present in the sensed signal. If the threshold is not exceeded at any of the F values, for $-\frac{1}{2} \leq F \leq \frac{1}{2}$, then we may claim that the signal is not present in the sensed observation.

(a)

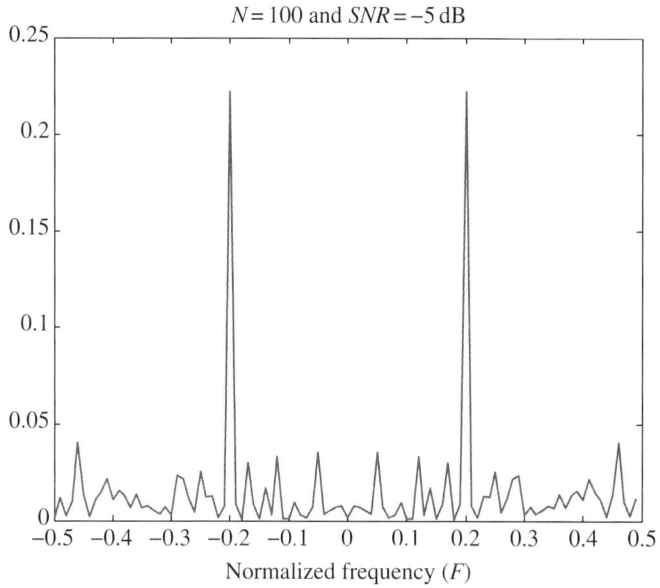

(b)

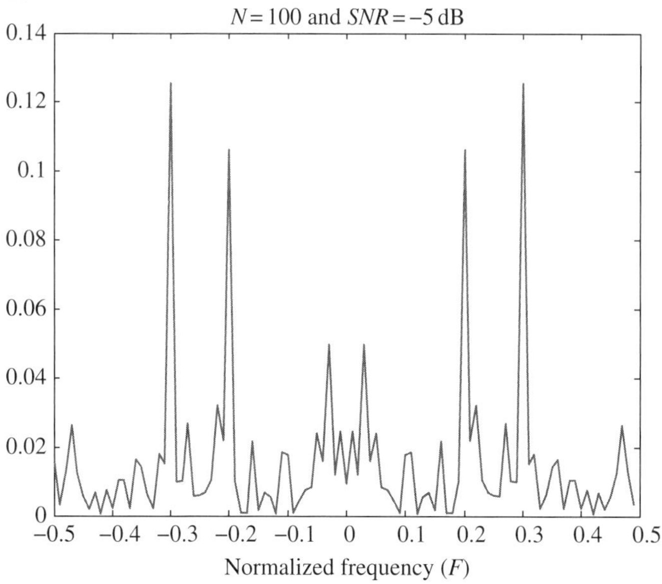

FIGURE 10.2 In low signal-to-noise ratio region, an approximately optimal framework for spectral activity detection consists of thresholding the output of correlating the observation periodogram with the power spectrum of the signal. When the signal is a sum of sinusoidal tones such that $x[n] = \sum_{k=1}^{K} \sqrt{2P_k} \cos 2\pi F_k n$, this is equivalent to simply thresholding the observation periodogram. (a) Observation periodogram $\hat{S}_y(F)$ when $K = 1$ with $F_1 = 0.2$. Other parameters are $N = 100$ and $SNR = -5$ dB. (b) Observation periodogram $\hat{S}_y(F)$ when $K = 2$ with $F_1 = 0.2$ and $F_2 = 0.3$. Other parameters are $N = 100$ and $SNR = -5$ dB.

Now, suppose $K = 2$ in (10.17). Again, if we did not know the two frequencies F_1 and F_2 at which the signals must be present, then we can compute the periodogram $\hat{S}_y(F)$ for all F and check at which two frequencies it exceeds the desired threshold. Figure 10.2(b) shows $\hat{S}_y(F)$ when $K = 2$ with $F_1 = 0.2$ and $F_2 = 0.3$ (both signals have $SNR = -5$dB and $N = 100$). When the two signals $x_1[n]$ and $x_2[n]$ are present in the sensed observations, this will happen at the two frequencies F_1 and F_2. This allows us to determine not only whether these signals are present but also their frequencies. On the other hand, if the periodogram does not exceed the threshold at two frequencies, then it must be the case that these signals are not present in the sensed observation. If the threshold is computed correctly under ideal conditions and $\hat{S}_y(F)$ exceeds it only at one frequency, then it must be the case that that is the only signal present in the observation. Of course, we can extend the same method to detect any K number of sinusoidal signals.

From the previous discussion, an approximately optimal approach for spectral activity detection is then to compute the periodogram $\hat{S}_y$ of the sensed observations and to check at what frequencies it exceeds a certain threshold. Those frequencies are the frequencies of the sinusoidal signals that are present in the sensed signal. Of course, the threshold must be chosen according to a certain optimality criterion as we have discussed in Chapter 4. Since the only approximation we have made is (10.10), this method is approximately optimal when SNR is sufficiently low and as long as signals and noise are Gaussian.

10.3 WIDEBAND SPECTRAL ACTIVITY DETECTION

In the previous section, we motivated an approximately optimal approach to perform spectral activity detection under ideal assumptions: Signals and noise were assumed to be Gaussian, and the approximate decision statistic (10.17) was based on sinusoidal signals in low SNR region. Of course, in practice, these may not be valid assumptions. Hence, in the following, we generalize the earlier approach to develop a suboptimal method to detect arbitrary signals that are not necessarily sinusoids.

Let us revisit the spectral activity detection problem (10.1) in a wideband cognitive radio based on sensed observations in a selected subband:

$$\begin{aligned} &H_0 \text{ (signal absent)} : y[n] = w[n] \quad \text{for } 0 \le n \le N-1 \\ &\qquad\qquad\qquad \text{versus} \\ &H_1 \text{ (signal present)} : y[n] = \sum_{k=1}^{K} x_k[n] + w[n] \quad \text{for } 0 \le n \le, N-1 \end{aligned} \tag{10.18}$$

where x_k, for $k = 1, \ldots, K$, are arbitrary signals located at arbitrary carrier frequencies and noise w is assumed to be zero-mean white Gaussian with $\mathscr{S}_W(F) = N_0$.

Motivated by the discussion in the previous section, in order to detect spectral activities in the sensed subband, we first compute the periodogram estimate of PSD of the received signal $y[n]$:

$$\begin{aligned}\hat{S}_y(F) &= \frac{1}{N}\left|\sum_{n=0}^{N-1} y[n]e^{-j2\pi Fn}\right|^2 \\ &= \frac{1}{N}\left|Y(F;N)\right|^2,\end{aligned} \tag{10.19}$$

where $Y(F;N)$ is the DTFT of sensed observations y_n. Recall from (10.13) that the optimal detector decision statistic for Gaussian problem in low SNR region is of the form of

$$\frac{1}{N}T(\mathbf{y}) \approx \frac{1}{N_0^2}\int_{-\frac{1}{2}}^{\frac{1}{2}} \hat{S}_y(F)\,\mathscr{S}_X(F)dF. \tag{10.20}$$

If the received signal consisted of K signals with constant spectra in nonoverlapping spectral intervals $B_1,\ldots, B_K$, where $B_k \subseteq [-\frac{1}{2},\frac{1}{2}]$, (10.20) becomes

$$\frac{1}{N}T(\mathbf{y}) = \frac{1}{N_0^2}\sum_{k=1}^{K}\int_{B_k} \hat{S}_y(F)dF. \tag{10.21}$$

Hence, a reasonable suboptimal approach to detect spectral activities in the sensed subband is to integrate the energy in the periodogram of the observations over small suitably chosen bands as implied by (10.21) and compare with a suitably chosen threshold. Those frequencies at which the threshold is exceeded represent the spectral activities in the sensed subband. The threshold may be chosen according to different optimality criteria as discussed in Chapter 4. For example, if we desire to keep the probability of falsely detecting signals below a certain threshold, say, of P_f, our criterion for setting the threshold can be the Neyman–Pearson optimality. In that case, the threshold will be a function of the false-alarm probability P_f. When considering a wide frequency band, it is likely that noise and signal characteristics may vary from one frequency to another. Hence, the required false-alarm probability P_f can, in general, be allowed to be a function of frequency F. In this case, we may denote it as $P_f(F)$. On the other hand, if one desires a fixed false alarm regardless of the frequency F, the threshold τ may be allowed to be a function of frequency $\tau(F)$ when noise and signal characteristics are varying over frequencies. To compute the exact thresholds, however, we need to assume specific noise and/or signal distributions.

The earlier approach essentially reduces the problem of spectral activity detection to a problem of power spectrum estimation, in particular the periodogram, and then applies a threshold test to the estimated periodogram at each frequency F for $-\frac{1}{2} < F < \frac{1}{2}$ as shown in Figure 10.3.

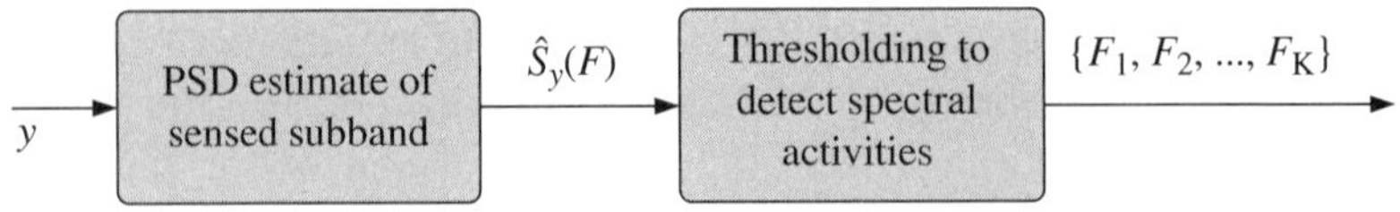

FIGURE 10.3 Wideband spectral activity detection in the presence of Gaussian noise.

Since the test statistic (10.21) is already suboptimal for non-Gaussian signals and noise as well as when the SNR is not sufficiently low, it then follows that there is no reason to limit the spectral estimator to the periodogram. Indeed, the periodogram emerged due to the Gaussian assumption. Although in the Gaussian case periodogram is a sufficient statistic for spectral activity detection, when this assumption is not valid, we may expect better overall performance in a suboptimal statistic such as (10.21) by having a better PSD estimator. In addition, we saw in Chapter 6 that the periodogram can be highly erratic even when N is large due to its nonzero asymptotic variance. Hence, a reasonable approach is to use a smoothed version of periodogram as discussed in Chapter 6.

Let us consider the frequency-domain smoothing of the periodogram using a rectangular smoothing window of (odd length) L. Let us suppose that the Fourier transform of the observations is computed using the DFT at a set of discrete frequency points $F_k = \frac{k}{N}$, for $k = 0, \ldots, N-1$:

$$\begin{aligned} Y[k] &= Y(F_k; N) \\ &= \sum_{n=0}^{N-1} y[n] e^{-j2\pi \frac{k}{N} n}. \end{aligned} \tag{10.22}$$

Let us assume that both signal and noise are such that $y[n]$'s are zero mean under both hypothesis. Then under both hypotheses, each $Y[k]$ is a sum of N iid zero-mean Gaussians. If we denote the variance of $y[n]$ under hypothesis H_j, for $j = 0, 1$, by σ_j^2, then for $k = 0, 1, \ldots, N-1$ under H_j

$$\begin{aligned} \mathbb{E}\left\{Y[k]Y^*[k']|H_j\right\} &= \mathbb{E}\left\{\sum_{n=0}^{N-1}\sum_{n'=0}^{N-1} Y[n]Y^*[n']e^{-j2\pi\frac{kn}{N}} e^{j2\pi\frac{k'n'}{N}} |H_j\right\} \\ &= \sum_{n=0}^{N-1}\sum_{n'=0}^{N-1} \sigma_j^2 \delta[n-n'] e^{-j2\pi\frac{kn}{N}} e^{j2\pi\frac{k'n'}{N}} \\ &= \sigma_j^2 \sum_{n=0}^{N-1} e^{-j2\pi\frac{(k-k')n}{N}} \\ &= \sigma_j^2 \begin{cases} N & \text{if } k = k' \\ 0 & \text{if } k \neq k' \end{cases}, \\ &= N\sigma_j^2 \delta_{kk'}, \end{aligned} \tag{10.23a}$$

where $\delta_{kk'}$ is the Kronecker delta function defined in (5.65). Hence, under both hypotheses, $Y[k]$'s are a set of iid Gaussian random variables with zero mean and a common variance $N\sigma_j^2$, for $j = 0, 1$.

To make a detection decision at frequency k, we form the following decision statistic by smoothing the periodogram with a length L rectangular window $\Pi\left(\frac{l-k}{L}\right)$ centered at frequency k:

$$T_3(\mathbf{Y}) = \frac{1}{LN} \sum_{l=-(L-1)/2}^{(L-1)/2} |Y[k+l]|^2. \tag{10.24}$$

Let $T_3'(\mathbf{Y}) - LT_3(\mathbf{Y})$. Under each hypothesis II_j, $T_3'(\mathbf{Y})$ is a sum of the squares of L-independent zero-mean complex Gaussian random variables $\frac{Y[k]}{\sqrt{N}}$'s. Since $\frac{Y[k]}{\sqrt{N}}$ is a zero-mean Gaussian with variance σ_j^2, then it follows that $T_3'(\mathbf{Y})$ is a sum of the squares of L-independent zero-mean complex Gaussian random variables with equal variance σ_j^2 under each hypothesis H_j. Equivalently, $T_3'(\mathbf{Y})$ is the sum of the squares of $2L$-independent zero-mean real Gaussian random variables with identical variance $\frac{\sigma_j^2}{2}$ each. From Appendix B, distribution of such a $T_3'(\mathbf{y})$ is a gamma distribution with scale parameter $\alpha = L$ and a shape parameter $\beta = \sigma_j^2$, denoted as $G(L, \sigma_j^2)$, having the corresponding probability density function (pdf):

$$f_{T_3'}(t \mid H_0) = \begin{cases} \dfrac{t^{L-1} e^{-\frac{t}{\sigma_j^2}}}{\sigma_j^{2L}\Gamma(L)} & \text{if } t > 0 \\ 0 & \text{if } t < 0 \end{cases}. \tag{10.25}$$

Hence,

$$f_{T_3}(t \mid H_0) = \begin{cases} \dfrac{L(Lt)^{L-1} e^{-\frac{Lt}{\sigma_j^2}}}{\sigma_j^{2L}\Gamma(L)} & \text{if } t > 0 \\ 0 & \text{if } t < 0 \end{cases}. \tag{10.26}$$

Hence, for a threshold τ_k at frequency $F = \frac{k}{N}$,

$$\begin{aligned} \Pr(T_3(\mathbf{Y}) \geq \tau_k | H_j) &= \int_{\tau_k}^{\infty} f_{T_3}\left(t \mid H_j\right) dt \\ &= 1 - \int_0^{\tau_k} f_{T_3}\left(t \mid H_j\right) dt \\ &= 1 - \int_0^{\tau_k} L \frac{(Lt)^{L-1} e^{-\frac{Lt}{\sigma_j^2}}}{\sigma_j^{2L}\Gamma(L)} dt \end{aligned}$$

$$= 1 - \int_0^{\frac{L\tau_k}{\sigma_j^2}} \frac{\theta^{L-1} e^{-\theta}}{\Gamma(L)} d\theta$$

$$= 1 - \frac{\gamma\left(L, \frac{L\tau_k}{\sigma_j^2}\right)}{\Gamma(L)}, \tag{10.27}$$

where $\Gamma(x)$ and $\gamma(x,t)$ are *gamma function* and *incomplete gamma function* defined, respectively, as

$$\Gamma(x) = \int_0^{\infty} \theta^{x-1} e^{-\theta} d\theta \tag{10.28}$$

and

$$\gamma(x,t) = \int_0^{t} \theta^{x-1} e^{-\theta} d\theta. \tag{10.29}$$

From (10.18), under hypothesis H_0, each $Y[n]$ is an independent zero-mean Gaussian random variable with variance $\sigma_0^2 = N_0$. Hence, to ensure a false-alarm probability of $P_f(k)$ at frequency F_k, we need

$$P_f(k) = \Pr(T_3(\mathbf{Y}) \geq \tau_k | H_0)$$

$$= 1 - \frac{\gamma\left(L, \frac{L\tau_k}{N_0}\right)}{\Gamma(L)}. \tag{10.30}$$

Hence, the threshold required to achieve a false-alarm probability of $P_f(k)$ is given by

$$\tau_k = \frac{N_0 \gamma^{-1}\left(L, \Gamma(L)(1 - P_f(k))\right)}{L}, \tag{10.31}$$

where $\gamma^{-1}(x,.)$ is the inverse of incomplete gamma function $\gamma(x,.)$ in its second argument.

We may also derive the detection probability at frequency $F = F_k$ with this $P_f(k)$-level Neyman–Pearson detector, if we assume that the signal is also zero mean Gaussian distributed. Indeed, note that under hypothesis H_1, at frequency F_k, the observation samples consist of signal plus noise. If variance of complex signal samples is assumed to be σ_s^2, then the decision statistic T_3 earlier is simply a $G(L, \sigma_1^2)$ random variable where $\sigma_1^2 = \sigma_s^2 + N_0$. Hence, the detection probability at frequency F_k is

$$\begin{aligned} P_D\left(\tilde{\delta}_{NP}\right) &= \Pr(T_3(\mathbf{Y}) \geq \tau_k | H_1) \\ &= 1 - \frac{\gamma\left(L, \frac{L\tau_k}{\sigma_1^2}\right)}{\Gamma(L)} \\ &= 1 - \frac{\gamma\left(L, \frac{N_0}{\sigma_1^2}\gamma^{-1}\left(L, \Gamma(L)(1 - P_f(k))\right)\right)}{\Gamma(L)}. \end{aligned} \tag{10.32}$$

It must be mentioned that smoothing as stated earlier has to be carefully designed because it degrades the spectral resolution. As we saw in Chapter 6, periodogram achieves the smallest possible minimum spectral resolution among all nonparametric approaches, which is on the order of $\frac{1}{N}$. Having more signal samples would improve the periodogram's spectral resolution. Hence, one may expect to compensate for the degradation in the spectral resolution due to smoothing by increasing N as needed. However, in wideband spectrum sensing applications, this may not always be possible. Recall that in a wideband spectrum sensing system N is determined not only by the sampling frequency and the A/D converter but also by the total number of subbands. The more the number of subbands N_b, the shorter dwell time T_D in each subband will be resulting in only a smaller number of samples N. This, in turn, will limit the minimum separation between two spectral components that the signal detector can distinctly detect. Hence, selection of the smoothing window must be carefully considered in order to ensure enough spectral resolution for not to miss closely separated, yet distinct, signals. Based on a required minimum spectral resolution and the tolerable estimator variance (which determines the performance of the earlier Neyman–Pearson detector), one needs to select the proper windowing.

Example 10.1 (Receiver Operating Characteristics of Wideband Spectral Activity Detector for a BPSK Signal in Gaussian Noise) Figure 10.4 shows simulated receiver operating characteristics (ROC) of a wideband spectral activity detector when there is just a single BPSK signal in the spectrum band of interest. The signal is centered at $F_1 = 0.125$ with a null-to-null bandwidth of 0.05 in normalized frequency. Note that the ROC in Figure 10.4 is evaluated for the specific frequency $F_1 = 0.125$. Figure 10.4 shows the effect of frequency-domain smoothing of the periodogram. As can be seen, smoothing improves the detection probability as expected. ■

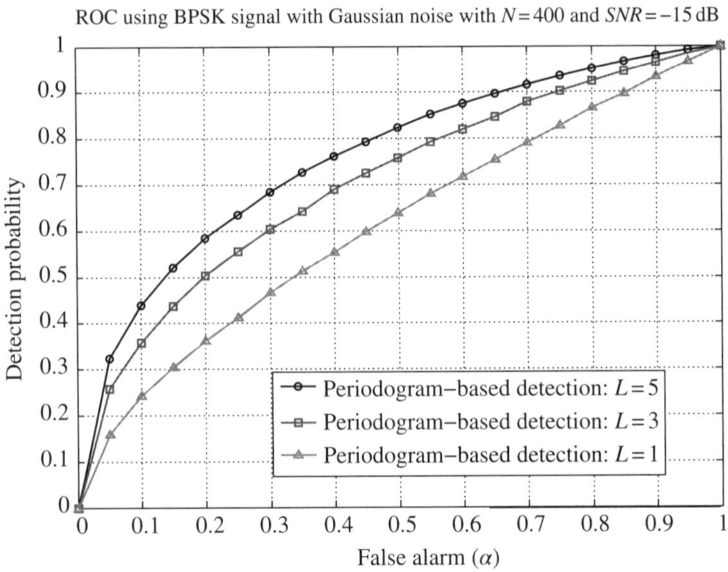

FIGURE 10.4 ROC of the spectral activity detection for a BPSK signal at $F_1 = 0.125$. Other parameters are $N = 400$ and $SNR = -15$ dB.

10.4 WAVELET TRANSFORM-BASED WIDEBAND SPECTRAL ACTIVITY DETECTION

Spectral activity detection approach in the previous section essentially boiled down to fist estimating the PSD $\hat{S}_y(F)$ of the received signal $y[n]$ and then thresholding. Since we did not know the center frequencies or bandwidths of the active signals over the subband, the only possibility was to employ a sliding window in the frequency domain. This can be computationally demanding and lack flexibility.

Recall from (10.18) that the sensed subband signal when there are active signals in it is of the form of

$$Y[n] = \sum_{k=1}^{K} X_k[n] + W[n] \quad \text{for } 0 \leq n \leq, N-1, \tag{10.33}$$

where X_k, for $k = 1, \ldots, K$, are arbitrary signals located at arbitrary carrier frequencies and noise W is assumed to be zero-mean white Gaussian with $\mathscr{S}_W(F) = N_0$. If noise and signals are all independent of each other, we have that

$$\mathscr{S}_Y(F) = \sum_{k=1}^{K} \mathscr{S}_{X_k}(F) + \mathscr{S}_W(F), \tag{10.34}$$

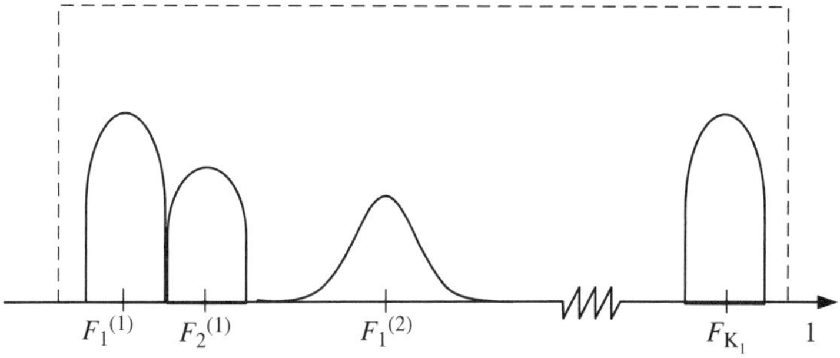

FIGURE 10.5 Wideband spectral activity detection in a subband can be reduced to a problem of detecting irregularities, discontinuities, or sharp edges in the estimated PSD of the sensed subband signal.

where $\mathscr{S}_{X_k}(F)$ is the PSD of the kth signal $X_k[n]$. The spectral activity detection problem is caused by the fact that these spectra $\mathscr{S}_{X_k}(F)$ of the active signals in the sensed subband are located at arbitrary frequencies and have arbitrary bandwidths, as in Figure 10.5, which are both unknown.

For the moment, assume that there is no noise and let us further suppose that the spectra $\mathscr{S}_{X_k}(F)$ do not overlap[1]. Then, essentially, spectral activities can be detected if we could detect the edges of the active signals in the spectrum $\mathscr{S}_Y(F)$ as shown in Fig. 10.5. Thus, spectral activity detection reduces to a problem of *edge detection* in spectrum $\mathscr{S}_Y(F)$, a real-valued function of the continuous variable F. Since we only have the estimate $\hat{S}_y(F)$ of the PSD $\mathscr{S}_Y(F)$, then our problem is edge detection in function $\hat{S}_y(F)$.

Let us assume that each $\mathscr{S}_{X_k}(F)$, for $1 \leq k \leq K$, is smooth over its support $\mathcal{B}_k \subseteq \left[-\frac{1}{2}, \frac{1}{2}\right]$. However, at the edges of the support set, there is an abrupt change. Then the spectrum $\mathscr{S}_Y(F)$ consists of this nonoverlapping set of smooth segments but discontinuities at the edges of the segments. In [141], it was observed that these edges in spectra can be detected by using the wavelet transform approach for singularity detection in functions developed in [142, 143]. The following discussion on wavelet transform-based spectral activity detection in a sensed subband signal is essentially an adaptation of the method proposed in [141] for our context. For more detailed discussion on the implementation of this approach and example performance results, we refer the reader to [141].

[1] Note that it is not possible to separate two signals that have overlapping spectra based simply on the PSD (i.e., energy detection) without further information on the signals. In wideband spectral activity detection that we consider here, such information is not presumed. If such information is available, then, in general, a multiuser detection or a source separation approach can be used to separate the signals with overlapping spectra (see e.g. [55]).

10.4.1 Wavelet Transform

Let us first briefly describe the wavelet transform of an arbitrary function $S(F)$ [142]. The wavelet transform of a function $S(F)$ can be defined as

$$WS(s,F) = S(F) \star \Psi_s(F), \tag{10.35}$$

where $WS()$ stands for the wavelet transform of function S, $\star$ denotes convolution, and $\Psi_s(F)$ is a *dilation* of a function $\Psi(F)$ defined as

$$\Psi_s(F) = \frac{1}{s}\Psi\left(\frac{F}{s}\right). \tag{10.36}$$

The possibly complex-valued function $\Psi(F)$ is called the *wavelet*. Note that since we are applying the wavelet transform to a function defined in the frequency domain F (although this discussion does not really depend on the meaning of these variables), we may consider $\Psi(F)$ as the Fourier transform of a function $\psi(\lambda)$ such that $\Psi(F) = \mathcal{F}\{\psi(\lambda)\}$. The function $\Psi(F)$ is a valid wavelet if and only if [144]

$$\int_{0}^{+\infty} \frac{|\psi(\lambda)|^2}{\lambda} d\lambda = \int_{-\infty}^{0} \frac{|\psi(\lambda)|^2}{|\lambda|} d\lambda = C_\Psi < +\infty, \tag{10.37}$$

where C_Ψ is some constant and we have used the fact that $\mathcal{F}\{\psi(F)\} = \psi(-\lambda)$. In most discussions involving transforms, the original functions are defined in time or space, and thus, Fourier transform will take them to frequency domain. However, in the case of wideband spectrum sensing, our original function is the PSD, which is already in the (normalized) frequency domain F.

If the wavelet is a valid wavelet, then the wavelet transform (10.35) is invertible. Indeed, the inverse transform is given by [142, 144]

$$S(F) = \frac{1}{C_\Psi} \int_{0}^{+\infty} \int_{-\infty}^{+\infty} WS(s,u)\psi_s^*(u-F)du\frac{ds}{s}. \tag{10.38}$$

The wavelet transform $WS(s,F)$ of $S(F)$ is a function of both the original variable F and the parameter s that is called the scale factor. Hence, $WS(s,F)$ is a function of two variables defined on the scale-frequency plane (S,F). For each scale factor $s = s_0$, the dilation is different, and we get the function $WS(s,F)$ evaluated along the line $s = s_0$ on the (S,F) plane.

There are several possible ways to compute the wavelet transform $WS(s,F)$ in (10.35). One approach is to observe that since convolution in the Fourier domain is the multiplication in the time domain, (10.35) can equivalently be written as

$$WS(s,F) = \mathcal{F}\left\{\mathcal{F}^{-1}\{S(F)\}\mathcal{F}^{-1}\{\Psi_s(F)\}\right\}. \tag{10.39}$$

But recall from Fourier transform properties that $\mathcal{F}\{\psi(s\lambda)\} = \frac{1}{s}\Psi\left(\frac{F}{s}\right) = \Psi_s(F)$. Hence, if we denote by $R(\lambda) = \mathcal{F}^{-1}\{S(F)\}$ the corresponding time-domain function of $S(F)$, then from (10.35),

$$WS(s,F) = \mathcal{F}\{R(\lambda)\psi(s\lambda)\}. \tag{10.40}$$

Since in our case $S(F)$ represents a PSD, then $R(\lambda)$ in fact corresponds to an autocovariance function, and λ represents the lag variable. For computational efficiency of Fourier transforms, it is convenient to select dyadic scale factors s such that $s = 2^j$ for some integer j.

10.4.2 Edge Detection with Wavelet Transform

Note that from the definition (10.35) of the wavelet transform, it is the convolution of the function $S(F)$ with the dilation $\Psi_s(F)$ of the wavelet $\Psi(F)$. Now, if we select $\Psi(F)$ to be the impulse response of a low-pass filter, then the wavelet transform $WS(s,F)$ is essentially a smoothed (low-pass filtered) version of function $S(F)$. The effect of low-pass filtering is to smooth out the sharp changes, or discontinuities, in $S(F)$. Of course, the extent to which a discontinuity in $S(F)$ will be smoothed out will depend on the particular $\Psi_s(F)$. For a fixed $\Psi(F)$, we will have different amounts of smoothing for different scales s. The wavelet transform $WS(s,F)$ will thus exhibit scale-dependent local variations at the discontinuity points of $S(F)$. Hence, detection of discontinuities, or edges, in $S(F)$ thus reduces to detecting local sharp variations in the wavelet transform $WS(s,F)$ [142].

These local sharp variations in wavelet transform $WS(s,F)$ of $S(F)$ at scale s can actually be detected by computing a slightly different wavelet transform of $S(F)$ at the first place. For analytical reasons in the following, we will assume that the wavelet $\Psi(F)$ has compact support and, for some m, it has m vanishing moments: That is,

$$\int_{-\infty}^{+\infty} F^k \Psi(F) dF = 0, \tag{10.41}$$

for all $k < m$. Let us define

$$\Psi^1(F) = \frac{d\Psi(F)}{dF} \tag{10.42}$$

and

$$\Psi^2(F) = \frac{d^2\Psi(F)}{dF^2}. \tag{10.43}$$

The wavelet transforms of $S(F)$ with respect to the wavelets $\Psi^1(F)$ and $\Psi^2(F)$ at scale s, by definition, are

$$W^1S(s,F) = S(F) \star \Psi_s^1(F) \tag{10.44}$$

and

$$W^2S(s,F) = S(F) \star \Psi_s^2(F), \tag{10.45}$$

where we have denoted the dilations of $\Psi^1(F)$ and $\Psi^2(F)$ at scale s, respectively, by $\Psi_s^1(F)$ and $\Psi_s^2(F)$:

$$\Psi_s^1(F) = \frac{1}{s}\Psi^1\left(\frac{F}{s}\right), \tag{10.46}$$

and

$$\Psi_s^2(F) = \frac{1}{s}\Psi^2\left(\frac{F}{s}\right). \tag{10.47}$$

The wavelet transforms $W^1S(s,F)$ and $W^2S(s,F)$ are easily seen to be related to the wavelet transform $WS(s,F)$. Indeed, from definitions (10.42) and (10.46),

$$\Psi_s^1(F) = \frac{1}{s}\left(\frac{d\Psi}{dF}\right)\left(\frac{F}{s}\right). \tag{10.48}$$

However,

$$\begin{aligned}\left(\frac{d\Psi_s}{dF}\right)(F) &= \frac{d}{dF}\left(\frac{1}{s}\Psi\left(\frac{F}{s}\right)\right)\\ &= \frac{1}{s^2}\left(\frac{d\Psi}{dF}\right)\left(\frac{F}{s}\right)\\ &= \frac{1}{s}\Psi_s^1(F),\end{aligned} \tag{10.49}$$

where the last step follows from (10.48). Hence, from (10.44) and (10.49), we have that [142]

$$\begin{aligned}W^1S(s,F) &= S\star\left(s\frac{d\Psi_s}{dF}\right)(F) = s\frac{d}{dF}\left(S\star\Psi_s\right)(F)\\ &= s\left(\frac{dWS(s,F)}{dF}\right)(F).\end{aligned} \tag{10.50}$$

Similarly, it can be shown that

$$\begin{aligned}W^2S(s,F) &= S\star\left(s^2\frac{d^2\Psi_s}{dF^2}\right)(F) = s^2\frac{d^2}{dF^2}\left(S\star\Psi_s\right)(F)\\ &= s^2\left(\frac{d^2WS(s,F)}{dF^2}\right)(F).\end{aligned} \tag{10.51}$$

Hence, $W^1S(s,F)$ and $W^2S(s,F)$ are in fact proportional to the first and second derivatives of the wavelet transform $WS(s,F)$, respectively. But then it follows that the local extrema of $W^1S(s,F)$ as a function of F for a fixed s must correspond to the inflection points of the wavelet transform $WS(s,F)$, which are an indication of the possible discontinuities and edges in $S(F)$. Similarly, the local extrema of $W^1S(s,F)$ also correspond to the zero-crossings of $W^2S(s,F)$. Hence, the edges and discontinuities in $S(F)$ can be detected by detecting the local extrema or zero-crossings, respectively, of $W^1S(s,F)$ and $W^2S(s,F)$ [142, 145]. However, it was pointed out in [142] that while slow and high variation points of $WS(s,F)$ correspond to local minima and maxima of $|W^1S(s,F)|$, respectively, they are both indicated as zero-crossings in $W^2S(s,F)$. Hence, $W^2S(s,F)$ does not provide a direct way to distinguish between slow and high local variations in $WS(s,F)$. Since the edges in $S(F)$ are reflected as high variation points in $WS(s,F)$, for edge detection, then it makes particular sense to focus on the local maxima of the wavelet transform modulus $|W^1S(s,F)|$ [142].

10.4.3 Spectral Activity Detection Based on Edge Detection

Following [141], we may thus perform spectral activity detection by computing the wavelet transform $W^1\hat{S}_y(s,F)$ of the PSD estimate $\hat{S}_y(F)$ and then finding the local maxima of the wavelet transform modulus $\left|W^1\hat{S}_y(s,F)\right|$. Hence, if we denote by $F_{k'}$, for $1 \le k' \le K'$, the edges present in the sensed subband signal $Y[n]$, where $F_1 < F_2 < \cdots < F_{K'}$, then they can be found as [141]

$$\{F_{k'}\}_{k'=1}^{K'} = \text{Local maxima of}_{F} \left|W^1\hat{S}_y(s,F)\right|. \tag{10.52}$$

Once these edge boundaries $\{F_{k'}\}_{k'=1}^{K'}$ are determined, then they can be mapped to spectral activity regions $\mathcal{B}_k$, for $1 \le k \le K$. This modified approach to wideband spectral activity detection is shown in Figure 10.6.

Since the PSD estimate $\hat{S}_y(F)$ contains noise PSD, in addition to the signal PSD, one may want to attempt to reduce the effect of noise on the above edge detection procedure. It was observed in [142] that for example, the average density of wavelet transform modulus maxima of white Gaussian noise is inversely proportional to the scale s. Hence, at large scales, one may expect to see less of modulus maxima due to white noise. In [142], based on this observation, it was pointed out that *denoising* approaches can be developed by exploiting the modulus maxima behavior across

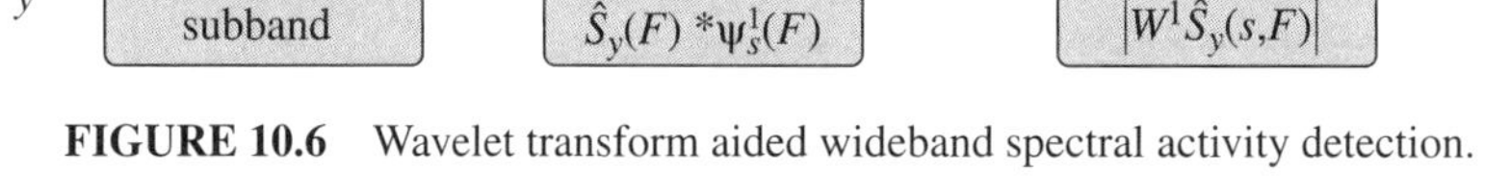

FIGURE 10.6 Wavelet transform aided wideband spectral activity detection.

multiple scales. This was used in [141, 146] to help reduce the effects of noise on edge detection by considering the multiscale wavelet product

$$W_J^1 \hat{S}_y(F) = \Pi_{j=1}^J W^1 \hat{S}_y(2^j, F), \tag{10.53}$$

where J dyadic scales of the form of $s = 2^j$, for $1 \leq j \leq J$ were employed. Since the noise-induced local maxima of $W^1 \hat{S}_y(2^j, F)$ will be randomly located for different scales $s = 2^j$, it can be expected that their effect will be reduced when one looks at the local maxima of the modulus of $W_J^1 \hat{S}_y(F)$:

$$\{F_{k'}\}_{k'=1}^{K'} = \text{Local maxima of}_F \left| W_J^1 \hat{S}_y(F) \right|. \tag{10.54}$$

However, in many applications of wavelet transform to detect singularities and edges, the original function is defined in the time or spatial (e.g., in image processing) domain. Hence, the assumption of white noise can at least be conceptually justifiable. On the other hand, in spectral activity detection, the noise in the original signal is the periodogram PSD estimate of noise. Hence, if time-domain signals were to be corrupted by white noise, then noise in $\hat{S}_y(F)$ will be constant noise. Hence, one may expect that the chance of noise-induced local maxima of wavelet transform to be less in this case compared to those in traditional applications of wavelet transform [141].

Determining that a local maximum of the wavelet transform modulus does indeed correspond to an actual spectral activity still needs a detection stage, which is a form of thresholding. This threshold can be selected based on an optimality criterion such as Neyman–Pearson as in the previous section, if it is possible to model the distribution of $|W^1 \hat{S}(s, F)|$. However, even if original signals and noise were to be Gaussian and if we assume $\hat{S}$ is the periodogram estimate of the PSD, then still, $W^1 \hat{S}(s, F)$ is not Gaussian but a linear transform of a chi-square random variable. The CFAR detectors of Chapter 4 appear to be particularly appealing in these scenarios due to possible lack of prior knowledge of signal distributions.

10.5 WIDEBAND SPECTRAL ACTIVITY DETECTION IN NON-GAUSSIAN NOISE

Thresholding of an estimate of the power spectrum of sensed data as discussed in the previous sections can be justified when noise is known to be Gaussian. In wideband spectrum sensing, however, the assumption of Gaussian noise may not always be true. With a wider spectrum band, it is more likely that the sensed signal Y may contain a mix of heterogeneous spectral activities from a variety of electromagnetic sources leading to outliers, jammers, and interferers that can better be modeled as non-Gaussian [147–149].

In a broad sense, there are two paradigms in statistic signal processing that handles non-Gaussian noise. First is the direct approach of non-Gaussian detection in which either a *known* non-Gaussian distribution or a nonparametric model is assumed [52, 148, 150]. The second approach, called the *robust detection*, is to allow the noise model to deviate from a certain nominal distribution, which usually is taken to be a Gaussian distribution. The deviation from the nominal noise distribution is modeled as being caused by an *unknown* contaminating distribution belonging to a certain known class of distributions [51, 151].

10.5.1 Arbitrary but Known Noise Distribution

Let us now modify the spectral activity detection problem in a wideband cognitive radio based on sensed observations in a selected subband we posed in (10.1) to allow for additive non-Gaussian noise:

$$\begin{aligned} H_0 \text{ (signal absent)} &: \mathbf{y}[n] = \mathbf{w}[n] \\ &\text{versus} \\ H_1 \text{ (signal present)} &: \mathbf{y}[n] = \theta\mathbf{x} + \mathbf{w}[n], \end{aligned} \tag{10.55}$$

where $\mathbf{x}[n] = \sum_{k=1}^{K} \mathbf{x}_k[n]$ with x_k's for $k = 1, \ldots, K$ are arbitrary signals located at arbitrary carrier frequencies. The signal vector $\mathbf{x}[n] = (x[0], \ldots, x[N-1])^T$ is assumed to be random with an arbitrary and unknown density function $f_X(\mathbf{x})$ except possibly for the mean and covariance matrix denoted, respectively, by $\bar{\mathbf{x}}$ and $\Sigma_{\mathbf{X}}$. The noise vector $\mathbf{w} = (w[0], \ldots, w[N-1])^T$ is allowed to have an arbitrary but known noise distribution characterized by the density function $f_{\mathbf{W}}(\mathbf{w})$. We have allowed the signal-present hypothesis H_1 to be a composite hypothesis by introducing the parameter θ. This is of interest since there will be inevitable fluctuations in the received signal level. It is also likely that this signal amplitude may not be known accurately during the spectral activity detection process. Hence, we model it as an unknown parameter $\theta > 0$. Obviously, $\theta = 0$ can be treated as the hypothesis H_0, and hence, it is a simple hypothesis, whereas H_1 is a composite hypothesis.

Given that signal $\mathbf{x}[n]$ is actually the sum of K different signals $\mathbf{x}_k[n]$'s, it may perhaps be more appropriate to introduce a different parameter θ associated with each of the signals $x_k[n]$. Indeed, this makes sense if one were to interpret θ as being caused by signal fading and attenuation encountered by the individual signals. As it will be seen later, the solution approach we develop will easily be amenable to this scenario.

With the previous definitions and assumptions, the spectral activity detection problem (10.55) is of the form of random signal detection in additive non-Gaussian noise we treated in Chapter 4, specifically in Section 4.3.4.3. As we saw in Chapter 4, the optimal tests for this problem will then be decision rules based on the following likelihood ratio:

$$\begin{aligned} L(\mathbf{y}, \theta) &= \frac{f_{\mathbf{Y}|\theta}(\mathbf{y}|\theta)}{f_{\mathbf{Y}|\theta=0}(\mathbf{y}|\theta = 0)} \\ &= \frac{\mathbb{E}\{f_{\mathbf{W}}(\mathbf{y} - \theta\mathbf{X})\}}{f_{\mathbf{W}}(\mathbf{y})}, \end{aligned} \tag{10.56}$$

where expectation is with respect to the distribution of $\mathbf{X}$.

We already know from Chapter 4 that uniformly most powerful optimal tests may or may not exist for this problem depending on the exact form of the densities $f_{\mathbf{W}}(\mathbf{w})$ and $f_{\mathbf{X}}(\mathbf{x})$. Recall that a uniformly most powerful test is a test that will ensure the maximum detection probability regardless of the actual value of θ among all decision rules that do not exceed a given false-alarm probability threshold. Even when the problem is known to have a UMP test, we may not be able to find or implement it easily. In Chapter 4, we put forth the notion of locally optimal, or locally most powerful, tests as an alternative in those situations. Here, we may pursue that approach.

From (4.77), the locally optimal decision rules for the spectral activity detection problem (10.55) will have the form

$$\delta^{LO}(\mathbf{y}) = \begin{cases} 1 & & > \\ \gamma & \text{if } \frac{\partial}{\partial\theta} L(\mathbf{y},\theta)\big|_{\theta=0} & = \ \eta, \\ 0 & & < \end{cases} \tag{10.57}$$

where $\eta \geq 0$ and $0 \leq \gamma \leq 1$ are to be chosen so that $P_F(\delta^{LO}) = \alpha$ for a given false-alarm rate α.

From (4.123c) in Chapter 4, the locally optimal decision statistic required in (10.57) is given by

$$\frac{\partial}{\partial\theta} L(\mathbf{y},\theta)\bigg|_{\theta=0} = -\frac{\bar{\mathbf{x}}^T \mathbf{f}'_{\mathbf{W}}(\mathbf{y})}{f_{\mathbf{W}}(\mathbf{y})}, \tag{10.58}$$

where $\mathbf{f}'_{\mathbf{W}}(\mathbf{w}) = \frac{\partial f_{\mathbf{W}}}{\partial \mathbf{w}} = \left(\frac{\partial f_{\mathbf{W}}}{\partial w_1}, \frac{\partial f_{\mathbf{W}}}{\partial w_2}, \ldots, \frac{\partial f_{\mathbf{W}}}{\partial w_N}\right)^T$.

Hence, if we knew the noise density function and the mean signal vector $\bar{\mathbf{x}}$, then we may possibly implement a locally optimal spectral activity detector of the form of (10.57) with the decision statistic (10.58). However, having the knowledge of $\bar{\mathbf{x}}$ may be questionable in the case of wideband spectrum sensing. After all, the reason we are modeling the signal as being random is due to the fact that we do not have much definitive knowledge about the signal. In such cases, the most common knowledge could be limited to that of a zero mean. Of course, when the signal mean is zero so that $\bar{\mathbf{x}} = \mathbf{0}$, the decision statistic (10.58) vanishes and the first-order derivative-based locally optimal detection does not work. The most likely quantity that one may justifiably be known is a second-order statistic related to signal energy since these can reasonably be estimated. As we saw in Chapter 4, in such cases, we may propose a locally optimal detector based on the second-order derivatives as

$$\delta^{LO}(\mathbf{y}) = \begin{cases} 1 & & > \\ \gamma & \text{if } \frac{\partial^2}{\partial\theta^2} L(\mathbf{y},\theta)\bigg|_{\theta=0} & = \ \eta. \\ 0 & & < \end{cases} \tag{10.59}$$

For the spectral activity detection problem at hand, it follows from (4.124d) in Chapter 4 that the relevant decision statistic is

$$\frac{\partial^2}{\partial \theta^2} L(\mathbf{y}, \theta)\bigg|_{\theta=0} = \frac{\mathrm{Tr}\left(\mathbf{F}''_{\mathbf{W}}(\mathbf{y})\Sigma_{\mathbf{X}}\right)}{f_{\mathbf{W}}(\mathbf{y})}, \tag{10.60}$$

where $\mathbf{F}''_{\mathbf{W}}(\mathbf{w}) = \frac{\partial}{\partial \mathbf{w}} \mathbf{f}'_{\mathbf{W}}(\mathbf{w})$ and we have assumed that indeed $\bar{\mathbf{x}} = \mathbf{0}$.

If one were to assume that the noise is independent over the samples,

$$f_{\mathbf{W}}(\mathbf{y}) = \prod_{n=1}^{N} f_{W_n}(y_n), \tag{10.61}$$

where $f_{W_n}(w)$ is the marginal noise distribution, then we saw in Chapter 4 that the decision statistic (10.60) can be written in the explicit form in (4.132b):

$$T_2^{(LO)}(\mathbf{y}) = \sum_{n=1}^{N} h_n^{(LO)}(y_n)\Sigma_{n,n}^{(x)} + \sum_{n=1}^{N} \sum_{n'=1,n'\neq n}^{N} g_{n'}^{(LO)}(y'_n) g_n^{(LO)}(y_n)\Sigma_{n',n}^{(x)} \tag{10.62a}$$

$$= \sum_{n=1}^{N} h_n^{(LO)}(y_n)\rho_0^{(x)} + \sum_{n=1}^{N} \sum_{n'=1,n'\neq n}^{N} g_{n'}^{(LO)}(y'_n) g_n^{(LO)}(y_n)\rho_{n'-n}^{(x)}, \tag{10.62b}$$

where in (10.62b) we have assumed that the signal X is wss so that the covariance matrix $\Sigma_{\mathbf{X}}$ is Toeplitz with $\Sigma_{n',n}^{(x)} = \rho_{n',n}^{(x)} = \rho_{n'-n,0}^{(x)} \triangleq \rho_{n'-n}^{(x)}$ for a covariance sequence $\{\rho\}$ and $g_n(.)$ and $h_n(.)$ are two instantaneous nonlinearities determined by the noise marginal densities $f_{W_n}(w)$ as follows:

$$g_n^{(LO)}(w) = \frac{f'_{W_n}(w)}{f_{W_n}(x)} \tag{10.63}$$

and

$$h_n^{(LO)}(x) = \frac{f''_{W_n}(x)}{f_{W_n}(n)}. \tag{10.64}$$

The similarity of (10.62b) to (10.7) and (10.8) can be better viewed by defining transformed observations

$$\tilde{y}_n \triangleq g_n^{(LO)}(y_n) \tag{10.65}$$

and

$$\hat{y}_n \triangleq h_n^{(LO)}(y_n), \tag{10.66}$$

so that (10.62b) becomes

$$T_2^{(LO)}(\mathbf{y}) = \sum_{n=1}^{N} \hat{y}_n \rho_0^{(x)} + \sum_{n=1}^{N} \sum_{n'=1, n' \neq n}^{N} \tilde{y}_{n'} \tilde{y}_n \rho_{n'-n}^{(x)} \tag{10.67a}$$

$$= \sum_{n=1}^{N} \sum_{n'=1}^{N} \tilde{y}_{n'} \tilde{y}_n \rho_{n'-n}^{(x)} + \sum_{n=1}^{N} \hat{y}_n \rho_0^{(x)} - \tilde{y}_n^2 \rho_0^{(x)} \tag{10.67b}$$

$$= \sum_{n=1}^{N} \sum_{n'=1}^{N} \tilde{y}_{n'} \tilde{y}_n \int_{-1/2}^{1/2} \mathcal{S}_x(F) e^{j2\pi F(n'-n)} dF + \sum_{n=1}^{N} \rho_0^{(x)} \left(\hat{y}_n - \tilde{y}_n^2 \right) \tag{10.67c}$$

where (10.67c) follows from the fact that autocovariance sequence $\{\rho^{(x)}\}$ of the signal is given by the inverse Fourier transform of its PSD function $\mathcal{S}_x(F)$. By denoting

$$T_{2,0}^{(LO)}(\mathbf{y}) = \rho_0^{(x)} \sum_{n=1}^{N} \left(\hat{y}_n - \tilde{y}_n^2 \right), \tag{10.68}$$

we have from (10.67c) that

$$T_2^{(LO)}(\mathbf{y}) = \int_{-1/2}^{1/2} \left(\sum_{n=1}^{N} \tilde{y}_n e^{-j2\pi Fn} \right) \left(\sum_{n'=1}^{N} \tilde{y}_{n'} e^{-j2\pi Fn'} \right)^* \mathcal{S}_x(F) dF + T_{2,0}^{(LO)}(\mathbf{y})$$

$$= \int_{-1/2}^{1/2} \left| \tilde{Y}(F) \right|^2 \mathcal{S}_x(F) dF + T_{2,0}^{(LO)}(\mathbf{y}) \tag{10.69a}$$

$$= \int_{-1/2}^{1/2} \hat{S}_{\tilde{y}}(F) \mathcal{S}_x(F) dF + T_{2,0}^{(LO)}(\mathbf{y}), \tag{10.69b}$$

where (10.69a) follows by defining $\tilde{Y}(F)$ as the DTFT of the observation sequence

$$\tilde{Y}(F) = \sum_{n=1}^{N} \tilde{y}_n e^{-j2\pi Fn}, \tag{10.70}$$

and in (10.69b), we have defined the periodogram of the transformed observation sequence $\{\tilde{y}_n\}_{n=1}^{N}$ to be $\hat{S}_{\tilde{y}}(F)$:

$$\hat{S}_{\tilde{y}}(F) = \left| \tilde{Y}(F) \right|^2. \tag{10.71}$$

Comparison of (10.69b) with (10.8) shows that aside from the second term $T_{2,0}^{(LO)}(\mathbf{y})$, the two decision statistics are both correlations of a periodogram estimate with the signal power spectrum $\mathscr{S}_x(F)$. In the non-Gaussian noise scenario, the periodogram, however, must be computed after first nonlinearly transforming the sensed observation sequence.

This motivates us to consider a similar framework to that of Section 10.3 to detect spectral activity in the presence of known, but non-Gaussian, noise. In particular, in wideband spectrum sensing, the signal covariance sequence and equivalently the signal spectrum $\mathscr{S}_x(F)$ are unknown, and the signal of interest X is actually a sum of K signals located at different frequencies. Then, we may make the same arguments as we did in Section 10.3 to justify essentially a decision statistic for each frequency F_0 of the form of

$$T_{F_0}(\mathbf{y}) = \int_{F_0-\Delta F}^{F_0+\Delta F} \hat{S}_{\tilde{y}}(F)dF + T_{2,0}^{(LO)}(\mathbf{y}), \tag{10.72}$$

where ΔF defines a window of frequencies.

It must, however, be pointed out that the second term $T_{2,0}^{(LO)}(\mathbf{y})$ in (10.69b) makes its form different from the Gaussian noise case. In the case of Gaussian noise, $T_{2,0}^{(LO)}(\mathbf{y})$ is a constant that does not depend on the observations. Hence, it can be absorbed into the threshold, and thus, direct thresholding of the (spectrum) correlator output is indeed optimal. However, for other noise distributions, $T_{2,0}^{(LO)}(\mathbf{y})$ can very well be a function of the observation $\mathbf{y}$. Then it must be included in the decision statistic (10.72).

Example 10.2 (Gaussian Mixture Noise) A commonly used non-Gaussian noise model is the Gaussian mixture noise. If all noise samples are to have the same marginal distribution of a Gaussian mixture with m mixture components, then the noise pdf will be of the form of

$$f_W(w) = \sum_{i=1}^{m} \kappa_i \mathcal{N}(\mu_i, \sigma_i^2)(w), \tag{10.73}$$

where $\mathcal{N}(\mu, \sigma^2)(.)$ denotes the pdf of a Gaussian distribution with mean μ and variance σ^2 and $\{\kappa_i\}_{i=1}^m$ are the mixing proportions. It is usually the case that these means and variances are modeled as $\mu_i = a_i\mu$ and $\sigma_i^2 = a_i^2\sigma^2$, for some set of coefficients a_i's [149]. For the noise model of (10.73), the nonlinear transformations $g_n^{(LO)}(w) = g^{(LO)}(w)$ and $h_n^{(LO)}(w) = g^{(LO)}(w)$, for $1 \le n \le N$, can be shown to be

$$g^{(LO)}(w) = \frac{\sum_{i=1}^{m} \frac{\kappa_i(w-\mu_i)}{\sigma_i^3} e^{-\frac{(w-\mu_i)^2}{2\sigma_i^2}}}{\sum_{i=1}^{m} \frac{\kappa_i}{\sigma_i} e^{-\frac{(x-\mu_i)^2}{2\sigma_i^2}}} \tag{10.74}$$

and

$$h^{(LO)}(w) = \frac{\sum_{i=1}^{m} \frac{\kappa_i}{\sigma_i^3} \left[\frac{(w-\mu_i)^2}{\sigma_i^2} - 1 \right] e^{-\frac{(w-\mu_i)^2}{2\sigma_i^2}}}{\sum_{i=1}^{m} \frac{\kappa_i}{\sigma_i} e^{-\frac{(w-\mu_i)^2}{2\sigma_i^2}}}. \tag{10.75}$$

Note that when $m = 1$, above model, of course, reduces to the usual iid Gaussian noise. Indeed, for $m = 1$ with zero-mean assumption, (10.74) and (10.75) simplify to

$$g^{(LO)}(w) = \frac{w}{\sigma^2} \tag{10.76}$$

and

$$h^{(LO)}(w) = \frac{1}{\sigma^2} \left[\frac{w^2}{\sigma^2} - 1 \right]. \tag{10.77}$$

where we have set $\kappa_1 = 1$. Hence,

$$\tilde{y}_n = \frac{y_n}{\sigma^2}, \tag{10.78}$$

so that

$$\tilde{Y}(F) = \frac{Y(F)}{\sigma^2}. \tag{10.79}$$

Similarly,

$$\hat{y}_n = \frac{1}{\sigma^2} \left[\frac{y_n^2}{\sigma^2} - 1 \right], \tag{10.80}$$

so that

$$T_{2,0}^{(LO)}(\mathbf{y}) = -\frac{N\rho_0^{(x)}}{\sigma^2}. \tag{10.81}$$

The decision statistic then also reduces to the familiar form of

$$T_2^{(LO)}(\mathbf{y}) = \frac{1}{\sigma^2} \int_{-1/2}^{1/2} \hat{S}_y(F) \mathscr{S}_x(F) dF - \frac{N\rho_0^{(x)}}{\sigma^2}, \tag{10.82}$$

where $\hat{S}_y(F) = |Y(F)|^2$ is the periodogram estimate of the observation power spectrum and the second term is a constant that can be absorbed into the threshold. ■

Example 10.3 (Gaussian–Laplacian Mixture Noise) Another non-Gaussian noise model that has been considered in literature is that of a mixture of Gaussian and Laplacian noise [150]. Again, let us assume that noise samples are iid with a common marginal distribution. If the common marginal distribution is a Gaussian–Laplacian mixture, then its pdf is of the form of

$$f_W(w) = (1-\epsilon)\frac{1}{\sqrt{2\pi\sigma^2}}e^{-\frac{w^2}{2\sigma^2}} + \epsilon\frac{\lambda}{2}e^{-\lambda|w|}. \tag{10.83}$$

The nonlinear transformations $g_n^{(LO)}(w) = g^{(LO)}(w)$ and $h_n^{(LO)}(w) = g^{(LO)}(w)$, for $1 \leq n \leq N$, corresponding to the case of Gaussian–Laplacian mixture noise model are given by [150, 152, 153]

$$g^{(LO)}(w) = \frac{\frac{\lambda^2\epsilon}{2}sgn(w)e^{-\lambda|w|} + \frac{(1-\epsilon)w}{\sigma^3\sqrt{2\pi}}e^{\frac{-w^2}{2\sigma^2}}}{\frac{\lambda\epsilon}{2}e^{-\lambda|w|} + \frac{1-\epsilon}{\sqrt{2\pi\sigma^2}}e^{-\frac{w^2}{2\sigma^2}}} \tag{10.84}$$

and

$$h^{(LO)}(w) = \frac{\frac{\lambda^2\epsilon}{2}\left(\lambda - 2\delta_D(w)\right)e^{-\lambda|w|} + \frac{1-\epsilon}{\sigma^3\sqrt{2\pi}}\left(\frac{w^2}{\sigma^2} - 1\right)e^{-\frac{w^2}{2\sigma^2}}}{\frac{\lambda\epsilon}{2}e^{-\lambda|w|} + \frac{1-\epsilon}{\sqrt{2\pi\sigma^2}}e^{-\frac{w^2}{2\sigma^2}}}, \tag{10.85}$$

where ϵ is the Laplace contamination rate, λ is the parameter of the Laplace noise, σ^2 is the variance of the zero-mean Gaussian mixture component, $\delta_D(.)$ is the Dirac delta function, and $sgn(.)$ is the *signum function* defined as

$$sgn(x) = \begin{cases} 1 & \text{if } x > 0 \\ -1 & \text{if } x < 0 \end{cases}.$$

When $\epsilon = 0$, (10.83) is simply the Gaussian noise, and (10.84) and (10.85) reduce to (10.76) and (10.77), respectively. On the other hand, when $\epsilon = 1$, (10.83) models iid Laplacian noise for which the two nonlinearities are given by

$$g^{(LO)}(w) = \lambda sgn(w) \tag{10.86}$$

and

$$h^{(LO)}(w) = \lambda^2 - 2\lambda\delta_D(w). \tag{10.87}$$

Thus, for Laplacian noise, the transformed observations are of the form of

$$\tilde{y}_n = \lambda sgn(y_n), \tag{10.88}$$

so that

$$\tilde{Y}(F) = \lambda \sum_{n=1}^{N} sgn(y_n) e^{-j2\pi Fn}, \tag{10.89}$$

and

$$\hat{y}_n = \lambda^2 - 2\lambda\delta_D(y_n), \tag{10.90}$$

so that

$$T_{2,0}^{(LO)}(\mathbf{y}) = -2\lambda\rho_0^{(s)} \sum_{n=1}^{N} \delta_D(y_n). \tag{10.91}$$

Since $\Pr\{T_{2,0}(\mathbf{y}) = 0\} = 1$ due to the continuous nature of the Laplacian noise pdf, in this case, essentially, the decision statistic can be taken to be

$$T_2^{(LO)}(\mathbf{y}) = \frac{1}{\sigma^2} \int_{-1/2}^{1/2} |\tilde{Y}(F)|^2 \mathscr{S}_x(F) dF \tag{10.92}$$

where $\tilde{Y}(F)$ is the DTFT of the sign sequence of the observations given by (10.89).

Figure 10.7 shows the conventional periodogram and the modified periodogram based on the above transformed variables according to the locally optimal detection in non-Gaussian noise when noise is indeed a Gaussian–Laplacian mixture distribution. Figures 10.7(a) and (b) show these periodograms when $\epsilon = 0.1$ and $\epsilon = 0.9$, respectively, with $N = 100$ samples. In all cases, a sinusoidal signal located at $F = 0.2$ normalized frequency is present at an $SNR = -5$ dB. These figures show that the conventional periodogram seems to be erratic in comparison to the modified periodogram that takes into account the non-Gaussian nature of noise distribution. It is thus reasonable to expect spectral activity detection based on the latter to provide better performance than that based on the former. ■

10.5.2 Robust Spectral Activity Detection

In practice, the periodogram estimate of the observation PSD will be computed based on a DFT $Y(k;N)$ in place of $Y(F;N)$, where $0 \leq k \leq N-1$ specifies the discrete frequencies corresponding to $F = \frac{k}{N}$. The discrete periodogram is then given by

$$\hat{S}_y(k) = \frac{1}{N} |Y(k;N)|^2. \tag{10.93}$$

Let us assume for the moment that the observation data is of the form of

$$y[n] = \mathbf{c}_n^H \mathbf{v} + w[n] \quad \text{for } n = 0, \ldots, N-1, \tag{10.94}$$

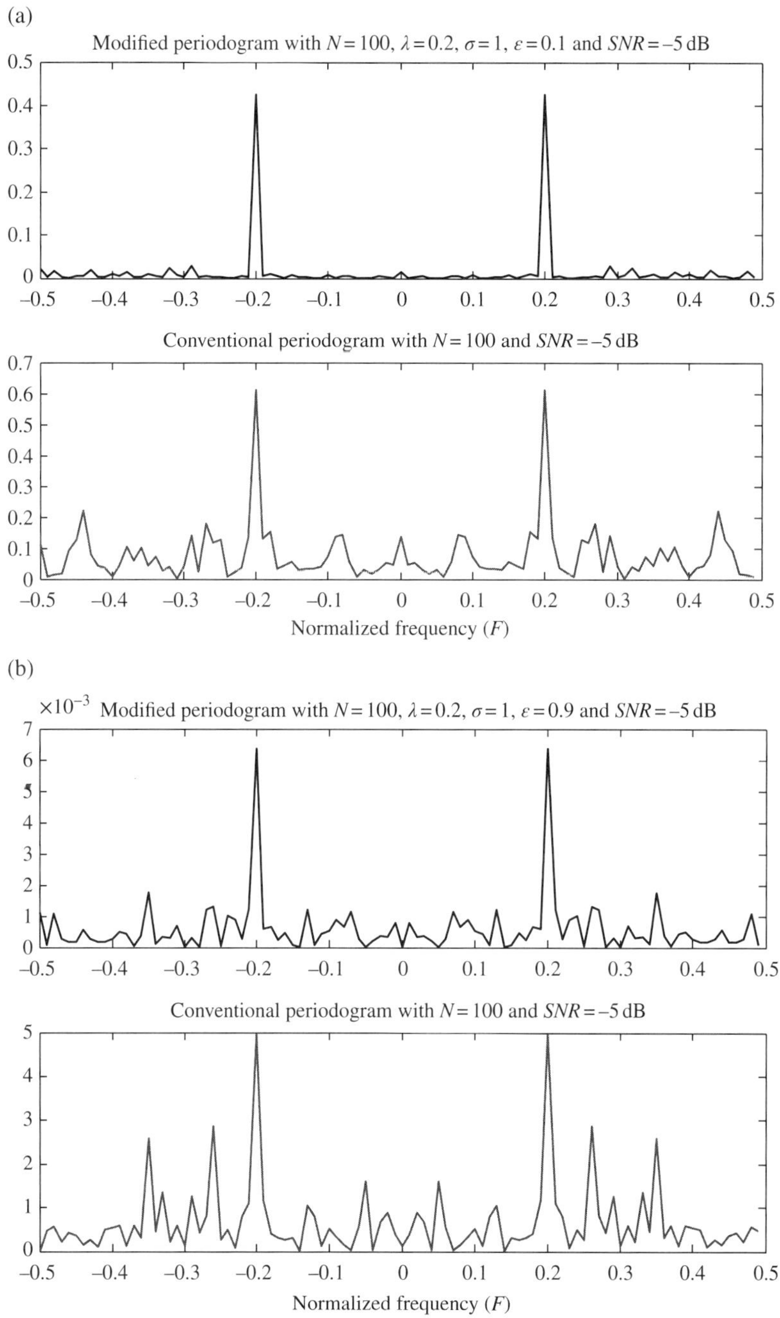

FIGURE 10.7 Conventional and modified periodograms in the presence of Gaussian–Laplacian mixture noise. (a) Conventional and modified periodograms in the presence of Gaussian-Laplacian mixture noise with $\epsilon = 0.1$. (b) Conventional and modified periodograms in the presence of Gaussian–Laplacian mixture noise with $\epsilon = 0.9$.

for some unknown vector $\mathbf{v}$, where $\mathbf{c}_n = \frac{1}{\sqrt{N}}[1, e^{-j2\pi\frac{n}{N}}, \ldots, e^{-j2\pi(N-1)\frac{n}{N}}]^T$, for $n{=}0, \ldots, N-1$. Let us denote $\mathbf{C} = [\mathbf{c}_1, \ldots, \mathbf{c}_N]$. Note that $\mathbf{C}$ is a DFT matrix and $\mathbf{C}\mathbf{C}^H = \mathbf{C}^H\mathbf{C} = \mathbf{I}_N$.

Essentially, (10.94) implies that $y[n]$ is a weighted sum of harmonic complex exponentials and the coefficient vector $\mathbf{v}$ specifies the strength of each of the harmonic frequency component contained in $y[n]$. But the strengths, or power, of the frequencies in a signal are what the PSD of a signal supposed to show. Hence, the periodogram (10.93) in this case will estimate the strengths of these coefficients specified by the vector $\mathbf{v}$. Not surprisingly, in fact, if we were to pose a *least squares estimation problem* for the coefficients $\mathbf{v}$ such that

$$\mathbf{v}^* = \arg\min_{\mathbf{v}\in\mathbb{C}^N} \|\mathbf{y} - \mathbf{C}^H\mathbf{v}\|^2, \tag{10.95}$$

then the solution $\mathbf{v}^* = (v_1^*, \ldots, v_N^*)^T$ to (10.95) is

$$\mathbf{v}^* = \mathbf{C}\mathbf{y}, \tag{10.96}$$

which is indeed the DFT of the observation sequence so that

$$\hat{S}_y(k) = |v_k^*|^2. \tag{10.97}$$

Hence, an alternative way to view the periodogram estimate is that it is the *least squares estimate* of the true power spectrum. Moreover, if noise $w[n]$ in (10.94) were to be Gaussian, clearly, it is also the ML estimate of the power spectrum.

This interpretation of the periodogram as the least squares estimate of the true PSD then immediately leads us to expect that the periodogram PSD estimate must suffer from all the well-known deficiencies of the least squares estimates. Among them is the lack of robustness against outliers and heavy-tailed noise distributions [51]. In the case of wideband cognitive radio spectrum sensing, it is very likely that noise may not always be Gaussian. In particular, deliberate jammers, unintentional interferers, and other electromagnetic radiation may cause noise to perhaps deviate around a certain nominal distribution. Often, we may be able to justify the nominal noise distribution to be a certain Gaussian but with possible deviations from it according to an unknown distribution with non zero probability. Other times, even the nominal distribution may be non-Gaussian with the deviations from it following some other unknown distributions as well. The least squares estimates, and thus the periodogram estimate of the PSD we used in earlier sections, will not be able to withstand such deviations from Gaussianity in that the performance will degrade severely. In other words, these are not robust PSD estimates against noise deviations.

There are various robust spectral estimation methods that can be found in literature. In the following, we may briefly introduce some of the commonly used approaches followed by a description of a *robust wideband spectral activity detection* framework based on one of them.

10.5.2.1 Regularization A commonly used approach to robustify, in the sense of improving the smoothness and the generalization capability, the least squares estimates is to modify the least squares cost function (10.95) by including a so-called regulation term:

$$\mathbf{v}^* = \arg \min_{\mathbf{v}\in\mathbb{C}^N} \|\Lambda\mathbf{v}\|^2 + \|\mathbf{y} - \mathbf{C}^H\mathbf{v}\|^2, \tag{10.98}$$

where Λ is called a regularization, or Tikhonov, matrix. Often, (10.98) is called *Tikhonov regularization*. The solution to (10.98) is called the *ridge regressor* and can be shown to be

$$\mathbf{v}^* = \left(\mathbf{C}\mathbf{C}^H + \Lambda^H\Lambda\right)^{-1}\mathbf{C}\mathbf{y}, \tag{10.99}$$

which of course reduces to the standard least squares solution when $\Lambda = \mathbf{0}$.

10.5.2.2 L_p-Norm Periodogram Since ridge regressor is based on a quadratic cost function of the estimation error residual, the ridge regressor-based spectral estimation is also only suitable for Gaussian noise models. An obvious alternative is to replace the quadratic, or L_2 norm, cost function by a general L_p-norm cost function to obtain what is called the *L_p-norm periodogram* [154]:

$$\mathbf{v}^* = \arg \min_{\mathbf{v}\in\mathbb{C}^N} \|\mathbf{y} - \mathbf{C}^H\mathbf{v}\|^p. \tag{10.100}$$

In particular, when $p = 1$ in (10.100), one obtains the so-called Laplace–periodogram that can be robust against heavy-tailed noise distributions such as Laplace and Cauchy [154, 155].

10.5.2.3 Robust Cost Functions Another approach that is based on certain desirable optimality criterion is to replace the quadratic error residual in the least squares problem with a so-called robust cost function (or a robustifying function) $\ell(x)$:

$$\mathbf{v}^* = \arg \min_{\mathbf{v}\in\mathbb{C}^N} \ell\left(\mathbf{y} - \mathbf{C}^H\mathbf{v}\right). \tag{10.101}$$

Suppose that noise $w[n]$ has a known nominal Gaussian distribution for the most part, except that some noise samples deviates from this according to some unknown distribution. We may model such noise W as having a distribution $P_w \in \mathcal{P}_\epsilon$ where $\mathcal{P}_\epsilon$ is a set of distributions of the form of

$$\mathcal{P}_\epsilon = \{P : P = (1-\epsilon)\Phi + \epsilon H, H \in \mathcal{H}\}, \tag{10.102}$$

for $\mathcal{H}$ is a set of unknown contaminating probability distributions that are completely arbitrary except for being symmetric, Φ is the known nominal Gaussian distribution, and $0 \le \epsilon < 1$ is a known parameter that specifies the rate of deviations from the nominal noise distribution [151]. The noise model (10.102) is called the *ϵ-contaminated*

Gaussian noise or simply ϵ-contaminated noise when the nominal distribution is not necessarily Gaussian.

It can be shown that the cost function $\ell(.)$ that is most robust against the ϵ-contaminated Gaussian noise in the sense of minimizing the *maximum asymptotic estimation error variance* (worst-case estimation error variance) when $P_w \in \mathcal{P}_\epsilon$ is [151]

$$\ell_H(x) = \begin{cases} x^2/2 & \text{if} \quad |x| \leq \delta_H \\ \delta_H(|x| - \delta_H/2) & \text{if} \quad |x| > \delta_H \end{cases}, \tag{10.103}$$

where δ_H is related to ϵ through

$$\frac{2\phi(\delta_H)}{\delta_H} - 2\Phi(-\delta_H) = \frac{\epsilon}{1-\epsilon} \tag{10.104}$$

with $\phi(.)$ denoting the Gaussian pdf corresponding to the Gaussian distribution function Φ.

Function $\ell_H(x)$ is referred to as the *Huber cost function*. Hence, with $\ell_H(.)$ of (10.103), the PSD estimate (10.101), is called the robust estimator against ϵ-contaminated Gaussian noise or the *Huber estimate*. It is easily seen that this is also the maximum likelihood estimate if noise W were to have a pdf of the form of $f_W(w) = Ke^{-\ell_H(w)}$ where $K = \frac{1-\epsilon}{\sqrt{2\pi}}$ [151].

10.5.2.4 Robust Spectral Activity Detection in Wideband Cognitive Radios In general, it is desirable to have a wideband spectral activity detection framework that is not only robust against possible deviations of noise from a nominal distribution due to contaminating distributions but also to retain the improved generalization and smoothness characteristics of the ridge regressor (10.98). Hence, in the following, we may seek a PSD estimator of the form of

$$\mathbf{v}^* = \arg\min_{\mathbf{v}\in\mathbb{C}^N} \left[\frac{1}{2}\|\Lambda\mathbf{v}\|^2 + \gamma L\left(\mathbf{y} - \mathbf{C}^H\mathbf{v}\right)\right], \tag{10.105}$$

where $\mathbf{v} = [v_1, \ldots, v_N]^T \in \mathbb{C}^N$, $\gamma > 0$, and $L : \mathbb{C}^N \to \mathbb{R}$ is a robust cost function defined as

$$L(\mathbf{v}) = \ell\left(\Re\{\mathbf{v}\}\right) + \ell\left(\Im\{\mathbf{v}\}\right), \tag{10.106}$$

for $\ell : \mathbb{R}^N \to \mathbb{R}$. For example, we may choose $\ell(.)$ to be the vector-generalized version of Huber cost function (10.103) so that $\ell(.) = \ell_H(.)$. By modifying γ and regularizating matrix Λ in (10.105), we can balance between robustness and smoothness of the derived PSD estimate. In particular, with $\ell(.) = \ell_H(.)$, the *Huber optimal* solution is obtained by setting $\Lambda = \mathbf{0}$ in (10.105). By defining $\mathbf{v} \triangleq \mathbf{v}_R + j\mathbf{v}_I$ where $\mathbf{v}_R = \Re\{\mathbf{v}\}$ and $\mathbf{v}_I = \Im\{\mathbf{v}\}$, (10.106) can be written as

$$\ell(\mathbf{v}) = \ell(\mathbf{v}_R) + \ell(\mathbf{v}_I). \tag{10.107}$$

In order to obtain an efficient numerical solution for the nonlinear optimization problem (10.105) over a complex domain, let us define

$$\begin{aligned}\mathbf{y} &\triangleq \mathbf{y}_R + j\mathbf{y}_I,\\ \mathbf{c}_k &\triangleq \mathbf{c}_{k,R} + j\mathbf{c}_{k,I},\end{aligned}$$

and

$$\mathbf{e} = \mathbf{y} - \mathbf{C}^H\mathbf{v} \triangleq \mathbf{e}_R + j\mathbf{e}_I,$$

where subscripts R and I denotes the real and imaginary parts, respectively. Similarly, letting $\mathbf{C} = \mathbf{C}_R + j\mathbf{C}_I$, we have that

$$\begin{aligned}\mathbf{e} &= \mathbf{y}_R + j\mathbf{y}_I - (\mathbf{C}_R - j\mathbf{C}_I)^T(\mathbf{v}_R + j\mathbf{v}_I)\\ &= \left[\mathbf{y}_R - \left(\mathbf{C}_R^T\mathbf{v}_R + \mathbf{C}_I^T\mathbf{v}_I\right)\right] + j\left[\mathbf{y}_I - \left(\mathbf{C}_R^T\mathbf{v}_I - \mathbf{C}_I^T\mathbf{v}_R\right)\right]\end{aligned}$$

so that

$$\mathbf{e}_R = \mathbf{y}_R - \left(\mathbf{C}_R^T\mathbf{v}_R + \mathbf{C}_I^T\mathbf{v}_I\right)$$

and

$$\mathbf{e}_I = \mathbf{y}_I - \left(\mathbf{C}_R^T\mathbf{v}_I - \mathbf{C}_I^T\mathbf{v}_R\right).$$

With this notation, computing the spectral estimate of a signal $\mathbf{y} \in \mathbb{C}^N$ that is the solution to (10.105) can be rewritten as

$$\min_{\mathbf{v},\mathbf{e}_R,\mathbf{e}_I} \frac{1}{2}\|\Lambda\mathbf{v}\|^2 + \gamma\left[\ell(\mathbf{e}_R) + \ell(\mathbf{e}_I)\right], \tag{10.108}$$

subject to

$$\begin{aligned}\mathbf{e}_R &= \mathbf{y}_R - \left(\mathbf{C}_R^T\mathbf{v}_R + \mathbf{C}_I^T\mathbf{v}_I\right)\\ \mathbf{e}_I &= \mathbf{y}_I - \left(\mathbf{C}_R^T\mathbf{v}_I - \mathbf{C}_I^T\mathbf{v}_R\right).\end{aligned}$$

Since $\|\Lambda\mathbf{v}\|^2 = \|\Lambda\mathbf{v}_R\|^2 + \|\Lambda\mathbf{v}_I\|^2$, the Lagrangian of (10.108) can be written as

$$\begin{aligned}\mathcal{L} &= \frac{1}{2}\|\Lambda\mathbf{v}_R\|^2 + \frac{1}{2}\|\Lambda\mathbf{v}_I\|^2 + \gamma\left[\ell(\mathbf{e}_R) + \ell(\mathbf{e}_I)\right] + \\ &+ \boldsymbol{\alpha}^T\left[\mathbf{y}_R - \left(\mathbf{C}_R^T\mathbf{v}_R + \mathbf{C}_I^T\mathbf{v}_I\right) - \mathbf{e}_R\right] + \\ &+ \boldsymbol{\beta}^T\left[\mathbf{y}_I - \left(\mathbf{C}_R^T\mathbf{v}_I - \mathbf{C}_I^T\mathbf{v}_R\right) - \mathbf{e}_I\right],\end{aligned}$$

where $\boldsymbol{\alpha}$ and $\boldsymbol{\beta}$ are the vector of Lagrange multipliers. A set of necessary conditions are obtained by setting the gradient of the Lagrangian to zero:

$$\frac{\partial\mathcal{L}}{\partial\boldsymbol{\alpha}} = 0 \Rightarrow \mathbf{e}_R = \mathbf{y}_R - \left(\mathbf{C}_R^T\mathbf{v}_R + \mathbf{C}_I^T\mathbf{v}_I\right)$$

$$\frac{\partial\mathcal{L}}{\partial\boldsymbol{\beta}} = 0 \Rightarrow \mathbf{e}_I = \mathbf{y}_I - \left(\mathbf{C}_R^T\mathbf{v}_I - \mathbf{C}_I^T\mathbf{v}_R\right)$$

$$\frac{\partial\mathcal{L}}{\partial\mathbf{e}_R} = 0 \Rightarrow \boldsymbol{\alpha} = \gamma\boldsymbol{\ell}'(\mathbf{e}_R)$$

$$\frac{\partial\mathcal{L}}{\partial\mathbf{e}_I} = 0 \Rightarrow \boldsymbol{\beta} = \gamma\boldsymbol{\ell}'(\mathbf{e}_I)$$

$$\frac{\partial\mathcal{L}}{\partial\mathbf{v}_R} = \mathbf{0} \Rightarrow \mathbf{v}_R = Q^{-1}\left(\mathbf{C}_R\boldsymbol{\alpha} - \mathbf{C}_I\boldsymbol{\beta}\right) \tag{10.109}$$

$$\frac{\partial\mathcal{L}}{\partial\mathbf{v}_I} = \mathbf{0} \Rightarrow \mathbf{v}_I = Q^{-1}\left(\mathbf{C}_I\boldsymbol{\alpha} + \mathbf{C}_R\boldsymbol{\beta}\right), \tag{10.110}$$

where we let Q denote the nonsingular matrix $Q \triangleq \Lambda^H\Lambda \in \mathbb{R}^{N\times N}$ and $\boldsymbol{\ell}'(\mathbf{u}) \triangleq [\ell'(u_1),\ldots,\ell'(u_N)]^T$ for all $\mathbf{u} = [u_1,\ldots,u_N]^T$.

It is easily seen that the solution characterized by the above equations can be obtained by solving the following fixed-point equations:

$$\begin{aligned}\boldsymbol{\alpha} &= \gamma\boldsymbol{\ell}'\left(\mathbf{y}_R - \mathbf{P}\boldsymbol{\alpha}\right) \\ &\triangleq h_R(\boldsymbol{\alpha})\end{aligned} \tag{10.111}$$

and

$$\begin{aligned}\boldsymbol{\beta} &= \gamma\boldsymbol{\ell}'\left(\mathbf{y}_I - \mathbf{P}\boldsymbol{\beta}\right) \\ &\triangleq h_I(\boldsymbol{\beta}),\end{aligned} \tag{10.112}$$

where we have defined $\mathbf{P} = \mathbf{C}^H\mathbf{Q}^{-1}\mathbf{C}$, $h_R : \mathbb{R}^N \to \mathbb{R}^N$ and $h_I : \mathbb{R}^N \to \mathbb{R}^N$. The weight vector $\mathbf{v}$ that is the PSD estimator can be then obtained by substituting the solutions to $\boldsymbol{\alpha}$ and $\boldsymbol{\beta}$ in (10.109) and (10.110).

Fixed-point equations (10.111) and (10.112) will admit unique solutions for $\boldsymbol{\alpha}$ and $\boldsymbol{\beta}$ if h_R and h_I are contraction mappings [156]. It has been established in [153] that if ℓ' is Lipschitz continuous[2] with a Lipschitz constant p and if $\gamma p\|\mathbf{P}\|_2 < 1$, then, indeed, h_R and h_I in (10.111) and (10.112) are contraction mappings on $\mathbb{R}^N$. Thus, it follows from the Banach fixed-point theorem that the solution of (10.111) and (10.112) can be obtained using fixed-point iterations [156]. Interestingly, since regularization parameter γ can be chosen arbitrarily, the condition $\gamma p\|\mathbf{P}\|_2 < 1$ can always be satisfied. Hence, it is possible to guarantee convergence of the fixed-point iteration by proper choice of the regularization parameter γ.

Example 10.4 (Robust Spectral Estimation in Gaussian–Laplacian Mixture Noise) Let us revisit the sinusoidal signal with normalized frequency $F = 0.2$ embedded in Gaussian–Laplacian mixture noise considered in Example 10.3. The noise pdf $f_W(w)$ was given in (10.83). In Figure 10.7 in Example 10.3, we saw the conventional periodogram and the periodogram based on the transformed variables according to the locally optimal detection in non-Gaussian noise for $\epsilon = 0.1$ and $\epsilon = 0.9$. In Figure 10.8, we show the same conventional PSD estimate but this time along with the robust periodogram derived previously with $\ell(.) = \ell_H(.)$. As before, we have assumed that $N = 100$ and $SNR = -5B$. Observe that especially when ϵ is relatively large, conventional periodogram may exhibit more erratic behavior than the robust periodogram. ■

10.6 WIDEBAND SPECTRAL ACTIVITY DETECTION WITH COMPRESSIVE SAMPLING

The total spectrum of interest to the cognitive radio spanned by all N_b subbands can, in general, be in the order of several GHz. While subband segmentation in conjunction with subband selection algorithms of Chapter 9 provides a way to handle the computational demand of wideband spectrum sensing, it may not still be enough given the fact that each subband may itself be of several hundred MHz wide. A promising approach is to follow subband segmentation with *compressive sampling* (CS): That is, first, the wide spectrum of interest is divided into a set of (still wide) subbands, and then each band is sensed via CS for the purpose of spectral activity detection.

CS is an efficient approach to reconstruct functions, or signals, that are *sparse* in a certain domain. Efficiency here refers to both the number of samples needed for the reconstruction as well as computational efficiency. In the case of wideband spectral activity detection, there may be situations in which spectral activities can be sparse within the frequency range of a subband. Then, the PSD of the sensed-subband signal, as a function F can be interpreted as a sparse signal. In certain frequency ranges, however, spectral activity may not be sparse. For example, in certain locations, ISM

[2] A function f is said to be Lipschitz continuous on the open subset $\mathcal{A} \subseteq \mathbb{R}^m$ if there exists a constant $p \geq 0$ such that $|f(\mathbf{x}) - f(\mathbf{y})| \leq p|\mathbf{x} - \mathbf{y}|$ for all $\mathbf{x} \in \mathcal{A}$, $\mathbf{y} \in \mathcal{A}$. The constant p is called the Lipschitz constant of f on $\mathcal{A}$.

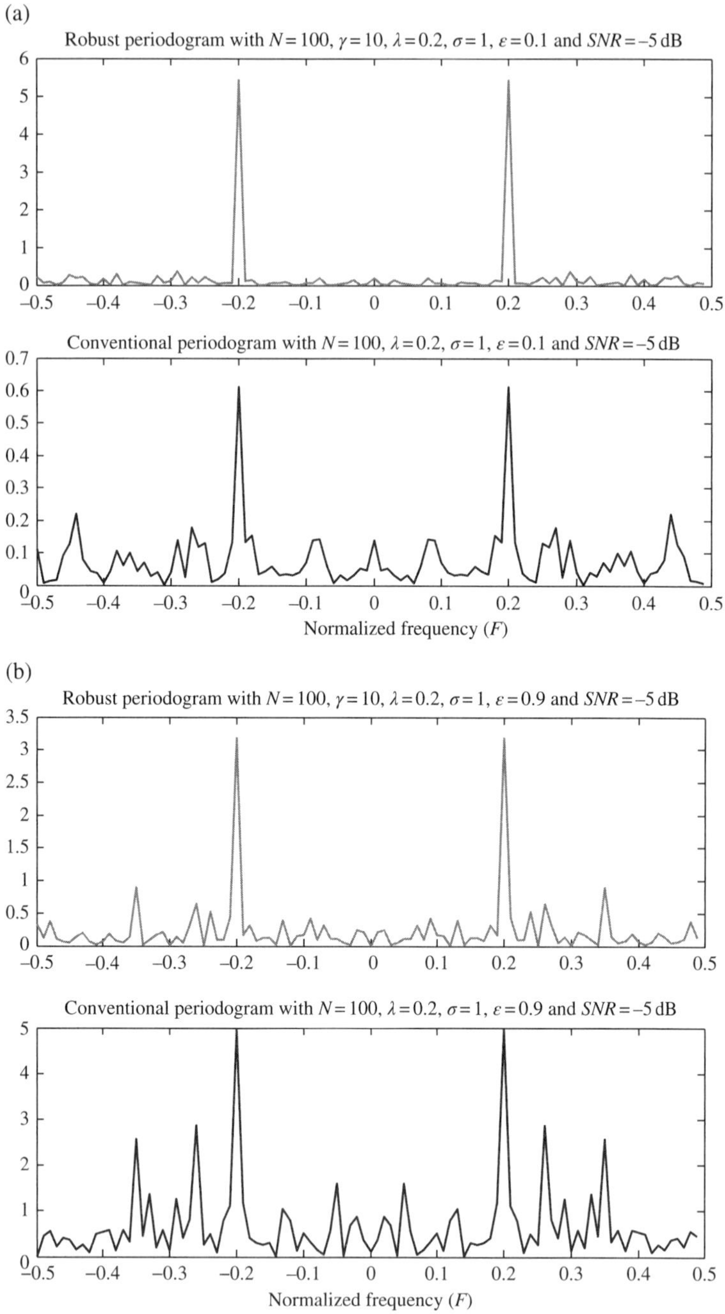

FIGURE 10.8 Conventional and robust periodograms in the presence of Gaussian–Laplacian mixture noise. (a) Conventional and robust periodograms in the presence of Gaussian–Laplacian mixture noise with $\epsilon = 0.1$.(b) Conventional and robust periodograms in the presence of Gaussian–Laplacian mixture noise with $\epsilon = 0.9$.

band frequencies may be highly congested. However, there are equally good cases in which at certain locales the spectrum subbands can be highly under utilized and thus sensed subband signal can be highly sparse in frequency domain.

10.6.1 Compressive Sampling

Let us first look at the concept of CS developed in [157–159] and summarized in [160]. For the moment, let us consider an arbitrary signal $x(t)$ for $0 \leq t \leq T$. Let us suppose that $\varphi_i(t)$, for $1 \leq i \leq N$, is an orthonormal basis for signals defined on $0 \leq t \leq T$. Then we know that we can express the signal $x(t)$ as

$$x(t) = \sum_{i=1}^{N} s_i \varphi_i(t), \tag{10.113}$$

where s_i, for $1 \leq i \leq N$, are the expansion coefficients that are given by the projection of $x(t)$ onto each of the basis vectors:

$$s_i =< x, \varphi_i > \quad \text{for } \; i = 1, \cdots, N.$$

Note that such expansions are abound. Perhaps the most familiar one is the Fourier series expansion in which the basis signals are chosen as (truncated) complex exponential signals with harmonic frequencies. In this case, the coefficients s_i are the Fourier series coefficients. Another familiar example is encountered in Shannon–Nyquist sampling in which the basis signals are the time-shifted $sinc(.)$ functions and s_i's are nothing but the sampled-signal values $x[n]$, for $n = 1, \ldots$. Once the orthogonal basis is specified, the set of expansion coefficients $\mathbf{s} = (s_1, \ldots, s_N)^T$ fully specifies the signal.

Indeed, our interest is in sampling the above signal $x(t)$. Suppose we sample the signal $x(t)$ to obtain the N-length vector $\mathbf{x} = (x[1], x[2], \ldots, x[N])^T = (x_1, x_2, \ldots, x_N)^T$. Then, we may write (10.113) as

$$\mathbf{x} = \Upsilon \mathbf{s}, \tag{10.114}$$

where Υ is the $N \times N$ matrix whose columns are the sampled basis signals:

$$\Upsilon = \left[\boldsymbol{\varphi}_1, \ldots, \boldsymbol{\varphi}_N\right], \tag{10.115}$$

with $\boldsymbol{\varphi}_i = (\varphi_i[1], \ldots, \varphi_i[N])^T$.

In general, if the basis signal set has dimension N as assumed above, then to fully characterize the signal, we do need to give the N-length vector $\mathbf{x}$. This is because all s_i coefficients in the signal $\mathbf{s}$ can be non zero. However, note from (10.114) that these coefficients depend on the chosen basis, or the matrix Υ. With different basis, vector $\mathbf{s}$ can look very different. In particular, it is possible that there may be a certain basis signal set $\{\varphi_i(t)\}_{i=1}^{N}$, or equivalently a matrix Υ, for which some of these coefficients

s_i may actually be zero. Indeed, for many signals, there exist such basis with respect to whom some of their expansion coefficients are zero.

If only $K < N$ expansion coefficients s_i are nonzero for a signal $\mathbf{x}$, with respect to some basis Υ, then the signal is called *K-sparse*. In this case, the basis Υ is called the *sparsity basis matrix*, or simply the *sparsity matrix*. Note that if a signal is K-sparse, then the signal is fully characterized by those K nonzero coefficients in vector $\mathbf{s}$. Hence, we can discard all but K of the N coefficients.

Let us now consider a signal that is K-sparse with respect to some (unknown) basis Υ. In this case, all but $K < N$ elements of the signal vector $\mathbf{s}$ are actually zero. The usual approach to sampling a signal, governed by the Shannon–Nyquist sampling theorem, however, does not pay attention to this extra information about the signal. All it considers are the spectral components present in the signal, and thus state that for reconstructing the signal $x(t)$ without any information loss, it is sufficient to have a sampling rate that is twice the maximum frequency in signal $x(t)$. Hence, if the dimension N of the signal is chosen to meet this criterion, then we must obtain the complete vector $\mathbf{x}$ in order to reconstruct the signal according to the Shannon–Nyquist sampling theorem. The reason is, in fact, that the Shannon–Nyquist sampling theorem is fixed on the Fourier basis. Hence, the required dimension of the basis signal set N is determined by the maximum frequency present in the signal $x(t)$. Once the basis dimension N is set by this criterion, we do need to have all N samples given by $\mathbf{x}$ in order to reconstruct the signal. It does not bother whether there is another basis with respect to whom the signal $\mathbf{s}$ is actually sparse. Worse yet, even if the sparsity basis was known and used to sample, since which coefficients are exactly zero is unknown, it is still required according to conventional sampling procedures to first obtain all N samples before actually throwing out all but the K nonzero samples.

Hence, the conventional approach to sampling will require us to specify the full signal $\mathbf{x}$ even if the signal $\mathbf{s}$ is actually K-sparse. Now, suppose that instead of conventional sampling, we perform a different type of sampling in which rather than the N-vector $\mathbf{x}$, we obtain an M-length vector $\mathbf{y}$, for $M < N$:

$$\mathbf{y} = \Phi\mathbf{x}, \tag{10.116}$$

where $M \times N$ matrix Φ is called the *sensing matrix*. If we denote rows of Φ by $\boldsymbol{\phi}_1^T, \ldots, \boldsymbol{\phi}_M^T$, then each element of $\mathbf{y}$ is a projection of $\mathbf{x}$ onto these row vectors of Φ. Note that usual Shannon–Nyquist sampling corresponds to $M = N$ and $\Phi = \mathbf{I}_M$. From (10.114) and (10.116),

$$\mathbf{y} = \Phi\Upsilon\mathbf{s}. \tag{10.117}$$

If $\mathbf{s}$ is K-sparse, is it possible to reconstruct $\mathbf{s}$ from above $\mathbf{y}$, where $\mathbf{y}$ is an M-length vector for some $M < N$? If so, what should be M, the sensing basis Φ, and the reconstruction approach? The CS of [157–159] develops an approach that answers these questions.

The answer to the first question is in the affirmative: It is indeed possible to reconstruct the N-length vector $\mathbf{s}$ that is K-sparse with respect to a basis Υ from an M-length vector **y** where $M < N$. This is not very surprising. For example, although Shannon–Nyquist sampling theorem states that one needs samples at the rate of twice the largest frequency in a signal, it is well known that passband signals may be reconstructed with samples obtained at a much lower rate than that. Indeed, versions of sampling theorem for passband signals state that it is only enough to sample at roughly about twice the bandwidth of the passband signal, which can be much lower than twice the highest frequency [161, 162]. Essentially, the signal is sparse in the frequency range from very low to the highest frequency in the signal. Only a small frequency interval equal to its bandwidth is really occupied.

The problem we need to solve is

$$(P_0) \quad \min_{\mathbf{s}\in\mathbb{R}^N} \|\mathbf{s}\|_{\ell_0} \text{ s.t. } \mathbf{y} = \Phi\Upsilon\mathbf{s}, \tag{10.118}$$

where $\|\mathbf{s}\|_{\ell_0}$ denotes the ℓ_0-norm of the vector **s** that is defined as the number of nonzero elements in **s**. The solution to (10.118) is the sparsest $\mathbf{s} \in \mathbb{R}^N$ that agrees with the samples **y**.

For $M < N$, the solution to (10.118), however, is not unique in general. Worse yet, the problem (10.118) is a combinatorial problem, and if the sparsity order K is unknown, it may require one to try out all subsets of all sizes of the set $\{1,\ldots,N\}$. One of the main ideas of CS is to instead solve the following convex optimization problem that replaces ℓ_0-norm by ℓ_1-norm:

$$(P_1) \quad \min_{\mathbf{s}\in\mathbb{R}^N} \|\mathbf{s}\|_{\ell_1} : \mathbf{y} = \Phi\Upsilon\mathbf{s}. \tag{10.119}$$

Note that ℓ_1-norm minimization (10.119) can be cast as a linear program whose time complexity is only polynomial. Thus, there are efficient algorithms to solve (10.119) for **s**. The justification for solving (10.119) instead of (10.118) is, in part, argued in the following theorem from [160, 163].

Theorem 10.1 (Compressive Sampling of Sparse Signals) If the N-vector **s** is K-sparse with respect to the sparsity basis Υ and if the M rows of Φ are drawn independently and uniformly from the vectors of an orthonormal basis $\{\boldsymbol{\phi}_i\}_{i=1}^N$, then the solution to (10.119) is the same as the solution to (10.118) with very high probability if

$$M \geq C\mu^2(\Phi,\Upsilon)K\log N, \tag{10.120}$$

where $\mu(\Phi,\Upsilon)$ is the coherence between the sensing basis Φ and the sparsity basis Υ defined as

$$\mu(\Phi,\Upsilon) = \sqrt{N} \max_{1\leq i\leq N, 1\leq j\leq M} \left| < \boldsymbol{\varphi}_i, \boldsymbol{\phi}_j > \right| \tag{10.121}$$

and C is some positive constant. In particular, if

$$M \geq C\mu^2(\Phi,\Upsilon)K\log\left(\frac{N}{\delta}\right), \tag{10.122}$$

then the probability that the solution to (10.119) is the same as the solution to (10.118) is greater than $1-\delta$.

Proof: See [163].

Thus, whether sparse $\mathbf{s}$ can be reconstructed by a considerably less number of samples than N is determined by the *coherence* $\mu(\Phi,\Upsilon)$ between the sensing basis Φ and the sparsity basis Υ and the sparsity order K. The smaller the coherence between the two bases, smaller the number of samples needed. Since $\frac{1}{\sqrt{N}} \leq \left| < \boldsymbol{\varphi}_i, \boldsymbol{\phi}_j > \right| \leq 1$ for orthonormal basis Φ and Υ, the coherence satisfies $1 \leq \mu(\Phi,\Upsilon) \leq \sqrt{N}$. If the sensing basis vectors Φ is chosen so that coherence is at the minimum possible value of 1, then according to Theorem 10.1, we only need samples on the order of $K \log N$ instead of N. The more sparse $\mathbf{s}$ is, we should be able to reconstruct $\mathbf{s}$ from less number of samples.

At this point, one may wonder how we should choose Φ so that it does have low coherence with the fixed basis Υ. It is well known that random vectors are highly incoherent with any fixed set of vectors [157, 160]. Hence, in theory, the most effective CS can be achieved by just picking a random matrix Φ: For example, Φ can be constructed by drawing its elements independently and randomly according to some distribution such as Bernoulli or Gaussian. Such a matrix Φ will be highly incoherent with any fixed sparsity basis Υ [160]. Hence, we do not even need to know Υ to design a good sensing matrix Φ that will guarantee that a small number of samples (as specified by Theorem 10.1) are enough to reconstruct $\mathbf{s}$. However, for actual reconstruction of the signal, the knowledge of Υ is needed.

The earlier discussion, and Theorem 10.1 in particular, essentially summarizes the main ideas of CS. However, in practice, it will be rare to find signals of interest to be exactly K-sparse for some K with respect to some basis. It is, however, very common that with respect to some basis, most signals have only very few coefficients that are dominantly large, while most of the other coefficients are comparatively very small. Hence, in the sense of the previous discussion, these signals are not sparse. In fact, they may have all their coefficients nonzero albeit most are very low. Hence, CS as discussed so far is not directly useful for these signals. Moreover, sampled-signal values may also be corrupted by, for example, additive noise:

$$\mathbf{y} = A\mathbf{s} + \mathbf{w}, \tag{10.123}$$

where A is an $M \times N$ matrix and $\mathbf{w}$ is additive noise. It is of interest to investigate and establish the relevance and effectiveness of the CS ideas to these situations: If most coefficients of $\mathbf{s}$ are relatively very small, but not identically zero, can we still expect to reconstruct $\mathbf{s}$ as earlier with a smaller number of samples than N? In particular, how good is the reconstruction of $\mathbf{s}$ obtained as below?:

$$\min_{\mathbf{s}\in\mathbb{R}^N} \|\mathbf{s}\|_{\ell_1} : \mathbf{y} = A\mathbf{s}. \tag{10.124}$$

It has been shown that indeed the solution to (10.124) can be a very good reconstruction of a *nearly K-sparse* **s** when matrix A satisfies certain conditions that are characterized in terms of the so-called restricted isometry property (RIP) [160].

Definition 10.6.1 (Isometry Constant of a Matrix) For each positive integer K, the smallest δ for which

$$(1-\delta)\parallel \mathbf{s} \parallel_{\ell_2}^2 \leq \parallel A\mathbf{s} \parallel_{\ell_2}^2 \leq (1+\delta)\parallel \mathbf{s} \parallel_{\ell_2}^2$$

for all K-sparse vectors **s** is called the order K isometry constant δ_K of matrix A.

Note that if isometry constant δ_K is small, essentially, matrix A preserves the ℓ_2-norm (Euclidean norm) of K-sparse vectors. Hence, when δ_K is sufficiently small, the matrix A is said to hold the RIP of order K.

The following theorem from [160] assures that it is indeed possible to obtain good reconstruction of nearly K-sparse signals using the CS approach (10.124) when order $2K$ isometry constant of the matrix A is sufficiently small.

Theorem 10.2 Let $\mathbf{s}_K$ denote the vector obtained by setting all but the K-largest elements of **s** to zero. Then, for some constant C_1, the solution $\mathbf{s}^*$ to (10.124) satisfies

$$\parallel \mathbf{s}^* - \mathbf{s} \parallel_{\ell_2} \leq C_1 \frac{\parallel \mathbf{s} - \mathbf{s}_K \parallel_{\ell_1}}{\sqrt{K}}$$

and

$$\parallel \mathbf{s}^* - \mathbf{s} \parallel_{\ell_1} \leq C_1 \parallel \mathbf{s} - \mathbf{s}_K \parallel_{\ell_1},$$

if the order $2K$ isometry constant of matrix A satisfies $\delta_{2K} < \sqrt{2}-1$.

Proof: See [160, 164, 165].

As in the case of K-sparse signals, it has also been shown that many random constructions of $M \times N$ matrices A in fact ensure small enough isometry constants as long as [160]

$$M \geq C_2 K \log \frac{N}{K},$$

for some constant C_2.

10.6.2 Compressive Sensing of Wideband Spectrum

When the subband of interest contains several signals located at unknown locations but the spectrum occupied by these signals is a small fraction of the overall bandwidth of the subband, the sensed subband signal is a sparse signal with respect to a spectral domain basis such as Fourier basis. Hence, it is clear that CS can be exploited to reduce the sampling rate requirements needed for wideband spectral sensing. In the following, we explain the approach developed in [166] based on this insight.

Let us consider the sensed subband signal (10.33) again. Suppose that this discrete-time signal is the analog-to-digital converter output of the received continuous-time subband signal $y(t)$. In forming (10.33), it has been implicitly assumed that the signal is sampled at least at the Nyquist rate. Hence, if the bandwidth of the subband is B, then the signal samples in (10.33) are obtained by sampling $y(t)$ at least at $2B$ samples per second. If the corresponding sampling period is $T_s = \frac{1}{2B}$, then signal $Y[n]$ corresponds to a time segment of NT_s. Let us denote by $\mathbf{y} = (y_1, \ldots, y_N)^T$ the Nyquist rate (or higher) sampled signal $y(t)$ in (10.33) and for the moment ignore noise.

Let us denote by $N \times N$ matrix $\mathbf{C}$ the N-point DFT matrix as assumed in (10.95). Then the frequency-domain representation of $\mathbf{y}$ can be written as

$$\mathbf{y}_f = \mathbf{C}\mathbf{y}. \tag{10.125}$$

Of course, $\mathbf{C}^H\mathbf{C} = \mathbf{I}_N$ and Fourier transform $\mathbf{C}$ is invertible:

$$\begin{aligned} \mathbf{y} &= \mathbf{C}^{-1}\mathbf{y}_f && (10.126)\\ &= \mathbf{C}^H\mathbf{y}_f. && (10.127) \end{aligned}$$

If the sensed subband signal $y(t)$ is sparse, however, $\mathbf{y}_f$ in (10.125) will only contain few large coefficients, while the rest of the coefficients are relatively very small. Hence, with respect to the Fourier basis $\mathbf{C}$, the signal $y(t)$ is sparse. Suppose that the bandwidth occupied by active signals in the subband in question is only a fraction $\frac{K}{N}$ of the total bandwidth B of the subband. Then, on average, we may expect only K out of the N elements of $\mathbf{y}_f$ in (10.125) to be nonzero or dominant compared to the other coefficients. Theorems 10.1 and 10.2 then assure us that we may reconstruct $\mathbf{y}_f$, the frequency-domain representation of the sensed continuous-time subband signal $y(t)$, with relatively small error by only having about $K \log N$ number of samples instead of N.

Hence, instead of sampling the signal $y(t)$ at the Nyquist rate of $2B$ samples per second over the time interval of length NT_s where $T_s = \frac{1}{2B}$, we may sample it at a much lower rate to just collect about $K \log N$ number of samples over the same time interval. From Theorem 10.1 and the discussion following Theorem 10.2, these samples do not have to be even carefully chosen: Over the time interval, NT_s just take random samples with an average rate of the order of $\frac{2BK \log N}{N}$ samples per second. If we denote this random sampling matrix by the $M \times N$ matrix Φ, then randomly compressive sampled version of $y(t)$ can be written as

$$\begin{aligned} \mathbf{y}_c &= \Phi\mathbf{y}\\ &= \Phi\mathbf{C}^{-1}\mathbf{y}_f. && (10.128) \end{aligned}$$

From Theorems 10.1 and 10.2, as long as $M > CK\log N$ for some C, we are guaranteed to obtain a very good (in the sense of Theorem 10.2) reconstruction $\hat{\mathbf{y}}_f$ of the frequency-domain representation $\mathbf{y}_f$ of the subband signal $y(t)$ by solving the following linear programming problem:

$$\hat{\mathbf{y}}_f = \arg\min_{\mathbf{y}_f \in \mathbb{R}^N} \left\| \mathbf{y}_f \right\|_{\ell_1} \quad \text{s.t.} \quad \mathbf{y}_c = \Phi \mathbf{C}^{-1} \mathbf{y}_f. \tag{10.129}$$

Once we have $\hat{\mathbf{y}}_f$, then we may perform spectral activity detection as in the previous sections. In particular, [166] combines the wavelet transform-based approach of [141] described in Section 10.4 with compressive sampled reconstruction $\hat{\mathbf{y}}_f$ to achieve spectral activity detection.

10.7 SUMMARY

- In general, a wideband cognitive radio needs to detect all spectral activities within its sensed spectrum subband. Detecting all signals located at unknown carrier frequencies over a possibly wide spectrum subband is called the wideband spectral activity detection problem.
- It can be shown that detecting spectral activities in a sensed signal is equivalent to an optimal test based on the statistic

$$\frac{1}{N} T(\mathbf{y}) = \int_{-\frac{1}{2}}^{\frac{1}{2}} \hat{S}_y(F) S_Q(F) dF,$$

  where $\hat{S}_y(F)$ is the periodogram estimate of the PSD of the sensed signal $\mathbf{y}$,

$$\hat{S}_y(F) = \frac{1}{N} \left| \sum_{n=1}^{N} y_n e^{-j2\pi Fn} \right|^2$$

  and $S_Q(F)$ is the DTFT of the sequence $\{q_m\}_m$ defined by any row of the matrix $Q = \Sigma_0^{-1} - \Sigma_1^{-1}$ with Σ_0 and Σ_1 denoting the covariance matrices of Y when signal is absent and present, respectively.
- An approximately optimal decision statistic for the low SNR region turns out to be

$$\frac{1}{N} T(\mathbf{y}) \approx \frac{1}{N_0^2} \int_{-\frac{1}{2}}^{\frac{1}{2}} \hat{S}_y(F) \mathscr{S}_X(F) dF.$$

- Hence, an approximately optimal approach for spectral activity detection is to compute the periodogram $\hat{S}_y$ of the sensed observations and to check at what frequencies it exceeds a certain threshold. In practice, this can be achieved by integrating the energy in the periodogram of the observations over small suitably chosen bands and comparing with a suitably chosen threshold. Those frequencies at which the threshold is exceeded represent the spectral activities in the sensed subband. The threshold can be chosen according to a desired optimality criterion.
- When spectra of the active signals in the sensed subband are located at arbitrary frequencies, have arbitrary bandwidths, and are nonoverlapping, spectral activities can also be detected by detecting the edges of the active signals in the spectrum $\mathscr{S}_Y(F)$. This reduces the spectral activity detection problem into a problem of edge detection in the spectrum estimate $\hat{S}_y(F)$. The wavelet transform is an effective tool to detect discontinuities in functions making it an ideal choice for detecting edges in $\hat{S}_y(F)$.
- For a possibly complex-valued function $\Psi(F)$ called a wavelet, the wavelet transform of a function $S(F)$ is defined as

$$WS(s,F) = S(F) \star \Psi_s(F),$$

where $WS()(.,.)$ stands for the wavelet transform of function S, $\star$ denotes convolution, and $\Psi_s(F)$ is a dilation of $\Psi(F)$ defined as

$$\Psi_s(F) = \frac{1}{s}\Psi\left(\frac{F}{s}\right).$$

The function $\Psi(F)$ is a valid wavelet if and only if

$$\int_0^{+\infty} \frac{|\psi(\lambda)|^2}{\lambda} d\lambda = \int_{-\infty}^{0} \frac{|\psi(\lambda)|^2}{|\lambda|} d\lambda = C_\Psi < +\infty,$$

for some constant C_Ψ.
- If the wavelet is a valid wavelet, then the inverse wavelet transform is given by

$$S(F) = \frac{1}{C_\Psi} \int_0^{+\infty} \int_{-\infty}^{+\infty} WS(s,u)\psi_s^*(u-F)du\frac{ds}{s}.$$

- It can be shown that the first and second derivatives of $WS(s,F)$ computed with a particular wavelet $\Psi(F)$ can be written as

$$\left(\frac{dWS(s,F)}{dF}\right)(F) = \frac{1}{s}W^1S(s,F)$$

and

$$\left(\frac{d^2 WS(s,F)}{dF^2}\right)(F) = \frac{1}{s^2} W^2 S(s,F),$$

where $W^1 S(s,F)$ and $W^2 S(s,F)$ are, respectively, the wavelet transforms of $S(F)$ with respect to the two wavelets $\Psi^1(F) = \frac{d\Psi(F)}{dF}$ and $\Psi^2(F) = \frac{d^2\Psi(F)}{dF^2}$ at scale s.

- Local extrema of $W^1 S(s,F)$ as a function of F for a fixed s correspond to the inflection points of the wavelet transform $WS(s,F)$, which are an indication of the possible discontinuities and edges in $S(F)$. They also correspond to the zero-crossings of $W^2 S(s,F)$. Hence, the edges and discontinuities in $S(F)$ can be detected by detecting the local extrema of $W^1 S(s,F)$ or the zero-crossings of $W^2 S(s,F)$. Since slow and high variation points of $WS(s,F)$ correspond to local minima and maxima of $|W^1 S(s,F)|$, respectively, and since the edges in $S(F)$ are reflected as high variation points in $WS(s,F)$, edges in $S(F)$ can be detected by finding the local maxima of the wavelet transform modulus $|W^1 S(s,F)|$.
- Spectral activity detection in a sensed subband can be achieved by computing the wavelet transform $W^1 \hat{S}_y(s,F)$ of the PSD estimate $\hat{S}_y(F)$ and then finding the local maxima of the wavelet transform modulus $\left|W^1 \hat{S}_y(s,F)\right|$:

$$\{F_{k'}\}_{k'=1}^{K'} = \text{local maxima of}_{F} \left|W^1 \hat{S}_y(s,F)\right|.$$

- Once these edge boundaries $\{F_{k'}\}_{k'-1}^{K'}$ are determined, then they can be mapped to spectral activity regions $\mathcal{B}_k$, for $1 \leq k \leq K$.
- It is known that the average density of modulus maxima due to white noise is inversely proportional to the scale s. Hence, the effect of noise on spectral activity detection can be reduced by comparing the modulus maxima behavior across multiple scales. One way to achieve this in detecting edges in the $\hat{S}_y(F)$ is to detect the edges as

$$\{F_{k'}\}_{k'=1}^{K'} = \text{local maxima of}_{F} \left|W_J^1 \hat{S}_y(F)\right|,$$

where $\left|W_J^1 \hat{S}_y(F)\right|$ is the modulus of the multiscale wavelet product

$$W_J^1 \hat{S}_y(F) = \Pi_{j=1}^{J} W^1 \hat{S}_y(2^j,F),$$

with J dyadic scales chosen so that $s = 2^j$, for $1 \leq j \leq J$.

- Possible non-Gaussianity in noise can be better handled by either replacing the noise density with an appropriate non-Gaussian density function or by resorting to robust detection approaches. Either way, many of these essentially lead to modifying the periodogram PSD estimate to obtain a robust periodogram.

- When non-Gaussian noise densities are assumed, it quite common to use locally optimal detectors. When signals are modeled as random, a locally optimal non-Gaussian spectral activity detector can be implemented, for each frequency F_0, as

$$T_{F_0}(\mathbf{y}) = \int_{F_0-\Delta F}^{F_0+\Delta F} \hat{S}_{\tilde{y}}(F)dF + T_{2,0}^{(LO)}(\mathbf{y}),$$

where ΔF defines a window of frequencies,

$$T_{2,0}^{(LO)}(\mathbf{y}) = \rho_0^{(x)} \sum_{n=1}^{N} \left(\hat{y}_n - \tilde{y}_n^2\right),$$

and $\hat{S}_{\tilde{y}}(F)$ is the periodogram of a transformed observation sequence defined as

$$\hat{S}_{\tilde{y}}(F) = \left|\tilde{Y}(F)\right|^2.$$

where

$$\tilde{y}_n \triangleq g_n^{(LO)}(y_n)$$

and

$$\hat{y}_n \triangleq h_n^{(LO)}(y_n),$$

with

$$g_n^{(LO)}(w) = \frac{f'_{W_n}(w)}{f_{W_n}(x)}$$

and

$$h_n^{(LO)}(x) = \frac{f''_{W_n}(x)}{f_{W_n}(n)}.$$

- Poor performance of periodogram in the presence of non-Gaussian noise can be related to the fact that it is the least squares estimate of the PSD and that least squares estimates lack robustness against outliers and heavy-tailed noise distributions. There are many modifications that one can do to least squares estimates in order to improve their robustness against noise outliers, including:
  - *Tikhonov regularization*: Regularization is an often used approach to improve the smoothness and the generalization properties of a least squares estimate.

It modifies the least squares cost function (10.95) by including a regulation term:

$$\mathbf{v}^* = \arg\min_{\mathbf{v}\in\mathbb{C}^N} \|\Lambda\mathbf{v}\|^2 + \|\mathbf{y} - \mathbf{C}^H\mathbf{v}\|^2,$$

where $\mathbf{C} = [\mathbf{c}_1,\ldots,\mathbf{c}_N]$, with $\mathbf{c}_n = \frac{1}{\sqrt{N}}[1, e^{-j2\pi\frac{n}{N}}\ldots, e^{-j2\pi(N-1)\frac{n}{N}}]^T$, for $n = 0,\ldots,N-1$, is a DFT matrix. The optimal solution turns out to be

$$\mathbf{v}^* = \left(\mathbf{C}\mathbf{C}^H + \Lambda^H\Lambda\right)^{-1}\mathbf{C}\mathbf{y}.$$

When $\Lambda = \mathbf{0}$, this reduces to the standard least squares solution $\mathbf{v}^* = \mathbf{C}\mathbf{y}$, which is essentially the periodogram.

- *L_p-norm periodogram*: An alternative to Tikhonov regularization is to replace the quadratic cost function by a general L_p-norm cost function to obtain what is called the *L_p-norm periodogram*:

$$\mathbf{v}^* = \arg\min_{\mathbf{v}\in\mathbb{C}^N} \|\mathbf{y} - \mathbf{C}^H\mathbf{v}\|^p.$$

When $p = 1$, this is called the Laplace periodogram, which is known to be robust against heavy-tailed noise distributions.

- Robust cost functions: Robust cost functions replaces the L_2- and L_p-norm cost functions used in both least squares and L_p-norm periodogram approaches by more general robust cost functions $\ell(x)$:

$$\mathbf{v}^* = \arg\min_{\mathbf{v}\in\mathbb{C}^N} \ell\left(\mathbf{y} - \mathbf{C}^H\mathbf{v}\right).$$

- A particularly interesting robust cost function is the Huber cost function $\ell_H(.)$ defined as

$$\ell_H(x) = \begin{cases} x^2/2 & \text{if} \quad |x| \le \delta_H \\ \delta_H(|x| - \delta_H/2) & \text{if} \quad |x| > \delta_H \end{cases},$$

where δ_H is related to ϵ through

$$\frac{2\phi(\delta_H)}{\delta_H} - 2\Phi(-\delta_H) = \frac{\epsilon}{1-\epsilon}$$

with $\phi(.)$ denoting the Gaussian pdf corresponding to the Gaussian distribution function Φ. It can be shown that the Huber cost functions is the most robust against the ϵ-contaminated Gaussian noise in the sense of minimizing the maximum asymptotic estimation error variance when noise distribution $P_w \in \mathcal{P}_\epsilon$ where

$$\mathcal{P}_\epsilon = \{P : P = (1-\epsilon)\Phi + \epsilon H, H \in \mathcal{H}\},$$

for $\mathcal{H}$ is a set of unknown contaminating probability distributions that are completely arbitrary except for being symmetric and Φ is the known nominal Gaussian distribution.

- A robust spectral activity detection approach for wideband cognitive radios can be benefitted by incorporating both a robust cost function as well as a regularization term to improve generalization capability as below:

$$\mathbf{v}^* = \arg\min_{\mathbf{v}\in\mathbb{C}^N}\left[\frac{1}{2}\|\Lambda\mathbf{v}\|^2 + \gamma L\left(\mathbf{y} - \mathbf{C}^H\mathbf{v}\right)\right],$$

where $L : \mathbb{C}^N \to \mathbb{R}$ is a robust cost function of the form of

$$L(\mathbf{v}) = \ell\left(\Re\{\mathbf{v}\}\right) + \ell\left(\Im\{\mathbf{v}\}\right),$$

for $\ell : \mathbb{R}^N \to \mathbb{R}$. It can be shown that the solution to above can be obtained by solving the following fixed-point equations:

$$\boldsymbol{\alpha} = \gamma\boldsymbol{\ell}'\left(\mathbf{y}_R - \mathbf{P}\boldsymbol{\alpha}\right)$$

and

$$\boldsymbol{\beta} = \gamma\boldsymbol{\ell}'\left(\mathbf{y}_I - \mathbf{P}\boldsymbol{\beta}\right),$$

where $\boldsymbol{\ell}'(\mathbf{u}) \triangleq [\ell'(u_1),\ldots,\ell'(u_N)]^T$ and $\mathbf{P} = \mathbf{C}^H\mathbf{Q}^{-1}\mathbf{C}$ with $Q = \Lambda^H\Lambda$. The weight vector $\mathbf{v} = \mathbf{v}_R + j\mathbf{v}_I$ that is the required PSD estimator is related to $\boldsymbol{\alpha}$ and $\boldsymbol{\beta}$ through

$$\mathbf{v}_R = Q^{-1}\left(\mathbf{C}_R\boldsymbol{\alpha} - \mathbf{C}_I\boldsymbol{\beta}\right)$$
$$\mathbf{v}_I = Q^{-1}\left(\mathbf{C}_I\boldsymbol{\alpha} + \mathbf{C}_R\boldsymbol{\beta}\right),$$

where $\mathbf{C} = \mathbf{C}_R + j\mathbf{C}_I$.

- Compressive sampling is an efficient approach to reconstruct functions that are known to be sparse in a certain domain. Its efficiency arises in two forms: using a smaller number of samples than that is required in standard sampling-based reconstruction approaches and using a computationally efficient reconstruction mechanism.
- An N-length signal vector $\mathbf{x} = \Upsilon\mathbf{s}$ where Υ is an $N \times N$ basis matrix and $\mathbf{s} = (s_1,\ldots,s_N)^T$ is its coefficients with respect to basis Υ is called K-sparse in this basis if only $K < N$ expansion coefficients s_i are nonzero. Then Υ is called the sparsity matrix.

- If an N-length signal vector $\mathbf{x}$ is K-sparse, for $K < N$, with respect to some basis Υ, then it can be reconstructed by using an M-length, for $M < N$, vector measurement $\mathbf{y} = \Phi\mathbf{x}$ as

$$\min_{\mathbf{s}\in\mathbb{R}^N} \|\mathbf{s}\|_{\ell_1} : \mathbf{y} = \Phi\Upsilon\mathbf{s}.$$

  In particular, if the number of samples is chosen so that

$$M \geq C\mu^2(\Phi,\Upsilon)K\log\left(\frac{N}{\delta}\right),$$

  where $\mu(\Phi,\Upsilon)$ is the coherence between the sensing basis Φ and the sparsity basis Υ defined as

$$\mu(\Phi,\Upsilon) = \sqrt{N}\max_{1\leq i\leq N, 1\leq j\leq M}\left|< \varphi_i, \phi_j >\right|$$

  and C is some positive constant, then the earlier reconstruction will agree with the true sparse signal with probability greater than $1-\delta$.
- The number of samples required for high accuracy in reconstruction can be reduced by selecting a sensing basis Φ that is highly incoherent with the sparsity basis Υ. Interestingly, this can be achieved by simply picking a random matrix for the sensing basis Φ due to the fact that random vectors are highly incoherent with any fixed set of vectors.
- If it is known that a signal is only nearly K-sparse and that the observations are corrupted by noise and are of the form of

$$\mathbf{y} = A\mathbf{s} + \mathbf{w},$$

  where A is an $M \times N$ matrix and $\mathbf{w}$ is the additive noise, then still it is possible to use the compressive sampling reconstruction of the form of

$$\min_{\mathbf{s}\in\mathbb{R}^N} \|\mathbf{s}\|_{\ell_1} : \mathbf{y} = A\mathbf{s}$$

  to obtain a highly accurate signal reconstruction. In this case, the accuracy will improve as the isometry constant δ_K of matrix $\mathbf{A}$ becomes smaller, where order K isometry constant δ_K of matrix A is the smallest δ for which

$$(1-\delta) \parallel \mathbf{s} \parallel_{\ell_2}^2 \leq \parallel A\mathbf{s} \parallel_{\ell_2}^2 \leq (1+\delta) \parallel \mathbf{s} \parallel_{\ell_2}^2,$$

  for all K-sparse vectors $\mathbf{s}$.
- A random $M \times N$ matrix will have small enough order K isometry constant if

$$M \geq C_2 K \log\frac{N}{K},$$

  for some constant C_2.

- Compressive sampling of subbands can be used to reduce the computational demand encountered in wideband spectrum sensing.
- Suppose that N-vector $\mathbf{y}$ denotes the subband signal collected over a time segment of NT_s with a sampling period $T_s = \frac{1}{2B}$ when the bandwidth of the subband is B. If $\mathbf{y}$ only occupies a small fraction of the total bandwidth B of the subband, then it is a sparse signal in the frequency domain.
- If $\mathbf{y}$ occupies only a fraction $\frac{K}{N}$ of the total bandwidth B of the subband, then on average, we may expect only K out of the N elements of the DFT $\mathbf{y}_f = \mathbf{C}\mathbf{y}$ of the signal $\mathbf{y}$ to be nonzero or dominant compared to the other coefficients. Hence, we may construct the sparse signal $\mathbf{y}_f$ by using the compressive sampling reconstruction

$$\hat{\mathbf{y}}_f = \arg\min_{\mathbf{y}_f \in \mathbb{R}^N} \|\mathbf{y}_f\|_{\ell_1} \quad \text{s.t.} \quad \mathbf{y}_c = \mathbf{\Phi}\mathbf{C}^{-1}\mathbf{y}_f.$$

  based only on an M-length observation vector $\mathbf{y}_c = \mathbf{\Phi}\mathbf{y}$. As long as $M > CK \log N$ for some C, we can expect a very good reconstruction $\hat{\mathbf{y}}_f$.

11

SIGNAL CLASSIFICATION IN WIDEBAND COGNITIVE RADIOS

11.1 INTRODUCTION

For a wideband cognitive radio, the ability to accurately infer RF environment is critical to achieving its performance objectives. Each of the spectrum subbands may contain multiple signals from multiple systems. Comprehending spectrum state requires not only the ability to detect RF signals at arbitrary carrier frequencies within each subband, as we already discussed in Chapter 10, but also being able to identify these signals.

In this chapter, we investigate how a cognitive radio can identify spectral activities detected in the sensed subband. Referring to the spectral knowledge acquisition architecture of Figure 2.4, at the first stage of spectrum sensing, sensed observations from a subband are used to detect spectral activities as discussed in the previous chapter. Once a set of spectral activities are detected, these are passed on to the signal classification module at the next stage of spectrum sensing chain in the cognitive engine as shown in Figure 2.4.

Note that spectrum sensing algorithms for DSA-based cognitive radios[1] scan a desired channel corresponding to one of the known (primary) systems to determine whether that particular signal is ON or OFF. For instance, a cognitive radio used for DSA over the idle TV bands treats the TV signals as the primary user signals. Hence, spectral characteristics (e.g., carrier frequency and bandwidth) of the signals

[1] We will discuss this topic in Chapter 12.

Signal Processing for Cognitive Radios, First Edition. Sudharman K. Jayaweera.

of interest can normally be taken as known. On the other hand, at any given time, a wideband cognitive radio will sense a spectrum subband that can be wide enough to contain several channels belonging to possibly different active systems. For example, in the 2.4 GHz ISM band, there may be WiFi (IEEE 802.11b), Bluetooth, as well as other unknown signals such as microwave oven radiation. Another example is the WMTS systems operating in lower and upper L bands (1395–1400MHz and 1427–1432MHz, respectively): These bands may contain an unknown mix of utility telemetry signals, government radar signals, as well as radio astronomy signals [39]. Hence, signals to be sensed are certainly not homogeneous and may have unknown center frequencies within the subband, posing a significantly more complicated problem.

Moreover, a wideband cognitive radio needs to detect and identify all spectral activities within the selected subband. There are several reasons for this. First, a spectrum opportunity of interest for a wideband cognitive radio may not necessarily be a white space as is the case with DSA-based cognitive radios. Indeed, a spectrum opportunity for a wideband cognitive radio will depend on its instantaneous performance objectives. For example, at one time, the objective may be to simply find the best channel or a combination of channels to transmit. At other times, the radio's objective may be to achieve covert or low probability of interception/detection (LPI/LPD) type of communications. Yet another objective may be to find unfriendly signals and possibly jam them. As the performance objective varies, so does what is meant by a spectrum opportunity. Second, a wideband cognitive radio can be sophisticated enough to make simultaneous use of more than one spectrum opportunity. It may have the capability to communicate over a combination of radio interfaces simultaneously or even pursue multiple objectives at the same time. Hence, the more comprehensive the spectrum understanding and awareness, the better may be the overall performance.

As a start, similar signals must be clustered together into common groups, thereby indicating the possibility they may belong to the same wireless system. A signal can then either be associated as belonging to one of the already known systems or may be assigned to a possibly new class of signals. Hence, the classification stage in a wideband cognitive radio will be a hybrid of classification, clustering, and association modules. At any given time, the number of active systems in a subband as well as the numbers of active signals belonging to each of these systems will likely not be known a priori. Hence, the number of active systems, the number of users in each system, and the number of extractable features may all have to be allowed to be variables to handle spectrum dynamism.

Since the scanned subband, as well as the spectrum of interest spanned by all the N_b subbands, may contain a number of different systems and at any given time there might be an arbitrary number of signals belonging to each active system, the Bayesian nonparametric classification approach of Chapter 8 seems to be an ideal framework for this problem. Often, classification is better achieved based on a set of suitable features extracted from observations rather than based on direct observations themselves. Hence, after formulating the spectral activity classification problem in a cognitive radio, we first discuss feature extraction before developing a nonparametric classification approach for spectral activity identification in wideband cognitive radios.

11.2 SIGNAL CLASSIFICATION PROBLEM IN A WIDEBAND COGNITIVE RADIO

The goal of the cognitive radio is to acquire as comprehensive a spectrum knowledge as possible. Referring to the cognitive radio wideband spectrum sensing architecture of Chapter 2, at the first stage, the radio will detect possible spectral activities present in the scanned subband. A subband may contain several signals possibly belonging to different systems. The wireless channel is a dynamic channel and signals may appear and disappear at arbitrary times. Hence, the third stage of the spectrum sensing chain of Figure 2.4 is tasked with identifying these detected spectral activities.

Spectral activity identification can be of two types: classification or clustering. In certain spectrum bands, the cognitive radio may have a good a priori knowledge on which type of systems are to be found. In that case, most of the detected signals may fall into one or another of the known systems. Hence, depending on the center frequencies of the detected signals, the radio may be able to extract certain predefined features and associate the signals with specific known systems. Hence, the task can be interpreted as either supervised classification or association. In other spectrum bands, the radio may not have much prior information thereby prompting it to take a more flexible approach to identification of detected spectral activities. For example, the radio may first attempt to group, or cluster, signals according to their common characteristics/features. Some of these clusters may in fact correspond to known systems, while others may not. Hence, the radio may follow this with an association step to attempt to associate certain clusters with known systems thereby identifying those signals. Signals in other clusters may tentatively be treated as signals belonging to unknown new systems. However, with more future observations from the same spectrum range, the radio may later be able to prove or disprove these hypotheses on new signal classes.

Be it classification or clustering, it is better achieved by identifying common characteristics of signals belonging to, for example, the same system. This is true in any classification problem. For instance, suppose a school administrator wants to assign students for a special ed class. A commonly extracted feature in this case is the students IQ numbers. In this case, the heights of the students may not be of much use as a feature. On the other hand, if the school's basketball coach wants to select players for the basketball team, the height of the students may very well be a feature of interest. This shows that depending on the anticipated groupings, the best-suited features may vary. In the first example, identifying a good feature is not that difficult although still one may argue whether it is sufficient as the sole, or the most important, feature. In some situations, however, identifying the most relevant features that will help distinguish among different classes may be difficult.

For instance, suppose we give a collection of earthly vegetables and fruits to an alien from another galaxy:

$$\{\text{banana, eggplant, apple, squash, beans, broccoli, mango}\}.$$

We ask the alien to classify them into two groups as *vegetables* and *fruits*. What should the alien extract as suitable features? Perhaps the difficulty lies in its inability at the first place to clearly define what is a vegetable and what is a fruit.

It is worth noting that not all relevant features for a classification task may directly be viewed as simple physical characteristics. Indeed, in many pattern recognition problems, the extracted features may be abstract features from transformed data. These features may only be viewed in a suitable abstract space. Still, it can be argued that fundamentally these features, or patterns, do have a physical origin. It is just that human sensory organs cannot directly see these patterns by merely observing raw data.

These same issues will have to be anticipated in signal classification. The basic signal classification problem, however, is at least a little more well defined than the aforementioned problem of classifying vegetables and fruits. For example, if one needs to separate detected signals in the 2.4GHz ISM band into either Bluetooth or WiFi, the problem is relatively easier. The reason is unlike vegetables and fruits, the definitions of WiFi and Bluetooth are much more specific. The signal classification problem in a wideband cognitive radio, on the other hand, can be more complicated than this.

Suppose, instead of asking to classify the given collection into two groups labeled as fruits and vegetables, that we merely require the alien group the given collection to any number of different classes as appropriate. We are, of course, anticipating the most suitable classification in this case to be two classes as vegetables and fruits. But depending on the extracted features, the alien may come up with any number of possible groupings. If the extracted feature is the COLOR, for example, the alien may come up with at least three different classes in which the (green) mango is grouped together with beans and broccoli. Note that the outcome here can be erroneous due not only to classifying things with wrong peers but also to either creating not enough or more than necessary distinct classes.

Signal identification problem encountered in a wideband cognitive radio, in general, can resemble this situation. Once the spectral activities are detected at certain frequencies (within a given subband), the radio needs to extract the most relevant features from each of the detected signals so as to help classification. The radio may have certain a priori knowledge of the type of signals that might be active in certain subbands. Hence, depending on at what frequencies it detects the spectral activities, it may have some guidance as to what features to look for. Once the appropriate features are extracted from detected signals, the radio will either classify them directly into known signal classes or cluster them into a set of data-defined classes. A final step of this procedure will then attempt to identify the classes with known systems.

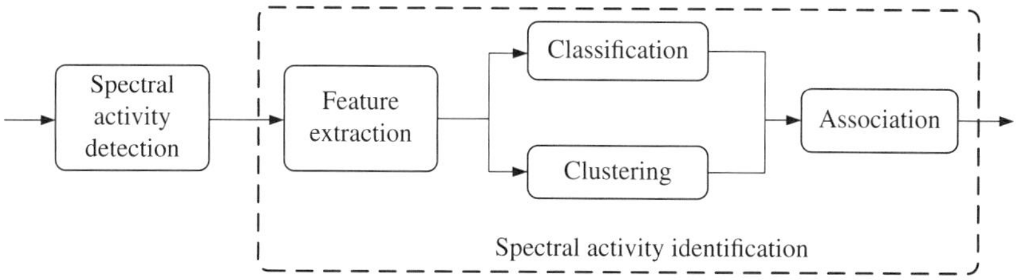

FIGURE 11.1 Spectral activity identification in a cognitive radio.

Thus, the signal identification problem in a wideband cognitive radio consists of the following three stages as shown in Figure 11.1:

1. Feature extraction
2. Classification and clustering
3. Signal and cluster association

The number of features extracted does not have to be the same at all frequencies. Based on prior information a cognitive radio may have about possible active systems at certain frequencies within a subband, the radio may know which type of features to look for. In practice, this is likely to be the case. Hence, the output from the feature extraction module in the first step can be allowed to be a variable-length feature vector (for each detected spectral activity).

The second stage may be either classification into some known classes or clustering into an arbitrary set of classes. The trigger may be the frequencies of the detected signals and some other basic features. The classification path is akin to a standard supervised classification problem in which the input feature vectors will be classified into a set of known classes. As we mentioned in Chapter 8, many algorithms that can do such classification already exist. Depending on the application scenario, the spectrum range, and other possible factors, the most suitable classification algorithm may vary. However, with little effort and modification, many of the available standard classification algorithms can solve the signal classification problem provided that suitable features are being extracted. Hence, in this chapter, we do not further discuss the signal classification path in the second stage of the spectral activity identification module shown in Figure 11.1.

The clustering path, on the other hand, comes with issues unique to the signal identification problem in a wideband cognitive radio, thus deserving a detailed look. Note that clustering path in Figure 11.1 may be applicable under several scenarios. One can argue that whenever a cognitive radio is first turned on in a given RF environment, it may always pursue the clustering path until it develops enough confidence in its prior knowledge so as to revert to the classification path later. Still, even when the radio has been operating in the classification mode for a while, sometimes, it may be triggered to pursue the clustering path if it encounters unexpected detected signal frequencies. If this was caused by an outlier, the radio may be able to verify this and falls back into the classification path after a few sensing cycles. The two parallel paths in the second stage will thus be interacting with each other.

As we have already noted, the clustering problem in a cognitive radio has several attributes that make it more challenging. First, we need to allow for an arbitrary number of active systems. Hence, the clustering algorithm must not presume, or require, the number of active clusters as a priori knowledge. It should be able to determine that from the input feature vectors themselves. Most clustering algorithms, however, require either the number of active clusters or the clusters themselves to be known a priori. For example, as we saw in Chapter 8, the commonly used K-means algorithm requires the number of clusters to be known: It can create the clusters from the data itself if the number of clusters were to be specified. On the other hand, a supervised classification algorithm such as k-nearest neighbor algorithm may not have to know the exact number of clusters represented in a given input data set as long as all possible clusters are already created through training. Hence, algorithms of both these sorts are not directly applicable to the problem of signal clustering in a wideband cognitive radio. One suitable approach, perhaps, is the X-means algorithm that we encountered in Chapter 8, which is a variation of the K-means algorithm that does not require knowledge of the number of clusters a priori. However, as we also discussed in Chapter 8, if the features do not have Gaussian likelihoods, the performance of the X-means algorithm may severely degrade. Hence, a more general alternative candidate is the DPMM-based Bayesian nonparametric classification approach introduced in Chapter 8. This framework not only allows the number of clusters to be arbitrary, but also it can grow as the complexity of the data increases. Moreover, the feature input likelihoods do not have to be limited to multivariate Gaussians. Indeed, extensions introduced in Chapter 8 allow the data inputs to have likelihoods belonging to different families of distributions.

In the final stage of signal identification process in Figure 11.1, the resulting signal classes are associated with known systems, be they the clusters created by clustering module or the already existing classes into which signals were classified by the classification module. This will be useful in later operational decision-making. The association step can also be interpreted as a form of supervised classification. Indeed, a cognitive radio may have knowledge of certain systems having certain characteristics in certain spectrum bands. It may thus be able to associate common features of signals in a certain cluster with features expected from such a known system thereby concluding that these signals are indeed from that particular system. There may also be situations in which the radio may not be able to do this. Then, the created cluster may possibly represent a new system unknown to the radio thus far. The cognitive radio may hold on to this cluster for some period of time to determine from additional future observations whether it deserves to be designated as a new system or whether that cluster was merely a mistake due to data outliers.

A cognitive radio does not necessarily have to follow these three steps sequentially in that order all of the time. The steps can be interweaved and the radio can go through them only as needed. Clearly, there is no need to perform unnecessary computations when radio can make a classification decision with sufficiently high confidence of being accurate. Hence, it is to be understood that there is a higher-level decision-making algorithm that directs and controls the sequence of processing within the spectral activity detection module. Indeed, this is the same module that will guide the selection of the classification path versus the clustering path that we already mentioned in the context of Figure 11.1.

11.3 FEATURE EXTRACTION FOR SIGNAL CLASSIFICATION

Extracting the most relevant features is critical to obtain correct classification results. The task of the classification algorithm can be that much easier if it is given the right feature vectors as input. In practice, it may not always be possible to determine the most relevant feature vectors. However, at least we may be able to extract reasonably suitable and relevant features. The number of features extracted as well as the type of features may be different from one spectrum subband to another as well as from one frequency to another depending on the type of prior information the radio may have about the spectrum environment. In the following, we discuss some basic candidates for features that might be useful in spectral activity identification. Clearly, these are not the only possible ones to be extracted nor they may be the most relevant in certain applications. They are used here merely to illustrate the steps of the spectral activity identification process in a cognitive radio.

11.3.1 Carrier/Center Frequency

Since spectral activities are detected in the frequency domain, the most obvious feature to be extracted is the carrier or center frequency of the detected spectral activity. Indeed, the number K of active signals within the sensed subband with their corresponding normalized center frequencies F_k's (or the center frequencies f_k's) is the basic output from the spectral activity detection step in the spectrum sensing architecture.

As we saw in Chapter 10, it is possible to use sophisticated spectral activity detectors to account for non-Gaussianity and other such effects. However, to keep the discussion simple, let us assume here that the spectral activities are detected by applying an energy detector implemented based on the periodogram as in Section 10.3 of Chapter 10. Let the sampled received signal sequence during the sensing time interval be denoted by $y[n] \triangleq y(t+nT_s)$, for $n = 0,\ldots,N-1$, where T_s is the sampling period and N is the number of samples during that time interval. As we saw in Section 10.3, the spectral activity detection module first computes the DFT of the signal sequence as in (10.22):

$$Y[k] = \sum_{n=0}^{N-1} y[n]e^{-j2\pi\frac{kn}{N}} \quad \text{for} \quad k = 0,\ldots,N-1. \tag{11.1}$$

Next, a smoothed periodogram power spectral estimator is obtained by applying a sliding window technique as in (10.24):

$$\hat{S}_{yy}[k] = \frac{1}{LN} \sum_{l=-(L-1)/2}^{(L-1)/2} |Y[k+l]|^2, \tag{11.2}$$

where we have assumed an odd window length L. Note that $Y[k] = Y[k + mN]$, for $m \in \mathbb{Z}$, from the properties of DFT.

Center frequencies of possible spectral activities are then determined by applying a Neyman–Pearson threshold test of the form of $\hat{S}_{yy}[k] > \eta$, where from (10.31) the threshold η for a false-alarm probability of α_F is [167, 168]

$$\eta = \frac{\gamma^{-1}\left(L; (1 - \alpha_F)\,\Gamma(L)\right) P_n}{L}, \tag{11.3}$$

where P_n denotes the noise power $\mathbb{E}\{|W[n]|^2\}$. Note that noise power can reasonably be estimated from the periodogram under hypothesis H_0 as [169, 170]

$$\hat{P}_n = \frac{1}{N}\sum_{k=0}^{N-1} \hat{S}_{yy}[k].$$

The output from this spectral activity detection approach is the number of active signals K contained in the sensed subband signal and their corresponding center frequencies. As we saw in Section 10.3, the frequency smoothing can be interpreted as a way of handling the nonzero bandwidth of the signals. Hence, the implementation of the earlier spectral activity detection can be combined with an edge detection procedure to determine a coarse estimate of the bandwidth of each of the detected signals as well. Let us suppose that the carrier/center frequencies of the detected spectral activities are estimated as the midpoints of the segments formed by the intersection between the PSD curve and a threshold line η_{PSD}, as shown in Figure 11.2. We may denote by $\mathcal{A}_c$ the set of detected center frequencies in the subband of interest. Since there may be some erroneous detections, let us denote by N_f the number of detected frequencies so that $|\mathcal{A}_c| = N_f$. Note that the number of feature vectors extracted will then be N_f.

11.3.2 Cyclostationary Features

It is well known that almost all man-made signals exhibit certain underlying periodicities due to, for example, their symbol rates, coding schemes, packet/frame header structures, and training symbol sequences [65, 66, 171]. As we saw in Chapter 6, second-order periodicities resulting from such periodic characteristics will lead to spectral correlation that can be unique to the type of signal. These spectral correlations are reflected in nonzero cyclic frequencies associated with the signal. Hence, cyclic frequencies can be a common feature shared by all signals belonging to a same system.

As we know from Section 6.6, cyclic frequencies of a signal are best viewed by looking at its spectral correlation function $S_{YY}^{\alpha}(F)$. Observe from (6.95) to (6.98) that an estimate of the spectral correlation function can be obtained by time smoothing the finite-time DTFT-based finite-time cyclic periodogram $\hat{S}_p^{\alpha}(F; N)$ where we have assumed $n_0 = 0$ for convenience. However, as established in (6.100), smoothing can also equivalently be done in the frequency domain as in (6.99). Since processing in

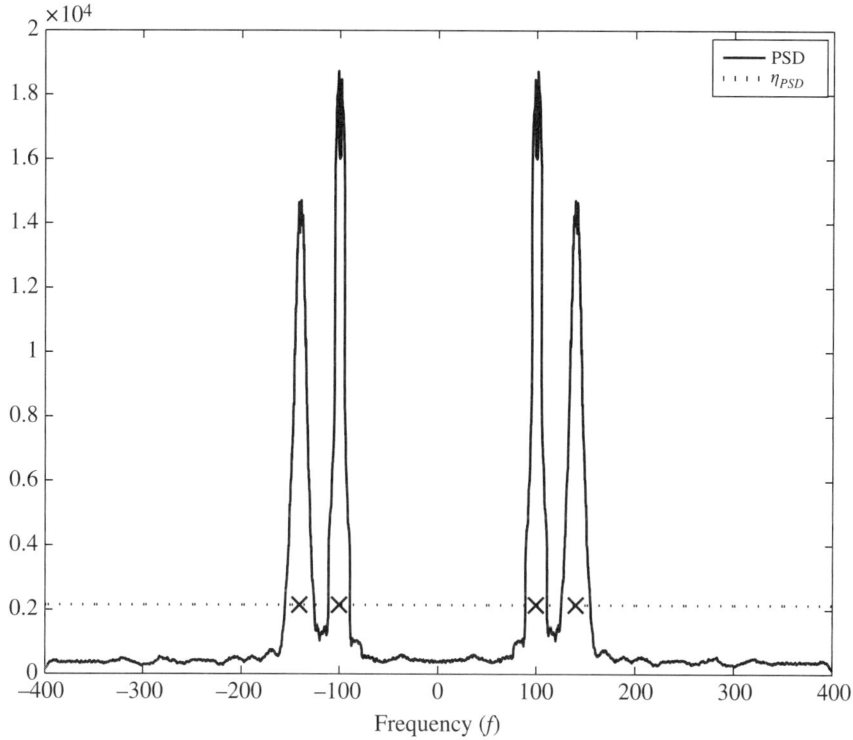

FIGURE 11.2 Carrier frequencies are estimated as the midpoints of the intersections between the PSD curve and the threshold line.

a cognitive radio will be done based on DFT, a frequency-smoothed estimate of the spectral correlation function thus can be obtained as

$$\hat{S}_{yy}^{\alpha}[k] = \frac{1}{LN} \sum_{l=-(L-1)/2}^{(L-1)/2} Y[k-l]Y^*[k+\alpha-l], \tag{11.4}$$

where $\alpha \in \left[-\frac{1}{2}, \frac{1}{2}\right]$ is the normalized cyclic frequency.

Example 11.1 (Extraction of Cyclic Frequencies Corresponding to Baud Rate of a Signal) Consider a passband signal at a carrier frequency $f_c = 10$ MHz. The signal is a 4-QAM signal with a symbol rate of 2 MBauds (Msymbols/s). Let us assume that sampling rate is $f_s = 32$ Msamples/s. Figure 11.3 shows the spectral correlation function magnitude when each sensing duration is of 20 μs long and the frequency-smoothing window length is $L = 101$. Note that for $\alpha = 0$ the SCF gives the power spectral density function. More importantly, Figure 11.3 shows that there are cyclic

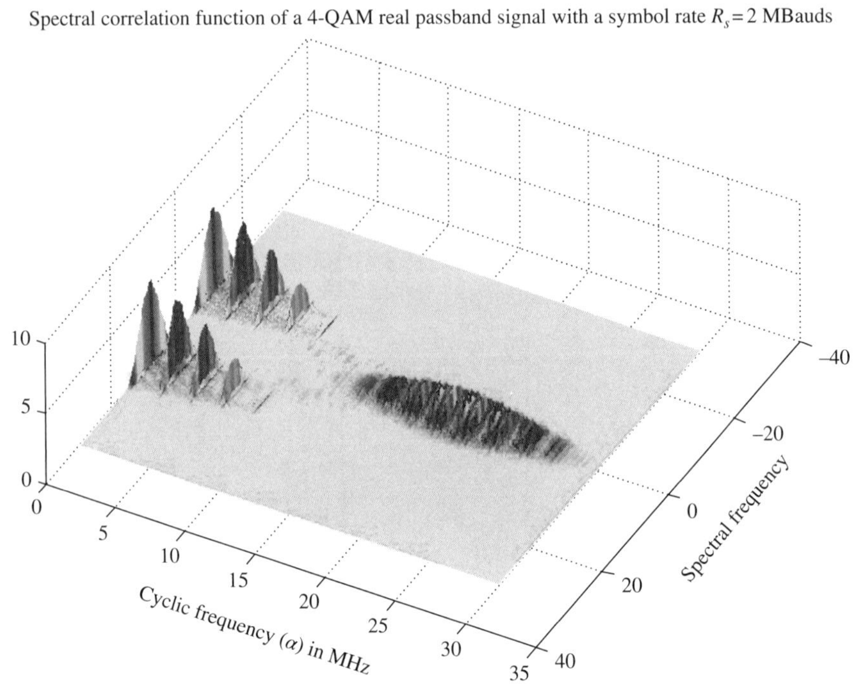

FIGURE 11.3 Spectral correlation function (SCF) magnitude of the real passband 4-QAM signal with a symbol rate *MBauds* at a center frequency $f_c = 10\,\text{MHz}$. Sensing duration is 20 μs long and the frequency-smoothing window length is $L = 101$.

frequencies at integer multiples of 2MHz corresponding to the symbol rate of the signal. ■

We may use the spectral correlation function $\hat{S}_{yy}^{\alpha}[k]$ to obtain an estimate of the spectral autocoherence function (or the magnitude of the complex autocoherence function):

$$|\hat{C}_{yy}^{\alpha}[k]| = \frac{|\hat{S}_{yy}^{\alpha}[k]|}{\sqrt{\hat{S}_{yy}[k]\hat{S}_{yy}[k+\alpha]}}. \tag{11.5}$$

The cyclic profile $I(\alpha)$ of the signal is then computed using the above spectral autocoherence function [171]:

$$I(\alpha) = \max_{k \in \{0,\cdots,N-1\}} |\hat{C}_{yy}^{\alpha}[k]|. \tag{11.6}$$

In wideband spectrum sensing, there might be multiple active signals in the sensed band simultaneously. The cyclic profile (11.6) does not help separately identify cyclic frequencies of different signals. However, it is easy to see that when the sensed signal is a superposition of multiple independent signals, then their spectral peaks also form a superposition in the (F,α) domain. To see this, suppose that the sensed signal is of the form of

$$Y[n] = \sum_{k=1}^{K} X_k[n] + W[n], \tag{11.7}$$

where $\{X_k[n]\}_{k=1}^{K}$ are independent zero-mean random processes denoting K superposed signals and $W[n]$ is an independent white noise process with a power spectral density of $\frac{N_0}{2}$. The autocorrelation function (6.78) of $Y[n]$ can be written as

$$\mathscr{R}_{YY}[n,m] = \sum_{k=1}^{K} \mathscr{R}_{XX}^{(k)}[n,m] + \frac{N_0}{2}\delta[m], \tag{11.8}$$

where $\mathscr{R}_{XX}^{(k)}[n,m]$ denotes the autocorrelation functions of $X_k[n]$, for $k = 1,\ldots,K$. From (6.85), then cyclic autocorrelation function of $Y[n]$ is given by

$$\begin{aligned}
\mathscr{R}_{YY}^{\alpha}[m] &= \lim_{N\longrightarrow\infty} \frac{1}{N} \sum_{n=-\frac{N-1}{2}}^{\frac{N-1}{2}} \mathscr{R}_{YY}[n,m]e^{-j2\pi\alpha n} \\
&= \lim_{N\longrightarrow\infty} \frac{1}{N} \sum_{n=-\frac{N-1}{2}}^{\frac{N-1}{2}} \left[\sum_{k=1}^{K} \mathscr{R}_{XX}^{(k)}[n,m] + \frac{N_0}{2}\delta[m]\right] e^{-j2\pi\alpha n} \\
&= \sum_{k=1}^{K} \mathscr{R}_{XX}^{(k),\alpha}[m] + \frac{N_0}{2}\delta[m]\delta_D(\alpha),
\end{aligned}$$

where $\mathscr{R}_{XX}^{(k),\alpha}[m]$ denotes the cyclic autocorrelation function of the kth signal $X_k[n]$. It then follows from (6.87) that the spectral correlation function of $Y[n]$ is indeed the superposition of the spectral correlation functions of the K-independent signals:

$$\begin{aligned}
\mathscr{S}_{YY}^{\alpha}(F) &= \sum_{n=-\infty}^{\infty} \mathscr{R}_{YY}^{\alpha}[m]e^{-j2\pi Fm} \\
&= \sum_{k=1}^{K} \mathscr{S}_{XX}^{(k),\alpha}(F) + \frac{N_0}{2}\delta_D(\alpha).
\end{aligned} \tag{11.9}$$

Since the spectral correlation function of the superposition of multiple signals will have peaks at their individual cyclic frequencies, an approach to separate their cyclic characteristics is then to partition the (F,α) plane into Voronoi cells whose point sites are located at the detected carrier frequency points $\mathcal{A}_c$ [172]. Thus, we may define the cyclic sub domain profile at spectral frequency $k_c \in \mathcal{A}_c$ as [167, 168]

$$I(\alpha, k_c) = \max_{k \in \{k_c - \Delta, \ldots, k_c + \Delta\}} |C_{yy}^{\alpha}[k]|, \tag{11.10}$$

where 2Δ is the estimated bandwidth of the signal transmitting at frequency k_c. Note that this bandwidth estimate only needs to be coarse and can be obtained from the periodogram as mentioned earlier.

A simple way to extract the strongest cyclic frequencies is to threshold the above cyclic sub domain profile:

$$\{\alpha \neq 0 : \mathcal{I}_E I(\alpha, k_c) \geq \zeta\}, \tag{11.11}$$

where $\mathcal{I}_E$ denotes the indicator function of the event

$$E = \{I(\alpha, k_c) \text{ is a local maximum}\}$$

and $\zeta \in (0,1)$ is a threshold for the peak detection in the cyclic subdomain profile (note from (6.94) that the autocoherence magnitude is always upper bounded by unity).

There are, of course, other ways to determine the relevant cyclic frequencies as well. For example, it is possible to use a threshold that is a function of the carrier frequency itself. Another alternative is to specify a specific number of cyclic frequencies to be extracted no matter what their strengths are. Yet another method may be to specify a minimum or a maximum number of cyclic frequencies. These choices need to be made based on the specific application considerations.

If we pursue the approach of (11.11) to extract the cyclic features, there may be different numbers of cyclic frequencies extracted from different types of signals. Hence, we may denote the set of cyclic frequencies selected at frequency k by $\{\alpha_1, \ldots, \alpha_{M_k}\}$.

In the following discussion, we will explicitly focus on cyclic properties induced by the symbol rates although wireless signals may be expected to exhibit cyclic frequencies due to many other underlying periodic properties. The same procedure can be modified to extract such cyclic frequencies. In an uncoded transmission system centered at k_c, normally, the symbol rate can be estimated from the location of the first peak in the cyclic subdomain profile $I(\alpha, k_c)$ (for $\alpha \neq 0$) [66, 167]. As can be seen from Figure 11.3 of the 4-QAM signal considered in Example 11.1, however, extracting the cyclic frequencies has to be done with care. Although ideally it should show an impulse-like spectral line at each of the expected α values (integer multiples of 2 MHz in this case), there is a nonzero spread along the α-axis. As we know from Chapter 6, this can be a result of estimating the spectral correlation function based only on finite-length data as well as any frequency smoothing used.

Example 11.2 (Effect of Finite Length Data and Windowing on Extracting Cyclic Frequencies) Consider again the same real passband 4-QAM signal with a symbol rate of 2 MBauds at the carrier frequency f_c = 10 MHz. Figure 11.4(a) shows the two-dimensional (2-D) contours of the spectral correlation function magnitude in the $f - \alpha$ plane with earlier 20 μs sensing duration and L = 101 frequency-smoothing window length. Figure 11.4(b), on the other hand, shows the difference in the sharpness (resolution) in the α dimension when a shorter sensing duration (5 μs) and a shorter frequency-smoothing window size (L = 31) are used. Thus, in using the spectral correlation function to find and extract cyclic features, one needs to carefully consider the suitable values for these parameters so that accurate enough estimates are obtained. ■

11.3.3 Modulation Type and Order Features

Cyclostationary features corresponding to symbol/baud rate or the bandwidth may not be able to distinguish between two signals having the same symbol rate but based on two different modulation orders: for example, the same baud rate signals with 8QAM versus 64PAM or 4PSK versus 8PSK. The standard approach to general modulation classification problem is to consider estimating the most likely modulation type within a set of modulations, or constellations, of interest. Under the assumption that all constellations of interest are equiprobable, the optimal solution is then the maximum likelihood estimate of the modulation order [173–176]. However, it has been argued that the ML approaches not only have high complexity but also are highly sensitive to the iid assumption, non-Gaussian noise, timing, phase, and frequency offsets as well as wireless channel effects [176–178]. A detailed recent survey of automatic modulation classification (AMC) approaches can be found in [179]. As an alternative to the drawbacks of the ML methods, it has been shown that an effective way to distinguish among different modulation types and orders is to use higher order statistics such as the fourth-order cumulants [177]. In the following, we discuss an approach developed in [177] for this purpose.

To see how higher-order statistics can be used for extracting features related to such signal properties, recall that the *moment generating function* $M_X(t)$ of an arbitrary random variable X is defined as

$$\begin{aligned} M_X(t) &= \mathbb{E}\{\exp(tX)\} \\ &= \sum_{n=1}^{\infty} \mu_n \frac{t^n}{n!}, \end{aligned}$$

where coefficient μ_n denotes the nth moment of X: $\mu_n = \mathbb{E}\{X^n\}$. The *cumulant generating function* $g_X(t)$ of an arbitrary random variable X is the natural logarithm of the moment generating function $M_X(t)$:

$$\begin{aligned} g_X(t) &= \log\left(M_X(t)\right) = \log\left(\mathbb{E}\{\exp(tX)\}\right) \\ &= \sum_{n=1}^{\infty} C_n \frac{t^n}{n!}. \end{aligned}$$

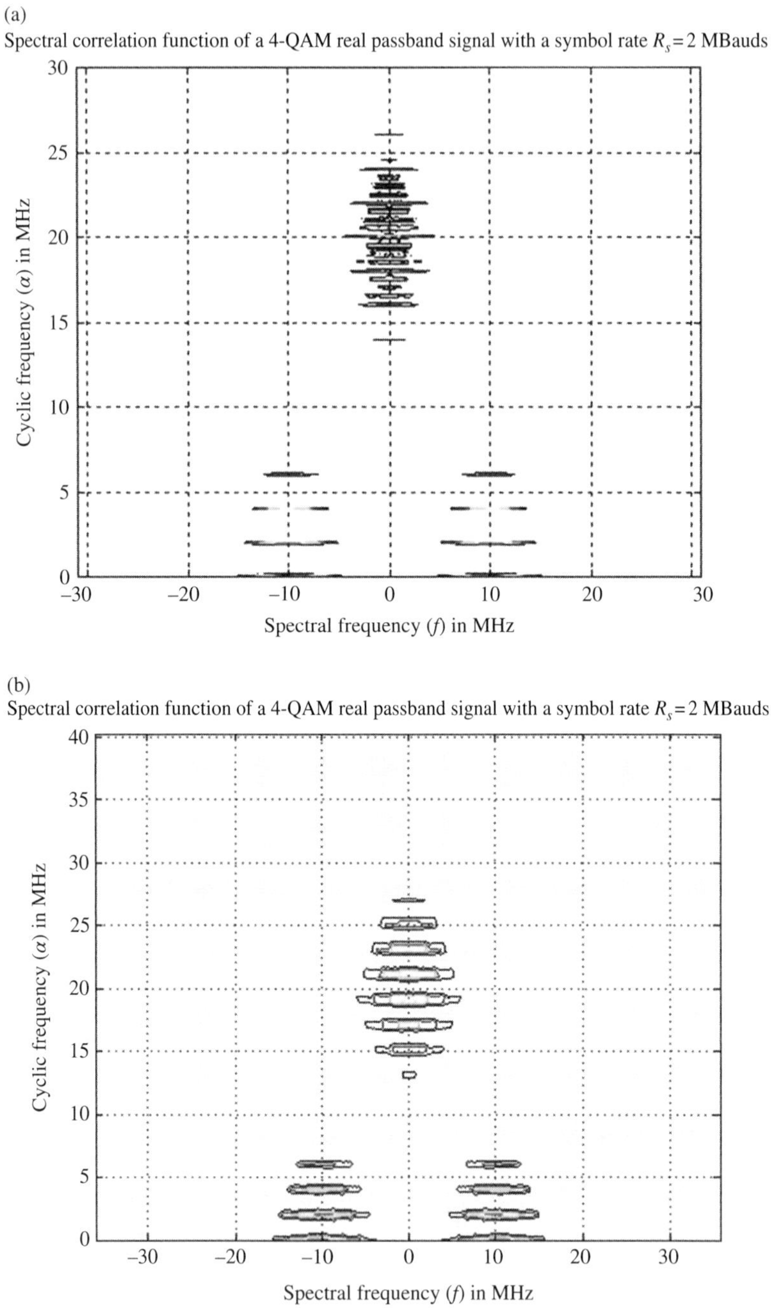

FIGURE 11.4 2-D contours of the spectral correlation function (SCF) magnitude of the real passband 4-QAM signal with a symbol rate 2MBauds at a center frequency $f_c = 10$ MHz. (a) Spectral correlation function (SCF) magnitude contours when sensing duration is 20 μs long and the frequency-smoothing window length is $L = 101$.(b) Spectral correlation function (SCF) magnitude contours when sensing duration is 5 μs long and the frequency-smoothing window length is $L = 31$.

Coefficient C_n is then called the nth order cumulant of X. It is not difficult to see that there is a relationship between the cumulants and the moments. Indeed, for the first few values of n, we may express the cumulants in terms of the moments as

$$\begin{aligned}
C_1 &= \mu_1\\
C_2 &= \mu_2 - \mu_1^2\\
C_3 &= \mu_3 - 3\mu_2\mu_1 + 2\mu_1^3\\
C_4 &= \mu_4 - 4\mu_3\mu_1 - 3\mu_2^2 + 12\mu_2\mu_1^2 - 6\mu_1^4.
\end{aligned}$$

As we go to the higher orders, this gets unwieldy, but there is a recursion to express any cumulant in terms of the moments:

$$C_n = \mu_n - \sum_{m=1}^{n-1} \binom{m-1}{n-1} C_m \mu_{n-m}.$$

Throughout, we will assume that the mean of the received signal is zero (if not, it can be estimated and subtracted out) so that the first 4 cumulants simplify to

$$\begin{aligned}
C_1 &= 0\\
C_2 &= \mu_2\\
C_3 &= \mu_3\\
C_4 &= \mu_4 - 3\mu_2^2.
\end{aligned}$$

When we consider complex baseband signal samples, however, the moments (and the cumulants) can be defined in several ways. For example, the second-order moment can be written either as $\mathbb{E}\{X^2\}$ or $\mathbb{E}\{|X|^2\}$. Hence, in the case of feature extraction from complex baseband signals, we have two second-order cumulants/moments as

$$C_{20} = \mu_{20} = \mathbb{E}\{X^2\} \tag{11.12a}$$

$$C_{21} = \mu_{21} = \mathbb{E}\{|X|^2\}, \tag{11.12b}$$

where the second subscript denotes how many terms have been conjugated. Similarly, there are three fourth-order cumulants denoted as

$$\begin{aligned}
C_{40} &= cum(X,X,X,X)\\
C_{41} &= cum(X,X,X,X^*)\\
C_{42} &= cum(X,X,X^*,X^*).
\end{aligned}$$

Since

$$cum(X,Y,Z,W) = \mathbb{E}\{XYZW\} - \mathbb{E}\{XY\}\mathbb{E}\{YZ\} - \mathbb{E}\{XZ\}\mathbb{E}\{YW\} - \mathbb{E}\{XW\}\mathbb{E}\{YZ\},$$

for four zero-mean random variables X, Y, Z, and W [177], we may express the three fourth-order cumulants in terms of moments as (note that from (11.12a) and (11.12b) the second-order cumulants are the same as the second-order moments)

$$C_{40} = \mathbb{E}\{X^4\} - 3C_{20}^2 \tag{11.13a}$$
$$C_{41} = \mathbb{E}\{X^2|X|^2\} - 3C_{20}C_{21} \tag{11.13b}$$
$$C_{42} = \mathbb{E}\{|X|^4\} - |C_{20}|^2 - 2C_{21}^2. \tag{11.13c}$$

Hence, a simple way to extract the fourth-order cumulants from a given signal is to first compute the sample estimates of the required second-order moments and then use them in sample estimates based on the last set of equations (11.13a)–(11.13c) as follows:

$$\hat{C}_{20} = \frac{1}{N}\sum_{n=1}^{N} x^2[n]$$
$$\hat{C}_{21} = \frac{1}{N}\sum_{n=1}^{N} |x[n]|^2$$

and

$$\hat{C_{40}} = \frac{1}{N}\sum_{n=1}^{N} x^4[n] - 3\hat{C}_{20}^2$$
$$\hat{C_{41}} = \frac{1}{N}\sum_{n=1}^{N} x^2[n]|x[n]|^2 - 3\hat{C}_{20}\hat{C}_{21}$$
$$\hat{C_{42}} = \frac{1}{N}\sum_{n=1}^{N} |x[n]|^4 - |\hat{C}_{20}|^2 - 2\hat{C}_{21}^2.$$

Many AMC approaches indeed use the earlier framework: Estimate the cumulant measures of the detected signals and compare them to a set of trained measures [177, 180–182]. These methods, however, presume that the detected signal is demodulated and sampled at exactly the symbol rate, which requires prior knowledge of the signal pulse shape [183]. Hence, the cumulant-based modulation techniques may not be applicable in the absence of such prior information.

An alternative feature extraction method for identifying the modulation scheme of a digital signal can be developed based on the kurtosis measure [182] defined as

$$\texttt{kurtosis}(X) = \frac{\mathbb{E}\left\{(X-\mu)^4\right\}}{\left[\mathbb{E}\left\{(X-\mu)^2\right\}\right]^2}.$$

The kurtosis can also be estimated from a given signal sequence $\{x[n]\}_{n=1}^{N}$ as

$$\texttt{kurtosis}(x) = \frac{N\sum_{n=1}^{N}\left(x[n]-\hat{\mu}\right)^4}{\left[\sum_{n=1}^{N}\left(x[n]-\hat{\mu}\right)^2\right]^2},$$

where $\hat{\mu}$ is the sample-mean estimate:

$$\hat{\mu} = \frac{1}{N}\sum_{n=1}^{N} x[n].$$

Unlike the cumulant-based method, the kurtosis-based approach does not require demodulation of the detected signal. A sensed signal of a wideband cognitive radio may include several signals at several different frequencies. In order to extract the kurtosis of each digital signal, we may filter each of the detected signals separately. This filtering can be made possible by using the results of the previous steps of energy and cyclo stationary detection. We can thus estimate the center frequencies and the bandwidths of the detected signals so that by applying bandpass filters at each estimated center frequency $\hat{F}_k$ (for $k = 1,\ldots,N_f$), we can generate the filtered output $x_k[n]$ (for $n = 1,\ldots,N$), which can be used to estimate the kurtosis of the kth squared-signal magnitude as $\texttt{kurtosis}\left(|x_k^2[n]|\right)$ [182].

11.4 A SIGNAL CLASSIFICATION ARCHITECTURE FOR A WIDEBAND COGNITIVE RADIO

Figure 11.5 shows a generic classification architecture for a wideband autonomous cognitive radio (features shown are only for illustration purpose and can be replaced/augmented by application-specific features). The input to the classification module is the observed signal with detected spectral activity information from the preceding spectral activity detection module. This architecture embeds the functional architecture of Figure 11.1. Indeed, the operation of the architecture in Figure 11.5 interweaves the feature extraction, classification/clustering, and identification steps of Figure 11.1.

The signal identification is performed at multiple stages, thus forming a sequential signal processing chain, as shown in Figure 11.5. However, it is not always necessary to go through all the steps. Indeed, an important design objective is to allow the radio to make an identification decision as soon as it is possible to do so with sufficiently high confidence. Thus, processing begins by extracting the simplest features that do not require any, or much, a priori knowledge about signals. The classification and identification modules then work together to classify the detected spectral activities into possibly known signal classes (i.e., systems) based on these features. Suppose there were N_f detected spectral activities at the detection stage. Then in this first stage of the classification process, $K_1 \leq N_f$ number of these might be successfully classified and identified based on these simplest form of features.

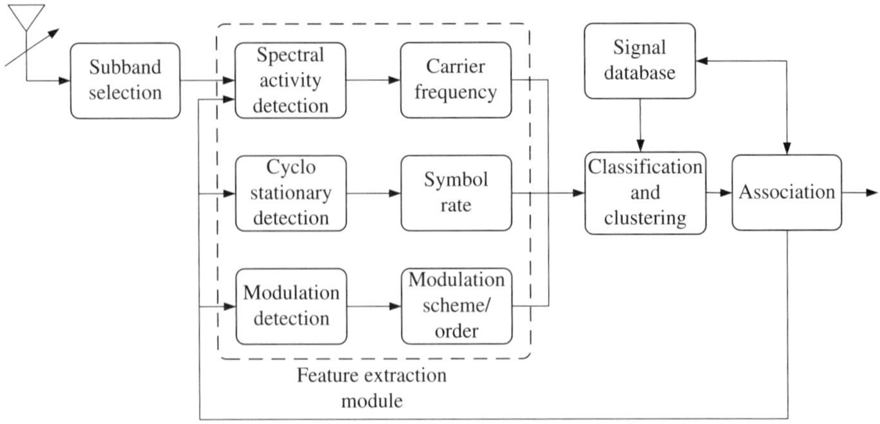

FIGURE 11.5 A signal classification and identification architecture in a cognitive radio.

For those remaining $N_f - K_1$ spectral activities, it may be that the classification and identification modules cannot make a final decision with sufficiently high confidence based on the given features. The classification/clustering module of Figure 11.1 then requests the feature extraction stage to extract more sophisticated features corresponding to these remaining spectral activities. In this round, the radio can extract features that it may not have been able to extract at the first stage by using those already extracted features as a priori information. Moreover, processing complexity is reduced since we only extract these sophisticated features (that may require more computationally demanding processing) for the remaining $N_f - K_1$ signals rather than for all N_f originally detected signals. Next, these remaining signals are again passed on to the classification/clustering and association modules with the modified, and extended, feature vectors. In this round, the classification module may be able to classify, say, K_2 number of spectral activities into known clusters by working with the identification module.

If $N_f - K_1 - K_2 = 0$, the process stops since the radio has identified all detected spectral activities. If, on the other hand, $N_f - K_1 - K_2 > 0$, then the feature extraction module may be revisited to extract perhaps even more sophisticated features for the remaining $N_f - K_1 - K_2$ signals. This iterative process continues until either all signals are successfully identified, or the number of features that can be extracted has reached the maximum possible value. If the signal algorithm terminates due to this latter reason, then the cognitive radio has a number of detected spectral activities that it has failed to associate with any of the known systems. The cognitive radio will hold these signals and their features in a temporary memory module. As time progresses, more sensing observations may be available so that the radio might attempt to confirm whether these signals indeed represent a new class of signals or whether they are just outliers. This is a form of self-learning. It may take several sensing durations before the radio can make a final decision on these signals with a predetermined confidence level. Here, we will not go into the details of this learning and confirmation aspects. However, in certain applications, such learning of new classes of signals may be one of the most important tasks of a wideband autonomous cognitive radio. For example, in military communications, these new signals may represent jamming or decoy signals that the radio needs to avoid or keep track of.

In the iterative spectral activity classification architecture of Figure 11.5, the simplest features extracted in the first stage constitute the carrier and certain cyclic frequencies. We may assume that energy detection implemented using power spectral estimation methods is used to identify the center frequencies of the detected signals. The estimated center frequencies constitute the first component of the feature vectors. Next, the radio proceeds to cyclo stationary analysis in order to identify the symbol rates and bandwidths of the detected signals. Hence, the cyclic frequencies corresponding to symbol rate or bandwidth will constitute the second component of the feature vectors. The first round of classification and identification may be attempted with this feature vector. If some of the detected signals cannot be successfully classified and associated with known systems during this first round, then the radio may request additional information about the detected signals from the feature extraction module. Certain features that may help radio discern the modulation methods and order may be extracted in this round as shown in Figure 11.5. As we discussed earlier, higher-order statistics such as kurtosis and cumulants may constitute such features as well as additional cyclic frequencies that may not have been used in the first stage. However, when cumulant-based approaches require prior knowledge about the pulse shapes of the detected signals, the kurtosis-based feature extraction approach might be preferable [182]. Furthermore, if the kurtosis-based approach is used, modulation classification may be performed independently of the cyclo stationary detection since it does not require knowledge about the symbol rates. This iterative signal classification process can easily be extended to include other more advanced features by introducing further iterative stages or by adding more features into the already existing stages.

11.5 BAYESIAN NONPARAMETRIC SIGNAL CLASSIFICATION

The extracted N_f feature vectors are supplied to the classification/clustering module at the next stage in Figure 11.1. As we discussed earlier, the classification branch of Figure 11.1 is similar to supervised classification encountered in many contexts: Associate each of the input feature vectors into the most appropriate class out of a set of signal classes that is already known to the radio. As we mentioned in Chapter 8, there are well-known supervised classification algorithms that can perform this. An example is the *nearest neighbor classification* algorithm that classifies a given feature vector into the same cluster that its nearest, in a certain distance sense, neighbor belongs to. One of the most common distance criterion used is the Euclidean distance, or some normalized version of it, between the feature vector and each of its neighbors. A simple variation of this algorithm is the k-nearest neighbor classification algorithm (kNN) in which the feature vector is classified into the cluster to which the majority of its k-nearest neighbors belongs to. More elaborate training-based classification algorithms have also been proposed based on, for example, neural networks [184] and support vector machines [185]. Overall, these do not differ much from the standard supervised classification algorithms except for the details on what type of features are used and how the system is trained. Hence, as we already mentioned earlier, we will not delve further into the classification path of the signal classification/clustering stage. Instead, our focus will be on the signal clustering path.

Certainly, there are situations in which supervised classification as earlier alone may not be sufficient to achieve signal identification task faced by a wideband cognitive radio. In practice, when the radio is turned on, it will first have to learn and create the signal clusters that can then be used in the supervised classification mode. This is also true when the radio is used in a new RF environment due to, perhaps, a new geographical location. In these situations, the signal identification will proceed by first clustering the signals into a set of clusters and then associating the clusters with some known systems. This allows for the possibility that some of the clusters may correspond to previously unknown systems, in which case the association step will leave these clusters as new clusters. Thus, this approach is desirable when the cognitive radio must be prepared to handle new signal environments.

The number of active systems and the number of signals belonging to each of the active systems in a sensed subband signal are both time varying. This makes many existing clustering algorithms inadequate for the signal clustering in a cognitive radio. For example, the unsupervised clustering algorithms, such as K-means algorithm we saw in Chapter 8, can allow the radio to learn the signal groups by itself based on the given feature vectors. However, it requires one to specify the number of clusters. Since this can be time varying in a wireless environment, it is not possible to use K-means directly for the signal clustering in a cognitive radio.

Ideally, we need a clustering approach that does not presume the number of clusters, or the number of signals belonging to each cluster. Suppose, for the moment, that the algorithm has been working for some time and thus it has a certain set of self-created distinct clusters in its memory. Still, given a new set of feature vectors as the input, the radio may not be able to presume that all these input feature vectors belong to the already known clusters. There is always the possibility that some feature vector inputs may correspond to new systems. Hence, a clustering algorithm for the signal identification in a cognitive radio should allow the number of clusters to be determined by the data itself.

As we saw in Chapter 8, one of the candidate algorithms for this scenario is the X-means algorithm [182]. It is based on the K-means algorithm, but rather than presuming the number of clusters, X-means algorithm approximates the number of clusters by maximizing a suitable information criterion such as Bayesian or the Akaike [91]. Similar to the K-means algorithm, however, it still assumes spherical Gaussian data, and its performance can also heavily depend on the choice of initialization [102].

The Bayesian nonparametric signal classification method based on the Dirichlet process mixture model (DPMM) developed in Chapter 8, on the other hand, offers flexibility to overcome these drawbacks. The DPMM-based clustering allows the distribution of classes to be drawn from a random distribution with infinite support. Each class (cluster) does have a parametric representation, but the number of classes are unbounded. Indeed, the number of classes is determined by the input data itself. Moreover, with the extension to allow for non-Gaussian likelihoods, the DPMM-based Bayesian nonparametric approach can effectively handle feature vectors with both Gaussian and non-Gaussian likelihoods in the same input data set. In particular, if we allow multiple types of different prior distributions for cluster parameters, this allows observations belonging to these different clusters to also have different likelihoods. Thus, it is able to naturally accommodate the non-Gaussian-distributed feature vector inputs.

The advantage of this approach is obvious: It does allow the radio to create new clusters by grouping signals that do not look sufficiently close to the already known signal classes. Moreover, it has the flexibility to separate even these new unknown signals into several distinct classes depending on similarities and dissimilarities among these signals.

The drawback of the approach is that it can be considerably more computationally complex than many other classification algorithms. In particular, if the features were to be indeed Gaussian distributed, the X-means algorithm may be simpler to use and provide good performance. It is, however, likely that extracted feature vectors corresponding to signals belonging to different systems may not always satisfy the Gaussian assumption. When some of the features are non-Gaussian, we may expect the X-means type of algorithms to perform poorly compared to the DPMM-based framework of Chapter 8. Another drawback of the Bayesian nonparametric approach is the need to assume a conjugate prior for the observation likelihoods. This again is imposed for the sake of computational efficiency. When the conjugate prior assumption is not held, the computation of the posterior conditional distributions, needed for Gibbs sampling, themselves can be computationally challenging. However, even with these drawbacks, the Bayesian nonparametric approach provides one of the most promising frameworks to address the unique challenges posed by the signal clustering problem in a wideband cognitive radio.

In Section 8.5.3.1, we explicitly developed a DPMM-based Bayesian nonparametric clustering approach that allows for two different types of observation likelihoods: Gaussian and lognormal. In anticipation of its use in the signal classification problem, in the following example, let us extend that framework to account for three distinct feature vector likelihoods. Of course, we will do so while preserving the conjugate prior property.

Example 11.3 (Bayesian Nonparametric Classification of a Mixture of Gamma, Lognormal, and Gaussian-Distributed Observations) Let us consider an example in which the number of prior distribution families allowed for the feature vectors is $L = 3$. Let us denote the observations $\mathbf{y}_i = [y_{i,1}, \ldots, y_{i,d}]^T \in \mathbb{R}^d$, so that each $\mathbf{y}_i|\theta_i$ has a likelihood of either Gaussian, lognormal, or gamma. It must be emphasized that the DPMM framework implies that even the observations having the same type of likelihood, say Gaussian, may belong to different clusters. This is so because the parameters of even the same type of observation likelihoods are drawn from a Dirichlet process. As has been the case throughout in this book, for analytical tractability, we will select a conjugate prior for each type of observation likelihood functions. Moreover, as in Section 8.5.3, we will continue to assume that in the case of non-Gaussian likelihoods, the elements of each feature vector $\mathbf{y}_i$ are independent.

The analytical expressions for the first two type of likelihoods, namely, the Gaussian and lognormal, were derived in Section 8.5.3.1. Indeed, in the case of Gaussian likelihoods, we allow the clusters to be distinct possibly due to both distinct mean vectors and covariance matrices. Hence, with the auxiliary variable $Z = 1$ denoting the Gaussian likelihoods in the mixture model of (8.55), we may let $\mathcal{S}_1 \triangleq \mathbb{R}^d \times \mathbb{R}^{d\times d}$ to be the set of possible parameters of the Gaussian likelihood function corresponding to $\theta_i|\{Z_i = 1\} \triangleq (\boldsymbol{\mu}_i, \mathbf{V}_i)$. As in Section 8.5.3.1, a suitable conjugate prior for the Gaussian likelihood with this parameter choice is the normal/inverse-Wishart prior. With this choice, the likelihood $f_{\theta_i}^{(1)}(\mathbf{y}_i)$, the posterior $f(\theta_i|\mathbf{y}_i, Z_i = 1)$, and the prior $G_0^{(1)}(\theta_i)$ are given, respectively, by (8.45), (8.49)–(8.50) in association with (8.51)–(8.52), and (8.47)–(8.48). The marginal $f(\mathbf{y}_i|Z_i = 1) = \int_{\theta\in\mathcal{S}_1} f_\theta^{(1)}(\mathbf{y}_i) g_0^{(1)}(\theta)d\theta$ can easily be evaluated using the fact that both likelihood $f_\theta^{(1)}(\mathbf{y}_i)$ and the prior $g_0^{(1)}(\theta)$ are Gaussian pdf's.

Similarly, allowing $Z = 2$ in the mixture model of (8.55) to represent lognormal likelihoods, we have the necessary analytical expressions already derived in Section 8.5.3.1 as well. We assume that the different clusters with lognormal likelihoods for observations are caused by having distinct mean parameters in the lognormal likelihood function, while shape parameters $\{\xi_k\}_{k=1}^d$ of these likelihoods are all fixed. Then we may define $\mathcal{S}_2 \triangleq \mathbb{R}^d$ where $\theta_i|\{Z_i = 2\} \triangleq \boldsymbol{\rho}$, where $\boldsymbol{\rho} = [\rho_1, \ldots, \rho_d]^T$ denotes the mean vector of a d-dimensional lognormal likelihood function. As in Section 8.5.3.1, we select the conjugate prior for the lognormal likelihood to be Gaussian so that each $\rho_k \sim \mathcal{N}\left(\rho_{0,k}, \xi_{0,k}^2\right)$, and is independent. Then, the likelihood $f_{\theta_i}^{(2)}(\mathbf{y}_i)$, the posterior $f(\theta_i|\mathbf{y}_i, Z_i = 2)$, the prior $g_0^{(2)}(\theta_i)$, and the marginal $f(\mathbf{y}_i|Z_i = 2) = \int_{\theta\in\mathcal{S}_2} f_\theta^{(2)}(\mathbf{y}_i) g_0^{(2)}(\theta)d\theta$ are given, respectively, by (8.59), (8.61), (8.60), and (8.62).

Next, we introduce the third type of observation likelihoods to be gamma distributions. We assume that the distinct clusters with the gamma likelihoods are caused by these distributions having distinct shape parameters, while rate/scale parameters $\{b_k\}_{k=1}^d$ stay fixed common to all (gamma) likelihood functions (see Appendix B for the definitions). Thus, letting $Z = 3$ for gamma likelihoods, we define $\mathcal{S}_3 \triangleq \mathbb{R}^d$ corresponding to $\theta_i|\{Z_i = 3\} \triangleq \mathbf{a}$ where $\mathbf{a} = [a_1, \ldots, a_d]^T$ are the shape parameters of a gamma-distributed likelihood function:

$$f_\theta^{(3)}(\mathbf{y}_i) = \prod_{k=1}^{d} \frac{b_k^{a_k}}{\Gamma(a_k)} y_{i,k}^{a_k - 1} e^{-b_k y_{i,k}}, \tag{11.14}$$

where $\theta = \mathbf{a}$. Note that in (11.14) we have assumed that each $y_{i,k}|\{\theta_i = \theta, Z_i = 2\} \sim G(a_k, \frac{1}{b_k})$ and is independent for $k = 1, \ldots, d$. In this case, a suitable prior for parameter $\theta = \mathbf{a}$ that preserves the conjugate property is shown to be (see Appendix C)

$$g_0^{(3)}(\theta_i) = g_0^{(3)}(\mathbf{a}) = \prod_{k=1}^{d} \frac{1}{J(a_0, b_0, b_k, c_0)} \cdot \frac{a_0^{a_k-1} b_k^{c_0 a_k}}{\Gamma(a_k)^{b_0}},$$

where a_0, b_0, and c_0 are the corresponding hyper-parameters and $J(a_0, b_0, b_k, c_0) \triangleq \int_0^\infty \frac{a_0^{x-1} b_k^{c_0 x}}{\Gamma(x)^{b_0}} dx$ is a normalization term. Then, it can be shown that the posterior distribution of θ_i is given by [119]

$$f(\theta_i | \mathbf{y}_i, Z_i = 3) = f(\mathbf{a}|\mathbf{y}_i) = \prod_{k=1}^{d} \frac{1}{J(a_0 y_{i,k}, b_0 + 1, b_k, c_0 + 1)} \cdot \frac{(a_0 y_{i,k})^{a_k - 1} b_k^{(c_0+1)a_k}}{\Gamma(a_k)^{b_0+1}}. \tag{11.15}$$

The marginal distribution of $\mathbf{y}$ can thus be computed as

$$f(\mathbf{y}_i | Z_i = 3) = \prod_{k=1}^{d} \int_0^\infty \frac{b_k^z}{\Gamma(z)} y_{i,k}^{z-1} e^{-b_k y_{i,k}} \frac{a_0^{z-1} b_k^{c_0 z}}{\Gamma(z)^{b_0}} \left[\int_0^\infty \frac{a_0^{t-1} b_k^{c_0 t}}{\Gamma(t)^{b_0}} dt \right]^{-1} dz. \tag{11.16}$$

Note that, in practice, the previous marginal distribution of $\mathbf{y}$ might have to be computed numerically. However, it only needs to be evaluated for a particular value of $\mathbf{y}_i$ ■

In the following example, we make use of the previous three observation likelihoods as possible models for features extracted from signals in the 2.4 GHz ISM band.

Example 11.4 (Bayesian Nonparametric Classification of Signals in the 2.4 GHz ISM Band) Let us consider a scenario in which there are 2 IEEE 802.11.b WiFi signals transmitting at 2.417 and 2.472 GHz (channels 2 and 13), respectively, and a Bluetooth signal transmitting at 2.45 GHz. The SNR at the receiver is 5 dB and each sensing window is $T_{ss} = 30\,\mu s$. We assume a fast Rayleigh fading channel with normalized fading coefficients, or *channel-state information* (CSI), h such that $\mathbb{E}\{h^2\} = 1$.

After each $30\,\mu s$ sensing time, three-dimensional (3-D) ($d = 3$) feature vectors (f_c, α, B) are extracted from the sensed signal, where f_c denotes the carrier frequency (down-converted to zero IF), α is the cyclic frequency component corresponding to the symbol rate, and B is the estimated signal bandwidth. The carrier frequencies f_c are extracted by applying the energy detection explained in Chapter 10, while corresponding cyclic frequencies α are then extracted through the cyclo stationary feature extraction explained earlier. The signal bandwidth B is estimated from the smoothed PSD estimate. Ideally, each WiFi signal has a bandwidth of 22 MHz, and the Bluetooth signal has a bandwidth of 1 MHz. Similarly, the Bluetooth signal has

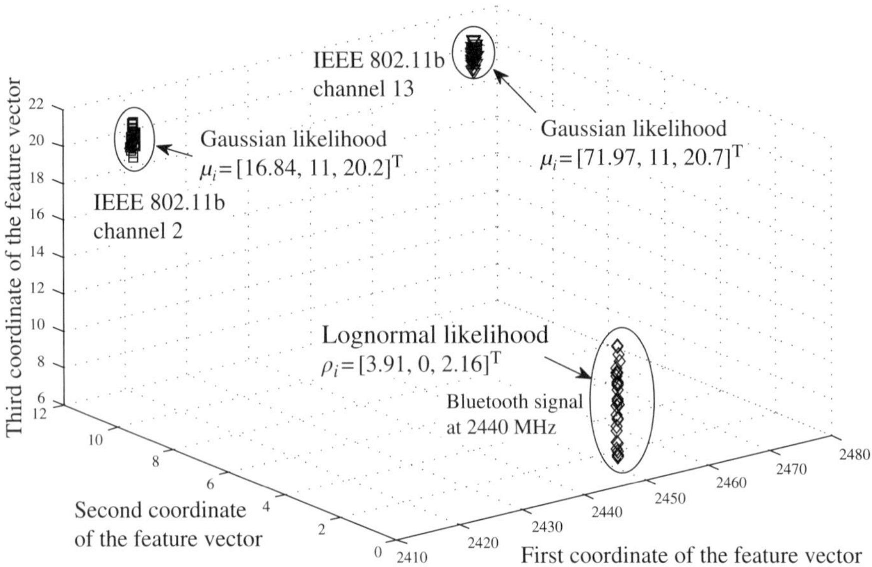

FIGURE 11.6 Signal classification of 2 WiFi signals and a Bluetooth signal. The feature point is denoted by (f_c, α, B), where f_c is the carrier frequency, α is the cyclic frequency component corresponding to the symbol rate, and B is the estimated bandwidth. Energy detection is applied for 30 μs at an SNR of 5 dB with Rayleigh fading (fast fading). The probability of correct classification is 100%.

a symbol rate of 1 MBaud, and the WiFi signals have a chip rate of 11 Mchips/s that is manifested in the α component of the feature points.

The DPMM-based feature classification algorithm is applied to features extracted over 50 such 30 μs-long sensing intervals (for a total dwell time of $T_D = 50 \times 30\,\mu s = 1.5$ ms). Figure 11.6 shows the results of the DPMM-based Bayesian nonparametric clustering in the 3-D feature space once Gibbs sampling has converged. The feature points that are marked with the same-shaped markers in Figure 11.6 are assigned to the same cluster. As can be seen, the two WiFi signals are estimated to have Gaussian observation models, while the Bluetooth signal is assigned a lognormal model. The classification accuracy, denoting the percentage of feature points classified into correct clusters, is estimated as 100% in this case.

It must be mentioned that in this example we have allowed a longer dwell time in the subband and a large number of Gibbs sampling in order to demonstrate the power of the Bayesian nonparametric classification approach to provide excellent clustering results under ideal circumstances. In practice, with less sensing time and lower SNR, of course, the classification accuracy can degrade. ■

The earlier example illustrates that, given the choice, the feature vectors of signals from some systems in 2.4 GHz ISM band (e.g., Bluetooth) may better be modeled as having non-Gaussian likelihoods. This then also implies that for such feature data, the DPMM-based classification algorithm may give better performance results than the X-means algorithm, which assumes spherical Gaussian feature vectors. In the following example, we may compare performance of the DPMM-based classification algorithm to that of the X-means algorithm-based classification approach proposed in [182].

Example 11.5 (Comparison of DPMM-Based and X-Means Algorithms in Classifying Signals in the 2.4 GHz ISM Band) Let us consider the same 2.4 GHz ISM band signal environment assumed in the previous example but with an additional 4-QAM digital signal transmitting at 2.44 GHz. For simplicity, we limit the feature vectors to be 2-D data (f_c, B). The same feature vectors are used as inputs to the X-means and DPMM-based clustering algorithms. Note that both these algorithms do not require the value of the number of clusters a priori.

Figure 11.7(a) and (b), respectively, show the performance of the X-means and the DPMM-based algorithms in clustering the same data. The X-means algorithm suffers by not being able to match the likelihoods of the observed data. As is seen from Figure 11.7(a), the feature vectors corresponding to the WiFi signals are split into multiple clusters by the X-means algorithm making it estimate the number of distinct clusters to be $X = 12$, as opposed to the true value of 4. This has led to the classification accuracy of the X-means algorithm to be only about 55.88%. On the other hand, the results of the DPMM-based classification in Figure 11.7(b) show perfect clustering due to its ability to better estimate the number of clusters.

The K-means algorithm, on which the X-means algorithm is based, on the other hand, may provide better performance than the X-means algorithm if the number of cluster K was known a priori. However, even if we were to use K-means algorithm with the correct value of $K = 4$ clusters, its performance can still be highly dependent on the initialization of the cluster centroids. Depending on the initial choice of centroid locations, the performance of the K-means algorithm can still be significantly inferior to the DPMM approach, even using the additional a priori knowledge of the exact value of K. For some initializations, it is observed that the K-means may perform as well as DPMM. Of course, the advantage of the DPMM-based method is that it does not require neither the number of clusters $K = 4$ nor their centroids as prior information. ■

So far, we have attributed all the classification errors to the signal classification algorithm, be it the DPMM-based or the X-means or the K-means algorithm. However, this can be misleading because the classification accuracy depends both on the feature extraction and signal classification performance. Indeed, in reference to the overall spectral knowledge acquisition framework of Figure 2.4 and the spectral activity identification architecture of Figure 11.1, the input to the classification algorithm is the features that were extracted at the carrier/center frequencies determined by the detection module as having active signals. If the detection algorithm makes an erroneous detection at a certain frequency, then the classification module will attempt to classify a feature vector that is actually not there in the observation. The feature

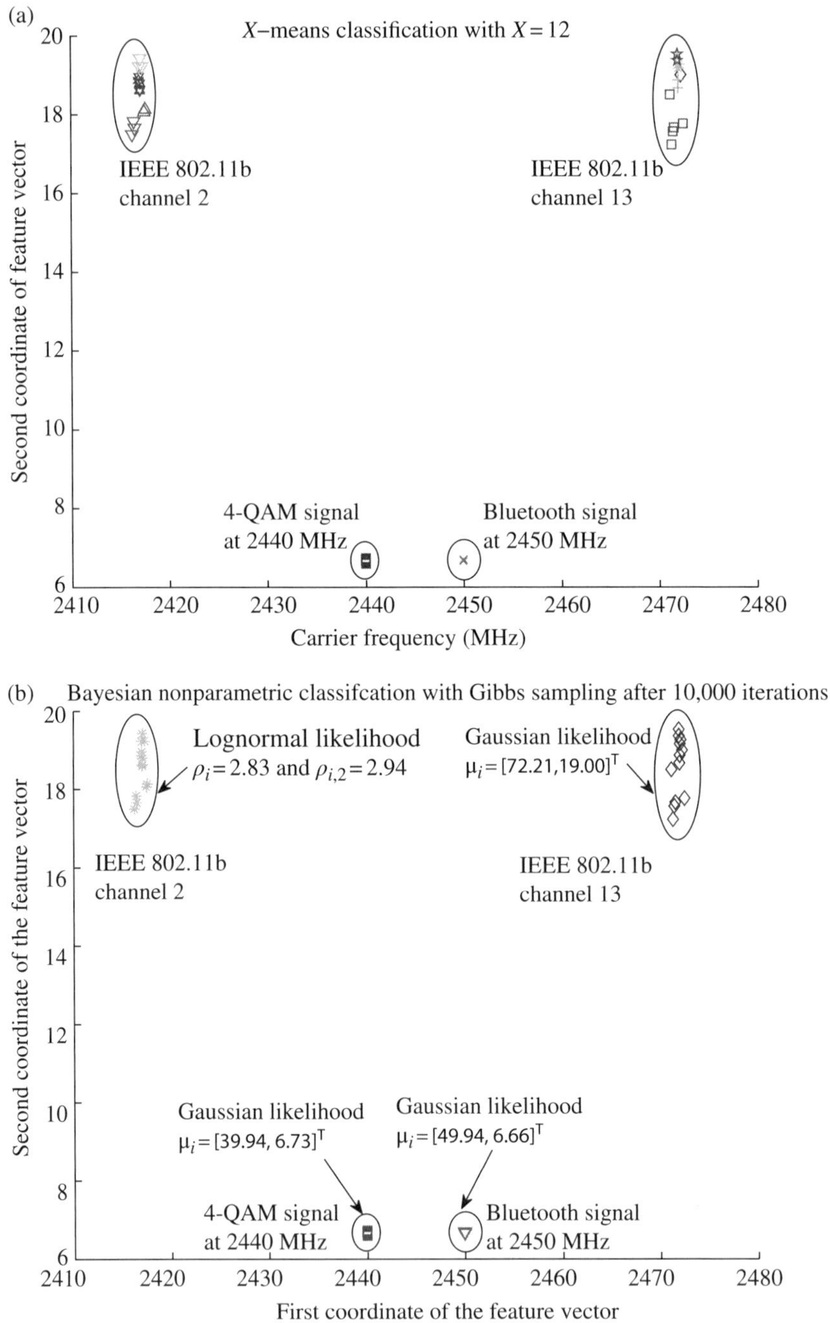

FIGURE 11.7 Comparison of nonparametric classification performance of the X-means algorithm and the DPMM-based algorithm. (a) The X-means algorithm, with an estimated number of clusters $X = 12$, gives a classification accuracy of about 55.88%. (b) The DPMM-based algorithm, with an estimated number of clusters $K = 4$, gives a classification accuracy of 100%.

extraction module will essentially extract a noise-only feature vector and pass on to the classification algorithm.

Hence, in detecting and classifying signals in the presence of noise, there can be two types of errors at the output of the classification module:

1. *Feature extraction errors*: These are the errors due to the input feature vectors corresponding to nonexistent signals. The origin of these errors is the false-alarm errors in the spectral activity detection algorithm.
2. *Signal classification errors*: These are the actual errors caused by the classification algorithm on feature inputs corresponding to the correctly detected signals. The probability of these errors may be called the *misclassification rate*.

In order to understand the effectiveness of a classification algorithm, we need to study the above two types of errors separately. For that, let us define the following types of error rates:

1. *Feature extraction error rate*: Defined as the proportion of feature vectors that do not correspond to existing signals. The complement of this can be called the feature extraction accuracy.
2. *Misclassification rate*: Defined as the proportion of feature vectors that are not classified into correct clusters. In this case, correct classification implies that feature vectors corresponding to different signals are classified into different clusters and erroneous feature vector inputs (due to false alarm) are classified into separate clusters.
3. *Overall feature extraction and signal classification error*: Defined as the proportion of feature vectors that do not correspond to existing signals and/or are misclassified. The complement of this is called the overall classification accuracy.

The overall classification accuracy is the proportion of feature vectors that are correctly detected and classified. Hence, a correct overall classification can result at the output of the signal classification stage only if a feature vector was extracted correctly for that particular signal at the detection/feature extraction stage. In the next example, we will take a closer look at the errors introduced by the actual signal classification step in Example 11.5 on classifying four different types of signals encountered in the 2.4 GHz ISM band.

Example 11.6 (Feature Extraction Error Bottleneck) Let us consider the same 2.4 GHz ISM band signal environment with four different types of signals assumed in Example 11.5: two IEEE 802.11.b WiFi signals transmitting at 2.417 and 2.472 GHz (channels 2 and 13), respectively, a Bluetooth signal transmitting at 2.45 GHz, and a 4-QAM signal transmitting at 2.44 GHz. All other assumptions are also as in Example 11.5. In particular, the feature vectors are the 2-D vectors of (f_c, B) corresponding to each detected spectral activity.

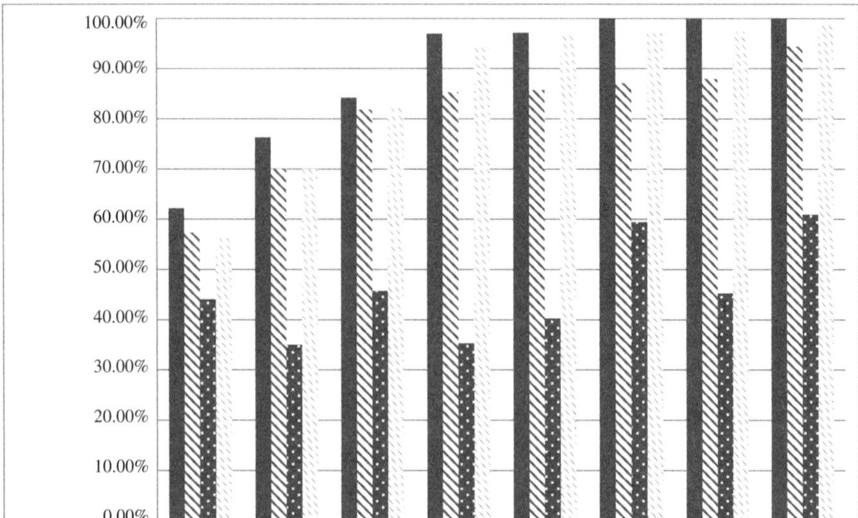

	$SNR = -4$ dB	$SNR = -2$ dB	$SNR = 0$ dB	$SNR = 2$ dB	$SNR = 4$ dB	$SNR = 6$ dB	$SNR = 8$ dB	$SNR = 10$ dB
Feature extraction	62.16%	76.32%	84.21%	96.97%	97.14%	100.00%	100.00%	100.00%
K-means	57.30%	70.00%	81.82%	85.28%	85.71%	87.06%	87.94%	94.41%
X-means	44.05%	35.00%	45.76%	35.28%	40.29%	59.41%	45.28%	60.88%
DPMM	56.22%	70.00%	82.42%	94.17%	96.86%	97.06%	97.35%	98.82%

FIGURE 11.8 The classification accuracy of the DPMM-based, *X*-means, and K-means (with $K = 4$ as given) algorithms in classifying a mixture of four different wireless signals in the 2.4 GHz ISM band along with the probability of correct detection/feature extraction at the detection stage.

In Figure 11.8, we show the classification accuracy of each of the three clustering algorithms (DPMM based, *X*-means, and *K*-means with $K = 4$ specified) along with the probability of correct detection/feature extraction at the spectral activity detection stage. In order to reduce the effect of initialization, the performance shown in Figure 11.8 for each SNR is obtained by averaging 10 independent runs of the classification experiments similar to those used in Figure 11.7.

Figure 11.8 verifies, as expected, that the classification accuracy of each of the three algorithms is upperbounded by the detection/feature extraction accuracy. As can be seen from Figure 11.8, the feature extraction accuracy improves as SNR improves. This indeed is as we have seen from the previous chapter as well as from general signal detection theory considered in Chapter 4. Except for the *X*-means algorithm, the other two algorithms give increasingly better classification accuracy as SNR improves as well. Note that the performance of the *X*-means algorithm seems to be highly erratic due to its inability to estimate the number of clusters accurately. Although *X*-means can provide highly competitive performance when feature vectors are indeed Gaussian distributed, it is conceivable that the non-Gaussian likelihoods in this feature vector set are likely causing it to go astray without careful fine-tuning (which we have not attempted).

The overall classification accuracy at the output of each of the classification algorithms is always less than that of the accuracy at the feature detection stage. For example, when $SNR = 4\,\text{dB}$, accuracy at the feature extraction stage is about 97.14%. Once the set of feature vectors extracted with this accuracy is input to the classification stage, the final (overall) classification accuracy of the signals can only be at most equal to this value, on average. Indeed, Figure 11.8 shows that the overall classification accuracy at the output of DPMM-based classification algorithm is about 96.86%, while that of the K-means algorithm is even further reduced to only about 85.71%. Note that, the K-means achieves this lower performance even after providing it with the additional information of the correct number of clusters (in this case $K = 4$). For higher SNR values, the feature extraction stage is able to perfectly detect spectral activities. Hence, each feature vector presented to the classification stage for these higher SNR values corresponds to a valid signal. Thus, any misclassifications at these SNR values are caused by the classification algorithm. For example, at $SNR = 6\,\text{dB}$, the DPMM-based classification algorithm achieves only about 97% accuracy, meaning that about 3% of the input feature vectors were misclassified although they all correspond to valid signals. On the other hand, at the same $SNR = 6\,\text{dB}$, the K-means misclassifies about 13% of the input feature vectors that indeed were corresponding to valid signals.

We may attribute the inferior classification accuracy of K-means algorithm primarily to its underlying reliance on the spherical Gaussian-distributed input vectors. It is conceivable that feature vectors corresponding to at least some of the signal types in this case have non-Gaussian likelihoods that degrade its performance. At the end, however, regardless of how good the classification algorithm is, the accuracy of feature extraction stage will be a performance bottleneck. ■

As good as the performance of the DPMM-based classification approach in the previous examples, its biggest drawback is its computational complexity due to the large number of Gibbs sampling iterations required to achieve convergence. The Gibbs sampling in the DPMM-based algorithm starts with arbitrary estimates of θ_i's, for all $1 \leq i \leq N_f$, and draws samples from the conditional distribution of each parameter θ_i, given the other parameters $\{\theta_j\}_{j \neq i}$, where $\{\theta_j\}_{j \neq i}$ take the values of their most recent estimates as was described in (8.64) in Chapter 8. Although under appropriate conditions these samples converge in probability to the actual posterior distribution of $(\theta_1, \ldots, \theta_{N_f})$ [97], the convergence can sometimes be unacceptably slow.

One cause of the time complexity of (8.64) is due to the fact that in each of its iterations, it samples all the parameters θ_i's sequentially (or, sometimes, randomly). Especially when N_f is large, this can make the process computationally prohibitive. However, an experimental observation is that some θ_i parameters do not change after a number of Gibbs iterations if they were assigned to large clusters. Hence, if an input feature vector $\mathbf{z}_i$, for $i = 1, \ldots, N_f$, is assigned to a certain cluster with a sufficiently large number of members, it is most likely to be reassigned to the *same cluster* if its corresponding parameter θ_i is sampled again according to (8.18). This is indeed a consequence of the underlying Chinese restaurant process property (8.14) of the DPMM in which the probability of assigning an observation to a certain cluster is proportional to the number of members already in that cluster. Hence, frequent resampling of parameters θ_i's belonging to large clusters may perhaps be unnecessary. On the other hand, the same Chinese restaurant process property implies that a feature vector input $\mathbf{z}_i$ that is assigned to a small cluster (i.e., a cluster with a relatively small number of members) is more likely to be reassigned to a different cluster when its parameter θ_i is resampled according to (8.18). Thus, perhaps, the convergence rate of the Gibbs sampling (8.64) can possibly be improved by nonuniformly selecting the parameters θ_i's for resampling. In particular, parameters belonging to small clusters may be resampled more frequently compared to those belonging to large clusters.

Algorithm 11.1

A biased selection strategy for Gibbs sampling in the DPMM-based classification algorithm.

(1) Initialize $\theta_i = z_i$ and $w_i = \frac{1}{N_f}$ for all $i \in \{1, \cdots, N_f\}$.

(2) Select θ_i from $\{\theta_i\}_{i=1}^{N_f}$ with probability w_i.

(3) Update θ_i by sampling from the posterior distribution in (8.18).

(4) Update the parameter selection policy $\{w_i\}_{i=1}^{N_f}$ so that $w_i \propto \frac{1}{m_k}$ for m_k the number of members in the cluster containing z_i.

(5) IF convergence condition is not satisfied, go to Step (2). Otherwise Stop.

Motivated by this, let us modify the DPMM-based classification algorithm slightly by incorporating a *parameter selection policy* to the Gibbs sampler (8.64). This policy that specifies the parameters θ_i's to be sampled during a particular iteration is designed to favor elements belonging to small clusters. Hence, we define w_i, for $1 \leq i \leq N_f$, to be the probability of selecting parameter θ_i to be resampled and let w_i to be inversely proportional to the number of members in the cluster containing $\mathbf{z}_i$ subject to the normalization condition $\sum_{i=1}^{N_f} w_i = 1$. Recalling that distinct values taken by θ_i's are denoted as ϕ_k's, if $\theta_i = \phi_k$, then $w_i \propto \frac{1}{m_k}$, where m_k is the number of elements of the cluster with parameter ϕ_k. This *biased selection policy*-based Gibbs sampling is given in Algorithm 11.1.

Note that according to the Gibbs sampling (8.64), initially, each feature vector $\mathbf{z}_i$ is assigned to its own cluster. Hence, initially, $m_k = 1$ for all k, implying that parameters θ_i's are selected uniformly. After several Gibbs sampling iterations, however, Algorithm 11.1 becomes biased toward selecting parameters θ_i's belonging to smaller clusters. Of course, there are other reasonable parameter selection methods one can develop as well.

In the next example, we will compare the convergence rate of this biased parameter selection policy with that of direct Gibbs sampling as in (8.64) when the observations have Gaussian likelihoods.

Example 11.7 (Convergence Performance of the Bayesian Nonparametric Classification of Gaussian Data with the Biased Parameter Selection Policy) Let us consider a set of $N_f = 50$ inputs drawn from four different clusters as shown in Figure 11.9(a). Here, we assume that the cluster parameters θ_i's have a bivariate normal/inverse-Wishart conjugate prior such that $\theta_i = (\boldsymbol{\mu}_i, \boldsymbol{\Sigma}_i)$ with components

$\boldsymbol{\Sigma}_i^{-1} \sim W_d(\mathbf{S}, s)$ and $\boldsymbol{\mu}_i | \boldsymbol{\Sigma}_i \sim \mathcal{N}(\mathbf{m}, \tau \boldsymbol{\Sigma}_i)$, for some mean $\mathbf{m}$ and a scale factor $\tau > 0$. Recall from Appendix B that the Wishart distribution $W_d(\boldsymbol{\Sigma}, n)$ is characterized by a positive definite matrix $\boldsymbol{\Sigma}$ and the degrees of freedom n.

The parameters of the model are chosen such that the mean of the observations $\mathbf{z}_i$'s is $\mathbf{m} = [2450, 15]^T$ (here $d = 2$), $\mathbf{S} = \mathbf{I}$, $s = 4$, and $\tau = 1$. From the definition of the Wishart distribution given in Appendix B, observe that small value of $s = 4$, corresponds to the case when there is little confidence in the prior $\mathbf{S}$ accurately estimating the covariance matrix $\boldsymbol{\Sigma}_i$. Hence, the model with small s allows the sampled $\boldsymbol{\Sigma}_i$'s to deviate from the prior estimate $\mathbf{S}$. Finally, τ represents the amount of dispersion of observations in the feature space. The parameter α_0 required for defining the Dirichlet process is also modeled as being random with a prior gamma distribution such that $\alpha_0 \sim G(a, b)$, where $a = 2$ (see Appendix B for the gamma density function). Thus, α_0 is also updated at each Gibbs iteration using Bayesian methods along with parameters θ_i's.

The rate of convergence of the Gibbs sampling with three different parameter selection policies, namely, round-robin, random, and the biased approach of Algorithm 11.1, is shown in Figure 11.9(b). From the number of clusters K at each iteration step shown in Figure 11.9(b), it can be observed that the biased selection policy converges in approximately 200 iterations, whereas the random selection policy requires more than 800 iterations for convergence. The convergence performance of the round-robin selection policy varies with each simulation run but, in general, does not achieve a faster convergence rate compared to the biased selection policy.

It is also interesting to note from Figure 11.9(b) that the number of clusters K remains constant for longer time durations in both the random and round-robin selection methods compared to the biased selection case. This is because the random and round-robin parameter selection policies require more iterations to find the desired parameter to be updated. The biased selection policy follows a more efficient procedure in selecting the appropriate parameters to be sampled at each iteration leading to its improved convergence rate. ■

The earlier example shows that at least under ideal Gaussian likelihoods, the biased parameter selection approach of Algorithm 11.1 can significantly improve the convergence performance of the Gibbs sampling required in the DPMM-based classification. Let us investigate its performance under not-so-ideal conditions in the next example in which we require the DPMM-based classification of QAM signals of different modulation orders.

Example 11.8 (Convergence Performance of the Bayesian Nonparametric Classification of QAM Signals with the Biased Parameter Selection Policy) Let us consider a signal environment in which wideband spectrum sensing picks up four different QAM signals (BPSK, 4-QAM, 16-QAM, and 64-QAM) that are transmitting simultaneously at baseband equivalent center frequencies of $-80, -40, 0$, and 50 MHz with respective symbol rates of 2, 2, 5, and 5 MBauds. The spectrum sensing is performed for $T_{ss} = 50\,\mu\text{s}$ in each sensing interval at a sampling rate of $f_s = 200$ MHz. The periodogram estimate is frequency smoothed with a smoothing window of length $L = 151$. We extract $N_f = 100$ feature vectors (whose components are the

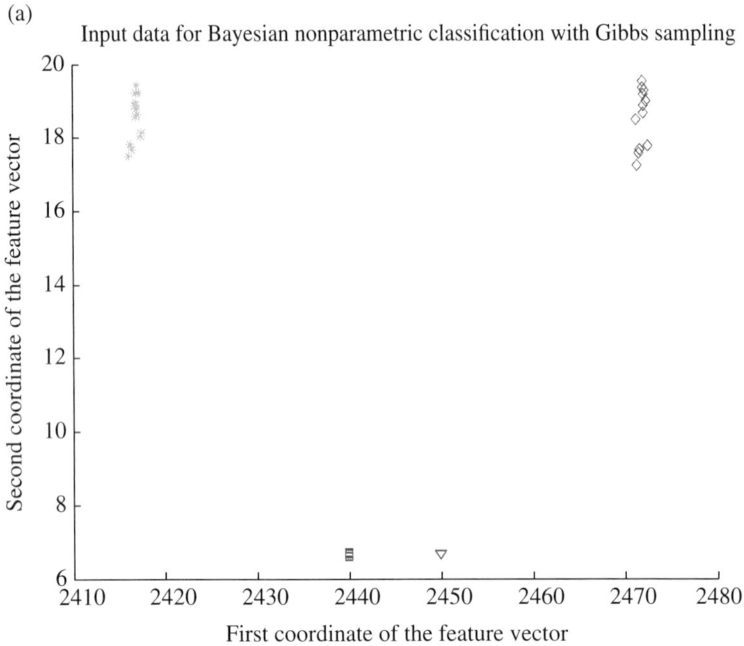

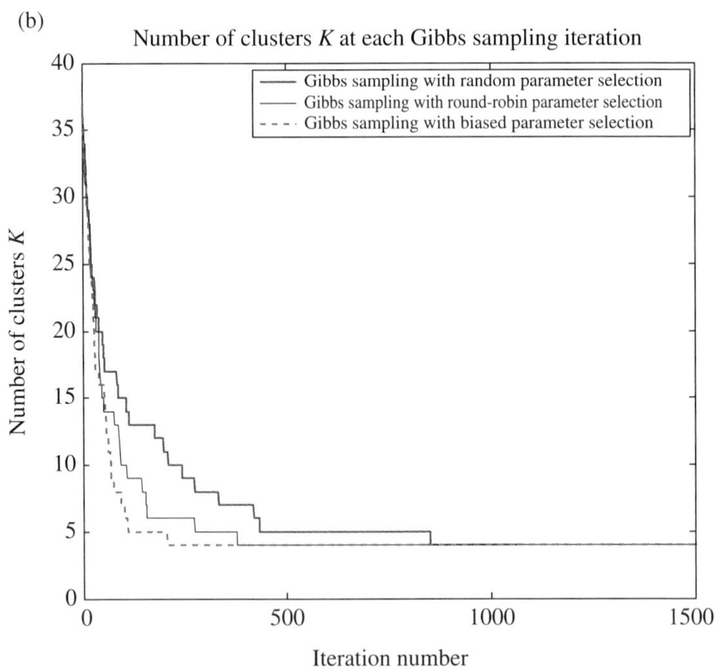

FIGURE 11.9 Comparison of round-robin, random, and biased parameter selection policies for the Gibbs sampler. (a) Input feature data to be classified using DPMM with different parameter selection policies.(b) The convergence rates of DPMM-based Gibbs sampling methods using either round-robin, random or biased parameter selection.

center frequencies, cyclo stationary feature, and kurtosis feature) after 25 such sensing intervals. Note that this means that during each sensing interval, exactly 4 spectral activities were detected from the estimated periodogram (giving $4 \times 25 = 100$). All channels are taken to be multipath Rayleigh fading channels with unity paths gains. The received *SNR* is 10 dB.

We again apply the DPMM-based classification algorithm using three different parameter selection policies for Gibbs sampling: random, round-robin, and biased parameter selection (i.e., simplified Gibbs sampling in Algorithm 11.1). To allow for the possibility of non-Gaussian-distributed feature vectors, we assume $L = 2$ number of different prior distributions on the clusters: specifically, Gaussian and lognormal priors with $\kappa_1 = \kappa_2 = 0.5$. Other hyper-parameters of the model are selected such that: $\mathbf{m} = [0, 0, 0]^T$, $\tau = 1$, $s = 5$, $\mathbf{S} = diag(0.1, 0.1, 1)$, $\boldsymbol{\rho}_0 = [\log(50), \log(2), \log(5)]^T$, $\boldsymbol{\xi}_0 = [1, 1, 1]^T$, and $\alpha_0 = 1$.

It is noted that after a sufficiently large number of Gibbs sampling iterations, all three parameter selection policies lead to perfect classification results. Indeed, Figure 11.10 shows that each of the parameter selection policies converges to the correct number of clusters (i.e., $K = 4$) after a sufficient number of Gibbs iterations.

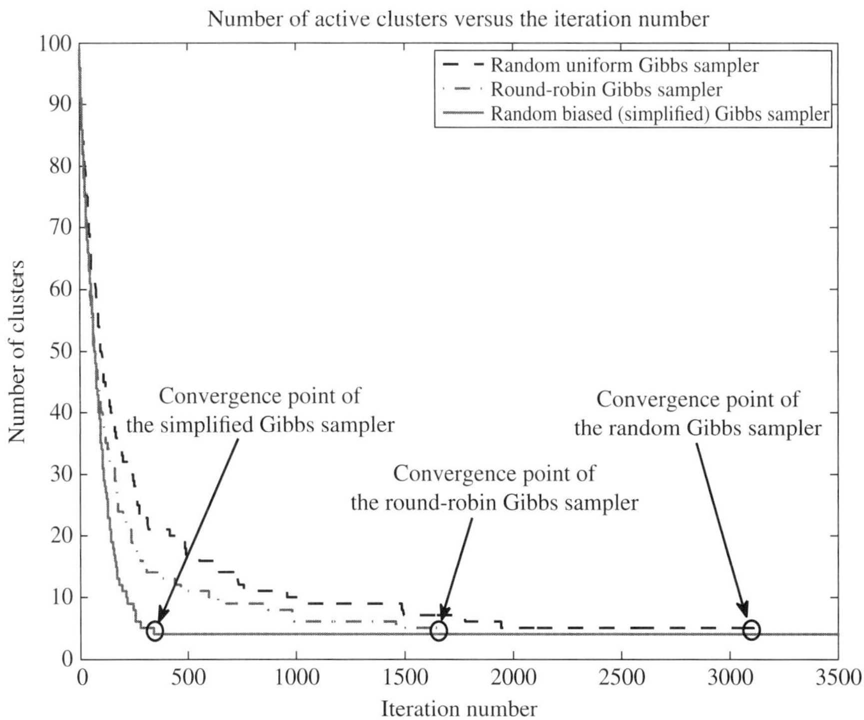

FIGURE 11.10 The convergence performance of the DPMM-based Bayesian nonparametric classifier with random uniform, round-robin, and random-biased parameter selection policies for Gibbs sampling.

However, it also shows that there is a notable difference in their convergence rates. The simplified Gibbs sampler with the biased parameter selection policy of Algorithm 11.1 requires only about 350 iterations for convergence, whereas round-robin and random uniform selection policies require, respectively, over 1600 and 3000 iterations. Thus, the simplified Gibbs sampler with the biased parameter selection can improve the convergence rate of the traditional Gibbs sampler by a factor of about 5 to 10 depending on whether parameters are selected round-robin or uniformly (randomly). The number of clusters stabilizes for longer durations in both uniform and round-robin selection policies before updating the proper parameters θ_i's to lead to convergence. On the other hand, by biasing the selection of parameters θ_i's, the simplified Gibbs sampler more efficiently selects the parameters θ_i's to be updated according to (8.18) leading to its faster convergence. ■

11.6 SEQUENTIAL BAYESIAN NONPARAMETRIC SIGNAL CLASSIFICATION

The DPMM-based classification algorithm of the previous section is a batch-processing algorithm in that it classifies the N_f input feature vectors as one group. Given the batch of N_f input feature vectors, the algorithm goes through Gibbs iterations where each iteration processes all N_f inputs together. Once the algorithm converges, we obtain the classification outputs corresponding to all N_f inputs. For instance, in the last example, it classifies the $N_f = 100$ feature vectors extracted from a sequence of 25 sensing intervals all at once.

A cognitive radio, however, needs the ability to perform classification of detected signals in real time as they are observed. In particular, given that the classifier has already classified feature vectors $\mathbf{z}_0 \ldots, \mathbf{z}_{N_f}$, it needs to classify a new feature vector input $\mathbf{z}_{N_f+1}$ into either one of the existing clusters or create a new cluster for itself. If we were to apply the formulation of the DPMM-based classification algorithm of the previous section, it will be wasteful, slow, and perhaps even redundant, since we will be classifying anew the whole feature vector set $\mathbf{z}_0 \ldots, \mathbf{z}_{N_f}, \mathbf{z}_{N_f+1}$.

What we would like to have is an algorithm that can, with high accuracy, cluster the new input $\mathbf{z}_{N_f+1}$ either to one of the existing clusters or put it in its own new cluster without having to reclassify all the previous inputs $\mathbf{z}_0 \ldots, \mathbf{z}_{N_f}$ and then update the cluster parameters accordingly. We should note that an optimal scheme is likely to require reclassifying all observations leading to possible new cluster assignments for any of the $\mathbf{z}_0 \ldots, \mathbf{z}_{N_f}$ as well. Thus, if we are to only classify $\mathbf{z}_{N_f+1}$ followed by cluster parameter updating, there is the possibility that we may never be able to correct some of the erroneous classifications in our recent past. This may lead to those feature points and their corresponding clusters inducing further errors. Hence, we should allow at least some of the most suspect past classifications to be adjusted after the clustering of the new input $\mathbf{z}_{N_f+1}$, so as to help reduce this error propagation.

These observations motivate us to modify the DPMM-based classification algorithm to create a sequential classification algorithm made of the following two steps:

1. Given a new input feature vector $\mathbf{z}_{N_f+1}$, the Gibbs sampler selects the parameter θ_{N_f+1} of $\mathbf{z}_{N_f+1}$ for sampling according to (8.18). Hence, the new observation $\mathbf{z}_{N_f+1}$ will be assigned to either an existing or a new cluster, and the parameter selection policy $\{w_i\}_{i=1}^{N_f+1}$ is updated accordingly.
2. A set of $J < N_f + 1$ parameters θ_i's is selected from $\{\theta_i\}_{i=1}^{N_f+1}$ with probabilities w_i's and is sampled according to (8.18). This ensures that all parameters θ_i's are being continuously sampled as time progresses (and new feature vector inputs are presented) to the algorithm.

A sequential DPMM-based classification algorithm based on above steps is given in Algorithm 11.2. Note that in step 1 above, the new input feature vector $\mathbf{z}_{N_f+1}$ is classified directly by itself rather than through batch-processing. Admittedly, this is not sufficient to ensure accurate classification. The resampling in step 2 attempts to overcome this by resampling a set of θ_i's. If N_f and J are both large enough, we may expect this to perform well although in general this sequential processing is suboptimal.

Algorithm 11.2

A sequential DPMM-based Bayesian nonparametric classification algorithm.

(1) **Training:** Apply off-line DPMM classification to $\mathbf{z}_{1:N_f}$.
(2) **Sequential classification:**
(3) FOR $i = N_f + 1, \cdots$
    Sample θ_i using the posterior distribution in (8.18).
    Update the parameter selection policy $\{w_{i'}\}_{i'=1}^{i}$.
    FOR $j = 1, \cdots, J$:
        Select $i_j = i'$, for $1 \le i' \le i$, with probability $w_{i'}$.
        Update θ_{i_j} by sampling from the posterior distribution in (8.18).
        Update the parameter selection policy $\{w_{i'}\}_{i'=1}^{i}$.
    END.
END.

The sequential DPMM-based classification approach of Algorithm 11.2 can especially be an effective low-complexity alternative when it is used in conjunction with an initialization step. During the initialization step, the batch-processing-based DPMM classification algorithm of previous section is to be used. Note that, unlike supervised learning approaches, however, this initialization step does not have to be based on training data. In fact, input data may be actual observations themselves, and the algorithm will classify them into clusters albeit incurring high computational complexity. However, after sometime, once the clusters have been stabilized, the cognitive radio can switch to the earlier sequential classification approach. The rationale for this lies in the fact that once a reliable set of clusters have been established, the sequential classification Algorithm 11.2 may perform satisfactorily at a lower computational complexity.

Indeed, under normal operating conditions, a cognitive radio may repeatedly be used in familiar signal environments. Hence, it will have stored in its long-term memory (signal database in Figure 11.5) previous clustering results of signals encountered in each of the subbands. In many situations, these can serve as the initialization stage of Algorithm 11.2. However, it is important to regularly update these established clusters in order to avoid error propagation.

The sequential DPMM-based classification approach of Algorithm 11.2 can be modified in different ways to better suit for a particular application. One obvious alternative implementation is to in fact combine the steps 1 and 2 above into one step so that we directly classify a set of $M < N_f + 1$ feature vectors out of $\mathbf{z}_0 \ldots, \mathbf{z}_{N_f}, \mathbf{z}_{N_f+1}$ where the new input $\mathbf{z}_{N_f+1}$ to be classified is included in the M selected feature vectors by default. If there are K already established clusters, then M is to be selected such that $K + 1 < M$. The remaining $M - 1$ feature vectors to be classified with the new input $\mathbf{z}_{N_f+1}$ can be selected in several possible ways: (i) First, uniformly select $M - 1$ clusters and then select a feature vector assigned to each of the selected clusters uniformly, or (ii) select $M - 1$ feature vectors according to the biased selection policy (then update the weights after classification), or (iii) uniformly select $M - 1$ feature vectors from among $\mathbf{z}_0 \ldots, \mathbf{z}_{N_f}$. Of course, there are other ways to perform these selections.

Example 11.9 (Sequential Bayesian Nonparametric Classification of a Mixture of Lognormal and Gaussian-Distributed Observations) In the following example, we consider a classification problem in a mixture of Gaussian and lognormal distributed observations corresponding to four different clusters. Note that details of posterior distribution updates in this model were discussed in Section 8.5.3.1. Two of the mixture components correspond to Gaussian models with respective means $\boldsymbol{\mu}_1 = [10, 10]^T$ and $\boldsymbol{\mu}_2 = [11, 11]^T$ and a common covariance matrix of $\boldsymbol{\Sigma}_i = diag(0.4, 0.025)$. The other two mixture components correspond to lognormal models with respective mean parameters $\boldsymbol{\rho}_3 = [\log(12), \log(12)]^T$ and $\boldsymbol{\rho}_4 = [\log(13), \log(13)]^T$ and shape parameters $\boldsymbol{\xi}_3 \in \mathbb{R}^2$ and $\boldsymbol{\xi}_4 \in \mathbb{R}^2$ that are the same. In this case, we denote $\boldsymbol{\xi}_l = [\xi_{l,1}, \xi_{l,2}]^T$, for $l = 3, 4$, where $\xi_{l,d}$ denotes the shape parameter of the dth coordinate of feature vector $\mathbf{y}_i$ that is drawn from the lth mixture model.

For the Gaussian likelihoods, we assume that the parameter $\theta_i = (\boldsymbol{\mu}_i, \boldsymbol{\Sigma}_i)$ admits a normal/inverse-Wishart conjugate prior with hyperparameters $\mathbf{m} = [9, 12]^T$, $\tau = 1$, $s = 4$, and $\mathbf{S} = diag(0.8, 0.05)$. In the lognormal case, on the other hand, we assume a Gaussian conjugate prior for the parameter $\theta_i = \boldsymbol{\rho}_i$ with hyperparameters $\rho_{0,k} = \log(13)$ and $\xi_{0,k} = 0.15$, for $k = 1, 2$ (see Section 8.5.3.1 for expressions for posterior distribution updates).

We use a set of $N_f = 100$ feature vectors $\mathbf{z}_i \in \mathbb{R}^2$ that are drawn from this mixture of Gaussian and lognormal likelihoods in an initial training stage of the sequential classifier as described in Algorithm 11.2. During this off-line training stage,

TABLE 11.1 Sequential Bayesian nonparametric classification of feature vectors that are drawn from a Gaussian and lognormal mixture using Algorithm 11.2

Classifier	Classification accuracy	
	$\underline{\mu}_5 = \underline{\mu}_1$ and Σ_i	$\underline{\mu}_5 = (9, 13)$ and Σ_i
DPMM-based classification	97%	82%
X-means classification	68.5%	68%

we allow a sufficiently large number of Gibbs sampling iterations to ensure proper convergence.

Next, new feature vectors are sequentially input to Algorithm 11.2 as well as X-means algorithm for online classification. In the first test case, a set of 100 new feature vectors are drawn from one of the clusters used in the training stage itself. Specifically, we draw these test feature vectors from a Gaussian model with mean $\boldsymbol{\mu}_5 = \boldsymbol{\mu}_1$ and $\boldsymbol{\Sigma}_i = diag(0.4, 0.025)$. Table 11.1 shows the classification accuracy during this sequential classification stage. The sequential DPMM-based Algorithm 11.2, with $J = 10$, achieves about 97% of classification accuracy, while X-means algorithm achieves only about 68% accuracy. Note that these performance rates are computed only during the sequential classification stage.

In the second case, a set of 100 new test feature vectors were drawn from a Gaussian distribution with mean parameter $\boldsymbol{\mu}_5 = [9, 13]^T$ and a covariance matrix $\boldsymbol{\Sigma}_i = diag(0.08, 0.005)$. Hence, these correspond to a completely new cluster. Table 11.1 shows that although performance of the DPMM-based sequential classification Algorithm 11.2 has significantly dropped from the earlier case, still it performs at a respectable 82% accuracy. The performance of the X-means classification seems to be almost insensitive to this new truth about test data. Clearly, Algorithm 11.2 far outperforms the X-means. ■

In the next example, we investigate the classification performance of the sequential algorithm when feature vectors correspond to QAM signals of different modulation orders.

Example 11.10 (Sequential Bayesian Nonparametric Classification of Digital Signals) Initially, we generate $N_f = 100$ feature vectors drawn from QAM signals with different modulation orders, including BPSK, 4-QAM, 16-QAM, and 64-QAM modulations. For each of these training signals, a 2-D feature vector whose components are the kurtosis of the squared-signal magnitude ($\texttt{kurtosis}(|s^2(t)|)$) and the fourth-order cumulant is extracted. All QAM signals are subjected to additive white Gaussian noise. We also assume Gaussian likelihoods for the extracted 2-D feature vectors with a normal/inverse-Wishart conjugate prior distribution for θ_i's. The parameters of the conjugate prior model are taken to be $\mathbf{m} = [0, 0]^T$, $\tau = 1$,

TABLE 11.2 Sequential Bayesian nonparametric classification of feature vectors corresponding to QAM signals from a mixture of four different modulation orders

Classifier	Classification accuracy	
	SNR = 3 dB	SNR = 10 dB
DPMM-based classification	98%	99%
X-means classification	50.5%	82%

$s = 4$, $\mathbf{S} = diag(0.002, 0.002)$. We assume $\alpha_0 = 1$ in the Dirichlet process model, corresponding to the rate of creating new clusters.

After the batch-processing-based initialization stage, 100 test feature vectors corresponding to the 64-QAM signals are generated and sequentially input to both Algorithm 11.2 and the X-means algorithm. Table 11.2 shows the classification performance of the two algorithms during the sequential classification stage for two different SNR values (SNR refers to that of the QAM signals when the feature vectors are extracted from them).

At the low SNR of 3 dB, the performance of the X-mean algorithm is almost useless. It seems to provide at least reasonable classification accuracy around SNR = 10 dB. However, the DPMM-based sequential classification algorithm performs extremely well even at the low SNR value of 3 dB. ■

A nonparametric classification algorithm, such as Algorithm 11.2 that allows for an arbitrary number of clusters can especially be relevant in dynamic signal environments. In practice, this is the situation one may encounter in wireless channels: Signals may come and go at arbitrary times. In particular, if a new signal belonging to a system unknown up to that instant enters the spectrum, the algorithm must be able to create a new cluster for it. Moreover, if it is indeed a signal of a new kind, the algorithm should still keep the classification assignments of all the previous inputs unchanged.

In the next example, we investigate the ability of the sequential DPMM classification algorithm to adapt to a changing RF environment.

Example 11.11 (Sequential Nonparametric Classification of Digital Signals When RF Environment Changes) Let us consider a sequential DPMM classifier trained in the QAM signal classification set up considered in Example 11.8. The signal scenario considered in Example 11.8 is used as the batch-processing-based initialization stage of the sequential classification Algorithm 11.2 in this signal environment. The classification space at the end of the initialization stage will look like as in Figure 11.11(a).

Next, we input new feature vectors into the sequential classifier given in Algorithm 11.2. All signals remain the same as during the initialization stage except for the signal

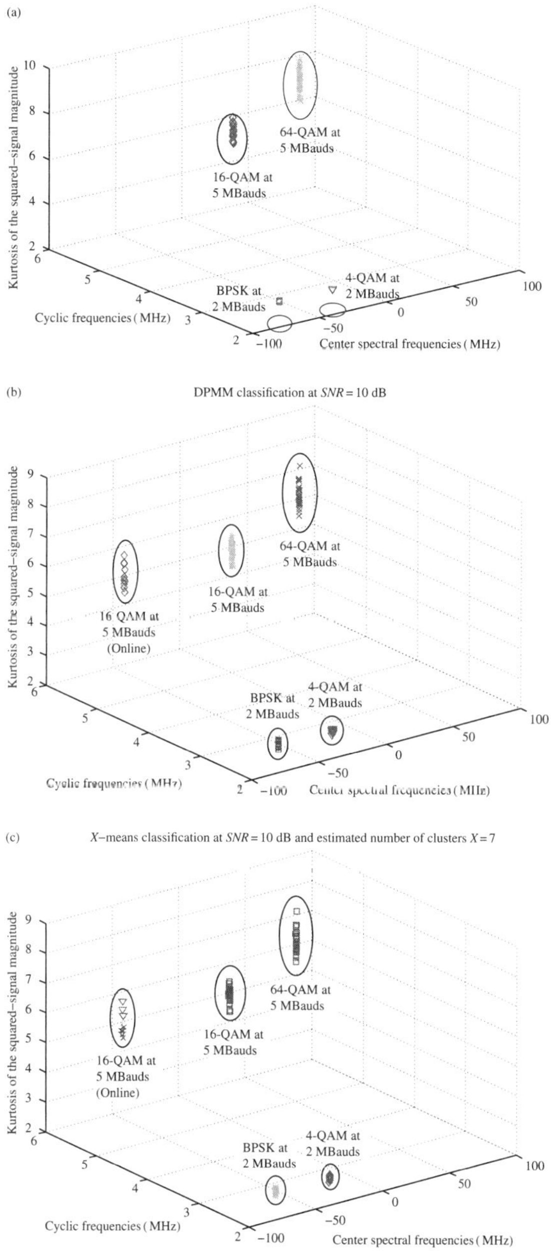

FIGURE 11.11 Sequential classification performance. (a) DPMM classification of four different QAM signals at a received $SNR = 10$ dB using Algorithm 11.2 at the end of the training period. Classification accuracy is equal to 100%. (b) Sequential DPMM classification of five different QAM signals at a received $SNR = 10$ dB using Algorithm 11.2. Classification accuracy is equal to 100%.(c) X-means classification of five different QAM signals at a received $SNR = 10$ dB. Classification accuracy is equal to 71%.

at -80 MHz. Recall that during the initialization stage, it was a 2MBauds BPSK signal. During real-time operation, the signal transmitting at -80 MHz is made to change its symbol rate from 2 to 5 MBauds and its modulation scheme from BPSK to 16-QAM. Figure 11.11(b) shows the resulting feature space, including feature vectors that were observed during both initialization stage and sequential processing stage, after 10 real-time sensing intervals that follow the initialization stage. Note that the 10 real-time inputs were processed by the sequential algorithm that classifies the detected features at the end of each sensing interval itself. Figure 11.11(b) shows that a new cluster has been created and it includes feature inputs corresponding to the new 16-QAM signal located at -80 MHz. All the other signals have still correctly joined the old clusters.

On the other hand, if we apply the X-means algorithm to the same set of data, it is seen that the classification accuracy is only about 71%. As shown in Figure 11.11(c), many feature vectors have not been classified correctly by the X-means algorithm. The primary reason is that the X-means algorithm failed to estimate the correct number of clusters in this case. Indeed, during the real-time operation, the number of clusters was estimated to be equal to 11, instead of 5. ■

Finally, let us look at the breakdown of the origin of errors that we observe at the output of the classification stage. Recall that in reality the errors can be already present in the feature vectors when they are input to the classification algorithm. As we saw in Example 11.6, it is unfair to attribute those errors to the classification algorithm. In the following example, we may do the same analysis for the sequential classification algorithms.

Example 11.12 (Overall versus Misclassification Error Rates of Sequential Non-parametric Classifiers of Digital Signals when RF Environment Changes) In this example, we consider the exact same scenario assumed in Example 11.11. An initialization stage over 25 sensing intervals is assumed, followed by 25 sensing intervals of sequential classification. In Figure 11.12, we show the breakdown of the two types of errors when we consider the whole 50 sensing intervals for both DPMM and X-means algorithms. It shows the feature extraction error rate at different SNR values and the overall classification error rates (denoted as "overall DPMM + feature extraction" and "X-means + feature extraction") at the output of the classification algorithm. Figure 11.12 also shows the misclassification rate itself, when the feature extraction errors were properly taken into account.

As expected, the feature extraction error drops as the SNR increases. Figure 11.12 shows that the overall classification accuracy of the DPMM is degraded mostly due to the high level of feature extraction errors at low SNR. At high SNR, when the feature extraction errors disappears, the DPMM classification errors have also decreased. Hence, as long as the rate of feature extraction errors is small, we may expect the performance of the DPMM classification to be very good.

On the other hand, the X-means algorithm shows extremely poor classification performance at low SNR. Worse yet, in contrast to the DPMM, it does not lead to perfect classification even when the SNR is high enough to eliminate all the feature

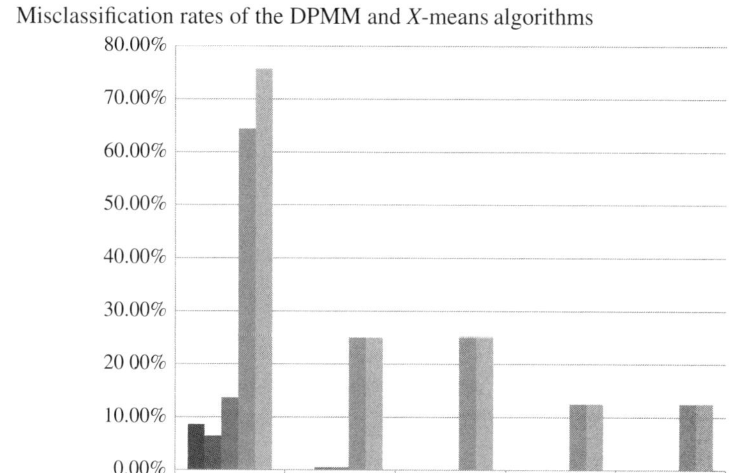

	SNR = −10 dB	SNR = −5 dB	SNR = 0 dB	SNR = 5 dB	SNR = 10 dB
■ Feature extraction (false alarm)	8.63%	0.00%	0.00%	0.00%	0.00%
■ Overall DPMM	6.47%	0.50%	0.00%	0.00%	0.00%
■ Overall DPMM + feature extraction	13.67%	0.50%	0.00%	0.00%	0.00%
■ X-means	64.41%	25.00%	25.00%	12.50%	12.50%
■ X-means + feature extraction	75.71%	25.00%	25.00%	12.50%	12.50%

FIGURE 11.12 Misclassification rates of the DPMM and X-means algorithms for different SNR levels in sequential mode. Overall DPMM misclassification rates are obtained over 50 sensing intervals, including both training and sequential stages. Misclassification rates of the X-means are also obtained over the whole 50 sensing intervals. Combined feature extraction errors and signal classification errors are computed using DPMM and X-means classifiers and are denoted as "overall DPMM + feature extraction" and "X-means + feature extraction," respectively. Four different QAM signals are being transmitted simultaneously and are sampled at a rate $f_s = 200$ MHz for a duration $T = 50\,\mu$s in each sensing interval.

extraction errors. This shows that X-means classification algorithm can lead to a significant amount of its own errors even when there are no erroneous feature vectors at its input. ■

11.7 SUMMARY

- Acquiring full spectrum knowledge requires that a cognitive radio be able not only to detect RF signals at arbitrary carrier frequencies within each subband but also to identify which systems these signals belong to.
- A spectrum opportunity of interest to a wideband cognitive radio will depend on its instantaneous performance objectives. In particular, a spectrum opportunity may not only be a white space as in the DSA-based cognitive radios.

- The signal identification problem in a wideband cognitive radio can be divided into three stages

    1. Feature extraction
    2. Classification and clustering
    3. Signal and cluster association

- In general, both classification and clustering are better achieved based on relevant features. Identifying the most suitable features, however, may be difficult without some guidance or prior knowledge. The number of features extracted does not have to be the same at all frequencies either.
- Signal classification task in a cognitive radio is not very different from the standard supervised classification problem: Given a set of input feature vectors, classify them into a set of known classes. There are many efficient classification algorithms available in literature that can perform this task.
- The signal clustering problem in a cognitive radio, however, includes challenges that are unique to this problem: The number of active systems in a subband is unknown and can vary with time. Moreover, the number of signals belonging to each of the active systems will also not be known a priori and can be time-varying. Finally, the likelihoods of feature vector inputs may not be multivariate Gaussians and can be distinct distributions from one signal to another.
- A promising candidate for solving the clustering problem in a wideband cognitive radio is the DPMM-based Bayesian nonparametric classification approach introduced in Chapter 8. This framework not only allows the number of clusters to be arbitrary, but also it can grow with the complexity of the data itself. Moreover, the feature input likelihoods do not have to be limited to multivariate Gaussians and can belong to different families of distributions.
- The final stage of signal identification process will associate the resulting signal classes with known systems, be they the clusters created by clustering module or the already existing classes into which signals were classified by the classification module. This association can also be posed as a supervised classification problem and be handled by using known algorithms.
- Extracting the most relevant signal features can be critical for their correct classification. In wideband spectrum sensing, possible features may include carrier/center frequency, signal bandwidth, cyclo stationary features corresponding to underlying periodicities in signals, and higher-order statistics such as cumulants and kurtosis. Which of these can be the best features will, of course, depend on the particular application scenario and available prior information.
- Given that computational requirements for wideband spectrum knowledge acquisition can be demanding, a multistage signal classification and identification architecture seems to be a realistic option for wideband cognitive radio. In such an architecture, the classification algorithm will first use only a small set of easily extractable features corresponding to each detected spectral activity. If

the algorithm cannot reliably classify and identify some signals based on these features, it will request for more detailed extra features corresponding to those signals. Even then, these additional features are extracted only for those signals that were not classified at the first stage. This process may repeat for several stages till either all signals are classified or the maximum number of extractable features is exhausted.

- Bayesian nonparametric signal classification method based on the Dirichlet process mixture model offers a promising approach to signal classification in a wideband cognitive radio. This method allows for the number of clusters to be unbounded a priori and be determined by the input data itself. Moreover, it can handle feature vectors with both Gaussian and non-Gaussian likelihoods in the same input data set. The drawback of the approach is that it can be considerably more computationally complex than some other alternative classification algorithms.
- Computational complexity of the DPMM-based nonparametric classification algorithm can somewhat be reduced by introducing a biased parameter selection policy during Gibbs sampling. Instead of uniformly and randomly selecting parameters θ_i's for updating, this method picks parameters according to weights or probabilities that are inversely proportional to the number of members in their respective clusters. Clearly, this policy favors those feature vectors belonging to smaller clusters. Since these are also the most likely candidates for cluster reassignment during subsequent Gibbs sampling, their frequent selection can speed up the convergence of the algorithm.
- DPMM-based Bayesian nonparametric classification algorithm can be modified so as to obtain a sequential algorithm suitable for real-time signal classification. While there are several ways one can do this, one (extreme) approach is to first use an initialization stage during which the batch-processing-based classification algorithm is used to establish a set of reliable clusters. After that, new inputs are classified sequentially one at a time. However, in order to ensure proper updating of the clusters so as to avoid error propagation, a randomly selected set of parameters corresponding to previously classified feature vectors is also resampled and updated after each new input.
- The sequential DPMM-based classification can have satisfactory performance in the presence of dynamic signal environments in which new signals may enter and old signals may exit. Regular updating of the clusters, however, will need to be performed in order to maintain good performance.

12

PRIMARY SIGNAL DETECTION IN DSA COGNITIVE NETWORKS

12.1 INTRODUCTION

As we discussed in Chapter 3, dynamic spectrum sharing (DSS) was perhaps the key application that drove the early interest in cognitive radios (CR). Dynamically sharing the spectrum has been around since the early days of wireless telecommunications. Indeed, any time multiple users share spectrum through some form of dynamic channel access, it does amount to DSS. For example, a communications device may use a random access protocol such as carrier sense multiple access (CSMA) to observe (or sense) whether there is an active carrier signal in a channel that it intends to transmit on (see Figure 12.1(a)). In a simpler random access protocol such as ALOHA, a radio will simply transmit without first verifying whether there is an active user in the channel or not, but it will still sense the channel to determine whether the transmission was successful or not. If it is not, it may follow up with specific protocol actions that make this also dynamically sharing the spectrum with other users. In centralized networks, for example, cellular networks, the radios will be assigned channels on the need-only basis in real time by a central controller facilitating dynamic sharing of the channels among radios.

DSS by a CR, however, differs from these usual spectrum sharing approaches in an important way as is shown in Figure 12.1(c). As we discussed in Chapter 3, CRs dynamically share the spectrum with users from other systems. In the case of DSS CRs, these other users in fact own the spectrum. CRs are supposed to access these already assigned spectrum bands only when they are unused by the licensed users.

Signal Processing for Cognitive Radios, First Edition. Sudharman K. Jayaweera.

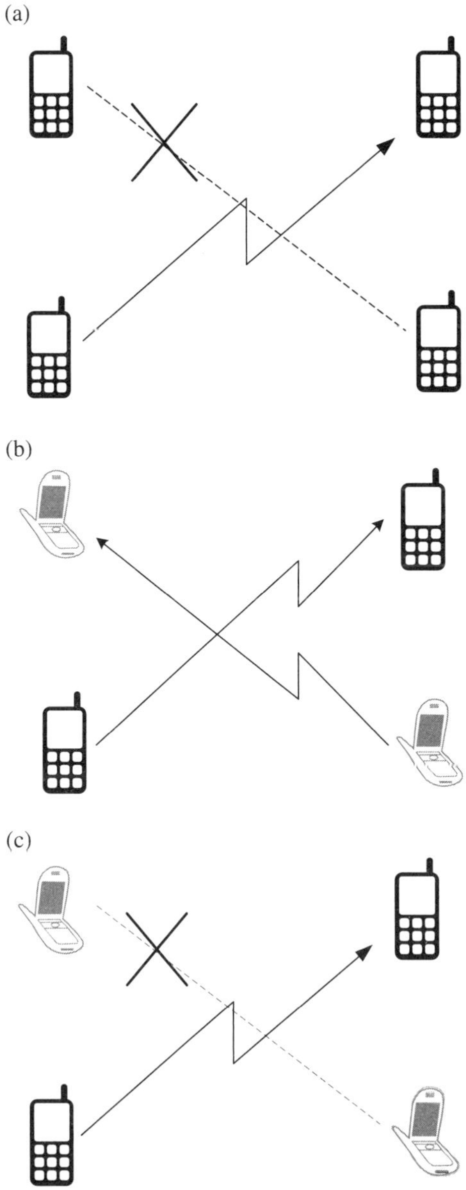

FIGURE 12.1 DSS in CR goes beyond traditional spectrum sharing seen in wireless networks. (a) Spectrum sharing among radios belonging to the same system is usually facilitated by channel-access and assignment protocols. (b) Spectrum sharing among radios belonging to different systems in ISM bands is primarily based on tolerating intersystem interference. (c) Spectrum sharing between cognitive secondary radios and primary radios is achieved by CR making a conscious effort not to interfere with primary transmissions.

A major requirement is that they do not harm licensed user communications. This is very different from spectrum sharing through dynamic channel allocation or random access protocols as seen in traditional wireless networks: Mostly, these schemes are dedicated for spectrum sharing among users belonging to the same system, as shown in Figure 12.1(a). Before the emergence of DSS by CRs, perhaps the only situation in which users from several different systems dynamically shared the same spectrum was in the ISM bands. For example, the IEEE 802.11 WLAN and Bluetooth systems both operate over the same 2.4–2.475GHz spectrum range. However, this is still different from DSS by CRs: In ISM bands, there are no licensed users. Hence, there is no need for a conscious effort from any user to not harm another user's communications. All systems sharing an ISM band have to live with each other by tolerating their mutual interference as shown in Figure 12.1(b).

While in this and the next two chapters we will be focusing exclusively on CRs dedicated for DSS purpose, to be explicit, we will sometimes refer to them as DSS CRs. The promise of CR is to allow radios belonging to a different system to dynamically share a spectrum band with the licensed, or primary, users in that spectrum by consciously making sure that the licensed user quality of service (QoS) is not compromised. For that, a DSS CR must continually monitor the spectrum before accessing it to proactively manage the *intersystem interference*. A CR is supposed to access the spectrum only if it can do so without harming the primary user transmission: that is, either when a channel is idle or when the interference to the primary user will be below an acceptable level. Recall from Chapter 3 that a CR that dynamically shares the spectrum with primary licensed users in this way is called a secondary user. In Chapter 3 we also observed that there are two broad paradigms for DSS between primary and secondary users:

1. Dynamic spectrum access (DSA) in which the DSS is managed all by the cognitive secondary radios themselves without any participation from the primary user network
2. Dynamic spectrum leasing (DSL) in which primary users may proactively facilitate the spectrum sharing in return for a reward from the secondary users.

As was discussed in Chapter 3, both DSA and DSL can be implemented on the basis of either spectrum underlay or spectrum interweave-based coexistence.

Just as in the case of a wideband CR, the key to the success of DSS is the CR's ability to gain spectrum knowledge. This is particularly important in the case of DSA-based spectrum sharing since CR network is solely responsible for managing the intersystem interference to the primary network. Even in a DSL-based spectrum sharing system, it is normally the CR that will have to bear most of the burden of interference management. A CR must be able to effectively sense all available primary channels and detect whether there is a primary signal present or not in a sensed channel. If the radio can do this well enough, then it can capitalize on idle spectrum opportunities to achieve its own communications needs.

While basic concepts and the underlying theory needed for spectrum sensing in DSA networks are not very different from those we have already seen in wideband

CR context, there are certain aspects that make specific discussion of the problem useful. Notably, the presence of primary users, and their priority status, imposes an important constraint on CRs. At the same time, having to only detect primary signals on licensed channels at least narrows the scope of what to look for as compared to that in the case of a wideband CR.

As we will see in the next section, the specific spectrum sensing problem encountered by a CR in a DSA network consists of three parts: deciding which channel to sense, detecting primary signals in the sensed channel, and deciding on which channel to transmit. In this chapter, we discuss the primary signal detection step. In the next chapter, we will address the problem of deciding which channel to sense and access.

12.2 SPECTRUM SENSING PROBLEM IN DYNAMIC SPECTRUM SHARING CR NETWORKS

All DSS approaches require a significant amount of spectrum awareness by the cognitive (secondary) users. The radios must be able to determine when a primary channel is idle and when it is not. To be specific, let us consider a primary–secondary coexistence scenario in which there are an M number of primary channels, K_p number of (licensed) primary users, and K_s number of secondary cognitive users as in Figure 12.2. Secondary radios are responsible for finding idle primary channels so that they can transmit on these channels without harming primary user QoS. Such idle channels represent spectrum opportunities. Recall from Chapter 3 that for a channel to be a *possible spectrum opportunity* at a given time instant, it must either be free of primary transmission activity or a secondary transmission on that channel should be able to achieve at a power level that will keep interference to the primary link on that channel below an acceptable threshold. However, as we saw in Chapter 3, this

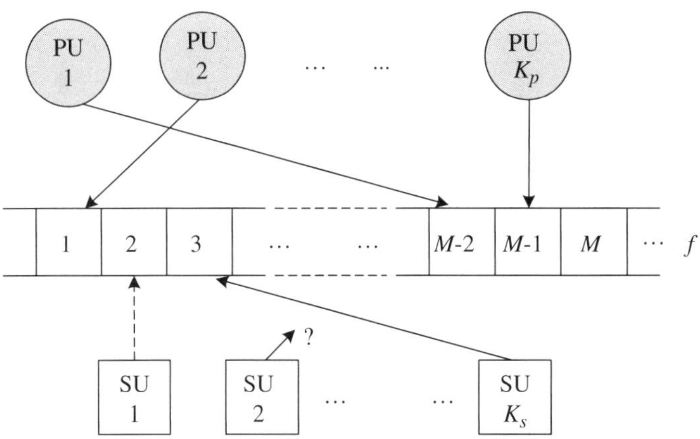

FIGURE 12.2 A DSS system contains K_s secondary users who are interested in accessing spectrum opportunities over M channels owned by a licensed system with K_p primary users.

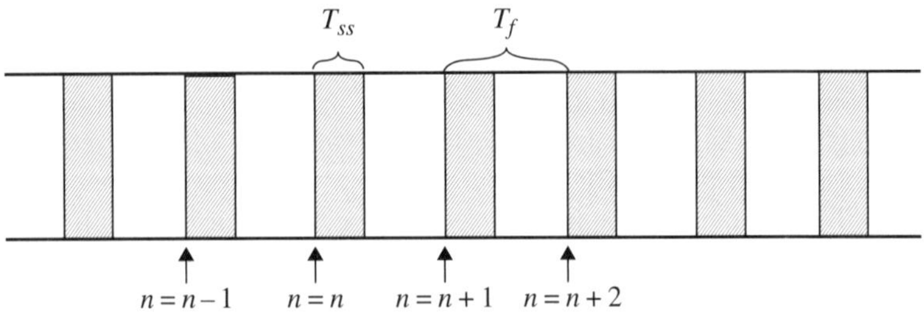

FIGURE 12.3 Slotted time horizon with sensing periods and transmitting periods.

alone may not be sufficient for a channel to be an *actual spectrum opportunity* for a cognitive secondary user. The missing part is that whether CR can indeed achieve successful communication over this channel in the sense that a transmission on this channel will successfully be received by the CR's intended receiver. This requires that the intended cognitive receiver does not receive interference from another primary or secondary transmission on this channel that the cognitive transmitter might not even be aware of.

For simplicity of discussion, let us assume that the primary transmissions are time slotted as in Figure 12.3. At the beginning of each time slot, there is a designated *sensing interval* during which all primary users will indicate whether they intend to use a particular channel for the duration of subsequent time slot. Let us denote by T_f and T_{ss}, respectively, the primary time slot and the spectrum sensing interval lengths. Note that this does not necessarily mean that the legacy primary systems need to do anything new. All we assume here is that primary users will always start their transmissions at the beginning of a time slot. It is the secondary CR that must agree among themselves to designate a small fraction of a time slot duration, which we have denoted as T_{ss}, at the beginning of each time slot as the common sensing interval. No secondary radio is supposed to transmit during this T_{ss} sensing duration. Having a designated sensing interval at the beginning of each primary time slot allows secondary radios to sense and detect primary signals without being misled by interference from other secondary transmitters.

This designated sensing interval assumption then implies that all secondary radios must terminate their transmissions at the end of each time slot. After each sensing interval, a CR only gets to transmit for a duration of $T_f - T_{ss}$, the remaining portion of the time slot after what is spent on spectrum sensing. Of course, a radio will get to transmit even for this duration only if the sensed channel was determined to be idle and a spectrum opportunity. At the beginning of the next time slot, it must perform spectrum sensing anew. Note that referring to Figure 3.10, T_f and T_{ss} may correspond to either one of the scenarios suggested in Figures 3.10(a) or 3.10(b).

It is worth noting that there is a trade-off on how long the sensing duration T_{ss} can be. Of course, we are assuming that $0 < T_{ss} \leq T_f$. Within this constraint, the spectrum sensing period T_{ss} has to be long enough so that the secondary CR can collect enough primary signal information, or energy, to make a reliable decision. If it is too short, these decisions may be erroneous with too high a probability. But if the sensing period T_{ss} is too long, then the time remaining in the time slot for transmission, that is, $T_f - T_{ss}$, will be small. Hence, T_{ss} should be long enough but not too long.

We now formalize the spectrum sensing problem in a DSA CR network under the aforementioned assumptions. We will limit the discussion specifically to spectrum interweave type of coexistence so that the objective is to find and use idle primary channels. The spectrum sensing problem in such a DSS network consists of three subproblems:

1. Decide which channel to sense (channel sensing decision-making problem)
2. Decide whether the channel is idle or busy based on observations of the sensed channel (idle/busy decision-making, primary signal detection or channel-state detection problem)
3. Decide whether to access the channel or not if it is determined to be indeed idle (channel-access decision-making problem)

The problem of *channel sensing decision-making* concerns with selecting a channel or a set of channels to be sensed at time $n=n$. The CR should select the channel, or the set of channels, that will most likely be idle during the sensing interval starting at time $n=n$ and also less likely to be sensed by another secondary user. The *channel-state detection problem* is the classical signal detection problem that can be modeled as a hypothesis testing problem: Based on sensed signals (observations), the cognitive secondary radio needs to decide which hypothesis (channel idle or channel busy) is true. Finally, the *channel-access decision problem* is to make a decision about whether to access or not a channel that is determined to be idle and can be similar to channel assignment problem in traditional wireless systems. We are implicitly assuming that a secondary CR will only access a channel that it has sensed and determined to be idle. Of course, this assumption makes sense in practice.

Solution approaches to these problems can look very different depending on the structure of the secondary CR network: whether each CR makes decisions autonomously or a centralized controller in the secondary network coordinates these decisions. All three above problems can be challenging problems when an individual CR has to make autonomous decisions by itself. For example, a channel that is going to be most likely idle during the next time slot might be sensed by more than one user, as in Figure 12.4. If two cognitive users sensed the same primary channel and determined it to be idle, then there will be a collision among their transmissions if both of them started to transmit (access) on that channel during the subsequent transmission duration. One or both of their transmissions may fail due to the interference from the other transmission. Hence, even after accurately predicting and sensing a channel that is going to be idle during the next time slot, one or both of them still may lose out. This shows that an autonomous CR faces the burden of not just having to predict an idle channel but also one that will least likely be sensed by another secondary CR. Moreover, in the case of an autonomous CR, the channel-access decisions may be the same as channel idle/busy (state) decisions since if a channel is found to be idle, then the radio may access it. As we will see in the following, determining whether a

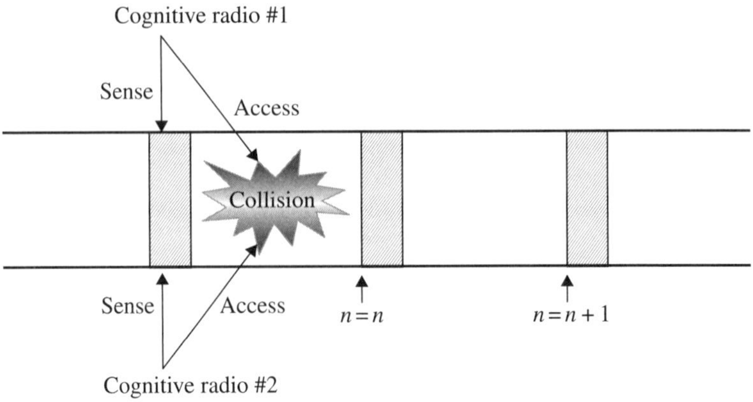

FIGURE 12.4 If two autonomous cognitive secondary radios sense and access the same primary channel, a collision will occur, causing possibly one or both of their transmissions to fail.

channel is idle or busy by a single CR may not be reliable either. Due to the so-called hidden terminal problems inherent in wireless communications, for example, a single CR may not be able to accurately detect a primary transmitter based on its own observations alone.

In a centralized system, on the other hand, both sensing and access decisions may be performed by a centralized unit, called the *secondary system decision center* (SSDC). All secondary users in such a network may forward their sensing results (called the local channel-state decisions) to the SSDC. The SSDC will make final decisions on which channels are indeed idle by combining these local data or decisions. Such combined decisions may possibly turn out to be more reliable compared to those of an individual CR. Similarly, the SSDC may also determine which secondary user should access which of those channels that were determined to be idle (channel-access decisions). This is similar to channel assignment problem in a cellular or a WiFi network. The SSDC may assign idle channels to CRs in need of channels based on various fairness, QoS, and priority criteria. Hence, collisions as in Figure 12.4 may be avoidable. In such a system, importantly, the SSDC can also instruct which secondary user should sense which primary channel during the next sensing period (sensing decisions). As we will see later, such centralized sensing decisions may be efficient in terms of both sensing several likely idle channels at once and making better decisions on whether a channel is indeed idle or not.

These two approaches (*autonomous* vs. *centralized*) are also called *noncooperative* versus *cooperative spectrum sensing*. However, the term autonomous seems better suited in this context since the opposite behavior to explicit cooperation may not necessarily be noncooperative behavior. Indeed, the radios may not cooperate in making decisions, but still their individual decisions may take into account each other's existence for it helps ensure their own transmission success. Hence, although they are not explicitly cooperative, they are certainly not noncooperative either. Yet another terminology for autonomous versus centralized is that of *distributed* versus *centralized*. The key difference, in any case, is where the decisions are made. If we presume that no secondary user will access a channel that has not been determined to be idle (by sensing it), in an autonomous CR system, then the decisions on which channel to sense and whether to access that channel or not are both made by the autonomous CRs themselves. The decisions are all local or distributed. On the other hand, in a centralized system, which radio should sense which channel and which radio can access which idle channel are both decided at the SSDC. Essentially, cognition in such systems is not really in the distributed radio units but at the SSDC.

In the remainder of this chapter, we discuss channel-state detection problem encountered in a DSS CR network under autonomous and cooperative assumptions. The sensing and access decision-making problems will be addressed in Chapter 13.

12.3 AUTONOMOUS SPECTRUM SENSING FOR DYNAMIC SPECTRUM SHARING

Cognitive wireless devices that may attempt to communicate over primary channels without being subjected to a central controller perform *autonomous* (or distributed) *spectrum sensing*. Such a CR will listen to a primary channel to see whether it is idle or busy, and if it finds the channel idle and if it has its own need to transmit, it will transmit over that channel. The decision on whether a channel is idle or busy is made by the (autonomous) radio itself, based only on its own sensing observation outcome. In that sense, the radio is autonomous. Moreover, whether to access the channel or not (channel-access decisions) as well as which channel to sense (channel sensing decisions) may also be made by the radio itself without regard to, or consulting, other secondary users in the network. However, for its own advantage, radio may take into account the existence of other secondary users in making these decisions although it may not explicitly be cooperating with them. In summary autonomous CRs may make decisions by themselves based only on their own sensing observations. Yet, they may take into account the existence of other secondary users in making these decisions because it might lead to their own better performance.

Let us consider a primary–secondary coexistence scenario in which the K_s cognitive secondary users have limited spectrum sensing capabilities in that each can only sense one primary channel at a given time (*narrowband spectrum sensing*). Note that, in general, all secondary users may not even belong to the same network. At any given sensing time instant n, the mth primary channel, for $m \in \{1, \ldots, M\}$, can be either *busy* or *idle*. As in Chapter 9, we call this the channel state. To avoid having to consider the situation in which a primary user may start or terminate transmission in the middle of a time slot, we assume that the state of a primary channel does not change within a single time slot. However, at least in this chapter, our real concern is that the channel state does not change during the T_{ss} length sensing interval within each time slot as shown in Figure 12.3. This would make the discussion of the hypothesis testing problem unnecessarily complicated. Although such changes may certainly be possible in an asynchronous network, under our assumption that each primary user is to start transmissions at the beginning of a time slot, this assumption on channel-state transitions is justifiable. Secondary users thus sense primary channels during the designated sensing interval at the beginning of each time slot as shown in Figure 12.3.

One of the main differences of spectrum sensing in DSA CR compared to wide-band spectrum sensing discussed in Chapter 9 is that secondary CRs have explicit knowledge of primary channels. In particular, they know the center frequencies and their (at least approximate) bandwidths. Hence, without loss of generality, we may assume that RF signals corresponding to a sensed channel are down-converted to an IF or baseband for processing. If a primary user intends to use a channel during a transmitting time slot, it will start to transmit from the beginning of that time slot so that a secondary user sensing that channel will likely pick it up during the sensing interval of the time slot. It is reasonable to assume that signals sensed on each channel will be processed separately.

Since a secondary user is assumed to sense only a single channel at a time, it then makes sense to assume that if a channel is determined to be idle, the user will also transmit on it subsequently. Hence, the following idle/busy (channel-state) decisions can be equivalent to access/don't-access decisions, as long as the radio has a payload to be transmitted.

12.3.1 Secondary User Sensing Observations

Assuming that at time n the ith secondary user, for $i \in \{1,\ldots,K_s\}$, senses the mth primary channel, for $m \in \{1,\ldots,M\}$, let us denote by $\mathbf{r}_{m,i}[n]$ the resulting L-length complex (baseband) observation vector:

$$\mathbf{r}_{m,i}[n] = (1 - S_m[n])h_{m,i}[n]\mathbf{x}_m[n] + \mathbf{w}_{m,i}[n], \tag{12.1}$$

where $S_m[n] \in \{0,1\}$ is the mth channel state in time slot n, $\mathbf{x}_m[n] \in \mathbb{C}^L$ is the complex-valued primary signal vector on channel m, and $\mathbf{w}_{m,i} \in \mathbb{C}^L$ is a complex random L-vector of i.i.d. zero-mean Gaussian random variables with independent real and imaginary parts each distributed as $\mathcal{N}(0,\sigma_w^2/2)$. Thus, each component of $\mathbf{w}_{m,i}$ is a circularly symmetric complex Gaussian random variable denoted as $\mathcal{N}_c(0,\sigma_w^2)$. We denote by $h_{m,i}[n] = \alpha_{m,i}[n]e^{j\theta_{m,i}[n]}$ the complex channel gain of the primary channel between the primary transmitter on the mth primary channel and the ith secondary user during time slot n, with amplitude $\alpha_{m,i}[n]$ and phase $e^{j\theta_{m,i}[n]}$. In this and the next two chapters, $S_m[n]=0$ and $S_m[n]=1$ will denote, respectively, channel-busy and channel-idle states.

Note that the L components of the observation vector $\mathbf{r}_{m,i}[n]$ in (12.1) may correspond to L time samples, L antenna measurements, or any other diversity domain (or a combination of them). We will assume that secondary user i has perfect knowledge of the channel-state information $h_{m,i}[n]$. Of course, this is not easy to justify under all conditions. However, we may assume that the channel fading is slow enough these channel gains do not vary much from symbol to symbol, and primary user, if active, periodically transmits training sequences intended for corresponding primary receivers to assist them in synchronization and channel estimation. Secondary users can overhear these transmissions and make use of them to estimate the fading coefficients between the primary transmitters and themselves. If fading is slow enough, these coefficients may not need fast updating.

In some situations, the received signal model (12.1) may not be completely adequate. For example, many wireless channels may exhibit frequency-selective fading that may require a multipath channel impulse response model as opposed to the frequency-flat fading assumed in (12.1). In general, the fading may have to also take into account shadowing and Doppler effects (caused by relative motion of transmitters and receivers). However, these are standard wireless communications issues that can be found in textbooks devoted to the subject. For our purpose, there is no loss in generality by assuming the simpler frequency-flat fading channel model of (12.1) since our objective is to develop the basic framework of spectrum sensing in a DSS CR. It is straightforward to modify the final decision rule as needed if one were to replace the channel model (12.1) with a more detailed channel model as required in a particular application. For details on specific and realistic wireless channel models, we refer the reader to [18, 186, 187].

It must also be mentioned that since we have assumed that the CR are autonomous and they make their channel-state decisions independently from the other CRs, we may have dropped the subscript i in the following discussion and denote the sensing observation on channel m in (12.1) as $\mathbf{r}_m[n]$. However, the notation of (12.1) will be necessary when we consider cooperative spectrum sensing later. Hence, in order to avoid having to use two different notations, we may adhere to notation of (12.1) in the case of autonomous spectrum sensing in this section as well.

12.3.2 Channel-State (Idle/Busy) Decisions

As we saw in Chapter 10, decision on whether a particular primary channel is idle or busy can be modeled as a binary hypothesis testing problem. Assuming that at time n the secondary CR senses the mth channel, the two hypotheses on the mth channel state are

$$\begin{aligned} H_0 &: S_m[n] = 0 \quad (m\text{th channel is busy}) \\ &\text{versus} \\ H_1 &: S_m[n] = 1 \quad (m\text{th channel is idle}). \end{aligned} \tag{12.2}$$

Note that in classical binary hypothesis testing literature, the alternative hypothesis (H_1) is taken to be the signal-present hypothesis. However, in (12.2), we have reversed this usual practice to denote the signal-absent hypothesis as the H_1 (alternative) hypothesis since our interest is to find channels with no primary signals. The hypothesis testing problem faced by an autonomous secondary user is to make a decision based only on its own sensing observation vector $\mathbf{r}_{m,i}[n]$. Hence, the decision-making problem of the ith CR at time n is

$$\begin{aligned} H_0 &: \mathbf{r}_{m,i}[n] = h_{m,i}[n]\mathbf{x}_m[n] + \mathbf{w}_{m,i}[n] \quad (m\text{th channel is busy}) \\ &\text{versus} \\ H_1 &: \mathbf{r}_{m,i}[n] = \mathbf{w}_{m,i}[n] \quad (m\text{th channel is idle}). \end{aligned} \tag{12.3}$$

Let $u_{m,i}[n] \in \{0, 1\}$ denote the ith secondary user's decision about the mth primary channel state at time n: $u_{m,i}[n] = 0$ means that the channel is busy (H_0 hypothesis is true), and $u_{m,i}[n] = 1$ means that the channel is idle (H_1 hypothesis is true). As we already know from Chapter 4, how to make this decision will depend on the optimality criteria and the assumptions on prior knowledge regarding the structure of the primary signal (e.g. deterministically known vs. random).

In DSS networks, whenever a channel is determined to be idle (H_1 is true) while actually the channel is busy (i.e., H_0 is true), this will lead to a collision with the primary user transmission since we have assumed that idle/busy decisions are equivalent to channel-access/channel-don't-access decisions. CRs are required to avoid such collisions with primary transmissions. Usually, in primary–secondary coexistence scenarios based on DSS, there is a constraint on how much collisions with primary transmissions will be tolerated. This is put in place so as not to degrade the primary license holders' QoS below an acceptable level. Recall from Chapter 4 that a false alarm is when H_1 is declared true while H_0 is actually true. Hence, the classical false-alarm probability (i.e., $P(H_1|H_0)$) of a decision rule is the same as the probability of collisions with the primary transmissions in the channel-state detection problem in a DSS system. Thus, it makes sense to require that the decision rule of the ith secondary user must be designed subjected to a constraint on the maximum allowed false-alarm (i.e., collision) probability. Since subject to this constraint the secondary user will naturally want to maximize its probability of correctly detecting idle channels as idle (i.e., declaring H_1 as true when indeed H_1 hypothesis is true), the Neyman–Pearson optimality of Chapter 4 seems to be a logical choice in this case:

$$u_{m,i}[n] = \arg\max_{\delta} P_D(\delta) \quad \text{subject to} \quad P_f(\delta) \leq \alpha \tag{12.4}$$

where $P_D = \Pr(H_1|H_1)$, $P_f = \Pr(H_1|H_0)$, and $\alpha \in [0, 1]$ denotes the maximum tolerable collision probability level.

From Chapter 4, the optimal Neyman–Pearson rule that solves the above problem will compare the likelihood ratio

$$L(\mathbf{r}_{m,i}[n]) = \frac{f_{\mathbf{r}|H_1}(\mathbf{r}_{m,i}[n] \mid S_m[n] = 1)}{f_{\mathbf{r}|H_0}(\mathbf{r}_{m,i}[n] \mid S_m[n] = 0)}, \tag{12.5}$$

to a threshold η chosen to satisfy the required collision probability, which in this case is the false-alarm probability. Although in general this optimal Neyman–Pearson rule can be a randomized test, under the assumption of a continuous-valued additive noise $\mathbf{w}_{m,i}[n]$ (e.g., Gaussian as in our case), the randomization turns out to be irrelevant. Hence, we may assume that $u_{m,i}[n] \in \{0, 1\}$.

The form of the conditional likelihood function $f_{\mathbf{r}|H_j}(.)$, for $j = 0, 1$, will depend on the assumptions on the knowledge of primary signal $\mathbf{x}_m[n]$ on channel m. The two extremes of this knowledge are:

1. Secondary system has no knowledge about the primary signaling waveform.
2. Secondary system has *perfect* knowledge about the primary signaling.

In the following, we will derive the local decision rules specifically in these two extreme cases.

12.3.2.1 Primary Signal Detection When Secondary System Has No Knowledge of Primary Signaling This assumption is justifiable when a secondary radio system may not know much about the primary signalling waveform except for perhaps its center/carrier frequency and bandwidth. Under these conditions, a common model is to assume that the primary signal is a stochastic process so that $\mathbf{x}_m[n]$ is a random vector. In this case, separately identifying the channel fading coefficient $h_{m,i}[n]$ from the signal $\mathbf{x}_m[n]$ as in (12.1) may not be of a much use. A more reasonable approach would be to assume that the noise power level is known (through estimation) and model the fading-effected primary signal as a realization of a random process with known overall covariance $\tilde{\Sigma}_{\mathbf{x}}$ so that:

$$\mathbf{r}_{m,i}[n] = (1 - S_m[n])\mathbf{x}_m[n] + \mathbf{w}_{m,i}[n], \tag{12.6}$$

where $\mathbf{x}_m[n]$ is a zero-mean complex Gaussian distribution with a (known) covariance matrix $\tilde{\Sigma}_{\mathbf{x}}$: $\mathbf{x}_m[n] \sim \mathcal{N}_c\left(\mathbf{0}, \tilde{\Sigma}_{\mathbf{x}}\right)$. In general, the elements of $\mathbf{x}_m[n]$ can be correlated, and thus, $\tilde{\Sigma}_{\mathbf{x}}$ may not necessarily be diagonal.

Of course, if we assume that $\mathbf{x}_m[n]$ in (12.1) has a covariance matrix $\Sigma_{\mathbf{x}}$, then essentially $\tilde{\Sigma}_{\mathbf{x}} = \mathbb{E}\{|h_{m,i}[n]|^2\}\Sigma_{\mathbf{x}}$. Hence, if we assume that fading coefficients are normalized so that $\mathbb{E}\{|h_{m,i}[n]|^2\} = 1$, then we may take $\tilde{\Sigma}_{\mathbf{x}} = \Sigma_{\mathbf{x}}$.

From (12.6), the hypothesis testing problem at the CR i is

$$\begin{aligned} &H_0 \text{ (busy)}: \quad \mathbf{r}_{m,i}[n] \sim \mathcal{N}\left(\mathbf{0}, \Sigma_{\mathbf{x}} + \Sigma_w\right) \\ &\qquad\qquad\text{versus} \\ &H_1 \text{ (idle)}: \quad \mathbf{r}_{m,i}[n] \sim \mathcal{N}\left(\mathbf{0}, \Sigma_w\right). \end{aligned} \tag{12.7}$$

The likelihood ratio needed for optimal tests is

$$\begin{aligned} L\left(\mathbf{r}_{m,i}[n]\right) &= \frac{f_{\mathbf{r}|H_1}(\mathbf{r}_{m,i}[n] \mid S_m[n] = 1)}{f_{\mathbf{r}|H_0}(\mathbf{r}_{m,i}[n] \mid S_m(n) = 0)} \\ &= \frac{\frac{1}{(2\pi)^{\frac{L}{2}} |\Sigma_w|^{\frac{1}{2}}} e^{-\frac{1}{2}\mathbf{r}_{m,i}^{\dagger}[n]\Sigma_w^{-1}\mathbf{r}_{m,i}[n]}}{\frac{1}{(2\pi)^{\frac{L}{2}} ||\tilde{\Sigma}_{\mathbf{x}}+\Sigma_w|^{\frac{1}{2}}} e^{-\frac{1}{2}\mathbf{r}_{m,i}^{\dagger}[n]\left(\tilde{\Sigma}_{\mathbf{x}}+\Sigma_w\right)^{-1}\mathbf{r}_{m,i}[n]}}. \end{aligned} \tag{12.8}$$

After some simplifications, the log-likelihood ratio turns out to be

$$\begin{aligned} \mathscr{L}\left(\mathbf{r}_{m,i}[n]\right) &= -\frac{1}{2}\mathbf{r}_{m,i}^{\dagger}[n]\left(\Sigma_w^{-1} - \left(\tilde{\Sigma}_{\mathbf{x}} + \Sigma_w\right)^{-1}\right)\mathbf{r}_{m,i}[n] + C \\ &= -\frac{1}{2}\mathbf{r}_{m,i}^{\dagger}[n]\mathbf{Q}_{m,i}\mathbf{r}_{m,i}[n] + C, \end{aligned} \tag{12.9}$$

where

$$\begin{aligned}\mathbf{Q}_{m,i} &= \Sigma_w^{-1} - \left(\tilde{\Sigma}_{\mathbf{x}} + \Sigma_w\right)^{-1} \\ &= \sigma_w^{-2} I_L - \left(\tilde{\Sigma}_{\mathbf{x}} + \sigma_w^2 I_L\right)^{-1} \qquad (12.10a) \\ &= \left(\sigma_w^{-2}\left(\tilde{\Sigma}_{\mathbf{x}} + \sigma_w^2 I_L\right) - I_L\right)\left(\tilde{\Sigma}_{\mathbf{x}} + \sigma_w^2 I_L\right)^{-1} \\ &= \sigma_w^{-2}\tilde{\Sigma}_{\mathbf{x}}\left(\tilde{\Sigma}_{\mathbf{x}} + \sigma_w^2 I_L\right)^{-1} \qquad (12.10b)\end{aligned}$$

and

$$\begin{aligned}C &= \frac{1}{2}\log\left(\frac{|\tilde{\Sigma}_{\mathbf{x}} + \Sigma_w|}{|\Sigma_w|}\right) \\ &= \frac{1}{2}\log\left(\frac{|\tilde{\Sigma}_{\mathbf{x}} + \sigma_w^2 \mathbf{I}_L|}{|\sigma_w^2 \mathbf{I}_L|}\right).\end{aligned} \qquad (12.11)$$

Hence, an optimal NP decision rule for primary signal detection can be written in the form of

$$u_{m,i}[n] = \begin{cases} 1 & \text{if} \\ & \mathbf{r}_{m,i}^{\dagger}[n]\mathbf{Q}_{m,i}[n]\mathbf{r}_{m,i}[n] \quad \begin{matrix} \leq \\ \\ > \end{matrix} \quad \eta_{m,i}^{*} \\ 0 & \text{if} \end{cases}, \qquad (12.12)$$

where $\eta_{m,i}^{*}$ is to be set so that the false-alarm probability is equal to the maximum allowed collision probability α.

If signal samples were to be iid so that $\tilde{\Sigma}_{\mathbf{x}} = \tilde{\sigma}_x^2 \mathbf{I}_L$, the above quadratic detector can further be simplified, due to (12.10b), becoming

$$\mathbf{Q}_{m,i} = \frac{\tilde{\sigma}_x^2}{\sigma_w^2\left(\tilde{\sigma}_x^2 + \sigma_w^2\right)} \qquad (12.13)$$

and

$$\begin{aligned}C &= \frac{1}{2}\log\frac{|\tilde{\sigma}_x^2 \mathbf{I}_L + \sigma_w^2 \mathbf{I}_L|}{|\sigma_w^2 \mathbf{I}_L|} \\ &= \frac{L}{2}\log\frac{\tilde{\sigma}_x^2 + \sigma_w^2}{\sigma_w^2},\end{aligned} \qquad (12.14)$$

so that the optimal NP detector turns out to be the following energy detector:

$$u_{m,i}[n] = \begin{cases} 1 & \text{if} \\ & \mathbf{r}_{m,i}^{\dagger}[n]\mathbf{r}_{m,i}[n] = ||\mathbf{r}_{m,i}[n]||^2 \quad \begin{matrix} \leq \\ \\ > \end{matrix} \quad \eta'_{m,i} \\ 0 & \text{if} \end{cases} \qquad (12.15)$$

where $\eta'_{m,i}$ is to be chosen to achieve the false-alarm (collision) probability of α.

Observe that under hypothesis $\mathcal{H}_0$, $\mathbf{r}_{m,i}[n]||^2$ is the sum of the squares of L independent zero-mean complex Gaussian random variables with common variance $\sigma_w^2 + \tilde{\sigma}_x^2$. Equivalently, it is a sum of squares of $2L$ independent zero-mean real Gaussian random variables with common variance $\frac{\sigma_w^2+\tilde{\sigma}_x^2}{2}$. Hence, as we also used in (10.26), from Appendix B, we note that $\mathbf{r}_{m,i}[n]||^2$ is a gamma random variable of the form of $G\left(L, \sigma_w^2 + \tilde{\sigma}_x^2\right)$ under hypothesis $\mathcal{H}_0$. Hence, following a similar procedure used to obtain (10.27), the false-alarm (collision) probability on channel m is

$$\begin{aligned} P_{F,i}^{(m)} &= \Pr(H_1|H_0) \ = \ \Pr\left(\text{channel is idle}|\text{channel is busy}\right) \\ &= \Pr\left(u_{m,i}[n] = 1|H_0\right) \\ &= \Pr\left(||\mathbf{r}_{m,i}[n]||^2 \leq \eta'_{m,i}|H_0\right) \\ &= \frac{\gamma\left(L, \frac{\eta'_{m,i}}{\sigma_w^2+\tilde{\sigma}_x^2}\right)}{\Gamma(L)}, \end{aligned} \tag{12.16}$$

where $\Gamma(x)$ is the gamma function and $\gamma(x,t)$ is the (lower) incomplete gamma function defined in (10.29).

To ensure that the maximum collision probability stays below α, the decision threshold in (12.15) must be

$$\eta'_{m,i} = \left(\sigma_w^2 + \tilde{\sigma}_x^2\right)\gamma^{-1}\left(L, \alpha\Gamma(L)\right), \tag{12.17}$$

where $\gamma^{-1}(L,.)$ denotes the inverse in the second argument of the incomplete gamma function.

We may also evaluate the probability of correctly detecting idle channels in closed form with this optimal Neyman–Pearson detector. Indeed, under H_1 (i.e., under the assumption that the channel is idle), $\mathbf{r}_{m,i}[n]||^2$ is a gamma random variable of the form of $G\left(L, \sigma_w^2\right)$. Hence, the probability of detection on channel m is

$$\begin{aligned} P_{D,i}^{(m)} &= \Pr(H_1|H_1) \ = \ \Pr\left(\text{channel is idle}|\text{channel is idle}\right) \\ &= \Pr\left(u_{m,i}[n] = 1|H_1\right) \\ &= \Pr\left(||\mathbf{r}_{m,i}[n]||^2 \leq \eta'_{m,i}|H_1\right) \\ &= \frac{\gamma\left(L, \frac{\eta'_{m,i}}{\sigma_w^2}\right)}{\Gamma(L)}. \end{aligned} \tag{12.18}$$

If the threshold is chosen to satisfy the α-level collision probability as in (12.17), then the detection probability can be written as a function of the tolerable collision probability as

$$P_{D,i}^{(m)} = \frac{\gamma\left(L, \left(1+\frac{\tilde{\sigma}_x^2}{\sigma_w^2}\right)\gamma^{-1}\left(L, \alpha\Gamma(L)\right)\right)}{\Gamma(L)}. \tag{12.19}$$

Hence, the receiver operating characteristic of the ith CR's primary signal detector is

$$P_{D,i}^{(m)} = \frac{\gamma\left(L, \left(1+\frac{\tilde{\sigma}_x^2}{\sigma_w^2}\right)\gamma^{-1}\left(L, P_{F,i}^{(m)}\Gamma(L)\right)\right)}{\Gamma(L)}. \tag{12.20}$$

*12.3.2.2 **Primary Signal Detection When Secondary System Has Perfect Knowledge of Primary Signaling*** If the CR i deterministically knows the primary signal to be $\mathbf{x}_m[n] = \mathbf{x}$, then from (12.1), we have that under hypothesis H_j, for $j = 0, 1$,

$$\mathbf{r}_{m,i}[n] \sim \mathcal{N}\left((1 - S_m[n])h_{m,i}[n]\mathbf{x}, \sigma_w^2 \mathbf{I}_L\right), \tag{12.21}$$

where $S_m[n] = 0$ or $S_m[n] = 1$, respectively, for H_0 and H_1. Assuming that the CR knows the channel-state coefficient between itself and the primary transmitter, the likelihood ratio (12.5) turns out to be

$$L\left(\mathbf{r}_{m,i}[n]\right) = \frac{\frac{1}{(2\pi)^{\frac{L}{2}}|\Sigma_w|^{\frac{1}{2}}} e^{-\frac{1}{2}\mathbf{r}_{m,i}[n]^\dagger \Sigma_w^{-1} \mathbf{r}_{m,i}[n]}}{\frac{1}{(2\pi)^{\frac{L}{2}}|\Sigma_w|^{\frac{1}{2}}} e^{-\frac{1}{2}(\mathbf{r}_{m,i}[n]-h_{m,i}[n]\mathbf{x})^\dagger \Sigma_w^{-1}(\mathbf{r}_{m,i}[n]-h_{m,i}[n]\mathbf{x})}}. \tag{12.22}$$

The corresponding log-likelihood ratio is

$$\mathscr{L}\left(\mathbf{r}_{m,i}[n]\right) = -\frac{\Re\left\{h_{m,i}[n]\mathbf{r}_{m,i}^\dagger[n]\mathbf{x}\right\}}{\sigma_w^2} + \frac{|h_{m,i}[n]|^2||\mathbf{x}||^2}{2\sigma_w^2}. \tag{12.23}$$

Hence, the optimal Neyman–Pearson rule can be implemented as a threshold test on a matched-filter output as below:

$$u_{m,i}[n] = \begin{cases} 1 & \text{if} \quad \Re\left\{h_{m,i}[n]\mathbf{r}_{m,i}^\dagger[n]\mathbf{x}\right\} \le \check{\eta}_{m,i}[n] \\ 0 & \text{if} \quad \Re\left\{h_{m,i}[n]\mathbf{r}_{m,i}^\dagger[n]\mathbf{x}\right\} > \check{\eta}_{m,i}[n] \end{cases}, \tag{12.24}$$

where $\Re(\cdot)$ denotes the real part and $\check{\eta}_{m,i}[n]$ is to be chosen to satisfy the α-level false-alarm probability.

It can be shown that under hypothesis H_0 (i.e., when the channel is busy),

$$\frac{\Re\left\{h_{m,i}[n]\mathbf{r}_{m,i}^\dagger[n]\mathbf{x}\right\}}{|h_{m,i}[n]|||\mathbf{x}||} \sim \mathcal{N}(|h_{m,i}[n]|||\mathbf{x}||, \sigma_w^2/2). \tag{12.25}$$

Hence, the false-alarm probability of declaring hypothesis H_1 (i.e., channel as being idle) when indeed hypothesis H_0 is true is given by

$$\begin{aligned} P_{F,i}^{(m)} = \Pr(H_1|H_0) = \Pr\left(\frac{\Re\left\{h_{m,i}[n]\mathbf{r}_{m,i}^{\dagger}[n]\mathbf{x}\right\}}{|h_{m,i}[n]|||\mathbf{x}||} \leq \frac{\check{\eta}_{m,i}[n]}{|h_{m,i}[n]|||\mathbf{x}||} \middle| H_0 \right) \\ = 1 - Q\left(\frac{\check{\eta}_{m,i}[n] - |h_{m,i}[n]|^2||\mathbf{x}||^2}{|h_{m,i}[n]|||\mathbf{x}||\sqrt{\sigma_w^2/2}} \right), \end{aligned} \tag{12.26}$$

where $Q(.)$ is the complementary cumulative distribution function (ccdf) of the standard normal distribution defined in (4.32). Since the false-alarm probability $\Pr(H_1|H_0)$ in a DSS CR is the collision probability with the primary users, in order to maintain maximum collision probability not exceeding α, we may choose the threshold in the optimal test (12.24) to be

$$\check{\eta}_{m,i}[n] = \sqrt{\frac{\sigma_w^2}{2}}|h_{m,i}[n]|||\mathbf{x}||Q^{-1}(1-\alpha) + |h_{m,i}[n]|^2||\mathbf{x}||^2. \tag{12.27}$$

Similarly, it can be shown that under hypothesis H_1 (i.e., when the channel is idle),

$$\frac{\Re\left\{h_{m,i}[n]\mathbf{r}_{m,i}^{\dagger}[i]\mathbf{x}\right\}}{|h_{m,i}[n]|||\mathbf{x}||} \sim \mathcal{N}(0, \sigma_w^2/2). \tag{12.28}$$

Hence, the probability of correctly detecting idle channels as being idle is given by

$$\begin{aligned} P_{D,i}^{(m)} = \Pr(H_1|H_1) = \Pr\left(\frac{\Re\left\{h_{m,i}[n]\mathbf{r}_{m,i}^{\dagger}[i]\mathbf{x}\right\}}{|h_{m,i}[n]|||\mathbf{x}||} \leq \frac{\check{\eta}_{m,i}[n]}{|h_{m,i}[n]|||\mathbf{x}||} \middle| H_1 \right) \\ = 1 - Q\left(\frac{\check{\eta}_{m,i}[n]}{|h_{m,i}[n]|||\mathbf{x}||\sqrt{\sigma_w^2/2}} \right). \end{aligned} \tag{12.29}$$

By substituting for threshold $\check{\eta}_{m,i}[n]$ from (12.27), we can relate the detection probability to the collision probability threshold as

$$P_{D,i}^{(m)} = 1 - Q\left(Q^{-1}(1-\alpha) + \frac{\sqrt{2}|h_{m,i}[n]|||\mathbf{x}||}{\sigma_w} \right), \tag{12.30}$$

where α is the allowed collision probability with primary users on channel m (which is assumed to be the same for all channels). Then the receiver operating characteristic of the ith CR's primary signal detector is

$$P_{D,i}^{(m)} = 1 - Q\left(Q^{-1}\left(1 - P_{F,i}^{(m)}\right) + \frac{\sqrt{2}|h_{m,i}[n]|||\mathbf{x}||}{\sigma_w}\right). \quad (12.31)$$

Example 12.1 (Fading-Averaged Probability of Detection) Probability of detection $P_{D,i}^{(m)}$ given in (12.31) is the probability of detection achieved with a particular realization $h_{m,i}[n]$ of the fading coefficient, assuming it is known. Hence, this is a random quantity that varies as a function of the random fading coefficient $h_{m,i}[n]$. The average probability of detection achieved by this detector will thus be given by

$$\begin{aligned}\bar{P}_{D,i}^{(m)} &= \mathbb{E}_{h_{m,i}}\left\{P_{D,i}^{(m)}\right\} \\ &= 1 - \mathbb{E}_{h_{m,i}}\left\{Q\left(Q^{-1}(1-\alpha) + \frac{\sqrt{2}|h_{m,i}[n]|||\mathbf{x}||}{\sigma_w}\right)\right\},\end{aligned} \quad (12.32)$$

where $\mathbb{E}_{h_{m,i}}\{.\}$ denotes the expectation with respect to the distribution of fading coefficient $h_{m,i}$.

If we assume that the fading coefficient $h_{m,i}$ is a complex Gaussian random variable with independent real and imaginary components such that average fading coefficient power $\mathbb{E}\left\{|h_{m,i}[n]|^2\right\} = 1$, then it is easily seen from Appendix B that $|h_{m,i}[n]|$ is a Rayleigh random variable with a pdf of the form of

$$f_{h_{m,i}}(h) = \begin{cases} 2he^{-h^2} & \text{if } x \geq 0 \\ 0 & \text{if } x < 0 \end{cases}.$$

Using an integration table,[1] it can be shown that then if the false-alarm probability is constrained to be α, the average detection probability (averaged over fading) is

$$\bar{P}_{D,i}^{(m)} = \alpha + \frac{e^{-\frac{Q^{-2}(1-\alpha)}{2(1+SNR)}}}{\sqrt{1+\frac{1}{SNR}}} Q\left(\frac{Q^{-1}(1-\alpha)}{\sqrt{1+\frac{1}{SNR}}}\right), \quad (12.34)$$

[1]For example, it can be shown that [55]

$$\int_0^{\infty} xe^{-x^2} Q\left(\frac{\sqrt{2}x+\mu}{\sigma}\right) dx = \frac{1}{2}\left[Q\left(\frac{\mu}{\sigma}\right) - \frac{e^{-\frac{\mu^2}{2(1+\sigma^2)}}}{\sqrt{1+\sigma^2}} Q\left(\frac{\mu}{\sigma\sqrt{1+\sigma^2}}\right)\right]. \quad (12.33)$$

where we have defined $SNR = \frac{||\mathbf{x}||^2}{\sigma_w^2}$ and $Q^{-2}(.) = \left(Q^{-1}(.)\right)^2$. It is important to note that in order to achieve the above average probability of detection, the threshold of the detector must be adjusted according to (12.27) for specific value of the fading coefficient so that the false-alarm probability is always maintained to be α. ■

The two optimal detectors (12.15) and (12.24) under the assumed two extreme scenarios are intuitively satisfying: When the signal is assumed to be nothing but a zero-mean Gaussian random process with a known covariance matrix embedded in zero-mean Gaussian noise, the optimal detector (12.15) determines whether the signal is present or not based on a threshold test on the sample energy of the observation. On the other hand, when the signal is a deterministically known vector $\mathbf{x}$ embedded in zero-mean Gaussian noise, the optimal detector (12.24) checks how closely the observation sample mean resembles the known signal $\mathbf{x}$.

Note that $u_{m,i}[n]$ denotes the channel-state decision (which can be the same as the channel-access decisions in this case) of the ith secondary user, for $i \in \{1, \ldots, K_s\}$, on the state of the mth primary channel, for $m \in \{1, \ldots, M\}$, that it sensed at time n. As we saw earlier, this channel-state detection is only one part of the overall spectrum sensing problem. Each radio must also make decisions on which channel to sense (sensing decisions) at each time. We will address the sensing decision-making problem in the next chapter.

12.4 LIMITATIONS OF AUTONOMOUS SPECTRUM SENSING

Autonomous spectrum sensing discussed in the previous section is relatively simple to implement since there is no need for any central coordination among secondary CRs. Schemes that involve central coordination require provision of control channels to exchange information between the secondary system's central controller, the SSDC, and the distributed CRs. Providing such control channels in DSS CR networks may not be possible, or efficient, since at the first place the assumption is that there is not much spectrum available for the secondary system to operate in. It makes sense to allow, as much as possible, distributed CRs to operate autonomously by making their own decisions about which channels to sense and access. However, there are several inherent limitations and inefficiencies to such autonomous spectrum sensing. In the following, we will look at some of these issues that motivate us to consider cooperative spectrum sensing in the next section.

Limitations of autonomous spectrum sensing can be seen in all three decision-making problems involved in spectrum sensing: sensing, channel-state, and channel-access decisions. Sensing and channel-access decisions get affected due to the absence of coordination among secondary radios on which channels to sense and access. Since each user makes its decisions on which channel to sense, there is the likelihood that more than one secondary user may sense the same primary channel m at a given time instant. If channel m were to be idle and if all these users determined it to be idle based on their sensing outcomes, then they may all attempt to access and transmit over channel m. This may result in collisions among secondary transmissions, perhaps leading to none of their signals getting through successfully. The spectrum opportunity on channel m would be wasted in this case. Since each user will likely optimize their decisions on which channel to sense based on what it believes is going to be the most likely idle channel during the next time slot, there is indeed a high likelihood that many users will sense the same channel that has a high likelihood of being idle.

Aside from the inefficiency of sensing and channel access due to no coordination, there is also a limitation in terms of how effectively a single radio can determine the true state of a channel based only on its own sensing observations. At the heart of this limitation is the well-known hidden terminal problems that are unavoidable in wireless communications systems. Hidden terminal problems can be traced back to either shadowing effects or the inherent difficulty in detecting a passive receiver. In the case of spectrum sensing for DSS, these issues may be exacerbated due to lack of intersystem knowledge.

To illustrate the hidden terminal problems, as in Chapter 3, let us denote by R_I and r_I the interference radii of primary and secondary signals, respectively, when their respective transmit powers are P_1 and P_2. Hence, a signal transmitted by a primary transmitter will interfere at any secondary receiver within an R_I-radius circle centered at the primary transmitter. Similarly, a signal transmitted by a secondary transmitter will interfere at any primary receiver within an r_I-radius circle centered at the secondary transmitter. For acceptable performance, the transmitter and receiver separation in primary and secondary systems must be less than R_Q and r_Q, respectively. However, r_Q does not have to be the maximum distance between a primary transmitter and a secondary CR for it to detect the primary signal. Thus, let us denote by r_D the *detection radius* of a secondary CR: That is, a secondary user can detect any (primary) transmitter within at most an r_D distance from it.

With these notations in place, we can identify at least three types of hidden terminal problems as shown in Figure 12.5 [12]:

1. *Hidden primary receiver*: The secondary transmitter S_{TX} in Figure 12.5(a) cannot detect the primary transmitter P_{TX} since it is located outside of the r_D-radius circle centered at S_{TX}. Hence, it may try to communicate with a possible secondary receiver. However, the intended primary receiver P_{RX} is located within r_I distance from S_{TX}. Hence, any transmission from S_{TX} will cause interference to this hidden primary receiver P_{RX} in Figure 12.5(a).
2. *Hidden primary transmitter*: The secondary transmitter S_{TX} in Figure 12.5(b) cannot detect the primary transmitter P_{TX} since it is located outside of the r_D-radius circle centered at S_{TX}. Hence, it may try to communicate with the secondary receiver S_{RX}. However, the intended secondary receiver S_{RX} is located within R_I distance from this primary transmitter P_{TX}. Hence, transmissions from this hidden primary transmitter P_{TX} will cause interference at the secondary receiver S_{RX}. A possible solution in this case is for the secondary receiver S_{RX} to also perform spectrum sensing and inform S_{TX} if it detects a primary transmitter. Since it may reasonably be assumed that $r_D \geq R_I$, in

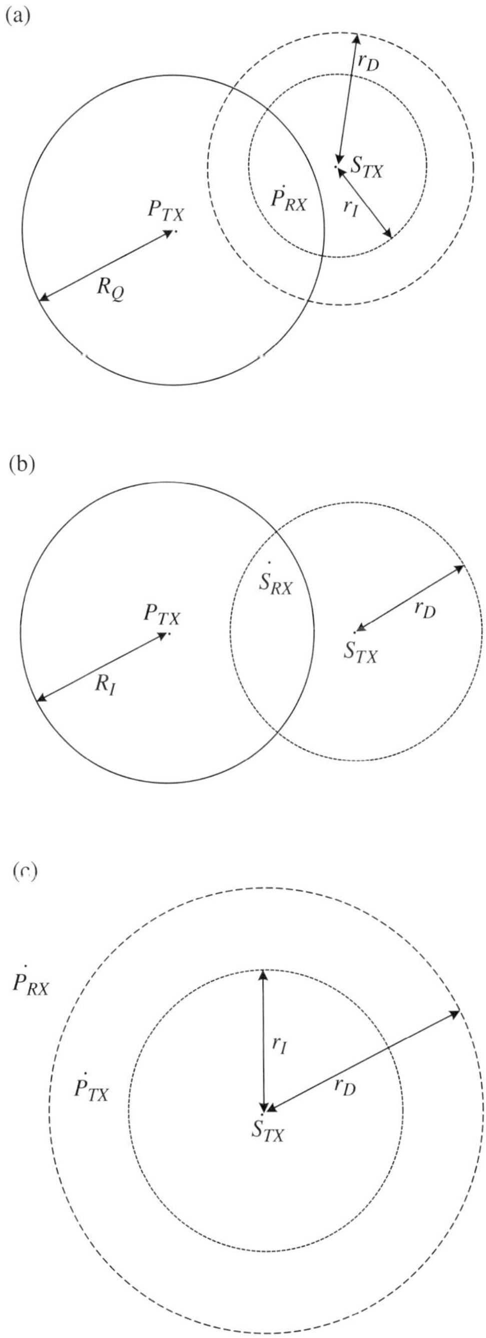

FIGURE 12.5 Hidden terminal problems encountered in DSS cognitive radio networks. (a) A hidden primary receiver. (b) A hidden primary transmitter. (c) An exposed primary transmitter.

Figure 12.5(b), the receiver S_{RX} should be able to detect the primary transmitter's presence.

3. *Exposed primary transmitter*: The secondary transmitter S_{TX} in Figure 12.5(c) detects the primary transmitter P_{TX} since they are separated by less than r_D distance. Hence, S_{TX} may not attempt to transmit. However, the intended receiver P_{RX} of P_{TX} is located outside the r_I-radius circle centered at S_{TX}. Hence, in this situation, the secondary node S_{TX} could have transmitted without causing any interference to the hidden primary receiver P_{RX}.

The above problems are exacerbated in a DSS CR network due to the lack of coordination with primary users. For example, both hidden primary receiver and the exposed primary transmitter problems could have been avoided if there was to be some coordination between the two systems. However, in DSA CR systems such coordination may not be present. On the other hand, in DSL CR systems, there may be an incentive for the primary users to facilitate the efficiency of the spectrum sharing process by providing some information to the secondary CRs so as to overcome the above limitations. Still, this may not resolve all possible forms of hidden terminal problems.

In the absence of coordination between the two systems, an alternative approach for the CR is to at least cooperate and coordinate among themselves. A hidden terminal problem for one CR may not be so for another, due to their spatial locations. Hence, CRs can share among themselves their local sensing results so that they can mitigate the effects of hidden terminal problems. Alternatively, we may introduce an SSDC that will collect local sensing results from all secondary users and make overall sensing and access decisions on behalf of them. Such cooperative spectrum sensing is the topic of next section.

12.5 COOPERATIVE SPECTRUM SENSING FOR DYNAMIC SPECTRUM SHARING

Let us now consider the spectrum sensing problem in a centrally controlled CR network. As before, we assume that each of the K_s secondary users can only sense one primary channel at a given time (narrowband spectrum sensing). However, now these radios are assumed to cooperate in finding and accessing idle spectrum opportunities available on multiple primary channels. A central unit in the CR network, called the SSDC, gathers all local channel sensing results from secondary CRs, possibly over a dedicated control channel. It then makes overall decisions regarding the states of each of the primary channels. All decisions on sensing and channel access are then made also by the SSDC based on these overall channel-state decisions. The SSDC is responsible for informing the distributed secondary CR, possibly over a dedicated control channel, these sensing and channel-access decisions so as to guide them in their actions. Such a combined optimal channel sensing and access policy can expect to maximize the total secondary system throughput accrued over time over all primary channels. However, it must be pointed out that such a network obviously requires a significant amount of control signal and information exchange between the SSDC and the CRs.

Specifically, let us consider time $n = n$ in Figure 12.3. Just before time slot n starts, that is, for $n = n^-$, the SSDC informs each CR i its sensing decisions. These sensing decisions $a_n^{(i)} \in \{1, \ldots, M\}$ instruct each radio i which primary channel it should sense during the sensing interval that is to begin at time $n = n$, as shown in Figure 12.6. Each secondary user may make a preliminary decision on the state of the primary channel $a_n^{(i)} = m$ it sensed based on its own observation $\mathbf{r}_{m,i}[n]$. As before, let us denote by $u_{m,i}[n]$ this local decision made by the ith CR about the state of channel $a_n^{(i)} = m$ at time instant n. As shown in Figure 12.6, each radio forward its local decision $u_{m,i}[n]$ to the SSDC who then in turn informs the radios its channel-access decisions (channel-access assignments) denoted by $d_n^{(i)}$'s.

Each secondary user is to forward the local sensing observation results, perhaps after some preprocessing, to the centralized SSDC over a dedicated control channel. The SSDC collects all channel sensing results from secondary users and makes decisions on whether a channel is idle or not. For the moment, let us assume that if the SSDC determines a particular channel to be idle at the end of the spectrum sensing interval T_{ss} at time $n = n$, then it will assign this channel to a secondary user for its access. We will not worry about how these channel-access assignments are made in this chapter. All that matters is that if a channel is determined to be idle by the SSDC, it will be accessed by a CR for the subsequent transmission period. In other words, idle/busy (channel-state) decisions are equivalent to channel-access/channel-don't-access decisions. The issue of which secondary user should access a given idle channel requires further network-specific considerations such as priority handling and fairness. The SSDC, for example, can be made to be responsible for balancing accessing opportunities among the secondary users (fairness) or assigning secondary users with any particular priorities.

Recall from our earlier discussion that at the end of the T_f length time period, each secondary user must terminate its transmission if it was assigned to transmit on a particular primary channel. At this time, the SSDC will make decisions regarding who should sense which channel at time $n = n+1$ based on the sensing reports it received at the end of the previous sensing interval, the overall channel-state decisions it made for time $n = n$, information it might have collected regarding the success or failure of CRs in accessing channels it assigned them during the just finished transmission period, as well as all its previous accumulated such knowledge (history). The SSDC will inform the CRs these sensing decisions $a_{n+1}^{(i)}$ at time $n = (n+1)^-$ and the process will repeat.

In this section, we will address the channel-state detection problem in a cooperative spectrum sensing CR network. The sensing and access decisions will be addressed in Chapter 13. Note that if we assume that the state $S_m[n]$, for $1 \leq m \leq M$, of the primary channel M is independent of those of the other primary channels, still the channel-state determination problems are decoupled over the primary channels. Hence, in the following, we may consider the problem for an arbitrary primary channel m.

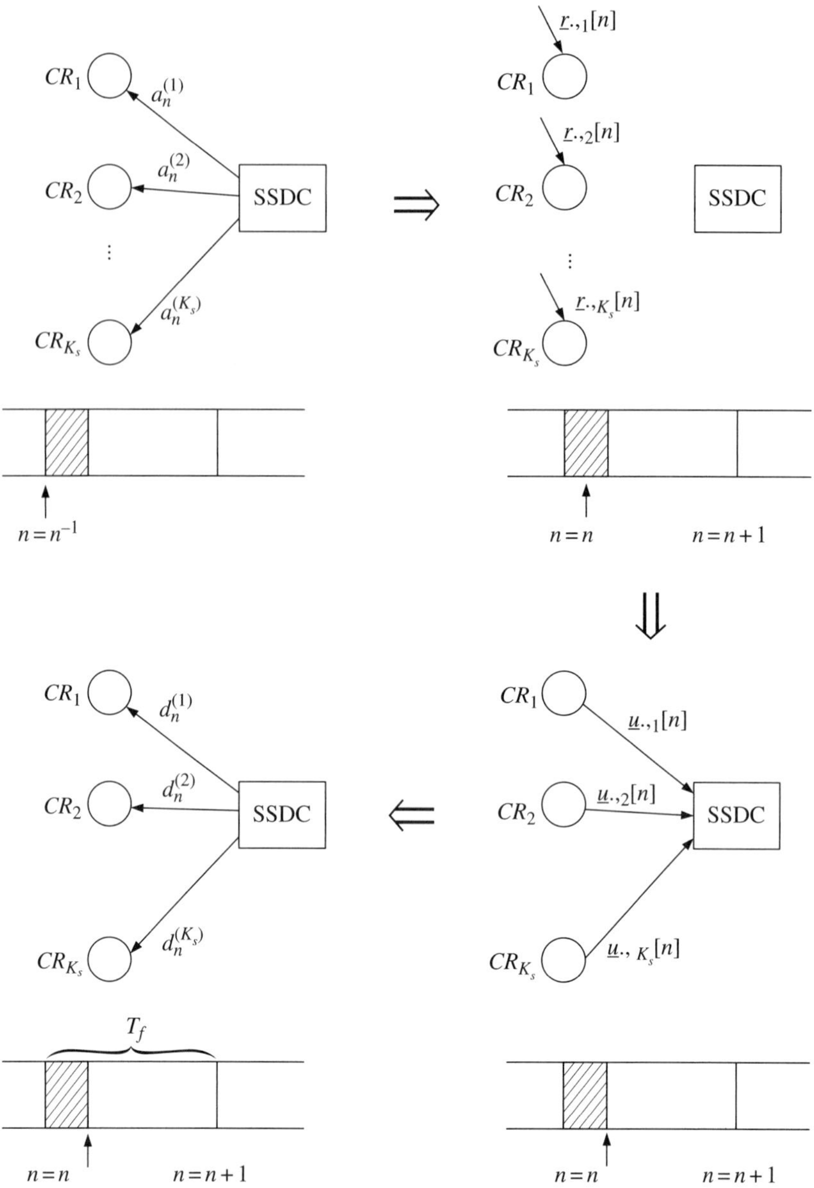

FIGURE 12.6 In a centrally controlled cognitive radio network, the SSDC is responsible for making both sensing and access decisions. Here, $a_n^{(i)}$ and $d_n^{(i)}$ are the sensing and access (channel index) decisions for the ith cognitive radio.

12.6 COOPERATIVE CHANNEL-STATE DETECTION

Cooperative channel-state detection is an application of the more general problem of distributed detection with fusion encountered in distributed sensor networks [188–193]. For example, in *wireless sensor networks*, usually there is a collection of distributed sensor nodes that observes a common phenomenon (or its absence) of interest. These sensor nodes inform their sensing results to a central *fusion center*, over a wireless network, that makes the final decision on the presence or the absence of the phenomenon. Since wireless channels are bandwidth and power constrained, an important aspect of the problem is to reduce the amount of information each sensor node sends to the fusion center. Hence, nodes perform some sort of local pre-processing before transmitting their observations to the fusion center. As shown in Figure 12.7, this preprocessing, in general, is a form of compression: In one extreme, the compression could be such that the node sends a binary decision. At the other extreme, there may not be any compression at all, and the node might directly send the observation itself to the fusion center. In the middle, one may get all possible schemes in between these two extremes.

The cooperative channel-state detection in a DSS CR network is very similar to this distributed detection problem with fusion considered in wireless sensor networks [194–199]. The distributed secondary users may perform some level of preprocessing to reduce the amount of information bits they need to send to the SSDC. One may think of the same two extremes as in the aforementioned: (i) no compression at all (Figure 12.7(a)) and (ii) compression to one bit (Figure 12.7(b)). In the first approach, the local observation $\mathbf{r}_{m,i}[n]$ of the ith CR is directly transmitted to the SSDC (or the fusion center, in wireless sensor networks terminology) perhaps with some gain adjustment. Hence, the local decision $u_{m,i}[n]$ from the ith radio is the same as its observation $\mathbf{r}_{m,i}[n]$. This is also similar to the so-called amplify and forward (AF) used in cooperative relaying. Since the task of the SSDC in this situation is to combine the original data itself to make a final decision, the problem at the SSDC is called *data fusion*. The other extreme is to compress the local observation to just a single bit so that $u_{m,i}[n] \in \{0, 1\}$. In the case of binary hypothesis testing, as in the case of primary channel-state detection, this is the same as local sensor making a binary decision on the state of the channel (busy or idle) and transmitting this $\{0, 1\}$ decision to the SSDC. Since the task of the SSDC now is to combine the local decisions, as opposed to data itself, to make a final decision, in this case, the problem at the SSDC is called *decision fusion*.

While the first approach requires the maximum amount of channel resources, the second approach uses the smallest. The purpose of spectrum sensing is to find spectrum opportunities for transmission. Hence, certainly, it does not make sense to assume that secondary users may directly be able to send their local observations to the SSDC. That would require too much bandwidth. Of course, one may optimize the number of bits to be used for compression to achieve a trade-off between the bandwidth requirement for sharing information between the nodes and the SSDC and the level of accuracy of the decisions made at the SSDC, as in Figure 12.7(c). Instead of delving into this optimization, however, in the following, we will simply consider

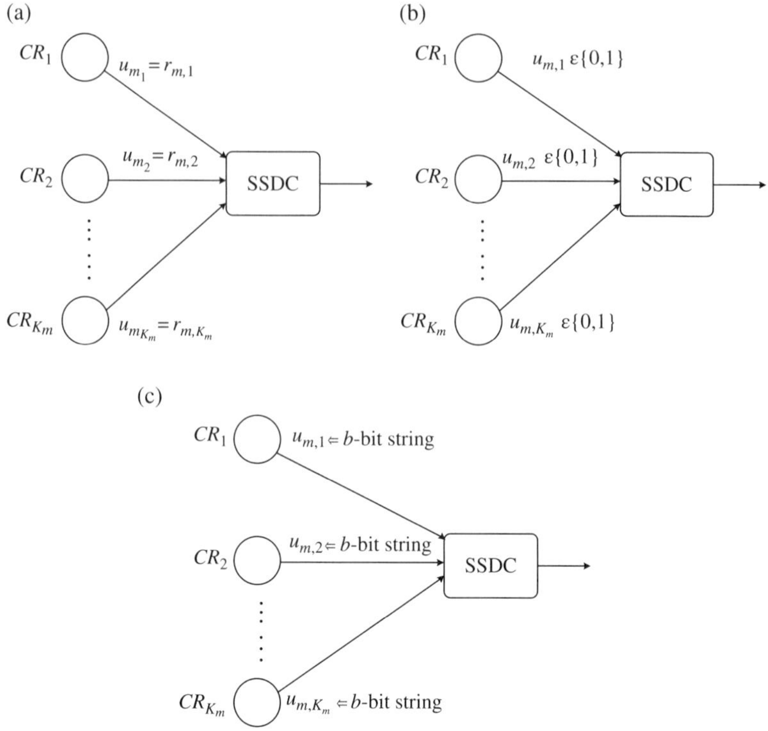

FIGURE 12.7 The preprocessing at the distributed secondary CRs is a form of compression that ranges from no compression at all at one extreme to compression down to a single bit at the other extreme. (a) In one extreme, the distributed secondary CR may send their observations themselves to the SSDC (no compression). (b) In the other extreme, the distributed secondary CR may compress their observations down to a single bit before sending to the SSDC (1-bit quantization). (c) In between, the distributed secondary CR may compress/quantize their observations into b-bit strings before sending to the SSDC (b-bit quantization). Note that K_m denotes the number of secondary CRs that sensed the primary channel m.

the extreme case of highest level of compression on $\mathbf{r}_{m,i}[n]$'s: The reports from secondary CRs to the SSDC are compressed, or quantized, to 0's and 1's, which can be considered as decisions of the state of the primary channel the radio sensed. Thus, in the following, we let $u_{m,i}[n] \in \{0, 1\}$ denote the local decision, or report, from the ith secondary user about the mth primary channel state at time n. It can be argued that the performance loss due to one-bit quantization can be negligible when the number of local reports from distributed CRs about the state of each primary channel m is sufficiently large.

Let us consider the channel-state detection problem on a particular primary channel m. The problem faced by the overall secondary system is to make a final decision at the SSDC based on the local (or intermediate) decisions forwarded from the sensing nodes. We may formulate this as a Neyman–Pearson hypothesis testing problem at the SSDC subject to a given collision probability constraint on channel m.

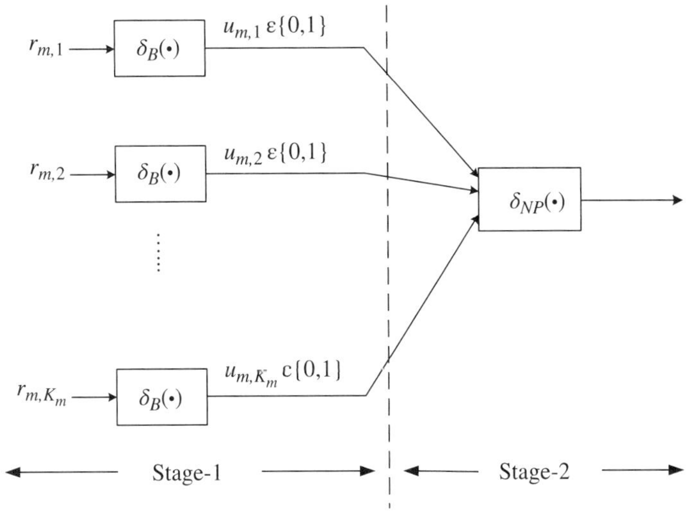

FIGURE 12.8 A practical approach to decision fusion in cooperative sensing is to separate the local processing and the fusion at the SSDC into two stages.

An optimal framework then will be to design all local decision rules and the decision rule at the SSDC jointly so as to maximize the probability of detection at the SSDC subject to this constraint. However, it is well known that this joint design can be very difficult in general [188, 191, 192]. For example, even under the assumption that conditional distributions at the sensor nodes are independent given the true hypothesis, the optimal local decision rules may not necessarily be simple threshold tests as we have seen in Chapter 4.

Thus, in terms of practicality, it makes sense to separate the optimization of local decisions and the final decision at the SSDC into two separate stages as shown in Figure 12.8. It is worth noting that in many cases of interest in practice, the global optimal design may not even lead to significant performance gain compared to its high complexity, justifying the above two-stage simpler approach. Hence, in the following, we will adapt this two-stage design approach as our cooperative channel-state detection framework.

In the first stage, the individual CRs make their local decisions $u_{m,i}[n] \in \{0, 1\}$ based on a detector that is designed to provide optimal decisions according to a local criterion. Since the local decisions are meant to be the information sent to the SSDC, it then makes sense to base these decisions under the Bayesian minimum probability of error criterion of Chapter 4. In the second stage, the SSDC makes its final decision based on the received local decisions under the Neyman–Pearson optimality in order not to exceed the maximum collision probability constraint.

When the local processing is such that $u_{m,i}[n] \in \{0, 1\}$, the true state $S_m[n]$ of the mth channel and the decision $u_{m,i}[n]$ of the ith secondary user can be viewed as the input and output, respectively, of a binary asymmetric channel (BAC) as in Figure 12.9. In the two-stage cooperative channel-state detection framework, all that matters

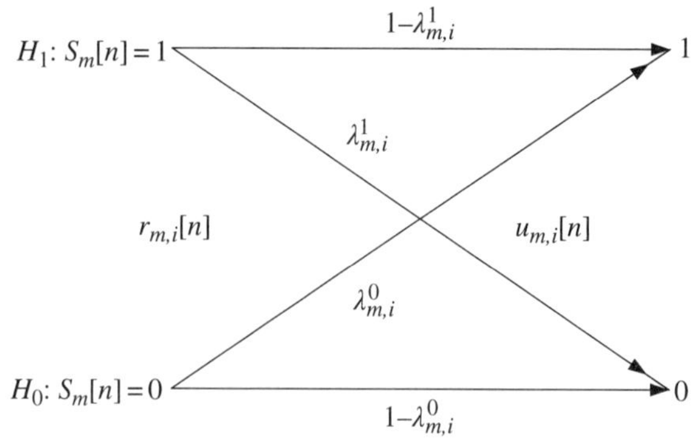

FIGURE 12.9 The ith secondary cognitive radio's decisions regarding the state of the mth primary channel can be modeled as a binary asymmetric channel (BAC).

to the second-stage processing is the crossover probabilities associated with this BAC representing the processing at the ith CR. Referring to Figure 12.9, these crossover probabilities are in fact nothing but the false-alarm (collision) and miss probabilities (missing idle states) of the (Bayesian) detector at the ith CR. In Figure 12.9, we use $\lambda_{m,i}^{0} = \Pr(H_1|H_0)$ and $\lambda_{m,i}^{1} = \Pr(H_0|H_1)$ to denote these crossover probabilities under $\mathcal{H}_0$ and $\mathcal{H}_1$, respectively.

12.6.1 Local Processing and Sensing Reports from Secondary Users

The L-length complex-valued observation vector on the mth channel of the secondary user i at time instant n is given by (12.1) with $m = a_n^{(i)}$ where $a_n^{(i)}$ is the sensing action taken by radio i at time n. The hypothesis testing problem at the ith CR regarding the state of the mth primary channel is still as given by (12.3).

The ith secondary radio makes its local decision $u_{m,i}[n]$ on channel m's state solely based on its observation $\mathbf{r}_{m,i}[n]$. This is then the same as the earlier case we discussed for the autonomous CRs. The only difference from the earlier case is that the local detector will be based on the Bayesian minimum probability of error optimality rather than the Neyman–Pearson optimality. As mentioned earlier, it makes sense that the objective of the secondary radios is to provide the SSDC with the most accurate decisions so that it can form the final decisions subject to the collision probability criteria. The more accurate the decisions sent from the local nodes, the better the overall decision the SSDC is able to make under whatever criteria it may adapt and, in particular, in this case, under the Neyman–Pearson optimality criterion.

From (4.15b) of Chapter 4, we know that the optimal $u_{m,i}[n]$ decisions that minimize the average probability of error are the MAP decisions:

$$u_{m,i}[n] = \arg\max_{j\in\{0,1\}} \Pr\{S_m[n]=j \mid \mathbf{r}_{m,i}[n])\}, \tag{12.35}$$

which is equivalent to the following likelihood ratio test:

$$\mathcal{L}(\mathbf{r}_{m,i}[n]) = \frac{f_{\mathbf{r}|H_1}(\mathbf{r}_{m,i}[n] \mid S_m[n]=1)}{f_{\mathbf{r}|H_0}(\mathbf{r}_{m,i}[n] \mid S_m[n]=0)} \underset{\mathcal{H}_0}{\overset{\mathcal{H}_1}{\gtrless}} \eta_m, \tag{12.36}$$

where $\eta_m = \frac{\Pr\{H_0\}}{\Pr\{H_1\}} = \frac{\Pr\{S_m[n]=0\}}{\Pr\{S_m[n]=1\}} = \frac{p_0^{(m)}}{p_1^{(m)}}$. Note that the likelihood ratio in (12.36) is the same as that in (12.5). We may assume that the required threshold $\eta_m = \frac{p_0^{(m)}}{p_1^{(m)}}$ is estimated at the SSDC and is provided to all nodes since it has access to more global information regarding the primary channel dynamics.

Again, the form of the likelihood function $f_{\mathbf{r}|H_j}(.)$ will depend on the assumptions on the knowledge of primary signal $\mathbf{x}_m[n]$ on channel m in (12.1). As we did in the case of autonomous spectrum sensing, we may consider the two commonly used assumptions on the primary signal $\mathbf{x}_m[n]$: (i) Secondary system has no knowledge about the primary signaling versus, and (ii) secondary system has perfect knowledge about the primary signaling. Note that the only difference compared to the autonomous sensing of the previous section here is in the optimality criterion being the Bayesian minimum error probability instead of the Neyman–Pearson. Hence, the only difference in the local detector is in its threshold. Thus, we only briefly summarize the results for these two assumptions on the primary signal.

12.6.1.1 Local Decisions When Secondary System Has No Knowledge of Primary Signaling When we model the primary signal on the mth channel to be a zero-mean random signal with a covariance matrix $\tilde{\Sigma}_{\mathbf{x}}$, the hypothesis testing problem at the ith CR is given by (12.7). The likelihood and log-likelihood ratios are given, respectively, by (12.8) and (12.9), so that the minimum probability of error optimal MAP decisions are given by

$$u_{m,i}[n] = \begin{cases} 1 & \text{if} \\ & \mathbf{r}_{m,i}^{\dagger}[n]\mathbf{Q}_{m,i}[n]\mathbf{r}_{m,i}[n] \quad \begin{matrix} \leq \\ \\ > \end{matrix} \quad \eta_{m,i}^{*} \\ 0 & \text{if} \end{cases}, \tag{12.37}$$

where $\mathbf{Q}_{m,i}[n]$ is as given in (12.10b) and $\eta_{m,i}^{*} = 2(C - \ln \eta_m)$ with C defined in (12.11).

Moreover, when the signal samples are iid so that $\tilde{\Sigma}_{\mathbf{x}} = \tilde{\sigma}_x^2 \mathbf{I}_L$, the above quadratic detector can further be simplified to an energy detector similar to that in (12.15):

$$u_{m,i}[n] = \begin{cases} 1 & \text{if} \\ & ||\mathbf{r}_{m,i}[n]||^2 \quad \begin{matrix} \leq \\ \\ > \end{matrix} \quad \eta'_{m,i} \\ 0 & \text{if} \end{cases}, \tag{12.38}$$

where $\eta'_{m,i} = \eta_{m,i}^{*} \frac{\sigma_w^2(\tilde{\sigma}_x^2+\sigma_w^2)}{\tilde{\sigma}_x^2} = 2(C - \ln \eta_m) \frac{\sigma_w^2(\tilde{\sigma}_x^2+\sigma_w^2)}{\tilde{\sigma}_x^2}$ with C as given in (12.14).

As we discussed earlier in the previous subsection, under H_0 and H_1, respectively, $\mathbf{r}_{m,i}[n]||^2$ is distributed as $G(L, \sigma_w^2 + \tilde{\sigma}_x^2)$ and $G(L, \sigma_w^2)$. Hence, we may easily compute the crossover probabilities of the equivalent BAC in Figure 12.9 for decisions at the ith CR about the state of the mth primary channel: Indeed,

$$\lambda_{m,i}^0 = \Pr(H_1|H_0)$$

$$= \frac{\gamma\left(L, \frac{\eta'_{m,i}}{\tilde{\sigma}_x^2 + \sigma_w^2}\right)}{\Gamma(L)} \tag{12.39a}$$

$$= \frac{\gamma\left(L, 2(C - \ln\eta_m)\frac{\sigma_w^2}{\tilde{\sigma}_x^2}\right)}{\Gamma(L)} \tag{12.39b}$$

where (12.39a) follows from (12.16). Similarly, from (12.18), we have that

$$\lambda_{m,i}^1 = \Pr(H_0|H_1)$$

$$= 1 - \frac{\gamma\left(L, \frac{\eta'_{m,i}}{\sigma_w^2}\right)}{\Gamma(L)}$$

$$= 1 - \frac{\gamma\left(L, 2(C - \ln\eta_m)\frac{\tilde{\sigma}_x^2 + \sigma_w^2}{\tilde{\sigma}_x^2}\right)}{\Gamma(L)}, \tag{12.40}$$

where $\eta_m = \frac{p_0^{(m)}}{p_1^{(m)}}$ and C is as given in (12.14).

12.6.1.2 Local Decisions When the Secondary System Has Perfect Knowledge of Primary Signaling When we assume that the primary signal on channel m is $\mathbf{x}_m[n] = \mathbf{x}$ and is deterministically known to the secondary user i, the relevant likelihood ratio for channel-state detection based on $\mathbf{r}_{m,i}[n]$ is given by (12.22). The corresponding log-likelihood ratio is similarly found in (12.23). Hence, the optimal MAP rule that minimizes the probability of errors at the CR i is

$$u_{m,i}[n] = \begin{cases} 1 & \text{if} \quad \Re\left\{h_{m,i}[n]\mathbf{r}_{m,i}^{\dagger}[n]\mathbf{x}\right\} \leq \check{\eta}_{m,i}[n] \\ 0 & \text{if} \quad \Re\left\{h_{m,i}[n]\mathbf{r}_{m,i}^{\dagger}[n]\mathbf{x}\right\} > \check{\eta}_{m,i}[n] \end{cases} \tag{12.41}$$

where $\check{\eta}_{m,i}[n] = \frac{|h_{m,i}[n]|^2||\mathbf{x}||^2}{2} - \sigma_w^2 \log\eta_m$, $\Re(\cdot)$ denotes the real part and $\eta_m = \frac{p_0^{(m)}}{p_1^{(m)}}$.

The distribution of the decision statistic in the above detector under hypotheses H_0 and H_1, respectively, is given in (12.25) and (12.28). From these, we have the crossover probabilities required in the simplified BAC model of state of the channel m to be

$$\begin{aligned}\lambda_{m,i}^{0} &= \Pr(H_1|H_0)\\ &= 1 - Q\left(\frac{\check{\eta}_{m,i}[n] - |h_{m,i}[n]|^2||\mathbf{x}||^2}{|h_{m,i}[n]|||\mathbf{x}||\sqrt{\sigma_w^2/2}}\right)\\ &= Q\left(\frac{|h_{m,i}[n]|^2||\mathbf{x}||^2 + 2\sigma_w^2 \ln(\eta_m)}{\sqrt{2}|h_{m,i}[n]|\sigma_w||\mathbf{x}||}\right)\end{aligned} \tag{12.42}$$

and

$$\begin{aligned}\lambda_{m,i}^{1} &= \Pr(H_0|H_1)\\ &= Q\left(\frac{\check{\eta}_{m,i}[n]}{|h_{m,i}[n]|||\mathbf{x}||\sqrt{\sigma_w^2/2}}\right)\\ &= Q\left(\frac{|h_{m,i}[n]|^2||\mathbf{x}||^2 - 2\sigma_w^2 \ln(\eta_m)}{\sqrt{2}|h_{m,i}[n]|\sigma_w||\mathbf{x}||}\right).\end{aligned} \tag{12.43}$$

The overall (minimum achievable) probability of error at the node i is given by the weighted sum of the above two error probabilities where weights are the prior probabilities $p_0^{(m)}$ and $p_1^{(m)}$ of the two hypotheses:

$$P_{e,i}^{(m)} = p_0^{(m)}\lambda_{m,i}^{0} + p_1^{(m)}\lambda_{m,i}^{1}. \tag{12.44}$$

Example 12.2 (Fading-Averaged Probability of Error) The minimum probability of error (12.44) is the probability of error at the ith CR for a particular realization of the fading coefficient value $h_{m,i}[n]$ assuming it is known to the CR. When fading coefficient is random, probability of error (12.44) is also random similar to the detection probability (12.31) earlier. Hence, the average probability of error achieved at the ith CR can be written as

$$\bar{P}_{e,i}^{(m)} = \mathbb{E}_{h_{m,i}}\left\{P_{e,i}^{(m)}\right\}, \tag{12.45}$$

where $\mathbb{E}_{h_{m,i}}\{.\}$ denotes the expectation with respect to the distribution of fading coefficient $h_{m,i}$. As in Example 12.1, we will assume Rayleigh fading in which case the fading coefficient $h_{m,i}$ is a complex Gaussian random variable with independent real and imaginary components and average fading coefficient power $\mathbb{E}\left\{|h_{m,i}[n]|^2\right\} = 1$.

If the two states were to be equally likely a priori, so that $p_0^{(m)} = p_1^{(m)} = \frac{1}{2}$ and $\eta_m = 1$, (12.45) simplifies to

$$\begin{aligned}\bar{P}_{e,i}^{(m)} &= \mathbb{E}_{h_{m,i}}\left\{Q\left(|h_{m,i}[n]|\sqrt{\frac{SNR}{2}}\right)\right\}\\ &= \int_0^{\infty} 2he^{-h^2}Q\left(h\sqrt{\frac{SNR}{2}}\right)dh\\ &= \left(1-\frac{1}{\sqrt{1+\frac{4}{SNR}}}\right)\end{aligned} \tag{12.46}$$

where, as before, $SNR = \frac{||\mathbf{x}||^2}{\sigma_w^2}$ and the last step follows by using the integral identity[2] (12.33) with $\mu = 0$.

It is, however, important to point out that the assumption of equally likely states may not be valid in many DSS scenarios. For example, it is known that the probability of primary channel occupancy can be significantly low in broadcast TV channels. When this is the case, (12.45) might have to be evaluated numerically. ■

12.6.2 Final Channel-State Decisions at the SSDC: Decision Fusion

In cooperative spectrum sensing for DSS, local decisions $u_{m,i}[n]$'s made at the secondary CRs are forwarded to the SSDC to make final decisions about the channel states. Note that individual sensing decisions $u_{m,i}[n]$'s are binary. In the following, we will also assume that these are transmitted to the SSDC using a binary phase-shift keying (BPSK) mapping.

Let us denote by $z_i[n]$ the received signal at the SSDC from the ith node at time n regarding the state of primary channel it sensed. Let us denote the set of secondary CR node indices who sensed channel m at time n by $\mathcal{A}_n^{(m)}$ and let $K_{m,n} = |\mathcal{A}_n^{(m)}|$ be the number of distributed CRs that sensed channel m at time n:

$$\mathcal{A}_n^{(m)} \triangleq \{i\ :\ \text{node } i \text{ senses channel } m \text{ at time } n\} \tag{12.47}$$

and

$$K_{m,n} \triangleq \left|\mathcal{A}_n^{(m)}\right|. \tag{12.48}$$

[2]We have the following integral identity [55]:

$$\int_0^{\infty} xe^{-x^2}Q\left(\frac{\sqrt{2}h}{\sigma}\right)dh = \frac{1}{2}\left(1-\frac{1}{\sqrt{1+\sigma^2}}\right).$$

To make notation a little simpler, let us define the variable-length vector $\mathbf{z}^{(m)}[n]$ as the vector whose elements are equal to $z_i[n]$'s for $i \in \mathcal{A}_n^{(m)}$. Note that $\mathbf{z}^{(m)}[n]$ contains the set of received channel sensing reports corresponding to the mth channel from all nodes that sensed channel m at time n.

As mentioned earlier, since individual channel states are assumed to be independent of each other, the decision fusion problem at the SSDC for individual channels is decoupled. Hence, let us consider an arbitrary channel m. The decision fusion problem at the SSDC subject to a constraint on the maximum collision probability with primary transmissions on the primary channel m is a Neyman–Pearson hypotheses testing problem of the form of

$$\begin{aligned} H_0\,(\text{channel busy, } S_m[n]=0):&\quad \mathbf{z}^{(m)}[n] \sim P_{m,0} \\ &\quad \text{versus} \\ H_1\,(\text{channel idle, } S_m[n]=1):&\quad \mathbf{z}^{(m)}[n] \sim P_{m,1}, \end{aligned} \tag{12.49}$$

where $P_{m,1}$ and $P_{m,0}$ denote the conditional distributions of $\mathbf{z}^{(m)}[n]$ given $S_m[n]=1$ and $S_m[n]=0$, respectively.

Note that in (12.49) we have already assumed that the channel-state decision regarding the state of the mth channel at time n is made solely based on the local decision vector $\mathbf{z}^{(m)}[n]$ from the CRs received at time n. Of course, it is possible to extend this to take into account previous sensing reports the SSDC had received from radios before time n as well. When channel occupancy dynamics exhibit temporal correlations, such historical information may indeed prove to be useful. However, to keep analysis tractable, in the following, we will assume the model (12.49), so that the likelihood ratio at the SSDC can be written as

$$L(\mathbf{z}^{(m)}[n]) = \frac{f_{m,1}(\mathbf{z}^{(m)}[n])}{f_{m,0}(\mathbf{z}^{(m)}[n])}. \tag{12.50}$$

The exact form of this likelihood ratio will depend on the form of the received signals $z_i[n]$ from the secondary radios as well as any interdependencies among them. Since local decisions at the distributed nodes were made based on conditionally independent observations $\mathbf{r}_{m,i}[n]$'s (conditioned on the true hypothesis), we may still assume that $u_{m,i}[n]$'s are also conditionally independent, so that (12.50) becomes

$$\mathcal{L}(\mathbf{z}^{(m)}[n]) = \prod_{i\in\mathcal{A}_n^{(m)}} \frac{f(z_i[n]|H_1)}{f(z_i[n]|H_0)}. \tag{12.51}$$

In addition, intuitively, we may expect that if the communications between the distributed nodes and the SSDC were to be reliable, then the local decisions will be received at the SSDC more accurately, leading to better fusion performance. In particular, if this communications were to be error-free, then the received $z_i[n]$'s at the SSDC will be exactly the same as the local decisions $u_{m,i}[n]$'s. If, on the other hand, communication from the secondary CRs to the SSDC is unreliable, then the local decisions may be received inaccurately at the SSDC, resulting in poor final decisions. In this case, $z_i[n]$'s are noisy versions of $u_{m,i}[n]$'s. Hence, depending on assumptions on communications from secondary CRs to the SSDC, the likelihood ratio (12.51) will be different. Then, so will be the optimal fusion detector at the SSDC.

12.6.2.1 Error-Free Communication between CRs and the SSDC Let us assume that transmitted $u_{m,i}[n]$'s are received error-free at the SSDC. Then we may as well assume that

$$z_i[n] = u_{m,i}[n], \tag{12.52}$$

where $1 \leq m \leq M$. Thus, the likelihood ratio at the SSDC (12.51) can be written as

$$\mathcal{L}\left(\mathbf{z}^{(m)}[n]\right) = \prod_{i \in \mathcal{A}_n^{(m)}} \frac{f(u_{m,i}[n]|H_1)}{f(u_{m,i}[n]|H_0)}. \tag{12.53}$$

Recall that under each of the hypothesis, $u_{m,i}[n]$'s are binary random variables. In particular, as can be seen from Figure 12.9, under H_1, $u_{m,i}[n]$ is a Bernoulli random variable with parameter $p = 1 - \lambda^1_{m,i}$ where $\lambda^1_{m,i} = \Pr(H_0|H_1) = \Pr(u_{m,i}[n] = 0|H_1)$. Similarly, from Figure 12.9, under H_0, $u_{m,i}[n]$ is a Bernoulli random variable with with the parameter $p = \lambda^0_{m,i}$ where $\lambda^0_{m,i} = \Pr(H_1|H_0) = \Pr(u_{m,i}[n] = 1|H_0)$. Hence, (12.53) can be written as

$$L\left(\mathbf{z}^{(m)}[n]\right) = \prod_{i \in \mathcal{A}_n^{(m)}} \left(\frac{\lambda^1_{m,i}}{1-\lambda^0_{m,i}}\right)^{1-u_{m,i}[n]} \left(\frac{1-\lambda^1_{m,i}}{\lambda^0_{m,i}}\right)^{u_{m,i}[n]} \tag{12.54}$$

and the corresponding log-likelihood ratio can be shown to be

$$\begin{aligned} \mathscr{L}(\mathbf{z}^{(m)}[n]) &= \sum_{i \in \mathcal{A}_n^{(m)}} u_{m,i}[n] c_{m,i} + d_m \\ &= T_m[n] + d_m, \end{aligned} \tag{12.55}$$

where we have defined the decision statistic at the SSDC for channel m to be

$$T_m[n] = \sum_{i \in \mathcal{A}_n^{(m)}} c_{m,i} u_{m,i}[n] \tag{12.56}$$

with

$$c_{m,i} = \ln\left(\frac{1-\lambda^0_{m,i}}{\lambda^1_{m,i}} \frac{1-\lambda^1_{m,i}}{\lambda^0_{m,i}}\right)$$

and

$$d_m[n] = \sum_{i \in \mathcal{A}_n^{(m)}} \ln\left(\frac{\lambda_{m,i}^1}{1-\lambda_{m,i}^0}\right) = K_m \ln\left(\frac{\lambda_{m,i}^1}{1-\lambda_{m,i}^0}\right).$$

Note that the decision statistic $T_m[n]$ in (12.56) is a weighted sum of error-free local decisions received at the SSDC regarding channel m. The weights are determined by the miss and false-alarm probabilities associated with these local decisions. To compute this weighted sum, thus, the SSDC must know these performance parameters associated with the local detectors at distributed CR nodes. Assuming these are available, the SSDC may then implement its optimal detector by setting a threshold according to the Neyman–Pearson criterion to keep the maximum collision probability with primary users below a given threshold α_m.

For a given set of local performance parameters $\left\{\lambda_{m,i}^1\right\}_i$ and $\left\{\lambda_{m,i}^0\right\}_i$, $T_m[n]$ is a discrete valued random variable taking only a finite set of discrete-values. Hence, in general, the optimal Neyman–Pearson channel-state detector at the SSDC on the state of the mth channel at time n is a randomized decision rule of the form of

$$\tilde{\delta}_{NP}(\mathbf{z}^{(m)}[n]) = \begin{cases} 1 & \text{if } T_m[n] > \tau_m \\ \gamma_m & \text{if } T_m[n] = \tau_m \\ 0 & \text{if } T_m[n] < \tau_m \end{cases}, \quad (12.57)$$

where τ_m and γ_m are to be determined so that the collision probability with the primary users on channel m does not exceed a given threshold α_m.

Example 12.3 (Decision Fusion Detector at the SSDC with Identical Channel Statistics) If all channels were to have identical channel statistics, decision statistic (12.56) at the SSDC simplifies to

$$\begin{aligned} T_m[n] &= c_m \sum_{i \in \mathcal{A}_n^{(m)}} u_{m,i}[n] \\ &= c_m K_{m,n}^{(1)} \end{aligned} \quad (12.58)$$

with

$$c_m = \ln\left(\frac{1-\lambda_m^0}{\lambda_m^1}\frac{1-\lambda_m^1}{\lambda_m^0}\right)$$

and

$$d_m[n] = K_{m,n} \ln\left(\frac{\lambda_m^1}{1-\lambda_m^0}\right),$$

where in (12.58) we have denoted by $K_{m,n}^{(1)}$ the number of $u_{m,i}[n]$'s in the set $\mathcal{A}_n^{(m)}$ that is equal to 1 (recall that each $u_{m,i}[n] \in \{0, 1\}$). Thus, $K_{m,n}^{(1)}$ denotes the number of distributed CRs that detected the channel m as being idle at time n. The decision fusion detector (12.57) at the SSDC for channel m can then be written as a threshold test on the number of local decisions that decided in favor of channel m being idle at each time instant n:

$$\tilde{\delta}_{NP}(\mathbf{z}^{(m)}[n]) = \begin{cases} 1 & \text{if } K_{m,n}^{(1)} > \tau_m \\ \gamma_m & \text{if } K_{m,n}^{(1)} = \tau_m \\ 0 & \text{if } K_{m,n}^{(1)} < \tau_m \end{cases}, \tag{12.59}$$

where τ_m and γ_m are to be determined so that the collision probability with the primary users on channel m does not exceed a given threshold α_m. ■

Let us denote by $\mathcal{C}_m$ the set of discrete values that $T_m[n]$ can take at time n. For $j=0,1$, we may denote by $p_{m,n}^j(t)$ and $F_{m,n}^j(t)$, respectively, the conditional probability mass function (pmf) and the conditional cumulative distribution function (cdf) of the decision statistic $T_m[n]$ conditioned on hypothesis $\mathcal{H}_j$ where $t \in \mathcal{C}_m$. Let us focus on the distribution of $T_m[n]$ under hypothesis H_0. Due to t taking values on a discrete set, the cdf $F_{m,n}^0(t)$, and correspondingly the ccdf $\Pr\{T_m[n] > t \mid H_0\} = 1 - F_{m,n}^0(t)$, will be piecewise constant with possible discontinuous jumps at discrete t values as shown in Figure 12.10. From the ccdf in Figure 12.10, it is easy to see that in order to ensure an α_m maximum collision probability, the threshold τ_m must satisfy

$$\Pr\{T_m[n] > \tau_m \mid H_0\} \leq \alpha_m < \Pr\{T_m > \tau_m^- \mid H_0\}, \tag{12.60}$$

or, equivalently in terms of the cdf $F_{m,n}^0(t)$,

$$1 - F_{m,n}^0(\tau_m) \leq \alpha_m < 1 - F_{m,n}^0(\tau_m^-), \tag{12.61}$$

where $\tau_m^- = \max\{t : t < \tau_m, t \in \mathcal{C}_m \bigcup \{-\infty\}\}$ is the maximum $t \in \mathcal{C}_m$ that is less than τ_m if τ_m is not the smallest element in $\mathcal{C}_m$ and is $-\infty$ otherwise. Given the monotonicity of the ccdf, as can be seen from Figure 12.10, there is always a unique τ_m as guaranteed by the Neyman–Pearson lemma (see [51] for a proof).

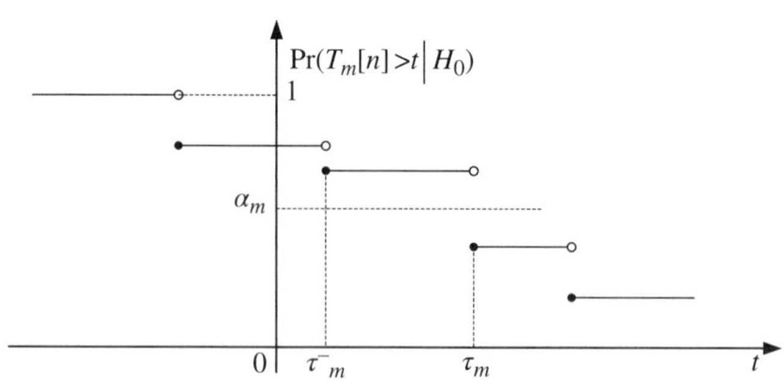

FIGURE 12.10 Selection of the Neyman–Pearson threshold τ_m for a given maximum collision probability α_m.

Once τ_m is chosen according to (12.61), the randomization variable γ_m needed in (12.57) can be chosen as [51]

$$\gamma_m = \frac{\alpha_m - \left(1 - F^0_{m,n}(\tau_m)\right)}{F^0_{m,n}(\tau_m) - F^0_{m,n}(\tau_m^-)}. \tag{12.62}$$

It is not difficult to verify that this choice for γ_m indeed ensures a collision probability that is exactly α_m. A derivation can be found in [51].

The probability of accurate detection of idle channels at the SSDC with the above detector is then given by

$$\begin{aligned} P_D^{(m,\mathcal{A}_n^{(m)})} &= \Pr\{T_m[n] > \tau_m[n] \mid H_1\} + \gamma_m[n]\Pr\{T_m[n] = \tau_m[n] \mid H_1\} \\ &= 1 - F^1_{m,n}(\tau_m) + \gamma_m[n]\left(F^1_{m,n}(\tau_m) - F^1_{m,n}(\tau_m^-)\right). \end{aligned} \tag{12.63}$$

Note that the above detection probability depends on which set of nodes sensed the channel m. Hence, the dependence on $\mathcal{A}_n^{(m)}$.

It is worth noting that the structure of the optimal channel-state decisions at the SSDC, given in (12.57), is independent of what types of local sensing rules were used by the distributed secondary users. All that matters is that the local decisions are binary and the associated crossover probabilities corresponding to the equivalent BAC. Hence, the decision rule (12.57) at the SSDC is valid under any assumptions on the knowledge of primary signals by the secondary users, including the two extreme cases we considered. As long as the local sensing decisions are quantized as 0 or 1 before transmitting to the SSDC, all that is needed for decision-making at the SSDC are the crossover probability parameters $\lambda^1_{m,i}$ and $\lambda^0_{m,i}$.

The decision statistic $T_m[n]$ used in the optimal channel-state detector (12.57) at the SSDC, when there are no communications errors as assumed earlier, is a weighted sum of the local decisions $u_{m,i}[n]$'s. The weights are determined by the crossover probabilities $\lambda^1_{m,i}$'s and $\lambda^0_{m,i}$'s. While these weights are important to ensure properly accounting for reliability of the individual decisions from CRs, determining them may require a certain amount of effort as well as perhaps regular updating. However, as we saw in Example 12.3, if all local decisions are on average equally reliable, at least approximately, then the weights may be taken to be equal so that the decision statistic reduces to the number of CRs that detected channel m as being idle[3] at time n, denoted as $K^{(1)}_{m,n}$.

12.6.2.2 Noisy Communication between CRs and the SSDC In general, the error-free communications between nodes and the SSDC may not always be possible. While classical distributed detection and data fusion have mostly considered the case of error-free communications of local decisions to a fusion center, recent interest

[3]The well-known *majority-vote decision fusion rule* is simply a Bayesian minimum probability of error detector based on this simplified fusion decision statistic $K^{(1)}_{m,n}$ when the prior probabilities are taken to be uniform.

on wireless sensor networks had motivated the need to take into account the communications errors. As a result, there are many formulations of decision/data fusion at a fusion center where local decisions/data are forwarded to a fusion center over noisy channels [194–197, 200]. In particular, there are optimal fusion rules and their performance analysis available for both AF types of local processing (i.e., sending direct local observations to the SSDC) and binary quantized local decisions (as we have assumed earlier). Different formulations have also considered different types of communications channels (e.g., AWGN, fading multipath, etc.), modulation schemes (BPSK, OOK, etc.), as well as multiple-access schemes (orthogonal, nonorthogonal, etc.). The important conclusion from these analysis is not surprising: When there are communications channel errors, the final fusion performance degrades. Essentially, each local decision is now corrupted not only by the local sensor observation noise but also by channel noise. In the following, we will discuss a simple formulation that shows the effect of communications noise on channel-state decisions made at the SSDC.

Let us assume that all CRs transmit there local decisions $u_{m,i}[n]$'s to the SSDC using BPSK, so that transmitted symbol values can be written as $x_{m,i}[n] = 2u_{m,i}[n] - 1$. For simplicity, let us also assume that nodes-to-SSDC channels are all orthogonal and noise to be additive Gaussian. Modifications to account for more realistic channel models, including fading and interchannel interference (which is normally called the multiple-access interference (MAI) or *multiuser interference*) can be done easily. Received signal at the SSDC from the ith CR corresponding to the channel state of the mth primary channel can then be written as

$$z_i[n] = g x_{m,i}[n] + w_i[n] \quad \text{for } i \in \mathcal{A}_n^{(m)}, \tag{12.64}$$

where g is the received power that is assumed to be the same for all nodes, for simplicity of the following analysis. We may still allow for the channels from different nodes to the SSDC to be independent but not necessarily identical by letting $w_i[n] \sim \mathcal{N}(0, \sigma_i^2)$. Note that, in light of this, the equal power assumption in the preceding text can still account for fading effects, except we need to assume that the fading coefficient for the channel between the ith node and the SSDC is known at the SSDC. We may write the combined received signal at the SSDC corresponding to the channel state of the mth primary channel in vector notation as

$$\mathbf{z}[n] = \mathbf{A}\mathbf{u}[n] + \mathbf{w}[n],$$

where we have dropped the primary channel index m for brevity of notation so that $\mathbf{z}[n] = \mathbf{z}^{(m)}[n] = [z_1[n], \ldots, z_{K_{m,n}}[n]]^T$, $\mathbf{A} = g\mathbf{I}_{K_{m,n}}$, and $\mathbf{w}[n] \sim \mathcal{N}(\mathbf{0}, diag(\sigma_1^2, \ldots, \sigma_{K_{m,n}}^2))$. Recall that $K_{m,n} = |\mathcal{A}_n^{(m)}|$ is the number of CRs that sensed the channel m and sent their decisions to the SSDC at time n.

The optimal detector must be a threshold test on the likelihood ratio (12.51) whose evaluation requires the conditional marginal pdf's $f(z_i[n]|H_0)$ and $f(z_i[n]|H_1)$. Since $u_{m,i}[n]$ is a binary Bernoulli random variable with parameter $p = 1 - \lambda_{m,i}^1$ under hypothesis H_1, the required conditional marginal density $f(z_i[n]|H_1)$ at the SSDC can be written as

$$f(z_i[n]|H_1) = \sum_{u_{m,i}[n]\in\{0,1\}} f(u_{m,i}[n]|H_1)f(z_i[n]|u_{m,i}[n])$$
$$= \frac{1}{\sqrt{2\pi}\sigma_i}\left[\left(1-\lambda^1_{m,i}\right)e^{-\frac{(z_{m,i}[n]-g)^2}{2\sigma_i^2}} + \lambda^1_{m,i}e^{-\frac{(z_{m,i}[n]+g)^2}{2\sigma_i^2}}\right], \qquad (12.65)$$

where we have used (12.64).

Similarly, since $u_{m,i}[n]$ is a binary Bernoulli random variable with parameter $p{=}\lambda^0_{m,i}$ under hypothesis H_0, we have that

$$f(z_i[n]|H_0) = \frac{\lambda^0_{m,i}e^{-\frac{(z_i[n]-g)^2}{2\sigma_i^2}} + (1-\lambda^0_{m,i})e^{-\frac{(z_i[n]+g)^2}{2\sigma_i^2}}}{\sqrt{2\pi}\sigma_i}. \qquad (12.66)$$

Substituting (12.65) and (12.66) in (12.51), and after some simplifications, we obtain that

$$L(\mathbf{z}[n]) - \prod_{i\in\mathcal{A}_n^{(m)}} \frac{\left(1-\lambda^1_{m,i}\right)e^{\frac{gz_i[n]}{\sigma_i^2}} + (1-\lambda^1_{m,i})e^{-\frac{gz_i[n]}{\sigma_i^2}}}{\lambda^0_{m,i}e^{\frac{gz_i[n]}{\sigma_i^2}} + (1-\lambda^0_{m,i})e^{-\frac{gz_i[n]}{\sigma_i^2}}}. \qquad (12.67)$$

There is little we can infer from the general likelihood ratio expression (12.67) corresponding to fusing noisy binary decisions. However, given the local performance parameters $\lambda^0_{m,i}$ and $\lambda^1_{m,i}$ along with g and σ_i^2, the exact likelihood ratio (12.67) can be evaluated. The threshold required not to exceed the collision probability α_m with primary transmissions on channel m, however, cannot be easily computed.

Intuitively, however, we see that it is natural to focus on the polarity of the received signals since antipodal signals with amplitude $\pm g$ are sent under the two hypothesis. Indeed, let us define the sets

$$S_0^m[n] = \left\{i : z_i[n] < 0, i \in \mathcal{A}_n^{(m)}\right\}$$

and

$$S_1^m[n] = \left\{i : z_i[n] > 0, i \in \mathcal{A}_n^{(m)}\right\}.$$

Then, the number of local sensing reports $z_i[n]$ corresponding to primary channel state m received as positive at the SSDC is simply

$$V_m[n] = \left|S_1^m[n]\right|.$$

Note that $V_m[n]$ is equivalent to $K_{m,n}^{(1)}$ when there is no channel noise.

If we define $X_i[n]$, for $i \in \mathcal{A}_n^{(m)}$, as

$$X_i[n] = \mathcal{I}_{S_1^m[n]}(i), \tag{12.68}$$

where $\mathcal{I}_A(.)$ is the indicator function of event A defined as

$$\mathcal{I}_A(z) = \begin{cases} 1 & z \in A \\ 0 & z \in A \end{cases},$$

then we may also write

$$V_m[n] = \sum_{i \in \mathcal{A}_n^{(m)}} X_i[n]. \tag{12.69}$$

From (12.68), we note that $X_i[n]$'s, for $i \in \mathcal{A}_n^{(m)}$, are a set of independent binary random variables that takes values in $\{0, 1\}$.

Similarly, the number of local sensing reports $z_i[n]$ corresponding to primary channel state m that are received as negative at the SSDC is $\left|S_0^m[n]\right| = K_{m,n} - V_m[n]$ where $K_{m,n}$ is the total number of radios that sensed the primary channel m at time n.

In terms of these, the likelihood ratio (12.67) can be rewritten as

$$L(\mathbf{z}[n]) = \prod_{i \in S_0^m[n]} \frac{\left(1-\lambda_{m,i}^1\right) + \lambda_{m,i}^1 e^{\frac{-2gz_i[n]}{\sigma_i^2}}}{\lambda_{m,i}^0 + \left(1-\lambda_{m,i}^0\right) e^{\frac{-2gz_i[n]}{\sigma_i^2}}} \prod_{i \in S_1^m[n]} \frac{\left(1-\lambda_{m,i}^1\right) e^{\frac{2gz_i[n]}{\sigma_i^2}} + \lambda_{m,i}^1}{\lambda_{m,i}^0 e^{\frac{2gz_i[n]}{\sigma_i^2}} + \left(1-\lambda_{m,i}^0\right)}. \tag{12.70}$$

Suppose further that all secondary CRs have the same average sensing performance for primary channel m so that $\lambda_{m,i}^0 = \lambda_m^0$ and $\lambda_{m,i}^1 = \lambda_m^1$ for all $i \in \mathcal{A}_n^{(m)}$ and the channels from them to the SSDC are good enough to justify the high SNR operation so that $\sigma_i^2 \longrightarrow 0$ for all $i \in \mathcal{A}_n^{(m)}$. Then, the likelihood ratio (12.70) can be simplified to

$$\begin{aligned} L(\mathbf{z}[n]) &= \prod_{i \in S_0^m[n]} \frac{\lambda_m^1}{(1-\lambda_m^0)} \prod_{i \in S_1^m[n]} \frac{(1-\lambda_m^1)}{\lambda_m^0}. \\ &= \left(\frac{\lambda_m^1}{1-\lambda_m^0}\right)^{K_{m,n}-V_m[n]} \left(\frac{1-\lambda_m^1}{\lambda_m^0}\right)^{V_m[n]}, \end{aligned}$$

so that the log-likelihood ratio at the SSDC can be written as

$$\mathscr{L}(\mathbf{z}[n]) = V_m[n] \log \frac{(1-\lambda_m^1)(1-\lambda_m^0)}{\lambda_m^0 \lambda_m^1} + K_{m,n} \log \frac{\lambda_m^1}{1-\lambda_m^0}. \tag{12.71}$$

Hence, under the above simplifying assumptions, the optimal decision fusion rule for primary channel m at the SSDC can be written as

$$\tilde{\delta}_{NP}(\mathbf{z}[n]) = \begin{cases} 1, & \text{if } V_m[n] > \tau_m \\ \gamma_m, & \text{if } V_m[n] = \tau_m \\ 0, & \text{if } V_m[n] < \tau_m \end{cases}. \tag{12.72}$$

To ensure that the collision probability with the primary users on channel m does not exceed the false-alarm probability α_m, the threshold τ_m must then be chosen to ensure that $\alpha_m = \Pr(V_m[n] \geq \tau_m | H_0)$. From (12.69), observe that the decision statistic $V_m[n]$ in the above optimal test (12.72) is essentially the sum of $K_{m,n}$ independent binary Bernoulli random variables $X_i[n]$'s.

Let $p_i^{(0)}$ denote the Bernoulli parameter that characterizes the conditional distribution of these $X_i[n]$'s under the H_0 hypothesis of channel being busy. We see that

$$\begin{aligned} p_i^{(0)} &= \Pr(X_i[n] = 1 | H_0) \\ &= \Pr(z_i[n] \geq 0 | H_0) \\ &= \int_0^\infty f(z_i[n] | H_0) dz_i[n] \\ &= \frac{1}{\sqrt{2\pi}\sigma_i} \int_0^\infty \left[\lambda_{m,i}^0 e^{-\frac{(z_i[n]-g)^2}{2\sigma_i^2}} + (1-\lambda_{m,i}^0) e^{-\frac{(z_i[n]+g)^2}{2\sigma_i^2}} \right] dz_i[n] \\ &= \lambda_{m,i}^0 + (1-2\lambda_{m,i}^0) Q\left(\frac{g}{\sigma_i}\right), \end{aligned} \tag{12.73}$$

where we have used (12.66) in the second last step.

Hence, under H_0, decision variable $V_m[n]$ is a sum of $K_{m,n}$ independent but non-identically distributed Bernoulli random variables with parameters $p_i^{(0)}$'s given by (12.73). At this point, one may design a Neyman–Pearson optimal decision rule by following the same procedure discussed under the error-free communications from CRs to the SSDC.

We may simplify the threshold computation, however, when the network is large enough so that each primary channel is sensed by many nodes. Then, we may employ asymptotic approximations to come up with simplified procedures. For example, it was shown in [196] that under certain, albeit restrictive, conditions, the asymptotic distribution of $V_m[n]$ can be evaluated in a useful form to facilitate this. In the following, we may assume a simplifying model in which the local detector performances and the channel SNR are all the same on average so that $p_i^{(0)} = p^{(0)}$ for all i in (12.73) where

$$p^{(0)} = \lambda_m^0 + (1 - 2\lambda_m^0) Q\left(\frac{g}{\sigma}\right), \tag{12.74}$$

where we have let $\sigma_i = \sigma$ for all i.

Then the conditional distributions of all the X_i's in (12.69) are the same under the hypothesis H_0. Hence, $V_m[n]$ is a binomial random variable with parameters $K_{m,n}$ and $p^{(0)}$. If $K_{m,n} \approx K_m$ is sufficiently large, the central limit theorem (CLT) can be invoked to claim that under H_0 hypothesis $V_m[n]$ is approximately Gaussian (in distribution) with mean $K_m p^{(0)}$ and variance $K_m p^{(0)} \left(1 - p^{(0)}\right)$ so that the false-alarm probability of the detector (12.72) is

$$\begin{aligned} P(H_1|H_0) &\approx \Pr\left(V_m[n] \geq \tau_m | V_m[n] \sim \mathcal{N}\left(K_m p^{(0)}, K_m p^{(0)}\left(1 - p^{(0)}\right)\right)\right) \\ &= Q\left(\frac{\tau_m - K_m p^{(0)}}{\sqrt{K_m p^{(0)}\left(1 - p^{(0)}\right)}}\right). \end{aligned} \tag{12.75}$$

Hence, for a collision probability not exceeding α_m on primary channel m, the threshold of the Neyman–Pearson test must be chosen as

$$\tau_m = \sqrt{K_m p^{(0)}\left(1 - p^{(0)}\right)} Q^{-1}(\alpha_m) + K_m p^{(0)}. \tag{12.76}$$

Similarly, under the hypothesis H_1 of channel being idle, the $X_i[n]$'s all have the same conditional Bernoulli distribution with a parameter $p^{(1)}$ given by

$$\begin{aligned} p^{(1)} = p_i^{(1)} &= \Pr(X_i[n] = 1|H_1) \\ &= \Pr(z_i[n] \geq 0|H_1) \\ &= \int_0^{\infty} f(z_i[n]|H_1) dz_i[n] \\ &= \frac{1}{\sqrt{2\pi}\sigma_i} \int_0^{\infty} \left[(1 - \lambda_{m,i}^1) e^{-\frac{(z_i[n]-g)^2}{2\sigma_i^2}} + \lambda_{m,i}^1 e^{-\frac{(z_i[n]+g)^2}{2\sigma_i^2}} \right] dz_i[n] \\ &= \left(1 - \lambda_{m,i}^1\right) + (2\lambda_{m,i}^1 - 1) Q\left(\frac{g}{\sigma_i}\right) \\ &= \left(1 - \lambda_m^1\right) + (2\lambda_m^1 - 1) Q\left(\frac{g}{\sigma}\right), \end{aligned} \tag{12.77}$$

where we have used (12.65).

Again, invoking the CLT to approximate the asymptotic distribution of $V_m[n]$, the detection probability of idle states of channel m at the SSDC can thus be approximated as

$$
\begin{aligned}
P_{D,m} &= \Pr(H_1|H_1) \approx \Pr\left(V_m[n] \geq \tau'_m | V_m[n] \sim \mathcal{N}\left(K_m p^{(1)}, K_m p^{(1)}\left(1-p^{(1)}\right)\right)\right) \\
&= Q\left(\frac{\tau_m - K_m p^{(1)}}{\sqrt{K_m p^{(1)}\left(1-p^{(1)}\right)}}\right) \\
&= Q\left(\frac{\sqrt{K_m p^{(0)}\left(1-p^{(0)}\right)}Q^{-1}(\alpha_m) + K_m p^{(0)} - K_m p^{(1)}}{\sqrt{K_m p^{(1)}\left(1-p^{(1)}\right)}}\right)
\end{aligned}
\tag{12.78}
$$

where in the last step we have substituted for τ_m from (12.76).

In general, we may expect final decisions at the SSDC to be less reliable when the quality of channels between the distributed CRs and the SSDC degrades. In particular, ultimately, there can be a performance floor induced by the quality of these channels regardless of how good the local decisions are. However, the effects of these channel errors will be diminished as the number of distributed nodes sensing any given channel is sufficiently large.

Example 12.4 (An Alternative Decision Fusion Detector When Links from CRs to the SSDC Are Noisy) A decision fusion detector similar to (12.72) can be motivated more directly by noting that the effect of noisy channel between each distributed CR node i and the SSDC is to essentially make the local decision $u_{m,i}[n]$ even more erroneous. When the channel was error-free, the SSDC directly received $u_{m,i}[n]$ and modeled it as a BAC as in Figure 12.9. The decision fusion in Section 12.6.2.1 was then based on these binary variables $u_{m,i}[n]$'s. The resulting decision fusion detector (12.57) at the SSDC was a threshold test on the weighted sum $T_m[n] = \sum_{i \in \mathcal{A}_n^{(m)}} c_{m,i} u_{m,i}[n]$ of these local decisions $u_{m,i}[n]$'s where the weights $c_{m,i}$'s were determined by the crossover probabilities of the BAC of Figure 12.9 that linked each $u_{m,i}[n]$ to the true state $S_m[n]$ of channel m at time n.

The noisy channel introduces its own errors to already erroneous $u_{m,i}[n]$. The SSDC may first detect, as best as it can, $u_{m,i}[n]$ from the received noisy signal $z_i[n]$ of (12.64). It may do so by using the minimum probability of error criterion on (12.64). Let us denote this minimum probability of error decoded value of $u_{m,i}[n]$ at the SSDC from the received noisy signal $z_i[n]$ as $\hat{u}_{m,i}[n] \in \{0,1\}$. It is straightforward to see that these are given by

$$
\hat{u}_{m,i}[n] = \begin{cases} 1 & \qquad \geq \\ & \text{if } z_i[n] \qquad \frac{\sigma_i^2}{2g} \log \hat{\eta}_{m,i} \\ 0 & \qquad < \end{cases}, \tag{12.79}
$$

where the required Bayesian threshold $\hat{\eta}_{m,i}$ is determined by the prior on $u_{m,i}[n]$ and is given by

$$\hat{\eta}_{m,i} = \frac{\Pr(u_{m,i}[n]=0)}{\Pr(u_{m,i}[n]=1)} = \frac{\Pr\left(p_1^{(m)}\lambda_{m,i}^1 + p_0^{(m)}\left(1-\lambda_{m,i}^0\right)\right)}{\Pr\left(p_1^{(m)}\left(1-\lambda_{m,i}^1\right) + p_0^{(m)}\lambda_{m,i}^0\right)}.$$

Similar to $\mathbf{z}^{(m)}[n]$, let $\hat{\mathbf{u}}^{(m)}[n]$ denote the variable-length vector whose elements are equal to $\hat{u}_{m,i}[n]$ for $i \in \mathcal{A}_n^{(m)}$. Now, instead of posing the decision fusion problem at the SSDC on the primary channel m as in (12.49), we may instead pose it as

$$\begin{aligned} H_0\,(\text{channel busy, } S_m[n]=0):&\quad \hat{\mathbf{u}}^{(m)}[n] \sim \hat{P}_{m,0} \\ &\quad\text{versus} \\ H_1\,(\text{channel idle, } S_m[n]=1):&\quad \hat{\mathbf{u}}^{(m)}[n] \sim \hat{P}_{m,1}, \end{aligned} \tag{12.80}$$

where $\hat{P}_{m,1}$ and $\hat{P}_{m,0}$ denote the conditional distributions of $\hat{\mathbf{u}}^{(m)}[n]$ given $S_m[n]=1$ and $S_m[n]=0$, respectively.

Now, if we treat $\hat{\mathbf{u}}^{(m)}[n]$'s as the local decisions, instead of $\mathbf{u}^{(m)}[n]$'s, then we may again use the formulation of error-free communication from nodes to the SSDC in Section 12.6.2.1 to design a decision fusion detector at the SSDC. The only caveat is that Figure 12.9 is no longer the BAC that links $\hat{\mathbf{u}}^{(m)}[n]$'s to the true state $S_m[n]$ of channel m. Instead, we have to append another BAC (in general) to that of Figure 12.9 to represent the channel errors introduced by going from $\mathbf{u}^{(m)}[n]$'s to $\hat{\mathbf{u}}^{(m)}[n]$'s, as shown in Figure 12.11(a). The serial concatenation of these two BAC is again a BAC, and that final BAC, shown in Figure 12.11(b), represents the crossover probabilities corresponding to the binary asymmetric channel linking $\hat{\mathbf{u}}^{(m)}[n]$'s to the true state of $S_m[n]$ of channel m. Let us denote these crossover probabilities of the overall BAC of Figure 12.11(b) that links $\hat{\mathbf{u}}^{(m)}[n]$'s to the true state of $S_m[n]$ of channel m by $\nu_{m,i}^0$ and $\nu_{m,i}^1$:

$$\begin{aligned} \nu_{m,i}^0 &= \Pr\left(\hat{\mathbf{u}}^{(m)}[n]=1|S_m[n]=0\right) \\ \nu_{m,i}^1 &= \Pr\left(\hat{\mathbf{u}}^{(m)}[n]=0|S_m[n]=1\right). \end{aligned}$$

We may easily write down these crossover probabilities of the equivalent final BAC in terms of the crossover probabilities of the two constituent BAC. Indeed, let us denote by $\mu_{m,i}^0$ and $\mu_{m,i}^1$ the crossover probabilities corresponding to the BAC that represents the channel effects:

$$\begin{aligned} \mu_{m,i}^0 &= \Pr\left(\hat{\mathbf{u}}^{(m)}[n]=1|\mathbf{u}^{(m)}[n]=0\right) \\ \mu_{m,i}^1 &= \Pr\left(\hat{\mathbf{u}}^{(m)}[n]=0|\mathbf{u}^{(m)}[n]=1\right). \end{aligned}$$

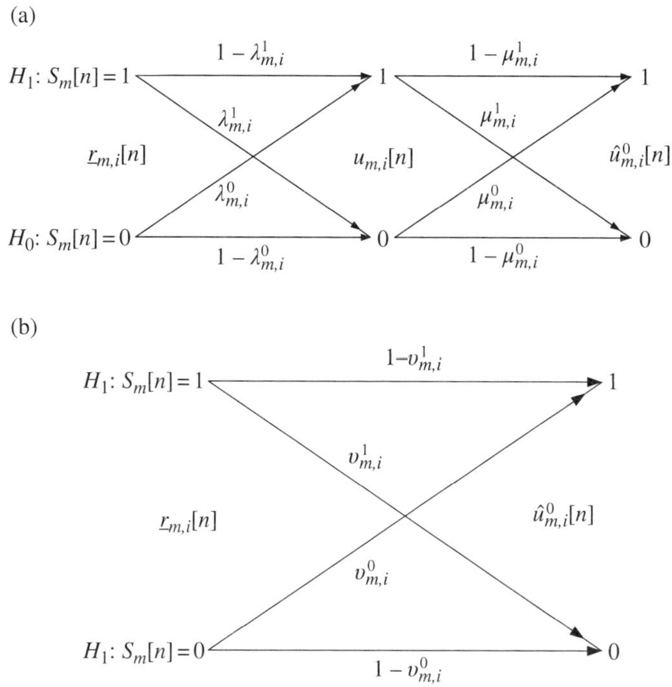

FIGURE 12.11 Decision fusion at the SSDC when links from nodes to the SSDC are noisy can be converted into an error-free link formulation by combining channel-induced errors with local decision errors. (a) The SSDC's estimates of the ith secondary cognitive radio's decisions regarding the state of the mth primary channel can be modeled as a concatenation of two binary asymmetric channels: The first-stage binary asymmetric channel represents the errors made by the local detector at the ith cognitive radio, while the second stage binary asymmetric channel represents the errors induced on those local decisions by the noisy channel from node i to the SSDC. (b) The concatenation of the two binary asymmetric channels can be represented by an equivalent single binary asymmetric channel.

Then it is straightforward to show that

$$\nu^0_{m,i} = \left(1-\lambda^0_{m,i}\right)\mu^0_{m,i} + \lambda^0_{m,i}\left(1-\mu^1_{m,i}\right)$$
$$\nu^1_{m,i} = \left(1-\lambda^1_{m,i}\right)\mu^1_{m,i} + \lambda^1_{m,i}\left(1-\mu^0_{m,i}\right).$$

With these crossover probabilities $\nu^0_{m,i}$ and $\nu^1_{m,i}$, respectively, replacing $\lambda^0_{m,i}$ and $\lambda^1_{m,i}$ and by replacing $\mathbf{u}^{(m)}[n]$'s in Section 12.6.2.1 by $\hat{\mathbf{u}}^{(m)}[n]$'s, then the decision fusion detector at the SSDC is again given by (12.57). Note that the decision statistic $T_m[n]$ in (12.56) is now computed by using $\hat{\mathbf{u}}^{(m)}[n]$'s in place of $\mathbf{u}^{(m)}[n]$'s and the weights $c_{m,i}$'s are determined by substituting $\nu^0_{m,i}$ and $\nu^1_{m,i}$, respectively, in place of $\lambda^0_{m,i}$ and $\lambda^1_{m,i}$.

Finally, it is worth noting from the detector (12.79) and (12.64) that

$$\mu_{m,i}^{0} = \mu_{m,i}^{1} = Q\left(\frac{g + \frac{\sigma_i^2}{2g}\log\hat{\eta}_{m,i}}{\sigma_i}\right).$$

Hence, the second-stage BAC in Figure 12.11 is in fact a binary symmetric channel (BSC). ■

12.7 SUMMARY

- CRs for dynamic spectrum sharing (DSS) is perhaps the most matured application of CRs. Unlike traditional spectrum sharing by users belonging to the same system or in ISM bands, DSS refers to the scenario of unlicensed CRs coexisting with users who own the spectrum without compromising their QoS. Hence, DSS requires the CRs to have the ability to gain spectrum awareness.
- CRs may dynamically share the spectrum with licensed primary users either based on dynamic spectrum access (DSA) or dynamic spectrum leasing (DSL) techniques. In DSA, the responsibility of intersystem interference management lies solely with the cognitive secondary radios without any participation from the primary users, while in DSL, primary users, may proactively facilitate the spectrum sharing to a certain extent in return for leasing payments from secondary users.
- Both DSA and DSL can be implemented on the basis of either spectrum underlay or spectrum interweave based coexistence.
- The spectrum sensing problem faced by a DSS cognitive radio is made of three subproblems:

  1. Decide which channel to sense (channel sensing decisions)
  2. Decide whether the channel is idle or busy based on observations of the sensed channel (idle/busy decision-making, primary signal detection or channel-state decision-making problem)
  3. Decide whether to access the channel or not if it is determined to be indeed idle (channel-access decisions)

- In cognitive radio networks formed by autonomous CRs, distributed CRs may make their own spectrum sensing decisions, while in networks formed by cooperative CRs, they may either coordinate among themselves in making these decisions or a central module in the network, called the secondary system decision center (SSDC), may make global decisions.
- Autonomous spectrum sensing refers to spectrum sensing performed by distributed CRs on their own without any centralized network control. Hence,

primary channel-state decisions, for example, are made by a radio based only on its own sensing observation outcome. Similarly, channel sensing and channel-access decisions are also made by the radio itself.

- In autonomous spectrum sensing, at any given time, a cognitive radio senses a particular primary channel and simply performs a hypothesis testing problem based on the sensed observation to determine whether a primary user is active in that channel or not. Since usually there is a threshold on the maximum tolerable probability of collision with primary users, Neyman–Pearson criterion seems to be a natural framework for making these channel-state decisions. Depending on what prior information the autonomous cognitive radio may have regarding the primary signals, thc form of the optimal detector may vary.
- Autonomous spectrum sensing has the advantage of implementation simplicity and low control overhead. However, its accuracy suffers from the hidden terminal problems that are inherent to all wireless networks. The lack of central coordination, however, can worsen the effects of these hidden terminal problems on the performance of autonomous spectrum sensing.
- In one form of cooperative spectrum sensing, all distributed CRs may forward their local decisions on primary channel states to a central unit in the network called the SSDC. The SSDC makes the final decision regarding the state of a particular primary channel by combining all local sensing reports it receives from distributed nodes who sensed that channel. Based on these global channel-state decisions and taking into account any other network-specific details, it also makes the channel-access decisions for the current time slot as well as the channel sensing decisions for the next sensing interval.
- While cooperative spectrum sensing can improve spectrum sensing performance and the overall spectrum utilization, these come at the price of increased implementation complexity and control overhead.
- Cooperative detection of channel idle/busy state is a problem of distributed detection with data/decision fusion. Joint optimal design of local detectors at the distributed cognitive radio nodes and the fusion detector at the SSDC, however, can be computationally too demanding. A reasonable alternative is to separate these into two stages in which the local detectors are optimized to achieve minimum error probability decisions and the SSDC is designed to maximize the detection of idle channels subjected to a given threshold on collisions with primary transmissions.
- In the case of distributed CRs sending their binary decisions $u_{m,i}[n] \in \{0, 1\}$ on the state of sensed channel to the SSDC, the problem at the SSDC is a decision fusion problem. If we were to assume a two-stage cooperative channel-state detection framework as shown earlier, then fusion performance at the SSDC only depends on the false-alarm $\lambda^0_{m,i} = \Pr(H_1|H_0)$ (collision) and miss $\lambda^1_{m,i} = \Pr(H_0|H_1)$ (missing idle states) probabilities of the local detectors at the distributed CRs. All that matters to the second-stage processing is the crossover probabilities associated with the binary asymmetric channel representing this processing at the ith cognitive radio.

- Local detectors at the distributed CRs can be designed following the same procedure as in the case of autonomous spectrum sensing except that the decision criterion may be the Bayesian minimum error probability.
- Decision fusion at the SSDC will be influenced by the effects of channels between the distributed CRs and the SSDC. In particular, if these channels were to be error-free, the optimal decision fusion rule at the SSDC can be a threshold test on the weighted sum of local decisions in the form of

$$\tilde{\delta}_{NP}(\mathbf{z}^{(m)}[n]) = \begin{cases} 1 & \text{if } > \tau_m \\ \gamma_m & \text{if } \sum_{i\in\mathcal{A}_n^{(m)}} c_{m,i}u_{m,i}[n] = \tau_m \\ 0 & \text{if } < \tau_m \end{cases},$$

where $\mathcal{A}_n^{(m)}$ denotes the set of secondary cognitive radio node indices who sensed channel m at time n,

$$c_{m,i} = \ln\left(\frac{1-\lambda_{m,i}^0}{\lambda_{m,i}^1}\frac{1-\lambda_{m,i}^1}{\lambda_{m,i}^0}\right),$$

$$d_m[n] = \sum_{i\in\mathcal{A}_n^{(m)}} \ln\left(\frac{\lambda_{m,i}^1}{1-\lambda_{m,i}^0}\right) = K_m \ln\left(\frac{\lambda_{m,i}^1}{1-\lambda_{m,i}^0}\right),$$

and the threshold τ_m and randomization constant γ_m are chosen to meet the constraint α_m on collision probability with the primary users on channel m.
- In the case of noisy communications channels between the distributed CRs and the SSDC, the design of the optimal decision fusion detector at the SSDC will need to take into account these effects. Essentially, one may expect the reliability of the final decisions at the SSDC to degrade as the quality of these channels becomes poor. In particular, there may be a performance floor induced by the noisiness of these channels.

13

SPECTRUM DECISION-MAKING IN DSA COGNITIVE NETWORKS

13.1 INTRODUCTION

As we discussed in Chapter 12, dynamic spectrum sharing (DDS) invariably requires the secondary cognitive radios to have a certain level of spectrum awareness: Cognitive radios must be able to determine when a primary user channel is idle and when it is not. In Chapter 12, we discussed how secondary radios can detect whether a channel is idle or busy based on sensing observations. However, this is only one part of the spectrum sensing problem for DSS. Indeed, recall from Chapter 12 that the overall spectrum sensing problem for DSS is made of three decision-making problems: channel-state detection, channel sensing decision-making, and channel-access decision-making.

Let us again consider the primary–secondary coexistence scenario in which there are an M number of primary channels in the system, K_p number of primary users, and K_s number of secondary users. For simplicity of this discussion, let us assume that the primary transmissions are slotted as in Figure 12.3 and at the beginning of each slot there is a (relatively) short sensing interval during which all primary users will indicate whether they intend to use a particular channel for the duration of the subsequent slot. Before this, secondary cognitive radios need to make decisions on which channel to sense during this sensing interval. This is the channel sensing decision-making problem.

Once the channel state is determined from sensing observations, cognitive radios will make channel-access decisions. The sensing and access decisions can be closely

Signal Processing for Cognitive Radios, First Edition. Sudharman K. Jayaweera.

coupled, especially in autonomous cognitive radios. In particular, if the cognitive radio can only sense a single channel at a time, then whatever the channel it senses is the one it can access. A joint sensing and access decision-making algorithm, however, can be complicated and perhaps lack the flexibility. In the following, we may primarily address the sensing decision-making problem while only briefly commenting on channel-access decision-making problem.

The problem of sensing decision-making is concerned with selecting the channel that is most likely to meet the desired performance objectives of the cognitive radio. In DSS, the objective is to achieve successful communications, and thus, the goal of the cognitive radio is to select a channel that will most likely be idle and be less likely to be sensed by another secondary user. Just as determining whether a channel is idle or busy based on a sensing observation, this also is a difficult decision to make by an individual cognitive radio alone because a channel that may most likely be idle might be sensed by more than one user. If, for example, two or more users sensed the same channel and determined it to be idle, there will be a collision among those secondary users if more than one of them started to transmit (access) on that channel. Inevitably, some users may not be able to get access to the channel even after sensing a channel that is to be idle during the next time slot. Thus, an autonomous cognitive radio needs to weight these issues in making a decision on which channel to sense and access.

The problem may somewhat be simplified in the case of centralized spectrum decision-making systems. In a centralized system, sensing and access decisions may be performed by the secondary system decision center (SSDC), just as in the case of cooperative idle/busy decisions in Chapter 12. Based on local sensing results forwarded by the distributed cognitive radios, the SSDC can also instruct which secondary user should sense which channel in the next sensing period. There may be a high likelihood of selecting the most likely idle channels for sensing under this scenario, and collisions among secondary cognitive radios may be avoidable if the SSDC also decides who get to access which channel.

In this chapter, we will discuss the remaining two subproblems of spectrum sensing for DSA: channel sensing and channel-access decision-making. We will, however, mostly be focusing on the sensing decision-making problem since access decisions may be highly dependent on the application context.

13.2 PRIMARY CHANNEL DYNAMIC MODEL

As we did in Chapter 12, let us assume a primary–secondary coexistence scenario in which there are multiple cognitive secondary users all with limited spectrum sensing capabilities in that each can only sense one primary channel at a given time. At the beginning of the sensing interval n, each cognitive radio needs to decide which channel to sense during the sensing period at time $n = n$. If all M primary channels are equally likely to be idle and busy, a priori, and no other information is available either, perhaps, there is nothing much one can do but to randomly pick a channel for sensing. However, this is not exactly the case. The cognitive radio has its past sensing observations, as well as results from previous attempts to access channels up to and including time n. This information may help it to do a better job in selecting the channel to be sensed at time $n = n$ than randomly picking a channel. Utilizing this information in a sensible way, however, would require knowing something about the idle and busy dynamics of the primary channel. For example, suppose that at time $n = n - 1$ the cognitive radio determined that the mth primary channel is free and was able to verify this by successfully accessing the channel during the access period corresponding to time $n = n - 1$. But how can the cognitive radio predict whether it will stay idle during the next time instant $n = n$ or not? Is it more likely or less likely for the channel m to be idle during the next time slot if it was idle at time $n - 1$?

As we mentioned in Chapter 9, a commonly used model for primary channel-idle/channel-busy dynamics is a first-order Markov chain similar to that in Figure 9.5 [129, 130, 132, 133, 201–203]. According to this model, each primary channel can be in one of two states: State 0 indicates that the channel is occupied by a primary user and is called the busy state and the state 1 indicates that there is no primary user transmission over that channel and is called the idle state. We will assume that primary channel occupancy dynamics are independent and identically distributed (iid) across channels. In this formulation, primary channels can also easily be generalized to be nonidentical in terms of their Markov parameters, although we will not attempt such generalizations here.

We denote the true state of the mth primary channel in sensing time slot n by $S_m[n] \in \{0, 1\}$. Note that unlike in Chapter 9, here, we do not bother to use double-subscript notation for the state. The reason is that most practical DSS proposals are limited to relatively smaller frequency bands and thus pose no need for subbanding as in the case of wideband spectrum sensing of Chapter 9. To keep this discussion simpler, we will assume that each secondary user can only sense one primary channel at any given time. However, this may be a too conservative assumption in practice. Generalizing the following approaches to such scenarios however is not that difficult.

For $m = 1, \ldots, M$, stationary transition probability of the mth primary channel Markov model from state s to state s' is defined as

$$p_{s,s'}^{(m)} = \Pr\{S_m[n+1] = s' \mid S_m[n] = s\}, \quad \forall s, s' \in \{0, 1\}. \tag{13.1}$$

The *transition probability matrix* of the mth primary channel Markov model is the 2×2 matrix:

$$P^{(m)} = \begin{pmatrix} p_{00}^{(m)} & p_{01}^{(m)} \\ p_{10}^{(m)} & p_{11}^{(m)} \end{pmatrix}.$$

We denote by vector $\boldsymbol{p}^{(m)} = \left[p_0^{(m)}, p_1^{(m)}\right]$ the stationary distribution of the mth primary channel state where $p_0^{(m)}$ and $p_1^{(m)}$ correspond to stationary probabilities of the channel being busy and idle, respectively. It is easily seen that $\boldsymbol{p}^{(m)}$ is given by the solution to $\boldsymbol{p}^{(m)} = \boldsymbol{p}^{(m)} P^{(m)}$.

Let us denote by $\mathbf{S}[n] = (S_1[n], \ldots, S_M[n])$ the state of the primary spectrum formed by the M primary channels. Since each $\mathbf{S}[n] \in \{0,1\}^M$, $\mathbf{S}[n]$ can take 2^M possible vector values. We will denote this set by $\mathcal{S}$. We may index these 2^M elements of the set of possible states $\mathcal{S}$ as $\mathcal{S} = \{\mathbf{s}_1, \mathbf{s}_2, \ldots, \mathbf{s}_{2^M}\}$. State transition probabilities of the overall spectrum are denoted by $p_{j,j'}$ where $p_{j,j'} = \Pr\left(S[n+1] = \mathbf{s}_{j'} | S[n] = \mathbf{s}_j\right)$ for $\mathbf{s}_j, \mathbf{s}_{j'} \in \mathcal{S}$.

13.3 SENSING DECISIONS IN DSS NETWORKS WITH AUTONOMOUS COGNITIVE RADIOS

Let us first consider autonomous cognitive radios that make their sensing decisions by themselves. As we already saw in Section 12.3, there is no cooperation or coordination among such secondary cognitive radios: Each secondary cognitive radio makes a sensing decision on which channel to sense at the next time instant $n = n$ based only on information available to it. Of course, its objective in making this decision has to be not only to determine channel that will most likely be idle during the next time instant but also will be less likely to be sensed by another cognitive radio. However, absent of any coordination among cognitive radios, the latter may be difficult to achieve in autonomous decision-making systems.

Since each cognitive radio i, for $i = 1, \ldots, K_s$, can observe only one out of the M primary channels at any given time n, from the perspective of each cognitive radio, the problem of sensing decision-making is one of partially observable type. Since decisions are based on state $\mathbf{S}[n]$ that evolves according to Markov dynamics and the state is not fully observable, the sensing decision-making problem thus is a partially observable Markov decision process (POMDP) similar to the subband selection problem of Chapter 9. The basic idea that the sensing decision-making in a cognitive radio can be formulated as a POMDP was first put forth in [129], which essentially developed the framework that we discuss in the following for an autonomous cognitive radio.

Recall from Chapter 7 that we may convert a finite-state POMDP in to an equivalent COMDP whose state takes values in a probability simplex. This completely observable new state is the belief state. Indeed, recall that belief $b_j[n]$ on the state at time n being $\mathbf{s}_j$, for $\mathbf{s}_j \in \mathcal{S}$, given all available evidence up to time n is the a posteriori probability of state $\mathbf{S}[n] = \mathbf{s}_j$ at time n given the complete information history $\underline{h}_n$ available to the cognitive radio up to time n:

$$b_j[n] = \Pr\left(\mathbf{S}[n] = \mathbf{s}_j \mid \underline{h}_n\right). \tag{13.2}$$

In (13.2), we have dropped the cognitive radio index since the same approach in the following applies to all radios. As in wideband spectrum sensing in Chapter 9, sensing decisions for time n in a DSS cognitive radio are also needed to be made before having access to the observation $Y[n]$ corresponding to state $S[n]$ at time n. Hence, similar to (9.31), the information history $\underline{h}_n$ available to the cognitive radio at time n is, for $n = 0, 1, \ldots,$

$$\underline{h}_n = (\underline{h}_{n-1}, a_{n-1}, y[n-1]) \tag{13.3}$$

with $\underline{h}_0 = \phi$. Note that both a_n and $y[n]$ depend on the particular cognitive radio although we have dropped their indices since the following approach is applicable to all cognitive radios.

The belief-state vector at time n is the 2^M-length vector defined as

$$\mathbf{b}[n] = \left(b_1[n], b_2[n], \ldots, b_{2^M}[n]\right)^T.$$

As we have already seen in Chapters 7 and 9, the belief-state vector $\mathbf{b}[n+1]$ is completely determined by $\mathbf{b}[n]$, the action a_n chosen at time n, and the resulting observation $y[n]$. Indeed, following the same steps as in deriving (9.37), we have that

$$\begin{aligned}
b_j[n+1] &= \Pr\left(S[n+1] = \mathbf{s}_j \mid \underline{h}_{n+1}\right) \\
&= \frac{\sum_{j'=1}^{2^M} \Pr\left(S[n+1] = \mathbf{s}_j | S[n] = \mathbf{s}_{j'}, \underline{h}_{n+1}\right) \Pr\left(S[n] = \mathbf{s}_{j'}, \underline{h}_{n+1}\right)}{\Pr\left(\underline{h}_{n+1}\right)} \\
&= \sum_{j'=1}^{2^M} \Pr\left(S[n+1] = \mathbf{s}_j | S[n] = \mathbf{s}_{j'}\right) \Pr\left(S[n] = \mathbf{s}_{j'} \left| \underline{h}_{n+1}\right.\right) \\
&= \sum_{j'=1}^{2^M} \Pr\left(S[n+1] = \mathbf{s}_j | S[n] = \mathbf{s}_{j'}\right) \Pr\left(S[n] = \mathbf{s}_{j'} \left| \underline{h}_n, a_n, y[n]\right.\right) \qquad (13.4) \\
&= \sum_{j'=1}^{2^M} p_{j',j} \Pr\left(S[n] = \mathbf{s}_{j'} \left| \underline{h}_{n+1}\right.\right) \\
&= \sum_{j'=1}^{2^M} p_{j',j} \frac{\Pr\left(y[n] \mid S[n] = \mathbf{s}_{j'}, a_n\right) b_{j'}[n]}{\sum_{j''=1}^{2^M} \Pr\left(y[n] \mid S[n] = \mathbf{s}_{j''}, a_n\right) b_{j''}[n]} \qquad (13.5) \\
&= \frac{\sum_{j'=1}^{2^M} p_{j',j} \Pr\left(y[n] \mid S[n] = \mathbf{s}_{j'}, a_n\right) b_{j'}[n]}{\sum_{j''=1}^{2^M} \Pr\left(y[n] \mid S[n] = \mathbf{s}_{j''}, a_n\right) b_{j''}[n]} \\
&= T_j\left(\mathbf{b}[n], a_n, y[n]\right), \qquad (13.6)
\end{aligned}$$

where (13.4) is due to the fact that the state transitions of the underlying channel dynamics are not affected by the actions chosen by the secondary cognitive radios or their observations, (13.5) follows from the same steps used in obtaining (9.36), and the $T_j(\mathbf{b}, a, y)$ in (13.6) is the same transformation in (9.37) that denotes the mapping from $\mathbf{b}[n]$ to $b_j[n+1]$.

Let us define the vector transformation

$$\mathbf{T} = (T_1, \ldots, T_{2^M}), \tag{13.7}$$

where T_j, for $j = 1, \ldots, 2^M$, is given by (13.6). Hence, we see that the belief vector $\mathbf{b}[n]$ is a Markov process:

$$\mathbf{b}[n+1] = \mathbf{T}(\mathbf{b}[n], a_n, y[n]). \tag{13.8}$$

As we already know from Chapter 7, $\mathbf{b}[n]$ is a sufficient statistic for the decision-making problem based on the partially observable state of the Markov process. From (13.8), thus, we have a COMDP based on the belief state $\mathbf{b}[n]$. However, when the core process state is $\mathbf{s}[n] = (s_1[n], \ldots, s_M[n])$ with each $s_m[n] \in \{0, 1\}$, the belief-state vector lies in the $(2^M - 1)$-dimensional probability simplex.

Observation probabilities $\Pr\left(y[n] \mid S[n] = \mathbf{s}_{j'}, a_n\right)$ needed for evaluating the transformation T_j in (13.5) can be written down easily by separating the channel-state detection problem from the sensing decision-making problem, as we did in wideband spectrum sensing in Chapter 9. Hence, although actual sensing observations are continuous-valued (e.g., Gaussian noise-corrupted signals $\mathbf{r}_{m,i}[n]$'s given in (12.1)), the observations in the decision-making problem of the ith cognitive radio are taken to be the binary decisions $u_{m,i}[n] \in \{0, 1\}$ already made by the Neyman–Pearson detectors discussed in Section 12.3. Hence, if we consider the decision-making problem faced by an arbitrary cognitive radio,

$$\begin{aligned}\Pr\left(y[n] \mid S[n] = \mathbf{s}_j, a_n = m\right) &= \Pr\left(u_m[n] \mid S_m[n] = s_{j,m}\right)\\ &= \begin{cases}\Pr\left(u_m[n] = 0 \mid S_m[n] = s_{j,m}\right) & \text{if} \quad y[n] = 0\\ \Pr\left(u_m[n] = 1 \mid S_m[n] = s_{j,m}\right) & \text{if} \quad y[n] = 1\end{cases},\end{aligned}$$

where $u_m[n]$ for the ith cognitive radio is given by the $u_{m,i}[n]$ in either (12.15) or (12.24). Hence,

$$\begin{aligned}\Pr\left(y[n] = 0 \mid S[n] = \mathbf{s}_j, a_n = m\right) &= \begin{cases}\Pr\left(H_0 \mid H_0\right) & \text{if} \quad s_{j,m} = 0\\ \Pr\left(H_0 \mid H_1\right) & \text{if} \quad s_{j,m} = 1\end{cases}\\ &= \begin{cases}1 - P_{F,i}^{(m)} & \text{if} \quad s_{j,m} = 0\\ 1 - P_{D,i}^{(m)} & \text{if} \quad s_{j,m} = 1\end{cases} &(13.9)\\ &= \begin{cases}1 - P_{F}^{(m)} & \text{if} \quad s_{j,m} = 0\\ 1 - P_{D}^{(m)} & \text{if} \quad s_{j,m} = 1\end{cases} &(13.10)\end{aligned}$$

and

$$\Pr\left(y[n]=1 \mid S[n]=\mathbf{s}_j, a_n=m\right) = \begin{cases} \Pr(H_1 \mid H_0) & \text{if} \quad s_{j,m}=0 \\ \Pr(H_1 \mid H_1) & \text{if} \quad s_{j,m}=1 \end{cases}$$
$$= \begin{cases} P_{F,i}^{(m)} & \text{if} \quad s_{j,m}=0 \\ P_{D,i}^{(m)} & \text{if} \quad s_{j,m}=1 \end{cases}, \tag{13.11}$$
$$= \begin{cases} P_{F}^{(m)} & \text{if} \quad s_{j,m}=0 \\ P_{D}^{(m)} & \text{if} \quad s_{j,m}=1 \end{cases}, \tag{13.12}$$

where $P_{F,i}^{(m)}$ for the ith cognitive radio is given by either (12.16) or (12.26) and $P_{D,i}^{(m)}$ for the ith cognitive radio is given by either (12.18) or (12.29). In (13.10) and (13.12), we have denoted these probabilities by dropping the cognitive radio index i since the equations apply to an arbitrary cognitive radio.

13.3.1 Optimal Sensing Policy Determination

The sensing decision-making problem is then to make a sensing decision $\pi_n(\mathbf{b}[n]) = m$, for $m \in \{1,\ldots,M\}$, at each time n based on the information available at time n, which is summarized in the form of the belief vector $\mathbf{b}[n]$. Hence, each $\pi_n(.)$ is a mapping $\pi_n : [0,1]^M \longrightarrow \{1,\ldots,M\}$. A policy $\underline{\pi}$ is a sequence of such decision functions for each time instant: $\underline{\pi} = (\pi_0(.),\ldots,\pi_{T-1})$. When $T \longrightarrow \infty$, we have an infinite-time-horizon problem. The policy $\underline{\pi}$ is said to be stationary if $\pi_n(.)$ does not depend on the time instant n.

Let us denote by $R(s,y,a)$ the (time-invariant) *immediate reward* to the secondary cognitive radio if when in state $S[n]=s$ it takes the sensing decision/action $a_n=a$ and observes $Y[n]=y$. Depending on the performance objectives of a radio, this reward function can vary. As we know from Chapter 7, the immediate reward cannot be evaluated by a cognitive radio that does not get to observe the state of the core process. Instead, similar to (7.73), it can only evaluate the expected immediate reward given the current belief state:

$$\begin{aligned} r(\mathbf{b}[n],a) &= \mathbb{E}\{R(S,Y,a)|\mathbf{b}[n]\} \\ &= \sum_{j=1}^{2^M} \mathbb{E}\left\{R(\mathbf{s}_j,Y,a)|\mathbf{b}[n]\right\} \Pr\left(\mathbf{S}[n]=\mathbf{s}_j|\mathbf{b}[n]\right) \\ &= \sum_{j=1}^{2^M} \mathbb{E}\left\{R(\mathbf{s}_j,Y,a)\right\} b_j[n] \\ &= \sum_{j=1}^{2^M} r_j(a) b_j[n] \\ &= \mathbf{r}(a)^T \mathbf{b}[n], \end{aligned} \tag{13.13}$$

where $\mathbf{s}_j = (s_{j,1}, \ldots, s_{j,M})$ for $j = 1, \ldots, 2^M$, and we have denoted $\mathbf{r}(a) = \left(r_1(a), r_2(a), \ldots, r_{2^M}(a)\right)^T$ with

$$r_j(a) = \mathbb{E}\left\{R(\mathbf{S}, Y, a) | \mathbf{S} = \mathbf{s}_j\right\}$$
$$= \mathbb{E}\left\{R(\mathbf{s}_j, Y, a)\right\}. \quad (13.14)$$

Example 13.1 (A Reward Function for Sensing Decision-Making in a DSS Cognitive Radio Network) As an example, we may assume that the immediate reward to the cognitive radio is of the following form:

$$R(s, y, a) = s_a B_a y_a, \quad (13.15)$$

where B_a is the bandwidth of the channel a and $y_a \in [0, 1]$ is the outcome from the channel-state detector. Note that the above $R(s, y, a)$ is an extension of that used in [129] and the building block of what we used earlier in Example 9.4:

$$R(S[n], Y[n], a_n) = S_{a_n}[n] B_{a_n} Y_{a_n}[n]$$
$$= \begin{cases} B_{a_n} Y_{a_n}[n] & \text{if} \quad S_{a_n}[n] = 1 \\ 0 & \text{if} \quad S_{a_n}[n] = 0 \end{cases}. \quad (13.16)$$

Thus, the immediate reward is zero if the radio chooses to sense a channel that happens to be busy during time interval n. On the other hand, if it chooses to sense a channel that will be idle during the time interval n, the radio receives an immediate reward proportional to both the bandwidth of the chosen channel and the outcome from the channel-state detector. If the channel-state detector was perfect in its decision so that $y_a = 1$, then the immediate reward will be the maximum of B_{a_n}. If however, channel-state detector erroneously declared $y_a = 0$, then the reward will be again zero since the radio will believe that the channel it sensed is busy and thus will not access it.

As we also discussed in Section 9.6.3, the reward (13.16) cannot be evaluated in a POMDP since the cognitive radio does not get to observe the actual state $\mathbf{S}[n]$ and it has to make decisions for time slot n before observations corresponding to time slot n are taken. Hence, the decisions will be based on the expected immediate rewards where expectation is with respect to both core state and observation processes.

If the state is observable, the expected immediate reward given that the current state $S[n] = \mathbf{s} = (s_1, \ldots, s_M)$ is

$$r(\mathbf{s}, a) = \mathbb{E}\{R(S, Y, a) | S = \mathbf{s}\}$$
$$= B_a s_a \mathbb{E}\{Y_a | S_a = s_a\}$$
$$= B_a s_a \mathbb{E}_{Y_a | S_a}\left\{Y_a | H_{s_a}\right\}$$

$$
\begin{aligned}
&= B_a s_a P(H_1|H_{s_a}) \\
&= \begin{cases} B_a P_D^{(a)} & \text{if} \quad s_a = 1 \\ 0 & \text{if} \quad s_a = 0 \end{cases} \\
&= B_a s_a P_D^{(a)},
\end{aligned}
\tag{13.17}
$$

where $P_D(a)$ is the probability of correct detection of idle hypothesis on channel a (which is $P_{D,i}^{(m)}$ for the ith cognitive radio and here we have dropped the index i).

Since expected immediate reward (13.17) cannot be evaluated by the cognitive radio since it does not have access to the true state $S[n]$, the radio may instead compute the expected immediate reward as in (7.73). Indeed, if action $a_n = a$ is to be taken from the belief state $\mathbf{b}[n]$, then we have that

$$
\begin{aligned}
r(\mathbf{b}[n], a_n = a) &= \mathbb{E}\{r(\mathbf{S}, a)|\mathbf{b}[n]\} \\
&= B_a P_D^{(a)} \mathbb{E}\{S_a|\mathbf{b}[n]\} \\
&= \sum_{j=1}^{2^M} B_a P_D^{(a)} s_{j,a} \Pr\left(\mathbf{S}[n] = \mathbf{s}_j|\mathbf{b}[n]\right) \\
&= \sum_{j=1}^{2^M} B_a s_{j,a} P_D^{(a)} b_j[n] \\
&= \sum_{j=1}^{2^M} r_j(a) b_j[n] \\
&= \mathbf{r}(a)^T \mathbf{b}[n],
\end{aligned}
\tag{13.18}
$$

where $\mathbf{s}_j = (s_{j,1}, \ldots, s_{j,M})$ for $j = 1, \ldots, 2^M$, and we have denoted $\mathbf{r}(a) = \big(r_1(a), r_2(a), \ldots, r_{2^M}(a)\big)^T$ with

$$
r_j(a) = B_a s_{j,a} P_D^{(a)}. \tag{13.19}
$$

■

We may then formulate the sensing decision-making problem as a POMDP in which the objective of the cognitive radio is to maximize the expected discounted total return. From Chapter 7, we know that an optimal POMDP decision policy that achieves this is characterized by

$$
\underline{\pi}^* = \arg\max_{\underline{\pi}} V^{\underline{\pi}}(\mathbf{b}), \tag{13.20}
$$

where the value function associated with the expected total discounted return under the policy $\underline{\pi}$ when the initial state is $\mathbf{b}[0] = \mathbf{b}$ is defined as

$$V_T^{\underline{\pi}}(\mathbf{b}) = \mathbb{E}_{\underline{\pi}}\left\{\sum_{n=0}^{T} \gamma^n r(\mathbf{b}[n], \pi_n(\mathbf{b}[n])) \mid \mathbf{b}[0] = \mathbf{b}\right\}, \tag{13.21}$$

with $\gamma \in (0, 1]$ as the discount factor. In the case of finite T, of course, one may drop the discount factor easily, while when $T \longrightarrow \infty$, the discount factor $0 < \gamma < 1$ can help ensure the finiteness of the expected reward for any bounded sequence of expected immediate rewards. Whether a finite- or an infinite-horizon formulation is better suited depends on the particular application context. In the case of an infinite-horizon formulation, however, we may particularly be interested in the existence and determination of a stationary policy. In Chapter 7, we discussed conditions that may ensure the existence of a stationary policy for a POMDP.

The key to obtaining an optimal policy is the optimality equation satisfied by the optimal value function corresponding to an optimal policy. In the case of a finite-horizon POMDP, Theorem 7.5 describes a procedure to find the optimal value function and thus an optimal policy. Indeed, if we denote by $V_T^*(\mathbf{b})$ and $v_m^*(\mathbf{b})$ the optimal value function and the optimal m-step value-to-go function, respectively, then from (7.84), we know that

$$V_T^*(\mathbf{b}) = v_T^*(\mathbf{b}).$$

The optimal T-step value-to-go function $v_T^*(\mathbf{b})$ can be computed in a backward induction thanks to the optimality equation given by Theorem 7.5:

$$v_m^*(\mathbf{b}) = \max_{a \in \mathcal{A}}\left\{\mathbf{b}^T \mathbf{r}(a) + \gamma \sum_{y \in \{0,1\}} v_{m-1}^*(\mathbf{b}^{(new)}) \Pr(y \mid \mathbf{b}, a)\right\}, \tag{13.22}$$

where $\mathbf{b}^{(new)} = (b_1^{(new)}, b_2^{(new)}, \ldots, b_{2^M}^{(new)})$ with

$$\mathbf{b}_j^{(new)} = T_j(\mathbf{b}, a, y), \tag{13.23}$$

for $j = 1, \ldots, 2^M$ and T_j is as given by (13.6). The recursion (13.22) is initialized for $m = 0$ by setting

$$v_0^*(\mathbf{b}) = \mathbf{b}^T \mathbf{r}_T \quad \text{for all } \mathbf{b}, \tag{13.24}$$

for some terminating reward vector $\mathbf{r}_T = (r(s_1), \ldots, r(s_M))^T$.

On the other hand, in the case of an infinite-horizon formulation of the sensing decision-making problem, the optimal value function $V^*(\mathbf{b})$ corresponding to a stationary policy satisfies the following fixed-point (optimality) equation given by Theorem 7.9:

$$V^*(\mathbf{b}) = \max_{a\in\mathcal{A}} \left\{ \mathbf{r}(a)^T\mathbf{b} + \gamma \sum_{y\in\{0,1\}} \Pr(y \mid \mathbf{b}, a) V^*(\mathbf{b}^{(new)}) \right\}, \tag{13.25}$$

where $\mathbf{b}^{(new)} = (b_1^{(new)}, b_2^{(new)}, \ldots, b_{2M}^{(new)})$ given by (13.6).

In both (13.22) and (13.25),

$$\begin{aligned}\Pr(y \mid \mathbf{b}, a) &= \mathbb{E}\{\Pr(Y[n] = y \mid S[n], a_n = a) \mid \mathbf{b}\}\\ &= \sum_{j=1}^{2^M} \Pr\left(Y[n] = y \mid S[n] = \mathbf{s}_j, a_n = a\right) b_j,\end{aligned} \tag{13.26}$$

where $\Pr\left(Y[n] = y \mid S[n] = \mathbf{s}_j, a_n = a\right)$ is given by either (13.9) or (13.11).

As we saw in Chapter 7, if the exact state were to be observable, the optimality equations of the form of (13.22) and (13.25) would be both solvable, at least in theory, although the computational complexity can be exponential in M, the number of channels. However, as we saw in Section 7.5, in the case of (13.22) and (13.25), this approach is not directly applicable since $\mathbf{b}$ is not a finite-state Markov process but takes a continuum of values in the $(2^M - 1)$-dimensional probability simplex $[0, 1]^M$.

For finite-horizon POMDPs, however, the backward induction algorithm can be made to be applicable, at least in theory, by noting that optimal value function of the finite-horizon POMDPs is always piecewise linear and convex. We saw in Chapter 7 that it is possible to have finite representations of the optimal value function due to this PWLC property of the optimal value function. The basic idea was to take advantage of the PWLC property of the optimal value function to partition the probability simplex into a finite number of convex regions $\mathcal{B}_1, \ldots, \mathcal{B}_{L_m}$ on each of which the value function $v_m(\mathbf{b})$ is piecewise linear. Hence, the optimal policy on each of the convex regions $\mathcal{B}_l$, for $l = 1, \ldots, L_m$, can be characterized by specifying the slope (gradient) vector of the optimal value function on this region. These slope vectors were called the β-vectors. Hence, at each step m, one has to only specify the optimal policy for each of the convex regions $\mathcal{B}_l$, for $l = 1, \ldots, L_m$, a finite collection, as opposed to an infinite continuum. In Section 7.5.4, we discussed several algorithms that can be used to compute an optimal finite-horizon policy for a POMDP exploiting this PWLC property of the optimal value function. These included, for example, the one-pass algorithm and the witness algorithm. Still, however, the number L_m of β-vectors at each time instant can, in general, be prohibitively large since it can grow exponentially with the number of observations, as can be seen from (7.100).

In the case of discounted infinite-horizon POMDPs, however, these algorithms that take advantage of the piecewise linearity of the optimal value function may not be applicable. As we discussed in Section 7.5.5, the reason is that as $T \longrightarrow \infty$, the optimal value function is no longer guaranteed to be linear. A situation in which the optimal infinite-horizon POMDP value function will still be piecewise linear is if it is known to have an optimal stationary policy that happens to be finitely transient. In this case, of course, we may again apply the algorithms developed for finite-horizon POMDPs with increasing horizon length T to obtain an optimal policy for the infinite-horizon POMDP. We do not discuss these solution algorithms specifically here since they are already given in Section 7.5.4.

13.3.2 Optimal Myopic Sensing Policy Determination

While in theory an optimal policy can be determined for either the finite- or the infinite-horizon POMDP formulation of the sensing decision-making problem faced by an autonomous cognitive radio using algorithms such as the one-pass algorithm and the witness algorithm of Section 7.5.4, the computational complexity is exponential in the number of channels M and the number of observations. Even then, the optimal policy has to be computed off-line assuming the full knowledge of the state transition dynamic probabilities. Thus, they preclude incorporating the actual observations to update these state transition probabilities and better fit the model to the data.

As we saw in Chapter 7, a myopic, or greedy, policy only considers the expected immediate reward to be achieved at current time. In selecting an action for time n, it does not care about its possible effect on future rewards. Such policies can be considerably easier to determine, compared to the optimal policies of the previous section. An additional advantage is that the state transition probabilities can be updated based on the most recent observations since the sensing decisions are computed at each time instant rather than off-line. This may allow adapting to time-varying nature of the spectrum, which is not possible in the case of an optimal policy determined off-line. Given the fact that wireless channels can be a highly dynamic system, myopic policies thus seem to be a natural fit for the sensing decision-making problem in an autonomous cognitive radio.

Suppose that at time n, the radio has computed the belief-state vector $\mathbf{b}[n]$ whose jth element $b_j[n]$, for $j=1,\ldots,2^M$, is the $\Pr(S[n]=\mathbf{s}_j|\mathbf{h}[n])$. An optimal myopic policy can be defined by

$$a_n^* = \arg\max_{a\in\mathcal{A}} \mathbb{E}\{R(S,Y,a)|\mathbf{b}[n]\}$$

$$= \arg\max_{a\in\mathcal{A}} \sum_{j=1}^{2^M} r_j(a)b_j[n] \tag{13.27}$$

$$= \arg\max_{a\in\mathcal{A}} \mathbf{r}(a)^T\mathbf{b}[n], \tag{13.28}$$

where $\mathbf{r}(a) = (r_1(a),\ldots,r_{2^M}(a))^T$ with $r_j(a)$ given by (13.14).

Determining the optimal myopic policy defined by (13.28) is considerably simpler than the optimal policy for the whole time horizon. In the case of an optimal POMDP solution for each time instant, we need to determine the optimal action (decision) for all possible state or belief vectors. Even in the case of an MDP, we need to evaluate the optimal policy for 2^M possible states as a contingency plan since we do not know which state we may encounter. In the case of a myopic policy, however, we do know the current belief state and simply pick the decision that maximizes the expected immediate reward given that belief state. No need to compute a contingency plan for all possible 2^M states of an MDP or a continuum of states of a POMDP.

Example 13.2 (An Optimal Myopic Sensing Policy for a DSS Cognitive Radio Network) We may define an optimal myopic sensing policy based on the immediate reward function introduced in Example 13.1. With the expected immediate reward $r(\mathbf{b}[n], a_n = a)$ for selecting action a when the belief state is $\mathbf{b}[n]$ given by (13.18) and (13.19), the optimal myopic channel sensing decision is

$$\begin{aligned} a_n^* &= \arg\max_{a \in \mathcal{A}} \mathbf{r}(a)^T \mathbf{b}[n] \\ &= \arg\max_{a \in \mathcal{A}} B_a P_D^{(a)} \sum_{j=1}^{2^M} s_{j,a} b_j[n], \end{aligned} \tag{13.29}$$

where $\mathbf{s}_j = (s_{j,1}, \ldots, s_{j,M})^T \in \mathcal{S}$. ■

Computation of an optimal myopic policy such as (13.28) only requires time updating of the belief-state vector $\mathbf{b}[n]$. This can be done by using (13.8) in conjunction with (13.6). An added benefit is that the model parameters such as transition probabilities can also be updated in parallel based on latest observations and be used in making future decisions. As we saw in the case of subband selection problem in Chapter 9, optimal myopic channel selection further simplifies when channel-state transitions are independent of each other as we show next.

13.3.2.1 Special Case: Independent Channel Dynamics A particularly interesting situation in practice is when the M primary channel occupancy dynamics are independent of each other. This can happen, and be justified, in various situations depending on the type of primary network in question. An example is a random-access network with independent user traffic. With independent channel dynamics, the joint system belief state can be characterized by an M-dimensional vector $\boldsymbol{\omega}[n] = (\omega_1[n], \ldots, \omega_M[n])^T$ instead of a 2^M-dimensional vector as assumed in the previous section. Each component $\omega_m[n]$, for $1 \le m \le M$, is simply the belief of the state of channel m defined as

$$\omega_m[n] = \Pr(S_m[n] = 1 | \underline{h}_n),$$

for $m = 1, \ldots, M$.

This reduced-dimensional belief vector $\boldsymbol{\omega}[n]$ can also be easily computed recursively from its previous value $\boldsymbol{\omega}[n-1]$, the action a_{n-1}, and the resulting observation $y[n-1]$ as was shown in (9.49b). Indeed, for $m = 1, \ldots, M$,

$$\omega_m[n] = \begin{cases} p_{11}^{(m)} \omega_m[n-1] + p_{01}^{(m)} (1 - \omega_m[n-1]) & \text{if} \quad a_{n-1} \neq m \\ \sum_{j \in \{0,1\}} p_{j1}^{(m)} \Pr(S_m[n-1] = j | \underline{h}_{n-1}, y_m[n-1]) & \text{if} \quad a_{n-1} = m \end{cases}$$
$$= \begin{cases} p_{11}^{(m)} \omega_m[n-1] + p_{01}^{(m)} (1 - \omega_m[n-1]) & \text{if} \quad a_{n-1} \neq m \\ p_{11}^{(m)} \Pr(S_m[n-1] = 1 | \underline{h}_{n-1}, y_m[n-1]) \\ \quad + p_{01}^{(m)} \Pr(S_m[n-1] = 0 | \underline{h}_{n-1}, y_m[n-1]) & \text{if} \quad a_{n-1} = m \end{cases},$$

where

$$\Pr\left(S_m[n-1] = 1 | \underline{h}_{n-1}, y_m[n-1]\right) = \frac{\Pr(y_m[n-1] | S_m[n-1] = 1)\, \omega_m[n-1]}{\Pr(y_m[n-1] | S_m[n-1] = 1)\, \omega_m[n-1] + \Pr(y_m[n-1] | S_m[n-1] = 0)\,(1 - \omega_m[n-1])}$$

with

$$\Pr(y_m[n-1] | S_m[n-1] = 1) = \begin{cases} P_D^{(m)} & \text{if} \quad y_m[n-1] = 1 \\ 1 - P_D^{(m)} & \text{if} \quad y_m[n-1] = 0 \end{cases}$$

and

$$\Pr\left(S_m[n-1] = 0 | \underline{h}_{n-1}, y_m[n-1]\right) = \frac{\Pr(y_m[n-1] | S_m[n-1] = 0)\,(1 - \omega_m[n-1])}{\Pr(y_m[n-1] | S_m[n-1] = 1)\, \omega_m[n-1] + \Pr(y_m[n-1] | S_m[n-1] = 0)\,(1 - \omega_m[n-1])}$$

with

$$\Pr(y_m[n-1] | S_m[n-1] = 0) = \begin{cases} P_F^{(m)} & \text{if} \quad y_m[n-1] = 1 \\ 1 - P_F^{(m)} & \text{if} \quad y_m[n-1] = 0 \end{cases}.$$

An optimal myopic policy for a DSS system made of independent primary channels can then be defined by

$$\begin{aligned} a_n^* &= \arg\max_{a \in \mathcal{A}} \mathbb{E}\{R(S_a, Y, a) | \boldsymbol{\omega}[n]\} \\ &= \arg\max_{a \in \mathcal{A}} \left[\mathbb{E}\{R(S_a = 1, Y, a)\}\, \omega_a[n] + \mathbb{E}\{R(S_a = 0, Y, a)\}\,(1 - \omega_a[n])\right]. \end{aligned} \quad (13.30)$$

Example 13.3 (An Optimal Myopic Sensing Policy for a DSS Cognitive Radio Network with Independent Primary Channel Dynamics) In the case of the immediate reward function introduced in Example 13.1, the optimal myopic policy (13.29) can be simplified when channel dynamics are independent of each other as

$$a_n^* = \arg\max_{a \in \mathcal{A}} \mathbb{E}\{S_a B_a Y_a | \omega_a[n]\}$$
$$= \arg\max_{a \in \mathcal{A}} B_a P_D^{(a)} \omega_a[n]. \tag{13.31}$$

■

13.4 SENSING DECISIONS IN COOPERATIVE DSS NETWORKS

Cooperative spectrum sensing may not only improve the final channel-idle/channel-busy decisions made at an SSDC by data/decision fusion, as we saw in the previous chapter, but also the likelihood of choosing a channel that will be idle during next time instant to sense at the first place, as we will see in the following. The reason is that the SSDC has access to more information than a single node. This allows it to make globally optimal, or at least globally better, decisions.

Sensing decisions at the SSDC determine which primary channel each secondary user should sense at each time instant. Since it makes no sense to spend resources (in this case secondary users) to find out that a particular channel is going to be busy, the SSDC must assign the nodes to channels that are highly likely to be idle. On the other hand, it is equally important that when SSDC decides that a particular channel is going to be idle during next time slot, it makes that decision with very high likelihood of being correct. Otherwise, it may lead to collisions with primary users wasting resources spent for sensing that particular channel.

Usually, reliability of the (optimal) channel-state decisions at SSDC with any reasonable fusion rule will monotonically improve with the number of local decisions received for a given channel. Hence, the reliability of channel-state decisions can be maximized by assigning, at each time instant, all secondary users to sense a single primary channel. Of course, that single channel then can be chosen to be the one that has the highest probability of being idle during the next time slot. But assigning all secondary users to sense a single channel may not be very efficient since it also leads to only one channel being available for secondary users to access (assuming no user will be assigned to access a channel that was not sensed and determined to be idle).

As an example, suppose there are $K_s = 12$ secondary users and $M = 6$ primary channels. If, at each time instant, all 12 secondary users were to sense a single primary channel, at most, only 1 secondary cognitive radio will get the opportunity to access a channel at each time instant. If, instead, the SSDC assigns 6 secondary users each to sense a channel, then at any given time, there may be reasonably good channel-state decisions about 2 primary channels. In this case, the two channels with highest likelihood of being idle may be selected for sensing by 6 cognitive radios each.

If, however, we try to push this argument to its extreme, then again, we end up by assigning the minimum possible number of secondary users to sense each channel. In this case, this will be to assign 2 secondary users each to sense each channel. Then the SSDC runs the risk of not receiving enough local decision reports for each channel to make reliable channel-state decisions. Not only that, there is the possibility that some of the channels might be busy during the next time slot with very high probability so that sensing them may be useless.

Hence, channel sensing decisions at the SSDC must consider several factors:

1. Assign nodes so that multiple channels are sensed.
2. Assign secondary users to sense only those channels that have a sufficiently high probability of turning out being idle during the next time slot.
3. Assign enough nodes to sense each channel that is going to be sensed, so that the final decision will be highly accurate.
4. Assign nodes to sense certain primary channels that have not been sensed for a long period so that their channel-state predictions can be made better.

Let us assume, as before, that each node can only sense a single primary channel at a time and each node is to be assigned to sense a primary channel. Hence, sensing decision-making problem at the SSDC is to determine a *partitioning* of the set of nodes $\{1,\ldots,K_s\}$ into at most M nonzero sets. Since neither the assigning all nodes to sense a single primary channel nor assigning nodes uniformly to sense all primary channels can, in general, be optimal, we need to formulate this problem in a way that will lead to decisions that may be considered as being optimal according to some performance criterion.

Let us denote by an $M \times K_s$ binary matrix $\mathbf{A}_n$ the sensing decisions made by the SSDC at time n, where $[\mathbf{A}_n]_{m,i} = 1$ or $[\mathbf{A}_n]_{m,i} = 0$. A sensing decision $[\mathbf{A}_n]_{m,i} = 1$ implies that the ith cognitive radio should sense the primary channel m at time n, while $[\mathbf{A}_n]_{m,i} = 0$ implies that the ith secondary user should not be sensing the primary channel m at time n. We may denote by $\mathcal{C}_n \subseteq \{1,\ldots,M\}$ the set of channel indices chosen to be sensed at time n according to the sensing decision matrix $\mathbf{A}_n$:

$$\mathcal{C}_n = \{m : \sum_{i=1}^{K_s} [\mathbf{A}_n]_{m,i} \neq 0, 1 \leq m \leq M\}.$$

As we did in (12.47), let $\mathcal{A}_n^{(m)}$ denote the set of node indices that were chosen to sense channel m at time n:

$$\mathcal{A}_n^{(m)} = \left\{i : [\mathbf{A}_n]_{m,i} \neq 0\right\}.$$

To simplify the discussion, let us specifically assume that the immediate reward of sensing channels according to the decision matrix $\mathbf{A}_n$ is

$$R(\mathbf{s}, \mathbf{A}_n, \mathbf{U}[n]) = \sum_{m \in \mathcal{C}_n} s_m B_m U_m[n], \tag{13.32}$$

where $U_m[n] \in [0,1]$ denotes the decision on the state of channel m made at the SSDC and $\mathbf{U}[n] = (U_1, \ldots, U_M)^T$. According to this reward function, when a secondary user successfully accesses a primary channel that is actually in idle state during a given time slot, it receives a reward proportional to the bandwidth of that channel if the SSDC in fact detected it to be idle. On the other hand, if a secondary user mistakenly accesses a primary channel that is in busy state, it causes a collision with the primary transmission and thus receives zero reward. The accumulated total reward of all secondary users can then be used as a measure of the secondary system throughput over the M primary channels. A reasonable objective for the SSDC can be to maximize this accumulated total throughput of the secondary system.

It is important to observe that the detection decision $U_m[n]$ at the SSDC in fact depends on the action matrix $\mathbf{A}_n$. Indeed, this is what makes the problem interesting as well as challenging. As mentioned earlier, assigning more nodes to sense channel m, the SSDC may be able to improve the reliability of the decision $U_m[n]$. However, this will come at the price of having to reduce the number of channels sensed so that $|\mathcal{C}_n|$ is decreased. Although we have not shown in the notation $U_m[n]$, it is important to remember that it depends on $\mathbf{A}_n$ or in turn on $\mathcal{A}_n^{(m)}$.

Since SSDC does not have access to the true spectrum state $\mathbf{S}[n]$, the immediate reward (13.32) cannot be evaluated. Instead, given the belief state $\mathbf{b}[n]$ on the state of the Markov system, it may make decisions based on the expected immediate reward defined as

$$
\begin{aligned}
r(\mathbf{b}[n], \mathbf{A}_n) &= \mathbb{E}\{R(S[n], \mathbf{A}_n, \mathbf{U}[n]) | \mathbf{b}[n]\} \\
&- \sum_{m \in \mathcal{C}_n} B_m \mathbb{E}\{S_m[n] U_m[n]\} \\
&= \sum_{m \in \mathcal{C}_n} \sum_{j=1}^{2^M} B_m s_{j,m} \mathbb{E}\left\{U_m[n] | S_m[n] = s_{j,m}\right\} \Pr\left(\mathbf{S}[n] = \mathbf{s}_j | \mathbf{b}[n]\right) \\
&= \sum_{m \in \mathcal{C}_n} \sum_{j=1}^{2^M} B_m s_{j,m} P_D^{(m), \mathcal{A}_n^{(m)}} \Pr\left(\mathbf{S}[n] = \mathbf{s}_j | \mathbf{b}[n]\right) \\
&= \sum_{m \in \mathcal{C}_n} \sum_{j=1}^{2^M} r_j(m) b_j[n] \\
&= \sum_{m \in \mathcal{C}_n} \mathbf{r}(m)^T \mathbf{b}[n],
\end{aligned} \tag{13.33}
$$

where $\mathbf{s}_j = (s_{j,1}, \ldots, s_{j,M})$ for $j = 1, \ldots, 2^M$, and we have denoted

$$
\mathbf{r}(m) = \left(r_1(m), \ldots, r_{2^M}(m)\right)^T, \tag{13.34}
$$

with

$$r_j(m) = B_m s_{j,m} P_D^{(m),\mathcal{A}_n^{(m)}}. \tag{13.35}$$

In the above, $P_D^{(m),\mathcal{A}_n^{(m)}}$ denotes the detection probability of idle states of channel m at the SSDC when the set of cognitive radios $\mathcal{A}_n^{(m)}$ is assigned to sense channel m. Recall from Chapter 12 that this detection probability is given by (12.63) when the communication from nodes to the SSDC is error-free.

At each time instant n, the decision rule at the SSDC is a mapping of the belief-state vector to a sensing decision matrix denoted as $\pi_n(\mathbf{b}[n]) = \mathbf{A}_n$, where $\mathbf{A}_n$ is an $M \times K_s$ matrix of binary elements satisfying the constraints

$$\sum_{m=1}^{M} [\mathbf{A}_n]_{m,i} = 1 \quad \text{for all } i, \tag{13.36}$$

where (13.36) is due to the constraint that each secondary user can only sense a single primary channel at a time. Note that the set of actions $\mathcal{A}$ available to the SSDC is the set of $M \times K_s$ binary matrices satisfying the constraint (13.36). Each such matrix represents a unique partitioning of the set $\{1, \ldots, K_s\}$ into at most M subsets subjected to (13.36).

Thus, we may cast the sensing decision-making problem at the SSDC as a POMDP optimization problem in which the objective is to maximize the following value function associated with the expected discounted total return:

$$V_T^{\underline{\pi}}(\mathbf{b}) = \mathbb{E}_{\underline{\pi}}\left\{\sum_{n=0}^{T} \gamma^n r(\mathbf{b}[n], \pi_n(\mathbf{b}[n])) \mid \mathbf{b}[0] = \mathbf{b}\right\}, \tag{13.37}$$

where $\underline{\pi} = (\pi_0(.), \ldots, \pi_{T-1}(.))$.

Hence, the decision-making problem at the SSDC is to determine an optimal policy π^* such that

$$\underline{\pi}^* = \arg\max_{\underline{\pi}} V_T^{\underline{\pi}}(\mathbf{b}). \tag{13.38}$$

In theory, finding an optimal policy (13.38) can be done following the same methods discussed in Section 7.5.4. We have already seen in Chapter 7 that optimal POMDP solution policies are prohibitively computationally demanding even in the case of autonomous nodes. The joint decision-making problem (13.38) at the SSDC is even more complicated since now the number of actions available at each time instant can be exponential in both M and K_s. Hence, in the following, let us consider independent channel dynamics.

13.4.1 Optimal SSDC Decisions for Independent Channel Dynamics

Observe from Figure 12.6 that at time $n = n^-$, the SSDC computes its sensing decision matrix $\mathbf{A}_n$ for time n and informs all nodes. In particular, the set of nodes in $\mathcal{A}_n^{(m)}$ will be informed that they are required to sense the primary channel m during the subsequent time slot. At the end of the sensing interval, the SSDC receives local decisions (reports) on the mth primary channel status. Recall from Chapter 12 that these reports, $u_{m,i}[n]$, are binary valued, where $u_{m,i}[n]$ is the report from the ith secondary cognitive radio about the mth channel. The complete observation corresponding to the channel m is the variable-length vector $\mathbf{z}^{(m)}[n]$ that we previously defined in Section 12.6.2. Hence, the observation process $Y[n]$ associated with the POMDP at the SSDC is made of these local report vectors $\{\mathbf{z}^{(m)}[n]\}_{m\in\mathcal{A}_n}$.

For simplicity of discussion, in the following, let us assume error-free communications from secondary cognitive radios to the SSDC as in Section 12.6.2.1. Hence, the elements of $\mathbf{z}^{(m)}[n]$ are directly the local decisions $u_{m,i}[n]$'s for $i \in \mathcal{A}_n^{(m)}$ as given in (12.52). Thus, we may perhaps make the notation easier by directly letting the vector $\mathbf{u}^{(m)}[n]$ denotes the observation at the SSDC at time n where $\mathbf{u}^{(m)}[n]$ is the variable-length vector with its elements being those $u_{m,i}[n]$'s for $i \in \mathcal{A}_n^{(m)}$. Thus, the complete history available at time n regarding channel m, for $1 \leq m \leq M$, is

$$\underline{h}_{m,n} = \left(\underline{h}_{m,n-1}, \mathbf{A}_{n-1}, \mathbf{u}^{(m)}[n-1]\right), \tag{13.39}$$

with $\underline{h}_{m,0} = \phi$ for all m.

Example 13.4 (An Alternative Observation Process for Sensing Decision-Making at the SSDC) Instead of considering the observation process associated with the POMDP at the SSDC to be the collection of variable-length binary vectors $\{\mathbf{u}^{(m)}[n]\}_{m\in\mathcal{A}_n}$, we could have chosen it to be the collection of final fused decisions at the SSDC for each channel $m \in \mathcal{A}_n$ given by either (12.57) (for error-free communications from secondary radios to the SSDC) or by (12.72) (noisy communications from secondary radios to the SSDC). In the case of error-free communications from distributed cognitive radios to the SSDC, this collection of final fused decisions is denoted by the vector $\mathbf{U}[n]$ as in (13.32).

The overall cooperative sensing decision framework in this section will still be applicable with appropriate modifications. ■

If we assume independent channel dynamics, then we may form an equivalent POMDP based on a reduced M-dimensional belief-state vector $\boldsymbol{\omega}[n] = [\omega_1[n], \ldots, \omega_M[n]]^T$, where

$$\omega_m[n] = \Pr\left(S_m[n] = 1 \mid \underline{h}_{m,n}\right), \tag{13.40}$$

where, as earlier, we have arbitrarily selected the probability of channel m being idle in defining the belief state because the state is assumed to be binary valued. Hence, $\Pr\left(S_m[n]=0 \mid \underline{h}_{m,n}\right) = 1-\omega_m[n]$.

The belief vector $\boldsymbol{\omega}[n]$ can easily be computed recursively. Due to the independence of channel dynamics, each component $\omega_m[n]$ of $\boldsymbol{\omega}[n]$ can be computed in its own recursion as in (9.49b). Indeed, while we could have used the complete history $\underline{h}_n$ in (13.40), the reason for denoting the history relevant for channel m as $\underline{h}_{m,n}$ was to emphasize that the belief states of each channel can be computed independent of each other. Hence, without loss of generality, we may focus on an arbitrary primary channel m, for $1 \leq m \leq M$, for the purpose of belief-state computation:

$$
\begin{aligned}
\omega_m[n+1] &= \Pr\left(S_m[n+1]=1 \;|\underline{h}_{m,n+1}\right)\\
&= \Pr\left(S_m[n+1]=1|S_m[n]=1,\underline{h}_{n+1}\right)\Pr\left(S_m[n]=1|\underline{h}_{n+1}\right)\\
&\quad +\Pr\left(S_m[n+1]=1|S_m[n]=0,\underline{h}_{n+1}\right)\Pr\left(S_m[n]=0|\underline{h}_{n+1}\right)\\
&= p_{11}^{(m)}\Pr\left(S_m[n]=1\left|\underline{h}_{m,n},\mathbf{A}_n,\mathbf{u}^{(m)}[n]\right.\right)+p_{01}^{(m)}\Pr\left(S_m[n]=0\left|\underline{h}_{m,n},\mathbf{A}_n,\mathbf{u}^{(m)}[n]\right.\right)\\
&= p_{11}^{(m)}\Pr\left(S_m[n]=1\left|\underline{h}_{m,n},\mathcal{A}_n^{(m)},\mathbf{u}^{(m)}[n]\right.\right)+p_{01}^{(m)}\Pr\left(S_m[n]=0\left|\underline{h}_{m,n},\mathcal{A}_n^{(m)},\mathbf{u}^{(m)}[n]\right.\right)\\
&= \begin{cases}
p_{11}^{(m)}\Pr\left(S_m[n]=1\left|\underline{h}_{m,n},\mathcal{A}_n^{(m)},\mathbf{u}^{(m)}[n]\right.\right) & \\
\qquad +p_{01}^{(m)}\Pr\left(S_m[n]=0\left|\underline{h}_{m,n},\mathcal{A}_n^{(m)},\mathbf{u}^{(m)}[n]\right.\right) & \text{if } \mathcal{A}_n^{(m)} \neq \phi\\
p_{11}^{(m)}\Pr\left(S_m[n]=1\left|\underline{h}_{m,n}\right.\right)+p_{01}^{(m)}\Pr\left(S_m[n]=0\left|\underline{h}_{m,n}\right.\right) & \text{if } \mathcal{A}_n^{(m)} = \phi
\end{cases}\\
&= \begin{cases}
p_{11}^{(m)}\Pr\left(S_m[n]=1\left|\underline{h}_{m,n},\mathcal{A}_n^{(m)},\mathbf{u}^{(m)}[n]\right.\right) & \\
\qquad +p_{01}^{(m)}\Pr\left(S_m[n]=0\left|\underline{h}_{m,n},\mathcal{A}_n^{(m)},\mathbf{u}^{(m)}[n]\right.\right) & \text{if } \mathcal{A}_n^{(m)} \neq \phi\,,\\
p_{11}^{(m)}\omega_m[n]+p_{01}^{(m)}(1-\omega_m[n]) & \text{if } \mathcal{A}_n^{(m)} = \phi
\end{cases}
\end{aligned}
\tag{13.41}
$$

where (13.41) follows from (13.40). Note that $\mathcal{A}_n^{(m)} = \phi$ implies that the mth primary channel was not sensed by any secondary cognitive radio at time n. Hence, belief state corresponding to the mth channel is updated by simply propagating the previous belief through the Markov state transition model probabilities. On the other hand,

$$
\Pr\left(S_m[n]=1\left|\underline{h}_{m,n},\mathcal{A}_n^{(m)},\mathbf{u}^{(m)}[n]\right.\right)
= \frac{\Pr\left(\mathbf{u}^{(m)}[n]|S_m[n]=1,\underline{h}_{m,n},\mathcal{A}_n^{(m)}\right)\Pr\left(S_m[n]=1,\underline{h}_{m,n},\mathcal{A}_n^{(m)}\right)}{\Pr(\mathbf{u}^{(m)}[n]|\underline{h}_{m,n},\mathcal{A}_n^{(m)})\Pr(\underline{h}_{m,n},\mathcal{A}_n^{(m)})}
$$

$$= \frac{\Pr\left(\mathbf{u}^{(m)}[n]|S_m[n]=1,\underline{h}_{m,n},\mathcal{A}_n^{(m)}\right)\Pr\left(S_m[n]=1|\underline{h}_{m,n},\mathcal{A}_n^{(m)}\right)}{\Pr(\mathbf{u}^{(m)}[n]|\underline{h}_{m,n},\mathcal{A}_n^{(m)})}$$

$$= \frac{\Pr\left(\mathbf{u}^{(m)}[n]|S_m[n]=1,\underline{h}_{m,n},\mathcal{A}_n^{(m)}\right)\Pr\left(S_m[n]=1|\underline{h}_{m,n}\right)}{\Pr(\mathbf{u}^{(m)}[n]|\underline{h}_{m,n},\mathcal{A}_n^{(m)})}$$

$$= \frac{\omega_m[n]\left[\prod_{i\in\mathcal{A}_n^{(m)}}\Pr\left(u^{(m,i)}[n]|S_m[n]=1\right)\right]}{\sum_{j=0}^{1}\Pr(\mathbf{u}^{(m)}[n]|S_m[n]=j,\underline{h}_{m,n},\mathcal{A}_n^{(m)})\Pr\left(S_m[n]=j|\underline{h}_{m,n}\right)}, \tag{13.42}$$

where (13.42) follows by assuming that local observations, and thus local decisions, are independent of each other conditioned on the true channel states. Similarly,

$$\Pr\left(S_m[n]=0\,\middle|\,\underline{h}_{m,n},\mathcal{A}_n^{(m)},\mathbf{u}^{(m)}[n]\right)$$

$$= \frac{(1-\omega_m[n])\left[\prod_{i\in\mathcal{A}_n^{(m)}}\Pr\left(u^{(m,i)}[n]|S_m[n]=0\right)\right]}{\sum_{j=0}^{1}\Pr(\mathbf{u}^{(m)}[n]|S_m[n]=j,\underline{h}_{m,n},\mathcal{A}_n^{(m)})\Pr\left(S_m[n]=j|\underline{h}_{m,n}\right)}. \tag{13.43}$$

In both (13.42) and (13.43), $\Pr\left(u^{(m,i)}[n]|S_m[n]=j\right)$, for $j=0,1$ are simply given by the crossover probabilities of the equivalent binary asymmetric channel characterizing the ith secondary radio's decisions on the state of channel m in Figure 12.9. Indeed,

$$\Pr\left(u^{(m,i)}[n]|S_m[n]=0\right) = \begin{cases} 1-\lambda_{m,i}^0[n] & \text{if} \quad u_{m,i}[n]=0 \\ \lambda_{m,i}^0[n] & \text{if} \quad u_{m,i}[n]=1 \end{cases} \tag{13.44}$$

and

$$\Pr\left(u^{(m,i)}[n]|S_m[n]=1\right) = \begin{cases} \lambda_{m,i}^1[n] & \text{if} \quad u_{m,i}[n]=0 \\ 1-\lambda_{m,i}^1[n] & \text{if} \quad u_{m,i}[n]=1 \end{cases}, \tag{13.45}$$

where $\lambda_{m,i}^0$ and $\lambda_{m,i}^1$ are given, respectively, by either (12.39b) and (12.40) when primary signals are modeled as random at the distributed cognitive radios or (12.42) and (12.43) when primary signals are modeled as being deterministically known to the distributed cognitive radios.

By substituting (13.42)–(13.45) in (13.41), we have that

$$\omega_m[n+1]$$
$$= \begin{cases} p_{11}^{(m)}\omega_m[n] + p_{01}^{(m)}(1-\omega_m[n]) & \text{if} \quad \mathcal{A}_n^{(m)} = \phi \\ \left(\begin{array}{l} p_{11}^{(m)} \dfrac{\left[\prod_{i\in\mathcal{A}_n^{(m)}} \Pr\left(u^{(m,i)}[n]|S_m[n]=1\right)\right]\omega_m[n]}{\sum_{j=0}^{1}\Pr(\mathbf{u}^{(m)}[n]|S_m[n]=j,\underline{h}_{m,n},\mathcal{A}_n^{(m)})\Pr\left(S_m[n]=j|\underline{h}_{m,n}\right)} \\ +p_{01}^{(m)} \dfrac{\left[\prod_{i\in\mathcal{A}_n^{(m)}} \Pr\left(u^{(m,i)}[n]|S_m[n]=0\right)\right](1-\omega_m[n])}{\sum_{j=0}^{1}\Pr(\mathbf{u}^{(m)}[n]|S_m[n]=j,\underline{h}_{m,n},\mathcal{A}_n^{(m)})\Pr\left(S_m[n]=j|\underline{h}_{m,n}\right)} \end{array} \right) & \text{if} \quad \mathcal{A}_n^{(m)} \neq \phi \end{cases} \tag{13.46}$$
$$= T_m^{(\omega)}(\omega_m[n], \mathbf{A}_n, \mathbf{u}^{(m)}[n]),$$

where we have denoted by $T_m^{(\omega)}(\omega_m[n], \mathbf{A}_n, \mathbf{u}^{(m)}[n])$ the mapping from $\omega_m[n]$ to $\omega_m[n+1]$.

With the reduced-dimensional belief-state vector $\boldsymbol{\omega}[n]$, the expected immediate reward (13.33) can be simplified as

$$r(\boldsymbol{\omega}[n], \mathbf{A}_n) = \sum_{m\in\mathcal{C}_n} r(m)\omega_m[n] \tag{13.47}$$

where from (13.32)

$$r(m) = B_m P_D^{(m,\mathcal{A}_n^{(m)})}, \tag{13.48}$$

with $P_D^{(m,\mathcal{A}_n^{(m)})}$ being the detection probability of channel m idle opportunities at the SSDC when the set of nodes that is assigned to sense channel m is the $\mathcal{A}_n^{(m)}$. An expression for evaluating this detection probability was given in (12.63).

With these expected immediate rewards, then the sensing decision-making problem at the SSDC can be cast as a POMDP optimization problem in which the objective is to maximize the achieved throughput over time. At each time instant, the action to be taken by the SSDC is $\pi_n(\boldsymbol{\omega}[n]) = \mathbf{A}_n$, where $\mathbf{A}_n$ is an $M \times K_s$ matrix of binary elements satisfying the constraint $\sum_{m=1}^{M}[\mathbf{A}_n]_{m,i} = 1$. A policy $\underline{\pi}$ for the finite-horizon sensing decision-making at the SSDC is then a sequence of decision rules $\underline{\pi} = (\pi_0(.), \ldots, \pi_{T-1}(.))$. The decision-making problem at the SSDC is now to determine an optimal policy $\underline{\pi}^*$ such that

$$\underline{\pi}^* = \arg\max_{\underline{\pi}} V^{\underline{\pi}}(\boldsymbol{\omega}), \tag{13.49}$$

where the expected discounted reward under the policy $\underline{\pi}$ when the initial state is $\boldsymbol{\omega}[0] = \boldsymbol{\omega}$ is defined as

$$V_T^{\underline{\pi}}(\boldsymbol{\omega}) = \mathbb{E}_{\underline{\pi}}\left\{\sum_{n=0}^{T} \gamma^n r(\boldsymbol{\omega}[n], \pi_n(.)) \mid \boldsymbol{\omega}[0] = \boldsymbol{\omega}\right\}. \tag{13.50}$$

When $T \longrightarrow \infty$, of course, we have an infinite-horizon problem.

The optimality equations for this decision-making problem can be written down straightforwardly from Theorem 7.5 (for finite T) or Theorem 7.9 (for $T \longrightarrow \infty$) of Chapter 7. Then optimal policy evaluation algorithms discussed in Section 7.5.4 can be used to obtain an optimal policy.

13.4.2 Optimal Myopic Sensing Decisions at the SSDC with Independent Channel Dynamics

As we already know, optimal policy algorithms can be computationally demanding. In the case of cooperative sensing decisions at the SSDC, even when channel dynamics are independent, the problem can still be demanding enough for a single time instant as we see in the following. Hence, unless the parameter space of the problem is very small, myopic decision rules can be particularly attractive in this case. In addition, optimal policy formulation in the previous section assumes that the channel-state dynamics are time invariant so that the transition probabilities are stationary and known to the SSDC. In practice, this may not also hold justifying the preference for optimal myopic policies even further.

An optimal channel sensing decision within the class of myopic policies (greedy policies) that only considers the instantaneous reward to be expected at the current moment can be obtained by maximizing the total secondary reward over all primary channels at each time step, that is, making the channel sensing decisions to obtain the highest instantaneous reward, rather than attempting to optimize the average (discounted) reward accrued over all times. This optimal myopic sensing decision $\mathbf{A}_n^*$ can be expressed as

$$\begin{aligned}\mathbf{A}_n^* &= \underset{\sum_{m=1}^{M}[\mathbf{A}_n]_{m,i}=1\ \forall i}{\arg\max} \quad r(\boldsymbol{\omega}[n], \mathbf{A}_n) \\ &= \underset{\sum_{m=1}^{M}[\mathbf{A}_n]_{m,i}=1\ \forall i}{\arg\max} \quad \sum_{m\in\mathcal{C}_n} r(m)\omega_m[n],\end{aligned} \tag{13.51}$$

where the constraint implies that each secondary user is assigned to sense one, and only one, primary channel and the expected immediate reward $r(m)$ is given, for example, by (13.48):

$$\mathbf{A}_n^* = \underset{\sum_{m=1}^{M}[\mathbf{A}_n]_{m,i}=1\ \forall i}{\arg\max} \quad \sum_{m\in\mathcal{C}_n} B_m P_D^{(m,\mathcal{A}_n^{(m)})}\omega_m[n]. \tag{13.52}$$

Myopic decision-making problem (13.52) is a constrained nonlinear integer optimization problem in which the objective is to assign each cognitive radio i, for $i \in \{1, \ldots, K_s\}$, to one and only one primary channel m, for $m \in \{1, \ldots, M\}$, so that the total reward $\sum_{m \in C_n} r(m)\omega_m[n]$ is maximized. Without further simplifying assumptions, there does not seem to be any efficient approach to solve it than standard techniques used for solving constrained nonlinear integer optimization problems. With certain simplifying assumptions, one may be able to transform the problem (13.52) into forms that indeed have efficient solution algorithms as have discussed in [133]. However, these modifications only provide suboptimal myopic policies.

Example 13.5 (Performance of Optimal Myopic Channel Sensing When $K_s = 1$ and $M = 2$) It is worth investigating the performance gain of making channel sensing decisions as in the preceding text. For example, compared to assigning all secondary users to sense a single channel (so that the idle/busy decisions are more reliable), do we gain anything by optimally (in the sense of maximizing (13.52)) allocating users to sense different channels? Indeed, assigning all secondary users to sense a single channel was considered in [203]. Alternatively, what if we simply assign the secondary cognitive radios as uniformly as possible across all available primary channels? These two approaches represent two extremes in sensing decision-making at the SSDC.

To gain insight into effects of these choices, let us consider a situation in which all primary channels have independent and identical Markov dynamics with a channel transition probability matrix of $\mathbf{P}^{(m)} = \mathbf{P}$, for all m, where

$$\mathbf{P}^{(m)} = \begin{bmatrix} 0.8 & 0.2 \\ 0.1 & 0.9 \end{bmatrix} \triangleq \mathbf{P}. \tag{13.53}$$

The corresponding stationary distribution of the busy and idle states is given by

$$\mathbf{p}^{(m)} = \left[p_0^{(m)}, p_1^{(m)}\right] = [0.3333, 0.6667] \triangleq \mathbf{p}. \tag{13.54}$$

Figure 13.1 shows the accumulated total return based on the reward function (13.32) in a simple system made of a single secondary user and two primary channels: $K_s = 1$ and $M = 2$. Admittedly, assuming a single cognitive radio $K_s = 1$ is not very useful in a discussion regarding cooperative decision-making. However, the primary goal of this example is indeed to show that what is best for a single cognitive radio can perform significantly poorly in a cooperative decision-making system as we will see later in Example 13.7.

We have normalized all primary channels to have bandwidth $B_m = B = 1$. The local binary decisions are made using either energy or matched filter detectors depending on the assumptions on the primary signal. These binary decisions are forwarded to an SSDC that makes the sensing decisions. In Figure 13.1, primary SNR denotes the average received SNR at the local secondary user in sensing the primary channel.

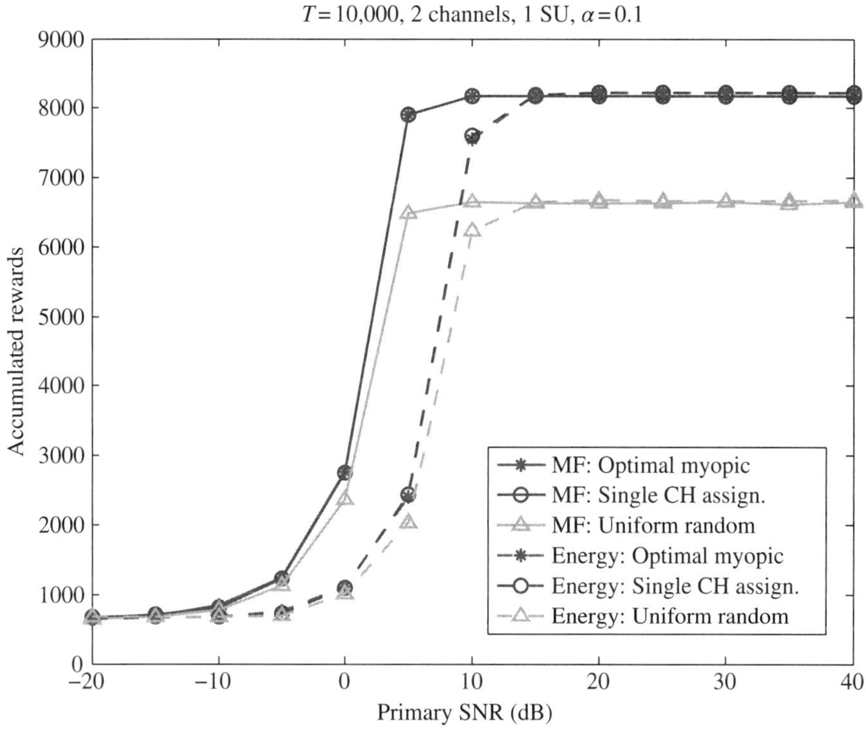

FIGURE 13.1 Accumulated reward in a centralized DSA system made of $K_s = 1$, a single secondary user, and $M = 2$, primary channels. Binary local decisions are forwarded to an SSDC for sensing decision-making.

In this setup, we have assumed there is no channel fading so that in the case of a primary signal that is deterministically known, the primary SNR is defined as

$$\text{Primary SNR} = \frac{||x||^2}{\sigma_w^2}$$

and the local detector is the matched filter, whereas in the case of a primary signal modeled as a random signal, the primary SNR is defined as

$$\text{Primary SNR} = \frac{\sigma_x^2}{\sigma_w^2}$$

and the local detector is the energy detector. In either case, the local detectors are Bayesian minimum error probability detectors. The final decision fusion at the SSDC is based on a Neyman–Pearson detector that meets the collision probability constraint with the primary users. Figure 13.1 corresponds to a collision probability of $\alpha = 0.1$.

In the case of Figure 13.1 with $K_s = 1$ and $M = 2$, the sensing decision is simply about which primary channel the only secondary user in the system should sense at any given time. In Figure 13.1, we have included three decision-making strategies: optimal myopic of (13.52), assigning all secondary users to sense a single channel (called the single channel assignment), and the uniformly random channel selection. Note that since there is only a single cognitive radio, optimal myopic scheme of (13.52) and the single channel assignment are equivalent: Both schemes assign the single available secondary user to sense the myopically best primary channel. The accumulated reward shown in Figure 13.1 indeed shows that.

As we would have expected, the performance at the SSDC improves with the increasing received primary SNR at the local secondary user regardless of whether the local detector is a matched filter or an energy detector, while in general the matched filter leading to better SSDC fusion decisions. However, when SNR is sufficiently large, local decisions from both detectors will mostly be correct. Thus, there may not be much difference in what is received at the SSDC. As a result, the final performance may not be different regardless of what local detector is used when the SNR is really large. This is also evident from Figure 13.1.

More importantly, Figure 13.1 justifies the computational burden of making optimal myopic (or, in this case, equivalently single channel assignment) decisions compared to simply picking one of the primary channels uniformly randomly. Indeed, the uniformly random channel sensing is always inferior to the optimal myopic except when SNR is really low so that both methods perform pretty badly.

When SNR is sufficiently large, however, there is a clear difference between their performance. The uniformly random sensing converges to an accumulated reward that is lower than the asymptotic accumulated reward achieved by the optimal myopic sensing. Indeed, the uniform random essentially converges, to what is predicted by the stationary-state distribution (13.54) of the channel Markov models. Since the two primary channels are identical and they both have a stationary-state distribution with idle-state probability of 0.6667, when SNR is large enough, the reward will be essentially to capture about 66.67% of the times that are idle. Note that the high SNR is needed at the secondary radio to achieve this because the reward is gained only if the radio (and subsequently the SSDC) can actually determine an idle channel as indeed idle. On the other hand, as Figure 13.1 shows, the optimal myopic sensing decisions can do better than this random guessing. By predicting which channel may have the better chance of being idle (using the belief-state information as in (13.52)) and sensing that particular channel, the optimal myopic scheme is able to capture a combination of idle opportunities from both channels that is larger than simply half of all idle opportunities across the two channels as with uniformly random sensing. ■

While accumulated total return is a meaningful criterion in making optimal decisions, a performance measure that is important in DSS systems is the efficiency of *primary channel utilization* by the cognitive radios. Essentially, this is the fraction (over time) of idle primary channels that the cognitive radio system actually detects and utilizes. Let us denote by $\hat{S}_m[n] \in \{0, 1\}$ the decision on the state of the mth primary channel according to any sensing policy with the understanding that $\hat{S}_m[n]=0$ for any channel m that was not sensed at time n as per the sensing policy. Hence, $\hat{S}_m[n]=1$ if and only if channel m was sensed at time n, and based on the sensing outcomes, it was determined to be idle by the SSDC using decision fusion as discussed in Chapter 12. The efficiency in primary channel utilization is then the *percentage of primary channel usage* U defined as

$$U = \frac{\sum_{m=1}^{M} \sum_{n=0}^{T} \hat{S}_m[n] S_m[n]}{\sum_{m=1}^{M} \sum_{n=0}^{T} S_m[n]}. \tag{13.55}$$

In the next example, we show the primary channel utilization measure corresponding to the same system that was considered in Example 13.5.

Example 13.6 (Efficiency in Primary Channel Utilization of Optimal Myopic Channel Sensing When $K_s = 1$ and $M = 2$) In this example, we assume the exact same system considered in Example 13.5. Figure 13.2 shows the efficiency of primary channel utilization U corresponding to the same system under the three sensing policies of interest for two different threshold values for collision probability with primary users.

Figure 13.2(a) shows the efficiency of primary channel utilization U corresponding to the accumulated reward shown in Figure 13.1 of Example 13.5 in which the collision probability threshold was assumed to be $\alpha = 0.1$. Figure 13.2(a) indeed verifies the fact that uniformly random sensing only uses exactly about half of the total available idle channel opportunities even when the local SNR is sufficiently high enough. When SNR is high, it accurately determines whatever the sensed channel as being idle or busy. The inefficiency is due to randomly picking a channel to sense. This causes it to on average choose either channel exactly about half the times when each channel is indeed idle, hence the about 50% utilization. On the other hand, optimal myopic and single channel assignment (which are equivalent since there is only a single cognitive radio in the system) can do better than the uniform random by always picking the myopically better channel. Hence, whenever these sensing policies sense either channel, that channel is more likely to be indeed idle, hence the better utilization than that with the uniform random.

Figure 13.2(b) shows the efficiency of primary channel utilization of exactly the same system when the collision probability constraint is more stringent at $\alpha = 0.01$. The overall behavior stays the same. ■

As we already mentioned, the case of $K_s = 1$ cognitive radio assumed in Examples 13.5 and 13.6, however, is not representative of a cooperative decision-making system. As we will see in the next example, although single channel assignment is the same as optimal myopic scheme (13.52) and outperforms the uniformly random sensing, it is actually the worst choice out of the three schemes when we have a cooperative decision-making system with more than one distributed cognitive radio.

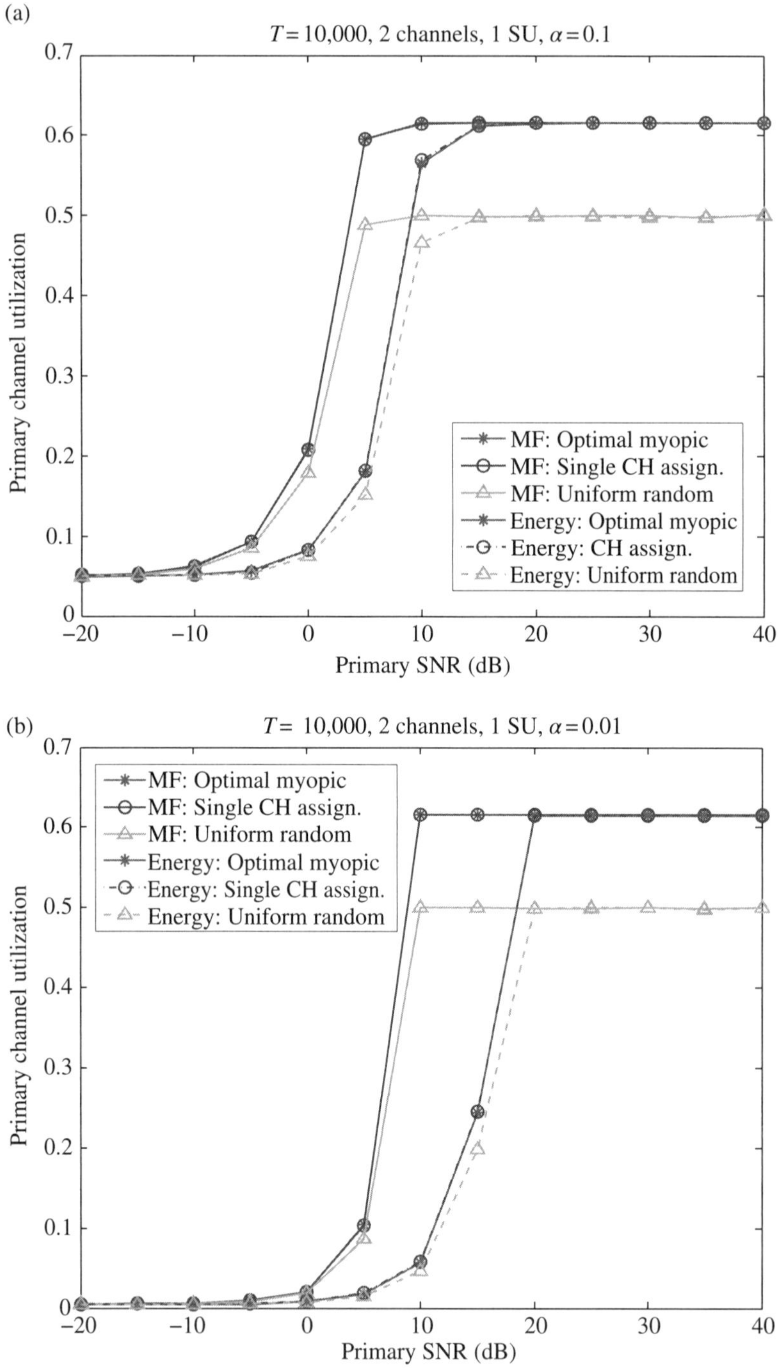

FIGURE 13.2 Efficiency of primary channel utilization of a sensing policy is the percentage of idle primary channels that were deemed accessible by the sensing policy. A centralized DSA system made of $K_s = 1$, a single secondary user, and $M = 2$, primary channels. (a) Efficiency of primary channel utilization when $\alpha = 0.1$. (b) Efficiency of primary channel utilization when $\alpha = 0.01$.

Example 13.7 (Inferior Performance of Single Channel Sensing Compared to Optimal Myopic Channel Sensing Decisions When $K_s = 2$ and $M = 2$) In this example, we consider exactly the same scenario as in the previous example with the only exception that the secondary system now has $K_s = 2$ secondary users. Figure 13.3 shows the efficiency of primary channel utilization achieved by the three sensing policies in this case.

First, let us consider the high SNR region of Figure 13.3. The policy of single channel sensing is clearly the worst option in this case. The reason is obvious: When SNR is high enough so that even a single node measurement of a primary channel is good enough to make an accurate decision on channel state, it is a waste to assign both cognitive radios to sense the same channel. In doing so, we constrain ourselves to be able to use only one primary channel at a time (since the SSDC will not assign a cognitive radio to access a primary channel unless it was sensed and verified to be idle.). Indeed, when SNR is high, the utilization achieved by the single channel assignment policy is independent of the number of secondary users in the cognitive radio system.

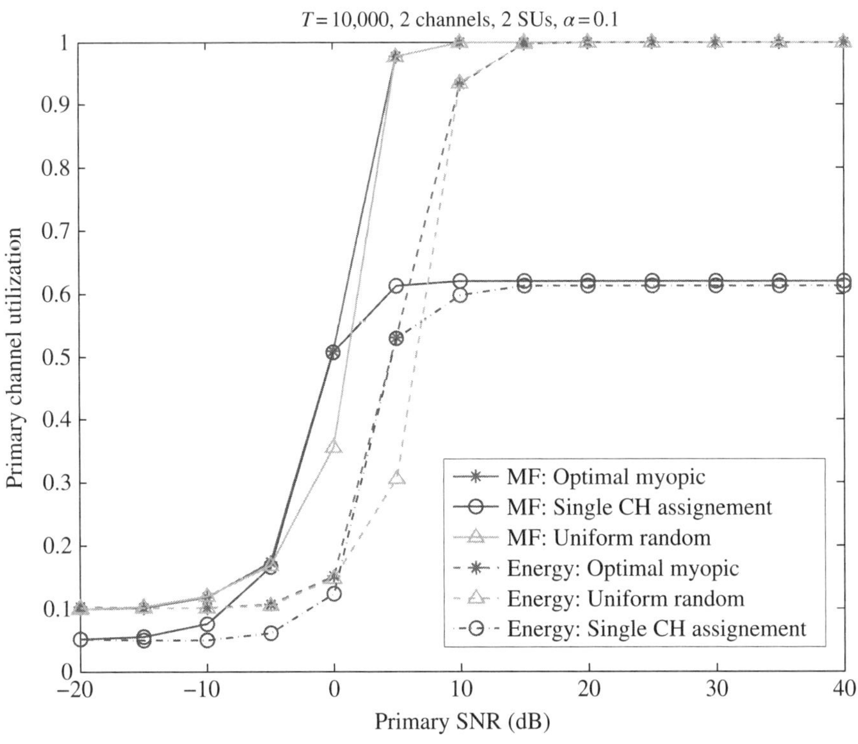

FIGURE 13.3 Efficiency of primary channel utilization of a sensing policy is the percentage of idle primary channels that were deemed accessible by the sensing policy. A centralized DSA system made of $K_S = 2$, a single secondary user, and $M = 2$, primary channels. The collision probability constraint is $\alpha = 0.1$.

Next important observation from the high SNR region of Figure 13.3 is that the uniformly random sensing does as well as the optimal myopic. This is again not surprising. When received SNR is sufficiently high, just a single user sensing a primary channel is enough to make a reliable decision on the state of that channel. Since uniformly random policy always senses one channel each in this case, it will never miss an actual idle opportunity on either channel, as long as SNR is high enough. Hence, it does achieve 100% utilization as confirmed by Figure 13.3. The optimal myopic decision-making scheme (13.52) also arrives at 100% utilization, but from a different direction: It may not always sense both channels as the uniformly random sensing. It may indeed assign both users to sense the same channel if it believes that channel has a high chance of being idle while it believes that the other channel has a very high chance of being busy. When SNR is high enough, these beliefs will mostly be accurate so that it will not miss any idle channel although it senses just the idle channel, while uniformly random scheme senses both channels. However, whenever these beliefs are not really clear-cut, the optimal myopic policy will also sense both channels by assigning one user each just as the uniformly random sensing scheme. The end result is, however, the same. It will also not miss an actual idle opportunity because it carefully makes the decision to sense the channel based on its belief on each channel being idle or not. Hence, uniform random and optimal myopic both lead to full utilization when SNR is large enough.

This observation also allows us to expect the situations in which uniformly random sensing may perform poorly compared to the optimal myopic sensing: One scenario is $K_s < M$ with nonidentical channels. In this case, when SNR is large, optimal myopic scheme will always assign the K_s users to sense the best M channels, whereas uniformly random scheme will only assign the K_s users to any M channels. Hence, it will only capture a $\frac{K_s}{M}$ fraction of average idle opportunities while optimal myopic sensing will capture the sum of the best K_s channels with highest average idle opportunities, when SNR is large enough.

Another scenario in which the uniformly random sensing may be inferior to optimal myopic sensing is when the SNR is not very large. Indeed, perhaps the more interesting region in terms of practical applications is not the high SNR region but the low-to-medium SNR range, say from −10 to 10 dB. When the received SNR at local nodes is in this medium range, the trouble of going through the computations needed for the optimal myopic policy over the simple uniformly random policy is again justified. Figure 13.3 shows that there is a noticeable gap in the primary channel utilizations with these two policies in this SNR region. Moreover, up to a certain SNR threshold, even the single channel sensing policy can be better than uniformly random policy as well. Of course, since the asymptotic utilization of single channel sensing policy is strictly inferior to that of the uniformly random policy, when SNR is larger than this threshold, the single channel sensing policy becomes worse off. This shows that for practical SNR values, one may not be able to rely on either single channel or uniformly random sensing to provide the performance consistently close to that of optimal myopic sensing scheme. ■

An important question is how good is the optimal myopic policy (13.52) compared to the optimal POMDP solutions such as (13.49). In [203], an upper bound for the optimal POMDP solution was derived for the special case when all secondary users are assigned to sense a single channel, which is the single channel assignment policy we have assumed earlier. The formulation of [203] assumes that the secondary user observations are directly sent to the SSDC rather than only local decisions are forwarded as we have assumed earlier. However, this is not crucial and the same approach can be used to obtain a similar upper bound for the case when only quantized local decisions are sent to the SSDC. In this case, the POMDP policy that one seeks is a policy to maximize the total discounted return by assigning all users to sense a particular channel at each time instant. Hence, the decision at each time instant is to pick this single channel that must be sensed by all distributed cognitive radios.

The upper bound of [203] was derived based on the important observation we made in Section 9.6.2.2 that the MDP corresponding to the sensing decision-making problem has a special simplifying structure compared to the general MDP problem: In the sensing decision-making problem, the SSDC decisions do not affect the channel dynamics. The primary user idle/busy characteristics have nothing to do with whether a secondary user senses that channel or not. This means that channel-state transitions are not affected by the decisions made (actions chosen) by the SSDC. This is a very significant simplification compared to the general MDP problem in which the chosen actions can steer the state transitions in certain directions. Indeed, if at each time instant we could observe the exact state of the whole system formed by M channels, then an optimal myopic policy similar to (13.52) but based on the exact state is indeed the optimal policy (13.49) for the spectrum sensing decision-making. There is no reason to look ahead into the future in making decisions, because future states (as shown by (9.30)), and thus future rewards, are completely independent of the current action.

In the sensing decision-making problem, it is the partial observability of the system state that makes us having to worry about the future rewards in making the current decisions. The dependence of current decisions on the future is, however, purely due to the effect of the current decision on the belief state. Since our decisions will indeed be based on the equivalent belief-state POMDP, this does make us to have a POMDP in which the decisions influence the future state as seen from the belief-state update (13.46): If we decide to sense a particular channel (or a set of channels), we only get to update the belief state of that channel based on an observation. The belief states of all other channels that were not sensed at the current time will simply be updated based on the state transition probability matrix. Hence, these belief states may become less reliable compared to those of the channels that were sensed at current time. If we pick the channel that gives the highest instantaneous reward based on this belief state of the system, due to partial observability, in time, our decisions may be erroneous unless we pick the channels in such a way so that all belief states also stay comparatively reliable.

It is that last observation that was used in [203] to derive an upper bound for the performance of the optimal POMDP solution for sensing a single channel each time. Indeed, according to this reasoning, in the case of sensing decision-making problem, a myopic policy based on full state observability should be better than any optimal policy that assumes partial observability. The optimal return with such a myopic policy can analytically be computed easily by assuming certain mild conditions on the state transition probability matrix. In [203], it was then shown that the optimal myopic policy (13.52) based on the belief state can provide performance close to this upper bound (in the case of single channel assignment). This then allows one to conclude that the performance of the optimal myopic policy (13.52) should also be close to that of the optimal POMDP policy (13.49).

The same reasoning can also be used to obtain an upper bound for the optimal POMDP policy that optimally assigns secondary users to the best set of channels (instead of assigning all to one single channel), at each time instant. In this case also, a simple upper bound for the performance of such an optimal policy is the performance of the optimal myopic policy based on full state observability. However, in this case, still, there is the issue that given the exact state, one needs to consider the best partition of users into channels to gain the highest instantaneous reward. In [133], this approach was used to show that the optimal myopic policy can be as close to the upper bound as in the single channel assignment considered in [203] or considerably inferior depending on the state transition probability matrix.

Normally, when state transition probabilities are such that it is highly likely for the states to flip from one to another often, then the optimal myopic policy seems to achieve performance closer to the upper bound. The reason is that in this case, the belief-state vector update purely based on the state transition probabilities is not considerably unreliable than that based on a current observation. Hence, the advantage of knowing the exact state, or making an observation related to the state, is reduced. Moreover, when SNR is sufficiently large, the performance gap also reduces. This is simply because the channel-state estimates are so reliable to start with, we may simply propagate them into future states purely based on the state transition probability matrix, and they will still be as good as belief updates based on an observation related to the state.

13.5 SUMMARY

- For successful dynamic spectrum sharing, cognitive radios operating as secondary users must be able to determine when a primary user channel is idle and when it is not. Since main objective of secondary users is to find idle spectrum opportunities, they need to make accurate decisions on which primary channel, or channels, to sense so that there is a high likelihood of sensing a channel that will in fact be idle. This is called the channel sensing decision-making problem.
- Channel sensing decisions can be made autonomously by individual cognitive radios based only on their individual observations or cooperatively by sharing information with other cognitive radios in the system. A common form of cooperative spectrum sensing decision-making involves having a secondary system decision center (SSDC) that assists all cognitive radios in the secondary system by collecting sensing reports from distributed radios and making

centralized sensing decisions. The SSDC then informs distributed cognitive radios who should sense which primary channel as well as possibly who should access which channel. Such cooperative spectrum sensing has the potential to not only find idle channels for sensing with high accuracy but also to help avoid collisions among secondary cognitive radios.

- Sensing decision-making requires a model for primary channel-idle/channel-busy dynamics. A commonly used model is a first-order Markov chain with two states denoted by busy (state 0) and idle (state 1). In general, these state dynamics on different primary channels may be nonidentical and dependent on each other.
- Since cognitive radios can only observe noisy primary channel states, decision-making problem encountered by them is a partially observable Markov decision process, or a POMDP. Moreover, an individual cognitive radio may not be able to sense all available primary channels due to physical limitations. An extreme end of this is when a cognitive radio can only sense a single primary channel at a time.
- In a secondary user system formed by autonomous cognitive radios, individual cognitive radios will make their sensing decisions independently. Decision-making problem can be transformed into an equivalent belief-state COMDP in which the belief $b_j[n]$ on the state at time n being $\mathbf{s}_j$, for $\mathbf{s}_j \in \mathcal{S}$, given all available evidence up to time n is defined as the a posteriori probability of state $\mathbf{S}[n] = \mathbf{s}_j$ at time n given the complete information history $\underline{h}_n$ available to the cognitive radio up to time n:

$$b_j[n] = \Pr\left(\mathbf{S}[n] = \mathbf{s}_j \mid \underline{h}_n\right),$$

where information history $\underline{h}_n$ available to the cognitive radio at time n is $\underline{h}_n = (\underline{h}_{n-1}, a_{n-1}, y[n-1])$ with $\underline{h}_0 = \phi$. The belief-state vector at time n is the 2^M-length vector $\mathbf{b}[n] = \left(b_1[n], b_2[n], \ldots, b_{2^M}[n]\right)^T$.

- Observation process corresponding to this POMDP can be taken as the binary output from the autonomous cognitive radio's own primary signal detector. The observation process probabilities are then characterized by the false-alarm and miss probabilities of this detector.
- Given a desired reward function, the sensing decision-making can then be formed as an optimal policy determination problem for a POMDP. Standard optimal policy computation algorithms for POMDPs such as modified exhaustive search, one-pass and witness algorithm can be used to obtain an optimal policy provided the problem size is not too large.
- A computationally less demanding alternative is to look for an optimal myopic policy that will sense the primary channel that promises the highest expected immediate reward without regard to its effect on future rewards. Such policies can be considerably easier to determine and offer the additional advantage of being able to easily adapt to time-varying nature of the spectrum.

- Cooperative spectrum sensing decisions made at an SSDC have the potential to improve the spectrum utilization performance compared to those of autonomous cognitive radios. In making sensing decisions, an SSDC needs to make a careful trade-off on how many users should sense each particular channel and altogether how many primary channels to sense. Allocating more users to sense a channel will increase the accuracy of channel-state detection at the expense of not being able to sense more primary channels. Hence, sensing decision-making problem at the SSDC is to determine a partitioning of the set of nodes $\{1, \ldots, K_s\}$ into at most M non-empty sets, so as to maximize a desired reward function.
- The cooperative sensing decision-making problem at the SSDC can also be posed as an optimal policy determination problem for a POMDP. For each time instant n, the policy will specify a decision rule that is a mapping of the belief-state vector at the SSDC to a sensing decision matrix denoted as $\pi_n(\mathbf{b}[n]) = \mathbf{A}_n$, where $\mathbf{A}_n$ is an $M \times K_s$ matrix of binary elements satisfying the constraint $\sum_{m=1}^{M} [\mathbf{A}_n]_{m,i} = 1$ for all $i = 1, \ldots, K_s$. If (m, i)th element of matrix $\mathbf{A}_n$ is 1, the radio i is assigned to sense channel m at time n, while it is not if the (m, i)th element is 0.
- If individual channel-state dynamics were to be independent of each other, the dimensionality of the belief vector can be reduced to be only linear in M rather than being exponential in M. In this case, the SSDC can essentially maintain a set of M belief values, one for each channel, affording considerable computational simplicity. However, such assumptions on independent channel dynamics may have to be justified in practice. In particular, if the primary system traffic is centrally allocated across channels, it is highly likely that channel-idle/channel-busy dynamics may show certain correlations.
- An SSDC may also resort to optimal myopic cooperative channel sensing policies in order to strike a balance between computational complexity and performance. In general, the optimal myopic policy can be posed as a constrained nonlinear integer optimization problem. Although, without additional structure, only possible solution approach seems to be the use of standard techniques for solving constrained nonlinear integer optimization problems, these decisions show superior performance compared to uniformly random channel sensing justifying their computational complexity. Moreover, it has also been shown that at least under certain assumptions on state transition probabilities, the performance of myopic policies can be close to optimal finite- or infinite-horizon POMDP policies.

14

DYNAMIC SPECTRUM LEASING IN COGNITIVE RADIO NETWORKS

14.1 INTRODUCTION

Dynamic spectrum leasing (DSL) refers to the dynamic spectrum sharing (DSS) approach in which primary users proactively facilitate the spectrum access by secondary cognitive radios. In contrast to DSA of the previous two chapters, wherein primary users were assumed to be completely oblivious to the spectrum sharing process, in DSL, there is an explicit incentive for the primary users to allow such secondary access.

Similar to DSA, DSL-based spectrum coexistence can also be achieved in the form of either spectrum underlay or spectrum overlay. In DSL, primary users dynamically adjust the extent to which they are willing to lease their spectrum (see Figure 3.8). In return, they receive a leasing payment, called a reward or a payback, that is proportional to the amount of allowed spectrum access. This reward can be considered as a monetary reward or an in-kind payment. An in-kind payment implies the primary users, or the system, receiving a certain cooperation from the secondary users that will help improve the primary communications performance. In the case of monetary rewards, it can be taken as a metric that can be converted into monetary terms. For our purpose, of course, we do not necessarily need to consider an exact monetary reward function. Instead, a certain technically and economically meaningful metric would be sufficient. In this chapter, we will discuss example DSL implementations based on both types of leasing paybacks. Both of them can be implemented either as spectrum underlay or spectrum interweave. Hence, we will discuss one type of

Signal Processing for Cognitive Radios, First Edition. Sudharman K. Jayaweera.

leasing paybacks in the context of an underlay network and the other in the context of a spectrum overlay network.

In the case of monetary (or direct) leasing payments, we will focus on a spectrum underlay DSL system in which the constraint on the secondary user/system is a maximum threshold on allowed interference to the primary users. This DSL interaction among the primary and secondary users can be modeled in a noncooperative game-theoretic formulation. Game theory has frequently been used in cognitive radio DSS formulations, starting with early DSA approaches [204–206]. However, unlike in a game model for DSA, here, both primary and secondary users are considered as the players of the game [207, 208]. We will look at the general structure of a suitable class of utility (reward) functions for both primary and secondary systems that reflects the demand for spectrum access by the secondary users, their payoffs in terms of a suitable performance measure, and the primary user quality of service (QoS) requirements. The challenge is to ensure that the system will operate at a stable and desirable operating point. In particular, when radios do not exactly know what other radios are doing, it is important that by making changes to their subsequent decisions, they arrive at such a desirable and predictable operating point. We will look at conditions to ensure such behavior within the game-theoretic formulation. Our discussion of spectrum underlay DSL will assume that the primary system is a centralized system, although this assumption can easily be dropped.

As we have already discussed, even if in DSL it is expected that primary system will have to bear responsibility for a certain amount of information sharing between the two systems, it is preferable in practice if this is kept to a minimum. Indeed, in practice, it may be desirable to minimize the awareness the primary system needs to have on the secondary operation. Hence, we will also look at how much interaction the primary system needs with the secondary system to enable this method of DSL. Perhaps surprisingly, information exchange between the two systems can indeed be kept to a few parameters. In the particular approach we discuss in this chapter, it will be seen that primary system only has to periodically broadcast just two parameter values that are readily available to them (or can easily be estimated).

Alternatively, rather than a metric that translates into monetary terms, we may model the primary reward to be the savings on their communications resources: A good example is the power savings enabled by the secondary users. In this chapter, we will discuss a primary–secondary interaction framework to achieve this type of DSL using an *asymmetric cooperative communications* architecture in the combined network (made of both primary and secondary systems) [33]. In effect, secondary users will spend a portion of their transmit power to *asymmetrically* relay the primary signals, to their destinations. The asymmetry results from having secondary users relay primary signals, while primary users transmit only their own signals. In return, however, primary users lease a certain portion of their spectrum resource to the secondary users. This could be interpreted as secondary users using their power as a form of currency to buy the primary bandwidth. Thus, in this formulation, the primary reward is the power saving they achieve due to *cooperative relaying*. In general, of course, one can replace relatively simple cooperative relaying by more sophisticated cooperative communications protocols [209–218].

The asymmetric cooperation-based DSL does not require a centralized primary system. Indeed, each *primary link* (i.e., transmitter and a receiver pair) can be allowed to act autonomously in making decisions on spectrum leasing. Acting as a unit, each transmitter–receiver pair can entertain bids from the secondary system specifying how much power they would be spending for relaying primary signals. This DSL interaction then turns out to be an *auctioning game* in which primary users act as the auctioneers. If the primary system is centralized, a central unit may act as the auctioneer. If each transmitter–receiver pair acts autonomously, then each pair will act as an autonomous auctioneer.

Secondary cognitive radios can also act either as a centralized system or as decentralized autonomous nodes. In the case of a centralized cognitive network model, we may assume, as in Chapters 12 and 13, that there is a secondary system decision center (SSDC) that is responsible for making spectrum leasing decisions for the whole secondary system: The SSDC decides which secondary user should cooperate with which primary user/link. Such decision-making by the secondary system may allow it to better negotiate with the primary users. However, each primary user may accept an offer of cooperation with a secondary user picked by the SSDC only if this would result in at least a certain minimum power saving. If the offer is too low, the corresponding primary user may simply decline the offer and the access to its channel would be denied until the next bidding interval. Each primary user may keep its threshold power level as private information so as to encourage bids as high as possible from the secondary system. Again, we will also discuss how such asymmetric cooperative communication can be realized in practice with very little intersystem information exchange.

While it gives the secondary system more control over its relaying power bids, implementation of centralized decision-making in secondary cognitive radio networks (CRN) may be costly. It requires dedicated (control) channels with enough bandwidth to support reporting of all spectrum sensing (in this case primary leasing offers) measurements at distributed cognitive radios to the SSDC, as well as channel and power allocation decisions from the SSDC back to the distributed radios. Thus, we may also consider a secondary CRN in which distributed cognitive radios make their own spectrum access decisions without centralized control. In such a decentralized secondary network, secondary users may compete with each other to gain access to available primary channels. This leads to what is called a *distributed dynamic spectrum leasing* (D-DSL) architecture [219]. A D-DSL system can support heterogeneous secondary radios who do not belong to the same system and do not have the same capabilities. Although an auctioning-based D-DSL framework can be realized either in spectrum interweave/overlay or underlay architecture, we will discuss it in the context of a spectrum overlay CRN.

14.2 DSL WITH DIRECT REWARDS TO PRIMARY USERS

Let us start by assuming that there is a single primary wireless communications system that owns the exclusive license rights to the spectrum band of interest. In this section, we will assume the primary system to be of a centralized system as far as DSL interaction with the secondary users is concerned, although relaxing this assumption is straightforward. The users in this primary system, however, may not be using its spectrum completely all the time, or they may be able to tolerate a certain amount of additional cochannel interference without compromising required QoS constraints. In both cases, there is an inefficient utilization of radio spectrum. Primary system/users can improve spectrum usage efficiency while earning extra revenue, by allowing cognitive radios from a secondary system (or systems) to access this spectrum band whenever it can tolerate. Of course, to maximize revenue, it makes sense to allow this access to the maximum possible extent.

To keep notation simple, let us assume that the centralized primary system is made of K_p users that communicate with a common receiver. It is not difficult to modify the following discussion to suit other situations. We index the primary transmitters by the index set $\mathcal{K}_p = \{0, 1, \ldots, K_p - 1\}$. Let us denote by K_s the number of secondary links in the secondary system. Note that we consider a pair of secondary transmitters and receivers as a single link. Thus, we will refer to the jth transmitter or the jth receiver to mean, respectively, the transmitter and receiver of the jth secondary link, for $j \in \mathcal{K}_s = \{K_p, K_p + 1, \ldots, K_p + K_s\}$.

Let us denote by h_{pl} the channel gain between the (common) primary receiver and the lth transmitter, be it a primary or a secondary transmitter so that $l \in \mathcal{K}_p \bigcup \mathcal{K}_s$. The channel gain between the kth transmitter (either primary or secondary) and the jth secondary receiver is denoted by h_{jk}. Throughout, we assume *quasistatic fading* in which fading coefficients stay fixed for a certain duration of time after which they change to a new set of values. Note that quasistatic fading model is frequently used in modeling many wireless communications environments. This can also be complemented with a channel estimation and tracking algorithm to cope with slowly time-varying situations. As we will see later, performance of the resulting DSL scheme is reasonably robust against such time-varying fading. We may use p_l to represent transmission power of the lth transmitter for $l \in \mathcal{K}_p \bigcup \mathcal{K}_s$.

An example of a CRN with $K_s = 3$ and $K_p = 1$ is shown in Figure 14.1. Note that, depending on the assumed type of wireless networks, the receivers of each link may or may not be physically distinct. For example, receivers of secondary Links 1 and 2 are different in the system shown in Figure 14.1(a), whereas in Figure 14.1(b), Links 1 and 2 share the same receiver. It is also possible that a particular node may act as both a transmitter and a receiver either on the same link or on two different links. Most of these situations can be handled by careful bookkeeping and notation. For simplicity of exposition, in the following discussion, all secondary links are assumed to belong to the same secondary system.

In spectrum underlay DSL networks, the primary system is assumed to adapt its *interference cap* (IC), denoted by Q_0, which is the maximum interference the primary system is willing to tolerate from all secondary transmissions at a given time. Dynamically adjusting the IC Q_0, the primary system can control the total transmit power secondary users impose on its spectrum band. In essence, the IC determines how much the primary system (or, in this case, the primary receiver) is ready to allow secondary user activity in its spectrum band at any given time. Thus, naturally, it makes sense to let its reward be a monotonically increasing function of the IC.

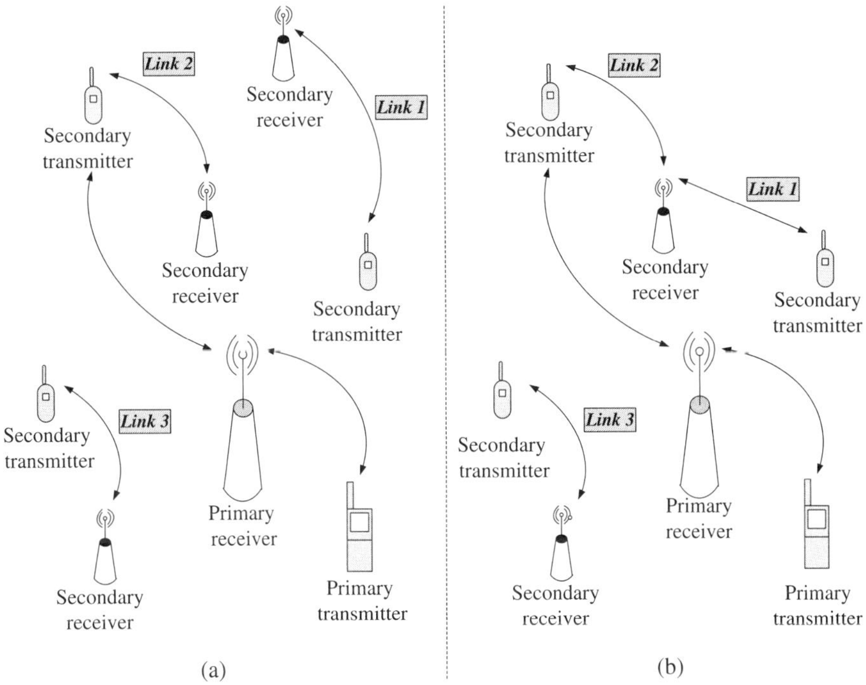

FIGURE 14.1 Cognitive radio system model with $K = 3$ secondary links. Depending on the type of network, some links may share nodes or links may all be made up of distinct nodes. (a) All secondary links are made of distinct nodes. (b) Secondary Links 1 and 2 share the same receiving node.

However, to ensure required QoS, each primary link should always maintain a target signal-to-interference-plus-noise ratio (SINR).

In general, a primary user can maintain a target SINR by increasing both Q_0 and its own transmit power, but this can of course be wasteful in terms of transmit power. Moreover, an unnecessarily large IC Q_0 by a primary user could potentially start a kind of an arms race among primary users since other primary users will also increase their transmit powers. This will degrade performance of both systems due to resulting high primary interference. Hence, IC Q_0 is not allowed to be increased arbitrarily. To achieve this, primary reward should not be allowed to increase simply monotonically with Q_0 but rather be also dependent on the fact that whether there is really a need for a high Q_0.

The goal of the secondary system, on the other hand, is to fully utilize the spectrum activity allowed by the primary system. Each secondary user may be assumed to act in its own interest to maximize its own utility. However, their transmission powers must be carefully self-regulated in order to ensure low interference to the primary user (within IC) as well as to other secondary users.

To simplify notation, let us consider all transmissions to be of BPSK. In the following, we may consider signals corresponding to a particular transmitted symbol and not worry about the symbol index. The continuous-time baseband signal received at the (common) primary receiver, during one symbol period, can be written as

$$r^{(p)}(t) = \sum_{i \in \mathcal{K}_p} A_{p,i} b_i s_i(t) + \sum_{j \in \mathcal{K}_s} A_{p,j} b_j s_j(t) + \sigma_p n(t), \tag{14.1}$$

where $A_{p,l} = h_{pl}\sqrt{p_l}$, for $l \in \mathcal{K}_p \bigcup \mathcal{K}_s$, is the received signal amplitude at the primary receiver; $s_l(t)$, for $l \in \mathcal{K}_p \bigcup \mathcal{K}_s$, is the lth user's transmit pulse waveform; $b_l \in \{+1, -1\}$ is the lth user's transmitted BPSK symbol (during the symbol period we are considering); $n(t)$ is zero-mean additive white Gaussian noise (AWGN) with unit spectral height; and σ_p^2 is the variance of the zero-mean additive noise at the primary receiver. Note that (14.1) is valid for t within an arbitrary symbol period.

It is perhaps easier to work with a finite-dimensional vector representation of continuous-time waveforms. To that end, let us assume that $\left\{\psi_1^{(p)}(t), \ldots, \psi_M^{(p)}(t)\right\}$ is an orthonormal basis signal set for the space spanned by the primary signals. We will call this space the *primary signal space* and the basis signal set $\left\{\psi_1^{(p)}(t), \ldots, \psi_M^{(p)}(t)\right\}$ the *primary basis* [30]. Every primary transmit signal can be represented by a linear combination of the M signals $\left\{\psi_1^{(p)}(t), \ldots, \psi_M^{(p)}(t)\right\}$, and thus they can equivalently be represented as an M-dimensional vector with respect to the primary basis. Hence, primary receiver processing can be based on the following M projections (or correlations) of the continuous-time received signal $r^{(p)}(t)$ onto each of the M orthonormal basis directions:

$$r_m^{(p)} = < r^{(p)}(t), \psi_m^{(p)}(t) > \text{ for } m = 1, \ldots, M. \tag{14.2}$$

Denoting

$$\mathbf{r}^{(p)} = \begin{bmatrix} r_1^{(p)} \\ \vdots \\ r_M^{(p)} \end{bmatrix},$$

we may then obtain the following finite-dimensional vector representation of the received signal (14.1) at the primary receiver:

$$\mathbf{r}^{(p)} = \sum_{i \in \mathcal{K}_p} A_{p,i} b_i \mathbf{s}_i^{(p)} + \sum_{j \in \mathcal{K}_s} A_{p,j} b_j \mathbf{s}_j^{(p)} + \sigma_p \mathbf{n}^{(p)},$$

where, for $l \in \mathcal{K}_p \bigcup \mathcal{K}_s$,

$$\mathbf{s}_l^{(p)} = \begin{bmatrix} s_{l1}^{(p)} \\ \vdots \\ s_{lM}^{(p)} \end{bmatrix}$$

is the M-vector representation of the lth radio's signaling waveform $s_l(t)$ with respect to the M-dimensional primary basis, where $s_{lm}^{(p)} =< s_l(t), \psi_m^{(p)}(t) >$, and

$$\mathbf{n}^{(p)} = \begin{bmatrix} n_1^{(p)} \\ \vdots \\ n_M^{(p)} \end{bmatrix} \sim \mathcal{N}(\mathbf{0}, \mathbf{I}_M)$$

with $n_m^{(p)} =< n_p(t), \psi_m^{(p)}(t) >$, for $m = 1, \ldots, M$.

Similarly, the received signal at the receiver of the jth secondary link, for $j \in \mathcal{K}_s$, can be written as

$$r_j^{(s)}(t) = \sum_{k \in \mathcal{K}_s} B_{j,k} b_k s_k(t) + \sum_{i \in \mathcal{K}_p} B_{j,i} b_i s_i(t) + \sigma_s n_j(t),$$

where $B_{j,l} = h_{jl}\sqrt{p_l}$, for $l \in \mathcal{K}_p \bigcup \mathcal{K}_s$, $n_j(t)$ is the zero-mean AWGN with unit spectral height at the jth receiver and σ_s^2 is the variance of secondary receiver noise.

Again, let us assume that all secondary waveforms lie in a space, referred to as the secondary signal space, spanned by an N-dimensional orthonormal basis signal set $\left\{\psi_1^{(s)}(t), \ldots, \psi_N^{(s)}(t)\right\}$ called the *secondary basis*. Then, the secondary receivers may base their processing on N-dimensional vectors obtained by projecting the received signals onto each of the basis signals in the secondary basis $\left\{\psi_1^{(s)}(t), \ldots, \psi_N^{(s)}(t)\right\}$. Thus, a finite-dimensional vector representation of $r_j^{(s)}(t)$ can be written as

$$\mathbf{r}_j^{(s)} = \sum_{k \in \mathcal{K}_s} B_{j,k} b_k \mathbf{s}_k^{(s)} + \sum_{i \in \mathcal{K}_p} B_{j,i} b_i \mathbf{s}_i^{(s)} + \sigma_s \mathbf{n}_j^{(s)}, \tag{14.3}$$

where

$$\mathbf{r}_j^{(s)} = \begin{bmatrix} r_{j1}^{(s)} \\ \vdots \\ r_{jN}^{(s)} \end{bmatrix} \quad \text{and} \quad \mathbf{n}_j^{(s)} = \begin{bmatrix} n_{j1}^{(s)} \\ \vdots \\ n_{jN}^{(s)} \end{bmatrix}$$

with, for $n = 1, \ldots, N$,

$$r_{jn}^{(s)} = < r_j(t), \psi_n^{(s)}(t) >$$

and

$$n_{jn}^{(s)} = < n_j(t), \psi_n^{(s)}(t) >,$$

respectively, being the projections of the received signal and noise at the secondary receiver j onto the nth orthonormal secondary-basis function, for $l \in \mathcal{K}_p \bigcup \mathcal{K}_s$,

$$\mathbf{s}_l^{(s)} = \begin{bmatrix} s_{l1}^{(s)} \\ \vdots \\ s_{lN}^{(s)} \end{bmatrix}$$

where, for $n = 1, \ldots, N$,

$$s_{ln}^{(s)} = < s_l(t), \psi_n^{(s)}(t) >$$

is the N-vector representation of $s_l(t)$ with respect to the N-dimensional basis employed by the secondary system and $\mathbf{n}_j^{(s)} \sim \mathcal{N}(\mathbf{0}, \mathbf{I}_N)$.

14.2.1 Interference at the Primary Receiver

For simplicity, let us assume that the detector at the primary receiver is based on the conventional matched filter (MF). Note that in order to decode signals of all K_p users, the primary receiver then employs a bank of MFs each matched to a signaling waveform $s_i(t)$ of a particular primary user i, for $i = 1, \ldots, K_p$. The output of the MF matched to the ith primary user's transmit signaling waveform can be shown to be

$$\begin{aligned} y_i^{(p)} &= \left(\mathbf{s}_i^{(p)}\right)^T \mathbf{r}^{(p)} \\ &= A_{p,i} b_i + \sum_{l \in \mathcal{K}_p \setminus \{i\}} \rho_{il}^{(p)} A_{p,l} b_l + \sum_{j \in \mathcal{K}_s} \rho_{ij}^{(p)} A_{p,j} b_j + \sigma_p \eta_i^{(p)}, \end{aligned} \tag{14.4}$$

where $\eta_i^{(p)} \sim \mathcal{N}(0,1)$ and, for $k, l \in \mathcal{K}_s \bigcup \mathcal{K}_s$,

$$\rho_{kl}^{(p)} = \left(\mathbf{s}_k^{(p)}\right)^T \mathbf{s}_l^{(p)}$$

denotes the cross-correlation between signaling waveforms of the kth and lth users.

It is, of course, well known that the MF receiver as above is suboptimal in the presence of the multiple-access interference (MAI) from other users [55]. Replacing the above primary receiver with a sophisticated *multiuser detector* (MUD), however, does not change the overall structure of the following DSL framework except to make notation and analysis more cumbersome. Generalization of this DSL architecture to more sophisticated receivers is indeed straightforward and has been discussed, for example, in [220]. Thus, since primary modulation is BPSK so that $b_i \in \{+1, -1\}$, the ith primary user symbols are detected as

$$\hat{b}_i = sgn\left(y_i^{(p)}\right).$$

It is straightforward to observe from (14.4) that the total secondary interference (SI) from all secondary transmissions to the ith primary user decisions is given by

$$\begin{aligned} I_i &= \sum_{j \in \mathcal{K}_s} \left(\rho_{ij}^{(p)}\right)^2 h_{pj}^2 p_j \\ &= \sum_{j \in \mathcal{K}_s} \tilde{A}_{i,j}^2 p_j, \end{aligned} \tag{14.5}$$

where we have let $\tilde{A}_{i,j} = \rho_{ij}^{(p)} h_{pj}$. Let us denote the maximum of this interference (during the symbol epoch under consideration) over all primary users i by I_0:

$$\begin{aligned} I_0 &= \max_{i \in \mathcal{K}_p} I_i \\ &= \sum_{j \in \mathcal{K}_s} \tilde{A}_j^2 p_j, \end{aligned} \tag{14.6}$$

where

$$\tilde{A}_j = \tilde{A}_{i^*j},$$

for some $i^* \in \mathcal{K}_p$. It is this worst-case total interference I_0 that needs to be below the primary IC Q_0:

$$I_0 \leq Q_0. \tag{14.7}$$

As we will see shortly, I_0 plays a key role in the spectrum underlay-based DSL systems.

Since IC Q_0 is the maximum interference the primary system is willing to tolerate from all secondary users, at any given time, the target SINR of the ith primary link can be defined in terms of Q_0 as

$$\bar{\gamma}_i = \frac{h_{pi}^2 p_i}{MAI(i) + Q_0 + \sigma_p^2}, \tag{14.8}$$

where

$$
\begin{aligned}
MAI(i) &= \sum_{l \in \mathcal{K}_p \setminus \{i\}} \left(\rho_{il}^{(p)}\right)^2 h_{pl}^2 p_l \\
&= \sum_{l \in \mathcal{K}_p \setminus \{i\}} \tilde{A}_{p,l}^2 p_l
\end{aligned}
$$

is the MAI from all other primary transmissions to the ith primary user. Note that $\bar{\gamma}_i$ represents the least acceptable transmission quality of the ith primary user.

On the other hand, actual instantaneous SINR of the ith primary link is determined by the actual SI to the primary user i:

$$
\begin{aligned}
\gamma_i &= \frac{h_{pi}^2 p_i}{\sum_{l \in \mathcal{K}_p \setminus \{i\}} \left(\rho_{il}^{(p)}\right)^2 h_{pl}^2 p_l + \sum_{j \in \mathcal{K}_s} \left(\rho_{ij}^{(p)}\right)^2 h_{pj}^2 p_j + \sigma_p^2} \\
&= \frac{h_{pi}^2 p_i}{MAI(i) + \sum_{j \in \mathcal{K}_s} \left(\rho_{ij}^{(p)}\right)^2 h_{pj}^2 p_j + \sigma_p^2} \\
&\geq \frac{h_{pi}^2 p_i}{MAI(i) + I_0 + \sigma_p^2} && \text{(14.9a)} \\
&= \bar{\gamma}_i \frac{Q_0 + MAI(i) + \sigma_p^2}{I_0 + MAI(i) + \sigma_p^2} && \text{(14.9b)} \\
&= \bar{\gamma}_i \left(1 + \frac{Q_0 - I_0}{I_0 + MAI(i) + \sigma_p^2}\right), && \text{(14.9c)}
\end{aligned}
$$

where (14.9a) and (14.9b) follow, respectively, from (14.7) and (14.8). Thus, as seen from (14.9c), each primary user's instantaneous SINR will be above the least acceptable SINR threshold $\bar{\gamma}_i$ as long as the worst-case total SI I_0 is below the primary system's IC Q_0 so that $Q_0 \geq I_0$.

Example 14.1 (A Single Primary User DSL System) As a special case, let us consider the case of a primary system with just a single primary user. This scenario is of particular interest for several reasons. First, if we consider each primary link to act autonomously inside its allocated channel, then the resulting DSL interaction between each autonomous primary user and the secondary system is equivalent to a single primary user primary system as assumed in this example. Second, if the primary system is a centralized system, the resulting DSL game is also essentially equivalent to a primary system with a single primary user.

When there is only a single primary user in the system, we may set $\mathcal{K}_p = \{0\}$ and $\mathcal{K}_s = \{1, \ldots, K_s\}$. Hence, the primary user is denoted as user 0, and the secondary users are labeled as users 1 through K. Moreover, here, we will assume that there is one primary receiver and one secondary receiver of interest.[1] Received signals at the primary and secondary receivers can, respectively, be written as

$$r_p(t) = A_0 b_0 s_0(t) + \sum_{k=1}^{K} A_k b_k s_k(t) + \sigma_p n_p(t) \tag{14.10}$$

and

$$r_s(t) = \sum_{k=1}^{K} B_k b_k s_k(t) + B_0 b_0 s_0(t) + \sigma_s n_s(t), \tag{14.11}$$

where A_k and B_k, defined as $A_k = h_{pk}\sqrt{p_k}$ and $B_k = h_{sk}\sqrt{p_k}$ for $k = 0, 1, \ldots, K$, represent the received signal amplitude at the primary and secondary receivers, respectively, h_{sk} is the fading coefficient between the kth user and the (common) secondary receiver, $n_s(t)$ is the secondary receiver noise that is assumed to be zero mean and of unit spectral height, and σ_s^2 is the noise variance at the secondary receiver. All other notations are the same as defined earlier.

Assuming M projections $r_m^{(p)} =< r_p(t), \psi_m^{(p)}(t) >$, for $m = 1, \ldots, M$, of the continuous-time received signal $r_p(t)$ onto a set of M orthonormal directions specified by $\left\{\psi_1^{(p)}(t), \ldots, \psi_M^{(p)}(t)\right\}$ that spans the primary signal space, we obtain the vector representation of the received signal at the primary receiver:

$$\mathbf{r}^{(p)} = A_0 b_0 \mathbf{s}_0^{(p)} + \sum_{k=1}^{K} A_k b_k \mathbf{s}_k^{(p)} + \sigma_p \mathbf{n}^{(p)},$$

where $\mathbf{s}_k^{(p)} = \left(s_{k1}^{(p)}, \ldots, s_{kM}^{(p)}\right)$, for $k = 0, 1, \ldots, K$, is the M-vector representation of $s_k(t)$ in the M-dimensional basis employed by the primary system, with $s_{km}^{(p)} =< s_k(t), \psi_m^{(p)}(t) >$, and $\mathbf{n}^{(p)} \sim \mathcal{N}(\mathbf{0}, \mathbf{I}_M)$.

Similarly, the finite-dimensional vector representation of $r_s(t)$ with respect to an N-dimensional orthonormal basis $\left\{\psi_1^{(s)}(t), \ldots, \psi_N^{(s)}(t)\right\}$ that spans the secondary signal space is

$$\mathbf{r}^{(s)} = \sum_{k=1}^{K} B_k b_k \mathbf{s}_k^{(s)} + B_0 b_0 \mathbf{s}_0^{(s)} + \sigma_s \mathbf{n}^{(s)},$$

where $\mathbf{r}^{(s)} = \left(r_1^{(s)}, \ldots, r_N^{(s)}\right)^T$, $r_n^{(s)} =< r_s(t), \psi_n^{(s)}(t) >$, for $n = 1, \ldots, N$, is the projection of the received signal at the secondary receiver onto the nth orthonormal basis function $\psi_n^{(s)}(t)$, $\mathbf{s}_k^{(s)} = \left(s_{k1}^{(s)}, \ldots, s_{kN}^{(s)}\right)$, for $k = 0, 1, \ldots, K$, is the N-vector representation of

[1] Generalization to more than one secondary receiver is straightforward and has been reported in [221].

$s_k(t)$ with respect to the N-dimensional basis employed by the secondary system with $s_{kn}^{(s)} =< s_k(t), \psi_n^{(s)}(t) >$, and $\mathbf{n}^{(s)} \sim \mathcal{N}(\mathbf{0}, \mathbf{I}_N)$.

Again, if we assume that the conventional MF detector is employed at both the primary and secondary receivers, then their decisions will be based on the following MF outputs:

$$\begin{aligned} y_0^{(p)} &= \left(\mathbf{s}_0^{(p)}\right)^T \mathbf{r}^{(p)} \\ &= A_0 b_0 + \sum_{k=1}^{K} \rho_{0k}^{(p)} A_k b_k + \sigma_p \eta^{(p)} \end{aligned}$$

and

$$\begin{aligned} y_k^{(s)} &= \left(\mathbf{s}_k^{(s)}\right)^T \mathbf{r}^{(s)} \\ &= B_k b_k + \sum_{j=1, j\neq k}^{K} \rho_{kj}^{(s)} B_j b_j + \rho_{k0}^{(s)} B_0 b_0 + \sigma_s \eta_k^{(s)} \end{aligned}$$

with $\rho_{0k}^{(p)} = \left(\mathbf{s}_0^{(p)}\right)^T \mathbf{s}_k^{(p)}$, $\rho_{kj}^{(s)} = \left(\mathbf{s}_k^{(s)}\right)^T \mathbf{s}_j^{(s)}$, for $j = 0, 1, \ldots, K$, and noise $\eta^{(p)}$ and $\eta_k^{(s)}$ are both distributed as $\mathcal{N}(0,1)$.

In this case, the worst-case total SI I_0 from all secondary transmissions to the primary user is simply the total SI given by

$$I_0 = \sum_{k=1}^{K} \tilde{A}_k^2 p_k, \tag{14.12}$$

where $\tilde{A}_k = \rho_{0k}^{(p)} h_{p_k}$ is the effective channel coefficient of the kth secondary user as seen by the primary receiver. Similarly, the total interference from all other secondary users to the kth (secondary) user signal, excluding that from the primary user, can be denoted by

$$i_k = \sum_{j=1, j\neq k}^{K} \left(\rho_{kj}^{(s)}\right)^2 h_{sj}^2 p_j.$$

In this scenario, the single primary user adapts its IC Q_0, which is the maximum total interference the primary user is willing to tolerate from secondary transmissions at any given time. By adjusting Q_0, the primary user can control the total transmit power the secondary users impose on its licensed channel. Thus, IC Q_0 determines how much secondary user activity the primary user is ready to allow in its spectrum. Hence, as we already mentioned, it is reasonable to make its reward an increasing function of the IC Q_0. However, we impose the realistic constraint that the primary user should always maintain its target SINR $\bar{\gamma}$ in order to ensure required QoS.

Recall also from our earlier discussion that an unnecessarily large IC by the primary user could hinder performance of both the secondary system and other primary receivers due to resulting high interference. ■

14.2.2 A Game Model for Dynamic Spectrum Leasing

In a DSL framework based on spectrum underlay, the primary system and secondary users interact with each other by adjusting their IC and transmit power levels, respectively, in order to maximize their own gains (which we will define in the following). Regardless of how many users are in the primary system, primary system's action in this interaction is to set the IC Q_0, which specifies the maximum combined interference it is willing to tolerate from all secondary users. This is true as long as primary system acts as a single centralized system. On the other hand, secondary users may adjust their transmit power levels to maximize a suitably defined gain of theirs without violating the primary IC constraint.

Noncooperative game theory provides a convenient way to model and analyze such interactions among rational agents [222]. In the following, we will assume a centralized primary system (or a single primary user, as in Example 14.1). We may model such a system as in the following noncooperative *DSL game* $(\mathcal{K}, \mathcal{A}_k, u_k(.))$:

1. *Players*: $\mathcal{K} = \{0, 1, 2, \ldots, K_s\}$, where we assume that the 0th user is the centralized primary system consisting of multiple primary users and $k = 1, 2, \ldots, K_s$ represents the kth secondary user.
2. *Action space*: $\mathcal{P} = \mathcal{A}_0 \times \mathcal{A}_1 \times \mathcal{A}_2 \cdots \times \mathcal{A}_{K_s}$, where $\mathcal{A}_0 = \mathcal{Q} = [0, \bar{Q}_0]$ represents the primary system's action set and $\mathcal{A}_k = \mathcal{P}_k = [0, \bar{P}_k]$, for $k = 1, 2, \ldots, K$, represents the kth secondary user's action set. Note that $\bar{Q}_0$ and $\bar{P}_k$ represent, respectively, the maximum IC of the primary system and the maximum transmission power of the kth secondary user. The lower limit of these action sets being zero indicates that at times, a secondary user may turn off its transmission or the primary system may not be willing to tolerate any interference from the secondary system at all. We may denote the action vector of all players in this noncooperative DSL game by $\mathbf{a} = [Q_0, p_1, \ldots, p_{K_s}]^T$, where $Q_0 \in \mathcal{Q}$ and $p_k \in \mathcal{P}_k$.
3. *Utility function*: We denote by $u_0(Q_0, \mathbf{a}_{-0})$ the primary system's utility function and by $u_k(p_k, \mathbf{a}_{-k})$, for $k = 1, 2, \ldots, K_s$, the kth secondary user's utility function where $\mathbf{a}_{-k}$, for $k = 0, 1, 2, \ldots, K_s$, denotes the action vector of all users excluding that of the kth user.

The motivation for DSL lies in the fact that providing the primary system with a reward for allowing secondary user access to its licensed spectrum can conceivably improve the spectrum utilization. It makes sense to consider a reward function for the primary system that, in general, increases with both the *demand* seen from the secondary network and the allowed SI. Hence, we may consider the following utility function for the primary system [29,30]:

$$\begin{aligned} u_0(Q_0, \mathbf{a}_{-0}) &= \left(\bar{Q}_0 - (Q_0 - I_0(\mathbf{a}_{-0}))\right) F(Q_0) \\ &= u_0(Q_0, I_0), \end{aligned} \tag{14.13}$$

where, as noted earlier, $\mathbf{a}_{-l}$ denotes the action vector of all users excluding the action of the lth user, for $l = 0, 1, \ldots, K_s$. Note that according to (14.13), utility of the primary system is directly proportional to demand and a suitable continuous reward function $F(.)$ of the IC Q_0. The demand is taken to be decreasing when extra interference margin $Q_0 - I_0$ increases. This discourages the primary system from swamping all other transmissions (both primary and secondary) by setting an unnecessarily too large an IC that will lead to higher transmission powers according to (14.8).

It is worth pointing out that the only dependence of the primary utility function (14.13) on the secondary system is through the single parameter I_0, which can be easily estimated (by the primary network) as we will see later. This avoids the need to have detailed channel-state information from secondary network.

Of course, the functional form (14.13) is not the only possibility for the primary utility function. One may make many other reasonable formulations. Even within the functional form of (14.13), there can be different utility functions corresponding to different choices of $F(.)$.

Example 14.2 (A Primary Utility Function) A simple example of a reward function $F(.)$ can be the linear function $F(Q_0) = Q_0$ so that the utility function (14.13) reduces to

$$\begin{aligned} u_0(Q_0, \mathbf{a}_{-0}) &= \left(\bar{Q}_0 - (Q_0 - I_0(\mathbf{a}_{-0}))\right) Q_0 \\ &= u_0(Q_0, I_0). \end{aligned} \tag{14.14}$$

■

In general, we may allow $F(.)$ in (14.13) to be an arbitrary function and establish conditions on $F(.)$ so that the proposed DSL game has desired equilibrium properties.

Example 14.3 (An Alternative Primary Utility Function) Another meaningful choice is $F(Q_0) = \log(1 + Q_0)$, so that the primary system utility is proportional to the capacity attained by the secondary system with respect to the primary receiver. This model for primary system utility is sensible in a DSL CRN when secondary system is concerned about the rate its users achieve rather than their transmission powers. Figure 14.2 shows the primary user utility $u_0(Q_0, I_0)$ with this choice for $F(.)$ as a function of the IC Q_0 for a fixed total SI I_0. Observe from Figure 14.2 that primary system utility u_0 is *quasiconcave* in IC Q_0. ■

As can be seen from (14.9c), as long as the total SI I_0 is below the IC Q_0 set by the primary system, the primary system QoS will be guaranteed for all its users. To ensure this, the utilities of secondary users must be fast decaying functions of $I_0 - Q_0$ when this difference is positive. Thus, if $f(p_k)$ is the reward achieved by the kth secondary

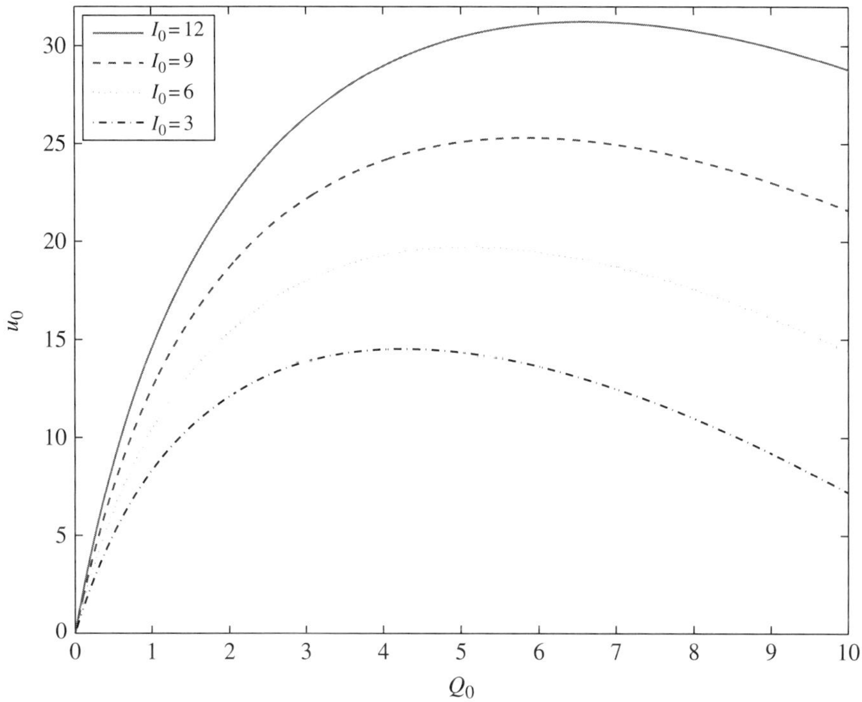

FIGURE 14.2 Primary user utility u_0 as a function of interference cap Q_0 for a fixed secondary interference I_0 assuming $F(Q_0) = \log(1 + Q_0)$ and a single-user secondary system. Parameter values are $\overline{Q} = Q_{max} = 10$, $h_{p1} = 1$, and $\rho_{01}^{(p)} = \rho_{10}^{(s)} = 1$.

user by transmitting at a power p_k, then a suitable utility function for it would ideally be $f(p_k)q(Q_0 - I_0)$ where $q(.)$ is the unit-step function. An undesirable property of the unit step function, however, is that it is discontinuous at the origin. To avoid that, we may use the following utility function for the kth secondary user, for $k \in \mathcal{K}_s$:

$$u_k(p_k, \mathbf{a}_{-k}) = \frac{f(p_k)}{1 + e^{\lambda(I_0 - Q_0)}} \tag{14.15}$$

where $f(.)$ is a suitable reward function chosen by the secondary system and the weighting term $\frac{1}{1+e^{\lambda(I_0-Q_0)}}$ in (14.15) is a *sigmoidal function* used to approximate the unit step with the property that (see Figure 14.3)

$$\lim_{(I_0 - Q_0) \longrightarrow \infty} \frac{1}{1 + e^{\lambda(I_0 - Q_0)}} = 0 \tag{14.16}$$

and

$$\lim_{(I_0 - Q_0) \longrightarrow -\infty} \frac{1}{1 + e^{\lambda(I_0 - Q_0)}} = 1. \tag{14.17}$$

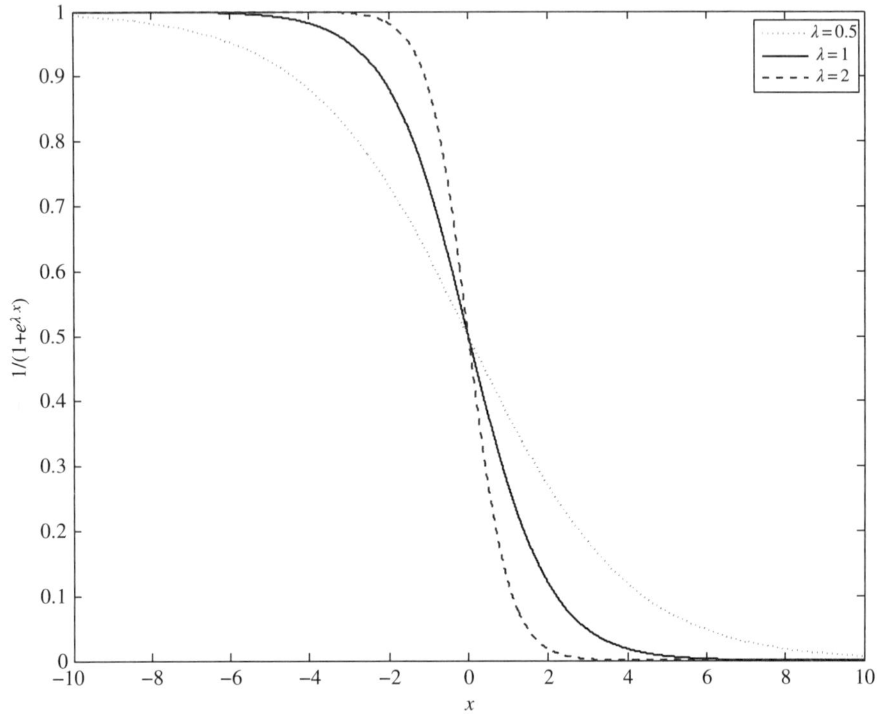

FIGURE 14.3 The sigmoidal function $\frac{1}{1+e^{\lambda x}}$.

Note that parameter λ can be used to adjust the steepness of the transition region as shown in Figure 14.3.

By using (14.6), or (14.12), we may express (14.15) also as

$$u_k(p_k, \mathbf{a}_{-k}) = \frac{f(p_k)}{1 + e^{\lambda(I_{0,-k}(\mathbf{a}_{-k}) - Q_0)} e^{c_k p_k}}, \tag{14.18}$$

where $c_k = \lambda \tilde{A}_k^2 \geq 0$, and we have defined the interference from all secondary transmissions to the worst-hit primary user excluding that from the kth secondary user as

$$I_{0,-k}(\mathbf{a}_{-k}) = I_0 - \tilde{A}_k^2 p_k. \tag{14.19}$$

The reward function $f(p_k)$ in (14.15), or (14.18), can be chosen depending on what is of interest to the secondary system as a whole or for individual secondary radios. An often justified objective of secondary systems in DSL networks is to maximize the sum capacity of all its users in the shared primary spectrum. However, from the perspective of a particular secondary user, it may be interested in gaining the maximum possible rate it can achieve for itself. Thus, each secondary user would like to maximize its own achieved capacity. If this is the case, then we may use $f(p_k) = W_k \log\left(1+\gamma_k^{(s)}\right)$ where $\gamma_k^{(s)}$ is the received SINR of the kth *secondary link* for $k \in \mathcal{K}_s$ and $W_k > 0$ is a scaling parameter that can be taken as proportional to the bandwidth.

Note that the received SINR $\gamma_k^{(s)}$ of the kth secondary link depends on the form of the receiver. In the following, we may assume that all secondary detectors are also based on the MFs. Since all secondary transmissions are BPSK, then the jth secondary link receiver detects the corresponding jth secondary transmitter's symbols as

$$\hat{b}_j = sgn\left(y_j^{(s)}\right),$$

where, for $j \in \mathcal{K}_s$,

$$y_j^{(s)} = \left(\mathbf{s}_j^{(s)}\right)^T \mathbf{r}_j^{(s)} \tag{14.20}$$

with $\mathbf{r}_j^{(s)}$ being the received signal corresponding to the jth secondary link given by (14.3). Substituting (14.3) in (14.20), we have that

$$\begin{aligned} y_j^{(s)} &= \left(\mathbf{s}_j^{(s)}\right)^T \mathbf{r}_j^{(s)} \\ &= B_{j,j}b_j + \sum_{l\in\mathcal{K}_s\setminus\{j\}} \rho_{jl}^{(s)} B_{j,l}b_l + \sum_{i\in\mathcal{K}_p} \rho_{ji}^{(s)} B_{j,l}b_l + \sigma_s\eta_j^{(s,j)}, \end{aligned}$$

where $\rho_{kl}^{(s)} = \left(\mathbf{s}_k^{(s)}\right)^T \mathbf{s}_l^{(s)}$ and $\eta_k^{(s,j)} = \left(\mathbf{s}_k^{(s)}\right)^T \mathbf{n}_j^{(s)} \sim \mathcal{N}(0,1)$. Hence, the jth secondary link's (average) SINR, with the MF receiver, is given by

$$\begin{aligned} \gamma_j^{MF} &= \frac{|h_{jj}|^2 p_j}{\sum_{l\in\mathcal{K}_s\setminus\{j\}} \left(\rho_{jl}^{(s)}\right)^2 |h_{jl}|^2 p_l + \sum_{i\in\mathcal{K}_p} \left(\rho_{ji}^{(s)}\right)^2 |h_{ji}|^2 p_i + \sigma_s^2} \\ &= \frac{|h_{jj}|^2 p_j}{i_j^{(j)} + \tilde{\sigma}_{s,j}^2} \\ &= \frac{p_j}{N_j}, \end{aligned} \tag{14.21}$$

where

$$i_k^{(j)} = \sum_{l\in\mathcal{K}_s\setminus\{k\}} \left(\rho_{kl}^{(s)}\right)^2 h_{jl}^2 p_l \tag{14.22}$$

is the total interference from all other secondary users to the kth secondary link signal at the jth secondary receiver,

$$\tilde{\sigma}_{s,j}^2 = \sum_{i \in \mathcal{K}_p} \left(\rho_{ji}^{(s)}\right)^2 |h_{ji}|^2 p_i + \sigma_s^2$$

is the effective primary interference plus noise seen by the jth link receiver, and $N_j = \frac{i_j^{(j)} + \tilde{\sigma}_{s,j}^2}{|h_{jj}|^2}$.

From (14.21), it is easily seen that when the secondary system employs the MF receiver,

$$\frac{\partial \gamma_j^{MF}}{\partial p_j} = \frac{\gamma_j^{MF}}{p_j}.$$

Thus, in the following, we may investigate equilibrium strategies of the DSL game $G = (\mathcal{K}, \mathcal{A}_k, u_k)$ where users are interested in maximizing the following utility functions:

$$\text{Primary system utility:} \quad u_0(Q_0, \mathbf{a}_{-0}) = \left(\bar{Q}_0 - (Q_0 - I_0(\mathbf{a}_{-0}))\right) F(Q_0) \tag{14.23}$$

$$\text{Secondary user utility:} \quad u_k(p_k, \mathbf{a}_{-k}) = \frac{W_k \log\left(1 + \gamma_k^{(s)}\right)}{1 + e^{\lambda(I_0 - Q_0)}} \quad \text{for } k \in \mathcal{K}_s. \tag{14.24}$$

Figure 14.4 shows the behavior of the secondary utility function (14.24) assuming a single-user secondary system and $\gamma_j^{(s)} = \gamma_j^{MF}$. Note that in Figure 14.4, we have assumed that $W_j = W = 1$, $h_{p1}^2 = h_{11}^2 = 1$, $\sigma_p^2 = \sigma_s^2 = 1$, and all cross-correlations are unity. Figure 14.4(a) shows secondary utility function (14.24) parameterized by the primary IC Q_0. Observe from Figure 14.4(a) that the secondary user utility function u_j is quasiconcave in p_j and that the unique maximum of u_j is an increasing function of primary IC Q_0. Hence, pushing the primary IC to a higher value encourages the secondary user to also transmit at a higher power level and thus allows the primary user to achieve higher leasing gains.

Figure 14.4(b) shows the effect of parameter λ on the utility function. As mentioned earlier, parameter λ can be used to adjust the steepness of the transition region of the secondary utility function. The higher the value of λ, the steeper the transition, as can be seen from Figure 14.4(b). Hence, higher values of λ indicate that the primary system expects the secondary users to strictly adhere to the $I_0 \leq Q_0$ requirement.

14.2.3 Nash Equilibria in Noncooperative Games

Game theory attempts to model, analyze, and predict interactions among rational entities called players of the game. An important aspect of such interactions is whether

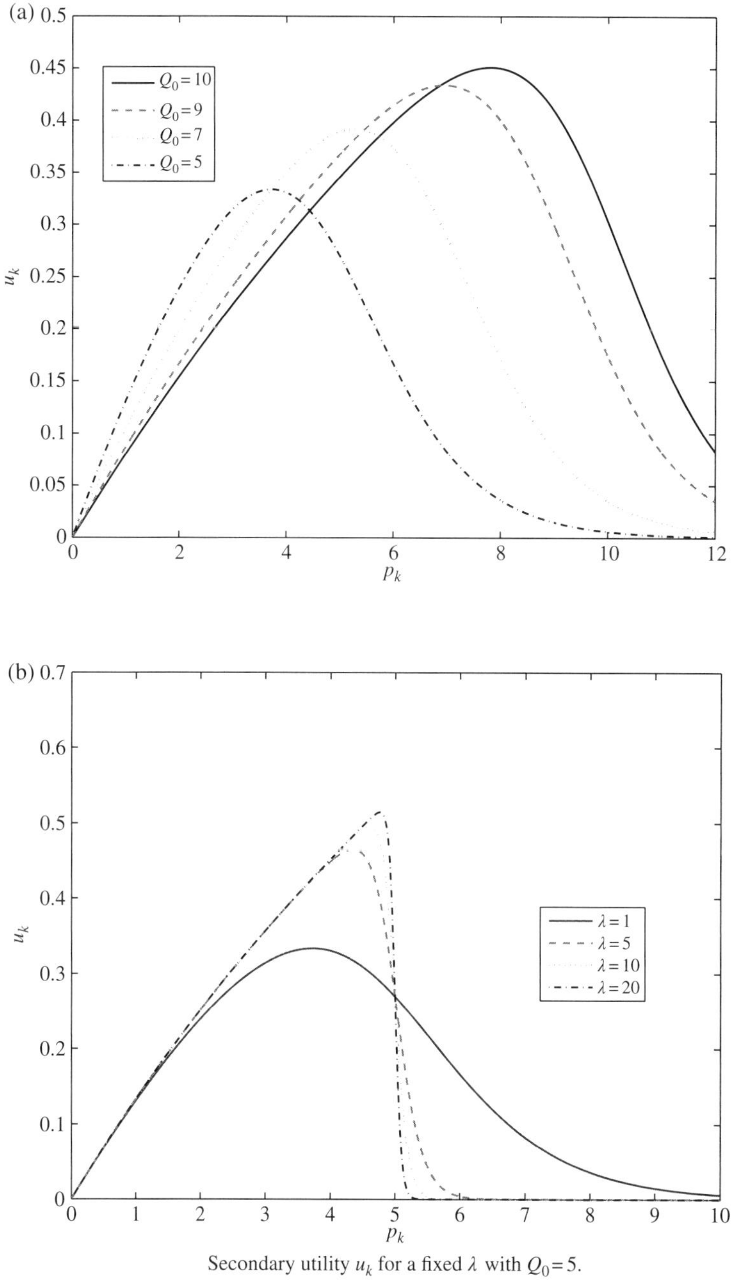

FIGURE 14.4 Secondary link utility u_k given by (14.24) with $W_k = W = 1$, $h_{p1}^2 = h_{11}^2 = 1$, $\sigma_p^2 = \sigma_s^2 = 1$, and all cross-correlation coefficients equal to unity. (a) Secondary utility u_k for a fixed interference cap Q_0. (b) Secondary utility u_k for a fixed λ with $Q_0 = 5$.

they lead to chaos or some sort of stable steady state or an equilibrium. The ability to predict under what conditions on the player actions and utilities things may or may not get out of control can be very helpful when such a model is used to capture interactions among players, or entities, in a real-world application. In the case of DSL between primary users and secondary cognitive radios, this is the reason why we are interested in modeling their interactions as a game. Game-theoretic analysis of DSL interactions may help us design a system that will perhaps ensure convergence of all user actions to certain desirable choices that will make the system to reach a certain steady state. Not only that, we may even be able to select the design parameters so that the behavior of the DSL interaction has desirable properties at the steady state or the equilibrium.

The most commonly used equilibrium concept in noncooperative game theory is the *Nash equilibrium*.

Definition 14.2.1 (Nash Equilibrium) A strategy vector $\mathbf{a} = (a_0, a_1, \ldots, a_k)$ is a Nash equilibrium of the primary–secondary user DSL game $G = (\mathcal{K}, \mathcal{A}_k, u_k)$ if, for every $k \in \mathcal{K}$, $u_k(a_k, \mathbf{a}_{-k}) \geq u_k\left(a'_k, \mathbf{a}_{-k}\right)$ for all $a'_k \in \mathcal{A}_k$.

In essence, when at a Nash equilibrium, no user has an incentive to unilaterally change its own strategy when all other users keep their strategies fixed. By definition, when at a Nash equilibrium, such unilateral changes do not help increase the player's utility from what it is right now. Hence, the Nash equilibrium is a stable outcome of a noncooperative game in the sense that if player, actions somehow arrive at a Nash optimal action vector, then we are certain all users may stick with these action choices as long as they are rational: Even though every user might be aiming at improving its own utility, no one has a reason to try to do anything differently.

The following well-known result due to Debreu, Glicksberg, and Fan gives a set of conditions that ensures the existence of a Nash equilibrium in a game [222] whose proof we omit.

Theorem 14.1 (Existence of a Nash Equilibrium) A Nash equilibrium exists in game $G = (\mathcal{K}, \mathcal{A}_k, u_k)$, if, for all $k \in \mathcal{K}$:

1. $\mathcal{A}_k$ is a nonempty, convex, and compact subset of some Euclidean space $\mathbb{R}^N$.
2. $u_k(\mathbf{a})$ is continuous in action vector $\mathbf{a}$.
3. $u_0(Q_0, \mathbf{a}_{-0})$ and $u_k(p_k, \mathbf{a}_{-k})$ are quasiconcave in Q_0 and p_k, respectively. ■

Proof: A proof can be found in p. 34 of [222].

But if all players are allowed to select their actions only caring about improving their own utility, how can we get to a Nash equilibrium? To answer this, we define the *best response* correspondence of a player in a game as the best reaction strategy a rational user would choose in order to maximize its own utility, in response to the actions chosen by other users.

Definition 14.2.2 (Best Response) The user k's best response $r_k : \mathcal{A}_{-k} \longrightarrow \mathcal{A}_k$ is the set

$$r_k(\mathbf{a}_{-k}) = \left\{ a_k \in \mathcal{A}_k : u_k(a_k, \mathbf{a}_{-k}) \geq u_k\left(a'_k, \mathbf{a}_{-k}\right) \quad \text{for all } a'_k \in \mathcal{A}_k \right\}.$$

It is reasonable to assume that a rational player's response to the actions chosen by all other players in a game will be one of the action choices from its best response correspondence set $r_k(\mathbf{a}_{-k})$. Of course, as soon as one user selects its action to be a best response to all the others' current actions, it may prompt one of the others to change its action to a new one that is a best response from the perspective of that second player. It is easy to see that this may keep repeating forever, unless each user's best response correspondence converges. Clearly, any such convergent point has to be a Nash equilibrium. We may characterize a sufficient condition for such convergence, in terms of the concept of *standard functions* [223].

Definition 14.2.3 (Standard Functions) A function $\mathbf{r}(\mathbf{a})$ is said to be a standard function if it satisfies following three properties:

I. *Positivity*: $\mathbf{r}(\mathbf{a}) > 0$.
II. *Monotonicity*: If $\mathbf{a} \geq \mathbf{a}'$, then $\mathbf{r}(\mathbf{a}) \geq \mathbf{r}(\mathbf{a}')$.
III. *Scalability*: For all $\mu > 1$, $\mu\mathbf{r}(\mathbf{a}) \geq \mathbf{r}(\mu\mathbf{a})$.

The notion of standard functions comes in handy in establishing the existence of a unique Nash equilibrium in a DSL game. Indeed, it has been shown that if the best response correspondences $r_k(\mathbf{a}_{-k})$ of a game are standard functions for $\forall k$, then the game has a unique Nash equilibrium [223].

14.2.4 Existence of a Nash Equilibrium in the DSL Game

Note that the primary system action set in the DSL game is of the form of $\mathcal{A}_0 = \mathcal{Q} = [0, \bar{Q}_0]$, where $\bar{Q}_0$ is the maximum IC that is determined by the required minimum QoS and the maximum possible transmit power of a primary user. Clearly, $\mathcal{A}_0$ is both compact and convex (being a closed and bounded interval on the real line). Similarly, for all $k \in \mathcal{K}_s$, the secondary user strategy sets are of the form of $\mathcal{A}_k = \mathcal{P}_k = [0, \bar{P}_k]$. Again, it is easy to observe that all secondary user action sets are convex and compact. Moreover, according to our construction, both $u_0(\mathbf{a})$ of (14.23) and $u_k(\mathbf{a})$ of (14.24) are continuous in the action vector $\mathbf{a}$.

Hence, the only remaining condition, as required in Theorem 14.1 to ensure the existence of a Nash equilibrium in the DSL game of Section 14.2.2, is the property of quasiconcavity of utility functions (14.23) and (14.24). The following proposition lists a set of sufficient conditions to achieve this and thus guarantees the existence of a Nash equilibrium in the DSL game (14.32)–(14.24).

Proposition 14.2.1 With $\mathcal{A}_k$'s and u_k's as defined earlier, the DSL game in (14.23)–(14.24) has a Nash equilibrium when $\gamma_k^{(s)} = \gamma_k^{MF}$ if $F(Q_0)$ satisfies the following conditions:

1. $F(Q_0)$ is continuous and strictly monotonic for $Q_0 > 0$.
2. $F(0) = 0, F'(0) > 0$, and $\lim_{Q_0 \to \infty} \frac{F(Q_0)}{F'(Q_0)} > -\infty$.
3. $\frac{F(Q_0)F''(Q_0)}{(F'(Q_0))^2} < 2$ for $Q_0 > 0$.
4. $0 \leq \bar{Q}_0 + I_0(\mathbf{a}_{-0}) < \infty$.

Proof: Since action sets $\mathcal{A}_k$ are both compact and convex for all $k = 0, 1, \ldots, K$ while both $u_0(\mathbf{a})$ and $u_k(\mathbf{a})$ are continuous in the action vector $\mathbf{a}$, the only remaining condition that we need to ensure for the existence of a Nash equilibrium according to Theorem 14.1 is the quasiconcavity of u_k's for all $k = 0, 1, \ldots, K$.

Let us define a function $\Phi(Q_0)$ as

$$\Phi(Q_0) = \frac{F(Q_0)}{F'(Q_0)} + Q_0. \tag{14.25}$$

It can be seen that u_0 has a local maximum that is indeed a global maximum, so that u_0 is quasiconcave, if $\Phi(Q_0) = \bar{Q}_0 + I_0(\mathbf{a}_{-0})$ has only one solution for $Q_0 \in \mathcal{Q}$. Clearly, this equation has a solution if $\Phi(Q_0)$ is continuous and $\lim_{Q_0 \to 0} \Phi(Q_0) \leq \bar{Q}_0 + I_0(\mathbf{a}_{-0}) < \lim_{Q_0 \to \infty} \Phi(Q_0)$. This solution would be a global maximum if in addition $\Phi'(Q_0) > 0$ for $Q_0 > 0$. It can be easily verified that $\Phi'(Q_0) > 0$ will be true if $F(Q_0)$ is such that $\frac{F(Q_0)F''(Q_0)}{(F'(Q_0))^2} < 2$. Note also that $\lim_{Q_0 \to \infty} \Phi(Q_0) = \infty$ if $\lim_{Q_0 \to \infty} \frac{F(Q_0)}{F'(Q_0)} > -\infty$. Hence, conditions 1–4 of the proposition indeed are a set of sufficient conditions on $F(Q_0)$ for u_0 to be quasiconcave.

Next, we can verify that u_k of (14.15), for each $k = 1, \ldots, K_s$, is in fact continuous and quasiconcave. Indeed, from (14.15), it is easy to see that $u_k(\mathbf{a})$ is continuous in $\mathbf{a}$. Consider the first-order derivative of $u_k(p_k) = \frac{W_k \log(1+\gamma_k^{MF})}{1+e^{\lambda(I_0 - Q_0)}}$ w.r.t. p_k where γ_k^{MF} is given by (14.21):

$$\begin{aligned}
\frac{\partial u_k(p_k)}{\partial p_k} &= \frac{W\left[\frac{\left(\frac{1}{N_k}\right)\left(1+e^{\lambda(I_{p,-k}(\mathbf{p}_{-k})-Q_0)}e^{c_k p_k}\right)}{1+\frac{p_k}{N_k}} - c_k e^{\lambda(I_{p,-k}(\mathbf{p}_{-k})-Q_0)}e^{c_k p_k}\log\left(1+\frac{p_k}{N_k}\right)\right]}{\left(1+e^{\lambda(I_{p,-k}(\mathbf{p}_{-k})-Q_0)}e^{c_k p_k}\right)^2} \\
&= \frac{We^{c_k p_k} g_k^{MF}(p_k)}{\left(1+\gamma_k^{MF}\right)\left(1+e^{\lambda(I_{p,-k}-Q_0)}e^{c_k p_k}\right)^2},
\end{aligned} \tag{14.26}$$

where

$$g_k^{MF}(p_k) = \frac{1}{N_k}\left(e^{-c_k p_k} + e^{\lambda(I_{0,-k}-Q_0)}\right) - c_k e^{\lambda(I_{0,-k}-Q_0)}\left(1+\gamma_k^{MF}\right)\log\left(1+\gamma_k^{MF}\right). \tag{14.27}$$

At an interior local extremum point of $u_k(p_k)$, for $p_k \in [0,\infty)$, we should have

$$g_k^{MF}(p_k) = 0. \tag{14.28}$$

Since

$$g_k^{MF}(0) = \frac{1}{N_k}\left(1 + e^{\lambda(I_{p,-k}-Q_0)}\right) > 0\,, \tag{14.29}$$
$$g_k^{MF}(\infty) \longrightarrow -\infty$$

and $g_k^{MF}(p_k)$ is continuous in p_k, it follows that $g_k^{MF}(.)$ must have at least one zero crossing. However, since

$$\frac{\partial g_k^{MF}(p_k)}{\partial p_k} = -\frac{c_k}{N_k}\left[e^{-c_k p_k} + e^{\lambda(I_{p,-k}-Q_0)}\left(1+\log\left(1+\frac{p_k}{N_k}\right)\right)\right]$$
$$< 0 \qquad \text{for } p_k \geq 0, \tag{14.30}$$

it can only have exactly one zero on $[0,\infty]$. From (14.26), this then implies that $u_k(p_k)$ has only one local extremum point on $p_k \in [0,\infty)$. If this extremum point is denoted as p^*, from (14.29) and (14.30), it follows that $g_k^{MF}(p_k) > 0$ for $0 \leq p_k < p^*$. From (14.26), then we have that $u_k(p_k)$ is monotonic increasing on $0 \leq p_k < p^*$. Similarly, since $g_k^{MF}(p_k) < 0$ for $p_k > p^*$, we have that $u_k(p_k)$ is monotonic decreasing on $p_k > p^*$. Hence, it follows that the local extremum point $p_k = p^*$ is indeed a global maximum of $u_k(.)$ on $[0,\infty)$, implying that $u_k(p_k)$ is quasiconcave in p_k, for each $k = 1,\ldots,K$.

From Theorem 14.1, then the above DSL game G has at least one Nash equilibrium.

In a noncooperative game that is guaranteed to have a Nash equilibrium, if all users are allowed to adapt their actions sequentially according to their best response correspondences, then all users are guaranteed to converge to a Nash equilibrium of the game. Since best response of a player in a game is a strategy that maximizes its utility given all the other players, actions, the unique interior solution for the best response of the primary system with utility (14.23) is given by

$$Q_0^*(I_0) = \left(\bar{Q}_0 + I_0\right) - \frac{F(Q_0^*)}{F'(Q_0^*)}. \tag{14.31}$$

Since $u_0(Q_0)$ is monotonic increasing for $Q_0 < Q_0^*$, the primary system best response is given by

$$r_0(\mathbf{a}_{-0}) = \min\{\bar{Q}_0, Q_0^*(I_0)\}, \tag{14.32}$$

where $Q_0^*(I_0)$ is the solution of (14.31).

Thus, to determine its best response for a power vector $\mathbf{a}_{-0}$ chosen by the secondary links, the only quantity primary system needs to know is the worst-case total SI I_0 experienced by the worst-hit primary user, as given in (14.6). This total interference I_0 can be estimated at the primary receiver without much difficulty.

One may, however, observe that making primary decisions based on the worst-case interference $I_0 = \max_{i \in \mathcal{K}_p} I_i$ may lead to not fully capitalizing on different channel conditions seen by different primary users. Indeed, if there were more than one primary receiver, different primary receivers may experience different interference from secondary users in general. However, in the context of our discussion, there is only a single primary receiver. In that case, the receiver needs to be able to ensure a minimum QoS for the worst-hit primary user and thus set the IC based on the effective interference to this worst-hit user.

On the other hand, the best response of the jth secondary link to the transmit powers of other secondary users as well as IC set by the primary user is given by the (unique) solution $p_j = p_j^*\left(Q_0, I_{0,-j}, i_j^{(j)}\right)$ to the equation $g_j^{MF}(p_j) = 0$ in (14.28), where $I_{0,-j}$ and $i_j^{(j)}$, given, respectively, by (14.19) and (14.22), are the interference from all secondary transmissions to the worst-hit primary user excluding that from the jth secondary user and the total interference from all other secondary users to the jth secondary link signal at the jth secondary receiver. Again, since u_j is quasiconcave in p_j, if $p_j^*\left(Q_0, I_{0,-j}, i_j^{(j)}\right) > \bar{P}_j$, then the best response of the jth secondary cognitive radio is to set its transmit power to $p_j = \bar{P}_j$. Hence, the best response of the jth secondary link, for $j \in \mathcal{K}_s$, is

$$r_j(\mathbf{a}_{-j}) = \min\{\bar{P}_j, p_j^*\left(Q_0, I_{0,-j}, i_j^{(j)}\right)\}. \tag{14.33}$$

Observe from (14.33) that, in general, the best response of the jth secondary link is a function of the residual interference $(Q_0 - I_{0,-j})$ of all other secondary users at the primary receiver and the total interference from all secondary and primary users to the jth link at its receiver. The jth secondary link receiver can easily estimate the latter quantity. However, to determine the residual interference $Q_0 - I_{0,-j}$, the secondary receiver needs to know the current IC Q_0 as well as the residual SI $I_{0,-j}$ at the primary receiver. To facilitate this, in noncooperative game theory-based DSL frameworks of [28–30, 220, 221], it was assumed that primary base station broadcasts both Q_0 and I_0, whenever it adjusts its IC to a new value. This is the only conscious interaction the primary system is assumed to be having with the secondary system. Knowing I_0, each secondary user can compute the residual interference $I_{0,-j} = I_0 - \tilde{A}_j^2 p_j$ since it knows its own transmit power and it may estimate the channel-state information $\tilde{A}_j$ if reverse-link signals are available on the same frequency band. When this is not possible, on the other hand, [30] has shown that even the simple approximation $I_{0,-j} \approx I_0$ may perform reasonably well.

Note that, in this formulation of the DSL game, all secondary links act autonomously. There is no central control or an SSDC, as in the previous chapters, involved. This avoids the need to have channel-state information of all secondary users at a central location and performing a centralized optimization to determine transmit powers of all secondary users. In addition, this also avoids the need to have a control channel dedicated to informing each individual secondary user its new transmit power p_k, which would cost bandwidth. However, it must be mentioned that autonomous secondary users may not always be able to utilize spectrum the most efficient way. This is the trade-off one has to consider in designing a centralized versus autonomous, or distributed, system.

Example 14.4 (The Nash Equilibrium of the DSL Game in a Single Secondary Link System) Perhaps the simplest DSL game of the aforementioned form is obtained when there is only a single secondary link. In this system, a single secondary user plays a DSL game against a single primary user (or the centralized primary system) so that $K = 1, I_{0,-1} = 0$ and $I_0 = \tilde{A}_1^2 p_1$ where $\tilde{A}_1 = \rho_{01}^{(p)} h_{p_1}$.

When $F(Q_0) = \log(1 + Q_0)$, the Nash equilibrium $\left(Q_0^*, p_1^*\right)$ for MF receiver is given by the intersection of the following two curves:

$$Q_0 = r_0(p_1) = \left(\text{Solution to equation } Q_0 + (1 + Q_0)\log(1 + Q_0) = \bar{Q}_0 + \tilde{A}_1^2 p_1\right) \tag{14.34}$$

and

$$p_1 = r_1(Q_0) = \left(\text{Solution to equation } g_1^{MF}(p_1) = 0\right), \tag{14.35}$$

where $g_1^{MF}(.)$ is given by (14.27) with $k = 1$.

Figure 14.5 shows the primary and secondary best response curves $Q_0 = r_0(p_1)$ and $p_1 = r_1(Q_0)$ for the system defined by (14.23) and (14.24) with MF receiver assuming $W = 1$, $\bar{Q} = 10$, $\bar{P}_1 = 20$, $\bar{\gamma}_0 = 1$, $\lambda = 5$, $h_{p1}^2 = h_{s1}^2 = h_{p0}^2 = h_{s0}^2 = 1$, $\sigma_s^2 = \sigma_p^2 = 1$. The intersection of these best response curves specifies the Nash equilibrium for this system, which in this case turns out to be $\left(Q_0^*, p_1^*\right) = (4.49, 6.01)$. ∎

Another situation in which one can characterize the best response correspondences of primary and secondary users in a way that the Nash equilibrium of the spectrum leasing game can be visualized graphically is when all secondary users are essentially identical in terms of channel coefficients and correlation coefficients. Although such a system may not be realized in a practical wireless channel, it however serves as a basis to draw useful conclusions and gain insight regarding DSL game behavior as can be seen from the next example.

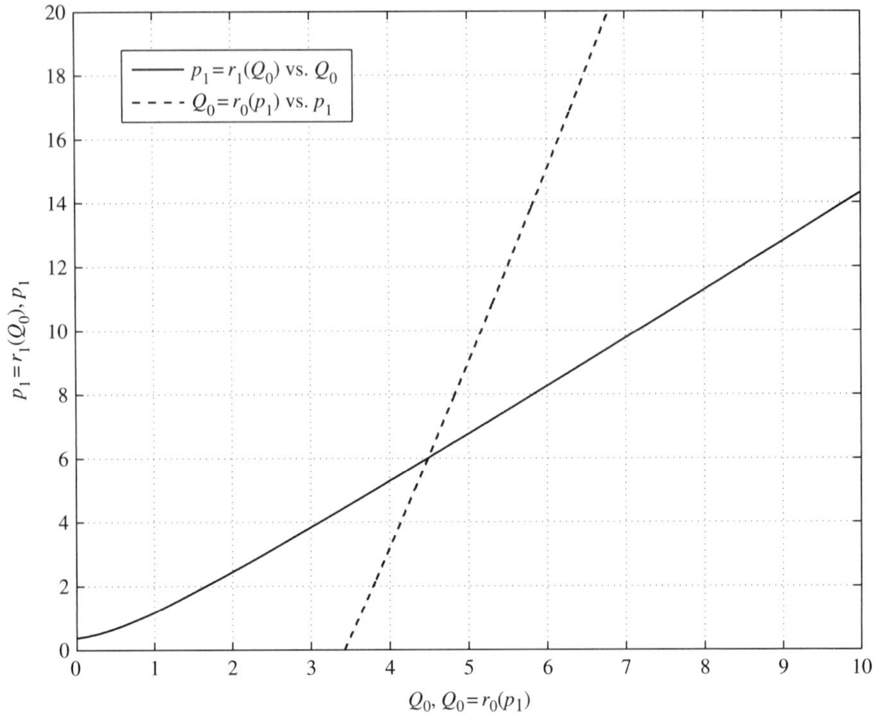

FIGURE 14.5 Primary and secondary user best response functions in a single secondary link dynamic spectrum leasing system with the MF receivers when $F(Q_0) = \log(1 + Q_0)$ and all cross-correlations are equal to 0.8.

Example 14.5 (The Nash Equilibrium of the DSL Game in a System with Identical Secondary Links) Consider a DSL game in which all secondary users have the same cross-correlation coefficients $\rho_{0k}^{(p)} = \rho_0^{(p)}, \rho_{k0}^{(s)} = \rho_0^{(s)}, \rho_{kj}^{(s)} = \rho^{(s)}$, for all $k,j = 1,\ldots,K$, and all channels are AWGN: $h_{sk} = h_{pk} = 1$ for all $k,j = 1,\ldots,K$. Then $\tilde{A}_k = \rho_0^{(p)} \triangleq \tilde{A}$ for all k.

By symmetry, it follows that all secondary users must have the same power $p_k = p^*$ at the Nash equilibrium (equivalently, the same SINR $\gamma_k = \gamma^*$). Thus, when the secondary system employs an MF receiver, with $F(Q_0) = \log(1+Q_0)$, the Nash equilibrium is characterized by the intersection $\left(Q_0^*, p^*\right)$ of the following two curves:

$$\begin{aligned} Q_0 &= r_0(p) \\ &= \left(\text{Solution to equation } Q_0 + (1+Q_0)\log(1+Q_0) = \bar{Q}_0 + K\tilde{A}^2 p\right) \end{aligned} \tag{14.36}$$

and

$$p = r_s(Q_0) = \left(\text{Solution to equation } g^{MF}(p) = 0\right), \tag{14.37}$$

where we have defined

$$g^{MF}(p) = \frac{1}{N}e^{-cp} + \frac{1}{N}e^{-\lambda Q_0}e^{(K-1)cp} - c\left(1+\frac{p}{N}\right)\log\left(1+\frac{p}{N}\right)e^{-\lambda Q_0}e^{(K-1)cp}.$$

Figure 14.6(a) shows the best response adaptations (14.36) and (14.37), in a $K=2$ secondary system with $W_k = W = 1$, $\bar{Q} = 10$, $\bar{P}_k = 20$, $\bar{\gamma}_0 = 1$, $h_{pk}^2 = h_{sk}^2 = 1$, for all k, $\sigma_s^2 = \sigma_p^2 = 1$, and all cross-correlations equal to 0.4. From the figure, we observe that the unique Nash equilibrium action profile in this case is $\left(Q_0^*, p_1^*\right) = (4.53, 12.5)$.

Figure 14.6(b) shows the first secondary link utility in this case as a function of its transmit power when the primary user and all other secondary users keep their actions fixed at the Nash profile. Of course, there is no incentive for this user to deviate from its Nash power value since its utility will only decrease if it does while others keep their actions fixed, as Figure 14.6(b) shows.

On the other hand, Figure 14.7 shows the equilibrium outcomes of the above (centralized) DSL game as a function of the secondary system size K_s. All parameter values are the same as in Figure 14.6 except for all cross-correlations being 0.5. From Figure 14.7, we see that the primary system can support only up to $K_s \leq 13$ secondary users before secondary system violates the primary IC.

Figure 14.8(a) shows the primary system utility at the Nash equilibrium of the system as a function of the secondary system size K_s, while Figure 14.8(b) shows both the sum rate $\sum_{k=1}^{K_s} \log\left(1+\gamma_k^{(s)}\right)$ and the per-user rate $\frac{1}{K_s}\sum_{k=1}^{K_s} \log\left(1+\gamma_k^{(s)}\right)$ achieved by the secondary system at the Nash equilibrium. As we observe from Figure 14.8(b), although the secondary sum utility increases slightly with the number of secondary users K_s in the system, the per-user utility monotonically decreases with increasing K_s. This is not surprising since the MF performance steadily worsens with the increasing interference. Depending on the application and the QoS requirement of the secondary system, each secondary user may have a minimum required rate (in bits per transmission) below which the transmissions would be useless. This QoS requirement will determine the maximum number of cognitive radios the secondary system would want to support at any given time. For example, if the minimum per-user rate required is 1 bit per transmission, according to Figure 14.8(b), the optimal K_s would be $K_s^* = 2$. If, on the other hand, the rate threshold was reduced to 0.5 bits per transmission, the secondary system might allow up to $K_s^* = 5$ users to operate simultaneously. In [220], it has been shown that by replacing the MF receiver by an advanced MUD such as *linear minimum mean squared error (LMMSE) receiver*, one may improve above performance. ■

In a realistic wireless channel with fading, the Nash equilibrium power profile of a DSL-based DSS system will depend on the channel-state realizations encountered by radios. Not surprisingly, the Nash equilibrium transmit powers of individual secondary users will also be different for each user. Hence, analytical solutions of the form of Examples 14.4 and 14.5 may not be possible. In the following example, we

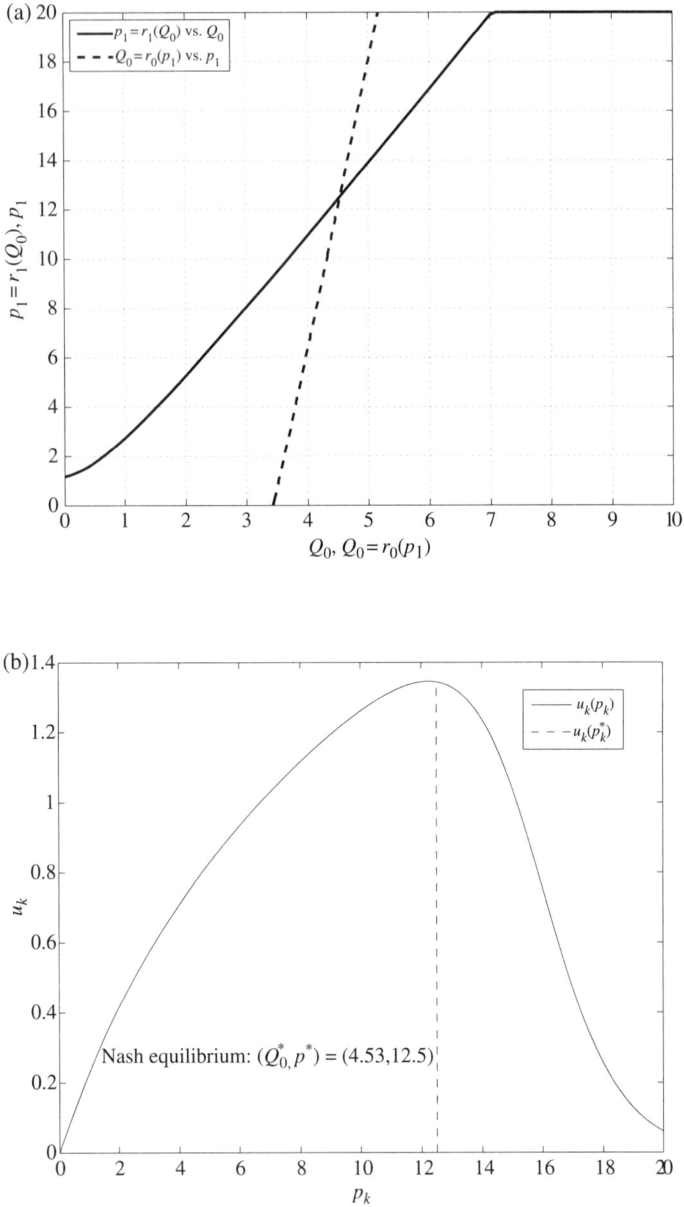

FIGURE 14.6 Best response functions and the secondary utility in a $K = 2$ (identical users) dynamic spectrum leasing network with secondary MF receiver, when $F(Q_0) = \log(1 + Q_0)$ and $\lambda = 5$. (a) Best response functions. (b) The utility $u_1(p_1)$ of the first secondary link when all other secondary users and the primary user keep their actions fixed at Nash equilibrium profile, that is, $Q_0 = Q_0^* = 4.53$ and $p_2 = p_2^* = 12.5$.

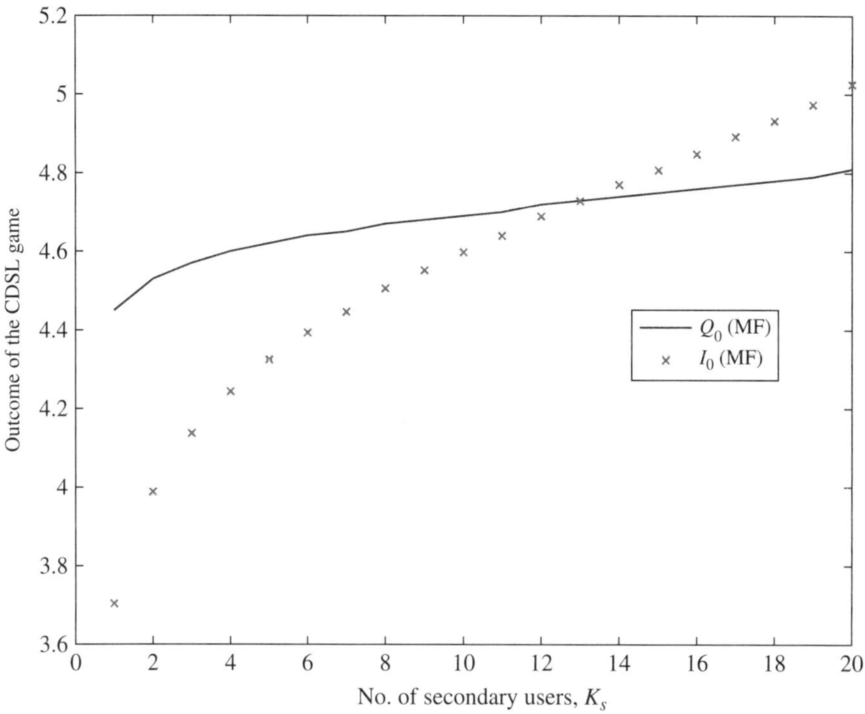

FIGURE 14.7 Outcome of the DSL game at the system NE with MF receiver, as a function of secondary system size K_s assuming identical secondary links, when $\lambda = 5$.

investigate the equilibrium behavior of a DSL game in the presence of fading channels based on Monte Carlo simulations.

Fading coefficients in wireless channels can be modeled in various ways depending on channel properties. Quasistatic fading model that we used in Examples 12.1 and 12.2 is one commonly used model. According to quasistatic fading model, each channel coefficient stays constant for the duration of a certain block but changes independently to a new value from one block to another. While this is a reasonable model that helps derive certain analytical error probabilities, it may not be entirely satisfactory in practice. In particular, it is likely that the fading coefficients may not stay constant over the duration of a time block but, at least slowly, time varying. In [220, 224], such temporally correlated fading has been modeled as a first-order Gauss–Markov process of the form of

$$h_{\cdot k}[n+1] = \sqrt{1-\epsilon^2} h_{\cdot k}[n] + \epsilon w_{\cdot k}[n], \tag{14.38}$$

where $w_{\cdot k}[n]$ is a complex white Gaussian random process of variance $\sigma^2_{\cdot k}$, independent of channel coefficients $h_{\cdot k}$'s, and ϵ is a parameter indicating the temporal

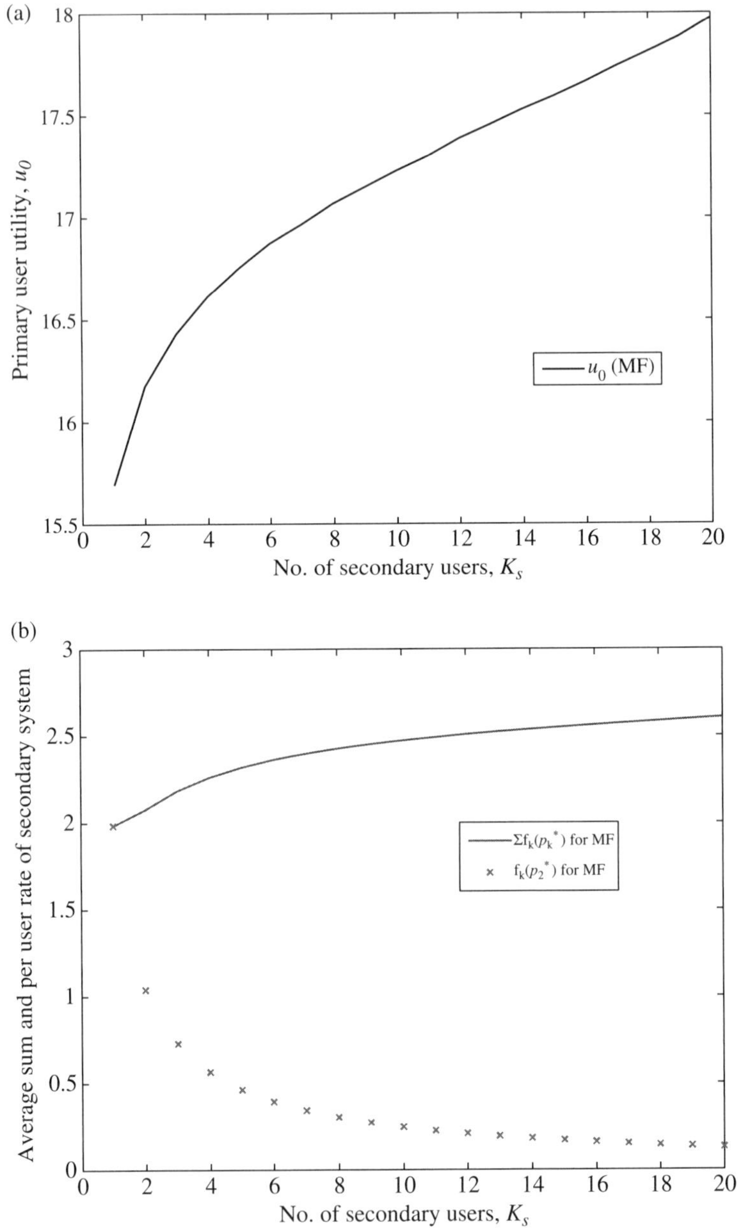

FIGURE 14.8 Primary and secondary utilities at the system NE of a DSL game with identical secondary users as a function of secondary system size K_S for $\lambda = 5$. (a) Primary system utility at the NE. (b) Sum rate and the per-user rate achieved by the secondary system at the NE.

variation rate of the channel. It is easy to verify that $\sqrt{1-\epsilon^2}$ represents the temporal correlation of the channel coefficients between two time instants.

In the following example, we investigate the performance of a DSL system with both quasistatic and time-varying fading models.

Example 14.6 (The Nash Equilibrium of the DSL Game in the Presence of Channels with Fading) Let us consider a primary–secondary DSL system in which, in the case of quasistatic fading, all channel fading coefficients are drawn from independent Rayleigh distributions normalized so that $\mathbb{E}\left\{|h|^2\right\} = 1$. Other parameters are $W_k = W = 1$, $\bar{Q}_0 = 10$, $\bar{P}_k = 20$, $\lambda = 5$, $\bar{\gamma}_0 = 1$, $\sigma_p^2 = \sigma_s^2 = 1$, and all cross-correlations equal to 0.5.

On the other hand, in the case of slowly time-varying fading, the Gauss–Markov fading model (14.38) with $\epsilon = 0.1$ is initialized with an $h_{\cdot k}^{(0)}$ drawn from a complex Gaussian distribution. In this case, we assume that the channel-state information is not instantaneously available to the receivers. Instead, each receiver updates channel-state information periodically every L samples, where the following simulations have assumed that $L = 10$.

In Figure 14.9(a), we have shown the DSL game outcome at the Nash equilibrium as a function of the number of secondary users K_s in the presence of both time-varying and quasistatic fading. It can be seen from Figure 14.9(a) that in quasistatic fading channels, the primary system can support up to $K_s \leq 13$ secondary users before the secondary system violates the primary IC. On the other hand, in the case of time-varying channels, the primary system can only support up to $K_s \leq 12$. Observe also from Figure 14.9(a) that in the case of time-varying fading scenario, values of Q_0 and I_0 are slightly higher than those for the quasistatic system.

Figure 14.9(b) shows that the secondary link utility is slightly decreased in time-varying channels compared to that with quasistatic channels. This is because of the incomplete channel information forcing the system to deviate from the actual Nash equilibrium. Note that the per-user utility decreases monotonically with K_s as before. ■

When $f(p_k) = \log\left(1+\gamma_k^{(s)}\right)$, the reward of a secondary link is the rate it can achieve assuming all other transmissions (both primary and secondary) as purely noise. The minimum transmission quality of the secondary system is defined as the average (over fading) minimum reward achieved by a link at the system equilibrium. Let us denote the minimum required QoS for secondary link k as $f_{min,k}$ and assume that $f_{min,k} = f_{min}$ for all secondary links. Thus, f_{min} represents the minimum rate a secondary user needs for successful communications. When the instantaneous rate falls below f_{min}, then the secondary user is considered to be in *outage*. We may thus define the *outage probability* of a typical secondary link as

$$\text{Outage probability} = \Pr\left(f_k\left(p_k^*\right) < f_{\min}\right).$$

Note that $f_k\left(p_k^*\right)$ is a random quantity as a result of user power levels and the IC being random quantities determined by the random channel fading coefficients.

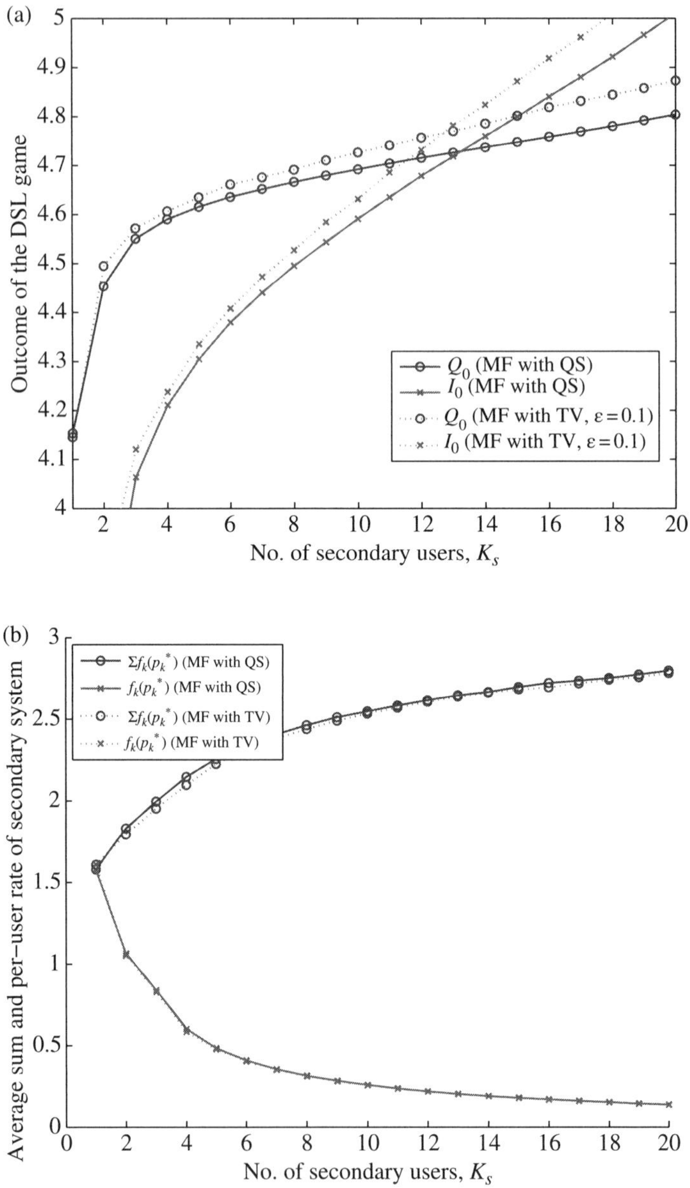

FIGURE 14.9 Equilibrium performance of a primary–secondary DSL game in the presence of channels with fading. (a) Outcome of the DSL game at the system NE as a function of secondary system size K_S. (b) Average sum rate and the per-user rate achieved by the secondary system at the NE as a function of secondary system size K_S.

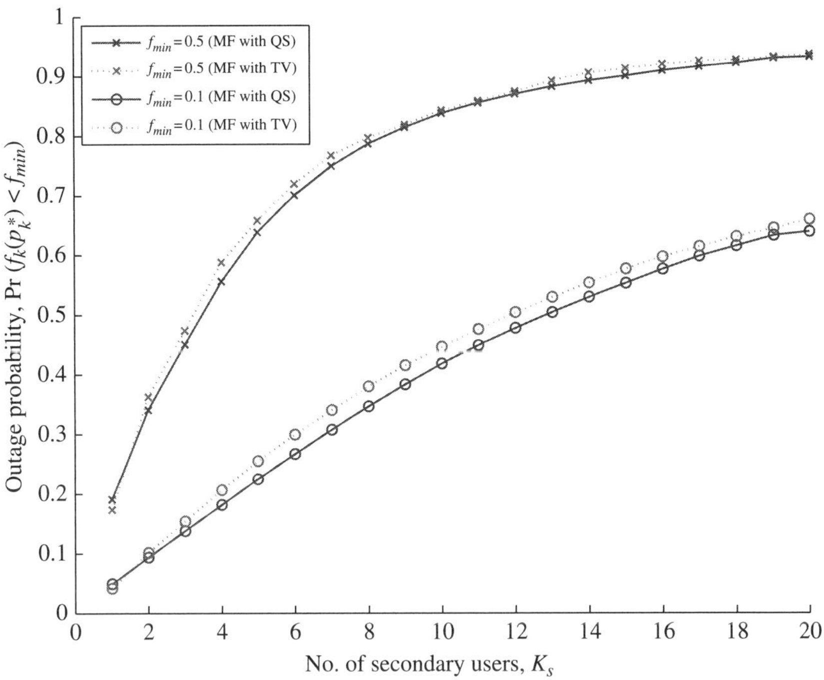

FIGURE 14.10 Outage probability $\Pr\left(f_k\left(p_k^*\right) < f_{min}\right)$ of a typical secondary user at the NE of the DSL game in fading channels as a function of secondary system size K_s for a required QoS requirement f_{min}.

Operating as secondary users causes secondary users to give up treating their own QoS as the first priority in favor of not violating the IC constraint. This forces them to select transmit power levels that will cause them to be in outage. In the following example, we use Monte Carlo methods to obtain the outage probability of a typical secondary user in a DSL game.

Example 14.7 (Outage Probability of a Secondary User in a DSL Game in the Presence of Channel Fading) In Figure 14.10, we show the outage probability of a typical secondary link in the same simulation setup considered in Example 14.6 as a function of K_s.

It can be seen from Figure 14.10 that, not surprisingly, the outage probability increases with K_s as well as with the minimum QoS requirement $f_{\min}$. The maximum secondary system size that can be supported according to Figure 14.9(b) thus needs to be interpreted in conjunction with the outage probabilities shown in Figure 14.10. For example, in quasistatic channels with a MF-based secondary system, Figure 14.9(b) shows that about four secondary links can (on average) meet the f_{min} = 0.5 bit per transmission QoS requirement. However, according to Figure 14.10, each of the these users may be in outage about 60% of time. This, of course, is the price of operating as

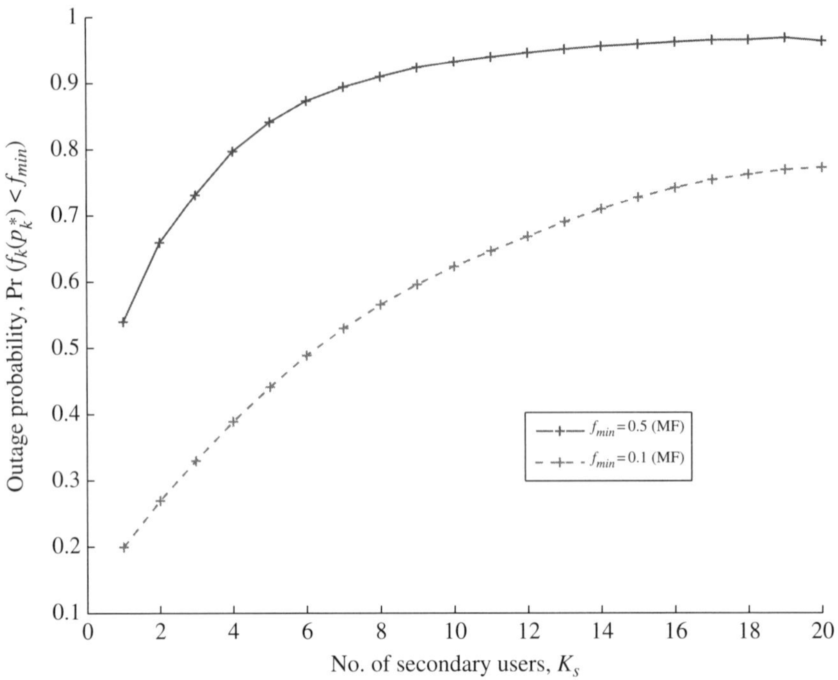

FIGURE 14.11 Outage probability $\Pr\left(f_k\left(p_k^*\right) < f_{min}\right)$ of a typical secondary user at the NE of the DSL game in fading channels with $K_p = 3$ for a required QoS requirement f_{min}.

a secondary system. Otherwise, radios can adjust their transmission powers according to the instantaneous fading coefficient values so that the transmission rate stays above the minimum required threshold.

Figure 14.11 shows the effect of more (primary) users in the primary system on the secondary user outage probability by considering a system with $K_p = 3$. In this case, the outage probability becomes worse due to increased total interference caused by the additional primary users in the system. ■

It must be emphasized that the above two forms of primary and secondary utility functions, however, are just two representative examples. Depending on the behavior one may want to impose on the resulting DSL interaction, these rewards and utility functions will have to be carefully chosen. That is a delicate issue to be addressed, taking into account not only the technical requirements of the form of the utility functions but also the practical meaning of them and the regulatory issues. We do not delve into such discussions since it is beyond the scope of our focus on signal processing aspects of cognitive radios. The above example scenario with the associated performance metrics is meant to provide reasonable insight into what issues need to be considered in evaluating a DSL system.

14.3 DSL BASED ON ASYMMETRIC COOPERATION WITH PRIMARY USERS

As radio systems evolve toward smart coexistence in the same spectrum bands, regardless of whether there is a license or not, it is likely that more systems will be designed to take advantage of cooperative behavior. The concept of cooperation can be applied at various levels (systems, intersystems, intersystem users, and intrasystem users). In this section, we may illustrate how cooperative communications gains can be used to enable DSL between a primary and a secondary system. To keep a balance, we will discuss this approach in the context of a spectrum interweave/overlay DSS network.

In the previous section, we considered DSL cognitive radios in which primary utility was of monetary nature. There, utility u_0 had no immediate consequence to the primary user's performance: The utility may be accrued over a time period and perhaps convert into monetary terms later. On the other hand, it is also possible to design DSL systems so that primary utility can be a resource that can be of immediate value to the radio. Indeed, rather than a monetary reward, the primary reward can be accrued in terms of savings on their communications resources, namely, the transmit power.

Such an approach can be realized by introducing asymmetric cooperative communications between primary and secondary radios: Secondary system, or secondary users, may facilitate cooperative communications gains for the primary signals, in return for getting access to the primary spectrum. For example, secondary users may spend a portion of their transmit powers to relay primary signals to their destinations. This is just one form of the so-called cooperative communications that we will discuss in more detail in Chapter 15. Note that we have an asymmetric cooperative communications architecture in the combined network consisting of primary and secondary systems since secondary users relay primary signals, while primary users do not reciprocate by relaying secondary signals. Instead, primary users lease a certain portion of their spectrum resource to secondary users in return for them relaying primary signals. One may think of this as secondary users utilizing their transmit power as a form of currency to buy bandwidth. The primary reward is the power savings achieved due to *cooperative relaying*. As we will see, such asymmetric cooperative communications can also be realized with very little intersystem information exchange.

Let us assume a decentralized primary system in which each primary link (i.e., a transmitter and a receiver pair) is allowed to act autonomously in making decisions on spectrum leasing. In a decentralized primary system, each primary link will entertain bids from the secondary system specifying how much power it would be willing to spend on relaying primary signals. The DSL game, thus, turns out to be an auctioning game in which primary users act as independent auctioneers. It is, however, straightforward to modify the following framework if the primary system acts as a centralized unit. Indeed, in that case, there is only one auctioneer and it may allow primary system to perhaps gain the best deal since there is no internal competition among primary users that may lead to some primary users accepting lower bids from secondary users.

The secondary cognitive radio system can also be either a centralized or a decentralized network. In a centralized CRN model, there may be an SSDC as in Chapters 12 and 13 that is responsible for making spectrum leasing decisions on behalf of the whole secondary system: The SSDC decides which secondary cognitive radio should cooperate with which primary user/link. Such coordinated decision-making as a single unit will allow secondary system to better negotiate with the primary users. However, each primary user may accept an offer of cooperation with a secondary user picked by the SSDC only if this would result in at least a certain minimum power saving. If the offer is too low, the corresponding primary user may simply decline the offer and access to its channel may be denied until the next round of bidding. Each primary user may keep its threshold power savings level as private information so as to encourage bids as high as possible from secondary system/users.

As we noted earlier also, in practice, it may be difficult to justify centralized decision-making in a CRN that operate as a secondary system. Such systems require dedicated (control) channels with enough bandwidth to support reporting of all spectrum sensing measurements or primary leasing offers at distributed cognitive radios to the SSDC, as well as channel and power allocation decisions from the SSDC back to the distributed radios. Thus, it is worth also considering a completely autonomous secondary CRN in which users make their own spectrum access decisions without any centralized control. In such a decentralized secondary network, secondary users may compete with each other in order to gain access to available primary channels. This leads to a D-DSL architecture that may particularly be justifiable when several heterogeneous (secondary) wireless networks are interested in leasing the same primary spectrum.

In a decentralized CRN, the secondary users who do not win a favorable channel at the beginning of a DSL auction can possibly learn over time to win a bid for a channel in subsequent bidding times. Each winning secondary node (one per each primary channel) may also use learning to revise its bid in subsequent bidding times to improve its own power savings. We may discuss a simple, yet robust, reinforcement learning mechanism to achieve distributed autonomous learning without any supervision in a D-DSL network. Interestingly, without any centralized control, both primary and secondary radios can still learn to arrive at an equilibrium in a completely distributed DSL framework.

14.3.1 A Primary–Secondary Coexistence Model

In almost all wireless systems, total available spectrum is usually divided into a multiple number of (primary) channels. For example, in familiar FM radios, the assigned spectrum from 88 to 108 MHz is divided into a set of 200 kHz wide FM radio channels. In early days of digital cellular communications, the spectrum allocated for global system for mobile communications (GSM) systems in Europe, 890–915 MHz (uplink) and 935–960 MHz (downlink), was also divided into a set of channels. Indeed, each 25 MHz spectrum allocated for uplink and downlink communications was divided into channels of 200 KHz each. In yet another example, various spectrum bands allocated for LTE systems are first divided into channels of 1.25–20 MHz width each.

Let us consider a spectrum band licensed to a primary system that is divided into a set of channels. A channel allocation scheme allocates these channels to primary users, as needed. This allocation can be static, as in FM radio, or it can be dynamic, as in cellular wireless systems. In the following, we assume that there are M primary users/links on M distinct primary channels. Thus, we will use the terms primary user, primary channel, and primary link interchangeably. A channel allocation to a primary user i implies that a certain channel resource W_i is allocated for that user. When a channel with bandwidth B_i is allocated to the primary user i, where $i \in \{1, \ldots, M\}$, then the channel resource W_i is the bandwidth B_i. It is this channel resource W_i that will be leased from primary users to the secondary users.

Channel resource, however, does not have to always correspond to channels in frequency. Thus, W_i does not have to be necessarily the bandwidth of a channel. For example, channels that are assigned to each primary user could be time division multiple access (TDMA) channels, in which case channel resource W_i would be the time slot length T_i of channel/user i. In an orthogonal frequency division multiple access (OFDMA)-based primary system, on the other hand, the ith primary user can be assumed to be allocated an L_i number of OFDMA subcarriers as in Figure 14.12.

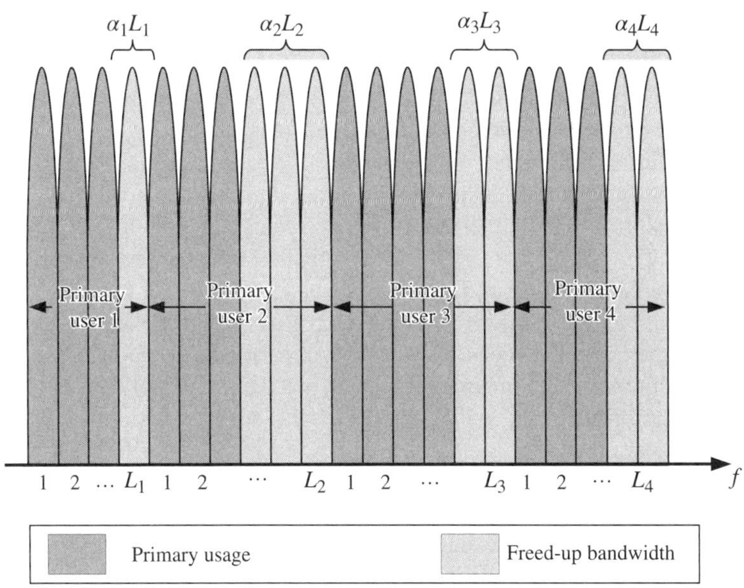

FIGURE 14.12 Distributed dynamic spectrum leasing (D-DSL) in an OFDMA-based wireless network. Channel resource to be leased can be considered as the OFDM subcarriers. Each primary user/link i, for $i = 1, \ldots, M$, may dynamically decide to lease an α_i fraction of its allocated subcarriers.

Hence, the channel resource W_i in this case is the L_i subcarriers. To be specific, in the following, we will assume primary channels to be distinct frequency division multiple access (FDMA) channels so that the channel resource W_i can indeed be thought of as bandwidth. However, the basic leasing framework is applicable to any type of channel definitions and resources W_i regardless of their physical interpretations.

For simplicity, we will also assume that each secondary user can transmit only over one channel at a time and that each primary transmitter–receiver pair (i.e., each primary link) i, for $i = 1, \ldots, M$, is allowed to lease its resource W_i to only one secondary user at a time. These assumptions can easily be relaxed in practice. They are assumed in order to keep the notation simple and discussion focused. Let us further assume that the primary system divides each time frame of duration T_f into a number of equal-length time slots. Suppose that channel fading coefficients vary only slowly within a time frame so that fading can be considered constant during one time frame justifying a quasistatic fading model. Let us denote the maximum transmit power of ith primary user by $\bar{P}_i$, for $i = 1, \ldots, M$.

As required QoS, interference, and the observed channel fading conditions change from one time frame to another, the ith primary user may be able to achieve its required QoS by using only $(1 - \alpha_i)$ fraction of its allocated bandwidth W_i, for $\alpha_i \in [0, 1]$. In DSS based on DSA, the primary users do not pay any attention to thus created spectrum white spaces. The secondary users are expected to sense the spectrum and detect these opportunities. If a secondary user successfully detects one of these seemingly random spectrum holes, it can access and use it until a primary user starts transmission on that channel again. As we already mentioned, there is no reason for the primary users to pay any attention to who accesses these white spaces, because they do not have anything to gain. The focus of DSA is on just utilizing the spectrum holes rather than efficient utilization of spectrum holes.

In contrast, in the following DSL framework, we assume that at the beginning of each time frame, the ith primary user can determine that it can achieve its required QoS by using only $(1 - \alpha_i)$ fraction, for $0 \leq \alpha_i \leq 1$, of its bandwidth W_i (or subcarriers L_i or time slot T_i). The ith primary user then broadcasts to the secondary users that it is making available up to an α_i fraction of its bandwidth W_i for secondary users to lease. If channel fading is frequency selective across bandwidth W_i, then ith primary user will have the freedom to decide which parts of its allocated bandwidth to be freed up. Although this may be an important aspect in practice, for simplicity, here, we will assume that each primary channel is frequency flat. Thus, we may arbitrarily assume that primary user i always frees the last $\alpha_i W_i$ portion of its channel for leasing.

As before, let us assume that there are K_s number of secondary users, each with maximum transmit power $\bar{P}_j$ for $j \in \mathcal{K}_s$ where $\mathcal{K}_s = \{1, \ldots, K_s\}$. We will call a primary user whose α_i announcements can be sensed by the jth secondary user a neighbor of the jth secondary user. Let us denote by $\Omega_j[n] \subseteq \{1, \ldots, M\}$ the set of neighboring primary channels/users of the jth secondary user at time n. Note that Ω_j sets, for $j = 1, \ldots, K_s$, are not necessarily disjoint.

At the beginning of each time frame, each secondary user $j \in \mathcal{K}_s$ receives all α_i's from primary users $i \in \Omega_j$ where we have dropped the time index n from $\Omega_j[n]$ to keep notation simple. For each $i \in \Omega_j$, the jth secondary user computes the portion $\beta_{j,i}$ of its transmit power $\bar{P}_j$, where $\beta_{j,i} \in [0,1]$, that can be allocated to relay the primary signals of the ith channel. Each secondary user computes this set of $\{\beta_{j,i}\}_{i\in\Omega_j}$ such that if it spends $\beta_{j,i}\bar{P}_j$ amount of its power to relay the ith primary user's signal, it can still achieve a minimum probability of error ϵ over a transmission bandwidth of $\alpha_i W_i$ using its remaining power $(1-\beta_{j,i})\bar{P}_j$. For this computation to be possible, the jth secondary user needs to know its channel-state information with respect to its intended receiver.

14.3.2 Asymmetric Cooperative Communications-Based DSL between Primary Users and a Centralized Secondary Network

Consider a secondary CRN that is centrally controlled. Suppose that at the beginning of each time frame, the ith primary transmitter determines it can free up an α_i fraction of its bandwidth (or channel resource) W_i and still meets its target QoS. The objective of the primary user is then to achieve a power saving by exploiting possible asymmetric cooperative communications facilitated by the secondary users in return for primary user i leasing them the α_i fraction of its bandwidth.

Let us denote by $h_{i,j}$ the fading coefficient from the jth secondary user (transmitter) to receiver of the ith primary link. As shown in Figure 14.13, the jth secondary user, for $j \in \mathcal{K}_S$, *coherently* relays the signal of the primary user $i \in \{1,\ldots,M\}$ over a link with fading coefficient $h_{i,j}$. For the sake of illustrating the method, we assume a *genie-aided cooperation* in that secondary user acting as the relay knows the primary message to be relayed instantaneously. In practice, this assumption can be realized when, for example, the two primary and secondary transmitters in question are close to each other compared to the distance between the primary transmitter and receiver [127]. In this case, the channel between these two transmitter nodes will have a relatively high gain allowing the secondary transmitter to successfully decode the message from the primary transmitter within a short fraction of the transmission

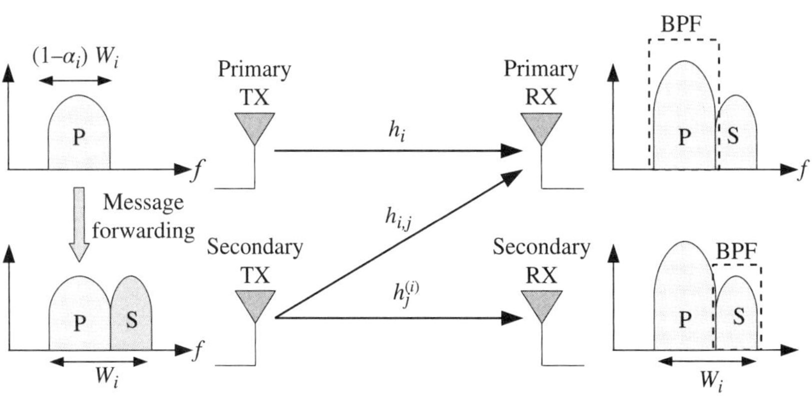

FIGURE 14.13 Asymmetric cooperative communication achieved by primary users with the help of secondary relays.

interval. Once the primary message is decoded, the jth secondary transmitter also starts transmitting the same primary message for the rest of the transmission interval simultaneously with the ith primary transmitter over the bandwidth $(1-\alpha_i)W_i$ that the primary user uses for its own transmissions. This genie-aided assumption, however, can easily be dropped by adapting a practical cooperative protocol at the expense of more elaborate notation [211, 225, 226] (see also Chapter 15).

The jth secondary user gets to transmit its own signal over the freed-up primary bandwidth $\alpha_i W_i$. We denote by h_i the fading coefficient between the primary transmitter i and the corresponding receiver, and $h_j^{(i)}$ the fading coefficient between the secondary transmitter j and its corresponding secondary receiver, when communicating over primary channel i.

14.3.2.1 *Primary and Secondary Actions* While transmitting at its nominal power level of $\bar{P}_i$, the rate that the ith primary user can achieve by freeing up α_i fraction of its bandwidth for leasing can be written as

$$R_i(\alpha_i) = (1-\alpha_i)W_i \log\left(1+\gamma_i(\alpha_i)\right), \tag{14.39}$$

where

$$\gamma_i(\alpha_i) = \frac{|h_i|^2 \bar{P}_i}{(1-\alpha_i)W_i N_i}$$

is the resulting signal-to-noise ratio (SNR) at the corresponding ith primary receiver. Suppose that the ith primary link needs a minimum rate of $R_i^{(min)}$ and that with the current realization of channel-state information h_i on the primary link, it can achieve a rate of $R_i^{(min)}$ with only using a minimum of $(1-\alpha_i^{(max)})$ fraction of its bandwidth (if it transmits at its nominal transmit power $\bar{P}_i$):

$$R_i(\alpha_i^{(max)}) = R_i^{(min)}.$$

Then, the ith primary link may free up an $\alpha_i \in [0, \alpha_i^{(max)}]$ fraction of its spectrum resource without degrading its QoS. All primary links select their α_i values according to their own preferences subject to this performance constraint and broadcast to the secondary radios.

Secondary user j receives all α_i's from its neighboring primary users $i \in \Omega_j$, as shown in Figure 14.14. Let us denote by $\gamma_{j,i}(\beta_{j,i})$ the SNR the jth secondary user can achieve (with respect to the corresponding jth secondary receiver) on the primary channel i (if it gets to transmit over that channel) using a transmit power level of $\bar{P}_j\left(1-\beta_{j,i}\right)$, where $0 \le \beta_{j,i} \le 1$:

$$\gamma_{j,i}(\beta_{j,i}) = \frac{\bar{P}_j\left(1-\beta_{j,i}\right)|h_j^{(i)}|^2}{\alpha_i W_i N_i}.$$

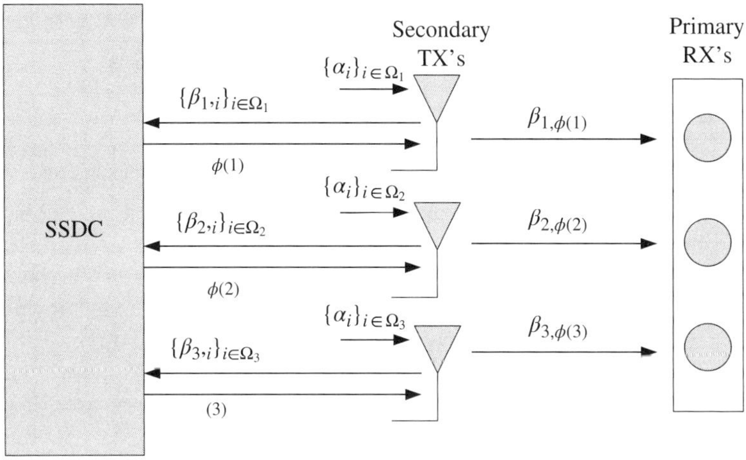

FIGURE 14.14 Primary channel and secondary cognitive radio assignment for cooperation at the secondary system decision center (SSDC).

Let us denote by $P_e^{(j,i)}(\beta_{j,i})$ the bit error rate (BER) of the jth secondary link when transmitting on primary channel i (if it gets to transmit over that channel) using the transmit power $\bar{P}_j\left(1-\beta_{j,i}\right)$.

Suppose that for successful communications, each secondary link needs to maintain a maximum BER of ϵ. For all primary channels $i \in \Omega_j$, then the jth secondary user computes the maximum power fractions $\beta_{j,i}^{(max)}$'s it can allocate to relaying the ith primary user signal while maintaining a maximum BER of ϵ with respect to its own receiver over the portion $\alpha_i W_i$ of channel i:

$$\beta_{j,i}^{(max)} = \arg \max_{\beta_{j,i}\in[0,1]} \beta_{j,i} \quad \text{s.t.} \quad P_e^{(j,i)}\left(\beta_{j,i}\right) < \epsilon. \tag{14.40}$$

For example, if we assume BPSK transmissions, then $P_{e,j,i}(\beta_{j,i}) = Q\left(\sqrt{\gamma_{j,i}(\beta_{j,i})}\right)$ for any $\alpha_i \neq 0$, so that $\beta_{j,i}^{(max)}$ can be obtained by numerically solving (14.40). If $\alpha_i = 0$, we may let $\beta_{j,i}^{(max)} = 0$ for all $j \in \mathcal{K}_s$, meaning that secondary users will not relay a primary user's signal who is not willing to lease any portion of its available bandwidth.

If the jth secondary user gets to transmit over channel i, then we may say that it receives a utility of

$$u_j(\beta_{j,i},\alpha_i) = \alpha_i W_i \log\left(1+\gamma_{j,i}(\beta_{j,i})\right) q\left(\epsilon - P_e(\beta_{j,i})\right), \tag{14.41}$$

where $q(.)$ is the unit-step function in (4.156).

Each secondary user's action is to determine the fraction of powers $\left\{\beta_{j,i}\right\}_{i\in\Omega_j}$ it would like to dedicate for relaying primary signals. Ideally, of course, each secondary user j would like to have $\beta_{j,i} = 0$ for $\forall j$ and $\forall i \in \Omega_j$. This will allow secondary transmitters to use all their power resources exclusively for the secondary signal transmission. Of course, primary users then will not have an incentive to lease the spectrum and thus would rather keep transmitting over the whole spectrum without freeing any portion.

Recall that each primary user will accept an offer of cooperation with a secondary user only if this would result in at least a certain minimum power saving to the primary user. Let us denote this minimum power savings threshold by P_i^{th}, where $P_i^{th} \leq \bar{P}_i$. In practice, each primary user expects to be able to reduce its transmit power below P_i^{th} by exploiting cooperative communications advantage in return for giving secondary users access to its channel. Recall also that P_i^{th} is the ith primary user's private information, and thus, in general, P_i^{th} of primary user $i \in \{1, \ldots, M\}$ is unknown to the secondary users. Hence, if the ith primary user does not receive an offer $\beta_{j,i}$ from the jth secondary user that will allow it to meet the target power reduction it expects, it may not accept the offer of cooperation with the jth secondary user. If the ith primary user does not receive an offer $\beta_{j,i}$ from any of the secondary users that is sufficient to meet its target power reduction, it may not lease the spectrum portion it advertised at all.

For that reason, the secondary users will attempt to choose their $\beta_{j,i}$ values large (just) enough to entice the ith primary user to agree to lease its spectrum portion while, of course, not exceeding $\beta_{j,i}^{(max)}$. Since they do not know P_i^{th}, this is a tricky situation for the secondary users. When $\beta_{j,i}$ is too small, there is no access at all. When $\beta_{j,i}$ is too large, there is unnecessarily wasting power to relay the primary signals.

There are many meaningful ways to strike a balance between these two extremes. Exact choice of best $\beta_{j,i}$ will depend on the particular application context and details of the secondary cognitive radio system. In particular, it will depend on how much power secondary radios have at its disposal. For example, if the secondary user j has a large battery life remaining, it can pick a value closer to $\beta_{j,i}^{(max)}$ and vice versa. Thus, one simple solution is to pick $\beta_{j,i}$ fraction using a particular distribution, or weighting, over $[0, \beta_{j,i}^{(max)}]$ (not necessarily uniform) where the weights are related to how concerned secondary radio is about its power availability:

$$\beta_{j,i} = \beta_{j,i}^{(max)} \left(1 - e^{-aT_j^{(res)}}\right), \tag{14.42}$$

where $T_j^{(res)}$ is the residual battery life of the jth secondary user.

14.3.2.2 Optimal Cooperative Channel Assignment at the SSDC Since the secondary cognitive radio system is a centralized system, however, it is not the individual secondary cognitive radio's utility that is of importance in making decisions but the overall gain to the secondary network as a whole. Hence, the SSDC is responsible for making final decisions on which secondary user should cooperate with which primary channel. A reasonable objective for the SSDC is to make bids for cooperation with primary users so as to maximize the secondary network sum rate.

At the beginning of each time frame, each secondary user computes the fractions $\{\beta_{j,i}\}_{i\in\Omega_j}$ of its power that it is willing to allocate for relaying primary signal as described above, for example, in (14.42). They then informs these $\{\beta_{j,i}\}_{i\in\Omega_j}$ values to the SSDC (over a control channel) as shown in Figure 14.14. The SSDC uses these $\{\beta_{j,i}\}_{i\in\Omega_j}$, for $j \in \mathcal{K}_s$, values and its knowledge of channel-state information (fading coefficients) to determine the cooperative channel assignment for each secondary user so as to maximize the secondary system's sum rate, as shown in Figure 14.14. The SSDC then broadcasts these decisions back to the secondary users through a control channel [133].

Formally, we may define the *scheduling function* at the SSDC that assigns each secondary user to a primary channel for possible cooperation as a mapping

$$\phi : \mathcal{K}_s \rightarrow \{0, 1, \ldots, M\},$$

where $\phi(j) = 0$ denotes that secondary user j is not assigned to cooperate with any primary channel.

Assuming that maximizing the secondary network sum rate is the objective of the SSDC, the optimal cooperative channel assignment ϕ^* of secondary users can be posed as the following optimization:

$$\phi^* = \arg\max_{\phi} \sum_{j=1}^{K_s} u_j(\beta_{j,\phi(j)}, \alpha_{\phi(j)}), \tag{14.43}$$

where $u_j(\beta_{j,i}, \alpha_i)$ is as given in (14.41). Note that we let $u_j(\beta_{j,\phi(j)}, \alpha_{\phi(j)})|_{\phi(j)=0} = 0$, meaning that the utility of a secondary user that is not assigned to any primary channel for cooperation is 0.

The problem (14.43) subjected to the constraints that each primary user i can be assigned to cooperate with only a single secondary cognitive radio and each secondary radio can cooperate with only a single primary user is an instant of the so-called assignment problem. Any assignment problem can be solved as a linear programming problem. However, they have special structure that allows for more efficient specialized solution algorithms. Indeed, the so-called Hungarian algorithm is known to find an optimal assignment at cubic complexity order [227, 228].

The Hungarian algorithm is best explained by assuming that the number of primary channels and secondary users is the same, say, M, and the objective is to minimize a given cost function. Then given a cost matrix C where $C_{i,j}$ denotes the cost of assigning i to j, for $i \in \mathcal{K}_p = \{1, \ldots, M\}$, and $j \in \mathcal{K}_s = \{1, \ldots, M\}$, the assignment problem is to find an assignment that minimizes the total cost $\sum_i \sum_j C_{i,j}$ subjected to the constraint that each $i \in \mathcal{K}_p$ can be assigned to only one $j \in \mathcal{K}_s$ and each $j \in \mathcal{K}_s$ can be assigned to only one $i \in \mathcal{K}_p$. This can be viewed as a matching problem on a *complete bipartite graph* as well. A *bipartite graph* is a graph whose vertices belong to two disjoint sets, and a complete bipartite graph is a bipartite graph that has the same number of elements in both vertex sets, similar to those in Figure 14.15. A matching in a complete bipartite graph is an assignment such that every vertex is connected to at most one vertex from the other set. Hence, the assignment problem is the same as finding a minimum cost matching in a complete bipartite graph.

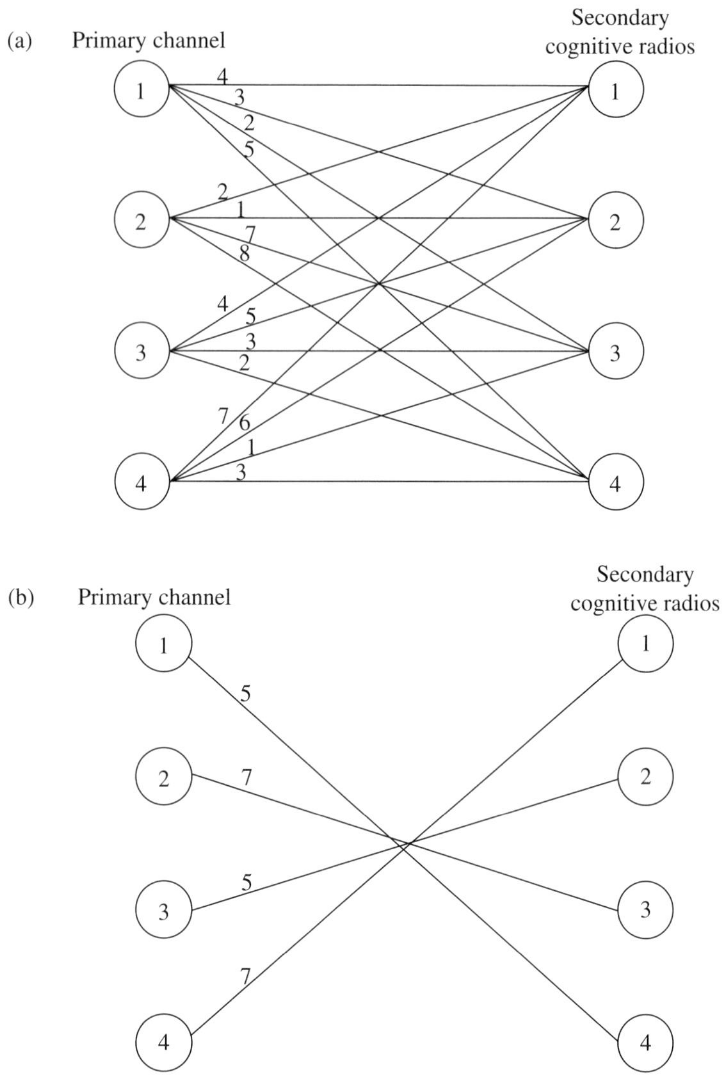

FIGURE 14.15 A bipartite graph representation of a cooperative channel assignment problem at an SSDC in a CRN. (a) The two vertex sets are formed by the primary channels and the secondary cognitive radios, respectively. The edge weights represent the utilities corresponding to assigning secondary user at one end of the edge to the primary channel at the other end of the edge. (b) The optimal assignment of cognitive radios to primary channels for the problem in Figure 14.15(a).

In the cooperative channel assignment problem of (14.43), the two disjoint sets of vertices are the primary channels and secondary users, while the edge weight between primary channel i and secondary cognitive radio j is taken to be the utility $u_j(\beta_{j,i},\alpha_i)$ given in (14.41):

$$C_{i,j} = u_j(\beta_{j,i},\alpha_i).$$

Figure 14.15(a) is an example of a weighted bipartite graph representation of a cooperative channel assignment problem made of $M = 4$ primary channels and $K_s = 4$ secondary cognitive radios.

Algorithm 14.1

```
The Hungarian Algorithm.
    Input: The M×M cost matrix C.
 (1) Subtract from each row of C the smallest entry in that row.
 (2) Subtract from each column of C the smallest entry in that
column.
 (3) Draw horizontal or vertical lines through appropriate rows and
columns of C so as to cover all zero entries of the obtained matrix
by using minimum number of lines.
 (4) If the minimum number of covering lines used in Step 3 is M,
go to step 6. Otherwise, go to Step 5.
 (5) Find the smallest entry not covered by any line. Subtract it
from each uncovered row of the current matrix. Then add it to each
covered column of the current matrix. Go to Step 3.
 (6)  Assign the zeros to obtain the optimal assignment.
```

Algorithm 14.1 gives the Hungarian algorithm that solves the assignment problem. Note that if we were to solve the general assignment problem of M nodes in each of the disjoint sets by trial and error, it will require trying out all $M!$ combinations of the pairs of nodes from the two sets. The Hungarian algorithm, however, solves the assignment problem (or the problem of finding an optimal matching in a weighted complete bipartite graph) with just cubic complexity order [227, 228].

The following example shows how Hungarian algorithm can be used to obtain an optimal assignment.

Example 14.8 (Finding a Minimum Cost Assignment: Hungarian Algorithm) Suppose we are given the following cost matrix:

$$\mathbf{C} = \begin{bmatrix} 5 & 8 & 7 \\ 8 & 10 & 6 \\ 4 & 8 & 5 \end{bmatrix}.$$

We apply Algorithm 14.1 to find a minimum cost assignment as below, with $M = 3$:

- *Step 1*: Subtract 5 from row 1, 6 from row 2, and 4 from row 3:

$$\mathbf{C} = \begin{bmatrix} 0 & 3 & 2 \\ 2 & 4 & 0 \\ 0 & 4 & 1 \end{bmatrix}.$$

- *Step 2*: Subtract 0 from column 1, 3 from column 2, and 0 from column 3:

$$\mathbf{C} = \begin{bmatrix} 0 & 0 & 2 \\ 2 & 1 & 0 \\ 0 & 1 & 1 \end{bmatrix}.$$

- *Step 3*: Cover all zeros in the matrix with the minimum number of lines. Note that an arrow at the beginning of a row or a column shows a line through that row or column:

$$\mathbf{C} = \begin{bmatrix} \rightarrow & 0 & 0 & 2 \\ \rightarrow & 2 & 1 & 0 \\ \rightarrow & 0 & 1 & 1 \end{bmatrix}.$$

- *Step 4*: Since the minimum number of lines in step 3 is 3, an optimal assignment of zeros can be done and we are finished:

$$\mathbf{C} = \begin{bmatrix} 0 & \boxed{0} & 2 \\ 2 & 1 & \boxed{0} \\ \boxed{0} & 1 & 1 \end{bmatrix}.$$

Thus, the minimum cost assignment in the original cost matrix looks like

$$\mathbf{C} = \begin{bmatrix} 5 & \boxed{8} & 7 \\ 8 & 10 & \boxed{6} \\ \boxed{4} & 8 & 5 \end{bmatrix}.$$

The minimum cost possible in this assignment problem is thus $c^* = 8 + 6 + 4 = 18$. ■

There are two modifications we need to do to use the Hungarian algorithm to solve the optimal cooperative channel assignment problem (14.43) at the SSDC, both of which are straightforward. First, the Hungarian algorithm solves a minimization problem, while our problem is a maximization problem. However, we can turn the problem (14.43) into an equivalent minimization problem by changing the cost matrix C to

$$C \Leftarrow c_m - C,$$

where

$$c_m = \max_{i,j} C_{i,j}.$$

We may then apply the Hungarian algorithm with this new cost matrix and find the optimal cooperative channel assignment.

Second, in the case of DSL cognitive radios, it may not always be true that $M = K_s$, as we have assumed in Figure 14.15 and as required for applying the Hungarian algorithm. However, any weighted bipartite graph can be turned into a complete weighted bipartite graph by setting the weights of the missing edges to be zero [228]. Thus, the Hungarian algorithm can still be used to find an optimal matching with this modification [229].

Example 14.9 (Finding a Maximal Matching in a Complete Weighted Bipartite Graph) Let us consider the problem of finding a maximal matching in the complete weighted bipartite graph shown in Figure 14.15(a). Note that, in this case, $M = 4$ and $K_s = 4$. The edge weights in Figure 14.15(a) correspond to, for example, utilities of pairing each primary channel with the corresponding secondary cognitive radio. These edge weights are used to define the following weight matrix W:

$$W = \begin{bmatrix} 4 & 3 & 2 & 5 \\ 2 & 1 & 7 & 8 \\ 4 & 5 & 3 & 2 \\ 7 & 6 & 1 & 3 \end{bmatrix}.$$

In order to apply the Hungarian algorithm, we first transform the above weight matrix into an equivalent cost matrix. By noting that $c_m = 8$ in matrix W, the equivalent cost matrix is obtained as

$$C = \begin{bmatrix} 4 & 5 & 6 & 3 \\ 6 & 7 & 1 & 0 \\ 4 & 3 & 5 & 6 \\ 1 & 2 & 7 & 5 \end{bmatrix}.$$

We apply Algorithm 14.1 to cost matrix C above to find a minimum cost assignment as below, with $M = 4$:

- *Step 1*: Subtract 3 from row 1, 0 from row 2, 3 from row 3, and 1 from row 4:

$$C = \begin{bmatrix} 1 & 2 & 3 & 0 \\ 6 & 7 & 1 & 0 \\ 1 & 0 & 2 & 3 \\ 0 & 1 & 6 & 4 \end{bmatrix}.$$

- *Step 2*: Subtract 0 from column 1, 0 from column 2, 1 from column 3, and 0 from column 4:

$$C = \begin{bmatrix} 1 & 2 & 2 & 0 \\ 6 & 7 & 0 & 0 \\ 1 & 0 & 1 & 3 \\ 0 & 1 & 5 & 4 \end{bmatrix}.$$

- *Step 3*: Cover all zeros in the matrix with the minimum number of lines. Note that an arrow at the beginning of a row or a column shows a line through that row or column:

$$C = \begin{bmatrix} \rightarrow & 1 & 2 & 2 & 0 \\ \rightarrow & 6 & 7 & 0 & 0 \\ \rightarrow & 1 & 0 & 1 & 3 \\ \rightarrow & 0 & 1 & 5 & 4 \end{bmatrix}.$$

- *Step 4*: Since the minimum number of lines in step 3 is 4, an optimal assignment of zeros can be done and we are finished:

$$C = \begin{bmatrix} 1 & 2 & 2 & \boxed{0} \\ 6 & 7 & \boxed{0} & 0 \\ 1 & \boxed{0} & 1 & 3 \\ \boxed{0} & 1 & 5 & 4 \end{bmatrix}.$$

Thus, the maximal assignment in the original weight matrix looks like

$$W = \begin{bmatrix} 4 & 3 & 2 & \boxed{5} \\ 2 & 1 & \boxed{7} & 8 \\ 4 & \boxed{5} & 3 & 2 \\ \boxed{7} & 6 & 1 & 3 \end{bmatrix},$$

which is shown in Figure 14.15(b). The maximum weight of the optimal assignment is thus $w^* = 5+7+5+7 = 24$. ■

Thus, with the above straightforward modifications, we use the Hungarian algorithm in Algorithm 14.1 to obtain the optimal cooperative channel assignment solution of (14.43). In this case, the two bipartite sets are $\mathcal{K}_p = \{1,\ldots,M\}$ and $\mathcal{K}_s$. The edge weights between primary $i \in \mathcal{K}_p$ and secondary $j \in \mathcal{K}_s$ are defined to be the utility $u_j(\beta_{j,i}, \alpha_i)$ given in (14.41).

14.3.2.3 Primary–Secondary DSL Auction with a Centralized Secondary System

At the beginning of each time frame, the SSDC informs the optimal channel assignment $\phi^*(j)$, computed as described in the previous section, to each secondary user $j \in \mathcal{K}_s$. In turn, each secondary user j sends, at its maximum transmit power $\bar{P}_j$, the value of power fraction $\beta_{j,\phi^*(j)}$ to its assigned primary user, as shown in Figure 14.14.

The receiver of the primary link $i = \phi^*(j)$ receives this bid from the jth secondary at a received power level of $P^R_{i,j} = |h_{i,j}|^2\bar{P}_j$. Using this knowledge, it computes the received SNR it will have if jth secondary user were selected to cooperate with it at the bid power level of $\beta_{j,i}\bar{P}_j$ to be

$$\gamma^{(p)}_{i,j} = \frac{\beta_{j,i}P^R_{i,j}}{(1-\alpha_i)W_iN_i}. \tag{14.44}$$

Note that $\gamma^{(p)}_{i,j}$ is the resulting SNR at the primary receiver i if the jth secondary user were to relay a signal over the $(1-\alpha_i)W_i$ bandwidth the primary link i decided to keep for its communications at a transmit power of $\beta_{j,i}P^R_{i,j}$.

Each primary user decides whether to accept or reject the offer of cooperation it receives from a secondary cognitive radio, depending on how much power saving it can achieve through cooperative communications. The achievable power savings may vary depending on the type of cooperative combining (e.g., the maximal-ratio combining (MRC), maximum SNR selection, or coherent relay detection)[2] the primary receiver employs. The ith primary receiver uses the particular combining rule to compute the resulting overall SNR it can achieve by combining the received signals from the two paths: direct path from the ith primary transmitter and the relayed path from the secondary node j with the SNR given by (14.44).

To be specific, let us consider coherent relay detection. Modifying the following steps for any other type of combining rule is straightforward. Let us denote by $R^f_i(P_i)$ the resulting final primary rate if the transmitter of the primary link i were to transmit at a power $P_i \leq \bar{P}_i$ while jth secondary user relays its signals at the agreed-upon bid power:

$$R^f_i(P_i) = (1-\alpha_i)W_i \log\left(1 + \frac{P_i|h_i|^2 + \bar{P}_j\beta_{j,i}|h_{i,j}|^2}{(1-\alpha_i)N_iW_i}\right). \tag{14.45}$$

For the primary link i to achieve a minimum transmission rate of $R^{(min)}_i$, if it were to accept the jth secondary user's bid for relaying, $R^f_i(P_i)$ should satisfy

$$R^f_i(P_i) \geq R^{(\min)}_i. \tag{14.46}$$

[2]We will illustrate some of these combining schemes in the next chapter.

From (14.45) and (14.46), primary user i can determine the minimum transmit power $P_i^{(min)}(\beta_{j,i})$ at which it will need to transmit if it were to accept the bid from the jth secondary user to be

$$P_i^{(min)}(\beta_{j,i}) = \min\left\{\bar{P}_i, \left[\frac{(1-\alpha_i)N_i W_i \bar{\gamma}_i(\alpha_i) - \bar{P}_j \beta_{j,i}|h_{i,j}|^2}{|h_i|^2}\right]^+\right\}, \tag{14.47}$$

where $[x]^+ = \max\{0, x\}$ and we have defined

$$\bar{\gamma}_i(\alpha_i) = 2^{\frac{R_i^{(min)}}{(1-\alpha_i)W_i}} - 1.$$

Thus, primary user i will accept the jth user's relaying power bid $\beta_{j,i}$ to cooperate with itself, only if

$$P_i^{(min)}(\beta_{j,i}) \leq P_i^{th} \leq \bar{P}_i, \tag{14.48}$$

where P_i^{th} is the ith primary user's target low-power threshold for accepting any cooperative communications requests.

In all the above computations, the primary receiver assumes that the channel from primary transmitter to the secondary relay is error-free. As we mentioned earlier, this can be a reasonable assumption under many scenarios [127]. However, when it cannot be justified, it is possible to modify the aforementioned method to take into account such errors at the expense of additional system complexity. Alternatively, observe that when the channel between the primary and secondary transmitters is not ideal, (14.45) can be considered as an upper bound to the actual primary transmission rate. Making the DSL cooperation decisions based on (14.45) is then still a valid criterion.

Finally, primary users who accept the bid offers from certain secondary users for cooperation will notify their decisions to those corresponding secondary users and start asymmetric cooperative communications-based DSL. Those primary users who decide not to accept possible offers from secondary users may keep transmitting over all of its licensed frequency band during the corresponding time frame.

14.3.2.4 Machine Learning in Centralized DSL Auctions Previous description only describes how DSL bidding and auctioning can be established at the beginning of a time frame. In practice, this interaction will continue over many time frames between these two systems. Being intelligent radio devices, cognitive radios thus have the ability to learn over time a best bidding strategy for each of the primary channels.

There are various ways to formulate such learning in cognitive radios participating in an asymmetric cooperative communications-based DSL. Many standard machine learning and adaptive algorithms can be used to implement such learning. Depending on how much interaction primary system is willing to have with the secondary system, the time scale of learning could, however, be different. For example, if primary system is only willing to entertain secondary power bids once every time frame, then any cognitive radio whose offer is not accepted will have to wait till the next time frame to place another bid. Since at the beginning of each time frame, primary radios may broadcast new α_i values, however, learning a strategy may take longer. On the other hand, if primary links are willing to consider revised bids during a time frame itself, this may give a cognitive radio the chance to learn a strategy faster.

In the following, we may consider a simple learning mechanism for a DSL environment. In this framework, secondary users who did not win access to a primary channel may send new offers $\beta_{j,i}$ to the SSDC in the following time slots. The SSDC computes new optimal channel assignments based on these new $\beta_{j,i}$ values. In particular, the secondary users in a centralized CRN may learn to increase their bid values $\beta_{j,i}$ within a time frame hoping that the primary users would accept the new offers. Moreover, a secondary user $j \in \mathcal{K}_s$ that has gained access to a channel $i \in \{1,\ldots,M\}$ might decrease its $\beta_{j,i}$ offer during the subsequent time slots in order to reduce the power spent on relaying the primary signal. We may refer to this model as the *centralized CRN with learning*. Note that compared to this model, the network model described in previous sections in which the secondary users keep their $\beta_{j,i}$ values fixed during a time frame may then be called a *static centralized CRN*.

In a centralized CRN with learning, at the beginning of each time frame, secondary users determine their bids $\beta_{j,i} \in [0, \beta_{j,i}^{(max)}]$ for $\forall i \in \{1,\ldots,M\}$. During subsequent time slots, secondary users update their $\beta_{j,i}$ values. The SSDC, in turn, computes the optimal assignment of secondary users to available primary channels (from (14.43)) based on these new $\beta_{j,i}$ values, for each of the time slots. The new bids are sent to the primary users who will decide whether to accept or reject those offers.

A simple reinforcement learning algorithm that secondary users may use to update their $\beta_{j,i}$ values in each time slot is given by

$$\begin{aligned} \text{Winning node: } \beta_{j,i}^{(new)} &= \beta_{j,i} - \mathcal{I}_{E(j,i)} \Delta\beta, \quad \text{for } \beta_{j,i}^{(new)} > 0 \\ \text{Losing node: } \beta_{j,i}^{(new)} &= \beta_{j,i} + \Delta\beta, \quad \text{for } \beta_{j,i}^{(new)} \leq \beta_{j,i}^{(max)}, \end{aligned} \tag{14.49}$$

where $\Delta\beta > 0$ is a suitable step size (or learning rate) and $\mathcal{I}_{E(j,i)}$ is the indicator function of the event

$$\begin{aligned} &E(j,i) \\ &\quad = \{\text{Secondary user } j \text{ has never lost a bid on channel } i \text{ in the current time frame}\}. \end{aligned}$$

In [33], it was shown that this simple learning algorithm indeed converges to fixed $\beta_{j,i}$ values after a sufficient number of time slots. As with any learning or adaptive system, the step size $\Delta\beta$ must be chosen carefully. A small step size could slow down the convergence, while a large step size will make $\beta_{j,i}$'s to have wild deviations around the final equilibrium. A common strategy is to start with a relatively large $\Delta\beta$ value and decrease it over the bidding iterations.

14.3.3 Asymmetric Cooperative Communications-Based DSL between Primary Users and Autonomous Cognitive Secondary Users

While having an SSDC to centrally compute the cooperative power bids for all secondary users may lead to maximizing the sum throughput achieved by the secondary CRN, communications overhead required for information exchange between the nodes and the SSDC may render it too expensive in practice. An alternative is to allow distributed secondary cognitive radios to make autonomous bid decisions and directly interact with primary users without any central coordination. Obviously, channel access in such a decentralized CRN is determined not only by the primary users but also by the competition among secondary users themselves.

This competition among secondary cognitive radios in a D-DSL system can also be formulated as an auctioning game in which each secondary user $j \in \mathcal{K}_s$ directly places a bid $\beta_{j,i}$ for each primary channel $i \in \Omega_j$. Recall that Ω_j, called the neighbor set of cognitive radio j, is the set of primary channels that the jth secondary user can sense. After computing its set of bids $\left\{\beta_{j,i}\right\}_{i\in\Omega_j}$, each secondary user j informs these values to corresponding primary receivers (could be the same receiver, if the primary system is centralized) at its maximum power level $\bar{P}_j$. There is no SSDC in the system, and cognitive radios do not forward their $\left\{\beta_{j,i}\right\}_{i\in\Omega_j}$ values to an SSDC as before. This D-DSL interaction is shown in Figure 14.16.

Receivers of each primary link will be responsible for determining the winning bid for that channel. If there are any ties among secondary users for winning access to a particular channel, then the corresponding primary user will randomly pick one of them. Once the winning bid is chosen, each primary user then informs its chosen winning secondary node of its bid being successful.

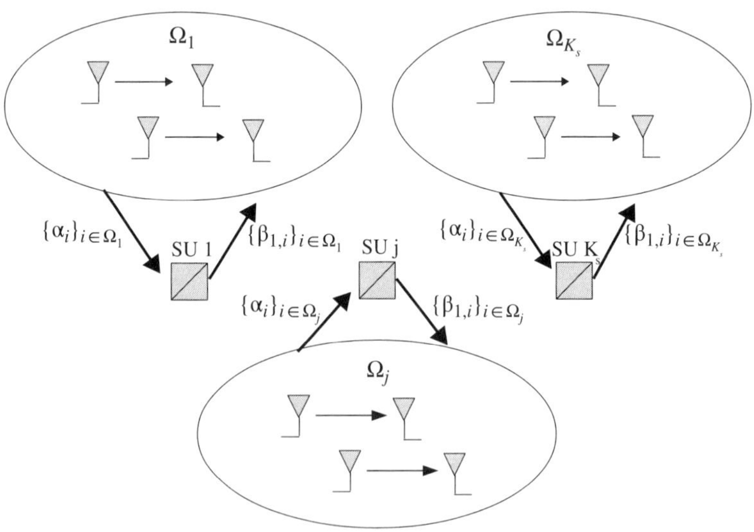

FIGURE 14.16 Distributed dynamic spectrum leasing (D-DSL) based on an auctioning game.

Unlike in the case of centralized secondary system, conflicts and competition among secondary cognitive radios are unavoidable in this scenario. Each secondary cognitive radio is interested in getting access to a primary channel, and the only way to achieve that is by beating out all other secondary radios in a game of bidding. While this can be beneficial for the primary users, it can, however, lead to inefficient spectrum usage.

Sometimes, a particular secondary user's bids may be selected by more than one primary channel as the winning bids. We will assume that each secondary user can cooperate on only one primary channel at a time. Then, this particular secondary user will decide to accept the invitation to cooperate with the primary channel that permits it to achieve the largest transmission throughput, assuming its main concern is getting channel access rather than how much power it will have to spend.[3] There are several ways the remaining primary users who accepted this secondary user's bids as the winning bid can react to the situation. Here, we assume that they will not lease their spectrum to any user during this time slot, thus encouraging the losing secondary users to increase their bids in the next time slot. Once the bid selection is done, then the primary and winning secondary users start to transmit based on the asymmetric cooperative communications, as is shown in Figure 14.13.

As before, each primary user computes its $\alpha_i^{(max)}$ value only at the beginning of a time frame since channel conditions are assumed to be constant over the duration of a frame. Secondary users are, however, free to place new bids at the beginning of each time slot within a given frame. This gives each secondary user a chance to revise its power bids in order to maximize its chance of getting access to the most favorable channel while minimizing the relaying power $\beta_{j,i}\bar{P}_j$ it has to spend. Thus, during each frame, the primary–secondary interaction can be modeled as a repeated auctioning game:

1. *Players*: M primary links (transmitter–receiver pairs) on M channels and K_s secondary users.
2. *Actions*: The action of the ith primary link is to choose $\alpha_i \in [0, \alpha_i^{(max)}]$ such that it satisfies the primary transmission rate requirements. The jth secondary user's action is to choose a set of power fractions $\beta_{j,i}$'s, for each $i \in \Omega_j$.

Each secondary user's aim is to relay primary signals at the lowest possible $\beta_{j,i}$ value. However, placing lower bids might reduce its chances of gaining channel access because primary users will prefer secondary users who are willing to spend a large amount of power for relaying their signals: The larger the relaying power from the secondary user, the smaller the power P_i primary user needs to spend. Since secondary radios act autonomously and compete among themselves, primary users have an edge in this type of systems compared to a centralized secondary system.

[3] Of course, if power is also a concern, then an acceptance criterion can be designed to take into account both throughput and power cost. It will, in general, be a trade-off between the two.

14.3.3.1 Selection of Winning Bids for Cooperative Communications in a D-DSL Network The objective of primary users is to minimize their own power expenditure by exploiting cooperative communications facilitated by secondary users. This objective can be achieved by maximizing the ith primary user utility function:

$$u_i\left(\alpha_i, \beta_{j^*(i),i}\right) = \frac{\bar{P}_i - P_i(\beta_{j^*(i),i})}{\bar{P}_i} q\left(R_i(\alpha_i) - R_i^{(\min)}\right), \tag{14.50}$$

where $\bar{P}_i$ and $P_i(\beta_{j^*(i),i})$ are, respectively, the nominal transmit power of the ith primary user without any cooperation and the transmit power of the ith primary user cooperating with the $j^*(i)$th secondary user and $j^*(i)$ is the index of the secondary radio who wins the chance to cooperate with the ith primary user. We let $P_i(\beta_{0,i}) = \bar{P}_i$ indicating that if primary user i does not reach an agreement with any secondary user, then it will be transmitting at its maximum power.

The utility (14.50) is maximized by selecting the secondary user that will allow the primary user to achieve its desired minimum transmission rate $R_i^{(min)}$ with the smallest transmit power $P_i(\beta_{j(i),i})$, which is $P_i^{(min)}(\beta_{j,i})$ given in (14.47). Hence, the ith primary receiver chooses the secondary user j that will lead to the smallest $P_i^{(min)}$ such that $R_i^f(P_i) \geq R_i^{(min)}$ as the winning bid for asymmetric cooperation on its channel:

$$j^*(i) \triangleq j^* = \arg\min_{j \in \mathcal{K}_s} P_i^{(min)}(\beta_{j,i}).$$

From (14.47), the above winning bid selection criterion simplifies to

$$j^*(i) \triangleq j^* = \arg\max_{j \in \mathcal{K}_s} \beta_{j,i} \bar{P}_j |h_{i,j}|^2. \tag{14.51}$$

Essentially, the primary user/link i will favor the secondary user j that will have the highest received power (or SNR, in general) at the receiver node of the ith link. Note that this is determined not only by the power $\beta_{j,i}\bar{P}_j$ secondary user j is willing to commit for relaying signals of the primary link i but also by the channel fading coefficient $h_{i,j}$ between the jth secondary user and the primary receiver i.

14.3.3.2 Repeated Auctioning-Based D-DSL with Reinforcement Learning The auctioning-based D-DSL game above assumes that both primary and secondary users can only choose their actions at the beginning of a time frame: Each primary user i picks its α_i and the winning secondary node for cooperation, while each secondary user picks its set of power commitments $\left\{\beta_{j,i}\right\}_{i \in \Omega_j}$. Once the winning nodes are selected, the primary users will cooperate with them for the duration of the whole time frame. Note that any primary channel that decided not to cooperate with a secondary user will then be not allocated to any secondary user for the whole time frame.

As in Section 14.3.2.4 in the case of a centralized secondary system, we may, however, modify this cooperative communications-based DSL structure so as to allow for better fairness as well as improved efficiency in spectrum usage. One way to achieve this is to allow primary users to consider new bids $\beta_{j,i}$ from secondary users at the beginning of each time slot as in Section 14.3.2.4. Similar to what we discussed in the case of a centralized secondary system, during the subsequent plays of this repeated auctioning game, secondary users can learn strategies of others, if the channel conditions stay fixed. This learning can help them win the auction for spectrum leasing during future time slots by revising their bids according to what they learn. As we discussed in Section 14.3.2.4, secondary users may update their bids using simple reinforcement learning strategy of (14.49). Note, however, that $\beta_{j,i}$ values adapted according to (14.49) may converge only if $\{\alpha_i\}$ stays fixed for a sufficiently long time.

Example 14.10 (A DSL Game with Asymmetric Cooperation) Let us consider a primary system with $M = 3$ equal bandwidth ($W_i = 10\,\text{kHz}$) channels and a secondary system with $K_s = 5$ cognitive radios. We will assume that all fading coefficients are Rayleigh with $\mathbb{E}\left\{|h|^2\right\} = 1$. The maximum transmit power of the primary users is $\bar{P}_i = 30$ mW, and the noise level at the receivers is $N_i = 0.1\,\mu\text{W/Hz}$.

The minimum transmission rate requirement of a primary user is assumed to be $R_i^{(min)} = 10$ kbps, while secondary users are assumed to require a BER smaller than $\epsilon = 0.05$. Each frame consists of 50 time slots, and the step size of the secondary learning algorithm is $\Delta\beta = 0.02$.

Figure 14.17(a) and (b) show the secondary sum rate and the average per-user primary power, respectively, as a function of maximum secondary transmit power $\bar{P}_j$, for both centralized and decentralized (i.e., with autonomous users) CRN, both with and without learning. Note that the no-learning (or static) scenario refers to setting α_i to $\alpha_i^{(\max)}$ and $\beta_{j,i}$ to $\beta_{j,i}^{(max)}\left(1 - e^{-aT_j^{(res)}}\right)$ for duration of the whole frame, where $a = \frac{1}{30*\text{number of slots/frame}}$ and $T_j^{(res)}$ is the total remaining number of slots at the beginning of a frame. Figure 14.17(a) shows that regardless of whether the secondary bid decisions are centralized or decentralized, learning permits the secondary network to achieve a higher sum rate, compared to the no-learning scenario.

In either case, the centralized CRN outperforms the decentralized CRN only if $\bar{P}_j \geq 50\,\text{mW}$. This is because when the secondary system is centralized, the ith primary user will agree to cooperate with a secondary user only if cooperation leads to it being able to transmit at a power level less than a certain threshold level of P_i^{th}. Since P_i^{th} is unknown to the secondary users, this forces them to allocate at least a minimum amount of power to relay the primary message, if they are to have any chance of winning access to the ith primary channel. Hence, secondary users will have an upper bound on the amount of power that they can use to transmit their own signals. However, if the secondary network is decentralized, the minimum power allocated to relay the primary message is only based on the competition among secondary users. In this case, the primary users do not have the expectation to reduce their own transmit power below a hard threshold (such as P_i^{th}). Instead, they accept the best bid from the competing secondary users irrespective of how small this bid is. This is seen by comparing the primary winning bid selection schemes (14.48) and (14.51). This difference makes both the primary power savings and the secondary sum rate

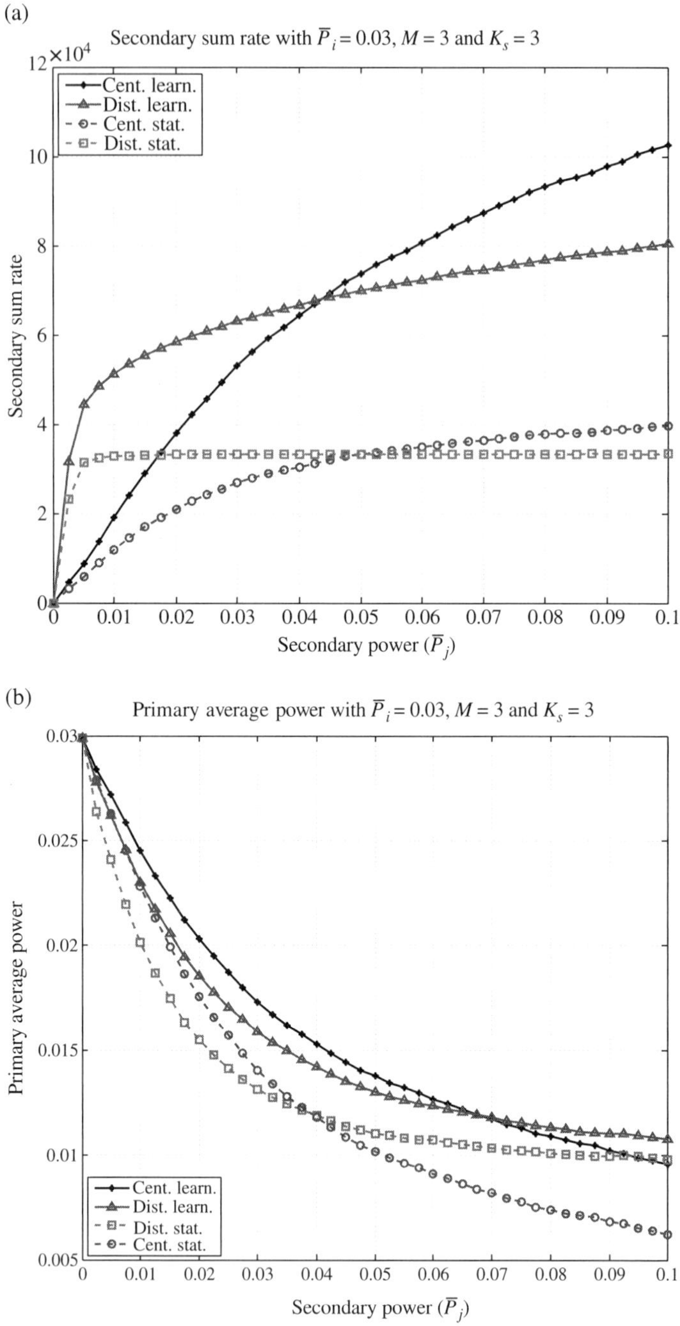

FIGURE 14.17 Performance of asymmetric cooperative communications-based DSL. (a) Secondary sum-rate performance versus $\bar{P}_j$. (b) Average per-user primary transmit power performance versus $\bar{P}_j$.

to be lower in the centralized case compared with the decentralized case when the secondary maximum power $\bar{P}_j$ is relatively small.

Figure 14.17(b) shows that, not surprisingly perhaps, having secondary users with learning ability is bad for the primary system. As can be seen from Figure 14.17(b), the average power spent by a primary user will be higher when secondary radios are allowed to learn compared to no-learning (static) case regardless of whether the system is centralized or decentralized. Of course, this is perhaps the reason why we are interested in cognitive radio at the first place: their ability to learn from experience. When there is no learning in the secondary cognitive radio system, secondary users will place higher $\beta_{j,i}$ bids at the beginning of each time frame since there is no second chance to revise these bids in case they are not accepted by the primary users. In a CRN with learning, on the other hand, secondary users learn to allocate just the minimum necessary power to relay primary signals while still gaining access to the primary channels. As a result, primary users will have to contend with smaller power commitments from the cooperative secondary radios, thereby leading to smaller savings in their own power expenditure. ■

The next example shows that the larger the number of cognitive radios in the secondary system, the better it is for both primary users and the overall secondary system performance.

Example 14.11 (Effect of the Secondary System Size on DSL Performance) Let us keep the same parameter values assumed as in Example 14.10 with the only exception that rather than assuming a fixed number of secondary users in the secondary system, now we investigate the performance of the asymmetric cooperation-based DSL game as a function of K_s. We will also fix the maximum transmit power of the primary and secondary users to be $\bar{P}_i = 10\,\text{mW}$ and $\bar{P}_j = 30\,\text{mW}$, respectively. Figure 14.18 shows the performance for two primary system sizes: $M = 2$ and $M = 4$.

Figure 14.18(a) shows that the sum rate of the secondary users increases with the number of secondary users K_s in the system. This is to be expected due to the increased *multiuser diversity*. It also shows that a significant gain can be achieved in the secondary sum rate when the secondary users employ learning.

Larger secondary system size is also beneficial to the primary system, as can be seen from Figure 14.18(b). Indeed, average per-user primary power decreases as K_s increases. Having more secondary users means there are more bidders in the DSL auction. Thus, primary users will have more options in selecting winning bids. ■

14.4 SUMMARY

- Dynamic spectrum leasing (DSL) is an alternative to DSA to implement dynamic spectrum sharing to improve spectrum utilization. The key difference between DSA and DSL is that while in DSA primary users are mostly passive spectators of the spectrum sharing process, in DSL, they are proactive participants in

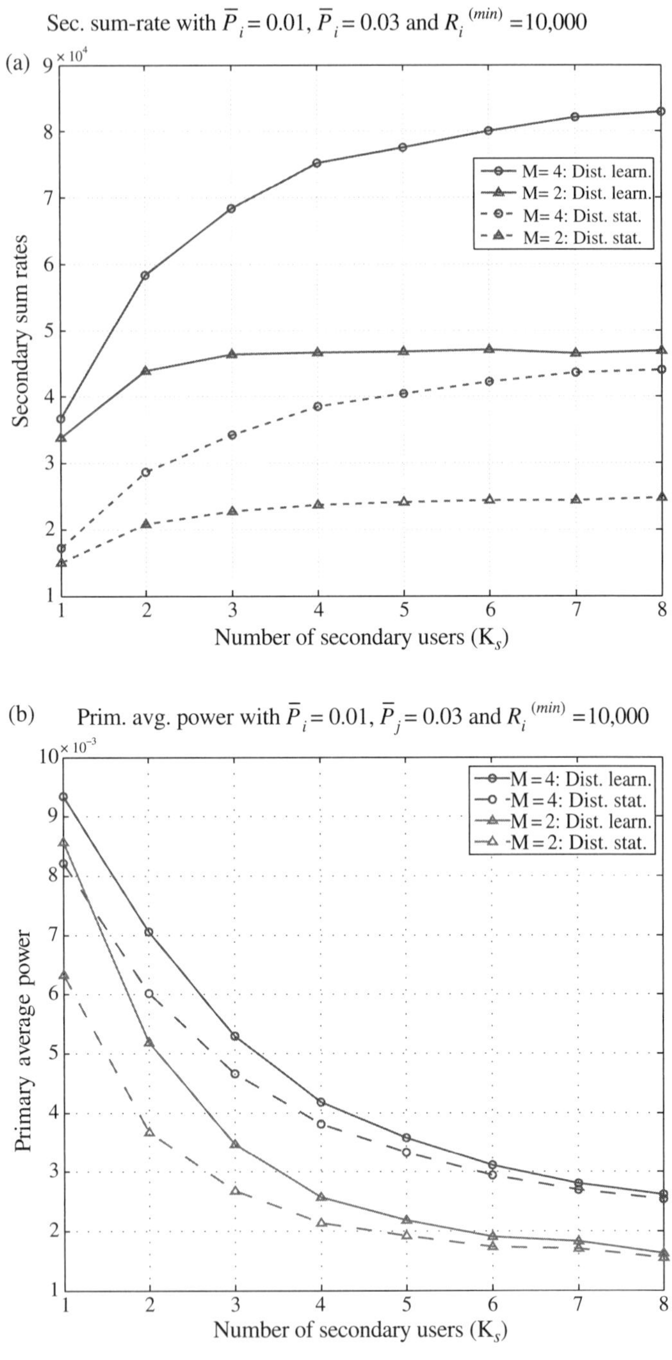

FIGURE 14.18 Performance of asymmetric cooperative communications-based DSL as a function of secondary system size. (a) Secondary sum rate performance versus K_S. (b) Average per-user primary power performance versus K_S.

it. In DSL networks, primary users are made to be interested in participating in the spectrum sharing process by providing them with an incentive to allow secondary spectrum access whenever it is feasible while maintaining their QoS above a minimum required threshold. This primary incentive can be proportional to the amount of secondary access it allows as well as to the demand from the secondary system to gain access to the primary spectrum.

- The primary system can lease its channels or spectrum under a centralized control or primary links may lease their resources autonomously. However, such autonomous spectrum leasing may not be justifiable since primary systems may not be cognitive systems.
- In a DSL game, secondary cognitive radio system can act as either a single centralized system or secondary users may be allowed to act as autonomous players. In a centralized secondary system, an SSDC may decide which secondary cognitive radio should attempt to gain access to which primary channel. As with all centralized decision-making systems, this will incur communications overhead to share information between the distributed cognitive radios and the SSDC. On the other hand, allowing cognitive radios to act autonomously may lead to certain inefficiencies but can be much simpler to implement.
- Similar to DSA, DSL can also be achieved with secondary system operating in either spectrum underlay or overlay/interweave modes.
- In general, in DSL, primary users dynamically adjust the extent to which they are willing to lease their spectrum. In return, they receive a leasing payment, called a reward or a payback, that is proportional to the amount of allowed spectrum access. This reward can be considered as a monetary reward or an in-kind payment that will help improve the primary communications performance.
- In a spectrum underlay DSL system, the primary users will indicate the amount of spectrum access they are willing to allow by setting a threshold on the maximum interference they are willing to tolerate from all secondary transmissions, called the interference cap. Whether acting as autonomous individuals or as a centralized system, secondary radios must honor this interference cap threshold. This DSL interaction among the primary and secondary users can be modeled as a noncooperative game. It is possible to implement such DSL schemes with minimal effort from primary users, in terms of the amount of control information they need to share with secondary system/users.
- In spectrum underlay-based DSL systems, managing interference can be the key to improve performance. In particular, if either system's receivers employ sophisticated multiuser interference suppression algorithms, this can lead to improved performance.
- When a DSL game is modeled as a noncooperative game, an equilibrium can be guaranteed by properly selecting the form of the utility functions. In particular, if conditions are established for the existence of a Nash equilibrium, then if all users are allowed to adapt their actions sequentially according to their best response correspondences, all user actions will converge to a Nash equilibrium of the game.

- In a spectrum overlay/interweave DSL system, the primary users will indicate the amount of channels, or spectrum, they are willing to leave idle for secondary users to access.
- Given the advanced capabilities of cognitive radios, DSL based on in-kind payments to the primary users can especially be implemented using asymmetric cooperative communications protocols. In these schemes, secondary users may spend a portion of their transmit power to asymmetrically relay primary signals to their destinations as an in-kind payment for getting access to primary spectrum. This could be interpreted as secondary users using their power as a form of currency to buy the primary bandwidth.
- Asymmetric cooperation between primary and secondary users need not be only limited to cooperative relaying. In general, many of the cooperative communications protocols such as distributed space time coding and virtual MIMO techniques can be used as the basis of cooperation.
- In an asymmetric cooperative relaying-based DSL framework, secondary users may spend a fraction of their power to relay primary signals in return for access to their spectrum. In general, secondary users will benefit by getting the primary users to accept a minimum amount of relaying power. Cognitive radios can learn over time to offer just the minimum relay power that a primary user will need to accept an offer of cooperation from them.
- The ability of learning in cognitive secondary radios can be beneficial to the cognitive radios, while it will be detrimental to the primary users. When cognitive radios learn, normally, primary users will end up having to accept lower relaying power offers.
- When the SSDC of the secondary cognitive radio system is responsible for determining which cognitive radio should cooperate with which primary user, the problem can be posed as finding a maximal matching in a bipartite graph or as an instant of the assignment problem. This problem can be solved efficiently using the so-called Hungarian algorithm.

15

COOPERATIVE COGNITIVE COMMUNICATIONS

15.1 INTRODUCTION

In Chapter 12, we discussed how secondary cognitive radios may make spectrum sensing decisions as a group with the aid of an SSDC. Such spectrum sensing frameworks are normally called cooperative spectrum sensing: Secondary cognitive radios cooperate among themselves to render spectrum sensing more accurate, effective, and efficient. In cognitive radio networks (CRNs), cooperation can, however, be used in a much broader sense than just for cooperative spectrum sensing. Indeed, initially cooperative communication was developed for improving performance of wireless communications systems independent of cognitive radios. The cognitive and intelligent abilities of a cognitive radio, or a CRN, of course provide an ideal platform to fully realize the potential gains of cooperation. Hence, naturally, it was later brought into CRN.

In general, cooperative strategies used among nodes in communications systems can be termed cooperative communications techniques. Cooperative spectrum sensing is just one form of cooperative communications. In fact, it is an incarnation of one of the traditional approaches used for improving the received signal-to-noise ratio (SNR) in wireless communications, namely, the *spatial diversity*. However, the more direct ancestor to cooperative communications as a broader theme arguably is the multiple-input multiple-output (MIMO), or dual antenna array, communication that was developed during the 1990s. Information theoretic characterization of enormous performance gains possible with MIMO communications in [230–232] spurred

Signal Processing for Cognitive Radios, First Edition. Sudharman K. Jayaweera.

a flurry of research activities and led to the development of a host of space–time coding [233–235] and signal processing [236, 237] algorithms that have already been incorporated into many emerging wireless communications systems.

When you have multiple transmit and receive antenna arrays, as in MIMO systems, it is possible to either increase the transmission rate (multiplexing gain) by using different data streams on different transmit antenna elements or improve the reliability (diversity gain) by essentially repeating data streams over different transmit antenna elements. Space–time coding and signal processing techniques essentially attempt to extract these gains possible with MIMO systems either in the form of diversity, multiplexing, or a combination of them. Of course, one has to come at the expense of the other and this obvious trade-off has been formalized in [238].

Realizing the potential of MIMO communications, by definition, requires both transmitter and the receiver to be equipped with multielement antenna arrays. Not all communications transceivers, however, are suited for this. Incidentally, a parallel research topic to MIMO was the wireless sensor networks that attempted to combine the flexibility of wireless communications with distributed sensor systems used in various remote sensing and monitoring applications. Wireless sensor nodes assumed in many of these WSN applications were to be constrained by limited battery power, making potential MIMO communications gains extremely attractive to these WSNs. In particular, it was shown in [213,215,217,218,239] that beyond a certain inter-node distance, power spent by a wireless sensor node on actual wireless transmissions can be much larger than that spent on actual operation of the sensor circuitry. However, wireless sensor nodes, as nodes in many other communications devices, were too small to host multielement antenna arrays precluding the direct use of MIMO techniques at very places they could have been extremely useful.

The modern cooperative communications emerged as a solution to this problem, albeit under various different terminologies such as *virtual MIMO*, *virtual antenna arrays*, *cooperative relaying*, or *distributed space–time coding* [209–212, 215, 240–243]. The common idea behind all of these techniques was to make a spatially close set of wireless nodes to cooperate to act as the elements of a virtual multiple antenna array.

Example 15.1 (Cooperative Space–Time Block Coding) Let us consider one of the simplest, and yet most elegant, forms of space–time block coding approaches and its direct cooperative implementation. It is well known that if there are multiple antenna elements at a receiver, by proper combining of the received signals at each element, the SNR can be improved in a fading channel. In particular, if there are M receiver antenna elements, received SNR can in fact be improved by a factor of M under ideal conditions. The so-called Alamouti scheme that can be considered as the forerunner to all space–time coding approaches for transmit antenna array systems was motivated by considering the situation in which the receiver only has a single antenna element but the transmitter is equipped with multiple antenna elements [233]. For example, in the context of wireless sensor networks, distributed nodes may be too small to support multiple receiver antennas, but a base station transmitter may be able to support multiple transmit antennas. Is it still possible to achieve an improved received SNR at the receiver by some clever processing, called space–time coding, of the signals transmitted from multiple antenna elements?

The Alamouti scheme is an elegant answer to this in the special case when there are only 2 antenna elements at the transmitter and a single antenna element at the receiver. Hence, it is also called the 2×1 space–time block code (STBC), where an $N \times M$ MIMO system means a system with N and M transmit and receive antennas, respectively. The clever approach of the Alamouti scheme allows one to achieve the same SNR improvement achievable in a 1×2 system with the 2×1 system.

Let us denote the possibly complex valued symbol sequence to be transmitted by $b[n]$, for $n = 0, 1, \ldots$. As shown in Figure 15.1(a), the Alamouti scheme takes two consecutive symbols, say $b[n]$ and $b[n+1]$, together at a time and transmits them by encoding across both space and time. Specifically, during the first transmission interval (or the symbol period), antennas 1 and 2 simultaneously transmit symbols $b[n]$ and $b[n+1]$, respectively. During the second transmission interval (or the symbol period), antennas 1 and 2 simultaneously transmit symbols $-b^*[n+1]$ and $b^*[n]$, respectively. This STBC is shown in Figure 15.1(b).

During each transmission interval, the received signal at the receiver antenna element (in baseband) can be written as

$$\text{Interval 1: } y[n] = h_1 b[n] + h_2 b[n+1] + w[n] \tag{15.1}$$

$$\text{Interval 2: } y[n+1] = -h_1 b^*[n+1] + h_2 b^*[n] + w[n+1], \tag{15.2}$$

(a)

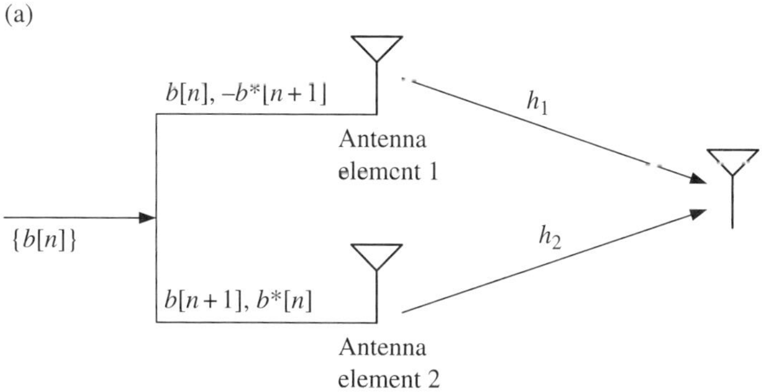

(b)

Space (antenna) \ Time	Symbol period 1	Symbol period 2
Antenna 1	b[n]	−b*[n+ 1]
Antenna 2	b[n+ 1]	b*[n]

FIGURE 15.1 Alamouti scheme for a transmitter with multiple antenna elements. (a) Alamouti scheme. (b) Alamouti scheme as a space–time block code.

where h_1 and h_2 denote the complex fading coefficients corresponding to the channel from transmit antenna elements 1 and 2, respectively, to the receiver antenna and $w[n]$ and $w[n+1]$ denote the additive zero-mean receiver noise that we will assume to be independent and have σ^2 variance. Let us assume that the fading coefficients are normalized so that $\mathbb{E}\{|h_1|^2\} = \mathbb{E}\{|h_2|^2\} = 1$.

Assuming that knowledge of channel-state information is available at the receiver, which is termed the receiver CSI case, the Alamouti scheme detects the symbols $b[n]$ and $b[n+1]$ based on the following decision statistics, respectively:

$$z_1 = h_1^* y[n] + h_2 y^*[n+1] \tag{15.3}$$

$$z_2 = h_2^* y[n] - h_1 y^*[n+1]. \tag{15.4}$$

Substituting (15.1) and (15.2), in (15.3), we get

$$z_1 = \left(|h_1|^2 + |h_2|^2\right) b[n] + \eta_1, \tag{15.5}$$

where $\eta_1 = h_1^* w[n] + h_2 w^*[n+1]$ is zero-mean noise with variance $\sigma_1^2 = \left(|h_1|^2 + |h_2|^2\right)\sigma^2$. Similarly, substitution of (15.1) and (15.2), in (15.4) shows that

$$z_2 = \left(|h_1|^2 + |h_2|^2\right) b[n+1] + \eta_2, \tag{15.6}$$

where $\eta_2 = h_2^* w[n] - h_1 w^*[n+1]$ is zero-mean noise with variance $\sigma_2^2 = \left(|h_1|^2 + |h_2|^2\right)\sigma^2$.

Let us denote the average SNR at the receiver when no multiple antennas were used at the transmitter and receiver by $\bar{\gamma}$:

$$\bar{\gamma} = \frac{\mathbb{E}\{|h|^2 |b[n]|^2\}}{\sigma^2} = \frac{\mathbb{E}\{|b[n]|^2\}}{\sigma^2}.$$

Let $\bar{\gamma}_1$ and $\bar{\gamma}_2$ denote, respectively, the average received SNR corresponding to detecting $b[n]$ and $b[n+1]$ from the decision statistics z_1 and z_2 above. Then, from (15.5) and (15.6), it follows immediately that

$$\bar{\gamma}_1 = \bar{\gamma}_2 = 2\bar{\gamma}, \tag{15.7}$$

which is the same SNR that would have resulted if a receiver array with 2 antenna elements were to be used. This shows that the Alamouti scheme indeed achieves the *full diversity* possible with two transmit antenna elements.

Now, instead of a single transmitter with a two-element antenna array, let us consider two transmitter nodes each equipped with a single antenna element as in Figure 15.2(a). Suppose both nodes attempt to communicate with the same receiver that is also equipped with a single antenna element. Let us denote the symbol sequence of nodes 1 and 2 by $b^{(1)}[n]$ and $b^{(2)}[n]$, respectively. In this situation, traditional approaches only allow achieving received SNR of $\bar{\gamma}$ for both node 1 and node 2.

(a)

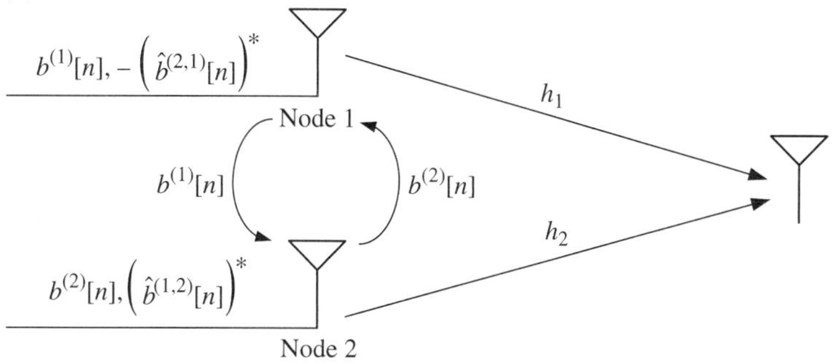

(b)

Time / Space (node)	Symbol period 1	Symbol period 2
Node 1	$b^{(1)}[n]$	$-\left(\hat{b}^{(2,1)}[n]\right)^* \approx -\left(b^{(2)}[n]\right)^*$
Node 2	$b^{(2)}[n]$	$\left(\hat{b}^{(1,2)}[n]\right)^* \approx \left(b^{(1)}[n]\right)^*$

FIGURE 15.2 Alamouti scheme and cooperative space–time block coding. (a) Cooperative implementation of the Alamouti scheme on two nodes each with a single transmit antenna element. (b) Distributed Alamouti scheme as a cooperative space–time block code.

A simple cooperative implementation of the Alamouti scheme, however, can allow both nodes to again achieve the same SNRs as $\bar{\gamma}_1$ and $\bar{\gamma}_2$ in (15.7), at least under ideal assumptions. Such cooperative schemes that implement MIMO transmission protocols developed for multiple-element transmit antenna arrays by exploiting node cooperation among nodes with single transmit antennas are also called the virtual MIMO techniques [215, 217, 218].

Indeed, suppose that the two transmit nodes are willing to cooperate. Let us assume a *genie-aided system* so that each node may know the other nodes symbol instantaneously without any error. In practice, this can indeed be achieved as follows: During the first transmission interval (or the symbol period), nodes 1 and 2 simultaneously transmit symbols $b^{(1)}[n]$ and $b^{(2)}[n]$, respectively. This is exactly the same as the first step in the Alamouti scheme above. However, during this transmission step, both node 1 and node 2 can also receive the other node's transmission. Hence, they may estimate each other's transmit symbols. Let us denote by $\hat{b}^{(1,2)}[n]$ the node 2's estimate of the node 1's symbol $b^{(1)}[n]$. Similarly, denote by $\hat{b}^{(2,1)}[n]$ the node 1's estimate of the node 2's symbol $b^{(2)}[n]$. During the second transmission interval (or the symbol period) of the Alamouti scheme, nodes 1 and 2 then simultaneously transmit symbols $-\left(\hat{b}^{(2,1)}[n]\right)^*$ and $\left(\hat{b}^{(1,2)}[n]\right)^*$, respectively. This cooperative STBC is shown in Figure 15.2(b).

If the two nodes are relatively closer, for example, we may assume that these estimates are error-free so that $\hat{b}^{(1,2)}[n] = b^{(1)}[n]$ and $\hat{b}^{(2,1)}[n] = b^{(2)}[n]$. Then the common receiver received signals are exactly the same as in (15.1) and (15.2). Hence, by forming decision statistics of the form of (15.3) and (15.4), the receiver can detect symbols $b^{(1)}[n]$ and $b^{(2)}[n]$ from nodes 1 and 2 both at an average received SNR of $2\bar{\gamma}$. ■

In general, there is an overhead involved in such distributed STBC or virtual MIMO approaches since nodes that act as the elements of an antenna array need to first share among themselves the data to be transmitted. For example, in the cooperative Alamouti coding in Example 15.1, if the two cooperating nodes cannot transmit and receive at the same time, then they need to first share their symbols before initiating the Alamouti scheme implementation. It was, however, shown in [213, 215, 239], albeit under simplifying assumptions, that even after accounting for such overhead, cooperative communications based on virtual MIMO or distributed space–time block coding can still offer significant performance gains. Moreover, various sophisticated cooperative communications protocols that go beyond simple and direct distributed implementation of then-existed space–time coding and signal processing algorithms have also been developed cementing cooperation as one of the most important resources in wireless communications [211, 212, 214, 226, 244].

In the context of cognitive radios, perhaps, cooperative communication receives its broadest meaning. Cognitive radios provide an ideal platform to realize the full potential of cooperation going beyond cooperative transmission protocols and combining schemes. The assumed abilities of reasoning, decision-making, and learning of cognitive radios can help realize sophisticated cooperative communications techniques. Interestingly, however, cooperation can in turn help boost these very same cognitive abilities of the radio, such as reasoning, decision-making, and learning, providing a perfect marriage.

As we already mentioned earlier, spectrum sensing was one of the first places where cooperation was used in cognitive radios. Simply combining either the local decisions or measurements collected by distributed secondary cognitive radios at a central location, that is, the SSDC, a CRN can improve the primary signal detection performance. As we saw in Chapter 13, however, spectrum sensing decisions can also benefit from cooperation: Making decisions as a single cooperative unit, cognitive radios can make more efficient decisions on which channels to sense at the first place to maximize global spectrum utilization performance of the network. This is one of the most unique advantages of cooperation in cognitive radios since spectrum sensing decisions are a unique feature of cognitive radios not encountered in traditional wireless networks. On top of all this, many sophisticated cooperative transmission protocols can also be integrated into cognitive radios, leading us to identify three ways cooperation is used in CRNs:

1. *Sensing decision fusion*: Combining distributed cognitive radio sensing measurements or decisions can improve received SNR. This is akin to received antenna array, or smart antenna, signal processing. Usually, this is called cooperative spectrum sensing.

2. *Spectrum sensing decision-making*: Making decisions on who should sense which channel or subband as a cooperative unit can greatly improve the chances of finding desired spectrum opportunities. This can lead to much better spectrum utilization. As mentioned, spectrum sensing decision-making is a function that is unique to cognitive radios. In many cases, these sensing decisions may also be made jointly with channel-access decisions.
3. *Virtual MIMO techniques*: Various cooperative communications protocols, distributed ST coding, and signal processing approaches can provide means of realizing the diversity and/or multiplexing gains offered by MIMO systems. These are collectively known as the cooperative communications. The cooperation can be among secondary cognitive radios or between the secondary cognitive radios and primary users in which case the protocols can be asymmetrical.

Literature on cooperative communications is extensive and there are already even several books dedicated to the topic [19, 245–247]. Taken in the context of virtual MIMO communications, the topic becomes even broader. Hence, in this book and in this chapter particularly, we do not attempt the impossible task of providing a comprehensive coverage of cooperative communications. Instead, our objective is to identify how cooperative communications, and node cooperation in general, can be incorporated into cognitive radios. Our discussion will be structured along the above three broad approaches in which cooperative behavior can be explored in a CRN.

15.2 COOPERATIVE SPECTRUM SENSING

In Chapter 12, we already discussed limitations of autonomous primary signal detection and the benefits of cooperation among secondary cognitive radios. In wireless channels, cooperative spectrum sensing can improve performance in several ways. First of all, consider secondary cognitive radios that cooperate in detecting spectral activities, or primary signals, by sharing their individual observations. Such sharing of distributed observations can be achieved under several different architectures: Among them are the sharing with one's own neighbors and sharing with a central fusion center such as the secondary system decision center we encountered in Chapter 12, shown, respectively, in Figures 15.3(a) and (b). Either way, if direct measurements are shared and combined in making detection decisions, it essentially leads to an improved effective SNR. If, instead, local decisions were shared among neighbors or with an SSDC, a similar improvement in performance can be gained. The performance improvement, or SNR improvement, can essentially be attributed to diversity afforded by distributed observations.

In Chapter 12, we also discussed the hidden terminal problems encountered in wireless communications and, in particular, in dynamic spectrum sharing cognitive radio networks. These effects are a manifestation of fading and shadowing effects inherent in wireless transmissions. Again, cooperation among distributed secondary

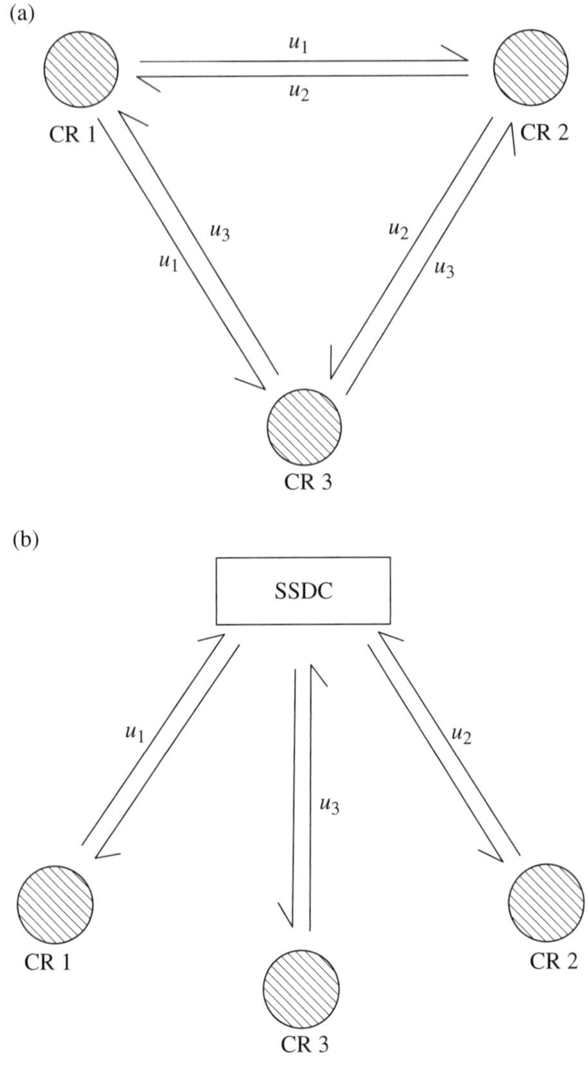

FIGURE 15.3 Two different frameworks for cooperative spectrum sensing. (a) Distributed CR nodes may share their local observations/decisions with their neighbors. (b) Distributed CR nodes may forward all their local observations/decisions to a common central node (such as the SSDC).

cognitive radios can allow them to observe the primary signals from several different vantage points, thus counteracting hidden terminal issues.

We do not go into details of cooperative spectrum sensing here since it was discussed in Chapter 12. Essentially, these techniques are very similar to distributed detection with data/decision fusion problems that have been investigated heavily in the context of wireless sensor networks (see, e.g., [194–196, 199]). A survey of cooperative spectrum sensing techniques can also be found in [248]. As noted in [248], perhaps the only difference in cooperative spectrum sensing for detecting primary users versus the plethora of work in the context of wireless sensor networks that proceeded it is in the assumptions on wireless channel properties and communications distances.

15.3 COOPERATIVE SPECTRUM SENSING AND CHANNEL-ACCESS DECISIONS

One of the unique functionalities found in cognitive radios that is absent in conventional radios, is the spectrum sensing decision-making by which we mean the decisions on which subband (Chapter 9) or channel (Chapter 13) to sense at any given time. These decisions are of extreme importance to cognitive radios—be they used as wideband devices or for the purpose of dynamic spectrum sharing. As we discussed earlier in Chapter 2, the intelligence and cognition relies heavily on radio's ability to be aware of its own environment and be able to respond to the perceived state of the environment. In light of unavoidable hardware limitations that prevent sensing the whole spectrum at once, a cognitive radio needs to make an educated guess as to which spectrum portion (be it a subband as in the case of a wideband cognitive radio or a channel as in the case of a DSS cognitive radio) it must sense in order to maximize the chance of finding spectrum opportunities that meet its performance objectives. The subband selection problem of Chapter 9 discussed the spectrum sensing decision-making in the context of a wideband cognitive radio. Its counterpart in DSS cognitive radios is the channel-sensing decision-making problem, which was also discussed earlier in Chapter 13. The previous two chapters on dynamic spectrum access and dynamic spectrum leasing, on the other hand, were examples of spectrum-access decision-making problems in the context of dynamic spectrum sharing cognitive radios. Cooperative techniques can be incorporated into all these decision-making problems.

In fact, in Section 13.4, we already discussed how cooperation can be beneficial in spectrum sensing decision-making problems. Recall that the basic idea of the approach in Section 13.4 was that final decisions are made at an SSDC on behalf of all secondary cognitive radios. As in the case of cooperative primary signal detection approaches discussed in the previous section, here also, distributed cognitive radios forward their local decisions, or observations themselves, to the SSDC. At the beginning of the sensing interval $n = n$, the SSDC is responsible for instructing each cognitive radio as to which primary channel it must be sensing during that sensing interval. As we saw in Section 13.4, making these decisions jointly at an SSDC avoids potential conflicts among nodes who may sense the same primary channel. This may lead to improved efficiency in idle primary channel utilization by the secondary CRN as a whole.

We discussed how such a cooperative spectrum decision-making problem at the SSDC can be formulated as a partially observable Markov decision process in

Section 13.4. The decisions are based on all available historical information about channel occupancy, dynamics, and actions at the SSDC. Similar to autonomous decisions, cooperative decisions can also be made to be optimal over a particular finite- or infinite-length time horizon or simply be myopic optimal.

The close similarity between the structure of the subband selection problem of Chapter 9 with the channel selection problem of Chapter 13 suggests that the cooperative decision-making framework of Section 13.4 can also be applied to the subband selection problem in wideband spectrum sensing when the network architecture permits it.

Recall that the wideband cognitive radios of Chapter 9 were treated as autonomous decision-making devices. However, this does not preclude them from being able to cooperate with other radios to reap benefits of cooperation. For example, regardless of how intelligent the radio itself is, it will belong to some wireless network. Hence, on top of the individual radio's own decisions on which subband to sense, a central control device, similar to the SSDC, at the network level may still provide at least additional decision support by taking into account historical information gathered by all nodes in the network. A more plausible and desirable approach in a wideband CRN, however, will be the direct cooperation among cognitive radios rather than relying on such centralized decision-making centers similar to the SSDC in a DSS network. We may refer to such cooperative approaches as *cooperative subband selection.*

As an example of cooperative subband selection, let us consider a set of K wideband cognitive radios operating over a certain spectrum range as in Chapter 9. In Chapter 9, each cognitive radio made its own subband selection decisions for sensing based entirely on its own historical information that consisted only of its own observation and action histories. Just as we discussed in Section 13.4, however, such an autonomous cognitive radio may be hampered by its lack of spatial diversity. If the K wideband cognitive radios are to cooperate among themselves, we may modify the subband selection decision-making framework of Chapter 9 by assuming that each cognitive radio periodically broadcasts a certain information regarding the spectrum. For simplicity, we may assume that this is done at the beginning of every decision-making epoch although it is easy to consider longer sharing intervals if so desired. Let us consider a particular cognitive radio denoted by index k, for $1 \leq k \leq K$. At time $n = n$, kth wideband cognitive radio receives information broadcast by its neighbors. As we did in Chapter 14, let us denote the set of cognitive radios that is the neighbors of the cognitive radio k at time n by $\Omega_k[n] \subseteq \{1, \ldots, K\}$. Note that the dependence of $\Omega_k[n]$ on time n is caused by possible relative node and/or channel dynamism. By definition, we will take node k to be a neighbor of itself so that $k \in \Omega_k[n]$ for all n and all k.

Recall from (9.31) in Section 9.6 that the cognitive radio k makes its selection of a subband for sensing at time n based on the information history $\underline{h}_n = (\underline{h}_{n-1}, a_{n-1}, y[n-1])$ available to it at time n.

For simplicity of this discussion, let us assume that the information that each cognitive radio shares with its neighbors is its observations $y_k[n]$ and the corresponding action choices $a_{k,n}$, for $1 \leq k \leq K$, where $a_{k,n}$ and $y_k[n]$ denote, respectively, the action and observation of the kth wideband cognitive radio at time n. Recall from Section 9.6

that each $Y_j[n]$ is binary since we assumed that the observations are the outputs of the channel-state detectors. In this case, the information the kth cognitive radio receives from its neighbors at time n is

$$\left\{\left(a_{j,n}, Y_j[n]\right)\right\}_{j\in\Omega_k[n]},$$

where we have assumed error-free information exchanges among cognitive radios. If the internode errors are to be taken into account, one may do so by modifying the crossover probabilities of the binary asymmetric channel in Figure 9.6 to incorporate the additional errors introduced by noisy communications channels.

Let us denote the information history available to the kth wideband cognitive radio for making decisions at time n by $\underline{h}_{k,n}$ where

$$\underline{h}_{k,n} = \left(\underline{h}_{k,n-1}, \left\{(a_{j,n-1}, y_j[n-1])\right\}_{j\in\Omega_k[n-1]}\right), \tag{15.8}$$

with the initialization $\underline{h}_{k,0} = \phi$ for all k. In Chapter 9, we saw that once we model the state of the spectrum at time n as $\mathbf{S}[n]$, optimal decisions of the kth cognitive radio within a POMDP model can be based on its belief of actual spectrum state given all available information. Hence, we may define a new belief state $b_{k,m}[n]$ at cognitive radio k given information history (15.8) as

$$b_{k,m}[n] = \Pr\left(\mathbf{S}[n] = \mathbf{s}_m | \underline{h}_{k,n}\right), \tag{15.9}$$

where $\mathbf{s}_m$, for $1 \le m \le 2^M$, denotes the possible states the spectrum can be in any given time. As in Section 9.6, we may define the belief-state vector of cognitive radio k at time n as

$$\mathbf{b}_k[n] = (b_{k,1}[n], \ldots, b_{k,2^M}[n])^T \in \mathcal{P}^{2^M-1}, \tag{15.10}$$

where, as before, $\mathcal{P}^{2^M-1}$ denotes the (2^M-1)-dimensional probability simplex $[0,1]^{2^M-1}$.

It is straightforward to see that if we now allow the subband selection decisions $\pi_{k,n}(.)$ of the kth cognitive radio at time n to be a function of the above belief-state vector $\mathbf{b}_k[n]$, so that

$$a_{k,n} = \pi_{k,n}\left(\mathbf{b}_k[n]\right), \tag{15.11}$$

where

$$\pi_{k,n} : \mathcal{P}^{2^M-1} \longrightarrow \{1, \ldots, N_b\}, \tag{15.12}$$

then decision-making strategies discussed in Section 9.6 can be applied verbatim. In particular, one may formulate either an optimal finite- or infinite-horizon POMDP decision-making framework as in (9.41) or simply make myopic optimal decisions as in (9.55). We omit details here since necessary modifications are straightforward and can be obtained simply by substituting the above belief-state vector (15.10) in place of the earlier one.

Clearly, it is also possible to implement a cooperative subband selection framework with the alternative reward function assumed in Section 9.7. Essentially, all we need to modify is the definition of the belief-state vector wherein we include the information received from cooperating neighbors in the information history, as in (15.8).

In either case, it is worth observing that these new cooperative subband selection frameworks are essentially equivalent to those of Chapter 9 but with a modified observation vector given by $\{Y_j[n]\}_{j\in\Omega_k[n]}$. The purpose of $\{a_{j,n}\}_{j\in\Omega_k[n]}$'s essentially is to clarify to which subband each $Y_j[n]$ corresponds. Shared observations from one's neighbors provide an extended observation vector that spans across several subbands. In some cases, this may result in the cognitive radio k having two or more observations corresponding to the same subband m at a particular time instant n. This will introduce diversity to its observations. On the other hand, since different cognitive radios may likely observe different subbands, now, the observation available to cognitive radio k may not be limited to the subband $a_{k,n}$ it observed at time n. Instead, it may have observations corresponding to a set of subbands. In the extreme case, when observations of the set of neighbors $\Omega_k[n]$ have spanned all the N_b subbands at time n, the cognitive radio k in fact will have a complete observation vector corresponding to all subbands. However, note that still the problem will in general remain a partially observable Markov decision process due to two reasons: First, the set of neighbors may not sense all subbands for each time instant n. Second, the observations on each subband are noisy and not perfect as captured by the binary asymmetric channel model of Figure 9.6.

The increased dimensionality of the effective observation vector at the cognitive radio k, for $1 \leq k \leq K$, can lead to conceivably better subband selection decisions. Of course, the computational complexity of cooperative spectrum decision-making can be significantly higher compared to autonomous decision-making. Moreover, information exchange among neighbors will require both transmit power and additional bandwidth. These are overhead costs inherent to most cooperative detection and decision-making architectures that need to be carefully weighed against gains in performance improvement.

15.4 COOPERATIVE COMMUNICATIONS STRATEGIES IN COGNITIVE RADIO NETWORKS

The two types of user cooperation discussed in the previous two sections are instances of cooperation that are either unique to CRNs or specifically directed toward a unique problem encountered in CRNs: spectrum sensing decision-making and spectral activity/primary signal detection. On the other hand, general cooperative communications include many adaptations of MIMO and space–time–frequency diversity techniques to virtual antenna arrays created by user cooperation. There are also cooperative communications strategies that may not explicitly seem to be direct generalizations of corresponding MIMO techniques to virtual MIMO techniques. Either way, in many cases, cooperation among distributed radios leads to spatial diversity that can be a useful resource against inherent wireless channel anomalies such as fading and shadowing.

As mentioned earlier, cognitive radios provide an ideal platform to implement these cooperative communications protocols. Although cooperative communications were developed independent of cognitive radios, there is an underlying assumption of somewhat sophisticated coordination and processing abilities in many of these protocols. Thus, cooperative communications strategies are a natural fit for cognitive radios.

Cooperative communications protocols usually involve two or more radio devices that are willing to cooperate in transmitting their payloads. We already saw an example of this in cooperative space–time block coding based on the distributed Alamouti scheme discussed in Example 15.1. The two cooperative radios transmit their two symbols, say, $b^{(1)}$ and $b^{(2)}$, during the first symbol duration simultaneously. If both of them are full duplex, for instance, each of the radio will be able to hear and decode the other's symbol transmission while it is transmitting its own symbol. Assuming this decoding is error-free, during the second symbol duration, the two radios will transmit the two symbols $-\left(\hat{b}^{(2,1)}\right)^* \approx -\left(b^{(2)}\right)^*$ and $\left(\hat{b}^{(1,2)}\right)^* \approx \left(b^{(1)}\right)^*$ simultaneously. By proper combining, a common receiver may decode both symbols $b^{(1)}$ and $b^{(2)}$ with a diversity gain of 2.

The simple set-up of Example 15.1 highlights usual assumptions in cooperative communications. One of them is the *symmetry in cooperation*: Both radios invest the same amounts of the same type of resources and reap the same benefits out of cooperation, on average. Usually this also means that both radios belong to the same wireless system. In the case of cognitive radios, however, this may not be always the case. In fact, as we have already seen in earlier chapters, many of the initial cognitive radios were targeted for primary–secondary coexistence. Hence, cooperative communications schemes proposed for such DSS CRNs can be inherently asymmetric between primary and secondary radios. Thus, in the context of CRNs, we may recognize two types of cooperative communications protocols:

1. *Protocols for cooperation among cognitive radios themselves*: The objective of these cooperative communications strategies is usually to take advantage of spatial diversity afforded by cooperation among distributed wireless nodes. Hence, almost any cooperative communications, distributed space–time coding, virtual MIMO, or virtual antenna array technique can be used in this category. Cooperative communication protocols used in this way among cognitive radios have nothing unique compared to their use in any traditional radio system. However, many cooperative communications protocols normally presume radios with sophisticated decision-making, coordination, and computing abilities. Hence, the inherent advanced reasoning, decision-making, and learning abilities of cognitive radios may greatly justify the feasibility of

implementing cooperative communications on CRNs than on traditional radio networks.

Moreover, such cooperative communications among cognitive radios themselves can be implemented regardless of whether they are wideband or aimed at DSS. When used in a secondary cognitive radio system, however, they may perhaps lead to a new interpretation of cooperative gains other than simply as improving the spatial diversity. Indeed, as was shown in [249], when there are more than one secondary system competing to get access to the same primary channels, a secondary system with cooperating users may be able to win out the competition due to their ability to achieve successful transmissions at lower individual transmit powers. Still, the underlying reason is the diversity gain.

2. *Protocols for cooperation between primary and secondary systems*: Dynamic spectrum sharing cognitive radios give rise to a unique type of cooperative communications paradigm in which two cooperating nodes can be asymmetric by design. This is in direct contrast to many cooperative protocols that have been developed independent of cognitive radios in which cooperating nodes are always treated symmetrically.

   In the case of primary–secondary coexistence, secondary cognitive radios can utilize cooperative protocols to achieve several different objectives: For instance, they may use asymmetric cooperative communications protocols as in Chapter 14 to provide primary users with an in-kind leasing payment. As we saw in Chapter 14, in this case, the amount of spectrum access granted to secondary cognitive radios will be determined by the power level at which they will relay primary signals. Though asymmetric, the primary users are usually proactive in facilitating such dynamic spectrum leasing and are fully aware of the secondary access to their spectrum.

   On the other hand, it is also possible that secondary cognitive radios may leverage asymmetric cooperative communications to gain better access to primary-owned spectrum without the explicit knowledge of primary users. In this scenario, secondary radios may, for example, relay primary signals without their explicit knowledge so that cognitive radios can also transmit their own secondary signals at higher power levels. By properly setting the relay and transmit power levels, secondary cognitive radios may prevent performance degradation at the primary receivers due to their high-power secondary signal transmissions.

The first type of cooperative communications strategies above is not unique to CRNs. Almost any cooperative communications protocol can be integrated into cognitive radios when the cooperation is assumed to be among nodes belonging to the same network. The second type of cooperative protocols above is, however, unique to DSS CRNs due to their explicit assumption of primary–secondary asymmetry. In Section 14.3.1 of the previous chapter, we already saw one example of such asymmetric cooperation between primary and secondary users in which primary users were aware of the cooperative communications attempts by the secondary cognitive radios. Indeed, in the method discussed in Section 14.3.1, primary users were instrumental in implementing the cooperative communications protocol. Specifically, they had the control over when to cooperate and to what extent.

Cognitive radios, however, may implement asymmetric cooperative communications protocols possibly without the explicit knowledge of primary users as well. In the next section, we describe an asymmetric cooperative communications architecture, proposed in [250], that is different from that of Section 14.3.1. Here, secondary cognitive radios relay primary signals possibly without their explicit awareness in order to make them gain better access to primary channels.

15.5 ASYMMETRIC COOPERATIVE RELAYING IN DSA COGNITIVE RADIOS

Let us consider a dynamic spectrum sharing cognitive radio system made of K_p primary users and K_s secondary cognitive transmitters. For simplicity, let us assume that all primary users communicate with a common primary receiver while all secondary users communicate with a common secondary receiver. We will implicitly assume that each of the K_p primary users is assigned a separate orthogonal channel, and thus, we may use primary channels and primary users interchangeably. Of course, these assumptions can easily be modified. Let $\mathcal{K} = \mathcal{K}_p \cup \mathcal{K}_s$ denote the set of user indices where $\mathcal{K}_p = \{1, \ldots, K_p\}$ and $\mathcal{K}_s = \{K_p + 1, \ldots, K_p + K_s\}$ are the sets of indices of the primary and secondary users, respectively.

Each primary user communicates with the common primary receiver over its own separate channel. Each secondary cognitive radio is interested in communicating with the common secondary receiver on any one of the K_p primary channels. To be specific, suppose that the jth secondary cognitive radio, for $j \in \mathcal{K}_s$, attempts to transmit its own message over the primary channel $i \in \mathcal{K}_p$. This transmission may likely cause interference at the primary receiver, thereby degrading the detection performance of the ith primary signal. However, cognitive radio j might be able to offset this interference that it causes at the primary receiver if it can ensure that the primary received signal power also increases accordingly. Indeed, it can achieve this by resorting to an asymmetric cooperative communications scheme in which it will relay the ith primary user's signal.

For simplicity, in the following, we will assume that each secondary user is capable of transmitting on at most a single primary channel at a time. Hence, at any given time, a cognitive radio only cooperates with at most a single primary user.

The cooperative communications scheme we discuss here is asymmetric, as in Section 14.3.1, since only secondary users will be relaying signals. However, this is different from the cooperative communications scheme in Section 14.3.1 since there is no explicit dynamic spectrum leasing involved. Rather, the dynamic spectrum sharing assumed here is more closer to dynamic spectrum access discussed in Chapters 12 and 13: Secondary cognitive radios voluntarily relay primary signals without explicit involvement of primary users who may in fact be oblivious to the fact that secondary radios are relaying their signals.

As usual, cognitive radios may act either autonomously or they may be directed centrally by a secondary system decision center. In the case of an autonomous cognitive radio, each radio needs to make two decisions:

1. Which primary channel to access and implement cooperative relaying
2. What should be the transmit and relay power levels

The first of these problems can be formulated as a decision-making problem similar to sensing decision-making problems discussed in Chapter 13. Indeed, if we assume that the autonomous cognitive radio can only sense a single channel at a time and it will only access a channel it has sensed and verified its state, then this access decision problem is the same as the sensing decision problem we already discussed in Chapter 13. Ideally, the autonomous cognitive radio may want to access a channel that will be free of primary signals or at least has very little primary signal power. Hence, one may develop a certain channel dynamics model for the primary activities and based on that make optimal (e.g., either for a certain decision horizon or just myopic) decisions as discussed in Chapter 13. Hence, we do not go into details of this process here.

The second problem, on the other hand, is new. Suppose that cognitive radio j is to cooperatively transmit on primary channel i. Since any transmission by cognitive radio j may introduce interference at the primary receiver, it will relay the primary signal simultaneously so as to offset this interference. However, its total power is constrained. Thus, the radio j needs to determine what amount of power it should allocate for relaying and what amount it should use for its own signal transmission. This is called the *power allocation* problem (in asymmetric cooperative relaying).

Before we discuss the solution to the power allocation problem in the next section, let us consider a centrally controlled secondary system. In this case, we allow the SSDC to make the decisions and inform the distributed cognitive radios. Still, the SSDC needs to make two decisions similar to an autonomous cognitive radio:

1. Which cognitive radio should access and cooperatively transmit on which primary channel
2. What should be the transmit and relay power levels of each cognitive radio on the channel it is assigned to

The first of these problem is one of scheduling problems. Unlike in the case of an autonomous cognitive radio, the SSDC has the opportunity to make access decisions to optimize a certain global performance metric. As we did in Section 14.3.2.2, we may use a scheduling function ϕ, where

$$\phi : \mathcal{K}_s \longrightarrow \mathcal{K}_p \bigcup \{0\}, \tag{15.13}$$

to indicate this cooperative primary channel assignment for a given cognitive radio. Note that $\phi(j) = 0$ indicates that secondary cognitive radio j is not assigned to any primary channel for transmission or cooperation.

We also assume that only a single secondary radio is assigned to cooperate with a primary user at any given time, so that for any $(u, v) \in \mathcal{K}_s \times \mathcal{K}_s$ with $u \neq v$ and $\phi(u)\phi(v) \neq 0$, we always have

$$\phi(u) \neq \phi(v).$$

Both these assumptions can also be dropped, in general, at the expense of notational complexity in the discussion and computational complexity in implementation. The scheduling function ϕ can equivalently be characterized in terms of the assignment vector

$$\Phi = \left[\phi(K_p + 1), ..., \phi(K_p + K_s)\right]^T, \tag{15.14}$$

where, under our assumptions, Φ will have $\left[K_s - K_p\right]^+ = \max\{0, K_s - K_p\}$ number of zero elements representing secondary cognitive radios that were not assigned to any primary channel because there are more secondary cognitive radios than the available number of primary channels. In Section 15.5.2, we will discuss in detail the optimal scheduling of cognitive radios for asymmetric cooperation with primary users.

The second problem faced by the SSDC is not very different from that of an autonomous cognitive radio. For each cognitive radio j, the SSDC needs to determine its transmit and relay power levels. Since we assume orthogonal primary channels, these power allocation problems are decoupled from one channel to another, and thus, the problem and the solution are the same as that in the case of an autonomous cognitive radio. Hence, in the next section, we will first discuss the optimal power allocation problem faced by an arbitrary cognitive radio on an arbitrary primary channel. The solution will apply to both autonomous cognitive radios and to the SSDC. Following that, in Section 15.5.2, we will discuss the optimal centralized scheduling of cognitive radios to primary channels for asymmetric cooperation. In Section 15.5.3 we will modify these schemes to obtain a more realistic implementation of the primary-secondary cooperative coexistence.

15.5.1 Secondary User Optimal Power Allocation for Asymmetric Cooperative Relaying

Let us denote by P_k the power used by any user k, for $k \in \mathcal{K}$, to transmit its own signal. However, transmissions of cognitive radios will not be only their own signals. Instead, it will be its own signal superposed with the primary signal on the channel it is transmitting. For example, if secondary cognitive radio j is transmitting on primary channel $i = \phi(j)$, it will transmit its own signal at a power level of P_j superposed with the ith primary signal at a power level of $q_{j,i}$. Hence, the total transmit power spent by the jth secondary cognitive radio in this case is $P_j + q_{j,i}$. If we denote by $\bar{P}_j$ the maximum total transmit power of secondary user $j \in \mathcal{K}_s$, then the jth cognitive radio's power allocation must always satisfy

$$P_j + q_{j,i} \leq \bar{P}_j. \tag{15.15}$$

As we did in Example 15.1, let us assume a system based on genie-aided cooperation. Recall that this implies that secondary cognitive radio may instantly know the primary message. For $k \in \mathcal{K}$, let $b_k[n]$ denote the nth symbol of user k. Then, the asymmetric cooperative relaying scheme between the secondary radio j and the primary $\phi(j) = i$ can be described as follows: The ith primary user transmits its nth symbol $b_i[n]$ at a power level of P_i. The jth secondary cognitive radio also transmits simultaneously with the primary user i. Thanks to the genie-aided assumption, the jth secondary user who is assigned to cooperate with primary user i (on channel i) also knows this primary symbol $b_i[n]$. Hence, as shown in Figure 15.4, the jth cognitive radio's transmission is a superposed signal made of symbol $b_i[n]$ at power level of $q_{j,i}$ and its own symbol $b_j[n]$ at a power level of P_j. Of course, $q_{j,i}$ and P_j are selected so that the maximum power constraint (15.15) is satisfied.

The *genie-aided assumption* can be justified by noting that if the primary transmitter is much closer to the secondary cognitive relay than the intended primary receiver, then the cognitive radio may be able to accurately decode the primary symbol $b_i[n]$ within only a small fraction of the symbol period [127,251]. Hence, one may assume that the primary message is known to the cognitive transmitter instantaneously. As soon as secondary cognitive radio finishes decoding the primary message $b_i[n]$, it may start to relay it at a power level of $q_{j,i}$ while also transmitting its own symbol $b_j[n]$ at a power level of P_j. More details on genie assumption and its justification can be found in [251].

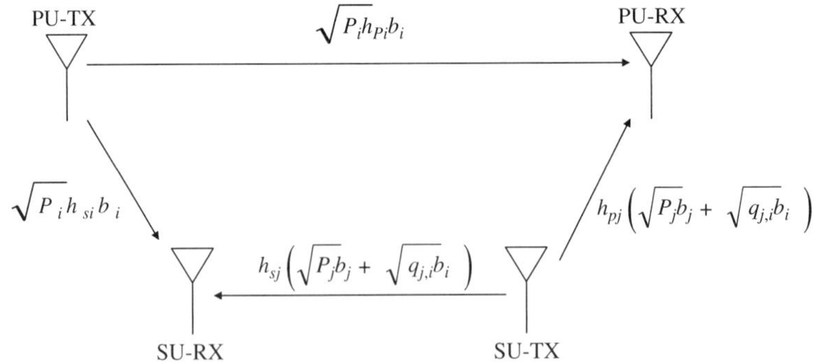

FIGURE 15.4 Asymmetric cooperative relaying: The jth secondary user transmits its own signal at power P_j superposed with primary signal at power $q_{j,i}$.

In practice, asymmetric cooperative relaying of primary signals can be implemented by assuming that every primary symbol is transmitted in two consecutive stages: During the first stage, the ith primary user sends its nth symbol $b_i[n]$ at a power level of αP_i, where $\alpha \in [0,1]$, to the jth secondary user. At the end of the first stage, the secondary cognitive radio j estimates the primary symbol $b_i[n]$ as $\hat{b}_i[n]$. During the second stage, the ith primary user retransmits the same symbol $b_i[n]$ but this time at a power level of $(1-\alpha)P_i$. During this stage, jth secondary cognitive radio also transmits simultaneously with the primary user i. The jth cognitive radio's transmission is a superposed signal made of symbol $\hat{b}_i[n]$ at power level of $q_{j,i}$ and its own symbol $b_j[n]$ at a power level of P_j. Implemented this way, the asymmetric cooperative protocol is a slight variation of a (symmetric) cooperative communications protocol proposed in [211,212]. It has been shown in [211,212] that the decrease in (primary) transmission rate due to this two-stage approach can be offset by the gain in improved symbol error probability.

Suppose that secondary user j is assigned to cooperate with primary user i (i.e., $\phi(j) = i$). Then, assuming a genie-aided system, the primary receiver receives two signals corresponding to the same symbol $b_i[n]$ of the ith primary user: one from the primary transmitter and the other from the jth secondary cognitive radio as shown in Figure 15.4. The received superposed signal y at the primary receiver can then be written as

$$y = \left(\sqrt{P_i}h_{pi} + \sqrt{q_{j,i}}h_{pj}\right) b_i[n] + \sqrt{P_j}h_{pj}b_j[n] + w \tag{15.16}$$

where h_{pk} is the channel fading coefficient between user $k \in \mathcal{K}$ and the (common) primary receiver and w is zero-mean additive noise with average power N_0. If, for the moment, we were to assume that the primary receiver knows all fading coefficients, the resulting signal-to-noise-plus-interference-ratio (SINR) at the primary receiver for detecting signals on the i-th channel is

$$\gamma_{pi} = \frac{\left|\sqrt{P_i}h_{pi} + \sqrt{q_{j,i}}h_{pj}\right|^2}{P_j\left|h_{pj}\right|^2 + N_0}. \tag{15.17}$$

Let us assume that the minimum required primary QoS (for the i-th primary user) is characterized in terms of a minimum required SINR at the primary receiver for detecting signals on the i-th channel, denoted as $\bar{\gamma}_{pi}$. Hence, the j-th cognitive radio is required to choose secondary transmit and relay powers, P_j and $q_{j,i}$ respectively, such that

$$\gamma_{pi} \geq \overline{\gamma}_{pi}. \tag{15.18}$$

However, since secondary system may not know instantaneous CSI associated with the primary system, it can only meet the constraint (15.18) on $\bar{\gamma}_{pi}$ on an average sense. A reasonable assumption for the secondary system to make will be that $\mathbb{E}\{h_{pi}\} = 0$

and $\mathbb{E}\{|h_{pi}|^2\} = G_{pi}$ for some known G_{pi}, so that under these and the genie-aided assumptions, we have that

$$\gamma_{pi} \approx \frac{P_i G_{pi} + q_{j,i}|h_{pj}|^2}{P_j|h_{pj}|^2 + N_0}. \tag{15.19}$$

Example 15.2 (Diversity Combining: Equal-Gain and Maximal-Ratio Combining) Note that, the received signal (15.16) assumes that two signal paths corresponding to the same primary symbol from the primary and secondary transmitters are received on the same channel as a single superposed signal. However, in practice, it may be the case that primary receiver may be able to receive these two signals as two distinct signal paths. In that case, if we denote the two received signals from the primary and secondary transmitters, respectively, by y_1 and y_2, the primary receiver may be able to employ so-called diversity combining schemes to achieve an SINR different from that in (15.17). Although, we will not be assuming such diversity combining here, in the following let us briefly discuss how such combining is performed.

Suppose that corresponding to the same symbol b, an L number of different signals are received at a receiver:

$$y_l = \sqrt{P} h_l b + w_l, \quad \text{for} \quad l = 1, \ldots, L \tag{15.20}$$

where b is the symbol to be decoded (assumed to be normalized so that $\mathbb{E}\{|b|^2\} = 1$), h_l is the channel fading coefficient of the lth received signal, and w_l is noise with noise power N_0. For example, y_ls may correspond to received signals from different multipaths or received signals at multiple antenna elements.

Diversity combining refers to appropriately combining such received signals in order to achieve improved receiver performance. In general, a combiner output can be written in the form of

$$\begin{aligned} y &= \sum_{l=1}^{L} G_l^* y_l \\ &= \sum_{l=1}^{L} \sqrt{P} G_l^* h_l b + \sum_{l=1}^{L} G_l^* w_l, \end{aligned}$$

where G_l denotes the weight given to the lth signal.

Depending on different choices for G_l's, we may obtain different types of combining schemes. For example, *equal-gain combining* refers to the choice of combiner weights

$$G_l = 1 \quad \text{for all } l = 1, \ldots, L. \tag{15.21}$$

On the other hand, assuming that noise w_l are independent, the SNR at the output of the combiner is given by

$$SNR_0 = \frac{\left|\sum_{l=1}^{L} G_l^* h_l\right|^2}{\sum_{l=1}^{L} |G_l|^2} \frac{P}{N_0}. \tag{15.22}$$

Maximal-ratio combining refers to the diversity combining scheme in which the weights G_l's are chosen to maximize the output SNR SNR_0 of (15.22). It can be shown easily that SNR_0 is maximized by choosing the weights to be

$$G_l = h_l \quad \text{for all } l = 1, \ldots, L. \tag{15.23}$$

With these weights, the output SNR of the maximal-ratio combiner is then turn out to be

$$SNR_0 = \frac{P}{N_0} \sum_{l=1}^{L} |h_l|^2. \tag{15.24}$$

Of course, maximizing the output SNR is justified in Gaussian noise environments due to the fact the probability of error is minimized by maximizing the SNR. Hence, maximal-ratio combining turns out to be the optimal combining scheme in such a channel in the sense of minimizing the probability of error. However, when signals contain interference in addition to noise, as in (15.16), the optimal combining scheme is different from (15.23). We refer the interested readers to texts such as [18, 187] for details. ■

In light of (15.15) and (15.19), the minimum required primary SINR constraint (15.18) leads to the following constraint on the transmit power P_j that the secondary cognitive radio j may allocate for its own message transmission:

$$\begin{aligned} P_j &\leq \min\left\{\bar{P}_j - q_{j,i}, \left[\frac{P_i G_{pi} - \bar{\gamma}_{pi} N_0 + q_{j,i}\left|h_{pj}\right|^2}{\bar{\gamma}_{pi}\left|h_{pj}\right|^2}\right]^+\right\} \\ &= \min\left\{\bar{P}_j - q_{j,i}, \left[\frac{q_{j,i}}{\bar{\gamma}_{pi}} + \tau_{j,i}\right]^+\right\}, \end{aligned} \tag{15.25}$$

with $[x]^+ \triangleq \max\{0, x\}$ and

$$\tau_{j,i} = \frac{P_i G_{pi} - \bar{\gamma}_{pi} N_0}{\bar{\gamma}_{pi}\left|h_{pj}\right|^2}, \tag{15.26}$$

for $i \in \mathcal{K}_p$ and $j \in \mathcal{K}_s$.

On the other hand, when $\phi(j) = i$, the instantaneous SINR at the (common) secondary receiver for detecting signals from the secondary cognitive radio j on channel i, as shown in Figure 15.4, can be written as

$$\gamma_{sj} = \frac{P_j \left|h_{sj}\right|^2}{P_i \left|h_{si}\right|^2 + q_{j,i} \left|h_{sj}\right|^2 + N_0}, \tag{15.27}$$

where h_{sk} denotes the channel fading coefficient between user $k \in \mathcal{K}$ and the secondary receiver.

A reasonable criterion that can be justified in many situations is for the cognitive radio j to select the power allocation $\left(q_{j,i}, P_j\right)$ to maximize its SINR γ_{sj} in (15.27) without violating the primary QoS constraint (15.18):

$$\left(q^*_{j,i}, P^*_j\right) = \arg \max_{\left(q_{j,i}, P_j\right)} \gamma_{sj} \tag{15.28}$$

subject to

$$\begin{aligned} P_j &\le \bar{P}_j - q_{j,i} \\ P_j &\le \frac{q_{j,i}}{\bar{\gamma}_{pi}} + \tau_{j,i} \\ P_j &> 0 \\ q_{j,i} &\ge 0, \end{aligned} \tag{15.29}$$

where $\tau_{j,i}$ is as given in (15.26). The shaded area in Figure 15.5 shows the feasibility region characterized by the above set of constraints (15.29) for power allocation for cooperative relaying. The solution to this power allocation problem can be written as [250]

$$\left(q^*_{j,i}, P^*_j\right) = \begin{cases} \left(0, \min\{\bar{P}_j, \tau_{j,i}\}\right) & \text{if} \quad \tau_{j,i} \ge \min\{\bar{P}_j, \tilde{\tau}_{j,i}\} \\ \left(\lambda_i, \bar{P}_j - \lambda_i\right) & \text{if} \quad -\frac{\bar{P}_j}{\bar{\gamma}_{pi}} < \tau_{j,i} < \min\{\bar{P}_j, \tilde{\tau}_{j,i}\} \\ (0,0) & \text{if} \quad \tau_{j,i} \le -\frac{\bar{P}_j}{\bar{\gamma}_{pi}} \end{cases}, \tag{15.30}$$

where $\tilde{\tau}_{j,i} = \frac{P_i h_{si}^2 + N_0}{\bar{\gamma}_{pi} h_{sj}^2}$ and $\lambda_i = \frac{\bar{\gamma}_{pi}}{1+\bar{\gamma}_{pi}} \left(\bar{P}_j - \tau_{j,i}\right)$. In addition to the knowledge of $\bar{\gamma}_{pi}$, implementing this optimal power allocation requires the key quantity $\tau_{j,i}$ given in (15.26).

On the other hand, if the secondary cognitive radios were not to attempt any cooperative relaying of primary signals, then the optimal noncooperative power allocation of cognitive user j on primary channel i is given by

$$\left(q^*_{j,i}, P^*_j\right) = \left(0, \min\left\{\bar{P}_j, \left[\tau_{j,i}\right]^+\right\}\right) \tag{15.31}$$

where again $\tau_{j,i}$ is as given in (15.26).

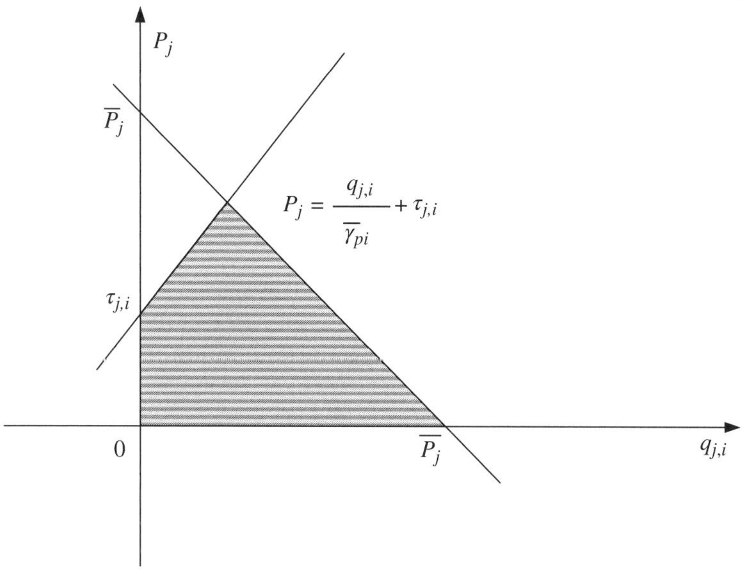

FIGURE 15.5 Feasibility region for power allocation for cooperative relaying.

15.5.2 Centralized Assignment of Cognitive Radios for Cooperation with Primary Users: An Ideal Approach

Power allocation problem is only one half of the asymmetric cooperative communications protocol design. As discussed earlier, the other half of the problem is to determine which cognitive radio is to cooperatively transmit over which primary channel. In the case of an autonomous cognitive radio, as pointed out earlier, these access and cooperate decisions are closely tied to the sensing decisions discussed in Chapter 13. Hence, the problem can be solved by adapting techniques discussed there.

In the case of a centralized cognitive radio system, however, the SSDC is responsible for making these scheduling decisions: Assign each secondary user $j \in \mathcal{K}_s$ to a primary channel $i \in \mathcal{K}_p$. This is akin to the optimal cooperative channel assignment problem at the SSDC we discussed in the context of DSL in Section 14.3.2.2.

The scheduling problem for cooperation is concerned with assigning each secondary cognitive radio to cooperate with one and only one primary user. The best assignment, of course, will depend on the objective of the network. Recall that the key motivation of DSS CRNs is to get access to the primary spectrum and improve its utilization. Hence, an ideal scheduling objective can be defined as to maximize the overall sum rate achieved jointly by the primary and secondary networks over the spectrum of interest.

To formalize this notion of optimality, consider the primary channel i, for $i \in \mathcal{K}_p$, and the secondary cognitive radio $j = \phi^{-1}(i)$ that is assigned to cooperate with the primary user i. For simplicity, let us assume that all receiver noise are Gaussian. Then, the achievable rate $R_{p,i}$ of the primary user i on its channel with respect to the (common) primary receiver is

$$R_{p,i}(\Phi) \triangleq \log_2\left(1+\gamma_{pi}\right), \tag{15.32}$$

where Φ is the scheduling function defined in (15.14) and γ_{pi} is given by (15.19). For $j=\phi^{-1}(i)$, similarly, let us denote by $R_{s,j}^{(i)}$ the achievable rate of the secondary cognitive radio j on primary channel i, with respect to the primary receiver:

$$R_{s,j}^{(i)}(\Phi) \triangleq \log_2\left(1+\gamma_{s\phi^{-1}(i)}\right), \tag{15.33}$$

with γ_{sj} given by (15.27). The sum rate $R_{i,j}$ on primary channel i when cognitive radio $j=\phi^{-1}(i)$ is assigned to cooperate with primary user i is the sum of these two rates:

$$R_{i,j} \triangleq R_{p,i}+R_{s,j}^{(i)}. \tag{15.34}$$

In [250], an ideal *cooperative channel assignment* was derived by assuming that the objective of the secondary system is to maximize these sum rates over all primary channels. However, in practice justifying such an objective for the secondary system may be difficult and we will discuss a more realistic objective in Section 15.5.3. For now, assuming that the common secondary receiver has access to all necessary channel-state information of relevant channels, it may determine an optimal cooperative channel assignment to maximize the transmission rate of the combined spectrum sharing system as follows:

$$\begin{aligned}\Phi^* &= \arg\max_{\Phi} R(\Phi)\\ &= \arg\max_{\Phi} \sum_{i\in\mathcal{K}_p} R_{i,\phi^{-1}(i)},\end{aligned} \tag{15.35}$$

where $R(\Phi)=\sum_{i\in\mathcal{K}_p} R_{i,\phi^{-1}(i)}$ is the sum rate of primary and secondary users over all primary channels.

Due to the assumptions that each primary user is allowed to share its channel with at most one secondary user at a time and each secondary user is allowed to cooperatively transmit over only one primary channel at a time, the channel assignment problem (15.35) is an instant of the assignment problem on a weighted bipartite graph similar to what we encountered in (14.43) in Chapter 14. In the assignment problem of (15.35), the two disjoint sets of vertices are the primary channels $\mathcal{K}_p$ and the secondary users $\mathcal{K}_s$, while the edge weight between primary channel i and secondary cognitive radio j is taken to be the sum rate $R_{i,j}$ of (15.34). Indeed, once we replace the edge weights with this new interpretation, Figure 14.15 is an example of the cooperative scheduling problem (15.35) at an SSDC of a cooperative cognitive radio system made of $K_p=4$ primary channels and $K_s=4$ secondary cognitive radios.

As we saw in Section 14.3.2.2, the Hungarian Algorithm given in Algorithm 14.1 efficiently solves the assignment problem in a complete bipartite graph (as a minimization problem) with cubic complexity in the number of vertices [227, 228].

In Section 14.3.2.2, however, we also pointed out that one may easily adapt the Hungarian Algorithm to solve an assignment problem posed as a maximization problem as in the case of (15.35). If $K_p = K_s$, the earlier bipartite graph has the same number of elements in both vertex sets.

Moreover, although in the case of a DSA cognitive radio system, it may not always be true that $K_p = K_s$, so that the assignment problem (15.35) is not defined on a complete bipartite graph, it is easy to turn any weighted bipartite graph into a complete weighted bipartite graph by setting the weights of the missing edges to be zero [228]. Indeed, in Section 14.3.2.2, we already used this fact.

In the next example, we consider the performance of a DSS system made of secondary cognitive radios that employ cooperative relaying with the scheduling assignments determined by the Algorithm 14.1, modified for this situation.

Example 15.3 (A DSS Cognitive Radio System with Asymmetric Cooperative Relaying) In this example, we consider a primary–secondary DSS cognitive radio system with $K_p = 3$ primary users and $K_s = 5$ secondary cognitive radios. In all cases, all fading coefficients are taken to be i.i.d. Rayleigh distributed with normalized power $\mathbb{E}[h^2] = 1$, $P_i = 1W$ for all $i \in \mathcal{K}_p$, $\bar{P}_j$ taken to be the same for all $j \in \mathcal{K}_s$ and the average noise power at the receivers is set to $N_0 = 0.1W$. All primary users are assumed to have the same minimum SINR requirement so that $\bar{\gamma}_{pi} = \bar{\gamma}_p$ for all $i \in \mathcal{K}_p$.

Figure 15.6 shows the average (averaged over fading) sum rate performance of this system both with and without cooperative communications. Note that, in both scenarios, we have used Algorithm 14.1 to optimally pair secondary cognitive radios to primary channels. In the case of noncooperative cognitive radios, however, this pairing is performed based on the edge weights computed using the optimal noncooperative power assignment (15.31).

Figure 15.6 shows the average sum rates with and without cooperative communications as a function of the minimum QoS requirement of the primary users. The average sum rate with cooperative communications is higher than that without cooperative communications especially in the mid-range of primary SINR requirements. When the minimum required primary SINR $\overline{\gamma}_p$ is very large (beyond 20 dB in the case of Figure 15.6), however, the average sum-rate improvement due to cooperative relaying becomes negligible. As can be seen from the optimal power allocation (15.30) and Fig. 15.5, for very large values of $\bar{\gamma}_p$, secondary cognitive radios do not get to transmit because $\tau \leq -\frac{\bar{P}_j}{\bar{\gamma}_{pi}}$. Indeed, as $\bar{\gamma}_p$ increases, the secondary cognitive radios need to spend more and more of their power budget to satisfy the QoS requirement of the primary user, thus leading to decreasing secondary rates as can be seen from Figure 15.6.

When $\bar{\gamma}_p$ is large enough so that the average primary rate boost due to cooperation is more than enough to offset the average secondary rate that has fallen to a very low level, the cooperative sum rate shows the largest spectrum utilization. However, as is clear from Figure 15.6, the primary system reaps the lion's share of this benefit. Of course, such large values of $\bar{\gamma}_p$ discourage secondary cognitive radios to cooperate. Ultimately when threshold $\tau \leq -\frac{\bar{P}_j}{\bar{\gamma}_{pi}}$ is reached, cognitive radios stop cooperative

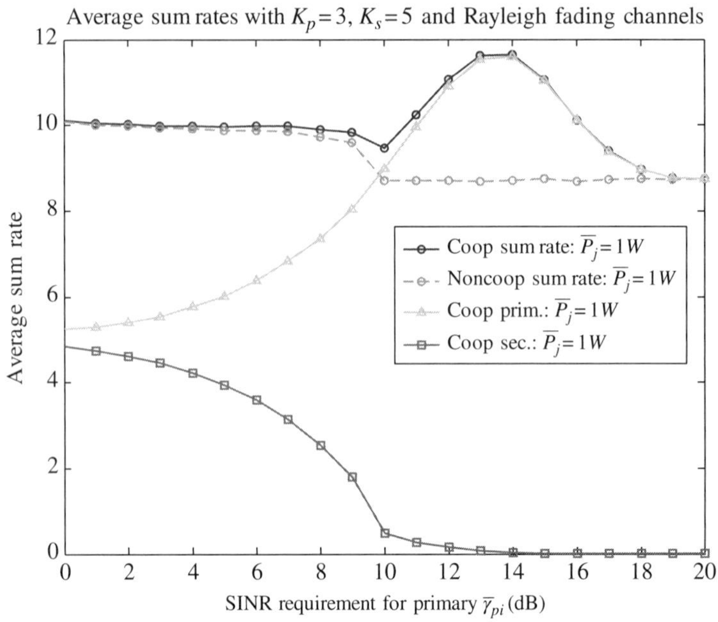

FIGURE 15.6 Average sum-rates in a DSS-based cognitive radio system with asymmetric cooperative relaying.

relaying and all that primary user gets is the average noncooperative rate close to $K_p \log_2\left(1+\bar{\gamma}_{pi}\right)$. ■

The Hungarian algorithm is able to find an optimal matching in a complete weighted bipartite graph with $2n$ vertices in $\mathcal{O}(n^3)$ arithmetic operations [228]. Hence, after adding zero-weight edges, the complexity order of Algorithm 14.1 in finding an optimal cooperative assignment of secondary users to the primary channels will be $\mathcal{O}\left(\max\{K_p^3, K_s^3\}\right)$. When this complexity is too large to afford, the simple heuristic algorithm detailed in Algorithm 15.1 (as proposed in [250, 252]) that has a quadratic complexity in $\max\{K_p, K_s\}$ can provide a fairly reasonable trade-off between the complexity and performance.

Algorithm 15.1

Heuristic Channel Assignment ($K_p \leq K_s$).
(1) Randomly pick a primary channel $i \in \mathcal{K}_p$.
(2) Find the secondary cognitive radio $j^*(i)$ such that
$j^*(i) = \arg\max_{j\in\mathcal{K}_s}\{\log_2\left(1+\gamma_{pi}\right)+\log_2\left(1+\gamma_{sj}\right)\}$.
(3) Remove both i from $\mathcal{K}_p$ and j^* from $\mathcal{K}_s$. IF $\mathcal{K}_p \neq \emptyset$, go to Step (1). Otherwise STOP.

Algorithm 15.1 assumes the case of $K_p \leq K_s$. When this is true, in each iteration of Algorithm 15.1, all currently unassigned secondary users simultaneously scan a randomly selected primary channel i that is not yet paired with a secondary cognitive radio. We will denote by $\mathcal{K}_p^{'}$ and $\mathcal{K}_s^{'}$ the set of primary channels and secondary users not yet assigned for cooperation. We assume that by scanning the primary channel i, the secondary cognitive radios can determine the corresponding CSI and primary transmit power P_i values. Based on these, each currently unassigned secondary cognitive radio j solves for the optimal $q_{j,i}$ and P_j using (15.30) and computes the resulting SINRs γ_{pi} and γ_{sj} as in (15.19) and (15.27), respectively. Next, the set of rates $\{R_{i,j}\}_{j\in\mathcal{K}_s^{'}}$ on channel i is computed where $\mathcal{K}_s^{'}$ denotes the currently unassigned set of secondary cognitive radios.

Algorithm 15.1 assigns the secondary cognitive radio $j^*(i)$ given by

$$j^*(i) = \arg\max_{j\in\mathcal{K}_s^{'}} R_{i,j} \tag{15.36}$$

to cooperate with the primary user in channel i. The algorithm proceeds to the next iteration after removing primary channel i and the secondary cognitive radio $j^*(i)$ from the sets $\mathcal{K}_p^{'}$ and $\mathcal{K}_s^{'}$, respectively. Note that the Algorithm 15.1 can easily be made to be applicable for the case of $K_p > K_s$ by simply modifying the termination condition in Step (3) to read as "IF both $\mathcal{K}_p \neq \emptyset$ and $\mathcal{K}_s \neq \emptyset$, go to Step (1). Otherwise STOP."

Another way Algorithm 15.1 can be modified for the case of $K_p > K_s$ is as follows: In each iteration of the algorithm, a secondary cognitive radio j is selected randomly from the set of still unassigned secondary nodes denoted by $\mathcal{K}_s^{'}$. The cognitive radio j will then scan all yet-to-be assigned primary channels, denoted by the set $\mathcal{K}_p^{'}$, and selects the channel $i^*(j)$ given by

$$i^*(j) = \arg\max_{i\in\mathcal{K}_p^{'}} R_{i,j} \tag{15.37}$$

as its cooperative primary channel assignment. The algorithm will proceed to the next iteration after removing $i^*(j)$ and j from the sets $\mathcal{K}_p^{'}$ and $\mathcal{K}_s^{'}$, respectively.

Example 15.4 (The Performance of the Heuristic Cooperative Assignment Algorithm) In this example, let us consider the same DSS system as in Example 15.3. Figure 15.7 shows the average sum-rates averaged over iid Rayleigh fading, versus minimum primary SINR requirement $\bar{\gamma}_p$ for two different maximum secondary transmit power levels $\bar{P}_j$. For any given $\bar{\gamma}_p$, cooperative communications-based system leads to a higher overall average sum rate compared to that of noncooperative systems. More importantly, however, Figure 15.7 also shows that the performance of the cooperative cognitive radio systems with heuristic cooperative assignments based on Algorithm 15.1 is reasonably close to that of cooperative systems with the optimal assignments based on Algorithm 14.1. ■

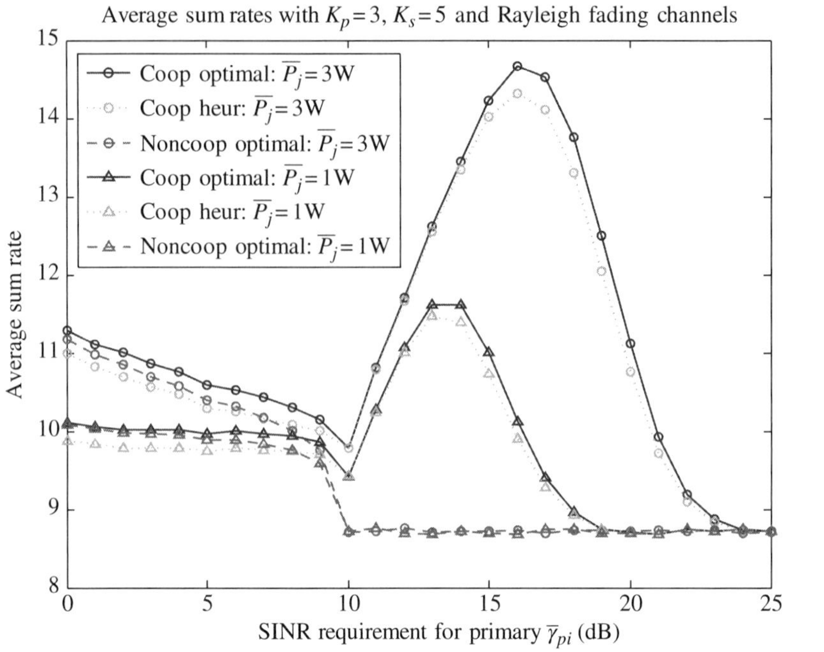

FIGURE 15.7 Average system sum rate in the presence of iid Rayleigh fading for a fixed maximum secondary transmit power limit $\overline{P}_j$.

15.5.3 Centralized Assignment of Cognitive Radios for Cooperation with Primary Users: A Realistic Approach

Admittedly, there are several obvious drawbacks in the ideal approach suggested in Section 15.5.2. First, (15.35) is not the only scheduling criterion one may use. Indeed, as already pointed out, (15.35) is an ideal objective that may not be implementable, or justifiable, in practice. Since only the secondary cognitive radio system is proactive in realizing this asymmetric cooperative relaying between primary and secondary radio systems, it is the secondary system that attempts to maximize the sum rate of both primary and secondary systems over all channels. In practice, this may be hard to justify. A more realistic criterion is that the secondary cognitive radio system maximizing its own sum rate. In addition, even if the secondary system wants to maximize the sum rate of the two systems, it cannot do so without having additional channel-state information regarding the primary system. In particular, as we see from the optimal power allocation scheme (15.30), the SSDC needs to know $\tau_{j,i}$ given by (15.26) for each $i \in \mathcal{K}_p$ and $j \in \mathcal{K}_s$.

A more realistic assignment of secondary radios for cooperation with primary users is thus based on maximizing only the sum rate of the secondary system:

$$\Phi^* = \arg\max_{\Phi} \sum_{i \in \mathcal{K}_p} R^{(i)}_{s,\phi^{-1}(i)}, \tag{15.38}$$

where $R^{(i)}_{s,\phi^{-1}(i)}$ is the rate of the secondary user $j = \phi^{-1}(i)$ on channel i with respect to the secondary receiver as given by (15.33).

It is important, however, to note that modifying the criterion for cooperative channel assignment as in (15.38) is not sufficient to obtain a practical scheme. Recall that the cooperative scheduling is one half of the problem. For each channel, we still need the power allocation values $(P_j, q_{j,i})$ as well. Indeed, rates $R^{(i)}_{s,\phi^{-1}(i)}$ given by (15.33) are determined by the SINR γ_{sj} of the secondary cognitive radio $j = \phi^{-1}(i)$ at the secondary receiver, which in turn depends on the power allocation $(P_j, q_{j,i})$. However, the optimal power allocation (15.30) depends on the channel-state information statistics $\{G_{pi}\}_{i\in\mathcal{K}_p}$ of the primary system through the dependence of λ_i on the parameter $\tau_{j,l}$. Hcncc, in addition to thc primary QoS constraint $\bar{\gamma}_{pi}$, thc SSDC nccds to know at least the quantity $\tau_{j,i}$ in order to be able to allocate power according to the optimal power allocation (15.30), even if we were to use the criterion (15.38) as the basis for cooperative scheduling of cognitive radios to primary channels.

Perhaps the most undesirable aspect of the asymmetric cooperative relaying scheme of Section 15.5.2, however, is that it does not really provide much benefit for the secondary users apart from allowing them to access the primary spectrum. Indeed, as Figures 15.6 and 15.7 show, when secondary cognitive radios get to transmit at any significant rate their rates with cooperative relaying is only slightly better than those without cooperative relaying. The combined sum rate of the two systems is also very close in this region with and without cooperation. In the region in which combined sum rate is significantly better with cooperative relaying than that without cooperative relaying, this gain is completely due to the rate boost received by the primary users. Secondary rates are almost negligible in this region.

This serious anomaly, however, was due to one unrealistic assumption made in Example 15.4: The fixed primary power P_i that is independent of primary target SINR. The more realistic scenario is to assume that the primary transmitter will adjust its transmit power according to its required SINR. Indeed, let us assume that in the absence of secondary cognitive radios, the primary transmit power levels are always tuned so that the primary SNR is equal to the required minimum QoS $\bar{\gamma}_{pi}$:

$$\bar{\gamma}_{pi} = \frac{P_i |h_{pi}|^2}{N_0}. \tag{15.39}$$

When secondary cognitive radio j transmits on the primary channel i under asymmetric cooperative communications, the resulting SINR of the ith primary user at the primary receiver was given in (15.19). If, however, secondary cognitive radios were to assume that primary transmit power levels are already tuned so that (15.39) is satisfied, then it can be shown that ensuring the primary QoS constraint $\gamma_{pi} \geq \bar{\gamma}_{pi}$ of (15.18) is equivalent to ensuring that

$$P_j \leq \frac{q_{j,i}}{\bar{\gamma}_{pi}}. \tag{15.40}$$

The optimal power allocation problem of the jth cognitive radio for cooperation on the ith primary channel thus can be posed as

$$\left(q^*_{j,i}, P^*_j\right) = \arg \max_{(q_{j,i}, P_j)} \gamma_{sj} \tag{15.41}$$

subject to:

$$\begin{aligned} P_j &\le \bar{P}_j - q_{j,i} \\ P_j &\le \frac{q_{j,i}}{\bar{\gamma}_{pi}} \\ P_j &> 0 \\ q_{j,i} &\ge 0, \end{aligned} \tag{15.42}$$

where γ_{sj} is the jth cognitive radio's SINR at the secondary receiver given in (15.27). It can be seen that the solution to this problem is

$$\left(q^*_{j,i}, P^*_j\right) = \left(\frac{\bar{\gamma}_{pi}}{1+\bar{\gamma}_{pi}}\bar{P}_j, \frac{1}{1+\bar{\gamma}_{pi}}\bar{P}_j\right), \tag{15.43}$$

which can readily be evaluated by the secondary system having only the knowledge of the minimum primary QoS requirement constraint $\bar{\gamma}_{pi}$.

Thus, we may implement an asymmetric cooperative relaying-based primary–secondary cooperative communications protocol by assigning secondary radios to cooperate with primary users according to (15.38) where secondary rates $R^{(i)}_{s,j}$ are computed based on the power allocation (15.43). With these rates as the edge weights of a bipartite graph, we may solve (15.38) by using the same assignment algorithms discussed in the previous section.

Example 15.5 (The Performance of a More Realistic DSS Cognitive Radio System with Asymmetric Cooperative Relaying) In this example, we consider the same DSS system as in Examples 15.3 and 15.4 but under the assumptions that primary power P_i is not fixed but is adjusted according to (15.39) and the secondary channel assignment is based on the objective function (15.38). As a result, the power allocation of secondary users will be according to (15.43). Figure 15.8 shows the average sum-rates of this system with and without cooperative relaying by the secondary cognitive radios. As expected, Figure 15.8(a) shows that without cooperative relaying, indeed, secondary radios do not get to access the primary channels at all. The reason is that since primary transmit power P_i is controlled according to (15.39) to exactly achieve the required primary SINR target of $\bar{\gamma}_{pi} = \bar{\gamma}_p$, any non-zero interference from secondary radios to primary receiver will lead to violating its SINR target. As a result, non-cooperative secondary sumrate is identically zero regardless of the primary target SINR $\bar{\gamma}_{pi}$. On the other hand, secondary users get to transmit at a non-zero rate when they cooperatively relay primary signals. Figure 15.8(a) shows that secondary radios can achieve significantly better rates without violating the primary SINR target $\bar{\gamma}_{pi}$ constraint especially when this SINR target is relatively smaller. As $\bar{\gamma}_{pi}$ increases, the secondary sum rate monotonically decays to zero.

Figure 15.8(a) perhaps shows the unexpected peculiarity of cooperative primary rate being slightly smaller than that without cooperation. This, however, is a

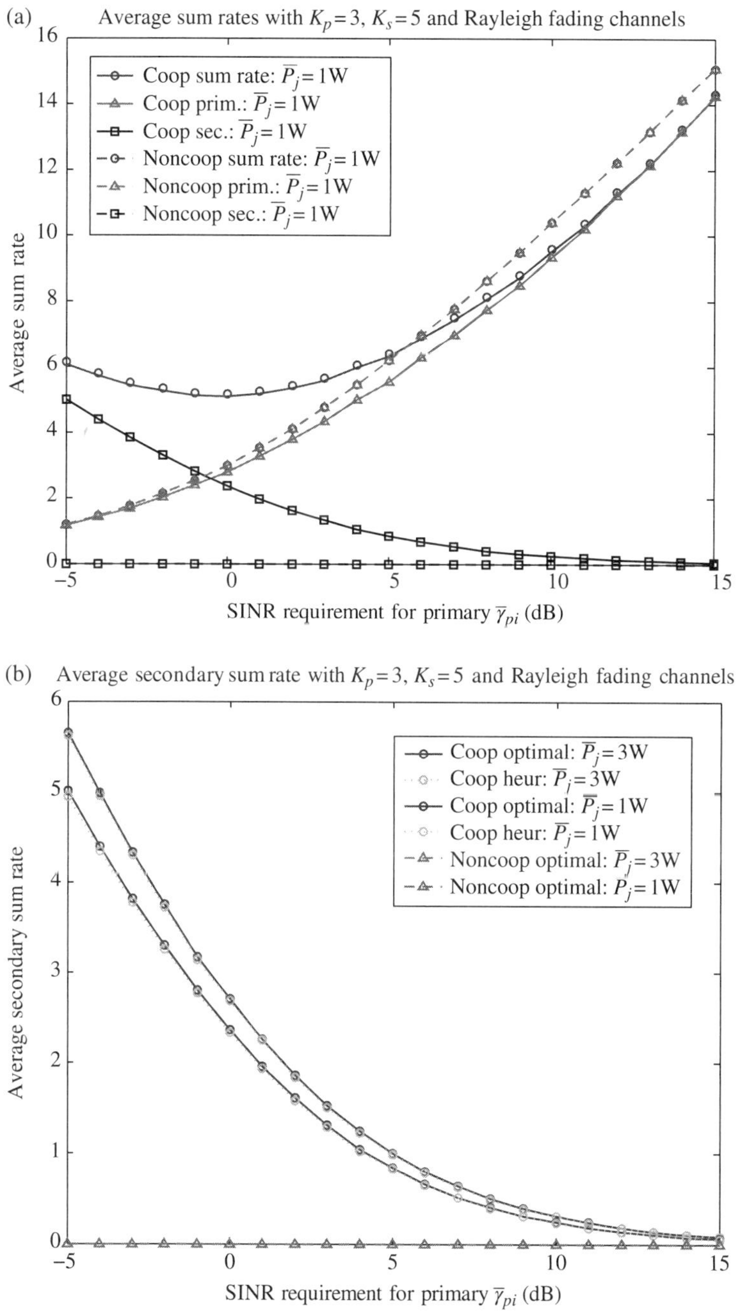

FIGURE 15.8 Benefits of asymmetric cooperative communications in a DSS based cognitive radio system with primary power control. (a) Average sum-rates with asymmetric cooperative relaying when primary system uses power control. (b) Performance of optimal versus heuristic channel assignment algorithms.

consequence of the approximation (15.19) used by the secondary cognitive radios in deriving the optimal power allocation scheme (15.43), instead of the true SINR expression of (15.17). As a result, it can indeed be verified that the constraint set (15.42) is not sufficient to exactly guarantee the true primary SINR target of $\bar{\gamma}_{pi}$. Instead, the solution (15.43) allows the primary SINR to have small variations around the target $\bar{\gamma}_{pi}$. Averaged over fading, these variations give rise to the small drop in the primary rate compared to the non-cooperative rate (which corresponds to exact $\bar{\gamma}_{pi}$) seen in Figure 15.8(a). As long as the primary target SINR $\bar{\gamma}_{pi}$ is less than a certain threshold value, this small drop in the primary rate is offset by the non-zero rate achieved by the secondary cognitive radios. Hence, the overall sum rate of the combined system is still better than that without cooperative relaying. When primary target SINR $\bar{\gamma}_{pi}$ exceeds this threshold value, however, secondary rate has dropped to such a smaller value that it is not large enough to offset the primary rate loss. In Figure 15.8(a), this threshold is around $\bar{\gamma}_{pi} \approx 5$dB. The important region of operation, however, is the small enough $\bar{\gamma}_{pi}$ region. Clearly, Figure 15.8(a) shows that with asymmetric cooperative relaying, secondary cognitive radios can get access to primary channels and transmit at significant rates without compromising the primary QoS.

Finally, Figure 15.8(b) shows that we may use the low-complexity heuristic channel assignment Algorithm 15.1, instead of the exact Hungarian Algorithm 14.1, within this more realistic primary-secondary cooperative coexistence framework. As Figure 15.8(b) shows, the loss due to using the heuristic channel assignment Algorithm 15.1 is almost negligible. ■

15.6 SUMMARY

- Cooperation among distributed wireless nodes emerged as an alternative solution to improving the diversity gain of wireless receivers, when multiple transmit/receive antenna elements were not possible due to physical constraints. Essentially, a set of cooperating wireless nodes acts as a virtual antenna array allowing the promised realizations of MIMO communications. This is called virtual MIMO or, in general, cooperative communications.
- MIMO, or dual antenna array, communication systems allow one to either improve the reliability of the communications by increasing the diversity or to improve the rate of communications by increasing the multiplexing. It is possible to trade off a certain amount of one of these gains to increase the other, and this characterization is called the diversity-multiplexing trade-off in MIMO systems.
- Cooperative communications include all techniques that exploit node cooperation to improve performance and come under various terms such as cooperative relaying, virtual antenna arrays, virtual MIMO, and distributed space–time coding.
- The genie-aided assumption is often invoked in cooperative communications systems in order to simplify the analysis. It refers to a cooperating node being

able to instantaneously know the message of the another node (its cooperating partner). This assumption can be made to hold at least approximately in practice under various conditions.

- Implicit in many cooperative communications protocols is the ability of cooperating nodes to perform advanced processing and computations. Hence, cognitive radios provide an ideal platform to realize the full potential of cooperative communications. Moreover, cooperation with other cognitive radios can in turn help improve reasoning, decision-making, and learning abilities of a cognitive radio.
- There are at least three ways node cooperation can be exploited in cognitive radio networks (CRN):

  1. Decision/data fusion for spectrum sensing
  2. Spectrum sensing decision-making
  3. Cooperative communications protocols

- In cooperative spectrum sensing, cognitive radios may share their local observations/decisions on spectrum conditions with their neighbors or an SSDC. Either way, fusion of these local decisions/data together can, in general, improve the primary signal detection performance in DSS CRNs and the spectral activity detection performance in wideband cognitive radio systems.
- In DSS cognitive radio systems, cognitive radios may cooperate with their neighbors in channel sensing and access decision-making as well. In many proposals, it is assumed that all nodes abide by sensing and access decisions made at a secondary system decision center.
- Wideband cognitive radios can also cooperate with their neighbors in making subband selection decisions. The optimal decision-making algorithms can be modified to take into account additional information available to a wideband cognitive radio due to such information sharing.
- While cooperative schemes related to spectrum sensing are unique to CRNs, they can also benefit from usual cooperative communications strategies that aim to improve diversity and/or multiplexing gain of communications systems. These cooperative protocols can broadly be categorized into two as:

  - Symmetric cooperative communications: These refer to cooperative communications techniques implemented among cognitive radios themselves—be they wideband or DSS cognitive radios.
  - Asymmetric cooperative communications: These refer to cooperative communications techniques implemented between cognitive radios and primary users/systems and can be unique to DSS CRNs.

- Symmetric cooperative communications among cognitive radios themselves are no different from cooperative communications encountered in regular wireless networks. Almost all cooperative techniques developed for such networks can be adapted verbatim to cognitive radios.

- Asymmetric cooperative communications schemes between primary and secondary users are unique to DSS systems. Cognitive secondary users may essentially either improve their chances of getting access to a primary channel or improve their performance in a primary channel by employing such cooperative communications protocols.

16

MACHINE LEARNING IN COGNITIVE RADIOS

16.1 INTRODUCTION

In Chapter 2, we defined a cognitive radio as a software-defined radio-based communications device that is cognitive and intelligent. Being cognitive requires the radio to be able to acquire knowledge and self-comprehend. To be considered as intelligent, it must not only be able to acquire knowledge and comprehend but also be able to apply acquired knowledge. Hence, acquiring knowledge and comprehension is at the heart of a cognitive radio's identity. However, acquiring knowledge and comprehension is learning. In the case of cognition, such learning necessarily needs to include conscious self-learning, while in the case of intelligence, it can be any form of learning. Regardless, it is the ability of learning that is at the heart of a cognitive radio.

Referring to Figure 2.2, learning is the process that converts information into actual knowledge. An entity that has the ability to gather information will not be an intelligent device if it does not have the ability to convert that information into knowledge (i.e., learn from information) and then apply that knowledge in making reasoned decisions, as shown in Figure 2.2. As such, a cognitive radio is expected to learn knowledge by processing information gathered through spectrum sensing and by executing actions.

The basic operation of a cognitive radio can be abstracted in a so-called cognitive cycle that highlights this process as shown in Figure 16.1: Already learned knowledge and current observations are used in reasoning to arrive at decisions that drive radio's

Signal Processing for Cognitive Radios, First Edition. Sudharman K. Jayaweera.

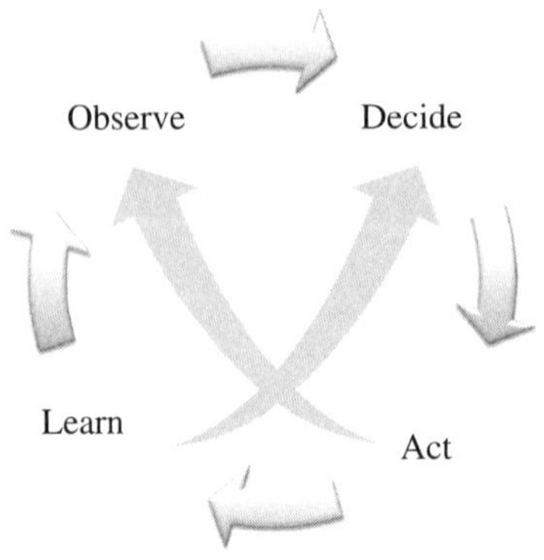

FIGURE 16.1 The cognitive cycle of a cognitive radio [16].

actions. The impacts of these actions on achieving performance objectives and environment state are observed and used for learning new knowledge that will be taken into account in future decision-making.

Thus, as we mentioned in Chapter 2, learning constitutes one of the three key functionalities a cognitive radio must have:

1. Spectrum awareness
2. Decision-making and execution
3. Learning

In previous chapters, we discussed spectrum awareness and decision-making aspects of cognitive radios mainly from a signal processing perspective. In the context of nonbiological entities, or agents, learning is often referred to as machine learning, which in signal processing terminology amounts to estimation problems. Although in biological entities, such as human brain, knowledge can be learned from information in a broader sense that is not confined to a single purpose or a task, in machines, learning is best achieved in the context of a particular objective. Perhaps, it may be that in the case of human brain, information is not discarded immediately after learning, or extracting, certain desired knowledge. Rather, original information itself is stored in memory for future use, whereas in many machine learning paradigms, information is used to extract certain knowledge and then the original information is discarded. As a result, human brain can any time recall many pieces of information gathered at different times in the past to combine in different ways to generate new knowledge or revise and modify current knowledge, whereas a machine may not be able to do that. This is a limitation of machine learning that is beyond the scope of this chapter. Our focus here is to identify value of learning ability to a cognitive radio and investigate various machine learning algorithms and their uses in achieving the objectives of a cognitive radio.

Figure 2.3 shows the role of learning in a cognitive radio: It links the spectrum awareness and communications decision-making. Learning ability is important for both of these. Essentially, learning algorithms are implicitly built into all spectrum knowledge acquisition and decision-making algorithms that we discusses earlier in the book, in the sense that they convert information (current and past observations) into decisions and actions. Hence, somewhere within these schemes, information is transformed into knowledge that then is used to derive these decisions and actions. To recognize what this place is, consider a simple example of a threshold-based decision rule for hypothesis testing. The threshold that specifies the rule is the useful knowledge that the radio has gained. Hence, actual learning takes place in deriving the correct threshold of the decision rule. Once this threshold learning is achieved, giving a certain output for a given input observation is simply reasoning and applying that learned knowledge to make decisions. This shows us the role of learning in any radio, let alone a cognitive radio: Learning is what generates the knowledge needed to define the structure of decision-making algorithms. In the signal processing terms, it is learning that allows us to obtain the values of the parameters that define a particular algorithm.

In many of our discussions of decision-making and spectrum awareness algorithms in earlier chapters, we conveniently ignored how to obtain the required parameters that specify the developed algorithms, be they thresholds in decision rules, transition probabilities of channel dynamic models, or fading coefficients of channels. In some other places, we proposed perhaps a specific approach to obtain a required parameter as it was the only possibility. This was the case with specifying the number of signal classes required for classification or obtaining Q-values to make decisions in a POMDP. The purpose of this chapter is to address the issue of learning knowledge needed for implementing these algorithms. As we will see in the following, there are various ways to achieve such learning depending on the assumptions. In many cases, there are more than one effective way to learn. However, there are certain learning approaches that might be particularly suitable for certain types of learning tasks.

Recall that in Chapter 2 we noted that learning algorithms can broadly be categorized as either supervised or unsupervised. These can be identified as training-based or blind approaches in the signal processing terminology. In cognitive radios, we may encounter both these types of learning problems. In some cases, even the same problem might have to be learned as an unsupervised learning task in one instant, while at another instant, there might be training data available. Hence, a cognitive radio has to be equipped with both these types of learning algorithms. Thus, in the following, we may discuss both supervised and unsupervised learning approaches.

It is almost impossible to list all learning problems that might possibly come up within a cognitive radio. Hence, rather than discussing specific learning problems and their solutions, we will discuss some of the most commonly used learning algorithms themselves and point out certain representative examples of their use in cognitive radio learning tasks. These learning algorithms will, of course, have much broader applicability in cognitive radios than just the given examples.

16.2 ARTIFICIAL NEURAL NETWORKS

One of the commonly used learning frameworks in various application contexts is the artificial neural networks (ANNs). An ANN, or simply a neural network, is a network of interconnected nonlinear decision elements with adjustable weights on the connections.

As the name suggests, ANNs are inspired by how learning and decision-making are performed within biological entities, in particular in human brain, by networks of interconnected neurons. In developing ANNs, one observes that biological neurons are best modeled as being nonlinear decision devices. As a result, the basic building block of ANNs is a non-linear decision module called the *linear threshold unit* (LTU) as shown in Figure 16.2. LTU is also referred to as the adaptive linear element or Adaline, perceptron, or threshold linear unit [253–255]. In addition, when used in an ANN, it is also simply referred to as the neuron. The following discussion of ANNs is mostly based on [253, 254], and the reader is referred to them for more details.

LTU is simply a *linear combiner* followed by a threshold operation: Indeed, if the input is $\mathbf{x} = (x_1, \ldots, x_d)^T$ and the weights of the linear combiner are $\mathbf{w} = (w_1, \ldots, w_d)^T$, then the output of the linear combiner is given by $\sum_{i=1}^{d} w_i x_i = \mathbf{w}^T\mathbf{x}$. If the LTU output is denoted by $f(.)$, then it is given by

$$\begin{aligned} f(\mathbf{x}) &= \begin{cases} 1 & \text{if } \mathbf{w}^T\mathbf{x} \geq \tau \\ 0 & \text{if } \mathbf{w}^T\mathbf{x} < \tau \end{cases} \\ &= q\left(\mathbf{w}^T\mathbf{x} - \tau\right), \end{aligned}$$

where τ is the threshold of the unit and $q(.)$ is the unit-step function.

It is convenient to redefine $\mathbf{x} \Leftarrow (x_1, \ldots, x_d, -1)^T$ and $\mathbf{w} \Leftarrow (w_1, \ldots, w_d, \tau)^T$ so that the above LTU output can be rewritten as [253]

$$f(\mathbf{x}) = q\left(\mathbf{w}^T\mathbf{x}\right). \tag{16.1}$$

We may mostly assume that the LTU output is as given by (16.1).

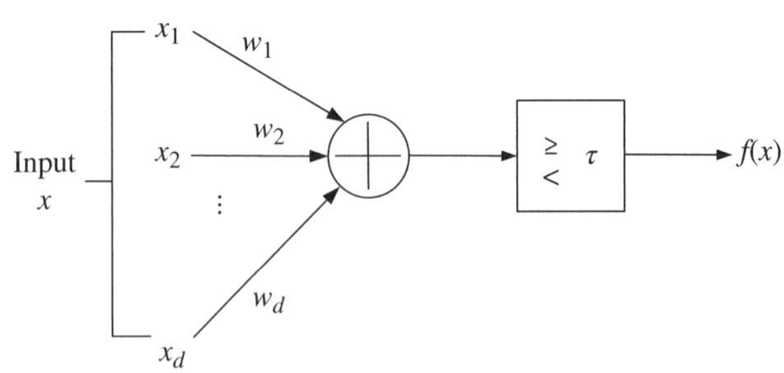

FIGURE 16.2 Linear threshold unit (LTU).

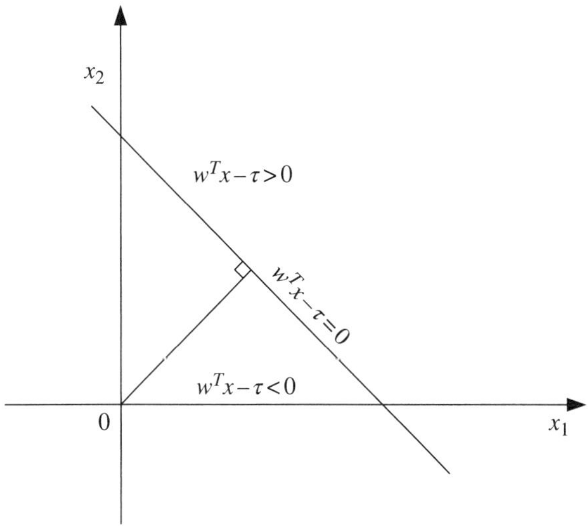

FIGURE 16.3 The linear threshold unit divides the space of **x** values into two regions by a separating hyperplane $\mathbf{w}^T\mathbf{x} - \tau = 0$.

Geometrically, the output of the linear combiner stage, $\mathbf{w}^T\mathbf{x}$, defines a hyperplane in the space of **x** values in the form of $\mathbf{w}^T\mathbf{x} = 0$. Thus, the LTU divides the space of **x** values, say, $\mathbb{R}^N$, into two by the hyperplane $\mathbf{w}^T\mathbf{x} - \tau = 0$, as shown in Figure 16.3. By changing the weight vector **w**, we may obtain a differently oriented hyperplane, and $\frac{\tau}{||\mathbf{w}||}$ denotes the distance from the origin to the hyperplane. This implies that the only input vectors that the LTU can perfectly separate into two classes are the ones that indeed can be separated into two groups using a hyperplane. Such input vectors are called *linearly separable*. Hence, the LTU can achieve perfect classification performance only on input vectors belonging to linearly separable classes [256].

16.2.1 Learning Algorithms for LTUs

The LTU can be made to learn by making it update its weight vector accordingly. There are several different methods one can use to adapt these weight vectors. They all, however, can be viewed as based on the basic idea of changing the weight vector along a direction that decreases some error criterion. To be specific, consider an LTU in a supervised learning framework. There is a set of training (feature) vectors $\mathbf{x}_l$, for $l = 1, \ldots, L$, given with their corresponding true class labels y_l's. We may denote such a training data set as $\{(\mathbf{x}_l, y_l)\}_{l=1}^{L}$. Suppose there are only two classes so that each $y_l \in \{0, 1\}$, for $l = 1, \ldots, L$. We initialize the weight vector $\mathbf{w} = \mathbf{w}_0$ arbitrarily and cycle through the given training data set using them as the inputs $\mathbf{x}_n$ to the LTU, for $n = 0, 1 \ldots,$. In the following, we may identify two simple methods to update the weight vectors [253]:

1. *Error correction training*: Suppose we present the LTU input vectors drawn from the assumed training set. Let us denote the output of the LTU to the nth input vector $\mathbf{x}_n \in \{\mathbf{x}_1, \ldots, \mathbf{x}_L\}$ as $f(\mathbf{x}_n)$. If the true class label corresponding to $\mathbf{x}_n$ is y_n, then the error corresponding to input $\mathbf{x}_n$ denoted by ϵ_n is

$$\epsilon_n = y_n - f(\mathbf{x}_n).$$

Note that since $y_n \in \{0, 1\}$ and $f(\mathbf{x}_n) \in \{0, 1\}$, it follows that $\epsilon_n \in \{-1, 0, 1\}$. Of course, if $\epsilon_n = 0$, then the current weight vector classifies the input $\mathbf{x}_n$ correctly. Then there is no need to do any changes to the weight vector. Indeed, we should modify the current weight vector only if it gives an erroneous output $f(\mathbf{x}_n)$ for a particular input $\mathbf{x}_n$. This is exactly what the *error correction training* approach (sometimes also called the *perceptron convergence procedure* [254]) to weight adaptation of an LTU does: The weight vector is changed only if the current weight vector produces an error. Hence, when a particular $\mathbf{x}_n$ leads to an erroneous output classification $f(\mathbf{x}_n)$, the weight vector is modified along the direction of input vector $\mathbf{x}_n$ as

$$\mathbf{w}_n = \mathbf{w}_{n-1} + \alpha \epsilon_n \mathbf{x}_n, \tag{16.2}$$

where $\mathbf{x}_n \in \{\mathbf{x}_1, \ldots, \mathbf{x}_L\}$ and $0 \leq \alpha \leq 1$ is called the learning rate or the step size. The learning rate α plays an important role in any learning method. The larger values of α may help initially to adapt the weight vector faster, but it may result in large steady-state errors. On the other hand, small α will lead to slow learning.

Observe that if the current weight vector accurately classifies the input vector so that $y_n = f(\mathbf{x}_n)$, then the weight vector update is zero. If, on the other hand, true class is $y_n = 1$ while LTU output is $f(\mathbf{x}_n) = 0$, this implies that the projection of the current weight vector $\mathbf{w}_{n-1}$ along the direction of $\mathbf{x}_n$, given by $\mathbf{w}_{n-1}^T \mathbf{x}_n$, is not large enough. Hence, the error correction training approach of (16.2) modifies the weight vector to be $\mathbf{w}_n = \mathbf{w}_{n-1} + \alpha \mathbf{x}_n$ so that new projection along $\mathbf{x}_n$ direction will be

$$\begin{aligned}\mathbf{w}_n^T \mathbf{x}_n &= \mathbf{w}_{n-1}^T \mathbf{x}_n + \alpha \mathbf{x}_n^T \mathbf{x}_n \\ &= \mathbf{w}_{n-1}^T \mathbf{x}_n + \alpha ||\mathbf{x}_n||^2 \\ &\geq \mathbf{w}_{n-1}^T \mathbf{x}_n.\end{aligned}$$

A similar argument applies when the LTU output is 1 while true label is 0.

It has been shown that the error correction training (16.2) indeed converges to the correct weight vector, if we cycle through the training data set sufficiently large number of times, provided that the training data set is indeed linearly separable [254, 257].

2. *Least mean squares* (*LMS*) *adaptation*: The LMS adaptation is a widely used adaptive, or learning, approach. The *LMS algorithm* can be motivated by the

problem of designing a linear combiner to minimize the mean squared error (MSE) of its output. It is simply a gradient descent algorithm in that the weight vector is updated along the negative gradient direction. Since the actual gradient of the MSE may not be known, the LMS algorithm simply replaces it by the noisy, sample value-based instantaneous gradient, or the *stochastic gradient*, of the MSE. For this reason, LMS is also called a stochastic gradient descent algorithm.

Note that, the LMS algorithm minimizes the output error of the linear combiner without assuming a threshold device. Hence, let us denote by e_n the linear combiner output error if we were to use the same weight vector we have at time $n-1$ with the new input $\mathbf{x}_n$:

$$e_n = y_n - \mathbf{w}_{n-1}^T \mathbf{x}_n. \tag{16.3}$$

The MSE of the linear combiner output at iteration n is

$$\begin{aligned} MSE &= \mathbb{E}\left\{e_n^2\right\} \\ &= \mathbb{E}\left\{\left(Y_n - \mathbf{w}_{n-1}^T \mathbf{X}_n\right)^2\right\}. \end{aligned}$$

The gradient of the MSE with respect to the weight vector $\mathbf{w}_{n-1}$ is given by

$$\begin{aligned} \nabla_{\mathbf{w}} MSE &= \frac{\partial}{\partial \mathbf{w}_{n-1}} \mathbb{E}\left\{Y_n^2 - 2\mathbf{w}_{n-1}^T \mathbf{X}_n Y_n + \mathbf{w}_{n-1}^T \mathbf{X}_n \mathbf{X}_n^T \mathbf{w}_{n-1}\right\} \\ &= -2\mathbb{E}\left\{\mathbf{X}_n Y_n\right\} + \mathbb{E}\left\{\mathbf{X}_n \mathbf{X}_n^T\right\} \mathbf{w}_{n-1}. \end{aligned} \tag{16.4}$$

Since the gradient $\nabla_{\mathbf{w}} MSE$ represents the direction at which the MSE grows the fastest, a gradient descent algorithm will apply a change to the current weight vector in the opposite direction to $\nabla_{\mathbf{w}} MSE$ so as to gain the largest slowdown in MSE:

$$\mathbf{w}_n = \mathbf{w}_{n-1} + \alpha\left(\mathbb{E}\left\{\mathbf{X}_n Y_n\right\} - \mathbb{E}\left\{\mathbf{X}_n \mathbf{X}_n^T\right\} \mathbf{w}_{n-1}\right), \tag{16.5}$$

where, for convenience, we have chosen the learning rate or the step size to be $\frac{\alpha}{2}$.

Since the required statistical averages $\mathbb{E}\left\{\mathbf{X}_n Y_n\right\}$ and $\mathbb{E}\left\{\mathbf{X}_n \mathbf{X}_n^T\right\}$ may not be known in practice, the LMS algorithm proposes to replace them directly by their noisy stochastic versions $\mathbf{x}_n y_n$ and $\mathbf{x}_n \mathbf{x}_n^T$ so that weight vector can be updated as

$$\begin{aligned} \mathbf{w}_n &= \mathbf{w}_{n-1} + \alpha\left(\mathbf{x}_n y_n - \mathbf{x}_n \mathbf{x}_n^T \mathbf{w}_{n-1}\right) \\ &= \mathbf{w}_{n-1} + \alpha\left(y_n - \mathbf{x}_n^T \mathbf{w}_{n-1}\right) \mathbf{x}_n. \\ &= \mathbf{w}_{n-1} + \alpha e_n \mathbf{x}_n, \end{aligned} \tag{16.6}$$

where e_n is given by (16.3).

The LMS adaptation (16.6) has many desirable properties and justifications for its use. It is easy to observe that the average of the LMS-adapted weight vector is clearly the optimal MMSE weight vector implied by (16.4):

$$\mathbf{w}^*_{MMSE} = \left(\mathbb{E}\left\{\mathbf{X}_n\mathbf{X}_n^T\right\}\right)^{-1}\mathbb{E}\{\mathbf{X}_nY_n\}.$$

Moreover, if, indeed, the given data is linearly separable (i.e., there exists a weight vector $\mathbf{w}^*$ such that each $y_n = (\mathbf{w}^*)^T\mathbf{x}_n$), then it can be shown that the squared error $|\tilde{\mathbf{w}}_n|^2$, where $\tilde{\mathbf{w}}_n = \mathbf{w}^* - \mathbf{w}_n$, vanishes as $n \longrightarrow \infty$ as long as the step-size parameter α satisfies

$$\alpha < \frac{2}{\mathbf{x}_n^T\mathbf{x}_n} \quad \text{for all } n.$$

Example 16.1 (The LMS Algorithm: An Alternative Derivation) The LMS algorithm can also be motivated directly as the adaptation that attempts to minimize the instantaneous squared error e_n^2 at each time instant, where e_n is given by (16.3). Recall that e_n is the resulting error if we were to use the same weight vector we have at time $n-1$ with the new input $\mathbf{x}_n$. The gradient of the above instantaneous squared error with respect to $\mathbf{w}$ is

$$\nabla_{\mathbf{w}} e_n^2 = \begin{bmatrix} \frac{\partial e_n^2}{\partial w_1} \\ \frac{\partial e_n^2}{\partial w_2} \\ \vdots \\ \frac{\partial e_n^2}{\partial w_d} \end{bmatrix}$$

where, for $i = 1, \ldots, d$,

$$\frac{\partial e_n^2}{\partial w_{n-1,i}} = -2e_n x_{n,i}$$

with $\mathbf{x}_n = (x_{n,1}, \ldots, x_{n,d})^T$. Hence, we have that

$$\nabla_{\mathbf{w}} e_n^2 = -2e_n\mathbf{x}_n.$$

Since the gradient $\nabla_{\mathbf{w}} e_n^2$ represents the direction at which the squared error e_n^2 grows the fastest, the LMS algorithm applies a change to the current weight vector in the opposite direction to $\nabla_{\mathbf{w}} e_n^2$ so as to gain the largest slowdown in e_n^2:

$$\begin{aligned}\mathbf{w}_n &= \mathbf{w}_{n-1} + \alpha\left(y_n - \mathbf{w}_{n-1}^T\mathbf{x}_n\right)\mathbf{x}_n \\ &= \mathbf{w}_{n-1} + \alpha e_n\mathbf{x}_n,\end{aligned}$$

where, again, $\frac{\alpha}{2}$ is the learning rate or the step size. ■

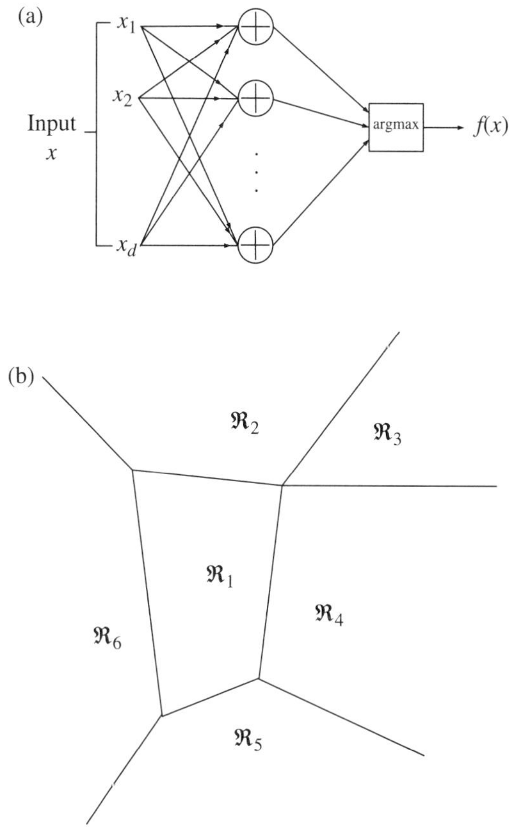

FIGURE 16.4 A parallel bank of linear combiners can be used to classify inputs belonging to more than two classes. (a) A parallel bank of linear combiners implementing a multiclass classifier. (b) Corresponding decision regions are polytopes.

16.2.2 Layered Neural Networks

As mentioned earlier, a single LTU is only capable of classifying inputs that can be divided into two classes by a separating hyperplane. If there are more than two classes, we may use a parallel bank of linear combiners and classify the given data input as belonging to the class with the largest linear combiner output, as shown in Figure 16.4(a) [253]. Such a parallel bank of linear combiners followed by a decision device can still only separate the **x** space into a set of polytopes defined by a set of hyperplanes as in Figure 16.4(b).

However, in many situations, the inputs may not be separated into classes by such hyperplanes. The separating boundaries may have to be more complex nonlinear surfaces. ANNs are an attempt to solve this limitation of LTUs by using networks of interconnected LTUs, as shown in Figure 16.5.

The structure of the ANN in Figure 16.5 is made of a set of layers of LTUs. The first layer (the leftmost layer) is called the *input layer*, and the last layer (the rightmost

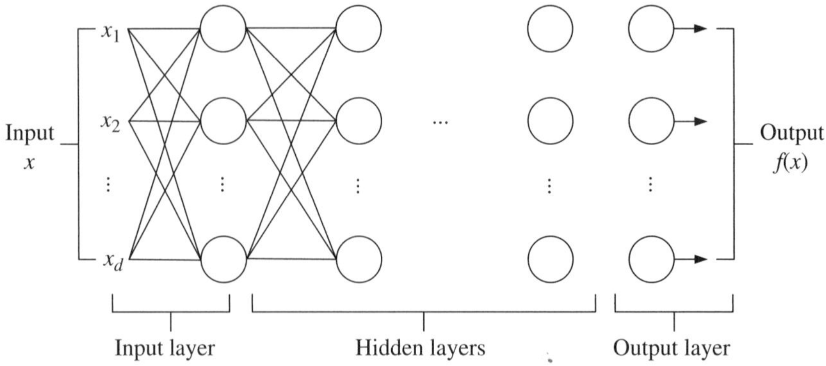

FIGURE 16.5 An artificial neural network is a network of interconnected linear threshold units.

layer) is called the *output layer*. The input vector **x** is fed into the LTUs in the input layer with certain associated weights. The output of the ANN is taken to be the outputs of the LTUs in the last layer of the network. In between the input and output layers, there are an arbitrary number of layers called *hidden layers*. The power of ANNs in learning complex nonlinear functions arises from these hidden layers. In general, each of these layers can have any number of LTUs.

The ANN shown in Figure 16.5 is, in fact, one particular type of possible networks called a *feed-forward ANN*. In feed-forward ANNs, no input of any of the LTUs depends on its output. In other words, there are no feedback connections. ANNs with feedback connections are called *recurrent networks*.

However, there is another way that the ANN in Figure 16.5 is a special type of an ANN. Indeed, in general, there is no reason to arrange the LTUs as layers as in Figure 16.5. In the layered architecture of Figure 16.5, the inputs of the LTUs in the kth layer are formed by the outputs of the LTUs in the previous layer (i.e., $(k-1)$th layer). Such an ANN is called a *layered neural network*, and Figure 16.5 is, thus, specifically a *layered feed-forward network* [253].

Depending on whether the network is layered or not and feed-forward or recurrent, one may obtain several different types of ANNs with differing behavior. Add to that different numbers of layers and different number of LTUs in each of the layers, we have a rich enough framework for approximating complex nonlinear functions. Indeed, it can be shown that an ANN of the form of Figure 16.5 consisting of just three layers can implement any arbitrary decision/classification boundary provided that the first two layers can have any number of LTUs in them [254].

16.2.3 Learning in Layered Feed-Forward Networks: Back-Propagation Algorithm

In section 16.2.1, we discussed how to adapt the weight vector of a single LTU to learn to classify inputs into one of two possible classes. In the case of an ANN made of several layers of LTUs, one needs a systematic way of adjusting weight vectors.

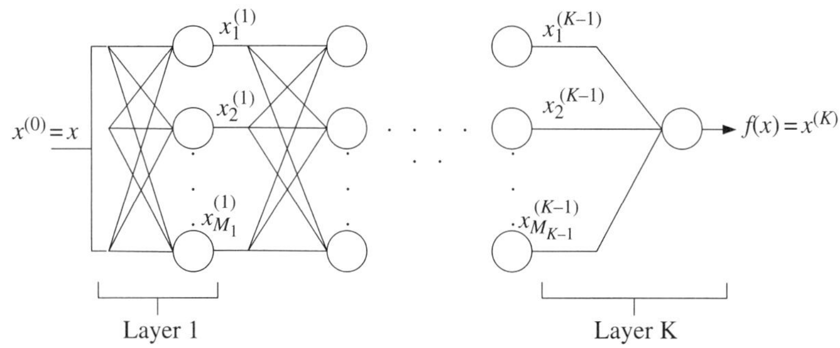

FIGURE 16.6 A layered feed-forward neural network.

In particular, consider a particular output LTU. If, for a given input $\mathbf{x}_l$, for $l = 1, \ldots, L$, its output differs from the desired output y_l specified by the training data, then it is not easy to see which place in the network is causing the error. In fact, it is possibly not one particular LTU that is causing this error, but all the LTUs in the whole network must in some way be responsible. Hence, for each error, we must update the weight vectors of all LTUs in all layers. One of the most widely used systematic approaches for achieving this in a layered feed-forward neural network is based on the criterion of minimizing the squared output error. This algorithm, based on the same basic idea of stochastic gradient descent, is called the *back-propagation algorithm* [258, 259].

Let us consider a layered feed-forward neural network made of K layers, as shown in Figure 16.6. Layer 1 is the input layer, whereas layer K is the output layer. Note that each circle in Figure 16.6 represents an LTU of the form of Figure 16.2. We may allow the kth layer, for $k = 1, \ldots, K$, to have M_k number of LTUs. For notational simplicity, in Figure 16.6, we have let the number of LTUs in the output layer K to be just one so that $M_K = 1$. However, allowing for an arbitrary number of output LTUs in the following discussion can be done easily.

Let us denote the output of the mth LTU in the kth layer as $x_m^{(k)}$, for $1 \leq m \leq M_k$ and $1 \leq k \leq K$. Recall from the definition of a layered feed-forward network that the inputs to the kth layer LTUs are the outputs of the LTUs of the $(k-1)$th layer. Hence,

$$\mathbf{x}^{(k-1)} \triangleq \begin{bmatrix} x_1^{(k-1)} \\ \vdots \\ x_{M_{k-1}}^{(k-1)} \end{bmatrix}$$

is the M_{k-1}-length input vector to the kth layer, for $k = 1, \ldots, K$, with $\mathbf{x}^{(0)} = \mathbf{x}$ where $\mathbf{x}$ is the external input vector, as shown in Figure 16.6.

As shown in Figure 16.7, input to the mth LTU in the kth layer is a weighted sum of the elements of this M_{k-1}-length input vector $\mathbf{x}^{(k-1)}$. Let us denote the (input) weight vector of the mth LTU in the kth layer by $\mathbf{w}_m^{(k)} \triangleq \left(w_{m,1}^{(k)}, \ldots, w_{m,M_{k-1}}^{(k)}\right)^T$. Then, by the

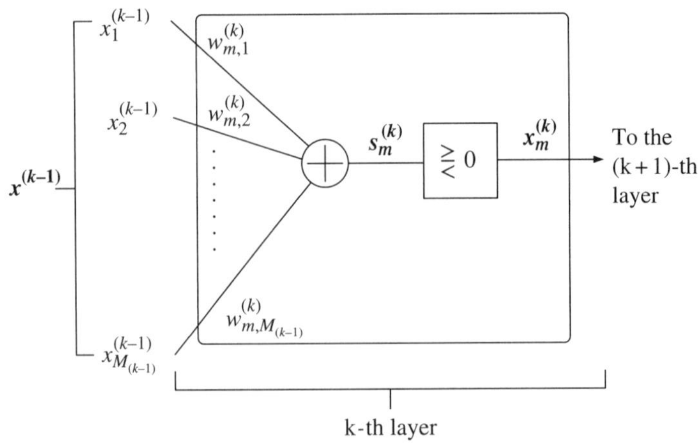

FIGURE 16.7 The mth linear threshold unit in the kth layer.

definition of the LTU function (16.1), it follows that the output of this mth LTU in the kth layer, $x_m^{(k)}$, for $1 \le m \le M_k$ and $1 \le k \le K$, is given by

$$\begin{aligned} x_m^{(k)} &= q\left(\left(\mathbf{w}_m^{(k)}\right)^T \mathbf{x}^{(k-1)}\right) \\ &= q\left(\mathbf{s}_m^{(k)}\right), \end{aligned} \tag{16.7}$$

where we have let

$$\mathbf{s}_m^{(k)} = \left(\mathbf{w}_m^{(k)}\right)^T \mathbf{x}^{(k-1)}. \tag{16.8}$$

Since we assume a single LTU in the output layer, we may let

$$\mathbf{s}_1^{(K)} = \mathbf{s}^{(K)}. \tag{16.9}$$

Let us consider a training input $\mathbf{x}$ with the associated true label y. According to our notation, the final output of the neural network to the input $\mathbf{x}$ is $x_1^{(K)}$. Again, since there is only a single output LTU, we may denote the final output of the network by $x^{(K)}$, so that

$$\begin{aligned} x^{(K)} &\triangleq x_1^{(K)} \\ &= q\left(\mathbf{s}^{(K)}\right). \end{aligned}$$

Let us denote by ϵ the final output error of the neural network:

$$\epsilon = y - x^{(K)} = y - q\left(\mathbf{s}^{(K)}\right). \tag{16.10}$$

The back-propagation algorithm is designed to adapt all weight vectors of the network $\mathbf{w}_m^{(k)}$, for $1 \le m \le M_k$ and $1 \le k \le K$, so as to minimize the squared output error ϵ^2. Similar to the LMS algorithm, the weight vector $\mathbf{w}_m^{(k)}$ is modified along the direction of negative gradient of ϵ^2 with respect to $\mathbf{w}_m^{(k)}$. The required gradient of ϵ^2 with respect to $\mathbf{w}_m^{(k)}$ is

$$\begin{aligned}
\frac{\partial \epsilon^2}{\partial \mathbf{w}_m^{(k)}} &= \frac{\partial \epsilon^2}{\partial \mathbf{s}_m^{(k)}} \frac{\partial \mathbf{s}_m^{(k)}}{\partial \mathbf{w}_m^{(k)}} \\
&= \frac{\partial \epsilon^2}{\partial \mathbf{s}_m^{(k)}} \mathbf{x}^{(k-1)} && (16.11a) \\
&= -2\epsilon \frac{\partial q\left(\mathbf{s}^{(K)}\right)}{\partial \mathbf{s}_m^{(k)}} \mathbf{x}^{(k-1)} && (16.11b) \\
&= -2\epsilon \delta_m^{(k)} \mathbf{x}^{(k-1)}, && (16.11c)
\end{aligned}$$

where (16.11a) and (16.11b) follow, respectively, from (16.8) and (16.10), and in (16.11c), we have defined

$$\delta_m^{(k)} = \frac{\partial q\left(\mathbf{s}^{(K)}\right)}{\partial \mathbf{s}_m^{(k)}}. \tag{16.12}$$

Note that $\delta_m^{(k)}$ is the partial derivative of the final output $x^{(K)} = q\left(\mathbf{s}^{(K)}\right)$ of the neural network with respect to the input $\mathbf{s}_m^{(k)}$ to the threshold device within the mth LTU in the kth layer.

From (16.11c), we note that the weight vector $\mathbf{w}_m^{(k)}$ of the mth LTU in the kth layer should be updated as follows:

$$\mathbf{w}_m^{(k)} \Longleftarrow \mathbf{w}_m^{(k)} + \alpha_m^{(k)} \epsilon \delta_m^{(k)} \mathbf{x}^{(k-1)}, \tag{16.13}$$

where we have let the learning rate to be $\frac{\alpha_m^{(k)}}{2}$ and allowed it to be possibly different for different LTUs and different layers.

For weight adaptation using (16.13), we need to compute $\delta_m^{(k)}$, for $1 \le m \le M_k$ and for each $1 \le k \le K$. However, note that the unit-step function $q(.)$ is not differentiable everywhere due to the discontinuity at the origin. The back-propagation algorithm gets around this problem by replacing the unit-step function by a differentiable function that approximates it. One of the best known replacements is the sigmoidal function defined as

$$g(s) = \frac{1}{1+e^{-s}}, \tag{16.14}$$

whose derivative can be written as

$$\frac{\partial g}{\partial s} = g(1-g). \tag{16.15}$$

With $q(s)$ replaced by $g(s)$, the output $x_m^{(k)}$ of the mth LTU in the kth layer in (16.7) becomes, for $1 \leq m \leq M_k$ and $1 \leq k \leq K$,

$$x_m^{(k)} = g\left(\mathbf{s}_m^{(k)}\right), \tag{16.16}$$

with $x^{(K)} = g\left(\mathbf{s}^{(K)}\right)$ for $k = K$. Similarly, (16.12) becomes

$$\delta_m^{(k)} = \frac{\partial g\left(\mathbf{s}^{(K)}\right)}{\partial \mathbf{s}_m^{(k)}}. \tag{16.17}$$

Using (16.14) in (16.17), we may easily obtain the $\delta_m^{(k)}$ corresponding to the output layer $k = K$. Indeed, since there is only one LTU in the output layer, from (16.17) and (16.9), we have that, for $m = 1$,

$$\begin{aligned}
\delta_m^{(K)} &= \frac{\partial g\left(\mathbf{s}^{(K)}\right)}{\partial \mathbf{s}_1^{(k)}} \\
&= \frac{\partial g\left(\mathbf{s}^{(K)}\right)}{\partial \mathbf{s}^{(k)}} \\
&= g\left(\mathbf{s}^{(K)}\right)\left(1 - g\left(\mathbf{s}^{(K)}\right)\right).
\end{aligned} \tag{16.18}$$

Since $m = 1$ for $k = K$, we may in fact let

$$\begin{aligned}
\delta^{(K)} &= \delta_1^{(K)} \\
&= g\left(\mathbf{s}^{(K)}\right)\left(1 - g\left(\mathbf{s}^{(K)}\right)\right) \\
&= x^{(K)}\left(1 - x^{(K)}\right).
\end{aligned} \tag{16.19}$$

Now, let us consider the mth LTU in the kth layer, for $k = K-1, \ldots, 1$. Observe from (16.12) and Figure 16.6 that the dependence of the output $g\left(\mathbf{s}^{(K)}\right)$ of the neural network on the input $\mathbf{s}_m^{(k)}$ to the mth LTU of the kth layer will be through the effect of $\mathbf{s}_m^{(k)}$ on the output $x_m^{(k)} = g\left(\mathbf{s}_m^{(k)}\right)$ of this LTU. This output $x_m^{(k)}$ of the mth LTU of the

kth layer affects $g\left(\mathbf{s}^{(K)}\right)$ through its effect on each of the inputs $\mathbf{s}_l^{(k+1)}$ of the lth LTU, for $1 \le l \le M_{k+1}$, of the $(k+1)$th layer. Hence, from (16.12), we have that

$$\begin{aligned}
\delta_m^{(k)} &= \frac{\partial g\left(\mathbf{s}^{(K)}\right)}{\partial \mathbf{s}_m^{(k)}} \\
&= \frac{\partial g\left(\mathbf{s}_1^{(k+1)}, \mathbf{s}_2^{(k+1)}, \ldots, \mathbf{s}_{M_{k+1}}^{(k+1)}\right)}{\partial \mathbf{s}_m^{(k)}} \\
&= \frac{\partial g}{\partial \mathbf{s}_1^{(k+1)}} \frac{\partial \mathbf{s}_1^{(k+1)}}{\partial \mathbf{s}_m^{(k)}} + \frac{\partial g}{\partial \mathbf{s}_2^{(k+1)}} \frac{\partial \mathbf{s}_2^{(k+1)}}{\partial \mathbf{s}_m^{(k)}} + \cdots + \frac{\partial g}{\partial \mathbf{s}_{M_{k+1}}^{(k+1)}} \frac{\partial \mathbf{s}_{M_{k+1}}^{(k+1)}}{\partial \mathbf{s}_m^{(k)}} \\
&= \sum_{l=1}^{M_{k+1}} \frac{\partial g}{\partial \mathbf{s}_l^{(k+1)}} \frac{\partial \mathbf{s}_l^{(k+1)}}{\partial \mathbf{s}_m^{(k)}} \\
&= \sum_{l=1}^{M_{k+1}} \delta_l^{(k+1)} \frac{\partial \mathbf{s}_l^{(k+1)}}{\partial \mathbf{s}_m^{(k)}},
\end{aligned} \tag{16.20}$$

where (16.20) follows by the definition of $\delta_m^{(k)}$. But from (16.8), observe that $\mathbf{s}_l^{(k+1)}$, for $l = 1, \ldots, M_{k+1}$, is a weighted sum of the outputs $x_m^{(k)} = g\left(\mathbf{s}_m^{(k)}\right)$, for $m = 1, \ldots, M_k$ of the kth layer:

$$\begin{aligned}
\mathbf{s}_l^{(k+1)} &= \left(\mathbf{w}_l^{(k+1)}\right)^T \mathbf{x}^{(k)} \\
&= \sum_{p=1}^{M_k} \mathbf{w}_{l,p}^{(k+1)} x_p^{(k)} \\
&= \sum_{p=1}^{M_k} \mathbf{w}_{l,p}^{(k+1)} g\left(\mathbf{s}_p^{(k)}\right).
\end{aligned}$$

Hence, for $l = 1, \ldots, M_{k+1}$ and $k = K-1, \ldots, 1$,

$$\begin{aligned}
\frac{\partial \mathbf{s}_l^{(k+1)}}{\partial \mathbf{s}_m^{(k)}} &= \frac{\partial}{\partial \mathbf{s}_m^{(k)}} \left[\sum_{p=1}^{M_k} \mathbf{w}_{l,p}^{(k+1)} g\left(\mathbf{s}_p^{(k)}\right) \right] \\
&= \sum_{p=1}^{M_k} \mathbf{w}_{l,p}^{(k+1)} \frac{\partial g\left(\mathbf{s}_p^{(k)}\right)}{\partial \mathbf{s}_m^{(k)}} \\
&= \mathbf{w}_{l,m}^{(k+1)} \frac{\partial g\left(\mathbf{s}_m^{(k)}\right)}{\partial \mathbf{s}_m^{(k)}} \\
&= \mathbf{w}_{l,m}^{(k+1)} g\left(\mathbf{s}_m^{(k)}\right) \left(1 - g\left(\mathbf{s}_m^{(k)}\right)\right).
\end{aligned} \tag{16.21}$$

Substituting (16.21) in (16.20), we have the following recursion for computing $\delta_m^{(k)}$: For $k = K-1, \ldots, 1$ and $m = 1, \ldots, M_k$,

$$\begin{aligned}\delta_m^{(k)} &= g\left(\mathbf{s}_m^{(k)}\right)\left(1 - g\left(\mathbf{s}_m^{(k)}\right)\right)\sum_{l=1}^{M_{k+1}} \delta_l^{(k+1)}\mathbf{w}_{l,m}^{(k+1)} \\ &= x_m^{(k)}\left(1 - x_m^{(k)}\right)\sum_{l=1}^{M_{k+1}} \delta_l^{(k+1)}\mathbf{w}_{l,m}^{(k+1)},\end{aligned} \tag{16.22}$$

with the initialization $\delta^{(K)}$ given by (16.19). By substituting $\delta_m^{(k)}$ in (16.13), then we have the back-propagation algorithm for updating the weight vectors of the layered feed-forward neural network.

The ANN is trained with the training data set $\{(\mathbf{x}_l, y_l)\}_{l=1}^{L}$ using the back-propagation algorithm until the changes in weight vectors are negligible from one iteration to the next. At this time, the network is considered to have learned to classify inputs of this type.

16.2.4 Neural Networks in Cognitive Radios

Neural networks have been applied to implement various decision and learning algorithms in cognitive radios. Depending on the problem, the choices of the types of neural networks used seem to vary. For example, both feed-forward and recurrent networks have been applied to implement certain decision-making and learning schemes. Not surprisingly, even within any of these classes, the structure of the specific neural network chosen can vary dramatically from one scenario to another.

In Chapter 11, we discussed a Bayesian nonparametric approach for classification and identification of detected signals over the sensed wide spectrum band. An alternative approach, albeit under different assumptions on prior knowledge, based on ANNs has been proposed in [260]. The objective is to identify the different types of signals by identifying their communications modes. This problem is called *mode identification*. For this, [260] extracts certain features from the time–frequency distributions of the sensed signal.

A time–frequency distribution of a signal is a two-dimensional (2-D) representation of the signal in the space of time and frequency. We have already seen such representations for cyclostationary signals in Chapter 6. However, there are other possible time–frequency representations as well. For example, one of the time–frequency distributions used in [260] to extract features from a signal $x(t)$ is the so-called Wigner–Ville distribution:

$$W(t,\omega) = \frac{1}{2\pi}\int x^*\left(t - \frac{\tau}{2}\right)x\left(t + \frac{\tau}{2}\right)e^{-j\omega t\tau}\,d\tau.$$

A discussion on the relationship of the Wigner–Ville distribution to cyclic spectrum and autocorrelation discussed in Chapter 6 can be found, for example, in [60].

From the computed time–frequency distribution, two features are extracted as relevant features for mode identification: (a) the standard deviation of *instantaneous frequency* σ_ω and (b) the maximum duration of a signal T_M. These particular features were selected based on the type of signals assumed in the sensed signal. In particular, [260] focused on mode identification of *spread spectrum* (SS) signals: direct-sequence code-division multiple-access (DS-CDMA) and frequency-hopping code-division multiple-access (FH-CDMA). Thus, it was assumed that the spectrum range of interest is the 2.4–2.4835 GHz ISM band and two systems are active in this spectrum: IEEE 802.11b WLAN (WiFi) employing DS-CDMA over 22 MHz width channels and Bluetooth that hops a 1 MHz bandwidth signal over 79 center frequencies across the whole ISM band at a rate of 1600 hops/s. The time–frequency distribution of the sensed signal is computed from the A/D converter output over a time duration corresponding to N samples. This window length N is chosen large enough so that it contains 10 Bluetooth frequency hops.

The instantaneous frequency of the signal $x(t)$ is computed from a given time–frequency distribution, for example, $W(t, \omega)$, as the average frequency at a fixed time:

$$\omega_i(t) = \frac{1}{W(t)} \int_{\infty}^{\infty} \omega W(t, \omega) d\omega,$$

where $W(t)$ is the marginal time distribution. The standard deviation σ_ω of the instantaneous frequency is then computed over a time duration of N samples as

$$\sigma_\omega = \sqrt{\frac{1}{T} \sum_{n=1}^{N} (\omega_i[n] - \overline{\omega_i}[n])^2},$$

where

$$\overline{\omega_i}[n] = \frac{1}{T} \sum_{n=1}^{N} \omega_i[n]$$

is the average instantaneous frequency.

The maximum duration of a signal is extracted from the time–frequency distribution by noting that while in DS-CDMA signals each frequency is continuous over the window length N, in FH-CDMA, the frequencies will be discontinuous due to frequency hopping. From these discontinuities, the duration of a signal is computed at each frequency, and then the maximum duration T_M is derived from these values.

A feed-forward neural network was used to determine the mode of the sensed signal as one of the following four possible choices: noise only, WLAN only, Bluetooth only, or WLAN and Bluetooth. Hence, both input and output layers were chosen to contain two LTUs each. The four hidden layers were to have 5, 5, 4 and 2 LTUs, respectively. The feed-forward neural network was trained using the back-propagation algorithm. The results obtained had shown that indeed the feed-forward neural network can provide promising results in identifying the mode of the signal.

On the other hand, [261] provides an example of the use of a recurrent neural network to implement learning algorithms in cognitive radios. In particular, [261] has discussed the possibility of using a recurrent ANN to make the cognitive radio be able to predict its capabilities in different radio configurations. The neural network-based learning method of [261] is specifically aimed at predicting achievable data rate in each of the radio configurations. However, it also discusses the possibility of extending the same framework for predicting all relevant PHY and MAC layer QoS parameters so that the cognitive radio can select the optimal mode of operation for a given state of the RF environment and user needs. This can be achieved by either designing a larger neural network with additional feature vector inputs and outputs denoting not just the data rate levels but also other relevant PHY and MAC QoS parameters or by having multiple parallel constituent neural networks for other relevant parameters, beyond data rate, that characterize the performance of a configuration modes such as bit error rate, transmit power, latency, etc.

One last example of neural network-based implementation of an unsupervised learning algorithm for signal classification is the use of *self-organizing maps* (SOMs) in [182]. These are a special type of neural networks that are particularly suited for unsupervised learning. Perhaps, the most common structure of SOMs are either as one-dimensional or 2-D grids made of decision devices, again, called the neurons. Each neuron is associated with a weight vector similar to the LTUs. These weight vectors are of the same dimension as the input vectors. Each neuron is, however, characterized by not only its associated weight vector but also by its location on the grid. A given input vector is classified by associating it with the neuron that has the closest weight vector to the input vector. This neuron is called the *best matching cell* (BMC).

For a given training input, weights of all neurons are modified by an update term that depends not only on the distance of its weight vector from the input vector but also on its distance from the best matching neuron in terms of their relative locations on the grid [262]. Note that there is no need for training examples and the weight updates are unsupervised. It was shown in [182] that a properly designed SOM with appropriate feature vectors can be used to classify 8 vestigial sideband (VSB) and BPSK signals successfully.

16.3 SUPPORT VECTOR MACHINES

The success of ANNs as a machine learning structure can be traced to it effectively addressing the key challenges encountered in learning problems. First, since the required decision rule can be nonlinear, the learning structure must have the capability to learn nonlinear separating surfaces (decision boundaries). The ANN addresses this representation issue by using layered neural networks. As we discussed earlier, having just a 3-layer feed-forward network is enough to represent any nonlinear decision boundary as long as the first two layers are allowed to have any number of LTUs. Second, once a network of interconnected LTUs is used, adaptation of weights becomes complicated compared to that in a single LTU. This computational issue is resolved by the back-propagation algorithm that gives a simple and efficient method to systematically update the weights. Third, and final, is that the trained ANN should have sufficient generalization performance. This generalization issue is addressed by proper choice of the network size. Indeed, sufficient generalization performance can be achieved by ensuring that the number of layers in the network and the number of LTUs in each layer are not too large [254].

Support vector machines (SVMs) are an alternative machine learning structure that also addresses these concerns effectively albeit using different approaches. In particular, as we will see in the following, a SVM addresses the representation issue by first non-linearly transforming the input data into a higher-dimensional space (compared to the dimension of the space of original input data) and then designing linear decision boundaries to classify this transformed data. The resulting decision rules in the higher-dimensional space correspond to nonlinear separating surfaces in the original lower-dimensional input vector space. The computational issues are resolved by formulating the problem of computing the optimal hyperplane in the higher-dimensional space as a convex optimization problem for which there are efficient solution algorithms [254]. Third, and final, issue of generalization performance of the classifier is achieved by looking for the hyperplane that separates the two classes with the largest margin. The justification behind this particular choice for achieving generalization capability, indeed, has an elegant theory of its own.

16.3.1 Statistical Learning Theory

Let us consider a classification problem in which each data $\mathbf{x} \in \mathcal{X} \subseteq \mathbb{R}^n$ is associated with a certain class label $y \in \mathcal{Y}$. A classifier function for this scenario should be of the form of

$$f : \mathcal{X} \longrightarrow \mathcal{Y}.$$

Our interest is in selecting the best (in some sense) classifier from among a class of classifiers. Hence, we may parameterize the classifier function $f(\mathbf{x})$ by a set of parameters, or a parameter vector, α. The form of the parameter set α will define the class of classifiers, or decision rules, of interest, while different parameter values assumed by α will characterize the individual classifiers from this class: By allowing for different $\alpha, f(\mathbf{x}, \alpha)$ represents a class of classifier functions. We may write $f(\mathbf{x}, \alpha)$ to imply the class of decision rules of interest as well as to explicitly denote the dependence of the classifier f on the parameters α.

Example 16.2 (The Class of Classifiers Represented by an LTU) As an example, recall that the class of classifiers that can be represented by a single LTU are those that linearly separate the two classes by a hyperplane. This class of classifiers thresholds a linear mapping of data $\mathbf{w}^T\mathbf{x}$. Any individual classifier within this class is essentially characterized by the weight vector $\mathbf{w}$, which defines the hyperplane as we saw in Figure 16.3. Hence, in this case, $\alpha = \mathbf{w}$, assuming a threshold of zero. For different weight vectors, we obtain a different classifier from the LTU. ■

In a probabilistic setting, we may assume that data $\mathbf{x}$ and their corresponding classes y are distributed according to some joint distribution $P_{\mathbf{X}Y}(\mathbf{x}, y)$. Then, for any assumed classifier $f(\mathbf{x}, \alpha)$, the expected value of the absolute error, given by

$$\begin{aligned} R(\alpha) &= \mathbb{E}\left\{\frac{1}{2}\left|Y - f(\mathbf{X}, \alpha)\right|\right\} \\ &= \frac{1}{2}\int \left|y - f(\mathbf{x}, \alpha)\right| dP_{\mathbf{X}Y}(\mathbf{x}, y), \end{aligned} \tag{16.23}$$

is called the *expected risk*. If we knew the joint distribution $P_{\mathbf{X}Y}(\mathbf{x}, y)$ of $\mathbf{X}$ and Y, we may possibly compute the expected risk $R(\alpha)$ associated with the classifier $f(\mathbf{x}, \alpha)$. A natural optimality criterion is to learn the classifier $f(\mathbf{x}, \alpha)$ that leads to the minimum expected risk $R(\alpha)$.

Of course, in many situations, this joint distribution $P_{\mathbf{X}Y}(\mathbf{x}, y)$ is not known, precluding this approach. Instead, what we have is a set of possible input data $\mathbf{x}_i \in \mathcal{X}$, for $i = 1, \ldots, L$, with their associated correct classification labels $y_i \in \mathcal{Y}$. As before, this set of labeled data $\{(\mathbf{x}_i, y_i)\}_{i=1}^{L}$ is called the training data set. For a given classifier $f(\mathbf{x}, \alpha)$, we may compute the sample average of errors on this training data set as

$$R_{emp}(\alpha) = \frac{1}{2L}\sum_{i=1}^{L}\left|y_i - f(\mathbf{x}_i, \alpha)\right|. \tag{16.24}$$

This sample average error $R_{emp}(\alpha)$ is called the *empirical risk*. Given a training data set, then a reasonable alternative is to learn the classifier $f(\mathbf{x}, \alpha)$ that minimizes the empirical risk $R_{emp}(\alpha)$.

A basic objective of any learning procedure, however, is that once the learning is completed, the resulting classifier, or the decision rule, should be able to perform well enough on, previously unseen, new test data. This ability of a classifier to do well on unseen test data is, indeed, what is called the generalization ability of the classifier. A classifier $f(\mathbf{x}, \alpha)$ that is designed to minimize the empirical risk $R_{emp}(\alpha)$ over a given training data set $\{(\mathbf{x}_i, y_i)\}_{i=1}^{L}$ may not necessarily perform well on all possible data $\mathbf{x}$ that is drawn according to the joint distribution $P_{\mathbf{X}Y}(\mathbf{x}, y)$. The issue at hand is that it is possible to have many classifier functions $f(\mathbf{x}, \alpha)$ that might minimize the empirical risk $R_{emp}(\alpha)$ over a given training data set $\{(\mathbf{x}_i, y_i)\}_{i=1}^{L}$, yet their performance on new test data may vary among themselves. Hence, while minimizing the empirical risk, it is also necessary to somehow ensure that the performance of the classifier on new unseen data will also be sufficiently good.

The overall cost of errors of a classifier $f(\mathbf{x}, \alpha)$ over all possible data $\mathbf{x}$ is given by the expected risk $R(\alpha)$. One way we can be confident that the classifier $f(\mathbf{x}, \alpha)$ may not perform too badly on unseen data is if we can, at least, have some idea that the empirical risk $R_{emp}(\alpha)$ on the training data set $\{(\mathbf{x}_i, y_i)\}_{i=1}^{L}$ is not too far from the overall risk $R(\alpha)$. We cannot, of course, know this directly, because we do not know $R(\alpha)$. However, if we consider a two-class classifying problem with $y_i \in \{+1, -1\}$ so that the risk $\frac{1}{2}\left|y_i - f(\mathbf{x}_i, \alpha)\right|$ also only takes the two possible values 0 and 1, then it has been shown that with probability $1-\eta$ the expected risk is upper bounded as [263]

$$R(\alpha) \leq R_{emp}(\alpha) + \sqrt{\frac{h\left(\log\left(\frac{2L}{h}\right)+1\right) - \log\left(\frac{\eta}{4}\right)}{L}}, \tag{16.25}$$

where $0 \leq \eta \leq 1$ and h is an integer-valued quantity called the *Vapnik–Chervonenkis (VC) dimension* [264].

Before we define VC dimension, it is worth pointing out that the bound (16.25) is valid as long as the data are drawn iid, regardless of $P_{\mathbf{X}Y}(\mathbf{x},y)$. The importance of this bound is that the VC dimension is a quantity that only depends on the class of functions $f(.,\alpha)$. Hence, if we knew h corresponding to the class of functions in question, then the upper bound (16.25) is easily computed. Moreover, it can easily be verified that the second term in the upper bound, called the *VC confidence*, is a monotonic increasing function of h for a fixed η and sample size L.

The VC dimension is defined in terms of the notion of *shattering*, which is defined below for the two-class classifier functions $f(\mathbf{x},\alpha) \in \{-1,+1\}$ [254, 264].

Definition 16.3.1 (Shattering) A set of vectors $\{\mathbf{x}_1,\ldots,\mathbf{x}_L\}$ is said to be shattered by a class of classifier functions $f(\mathbf{x},\alpha)$ if all the 2^L possible labelings of $\{\mathbf{x}_1,\ldots,\mathbf{x}_L\}$ can be generated by using classifier functions from the class $f(\mathbf{x},\alpha)$.

The 2^L possible labelings above refers to 2^L possible ways the set of vectors $\{\mathbf{x}_1,\ldots,\mathbf{x}_L\}$ may be assigned to the two classes denoted by -1 and $+1$. Hence, if the class $f(\mathbf{x},\alpha)$ of classifier functions shatters the set of vectors $\{\mathbf{x}_1,\ldots,\mathbf{x}_L\}$, then $f(\mathbf{x},\alpha)$ must be rich in its capacity to classify these vectors in the sense that we can always find a classifier function from this class that can correctly classify these vectors to the two classes. In other words, the set of vectors $\{\mathbf{x}_1,\ldots,\mathbf{x}_L\}$ can be separated into the two classes in all possible ways by decision rules from $f(\mathbf{x},\alpha)$.

However, this richness of the class of classifiers $f(\mathbf{x},\alpha)$ can be a curse when having to learn a decision rule from a given set of training data $\{(\mathbf{x}_l,y_l)\}_{l=1}^{L}$ that will also have good generalization performance: If $f(\mathbf{x},\alpha)$ shatters $\{\mathbf{x}_1,\ldots,\mathbf{x}_L\}$, then there will be a decision rule in the class $f(\mathbf{x},\alpha)$ that will be able to exactly match the given true classification $\{y_1,\ldots,y_L\}$. In other words, the empirical risk $R_{emp}(\alpha)$ can be made zero. There is, however, no guarantee how this decision rule will perform on new input data vectors. The VC dimension of a class of decision rules is the maximum number of vectors that can be shattered by that class of decision rules.

Definition 16.3.2 (VC Dimension) The VC dimension of a class of classifier functions $f(\mathbf{x},\alpha)$ is the maximum number of input vectors that can be shattered by $f(\mathbf{x},\alpha)$.

It is important to observe that if the VC dimension of the class of classifiers $f(\mathbf{x},\alpha)$ is h, it does not mean that $f(\mathbf{x},\alpha)$ will shatter any given set of h input vectors. It only means that there is at least one set of h input vectors $\{\mathbf{x}_1,\ldots,\mathbf{x}_h\}$ that it can shatter.

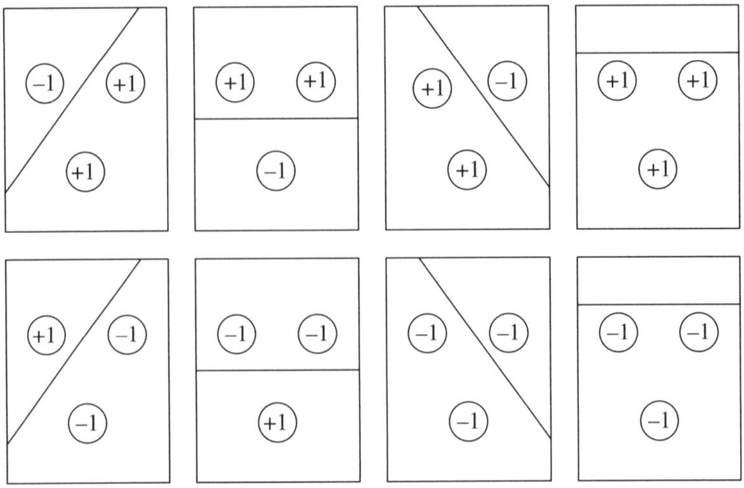

FIGURE 16.8 Lines on the 2-D plane can shatter three points on the planes that are not colinear.

Example 16.3 (VC Dimension of Hyperplanes in the Plane) As an example, let us suppose $d = 2$, so that the input data $\mathbf{x}$ are vectors on the 2-D plane. Figure 16.8 shows three input vectors $\mathbf{x}$. A hyperplane in the 2-D plane is a line, and as can be seen from Figure 16.8, hyperplanes can certainly shatter (i.e., gives all possible labelings) these three input vectors $\mathbf{x}$. Indeed, although not necessary, lines can shatter any three points on the 2-D plane as long as they are not colinear. It can, however, be shown that lines cannot shatter any four points on the 2-D plane (see, e.g., [254] for a simple proof). Hence, the maximum number of 2-D input vectors that can be shattered by the hyperplanes on the 2-D is 3. Thus, the VC dimension of 2-D hyperplanes, that is lines on the plane, is $h = 3$. ■

Example 16.3, in fact, is just a special case of a general result on the VC dimension of hyperplanes. Indeed, it has been shown that the VC dimension of hyperplanes in the d-dimensional Euclidean space $\mathbb{R}^d$ is $h = d + 1$ [264, 265].

The VC dimension of a class of functions is thus a measure of their richness in terms of the ability to classify. The richer the class of functions $f(., \alpha)$, the larger h will be. As mentioned earlier, a richer class of functions $f(\mathbf{x}, \alpha)$ has the ability, or capacity, to fit in to a given set of training data $\{(\mathbf{x}_i, y_i)\}_{i=1}^{L}$ very well so that the empirical risk could be made small. However, since h is large for such a class of functions, the VC confidence term (the second term in the upper bound (16.25)) will be large, resulting in a large overall upper bound (16.25). If we have reason to believe that the upper bound is tight in a given scenario, then this large upper-bound value may serve as an indicator of possible poor generalization ability of the classifier. On the other hand, a not so rich class of functions $f(., \alpha)$ will have a smaller VC dimension h and thus a smaller VC confidence. But they may not have the ability to fit very well in to the given training data set $\{(\mathbf{x}_l, y_l)\}_{l=1}^{L}$, resulting in a higher empirical risk. Again, the overall upper bound (16.25) can be large.

Hence, at least when the upper bound (16.25) is indeed tight, it makes sense to select a function $f(\mathbf{x}, \alpha)$ that has large enough VC dimension, but not too large. In particular, if the choice is among a class of functions $f(\mathbf{x}, \alpha)$ all giving zero empirical risk $R_{emp}(\alpha)$, then it is reasonable to prefer the one with the smallest VC dimension since VC confidence term in the upper bound (16.25) is monotonic increasing in h.

We now have a reasonable approach to select a decision rule that may result in good generalization performance: Given a class of functions $f(\mathbf{x}, \alpha)$, rather than selecting the one that minimizes just the empirical risk $R_{emp}(\alpha)$, we may select the one that minimizes the upper bound (16.25) on the expected risk $R(\alpha)$. This approach is called the *structural risk minimization* as opposed to *empirical risk minimization* that looks for the decision function $f(\mathbf{x}, \alpha)$ that minimizes simply the empirical risk $R_{emp}(\alpha)$.

In general, structural risk minimization considers classes of functions with different VC dimensions [264]. For each class, we may find the function $f(\mathbf{x}, \alpha)$ that minimizes the empirical risk $R_{emp}(\alpha)$. Out of these empirical risk minimizers, we pick the one that minimizes the overall upper bound (16.25) as the best classifier.

16.3.2 Structural Risk Minimization with Support Vector Machines

When it is not possible to find the exact VC dimensions of classes of functions, we may instead use a computable upper bound to the VC dimension. An SVM for a given classifier design problem is the solution to the associated structural risk minimization problem obtained by assuming a specific class of decision functions and a specific upper-bound characterization for the VC dimension of this class of functions.

The class of functions assumed in SVM is the hyperplanes. Suppose that data is d-dimensional real vectors $\mathbf{x} \in \mathbb{R}^d$ and the two classes are labeled as $y \in \{+1, -1\}$. We are given a training data set of $\{(\mathbf{x}_l, y_l)\}_{l=1}^{L}$. The SVM approach to structural risk minimization seeks to find a hyperplane $\mathbf{w}^T\mathbf{x} + b = 0$ that has the smallest VC dimension while also correctly classifying all training data (i.e., makes $R_{emp}(\alpha) = 0$).

Without loss of generality, let us assume that we normalize the weight vector $\mathbf{w}$ so that

$$\min_{1 \le l \le L} \left| \mathbf{w}^T \mathbf{x}_l + b \right| = 1. \tag{16.26}$$

Since distance D_l from an arbitrary point $\mathbf{x}_l$ to the hyperplane $\mathbf{w}^T\mathbf{x} + b = 0$ is equal to $D_l = \frac{|\mathbf{w}^T\mathbf{x}_l + b|}{||\mathbf{w}||}$, normalization (16.26) means that

$$\min_{1 \le l \le L} D_l = \frac{1}{||\mathbf{w}||}. \tag{16.27}$$

In other words, the closest data point to the hyperplane is at a distance of $\frac{1}{||\mathbf{w}||}$.

Let us denote by R the radius of a d-dimensional sphere centered at origin that encloses all the given data $\mathbf{x}_1, \ldots, \mathbf{x}_L$. Then, if $||\mathbf{w}|| < A$, for some $A > 0$, it is known that the VC dimension h of these hyperplanes is upper bounded as [266]

$$h \leq \min\left\{\lceil R^2 A^2 \rceil, d\right\} + 1. \tag{16.28}$$

This shows that minimizing the VC dimension of the hyperplanes is equivalent to minimizing the norm of the weight vector $||\mathbf{w}||$. This is the approach that SVM takes for structural risk minimization: Find the hyperplane defined by the weight vector with the smallest $||\mathbf{w}||$ that correctly classifies all training data.

Hence, SVM solution to the structural risk minimization problem is given by the solution $\mathbf{w}^*$ to the following optimization problem:

$$\mathbf{w}^* = \arg\min \frac{1}{2}||\mathbf{w}||^2 \text{ s.t. } sgn(\mathbf{w}^T\mathbf{x}+b) = y_l \quad \text{for all } l = 1, \ldots, L. \tag{16.29}$$

In light of (16.27), this is the same as finding the hyperplane that will maximize the minimum distance from the hyperplane to a given data point while also correctly classifying all training data. Hence, designing an SVM that will have good generalization performance to classify data, based on a given set of training examples, simplifies to finding this *optimal hyperplane* (optimal in the sense of maximizing the minimum distance from the hyperplane to a given data point while also correctly classifying all training data).

This, however, assumes that given training data set can, in fact, be classified correctly by a hyperplane. Hence, if they are not linearly separable, (16.29) cannot be applied as it is. Indeed, as we mentioned earlier, in most applications, the true classification surface can be nonlinear. SVMs get around this problem by first nonlinearly transforming data into a higher-dimensional space in which the transformed data can be linearly separable. Then, an optimal hyperplane, as defined earlier, is found in this higher-dimensional space. This hyperplane will define a nonlinear classification surface in the original space of data.

16.3.3 Linear Support Vector Machines

Let us first consider a situation in which a given training data set of $\{(\mathbf{x}_l, y_l)\}_{l=1}^{L}$, where $\mathbf{x}_l \in \mathbb{R}^d$ and $y_l \in \{+1, -1\}$ for $l = 1, \ldots, L$, is indeed linearly separable into two classes with labels $y \in \{+1, -1\}$. Suppose that $\mathbf{w}^T\mathbf{x} + b = 0$ is such a hyperplane that separates the given data into the two classes correctly. Then, for $l = 1, \ldots, L$,

$$\text{If } y_l = 1 \Longrightarrow \mathbf{w}^T\mathbf{x}_l + b \geq +1 \tag{16.30a}$$

$$\text{If } y_l = -1 \Longrightarrow \mathbf{w}^T\mathbf{x}_l + b \leq -1. \tag{16.30b}$$

We may combine (16.30a) and (16.30b) into a single equation as

$$y_l\left(\mathbf{w}^T\mathbf{x}_l + b\right) - 1 \geq 0 \quad \text{for } l = 1, \ldots, L. \tag{16.31}$$

Then the SVM optimization problem (16.29) for this case can be written as

$$\begin{aligned} &\min \frac{1}{2}||\mathbf{w}||^2 \\ &\text{s.t.} \quad y_l\left(\mathbf{w}^T\mathbf{x}_l + b\right) - 1 \geq 0 \qquad \text{for } l = 1,\ldots,L. \end{aligned} \tag{16.32}$$

The Lagrangian for this constrained optimization problem is

$$\begin{aligned} \mathcal{L}_p(\mathbf{w},b,\boldsymbol{\alpha}) &= \frac{1}{2}||\mathbf{w}||^2 - \sum_{l=1}^{L}\alpha_l\left(y_l\left(\mathbf{w}^T\mathbf{x}_l + b\right) - 1\right) \\ &= \frac{||\mathbf{w}^2||}{2} - \sum_{l=1}^{L}\alpha_l y_l\left(\mathbf{w}^T\mathbf{x}_l + b\right) + \sum_{l=1}^{L}\alpha_l, \end{aligned} \tag{16.33}$$

where $\alpha_l \geq 0$, for $l = 1,\ldots,L$, are the Lagrangian multipliers and $\boldsymbol{\alpha} = (\alpha_1,\ldots,\alpha_L)^T$.

The primal problem associated with (16.33) is

$$\begin{aligned} &\text{Primal:} \quad \min_{\mathbf{w},b} \mathcal{L}_p(\mathbf{w},b,\boldsymbol{\alpha}) \\ &\text{s.t.} \quad \frac{\partial}{\partial\alpha_l}\mathcal{L}_p = 0 \quad \text{for } l = 1,\ldots,L \\ &\qquad \alpha_l \geq 0 \quad \text{for } l = 1,\ldots,L. \end{aligned} \tag{16.34}$$

The Lagrangian dual function associated with $\mathcal{L}_p$ is given by

$$\mathcal{L}_d(\boldsymbol{\alpha}) = \inf_{\mathbf{w},b} \mathcal{L}_p(\mathbf{w},b,\boldsymbol{\alpha}). \tag{16.35}$$

The Lagrangian dual problem associated with (16.33) is then defined as

$$\begin{aligned} &\text{Dual:} \quad \max_{\boldsymbol{\alpha}} \mathcal{L}_d(\boldsymbol{\alpha}) \\ &\text{s.t. } \alpha_l \geq 0 \quad \text{for } l = 1,\ldots,L. \end{aligned} \tag{16.36}$$

Note that $\mathcal{L}_p(\mathbf{w},b,\boldsymbol{\alpha})$ is convex in $\mathbf{w}$ and b. Hence, the infimum required in (16.35) is achieved when the gradient $\nabla_{\mathbf{w},b}\mathcal{L}_p = \mathbf{0}$, giving us

$$\frac{\partial}{\partial\mathbf{w}}\mathcal{L}_p = 0 \Longrightarrow \mathbf{w} = \sum_{l=1}^{L}\alpha_l y_l \mathbf{x}_l \tag{16.37}$$

and

$$\frac{\partial}{\partial b}\mathcal{L}_p = 0 \Longrightarrow \sum_{l=1}^{L}\alpha_l y_l = 0. \tag{16.38}$$

Substituting (16.37) and (16.38) in (16.35), the Lagrangian dual function can be written as

$$\mathcal{L}_d(\boldsymbol{\alpha}) = \sum_{l=1}^{L} \alpha_l - \frac{1}{2}\sum_{l=1}^{L}\sum_{m=1}^{L} \alpha_l \alpha_m y_l y_m \mathbf{x}_l^T \mathbf{x}_m, \tag{16.39}$$

so that the Lagrangian dual problem (16.36) can be rewritten as

$$\begin{aligned}\text{Dual:} \quad & \max_{\boldsymbol{\alpha}} \sum_{l=1}^{L} \alpha_l - \frac{1}{2}\sum_{l=1}^{L}\sum_{m=1}^{L} \alpha_l \alpha_m y_l y_m \mathbf{x}_l^T \mathbf{x}_m \\ \text{s.t.} \quad & \sum_{l=1}^{L} \alpha_l y_l = 0 \\ & \alpha_l \geq 0 \quad \text{for } l = 1, \ldots, L.\end{aligned} \tag{16.40}$$

The problem (16.40) is a convex quadratic optimization problem that can be solved efficiently using specialized algorithms. By solving (16.40), we may obtain the optimal Lagrange multipliers α_l^*'s, for $l = 1, \ldots, L$. By back substituting these into (16.37), we obtain the optimal weight vector $\mathbf{w}^*$ to be

$$\mathbf{w}^* = \sum_{l=1}^{L} \alpha_l^* y_l \mathbf{x}_l. \tag{16.41}$$

The Karush–Kuhn–Tucker (KKT) conditions that must be satisfied by the primary problem give the following additional equations:

$$\alpha_l^* \left(y_l \left(\mathbf{x}_l^T \mathbf{w}^* + b^* \right) - 1 \right) = 0 \quad \text{for } l = 1, \ldots, L \tag{16.42a}$$

$$y_l \left(\mathbf{x}_l^T \mathbf{w}^* + b^* \right) - 1 \geq 0 \quad \text{for } l = 1, \ldots, L \tag{16.42b}$$

$$\alpha_l \geq 0 \quad \text{for } l = 1, \ldots, L. \tag{16.42c}$$

We may obtain the optimal b^* by solving (16.42a) for any $\alpha_l \neq 0$. However, for numerical accuracy, it is often better to compute b for all $\alpha_l \neq 0$ and then use their average as the final b^* [254, 264].

From (16.41), we can see that the optimal weight vector is determined by data points $\mathbf{x}_l$'s that correspond to $\alpha_l^* \neq 0$ in the solution. From (16.42a), on the other hand, we see that $\alpha_l^* \neq 0$ only if $y_l \left(\mathbf{x}_l^T \mathbf{w}^* + b^* \right) - 1 = 0$, or equivalently, $\left(\mathbf{x}_l^T \mathbf{w}^* + b^* \right) = \pm 1$. Hence, in light of (16.26), nonzero α_l^*'s correspond to those data vectors that are closest to the optimal hyperplane, and those data points are the only ones that matter in determining the optimal separating hyperplane. These data vectors corresponding to nonzero α_l^*'s are called the *support vectors*.

To see this geometrically, let us denote by d_+ and d_-, respectively, the shortest distance from the closest data point in the class $y = +1$ (i.e., $\mathbf{w}^T \mathbf{x} + b \geq 0$) and that in the class $y = -1$ (i.e., $\mathbf{w}^T \mathbf{x} + b < 0$), as shown in Figure 16.9. The closest points will correspond to those input vectors $\mathbf{x}_l$ that satisfy (16.30a) or (16.30b) with equality.

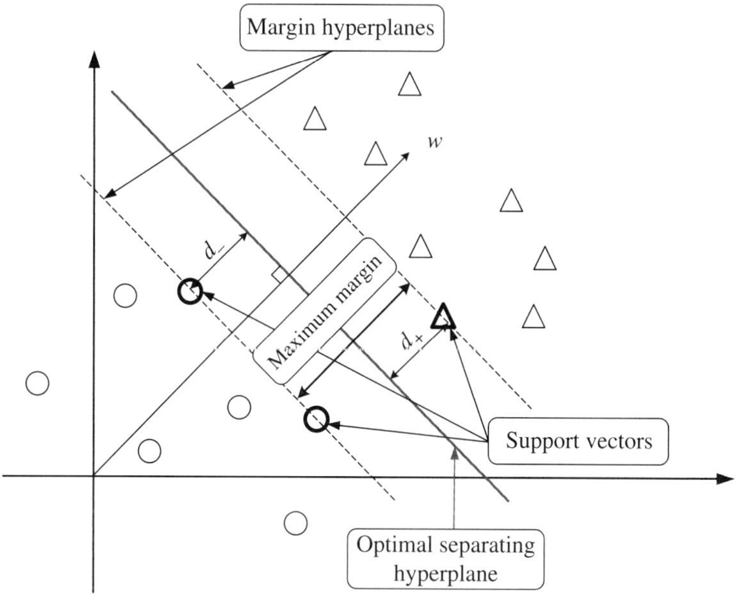

FIGURE 16.9 According to the support vector machine optimality criterion, the optimal hyperplane for classifying two-class data is the one that has the maximum margin.

Let us denote by $\mathbf{x}_+$ and $\mathbf{x}_-$ two such points satisfying, respectively, (16.30a) and (16.30b) with equality:

$$\mathbf{w}^T\mathbf{x}_+ + b = +1 \tag{16.43a}$$

$$\mathbf{w}^T\mathbf{x}_- + b = -1. \tag{16.43b}$$

Note that the normalization (16.26) guarantees that such two points $\mathbf{x}_+$ and $\mathbf{x}_-$ do exist.

The distance d_+ from a point $\mathbf{x}_+$ to the hyperplane $\mathbf{w}^T\mathbf{x}+b=0$, however, is given by

$$d_+ = \frac{|\mathbf{w}^T\mathbf{x}_+ + b|}{||\mathbf{w}||} = \frac{1}{||\mathbf{w}||}, \tag{16.44}$$

where (16.44) follows from (16.43a). Similarly, using (16.43b), distance d_- from a point $\mathbf{x}_-$ to the hyperplane $\mathbf{w}^T\mathbf{x}+b=0$ is given by

$$d_- = \frac{|\mathbf{w}^T\mathbf{x}_- + b|}{||\mathbf{w}||} = \frac{1}{||\mathbf{w}||}. \tag{16.45}$$

Hence, we note that the margin between the closest points on either side of the hyperplane $\mathbf{w}^T\mathbf{x}+b=0$ is equal to

$$d_+ + d_- = \frac{2}{||\mathbf{w}||}. \tag{16.46}$$

Thus, the SVM optimization problem (16.29) is also equivalent to finding the hyperplane that maximizes the margin between closest points to the hyperplane on the two sides. For this reason, sometimes, the SVM optimization problem is simply called finding the *maximum margin hyperplane*.

This shows that it is those points $\mathbf{x}_+$ and $\mathbf{x}_-$ on either side of the optimal hyperplane that satisfy (16.31) with equality that defines the optimal separating hyperplane. In fact, the rest of the data points in the training data set $\{(\mathbf{x}_l, y_l)\}_{l=1}^L$ are irrelevant for finding this optimal hyperplane. If one were to change those other data points in the training set $\{(\mathbf{x}_l, y_l)\}_{l=1}^L$ to some other arbitrary points while they are still not the closet points to the hyperplane, the optimal separating hyperplane will not change. On the other hand, if we change any of the data points $\mathbf{x}$ in the training data set $\{(\mathbf{x}_l, y_l)\}_{l=1}^L$ that are the closest points to the hyperplane (i.e., those points $\mathbf{x}_+$ and $\mathbf{x}_-$), the optimal hyperplane will change. Hence, these closest points $\mathbf{x}_+$ and $\mathbf{x}_-$ are the ones that support defining the optimal separating hyperplane, justifying them being called the support vectors.

16.3.4 Nonlinear Support Vector Machines

Linear SVMs of the previous section assume that the given training data set $\{(\mathbf{x}_l, y_l)\}_{l=1}^L$ is indeed linearly separable. If, however, data is linearly nonseparable, the approach of section 16.3.3 will not work because we may not be able to find any hyperplane that correctly classifies the given training data $\{(\mathbf{x}_l, y_l)\}_{l=1}^L$ into the two classes. Recall that an ANN resolves this same limitation encountered by a single LTU by introducing interconnected networks of LTUs. SVMs resolve this issue by transforming the original data $\mathbf{x}$ to a higher-dimensional space $\mathcal{H}$ using an appropriate nonlinear mapping Φ [267, 268]. The linear SVM approach of the previous section is then applied to the new training data set $\{(\Phi(\mathbf{x}_l), y_l)\}_{l=1}^L$, where $\Phi(\mathbf{x}_l) \in \mathcal{H}$ and $y_l \in \{+1, -1\}$ for $l = 1, \ldots, L$. Of course, the transformation $\Phi(.)$ is to be selected so that the transformed data in $\mathcal{H}$ is possibly linearly separable.

It is, however, possible even after the transformation the data may not be linearly separable. In general, this is handled by relaxing the constraint in (16.29) that all training data be correctly classified by the hyperplane. This means that we relax the constraints (16.30a) and (16.30b) to a certain extent by only requiring, for $l = 1, \ldots, L$,

$$\text{If } y_l = 1 \Longrightarrow \mathbf{w}^T\Phi(\mathbf{x}_l) + b \geq 1 - \xi_l \tag{16.47a}$$

$$\text{If } y_l = -1 \Longrightarrow \mathbf{w}^T\Phi(\mathbf{x}_l) + b \leq -1 + \xi_l, \tag{16.47b}$$

where $\xi_l \geq 0$, for $l = 1, \ldots, L$, are called (positive) slack variables and $\mathbf{w}$ is now a vector in the higher-dimensional space $\mathcal{H}$. The resulting separating hyperplane will give erroneous classifications of training data for those $\xi_l > 1$. Hence, we will impose a cost for giving such erroneous classifications. A simple upper bound for the number of training data points that are classified erroneously is given by $\sum_{l=1}^{L} \xi_l$ [264]. Hence, the modified cost function to be minimized in the higher-dimensional space $\mathcal{H}$ is

$$\min \frac{1}{2}||\mathbf{w}||^2 + C\sum_{l=1}^{L} \xi_l \tag{16.48a}$$

$$\text{s.t.} \quad y_l\left(\mathbf{w}^T \Phi(\mathbf{x}_l) + b\right) - 1 + \xi_l \geq 0 \quad \text{for } l = 1,\ldots,L. \tag{16.48b}$$

$$\xi_l \geq 0 \quad \text{for } l = 1,\ldots,L, \tag{16.48c}$$

where (16.48b) is obtained by combining the two types of constraints (16.47a) and (16.47b) into one equation. For convenience, let us denote $\boldsymbol{\xi} = (\xi_1,\ldots,\xi_L)^T$.

The Lagrangian for this new constrained optimization problem is

$$\begin{aligned}\mathcal{L}_p(\mathbf{w},b,\boldsymbol{\xi},\boldsymbol{\alpha},\boldsymbol{\mu}) &= \frac{1}{2}||\mathbf{w}||^2 + C\sum_{l=1}^{L}\xi_l \\ &\quad - \sum_{l=1}^{L}\alpha_l\left(y_l\left(\mathbf{w}^T\Phi(\mathbf{x}_l) + b\right) - 1 + \xi_l\right) - \sum_{l=1}^{L}\mu_l\xi_l \end{aligned} \tag{16.49}$$

where $\alpha_l \geq 0$ and $\mu_l \geq 0$, for $l = 1,\ldots,L$, are the Lagrangian multipliers and $\boldsymbol{\alpha} = (\alpha_1,\ldots,\alpha_L)^T$ and $\boldsymbol{\mu} = (\mu_1,\ldots,\mu_L)^T$.

The primal problem associated with (16.49) is

$$\begin{aligned}\text{Primal:} \quad & \min_{\mathbf{w},b,\boldsymbol{\xi}} \mathcal{L}_p(\mathbf{w},b,\boldsymbol{\xi},\boldsymbol{\alpha},\boldsymbol{\mu}) \\ \text{s.t.} \quad & \frac{\partial}{\partial \alpha_l}\mathcal{L}_p = 0 \quad \text{for } l = 1,\ldots,L \\ & \frac{\partial}{\partial \mu_l}\mathcal{L}_p = 0 \quad \text{for } l = 1,\ldots,L \\ & \alpha_l \geq 0 \quad \text{for } l = 1,\ldots,L \\ & \mu_l \geq 0 \quad \text{for } l = 1,\ldots,L. \end{aligned} \tag{16.50}$$

The Lagrangian dual function associated with $\mathcal{L}_p$ is given by

$$\mathcal{L}_d(\boldsymbol{\alpha},\boldsymbol{\mu}) = \inf_{\mathbf{w},b,\boldsymbol{\xi}} \mathcal{L}_p(\mathbf{w},b,\boldsymbol{\xi},\boldsymbol{\alpha},\boldsymbol{\mu}). \tag{16.51}$$

The infimum in (16.51) is achieved when the gradient $\nabla_{\mathbf{w},b,\boldsymbol{\xi}} \mathcal{L}_p = \mathbf{0}$, so that

$$\frac{\partial}{\partial \mathbf{w}} \mathcal{L}_p = 0 \Longrightarrow \mathbf{w} = \sum_{l=1}^{L} \alpha_l y_l \Phi(\mathbf{x}_l) \tag{16.52}$$

$$\frac{\partial}{\partial b} \mathcal{L}_p = 0 \Longrightarrow \sum_{l=1}^{L} \alpha_l y_l = 0 \tag{16.53}$$

and

$$\frac{\partial}{\partial \xi_l} \mathcal{L}_p = 0 \Longrightarrow \alpha_l + \mu_l = C \quad \text{for } l = 1, \cdots, L. \tag{16.54}$$

The Lagrangian dual problem is

$$\begin{aligned} \text{Dual:} \quad & \max_{\boldsymbol{\alpha},\boldsymbol{\mu}} \mathcal{L}_d(\boldsymbol{\alpha}, \boldsymbol{\mu}) \\ \text{s.t. } \alpha_l \geq 0 \quad & \text{for } l = 1, \cdots, L \\ \mu_l \geq 0 \quad & \text{for } l = 1, \cdots, L. \end{aligned} \tag{16.55}$$

Substituting (16.52)–(16.54) in (16.51), the Lagrangian dual function can be written as

$$\mathcal{L}_d(\boldsymbol{\alpha}) = \sum_{l=1}^{L} \alpha_l - \frac{1}{2} \sum_{l=1}^{L} \sum_{m=1}^{L} \alpha_l \alpha_m y_l y_m \Phi(\mathbf{x}_l)^T \Phi(\mathbf{x}_m), \tag{16.56}$$

so that the Lagrangian dual problem (16.55) becomes

$$\begin{aligned} \text{Dual:} \quad & \max_{\boldsymbol{\alpha}} \sum_{l=1}^{L} \alpha_l - \frac{1}{2} \sum_{l=1}^{L} \sum_{m=1}^{L} \alpha_l \alpha_m y_l y_m \Phi(\mathbf{x}_l)^T \Phi(\mathbf{x}_m) \\ \text{s.t.} \quad & \sum_{l=1}^{L} \alpha_l y_l = 0 \\ & 0 \leq \alpha_l \leq C \quad \text{for } l = 1, \ldots, L. \end{aligned} \tag{16.57}$$

Again, (16.57) turns out to be a standard convex quadratic optimization problem that can be solved efficiently for the optimal Lagrange multiplier vector $\boldsymbol{\alpha}^*$. The optimal weight vector is then obtained from (16.52):

$$\mathbf{w}^* = \sum_{l=1}^{L} \alpha_l^* y_l \Phi(\mathbf{x}_l) \tag{16.58}$$

The additional equations necessary for computing b are obtained by the KKT conditions of the primal problem:

$$\alpha_l^* \left(y_l \left(\Phi(\mathbf{x}_l)^T \mathbf{w}^* + b^* \right) - 1 + \xi_l \right) = 0 \quad \text{for } l = 1, \ldots, L \tag{16.59a}$$

$$y_l \left(\mathbf{x}_l^T \mathbf{w}^* + b^* \right) - 1 + \xi_l \geq 0 \quad \text{for } l = 1, \ldots, L. \tag{16.59b}$$

Again, for numerical reasons, it is normally better to compute b^* as the average of all b values obtained by solving (16.59a) for all $\alpha_l \neq 0$.

16.3.5 Kernel Function Implementation of Support Vector Machines

In the case of nonlinear SVMs, the transformation $\Phi(.)$ is very important. We have assumed that we can find a suitable $\Phi(.)$ and perform necessary computations. However, by definition, the range of $\Phi(.)$ is of much higher dimension than the actual input vectors $\mathbf{x}$, which can already be large. Hence, computing and storing such long vectors $\Phi(\mathbf{x})$ may not always be easy. In fact, sometimes the desirable $\Phi(.)$ may even be an infinite-dimensional transformation.

It is, then, worth observing that the convex quadratic optimization problem (16.57) depends on the nonlinear mapping $\Phi(.)$ only through inner product terms of the form of $\Phi(\mathbf{x}_l)^T \Phi(\mathbf{x}_m)$. Moreover, once the SVM has learned the optimal weight vector $\mathbf{w}^*$ in (16.58) from the training data set $\{(\mathbf{x}_l, y_l)\}_{l=1}^L$, decision-making on a new test input $\mathbf{x}$ is performed as

$$\begin{aligned} f(\mathbf{x}) &= sgn\left(\Phi(\mathbf{x})^T \mathbf{w}^* + b^* \right) \\ &= sgn\left(\sum_{l=1}^{L} \alpha_l^* y_l \Phi(\mathbf{x}_l)^T \Phi(\mathbf{x}) + b^* \right). \end{aligned} \tag{16.60}$$

Again, we see that the nonlinearity $\Phi(.)$ enters the computation through the inner product $\Phi(\mathbf{x}_l)^T \Phi(\mathbf{x})$. Inner products are scalars, and thus, the computational and storage requirements would greatly reduce if we could directly compute these inner products from data $\mathbf{x}$ rather than first computing the higher-dimensional vectors $\Phi(\mathbf{x})$. This would be possible if we could find a function $K : \mathcal{X} \times \mathcal{X} \longrightarrow \mathbb{R}$ such that $\Phi(\mathbf{x}_1)^T \Phi(\mathbf{x}_2) = K(\mathbf{x}_1, \mathbf{x}_2)$, for all $\mathbf{x}_1, \mathbf{x}_2 \in \mathcal{X}$. If this is possible, then the optimization problem (16.57) that needs to be solved during the learning phase of the SVM can be rewritten as

$$\begin{aligned} &\max_{\boldsymbol{\alpha}} \sum_{l=1}^{L} \alpha_l - \frac{1}{2} \sum_{l=1}^{L} \sum_{m=1}^{L} \alpha_l \alpha_m y_l y_m K(\mathbf{x}_l, \mathbf{x}_m) \\ &\text{s.t.} \sum_{l=1}^{L} \alpha_l y_l = 0 \\ &0 \leq \alpha_l \leq C \quad \text{for } l = 1, \ldots, L, \end{aligned} \tag{16.61}$$

and the computation (16.60) required for testing phase can also be rewritten as

$$f(\mathbf{x}) = sgn\left(\Phi(\mathbf{x})^T \mathbf{w}^* + b^*\right)$$
$$= sgn\left(\sum_{l=1}^{L} \alpha_l^* y_l K(\mathbf{x}_l, \mathbf{x}) + b^*\right). \quad (16.62)$$

Such a function $K(.,.)$ is called a *kernel function*. If we could find a such a function to replace the inner products, (16.61) and (16.62) show that the effect of the non-linear mapping $\Phi(.)$ on the performance of the SVM is completely determined by this function $K(.,.)$. Hence, in selecting the high-dimensional transformation, it is the properties of $K(.,.)$ defined as $K(\mathbf{x}_1, \mathbf{x}_2) = \Phi(\mathbf{x}_1)^T \Phi(\mathbf{x}_2)$ that we need to be concerned of. Then, it follows that we may in fact first select a desired kernel function $K(.,.)$ and directly solve (16.61) and (16.62), respectively, during the learning and testing phases. As long as our choice of the kernel function $K(.,.)$ satisfies the property that it can indeed be written as $K(\mathbf{x}_1, \mathbf{x}_2) = \Phi(\mathbf{x})^T \Phi(\mathbf{x})$ for some function $\Phi(.)$ and a (high-dimensional) space $\mathcal{H}$ such that $\Phi(\mathbf{x}) \in \mathcal{H}$, the effect will be the same as starting from the nonlinear mapping.

The replacement of the inner products in (16.57 and (16.60) by a kernel function as in (16.61) and (16.62) is often called the *kernel trick* in the literature on SVMs. Essentially, the kernel trick allows mapping input data vectors $\mathbf{x}$ into a high-dimensional space $\mathcal{H}$ and performing a computation (i.e., computing the inner product) in the space of $\mathcal{H}$ without actually having to transform data explicitly to $\mathcal{H}$.

In selecting a kernel $K(.,.)$ to be used directly in (16.61) and (16.62), however, we need to verify that indeed $K(.,.)$ satisfies the property that, for all $\mathbf{x}_1, \mathbf{x}_2 \in \mathcal{X}$,

$$K(\mathbf{x}_1, \mathbf{x}_2) = \Phi(\mathbf{x}_1)^T \Phi(\mathbf{x}_2) \quad (16.63)$$

for some function $\Phi(.)$ and space $\mathcal{H}$ such that $\Phi(\mathbf{x}) \in \mathcal{H}$. Under certain conditions, this is possible due to the following result.

Theorem 16.1 (Mercer's Theorem) Suppose that $\mathcal{X}$ is an interval and $K : \mathcal{X} \times \mathcal{X} \longrightarrow \mathbb{R}$ is a symmetric and continuous function that is square integrable:

$$\int\int K^2(t,u)dtdu \geq 0.$$

If for all square-integrable[1] functions $g(t)$

$$\int\int K(t,u)g(t)g(u)dtdu \geq 0,$$

[1] A real-valued function $g(t)$ is square integrable if $\int g^2(t)dt < \infty$.

then $K(t,u)$ can be expanded in a uniformly and absolutely convergent series as

$$K(t,u) = \sum_{k=1}^{\infty} \lambda_k \Phi_k(t)\Phi_k(u) \quad \text{for all } t,u \in \mathcal{X}$$

where $\lambda_k \geq 0$ and $\Phi_k(t)$ are called, respectively, the eigenvalues and eigenfunctions of $K(.,.)$ and are given by

$$\int K(t,u)\Phi_k(u)du = \lambda_k \Phi_k(t)$$

with for all $1 \leq k,j < \infty$

$$\int \Phi_k(t)\Phi_j(t)dt = \begin{cases} 1 & \text{if } k = j \\ 0 & \text{if } k \neq j \end{cases}.$$

Proof: See, for example, [269].

The property that $\int\int K(t,u)g(t)g(u)dtdt \geq 0$, for all square-integrable functions $g(t)$, is called the *Mercer's condition*. Thus, from Theorem 16.1, it follows that a kernel function $K(.,.)$ will indeed hold (16.63) if it satisfies the Mercer's condition.

Some of the commonly used kernel functions include:

- Polynomial kernels:

$$K(\mathbf{x}_1, \mathbf{x}_2) = \left(\mathbf{x}_1^T \mathbf{x}_2 + 1\right)^p \tag{16.64}$$

- Gaussian radial basis function kernels:

$$K(\mathbf{x}_1, \mathbf{x}_2) = e^{-\frac{1}{2\sigma^2}||\mathbf{x}_1 - \mathbf{x}_2||^2} \tag{16.65}$$

- Laplacian radial basis function kernels:

$$K(\mathbf{x}_1, \mathbf{x}_2) = e^{-\frac{1}{\lambda||\mathbf{x}_1 - \mathbf{x}_2||}}$$

- Sigmoidal kernels:

$$K(\mathbf{x}_1, \mathbf{x}_2) = \tanh\left(\kappa \mathbf{x}_1^T \mathbf{x}_2 - \delta\right)$$

16.3.6 SVMs in Cognitive Radios

In [270], an SVM approach was used to classify signals based on cyclostationary features similar to those we discussed in Chapter 11. In a simulated signal environment, six different types of modulation schemes were considered: AM, ASK, FSK,

PSK, MSK, and QPSK. The SVM was based on a Gaussian radial basis function kernel given in (16.65). The feature vector consisted of four features extracted from the spectral correlation function that were expected to reflect differences in modulation types. The results of [270] showed that the SVM outperforms two alternative classifiers based on a layered ANN and a binary decision tree (see, e.g., [79] for an introduction to binary decision trees).

These results were consistent with those of [260], which used time–frequency signatures extracted from a different time–frequency distribution instead of the spectral correlation function, as we discussed earlier. In addition to ANNs, [260] also considered SVMs for the same signal classification and identification problems in cognitive radios. The SVM used was again based on a Gaussian radial basis function kernel. It was observed in [260] also that SVM was somewhat better in comparison to the feed-forward neural network in terms of mode identification performance.

SVMs and general kernel-based learning formulations for several typical signal processing problems encountered in cognitive radios were also discussed in [271]. In general, as the name suggests, all kernel-based learning approaches are based on kernels that implicitly project data onto high-dimensional spaces in which classification can be achieved more efficiently. Classification, however, can be based on other cost functions instead of the SVM cost function that seeks for a maximum margin hyperplane. Indeed, there are various other cost function formulations based on transformed data that have been considered in the literature, including kernel-based K-means clustering [272], kernel-based principal component analysis (PCA) [273], kernel-based online learning [274], and Fisher discriminant analysis [275].

Both linear and nonlinear SVMs were considered for detecting primary coverage boundary in [271]. The problem scenario is that of a distributed set of secondary users who cooperate to determine the coverage area of a primary transmitter. It was noted in [271] that due to fading, shadowing, and attenuation effects in wireless channels, the coverage boundary of a transmitter can be an arbitrary nonlinear hyperplane on the 2-D terrain. Based on their local sensing observations, these K_p numbers of secondary users report their local decisions on whether the primary transmitter covers its location or not. For the nonlinear SVM implementation, [271] used both a polynomial kernel of the form of (16.64) with $p = 5$ and a quadratic kernel. Results showed that, as expected, the linear SVM was not effective in learning the nonlinear coverage boundary. The polynomial kernel nonlinear SVM, however, was able to learn the coverage boundary satisfactorily.

A different direction in which SVM ideas have been used in cognitive radios is found in [276]. The problem addressed in [276] is of spectral activity detection rather than classification. In particular, WiFi and Bluetooth signals in the 2.4 GHz ISM band were the focus. It was noted that due to various noise and interference phenomena encountered in the 2.4 GHz ISM band, the Gaussian noise assumption may not hold, and thus, non-Gaussian spectral estimation approaches may be needed. Thus, [276] used a robust cost function and an SVM-inspired regularization term in the regularized spectral estimator (10.98) discussed in Chapter 10. The robust cost function itself was a modified version of the Huber cost function $\ell_H(x)$ that we saw in (10.103). The modification was to add a third region with zero cost to the two regions considered in (10.103). The regularization term, on the other hand, was borrowed from the SVM optimality criterion so that it was a function of the norm of the weight vector, or the margin of the estimator weight vector. Hence, this approach can also be termed a regularized robust spectral estimator similar to the scheme we considered in Section 10.5.2.

This SVM-inspired regularized robust approach was applied for spectral activity detection in various noise environments in [276]: additive white Gaussian, impulsive, and block noise. The results reported in [276] showed, perhaps not surprisingly, that indeed regularized robust spectral estimator can perform better than the standard periodogram (or the LMS solution) in detecting Bluetooth and WiFi signals in the presence of non-Gaussian noise while providing comparable performance in the presence of Gaussian noise.

16.4 REINFORCEMENT LEARNING

In Chapter 9, when faced with having to learn a policy in a POMDP without knowing model parameters, we discussed a specific machine learning approach called Q-learning. Q-learning is one example of a broader paradigm of learning called reinforcement learning. Reinforcement learning is an unsupervised learning approach that attempts to learn by direct interaction with the environment. The information available for the learner is a feedback received from environment after the execution of a selected action. In general, reinforcement learning is an ideal learning methodology for autonomous learning and thus can be applicable to many cognitive radio learning problems.

Reinforcement learning is essentially learning by trial and error. The learner does not have to have any a priori knowledge about the environment. It may select actions randomly. After executing each action, however, the learner will receive a delayed reward, which is the feedback from the environment. This delayed reward will reflect how good or bad the executed action is. However, since learner does not know the absolute maximum or minimum rewards possible a priori, it may only evaluate how good or bad an action is, when in a certain state, with respect to rewards it has already seen. Hence, in order to learn a good policy, the learner needs to explore the space of states and actions as fully as it may. This, however, conflicts with its ability to act according to what it has already learned because it needs to try out state–action pairs it has not already tried. This is the trade-off of exploration and exploitation that we mentioned in Section 9.8.1: It is important that the learner has the ability, or the self-control, to strike the right balance between trying out new state–action pairs versus following an already learned policy that may be reasonably good.

Reinforcement learning has the flavor of how humans learn in general. There are no absolutely good or bad actions or guarantees on absolute best that can be achievable a priori. We mostly learn by trial and error. Think of when we have to decide between two routes to get to office from work. There are too many variables that might be out of our control. The only way to find what will take the shortest time is to try out both routes and see for ourselves.

Reinforcement learning can be particularly suitable for learning in Markov environments. Hence, to formalize, let us consider a Markov decision process (MDP) as we defined in Section 7.2. As a result of taking action $a_{n-1} = a$ when in state $S[n-1] = s$ at time $n-1$, the state of the environment changes to a new state $S[n] = s'$ while the learner receives delayed reward $r_n(s,a)$ at time n representing the reward corresponding to taking action $a_{n-1} = a$ when in state $S[n-1] = s$ at time $n-1$. In addition, the learner will be able to observe an observation $Y[n] = y$ that is related to the new state $S[n] = s'$ of the environment. In this formulation, the learner's goal is to use the observation $Y[n] = y$ and the delayed reward $r_n(s,a)$ to learn improved action choices for the future.

In the context of an MDP, $Y[n] = S[n]$ and the objective is to learn an improved (stationary) policy $\underline{\pi} \triangleq \pi \triangleq (\pi, \pi, \ldots)$ in the sense of maximizing the expected total discounted return defined in (7.31):

$$\overline{R_{\gamma}^{\underline{\pi}}} = \lim_{T \longrightarrow \infty} \mathbb{E}_{\underline{\pi}} \left\{ \sum_{n=0}^{T} \gamma^n r_n(S[n], \pi_n(.)) \right\}, \tag{16.66}$$

where $\gamma \in [0, 1)$ is a discount factor and the expectation is with respect to the state transitions induced by the policy π.

In Section 7.4, we showed that an optimal policy that maximizes (16.66) can be found by finding a policy that maximizes the value function $V^{\underline{\pi}}(s)$ defined in (7.33). Moreover, Theorem 7.3 guaranteed the existence of an optimal stationary policy under certain regularity conditions. Section 7.4 discussed several iterative algorithms that can then be used to compute an optimal stationary policy by computing the optimal value function $V^*(s) = \sup_{\pi} V^{\pi}(s)$. Of course, those algorithms require the full knowledge of the model parameters. In a learning environment, the objective is to learn an optimal policy.

Recall that one of the algorithms that we discussed in Section 7.4 for computing an optimal stationary policy for the MDP was the policy iteration in Section 7.4.2.1. The policy iteration algorithm learns an optimal policy by always deriving an improved policy from a given policy via the policy improvement step of (7.60):

$$\pi'(s) = \arg\max_{a} Q^{\pi}(s,a) \quad \text{for all } s, \tag{16.67}$$

where $Q^{\pi}(s,a)$ is the action-value function of policy π defined in (7.52) as

$$Q^{\pi}(s,a) = \mathbb{E}_{\pi} \left\{ \sum_{n=0}^{\infty} \gamma^n r(s[n], \pi(.)) \middle| S[0] = s, \pi_0 = a \right\}. \tag{16.68}$$

The action-value function $Q^{\pi}(s,a)$ denotes the expected total discounted reward starting from the initial state $S[0] = s$, but taking the action $\pi_0 = a$ at time $n = 0$, and then following the stationary policy $\pi(.)$ from there on. In Section 7.4.2.1, we showed that indeed the policy π' obtained as in (16.67) is better or at least as good as the current policy π.

One of the widely used reinforcement learning algorithms to learn an optimal policy for an MDP is based on the these observations. This algorithm, called the Q-learning, was discussed in Section 9.8.1 in order to learn an optimal subband selection policy for a wideband cognitive radio. Clearly, if we can learn the optimal action-value function $Q^*(s,a)$, then from (16.67) we may obtain an optimal policy. As we mentioned in Section 9.8.1, Q-learning can indeed converge to the optimal Q values and thus an optimal policy under fairly reasonable conditions on an MDP.

The Q-learning algorithm was discussed in detail in Section 9.8.1, and we may state it in the following for the sake of completeness. If as a result of taking action $a_{n-1} = a$ when in state $S[n-1] = s$ at time $n-1$, the learner observed the delayed reward $r_n(s,a)$ and the state transitioned to a new state $S[n] = s'$, then the Q value corresponding to the state-action pair (s,a) is updated as

$$Q(s,a) \leftarrow (1-\alpha)Q(s,a) + \alpha\left[r_n(s,a) + \gamma \max_a Q(s',a)\right], \tag{16.69}$$

where $0 < \alpha < 1$ denotes the learning rate. In order to achieve a desired level of exploration, the actions can be selected according to

$$a^* = \begin{cases} \arg\max_{a \in \mathcal{A}} Q(s,a) & \text{with probability } 1-\epsilon \\ \sim P(\mathcal{A}) & \text{with probability } \epsilon \end{cases}, \tag{16.70}$$

where $P(\mathcal{A})$ denotes a suitable probability distribution on the set of actions $\mathcal{A}$ and $0 \le \epsilon \le 1$ is the learning rate. We refer to Section 9.8.1 for details.

16.4.1 Temporal Difference Learning

Q-learning, however, is just one possible reinforcement learning algorithms. Indeed, Q-learning belongs to a class of reinforcement learning methods called the temporal difference (TD) learning methods. Recall that the basic idea of reinforcement learning is the learning by direct interaction with the environment and observing a certain feedbacks (rewards) after each action execution that depends on the executed action. A key element is that, unlike in supervised learning, these rewards are not right or wrong type of values. Hence, the learner can judge the merit of a certain action executed under certain conditions only relative to rewards it has encountered so far: There is no knowledge of the absolute best reward achievable. Essentially, TD methods use a function of the immediate reward r_n and the expected total return from the resulting new state as the correct outcome (or the best achievable reward) to update certain value functions. For example, rewriting (16.69) as

$$Q(s,a) \leftarrow Q(s,a) + \alpha\left[\left(r_n(s,a) + \gamma \max_a Q(s',a)\right) - Q(s,a)\right], \tag{16.71}$$

we may note immediately that Q-learning takes $r_n(s,a) + \gamma \max_a Q(s',a)$ as the correct or best outcome.

On the other hand, an alternative TD reinforcement learning algorithm called SARSA also learns an action-value function Q but using the following update equation:

$$Q(s,a) \leftarrow Q(s,a) + \alpha\left[\left(r_n(s,a) + \gamma Q(s',a')\right) - Q(s,a)\right], \tag{16.72}$$

where a' is the action choice for the new state s' according to the current policy. The difference between the two algorithms is that in Q-learning, $Q(s,a)$ is updated based on the possible best action, say, $a^*(s')$, choice for the new state s', whereas in SARSA, it is simply based on the action a' that would have been taken from state s' according to the actual action selection policy. In other words, although Q-learning updated $Q(s,a)$ based on the maximum $Q(s',a^*(s'))$ value possible from state s', it is possible that the actual action taken by the learner from state s' in the next step is something other than $a^*(s')$: For example, if the learner explores with probability ϵ, it will be an arbitrary action $a' \neq a^*(s')$ with probability ϵ. This small difference, however, can lead to important differences between the way learning progresses and the final learned policy. This is illustrated in terms of a particularly illuminating example of a mouse learning not to get killed by falling off a cliff while getting to where a piece of cheese is in [69]. If the policy that is used to select actions indeed converges to the optimal policy, then it is known that SARSA converges to the optimal Q function [69].

We refer an interested reader to [69], a textbook devoted to reinforcement learning, for a broader coverage on TD methods as well as other reinforcement learning algorithms.

16.4.2 Q-Learning in a POMDP: Replicated Q-Learning

The Q-learning algorithm (16.69) is aimed at learning a near-optimal policy for a COMDP when the state transition probabilities are unknown. However, many learning problems in cognitive radios that can be posed as MDPs will turn out to be POMDPs. The subband selection problem discussed in Chapter 9 and channel sensing decision-making problem in Chapter 13 are examples of such POMDP problems in cognitive radios. In Section 9.8.2, we discussed an ad hoc modification of the POMDP subband selection problem by defining a new observable state so that the standard Q-learning algorithm can be used to learn a subband selection policy. However, there are several modifications of Q-learning that can indeed be used to learn an optimal, or near-optimal, policy for a POMDP.

Perhaps the more pressing concern in computing an optimal policy in a POMDP is the computational complexity. Hence, most Q-learning-type algorithms for POMDPs are aimed at learning an optimal policy for a POMDP with less computational complexity rather than addressing the issue of unknown state transition probabilities. It is, of course, possible that state transition probabilities of the model may also be unknown. However, these transition probabilities can at least be estimated using standard estimation techniques such as those discussed in Chapter 5.

A modification of the Q-learning algorithm that can be used to learn an effective policy in a POMDP, if Markov model transition probabilities were to be known, is called the *replicated Q-learning* [277–279]. It can be an alternative to simply partitioning the state space of the belief vector and maintaining a Q-table for the *quantized belief state*. Although in the following we may assume that the transition probabilities of the model are known, it has been noted in [277–279] that, indeed, replicated Q-learning may be used with unknown transition probabilities so that these can also be learned while learning a policy.

First, note that we may rewrite the Q-learning update of (16.69), which is the same as that in (9.79), for an MDP as

$$\begin{aligned} Q(s[n-1],a_{n-1}) \leftarrow\ & Q(s[n-1],a_{n-1}) \\ & + \alpha\left[r_n(s[n-1],a_{n-1}) + \gamma \max_a Q(s[n],a) - Q(s[n-1],a_{n-1})\right]. \\ =\ & Q(s[n-1],a_{n-1}) + \alpha\Delta Q(s[n-1],a_{n-1}), \end{aligned} \quad (16.73)$$

where we have denoted the update term to the Q-values by

$$\Delta Q(s[n-1],a_{n-1}) = r_n(s[n-1],a_{n-1}) + \gamma \max_a Q(s[n],a) - Q(s[n-1],a_{n-1}). \quad (16.74)$$

If the model transition probabilities are known, then it is possible to define a COMDP in terms of a belief state $\mathbf{b}[n]$ as we have seen in earlier sections and in Chapter 7. Belief state $\mathbf{b}[n]$ is completely observable since it can be updated using (7.70) and (7.69). But recall that the jth element b_j of the belief state vector $\mathbf{b}$ is the probability (or the belief) of true state being the jth possible state (earlier we used $\mathbf{s}_j$ to denote the jth state, but here, we will simply use j since we only need the state label). Hence, given a Q-table $Q(s,a)$ we may define the *average Q-value* of taking action a when in belief state $\mathbf{b}$ as

$$Q(\mathbf{b},a) = \mathbb{E}\{Q(S,a)|\mathbf{b}\} = \sum_j b_j Q(j,a). \quad (16.75)$$

Essentially $Q(\mathbf{b},a)$ is a weighted sum of Q-values resulting from taking action a from all possible states, where weights are taken to be our belief that each of these is the true state.

Now suppose that the belief state at time $n-1$ is $\mathbf{b}[n-1]$ and the resulting state by taking action a_{n-1} from belief state $\mathbf{b}[n-1]$ at time $n-1$ is the state $\mathbf{b}[n]$ at time n. Since next state $s[n]$ required for computing the update term (16.74) is not available, the replicated Q-learning algorithm computes it by making use of the resulting belief state $\mathbf{b}[n]$ as follows:

$$\Delta Q(s[n-1],a_{n-1}) = r_n(s[n-1],a_{n-1}) + \gamma \max_a Q(\mathbf{b}[n],a) - Q(s[n-1],a_{n-1}).$$

Moreover, since we do not know whether the current state at time $n-1$ is $s[n-1] = j$ exactly, but only has the belief that it is the jth state with probability $b_j[n-1]$, the replicated Q-learning algorithm weights the update term in (16.73) by the belief of that state as follows:

$$Q(j,a_{n-1}) \leftarrow Q(j,a_{n-1}) + \alpha b_j[n-1]\left[r_n(j,a_{n-1}) + \gamma \max_a Q(\mathbf{b}[n],a) - Q(j,a_{n-1})\right]. \tag{16.76}$$

Hence, if the belief state at time n is $\mathbf{b}[n]$, then the equivalent decision rule to (9.80) becomes

$$a_n^* = \arg\max_a Q(\mathbf{b}[n],a), \tag{16.77}$$

while equivalent decision rule to (9.81) becomes

$$a_n^* = \begin{cases} \arg\max_{a\in\mathcal{A}} Q(\mathbf{b}[n],a) & \text{with probability } 1-\epsilon \\ \sim\mathcal{U}(\mathcal{A}) & \text{with probability } \epsilon \end{cases}. \tag{16.78}$$

If observations are sufficient to ensure that at each time n, $b_j[n] = 1$ for some j so that there is no uncertainty in the true state, it has been shown that (16.76) indeed reduces to the original Q-learning algorithm (9.79). Hence, when this is true, (16.76) is known to converge to the optimal Q-function, and thus an optimal policy, under sufficient regularity conditions similar to (9.79) [137,277]. The replicated Q-learning algorithm has been shown to be effective in learning optimal policies in many learning problems [277].

16.4.3 Reinforcement Learning in Cognitive Radios

TD reinforcement learning has been investigated in [280] to implement a cooperative sensing method in a cognitive radio network. It is assumed that a cognitive secondary user requests cooperation from its neighbors when it cannot determine the state of a particular primary channel on its own. However, given the number of available cooperative neighbors (the state), a radio is interested in learning to select the optimal number of neighbors for cooperation according to a certain reward function. Several TD algorithms, including the Q-learning and SARSA, were evaluated. It was noted that all methods finally converged to the optimal solution, although Q-learning exhibited somewhat desirable characteristics during the learning process.

In [138], replicated Q-learning was used to learn an optimal policy for aggregated interference control in a DSS CRN when the state was only partially observable. Specific scenario of interest was when multiple IEEE 802.22 WRAN cells are deployed around a digital TV (DTV) cell (Figure 16.10). Here, IEEE 802.22 WRAN base stations are the secondary cognitive radios and the DTV cell represents the primary user. The problem faced by each of these distributed cognitive radios (base stations) is to autonomously determine their transmit powers so that their aggregated secondary

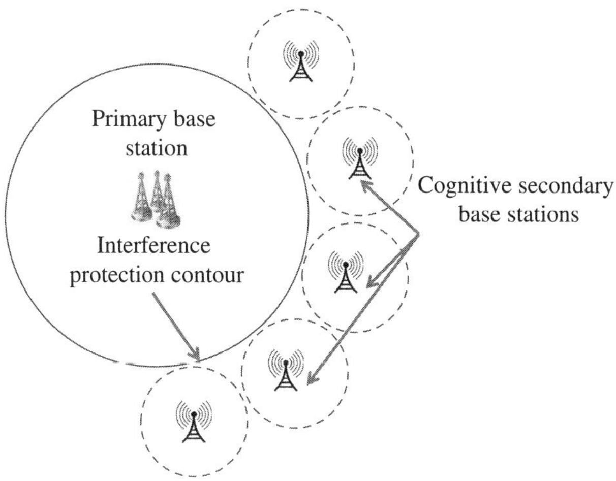

FIGURE 16.10 Multiple WRAN secondary base stations surrounding a primary digital TV cell are required to ensure their overall interference to the digital TV system is within acceptable levels [138].

interference at the primary receiver does not exceed a given threshold (which is similar to the interference cap we discussed in Section 14.2).

The approach in [138] defines a separate state for each of the WRANs. The state of the kth WRAN network at time n is defined as [138]

$$S_k[n] = \{I_k[n], d_k[n], p_k[n]\}, \tag{16.79}$$

where $I_k[n]$, $d_k[n]$, and $p_k[n]$ denote, respectively, a binary indicator that specifies whether the secondary interference to the primary network is above or below the given threshold, an estimate of distance between the secondary user and the protection contour of the primary system, and the current transmit power level of the secondary user. It is assumed that both $d_k[n]$ and $p_k[n]$ are allowed to take only a finite set of finite values, and these are known exactly to the kth base station.

Thus, the only part of the state $S_k[n]$ that the kth base station cannot determine exactly is the interference indicator $I_k[n] \in \{0, 1\}$. Note that $I_k[n] = 1$ indicates that the aggregated secondary interference is within the interference threshold of the primary, while $I_k[n] = 0$ indicates otherwise. Hence, the kth base station maintains a belief state $\mathbf{b}_k[n]$ on $I_k[n]$. Actions available to the kth secondary base station are the set of discrete power levels it can assign to the secondary user. Rather than rewards, [138] defines a cost function c and formulates the POMDP as a minimization problem. The cost is defined as a function of the difference between the instantaneous and required primary SINR threshold.

Results reported in [138] showed that replicated Q-learning can learn an optimal aggregated interference control policy effectively. The convergence performance of the replicated Q-learning was shown to be influenced by the form of the cost function.

By considering the absolute value and the quadratic and fourth powers of the difference between the instantaneous and required primary SINR threshold, [138] observed that the higher the order of the cost, the faster the convergence.

Perhaps the most important aspect of this is that the system assumed in [138] is in fact different from the system we have considered in Chapter 7: We always assumed that there is only one decision-maker in the system, whereas in the system of [138] all secondary base stations are independent distributed decision-makers. These two types of environments are called single-agent versus multiagent, and reinforcement learning applied to multiagent environments are called multiagent reinforcement learning (MARL). Not surprisingly, multiagent learning environments are exceedingly complicated and difficult to handle, as we will discuss in detail in the next section. However, reinforcement learning has been shown to converge to optimal, or good, policies even in such environments where multiple distributed decision-makers simultaneously attempt to learn optimal policies by interacting with the same system. In the case of [138], once each base station has learned an optimal policy during a learning phase, the secondary system made of multiple base stations was shown to reach at an efficient power allocation scheme in one iteration.

It must be mentioned that reinforcement learning algorithms such as Q-learning rely on underlying processes being Markov as well as only a single learner interacting with the environment for their convergence performance. When non-Markovian system dynamics are involved, environment is partially observable, or when there are multiple learners simultaneously interacting with the same environment, performance of reinforcement learning methods can degrade. There are various other machine learning approaches that have been developed that can be specifically suited for such learning problem. For example, some of these alternatives, such as evolutionary learning [281–283], learning by imitation, learning by instruction [284], and policy gradient methods [285, 286], have indeed been shown to outperform reinforcement learning on certain non-Markovian learning problems. A recent survey of various machine learning schemes that have been applied in the context of cognitive radios can be found in [287].

16.5 MULTIAGENT LEARNING

Recall from Chapter 2 that one of the major signal processing challenges encountered in cognitive radios is that the wireless spectrum is inherently a distributed multiagent system: Almost always, there are more than one wireless device operating in any spectrum band. Indeed, perhaps, the very goal of cognitive radios is to improve such spectrum coexistence of many wireless systems and devices. This may especially be the case when new spectrum allocation policies are put in place along the lines we discussed in Section 3.6 and CRNs become prevalent. Collectively, the spectrum of interest and all its users will then form a complex *distributed multiagent system.*

In general, individual cognitive radios, or agents, may attempt to learn on their own in this multiagent system while interacting with the same spectrum environment at the same. This scenario is referred to as multiagent learning, and the type of learning in which independent distributed agents attempt to learn their own decision policies simultaneously is called *concurrent learning*. Such multiagent learning is made difficult by the fact that changes in the state of the spectrum are due to the combined effect of actions executed by all distributed radios. Worse yet, a particular cognitive radio may not know the exact actions taken by other radios.

Perhaps the most widely investigated multiagent learning paradigm is based on (noncooperative) game theory. As we mentioned in Section 14.2.2, game theory provides a systematic way to model interactions among rational agents. The basic concept that allows predicting and understanding the behavior of interacting rational agents modeled as a game is the Nash equilibria of the game (see Definition 14.2.1). As multiple players of the game act independently and simultaneously attempt to learn what is best for each of them, often there is no cooperation among them. Each player is only concerned about what is best for itself. Under such conditions, it is reasonable to expect that many concurrent learning approaches will tend to converge to a Nash equilibrium of the game, if one exists. It is, however, possible that this Nash equilibrium may not be the globally optimal solution since, in general, a game may have multiple Nash equilibria and global performance measures are not a concern of individual players.

In Section 14.2.3, we noted that the best response strategy is a way to arrive at a Nash equilibrium of a game when distributed cognitive radios make their decisions independent of each other (see Definition 14.2.2). This already assumes that the game is what is termed a *repeated game*. Recall that in Section 14.2.2 we defined a game G as $G = (\mathcal{K}, \mathcal{A}_k, u_k(.))$. Such a game G is, in fact, what is called a *stage game* or a *one-shot game*. It defines the interaction among players in one round of playing the game G.

A repeated game is a game in which players repeatedly interact with each other over many rounds. In each round, their interaction is a stage game. We may denote the repeated game made of T repeated plays of a stage game G by G^T. If the number of rounds T is finite, the game is called a *finitely repeated game*, while if the number of rounds is infinite, the game is called an *infinitely repeated game*. The utility in a repeated game can be defined as the average, or the discounted average, utility over the course of all T rounds of the game (note that T can be infinite).

The aforementioned stage game G is also what is called a game in *strategic form*. This is also referred to as a game in *normal form* or in *matrix form* because, as the name suggests, the game can be represented as a matrix. The classic prisoner's dilemma game is shown in Table 16.1 in its strategic, or matrix, form [288].

Repeated games can also be represented in strategic form. However, it is more illuminating to represent a repeated game in a so-called extensive form, or tree form, since this representation has a time element built into it. Figure 16.11 shows the prisoner's dilemma in extensive form. Note that each node (except the leaf nodes) represents a particular player's decision-making points. The branches from a node show the action choices available to the player. Note that Figure 16.11 does not necessarily imply that players are acting sequentially. This is implied by the oval that groups nodes of player 2. It shows a so-called information set, which is a collection of nodes that are under the control of the same player and which the player is not

TABLE 16.1 Prisoner's dilemma: both prisoners loose and go to jail if they both defect and confess to the investigators. If, on the other hand, they both cooperate and do not confess to the investigators, they both will walk free. If one defects while the other does not, the one who defect not only walk free but will have the reputation intact too, while the other is completely condemned.

	Player 2	
Player 1	Cooperate	Defect
Cooperate	1, 1	−1, 2
Defect	2, −1	0, 0

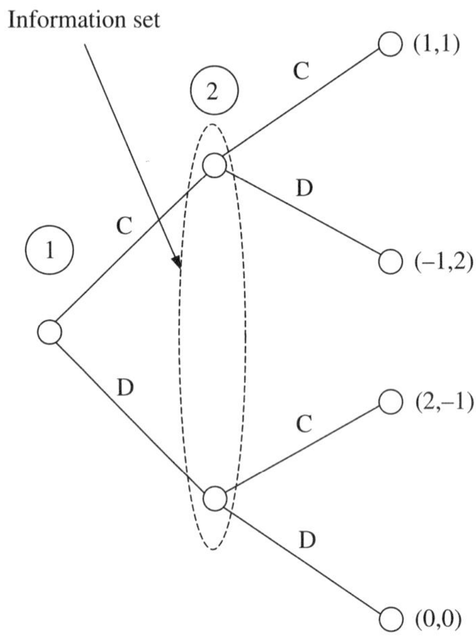

FIGURE 16.11 Prisoner's dilemma as a game in extensive form.

able to distinguish from one another [289]. In effect, nodes in an information set can be treated as a single node from the point of view of the player who is making decisions at these nodes. With this interpretation, Figure 16.11 allows us to model the same strategic form of prisoner's dilemma in Table 16.1, in which either they make decisions simultaneously or the action of first decision-maker is not revealed to the other before it makes its decision. If, however, we were to remove the information set in Figure 16.11, then the same extensive form representation can be interpreted as a game in which two players are taking actions sequentially and player 2 being able to know the action of player 1 before making its own action. Now, it is straightforward to extend the tree in Figure 16.11 if the game is repeated.

Repeated games, as described, assume that in every round the players play the same stage game G. This, however, can be restrictive in many real-world applications. The concept of *stochastic games* extends the concept of repeated games in a way that in each round the game played can be different. Specifically, the game played in each round is allowed to be drawn probabilistically from a collection of possible games. In each round, this collection of possible games is defined by the history of the games played in the previous rounds and the actions chosen by players in those rounds. In many cases, it is assumed that this complete history of the games being played so far and the corresponding players actions can be summarized by a state, and this state follows Markov dynamics. Such a game is called a *Markov game*. Thus, in a Markov game, the game played in the current round is determined by the current state.

Markov games provide a richer framework for modeling multiple interactions of distributed independent players and analyzing their behavior. Not only Markov games generalize the repeated games by allowing for stage games to be different in each round, but they are also a generalization of MDPs that we saw in Chapter 7. Indeed, observe that an MDP is a special case of a Markov game in which there is only one decision-maker (or one player). The single player interacts with an environment whose state dynamics are governed by Markov transitions. When multiple players interact, we thus need to resort to Markov games. On the other hand, a repeated game is a single-state Markov game.

16.5.1 Game-Theoretic Multiagent Learning

Importantly, repeated games and, in particular, Markov games may provide players with the possibility to learn to arrive at a better, or more efficient, equilibrium, whereas in a single-stage game, they may end up in a suboptimal Nash equilibrium for all players. To see the possibility of repeated games leading to more efficient equilibria, one needs to investigate other notions of the concept of equilibrium beyond the Nash equilibrium. One equilibrium concept that is more general than the Nash equilibrium is the *correlated equilibrium*. First, recall that a randomized strategy for player k is a probability distribution $P_k(a_k)$ over the action space $\mathcal{A}_k$ for the action a_k of the player k, for $k = 1, \ldots, K$. When action set $\mathcal{A}_k$ is finite, we may equivalently think of a randomized strategy for player k as a pmf $p_k(a_k)$.

Definition 16.5.1 (Correlated Equilibrium) A joint probability distribution $P(\mathbf{a})$, or equivalently a pmf $p(\mathbf{a})$, is called a correlated equilibrium if and only if, for all $k \in \mathcal{K}$, $a_k \in \mathcal{A}_k$, $\mathbf{a}_{-k} \in \mathcal{A}_{-k}$,

$$\sum_{a_{-k} \in \mathcal{A}_{-k}} p(a_k, \mathbf{a}_{-k}) \left[u_k(a'_k, \mathbf{a}_{-k}) - u_k(a_k, \mathbf{a}_{-k}) \right] \leq 0, \forall a'_k \in \mathcal{A}_k. \tag{16.80}$$

Clearly, every Nash equilibrium is a correlated equilibrium. In fact, a Nash equilibrium corresponds to the special case of a correlated equilibrium in which $p(\mathbf{a})$ is a product distribution of the form of $p(\mathbf{a}) = \prod_k p_k(a_k)$ (i.e., the decisions of the different players are independent). From the definition, playing the marginal distribution $p_k(a_k)$ derived from a correlated equilibrium $p(\mathbf{a})$ is a best response strategy (see Definition 14.2.2) for player k as long as all the other players also play their marginal distributions derived from the correlated equilibrium $p(\mathbf{a})$.

The best response adaptation discussed in Section 14.2.3 is, however, just one way of learning in a repeated game played by multiple decision-makers. An even simpler game-theoretic learning strategy for a multiagent system is the *random better response strategy* that can be defined as follows [289].

Definition 16.5.2 (Random Better Response) Suppose that the current action of player k according to the joint action vector $\mathbf{a} = (a_k, \mathbf{a}_{-k})$ is $a_k \in \mathcal{A}_k$. A random better response learning strategy for user k is to select an action $a'_k \in \tilde{r}_k$ from the set $\tilde{r}_k$ where $\tilde{r}_k : \mathcal{A}_{-k} \longrightarrow \mathcal{A}_k$ is a correspondence defined as

$$\tilde{r}_k(\mathbf{a}_{-k}) = \left\{a'_k \in \mathcal{A}_k : u_k\left(a'_k, \mathbf{a}_{-k}\right) > u_k(a_k, \mathbf{a}_{-k})\right\}.$$

If the set $\tilde{r}_k$ is empty, then play $a'_k = a_k$.

It can be shown that if they converge, both best response and better response learning will converge to a Nash equilibrium [289]. However, they both assume that in every round, the actions of other players stay constant and there are situations in which these learning methods may not converge. A situation that these games are guaranteed to converge is when the game is a so-called finite ordinal *potential game* [289]. However, such games can be restrictive.

An alternative game-theoretic learning strategy that is widely used in many contexts, and in particular in cognitive radio applications, is the so-called regret-based learning. In game-theoretic learning, *regret* refers to the possible loss incurred by a player following a certain policy $\underline{\pi}_k$ compared to another, perhaps simpler, policy. Note that regret is always a measure with respect to an assumed alternative policy. Let us consider time $n = n$. The policy $\underline{\pi}_k$ of player k specifies a probability distribution over its action set $\mathcal{A}_k$. Note that for each $a \in \mathcal{A}_k$, $p_{k,n}(a)$ is the probability that according to the policy $\underline{\pi}_k$ player k will play the action $a_k = a$. For simplicity, let us denote this pmf using the $|\mathcal{A}|$-length vector $\mathbf{p}_{k,n}$. Similarly, denote by $\mathbf{u}_{k,n}$ the $|\mathcal{A}|$-length reward, or utility, vector of player k at time n where each element $u_{k,n}(a_k)$, for $a \in \mathcal{A}_k$, of $\mathbf{u}_{k,n}$ denotes the reward achieved if player k played action $a \in \mathcal{A}_k$ at time n. Then the expected reward, or utility, at time n for player k with the strategy $\mathbf{p}_{k,n}$ is

$$\begin{aligned} r_{k,n}(\mathbf{p}_{k,n}) &= \mathbb{E}\{u_{k,n}(a_k)\} \\ &= \mathbf{p}_{k,n}^T \mathbf{u}_{k,n}. \end{aligned}$$

The expected reward of player k over T rounds of the repeated game with policy $\underline{\pi}_k$ is $\sum_{n=1}^{T} r_{k,n}$. Now, suppose, instead of this deliberate policy $\underline{\pi}_k$, the player k follows the simpler strategy in which it always plays the fixed action $\tilde{a} \in \mathcal{A}_k$ at each time instant n, regardless of any information it may have. This policy can be characterized by a pmf vector $\mathbf{p}_{\tilde{a}}$ that has all zeros except the element that correspond to the action $\tilde{a}$, which is one. The expected reward at time n with this simple fixed action policy is simply $u_{k,n}(\tilde{a})$, which we may also write as

$$r_{k,n}(\mathbf{p}_{\tilde{a}}) = \mathbf{p}_{\tilde{a}}^T \mathbf{u}_{k,n}.$$

At time n, the regret of policy $\underline{\pi}_k$ of player k with respect to this simpler policy of playing the same fixed-action $\tilde{a}$ all the time is

$$\begin{aligned} l_{k,n}^{(\underline{\pi}_k)}(\tilde{a}) &= r_{k,n}(\mathbf{p}_{\tilde{a}}) - r_{k,n}(\mathbf{p}_{k,n}) \\ &= u_{k,n}(\tilde{a}) - \mathbf{p}_{k,n}^T \mathbf{u}_{k,n}. \end{aligned}$$

The *average regret per time step* (or average loss per time step) over the first T rounds of the repeated game of policy $\underline{\pi}_k$ of player k with respect to this simpler policy of playing the same fixed-action $\tilde{a}$ all the time is then

$$l_k^{(\underline{\pi}_k)}(\tilde{a}) = \frac{1}{T} \sum_{n=1}^{T} l_{k,n}^{(\underline{\pi}_k)}.$$

The *maximum average regret per time step* of policy $\underline{\pi}_k$ of player k with respect to any such fixed-action policy, defined as

$$L_k^{(\underline{\pi}_k)} = \frac{1}{T} \max_{\tilde{a} \in \mathcal{A}_k} l_k^{(\underline{\pi}_k)}(\tilde{a}),$$

is called the *external regret* of policy $\underline{\pi}_k$ [290].

The *external regret minimization* approach to learning is to seek a policy $\underline{\pi}_k$ such that the above maximum average regret $L_k^{(\underline{\pi}_k)}$ is minimized. A learning algorithm is called a *no-regret learning algorithm* if $L_k^{(\underline{\pi}_k)}$ decays to zero as the time horizon $T \longrightarrow \infty$. Importantly, it can be shown that such no-regret learning algorithms do exist, as summarized in the following theorem attributed to [291, 292].

Theorem 16.2 (Existence of Good No-Regret Learning Algorithms) There exist no-regret learning algorithms with expected external regret $O\left(\sqrt{\frac{\log|\mathcal{A}|}{T}}\right)$ with respect to every fixed-action policy.

Proof: See [290] for a proof.

The *randomized weighted majority algorithm* of [292] is the standard example of a no-regret learning algorithm that achieves the above upper bound for the expected external regret. However, there are various modifications of the randomized weighted majority algorithm that provide similar upper bounds. We refer an interested reader to [290] for details on such algorithms.

Instead of the fixed-action policies, if we were to compute the regret with respect to the class of policies in which we replace just one action choice of policy $\underline{\pi}_k$ with a different action choice in all its occurrences within policy $\underline{\pi}_k$, we obtain what is called *internal regret*. Specifically, given the policy $\underline{\pi}_k$, we form a new policy $\tilde{\underline{\pi}}_k(a, \tilde{a})$ by replacing every occurrence of action a within policy $\underline{\pi}_k$ by action $\tilde{a}$. For this, reason internal regret is also called *swap regret*. It has been shown that there also exist no-regret learning algorithms that can achieve decaying swap regret with square root of the time horizon T. In fact, it is possible to find no-regret algorithms such that the swap regret is $O\left(\sqrt{\frac{|\mathcal{A}| \log |\mathcal{A}|}{T}}\right)$.

From the definition of the swap (or internal) regret, it should be clear that it is closely related with the notion of a correlated equilibrium. Indeed, if a joint probability distribution $p(\mathbf{a})$ over the action space $\mathcal{A}_1 \times \cdots, \mathcal{A}_K$ is a correlated equilibrium, then every player will have zero or negative swap regret when playing it. This interpretation of a correlated equilibrium then allows us to define a weaker version of the correlated equilibrium called an ϵ-correlated equilibrium as follows.

Definition 16.5.3 (ϵ-Correlated Equilibrium) A joint probability distribution $P(\mathbf{a})$ over the action space $\mathcal{A}_1 \times \cdots, \mathcal{A}_K$ is an ϵ-correlated equilibrium, if the swap regret is less than or equal to ϵ for every player and for every action chosen to be swapped.

The following theorem from [290] assures that no-regret learning can converge to an ϵ-correlated equilibrium.

Theorem 16.3 (Swap Regret and ϵ-Correlated Equilibria) If every player in a repeated game follows a strategy that has swap regret of at most ϵ, then the empirical distribution $P^{(emp)}$ of the joint actions played by the players is an ϵ-correlated equilibrium.

Proof: See [290] for a proof.

It is possible to modify a given no-regret learning algorithm with good external regret performance to obtain a no-regret learning algorithm with good swap regret [290]. Hence, the existence of good no-regret learning algorithms, such as the randomized weighted majority algorithm for external regret minimization assured by Theorem 16.2, provides the possibility to find no-regret learning algorithms that may converge to an ϵ-correlated equilibrium in a multiagent learning environment.

16.5.2 Cooperative Multiagent Learning

Learning in multiagent environments is another task in which cooperation can be of use. In particular, when multiple cognitive radios attempt to learn a policy distributively, it is conceivable that any coordination and/or cooperation among them can be of benefit to everyone and the overall network. This is called *cooperative multiagent learning*. Cooperative learning can benefit nodes and the overall network in several ways: First, the learning time required to converge at a policy can be reduced. Second, cooperation may help arrive at a globally optimal policy or at least a policy better than what individual noncooperating radios may be able to achieve.

Many of the learning paradigms we discussed so far can also be made to take advantage of any cooperative sharing of knowledge or information that may exist among distributed nodes. In that sense, many of the cooperative communications strategies that we discussed throughout the book, and in Chapter 15 specifically, can be considered as cooperative learning. The general problem of cooperative multiagent learning, however, usually refers to learning optimal policies in multiagent systems with certain restrictions on communications among distributed nodes. Indeed, it has been observed that in the extreme case of unlimited and unrestricted communications among agents, the multiagent learning problem can essentially be reduced to a learning system that is isomorphic to a single-learner system [293, 294]. Hence, for a cooperative multiagent learning system to be meaningful, we may exclude systems with unrestricted communications among nodes.

In [294], cooperative multi-agent learning approaches were divided into two broad categories:

1. Team learning
2. Cooperative concurrent learning

In *team learning*, there is only a single learner in the multiagent system. This single learner attempts to learn policies for all agents in the system. This is akin to a cooperative CRN equipped with a secondary system decision center (SSDC). Clearly, the SSDC is the single learner. Team learning problems can essentially be handled by single-agent learning approaches.

Cooperative concurrent learning refers to the scenario in which multiple agents in the multiagent system attempt to learn their own decision rules. This is more complicated than the team learning. Each player has its own learning process and has no control over the actions of the other players. Moreover, since each learner's actions vary over time as they learn, the learning environment becomes nonstationary. However, the nonstationarity is not decoupled from any of the individual learners either since the changes in policies of other learners are in response to this individual learner's actions [294]. Subband selection decisions of autonomous cognitive radios or distributed dynamic spectrum leasing (DSL) systems we discussed earlier can be considered as concurrent learning situations. Concurrent learning problems can be exceedingly difficult due to the existence of multiple learners, even with cooperation.

A common approach to modeling cooperative concurrent learning is to use a game-theoretic setting in which there is a global reward function that is monotonic in each agent's reward. Then, each distributed learner's objective of learning an optimal policy for itself will also help the cooperative multiagent learning system to reach a certain global target (e.g., maximizing a global utility). However, this approach may not always ensure full cooperation among nodes unless all agent rewards are the same [294].

One may wonder why it is ever an issue to assign unequal rewards to different players if the objective is to maximize a global utility of the whole system. Note that, however, when the global utility improves by a certain amount, it is possible that amount of contributions of different players toward this improvement may be different. As an example, think of a secondary cognitive radio system in which the objective is to maximize the total throughput of the secondary system. When a particular secondary user k has a very good channel compared to another secondary user k' and if they both create the same interference at the primary receiver for the same transmit power, then it may make sense to allow user k to transmit at a higher power at the expense of user k' having to transmit at a lower power. Such assignment of unequal rewards to players is called the problem of *credit assignment*. As can be inferred from this example, such unequal credit assignment may, however, discourage players from full cooperation and thus need to be carefully considered [294].

Even with restrictions, the communication among cooperating learners can be helpful in multiagent systems. Cooperating learners can help each other by sharing their knowledge as permitted by the allowed level of communications. This include sharing one's own observation, information history, or their learned policies with other players [295]. The results reported in [295], not surprisingly, show that in general any of these type of sharing can be helpful but at the expense of certain costs, communications overhead being the obvious.

In a CRN, this may be equivalent to informing a radio's neighbors of its current sensing observation and a summary of past channel/subband access decisions and the observed rewards corresponding to them or, in some cases, directly their own learned policies. Cooperating neighbors may be able to utilize such shared knowledge to better and quickly learn policies. In particular, when a new radio joins a network, such shared knowledge received from already active cognitive radios in the network can help it to at least form a good initialization point for its own policy. Depending on the amount of communications allowed, the frequency of knowledge sharing can be different. Cooperating players may use either game-theoretic or reinforcement learning approaches, for example, in combination with such shared knowledge to achieve better learning performance.

16.5.3 Multiagent Learning in Cognitive Radio Networks

Game theory has been applied for modeling and analysis of cognitive radio behavior in a wide range of problem settings. DSL between primary users and secondary users that we discussed in Chapter 14 is just one example scenario. There, our model was a simple noncooperative game, and the assumed learning strategy was best response learning. A survey of cognitive radio literature reveals a wide variety of different game-theoretic models that have been used in cognitive radio context. For example, [31] used a Stackelberg game to model a cooperative spectrum leasing scenario in which the primary users act as the game leaders and secondary cognitive users act as the followers.

In the context of dynamic spectrum access CRNs, the so-called supermodular games[2] were used to model distributed power control, power and channel allocation, and rate adaption in cognitive radios, respectively, in [296], [297], and [298], among others. A repeated game model was used in [299] to model the dynamic spectrum access in which multiple secondary users share the same spectrum hole. An example of the use of a Markov game to model the greedy selfish behavior of cognitive radios in a CRN is [300]. Finally, just as in modeling asymmetric cooperative communications-based DSL in Chapter 14, auction games have also been applied for spectrum allocation problems in [301].

Game-theoretic learning concepts are also prevalent in the cognitive radio literature. The simple best response learning was already used in Chapter 14 for learning a Nash equilibrium in a DSL game. On the other hand, in [302], cognitive radios used no-regret learning to learn to simultaneously update both transmission power and frequency. No-regret learning was also used in [303] to detect malicious nodes in spectrum sensing. It was shown in [303] that the correlated equilibrium achieved by no-regret learning had better performance and fairness compared to those of non-cooperative Nash equilibrium. Game-theoretic stochastic learning algorithms were proposed for cognitive radio problems in [304] and [305]. In particular, in [304], stochastic learning has been used to get around the requirement of prior knowledge of the environment and the actions taken by the other users.

In [306, 307], cooperative learning was used to speed up the convergence performance of the distributed cognitive base stations in the aggregate interference control problem from [138] that we discussed earlier in Section 16.4.3. Since nodes who share knowledge with others act as experts who teach these nodes, this type of learning was called *docitive learning* in [306, 307]. The knowledge sharing methods for cooperative learning among nodes identified in [306] can, however, be applied in general. In fact, we may broadly recognize two ways of invoking docitive learning in a CRN:

1. Start-up docitive learning: Wireless networks are almost always dynamic. When a new node dynamically joins a CRN, it can greatly benefit from the cooperative knowledge sharing with others. Hence, in start-up docitive learning, radios that are already in the network share the policies or the knowledge they have already learned with the newly joined network members. For example, in the case of a policy that is based on learning a Q-table, this knowledge sharing may boils down to providing Q-tables of the incumbent radios to newcomers. Indeed, in the case of aggregate interference control problem of [138], this is what [306, 307] proposed to do. Such start-up cooperation may provide a new-coming radio at least a good initialization point for it to start learning its own Q-table and, thus, its own optimal policy. Indeed, with this initialization, new radios learn on their own. Often, but not necessarily, in many policy learning problems, it can be expected that optimal policies of neighboring nodes

[2]Supermodular games have the important property that they possess at least one pure strategy Nash equilibrium.

to be similar in a certain sense. Hence, when start-up knowledge sharing is among neighboring radios, this can provide a newcomer with a reasonably good initialization that can help speed up its learning convergence performance.

2. Iterative docitive learning: There is no reason why cooperative knowledge sharing has to be limited to provide initializations. Certainly, cognitive radios may benefit by sharing their knowledge and policies with others at regular intervals. Such cooperative learning is called iterative, or periodic, docitive learning [306].

Regardless of whether cooperation is at the start-up or at regular intervals, it is advantageous for a learner to know qualifications of a teacher. In particular, when there are several cognitive radio nodes that provide a learning node with their expert knowledge, the learner can benefit by knowing the quality of these expert nodes. There are various measures one may propose as indicators of the strength, or weakness, of a node as an expert and its knowledge as being superior to others. Common performance indicators include the variance of the steady-state oscillations (with respect to the target) and speed of convergence. Hence, when possible, the nodes may provide their cooperative learning partners with not just their knowledge but also the performance parameters associated with their own learning.

In [306, 307], these cooperative docitive learning strategies were used to improve the learning performance of cognitive base stations in controlling aggregate interference to the primary system. It was shown that proper docitive learning can indeed lead to superior performance in terms of both speed of convergence and accuracy. Moreover, perhaps not surprisingly, iterative docitive learning was shown to outperform the other approaches [307].

The drawback of cooperative learning approaches, of course, is the possible communications overhead involved. This is no different from any of the cooperative communications schemes discussed in the previous chapter. Hence, any policy improvement achieved by cooperative learning needs to be balanced against the overhead costs incurred for knowledge sharing.

16.6 SUMMARY

- Learning is acquiring knowledge and comprehension, which is an integral part of both cognition and intelligence.
- The operation of a cognitive radio can be abstracted as a cognitive cycle formed of four basic steps: observe, decide, act, and learn.
- An artificial neural network is a network of interconnected nonlinear decision elements with adjustable weights on the connections. The nonlinear decision element is normally a linear threshold unit (LTU).
- A single LTU can only implement decision rules that linearly separate the input data space by a hyperplane. Data that can in fact be separated by such linear hyperplanes are called linearly separable.

- An LTU, and the decision rule it implements, is characterized by the weight vector associated with it.
- Error correction training is a commonly used learning algorithm in LTUs. According to this algorithm, whenever the LTU makes an error in its classification of the training data, the weight vector **w** is modified in the direction of the input data vector. The size of the modification is determined by a parameter called the learning rate. When the classifications are correct, no changes are made to the weight vector. Simple as it may, this procedure in fact converges to the correct weight vector if the training data is indeed linearly separable and we cycle through the training data sufficiently large number of time.
- An alternative learning procedure that is used in a wide variety of learning and adaptive applications is the LMS algorithm. LMS algorithm can be derived in several different ways. One approach is to seek the weight vector that minimizes the mean squared error between the output of the LTU and the true classification. The weight vector must then be adjusted in the direction of the negative gradient of the mean squared error. The LMS algorithm emerges if we replace the required gradient by its noisy instantaneous approximation, which is called the stochastic gradient. For this reason, the LMS algorithm is also called the stochastic gradient descent algorithm.
- The LMS algorithm can also be obtained if one seek the weight vector that minimizes the instantaneous squared error. Adapting the weight vector along the direction of the negative gradient of this instantaneous squared error results in the same LMS adaptive/learning algorithm.
- Artificial neural networks can be of many different types depending on the architecture of the network. A commonly used popular structure is the layered feed-forward architecture. In a layered network, the first and last layers of LTUs are called, respectively, the input and output layers. The layers of LTUs in between them are called hidden layers. A feed-forward network is a network in which no LTU's input depends on its output. Such networks, thus, do not have feedback connections. Artificial neural networks with feedback connection are called recurrent networks. Both feed-forward and recurrent neural networks have been applied for signal classification and mode identification in cognitive radios.
- Although a single LTU can only implement decision rules characterized by hyperplanes, networks of interconnected LTUs can implement complex nonlinear decision rules. Indeed, it can be shown that a layered feed-forward network made of just three layers can implement any arbitrary decision rule as long as the first two layers can have any number of LTUs.
- Perhaps the reason for the popularity of layered feed-forward networks is the fact that there is a simple and effective learning algorithm for these algorithms. This algorithm, called the back-propagation algorithm, allows one to update the weight vectors of all LTUs in the network in response to an output error for a given training data sample.

- A class of artificial neural networks that are particularly suited for unsupervised learning is called the self-organizing maps. These networks are normally implemented as one- or two-dimensional grids made of decision devices called the neurons, which are in fact can be simply the LTUs. Each LTU in a self-organizing map is, however, characterized by not only its associated weight vector but also by its location on the grid. A given input vector is classified by associating it with the neuron that has the closest weight vector to the input vector, called the best matching neuron or the best matching cell.
- Self-organizing maps learn by updating their weight vectors by an update term that depends not only on the error between its weight vector and the input data vector but also on its distance from the best matching neuron in terms of their relative locations on the grid.
- Empirical risk minimization approach to classifier design selects the classifier that minimizes the error over the given training data set. However, this has the potential to lead to a classifier that may not have sufficient generalization performance against unseen new data.
- Structural risk minimization approach seeks a classifier that minimizes a linear combination of the empirical risk and a regularization term. A particularly interesting regularization term is a term that depends on the generalization ability, or the capacity, of a classifier.
- The VC dimension is a measure of the capacity of a class of decision rules. The higher the VC dimension, the more capable this class of classifiers to correctly classify a given training data set.
- Hence, one approach to structural risk minimization is to seek a decision rule that minimizes the VC dimension subject to the constraint that the empirical risk is made to be zero. When, instead of the VC dimension itself, a suitable upper bound to the VC dimension is used, the resulting structural risk minimization is called a support vector machine optimization problem. The chosen upper bound to the VC dimension is simply the norm of the weight vector. It can also be shown that the resulting structural risk minimization problem is then equivalent to looking for the hyperplane that separates the given data into two classes with the maximum separation between them. For this reason, support vector machine solution is also called the maximum margin hyperplane.
- Such a hyperplane may not exist if the given training data set is not linearly separable. In such cases, the input data is first transformed into a higher-dimensional space in which they may in fact be linearly separable. This, however, can be computationally demanding since transformed data can be of much higher dimension than the original data.
- Support vector machines overcome this computational hurdle by using the so-called kernel trick. Essentially, it is noted that the transformed high-dimensional data vectors always enter the computations as inner products. Hence, by choosing a function $K(.,.)$, called a kernel function, defined on pairs of data vectors that are always equal to the inner products of transformed data vectors, the need for ever having to compute the transformed data vectors can be avoided.

The kernel trick is to then go one step even further and directly start with a kernel function having the desired properties for a particular classifier design application.

- There are many possible kernel functions with differing properties. Among them are the various radial basis function kernels and polynomial kernels. Obviously, which one is better suited for a particular problem depends on the specific details of the problem.
- Reinforcement learning, or learning by trail and error, is an unsupervised learning approach that attempts to learn by direct interaction with the environment. Each time the learner executes an action, it receives a delayed feedback received from environment. This feedback is used as a coarse guide in rating the level of preference for the particular action under the same conditions after the execution of a selected action. In general, reinforcement learning is an ideal learning methodology for autonomous learning and thus can be applicable to many cognitive radio learning problems.
- In order to learn a good policy through reinforcement learning, it is essential that the learner explores the space of states and actions as fully as it can. This exploration has to be carefully balanced against the exploitation of an already learned policy so as both not to wonder around too much randomly and not to be trapped in a local extremum.
- A widely used reinforcement learning algorithm is the Q-learning, which is well suited for learning problems posed in Markov environments. In Q-learning, the objective is to learn the optimal action-value function $Q^*(s,a)$ without having knowledge of the model parameters. Section 9.8.1 discussed the Q-learning algorithm in detail.
- The Q-learning algorithm can be modified to learn an effective policy for a POMDP if the belief state of the POMDP can be computed. The replicated Q-learning algorithm updates the values of the Q-table as

$$Q(j,a_{n-1}) \leftarrow Q(j,a_{n-1}) + \alpha b_j[n-1] \times \left[r_n(j,a_{n-1}) + \gamma \max_a Q(\mathbf{b}[n],a) - Q(j,a_{n-1}) \right],$$

where

$$\Delta Q(j,a_{n-1}) = r_n(j,a_{n-1}) + \gamma \max_a Q(\mathbf{b}[n],a) - Q(j,a_{n-1})$$

with average Q-value of taking action a when in belief state $\mathbf{b}$ defined as

$$Q(\mathbf{b},a) = \sum_j b_j Q(j,a).$$

Then the actions are chosen according to either

$$a_n^* = \arg\max_a Q(\mathbf{b}[n], a),$$

or

$$a_n^* = \begin{cases} \arg\max_{a \in \mathcal{A}} Q(\mathbf{b}[n], a) & \text{with probability } 1 - \epsilon \\ \sim \mathcal{U}(\mathcal{A}) & \text{with probability } \epsilon \end{cases}.$$

- Multiagent learning refers to learning by distributed multiple players in a multi-agent system. In particular, when distributed agents attempt to learn their own decision policies simultaneously while interacting with the same system, it is called concurrent learning.
- A repeated game is a game in which players repeatedly interact with each other over many rounds. In each round, their interaction is a stage game. Depending on the number of rounds, a game can be either finitely repeated or infinitely repeated.
- When the stage game in each round of a repeated game is allowed to be drawn probabilistically from a collection of possible games, it is called a stochastic game.
- A Markov game is a stochastic game in which the collection of possible games is determined by a state variable that has Markov dynamics. A Markov game made of a single player is an MDP, while a Markov game with a single state is a repeated game.
- A joint probability distribution $P(\mathbf{a})$, or equivalently a pmf $p(\mathbf{a})$, is called a correlated equilibrium if and only if, for all $k \in \mathcal{K}$, $a_k \in \mathcal{A}_k$, $\mathbf{a}_{-k} \in \mathcal{A}_{-k}$:

$$\sum_{a_{-k} \in \mathcal{A}_{-k}} p(a_k, \mathbf{a}_{-k}) \left[u_k(a'_k, \mathbf{a}_{-k}) - u_k(a_k, \mathbf{a}_{-k}) \right] \leq 0, \forall a'_k \in \mathcal{A}_k. \tag{16.81}$$

- The external regret of a policy is its maximum average regret per time step with respect to fixed-action policies. The swap regret of a policy is its average regret per time step with respect to swapped-action policies in which an action choice of policy $\underline{\pi}_k$ is replaced with a different action choice in all its occurrences within policy $\underline{\pi}_k$.
- No-regret learning refers to learning algorithms that attempt to minimize either the external or internal regret. It is known that there exist good no-regret learning algorithms whose maximum average regret decays to zero as time horizon tends to infinity.
- A joint probability distribution $P(\mathbf{a})$ over the action space $\mathcal{A}_1 \times \cdots, \mathcal{A}_K$ is an ϵ correlated equilibrium, if the swap regret is less than or equal to ϵ for every player and for every action chosen to be swapped.

- If every player in a repeated game follows a strategy that has swap regret of at most ϵ, then the empirical distribution $P^{(emp)}$ of the joint actions played by the players is an ϵ-correlated equilibrium.
- When distributed agents in a multiagent system cooperate among themselves in learning, it is called cooperative multiagent learning.
- Cooperative multiagent learning problems can be either team learning or concurrent learning problems.
- Team learning refers to the situation in which there is only a single learner in the multiagent system, and this single learner learns policies for all agents in the system. Team learning problems can essentially be handled by single-agent learning approaches.
- Cooperative concurrent learning refers to the situation in which multiple agents in the system attempt to learn their own decision rules.
- One way to model cooperative concurrent learning is to use a game-theoretic setting in which there is a global reward function that is monotonic in each agent's reward.
- Another approach is to explicitly take into account communications among cooperating radios. Cooperating learners can help each other by sharing their knowledge as permitted by the allowed level of communications. The shared knowledge can be either their observations, information histories, or learned policies.
- Cooperative learning can be a very general approach to improving the efficiency of learning in cognitive radios. Cooperative learning can help a cognitive radio to both learn faster and achieve better learning accuracy.
- In the case of cognitive radio networks, cooperative learning can be evoked either when a new node joins a network or as an iterative procedure at regular intervals.
- As with any cooperative protocol, the price of cooperative learning is the cost of overhead communications needed to share the knowledge among distributed nodes.

APPENDIX A

NYQUIST SAMPLING THEOREM

Software-defined radios and cognitive radios are digital communications devices that hold and process information in digital domain. The first stage of interface between analog and digital worlds is the process called sampling. Given an analog information waveform $x(t)$, sampling generates a discrete-time sequence $x[n]$. The important objective is to ensure that there is no information loss in representing the analog waveform $x(t)$ by the sampled sequence $x[n]$. If this objective is indeed satisfied, then the sequence $x[n]$ must be equivalent to the analog waveform $x(t)$ in the sense that it is possible to reconstruct $x(t)$ from the sequence of samples $x[n]$.

Let f_s denote the sampling rate (in the number of samples per second) so that $T_s = \frac{1}{f_s}$ is the corresponding sampling period. Let $x_s(t)$ denote the sampled version of $x(t)$ as shown in Figure A.1:

$$
\begin{aligned}
x_s(t) &= \sum_{n=-\infty}^{\infty} x(nT_s)\delta_D(t-nT_s) \\
&= \sum_{n=-\infty}^{\infty} x(t)\delta_D(t-nT_s) &\text{(A.1)}\\
&= x(t)\left[\sum_{n=-\infty}^{\infty} \delta_D(t-nT_s)\right] &\text{(A.2)}\\
&= x(t)\delta_s(t), &\text{(A.3)}
\end{aligned}
$$

Signal Processing for Cognitive Radios, First Edition. Sudharman K. Jayaweera.

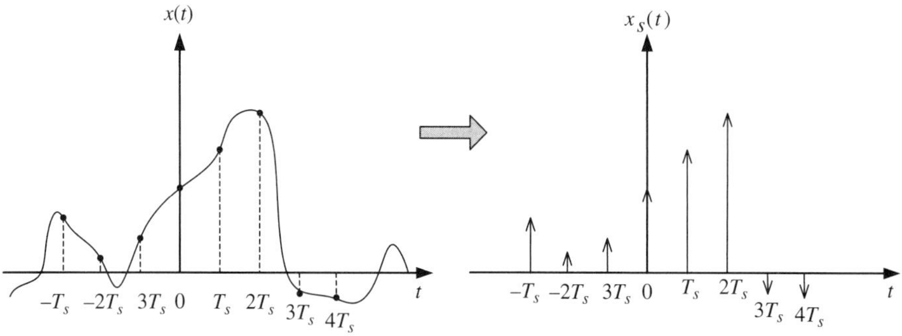

FIGURE A.1 A signal $x(t)$ and its sampled version $x_s(t)$.

where (A.1) and (A.2) follow from the properties of the Dirac delta function $\delta_D(t)$, and by $\delta_s(t)$, we denote an impulse-train signal:

$$\delta_s(t) = \sum_{n=-\infty}^{\infty} \delta_D(t - nT_s). \tag{A.4}$$

Let $X(f)$, $X_s(f)$, and $\Delta_s(f)$ denote the Fourier transforms of $x(t)$, $x_s(t)$, and $\delta_s(t)$. Applying the multiplication theorem for Fourier transforms to (A.3), we then have that

$$X_s(f) = X(f) \star \Delta_s(f). \tag{A.5}$$

Since impulse-train signal $\delta_s(t)$ is periodic with period T_s, its Fourier transform $\Delta_s(f)$ is also an impulse train in the frequency domain with period f_s:

$$\Delta_s(f) = f_s \sum_{m=-\infty}^{\infty} \delta(f - mf_s). \tag{A.6}$$

From (A.5) and (A.6), we see that the spectrum $X_s(f)$ of the sampled signal $x_s(t)$ is

$$\begin{aligned} X_s(f) &= X(f) \star \sum_{m=-\infty}^{\infty} f_s \delta(f - mf_s) \\ &= f_s \sum_{m=-\infty}^{\infty} X(f - mf_s). \end{aligned} \tag{A.7}$$

Thus, aside from a scaling factor f_s, the spectrum $X_s(f)$ of the sampled signal $x_s(t)$ is the sum of an infinite number of shifted versions of the spectrum $X(f)$ of the original signal $x(t)$ where shifts are equal to integer multiples of the sampling frequency.

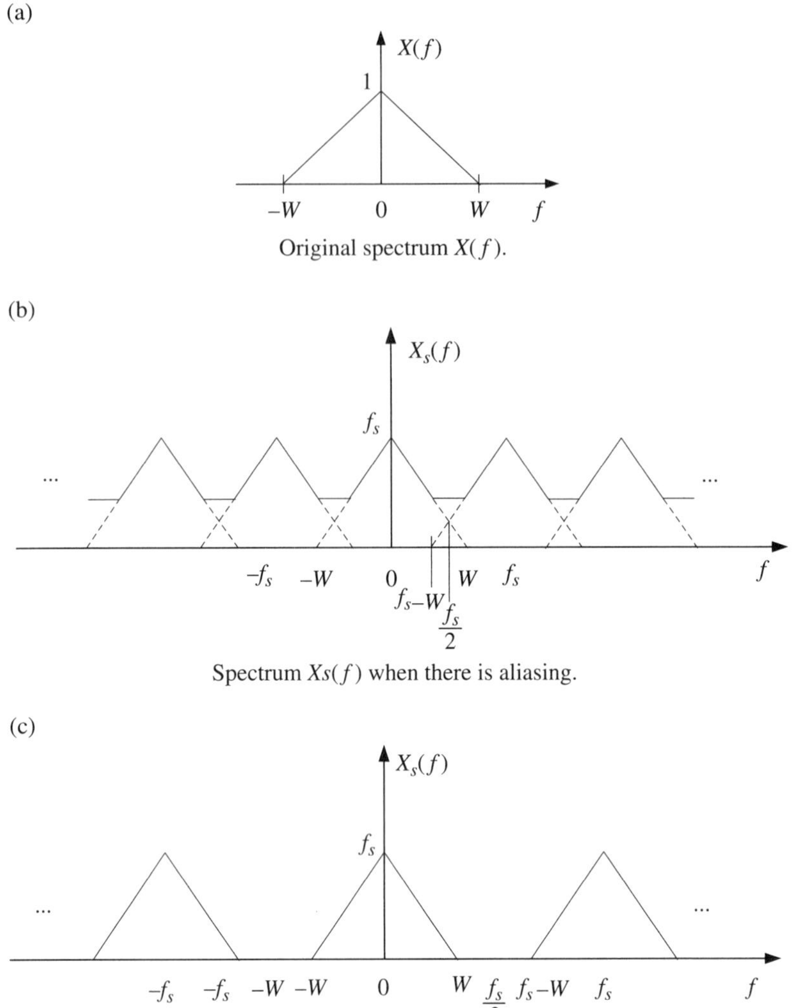

FIGURE A.2 Original spectrum $X(f)$ and the spectrum $X_s(f)$ of the sampled signal $x_s(t)$.

As can be seen from Figure A.2(b), in general, this spectrum $X_s(f)$ can be different from the original spectrum $X(f)$ due to possible overlapping of shifted versions of $X(f)$. However, as Figure A.2(c) shows, under certain special conditions on the original spectrum $X(f)$ and the sampling frequency f_s, it seems possible to recover the original spectrum $X(f)$ from $X_s(f)$. Indeed, Figure A.2(c) shows that if the original information signal $x(t)$ is bandlimited to a maximum frequency $f_{\max} = W$ and if the sampling frequency f_s is chosen to be greater than $2W$, then the shifted versions

$X(f - mf_s)$, for all m, that are added up to form the spectrum $X_s(f)$ do not overlap with each other. Hence, under these conditions, the spectrum $X_s(f)$ of the sampled signal $x_s(t)$ is exactly equal to the original signal spectrum $X(f)$ for $-\frac{f_s}{2} < f < \frac{f_s}{2}$ except for the scaling factor f_s. As can be seen by comparing Figs. A.2(a) and A.2(c), outside of the frequency interval $-\frac{f_s}{2} < f < \frac{f_s}{2}$, which is called the *fundamental frequency interval*, the original signal spectrum is zero. However, spectrum of the sampled signal repeats its spectrum within the fundamental frequency interval with a period of f_s.

Hence, if the sampling frequency $f_s \geq 2W$, where W is the maximum frequency of the analog signal $x(t)$, then we may recover the original signal spectrum $X(f)$ from the sampled signal spectrum $X_s(f)$ by passing the sampled signal $x_s(t)$ through an ideal low-pass filter having a transfer function

$$H(f) = \frac{1}{f_s} \prod \left(\frac{f}{2B} \right). \tag{A.8}$$

If the low-pass filter bandwidth B is such that $W \leq B \leq f_s - W$, then the output $Y(f)$ of the filter corresponding to the input $x_s(t)$ is

$$Y(f) = X_s(f)H(f) \ = \ X(f), \tag{A.9}$$

showing that indeed perfect recovery of the original analog signal is possible under these conditions. The above discussion is summarized in the following called the Nyquist sampling theorem.

Theorem A.1 (Nyquist Sampling Theorem) If a signal $x(t)$ contains no frequencies above $f = W$, then it is completely characterized by its sample values uniformly spaced in time with a period $T_s \leq \frac{1}{2W}$. The signal can be perfectly reconstructed by passing the sampled waveform $x_s(t)$ through an ideal low-pass filter having a bandwidth B if $W < B < f_s - W$.

Let us denote by $x[n]$ the discrete-time signal (sequence) obtained by sampling the analog signal $x(t)$: That is, $x[n] = x(nT_s)$ for $-\infty < n < \infty$. Note that given the sequence of sample values $x[n]$ for $-\infty < n < \infty$, we can directly compute the Fourier transform $X_s(f)$ of the sampled signal $x_s(t)$ as

$$\begin{aligned}
X_s(f) &= \int_{-\infty}^{\infty} x_s(t)e^{-j2\pi ft}dt \\
&= \int_{-\infty}^{\infty} \sum_{n=-\infty}^{\infty} x(nT_s)\delta_D(t - nT_s)e^{-j2\pi ft}dt \\
&= \sum_{n=-\infty}^{\infty} x(nT_s) \int_{-\infty}^{\infty} \delta_D(t - nT_s)e^{-j2\pi ft}dt
\end{aligned}$$

$$= \sum_{n=-\infty}^{\infty} x(nT_s)e^{-j2\pi f n T_s}$$

$$= \sum_{n=-\infty}^{\infty} x[n]e^{-j2\pi \frac{f}{f_s} n}$$

$$= \sum_{n=-\infty}^{\infty} x[n]e^{-j2\pi F n}, \tag{A.10}$$

which is, in fact, the discrete-time Fourier transform (DTFT) of the sequence $x[n]$ with $F = \frac{f}{f_s}$ denoting the normalized frequency. As is seen from (A.9), proper low-pass filtering of this spectrum allows recovering the spectrum $X(f)$ of the original analog signal $x(t)$.

We may also obtain a direct reconstruction formula for $x(t)$ in terms of the sampled sequence $x[n]$ by noting that the impulse response function $h(t)$ corresponding to the ideal low-pass filter (A.8) is given by

$$h(t) = 2BT_s sinc\,(2Bt)\,,$$

where $sinc(t) = \frac{\sin \pi t}{\pi t}$. Hence, the output of the ideal low-pass filter (A.8) corresponding to the input $x_s(t)$ is

$$y(t) = x_s(t) \star h(t)$$

$$= \left[\sum_{n=-\infty}^{\infty} x(nT_s)\delta_D(t - nT_s) \right] \star [2BT_s sinc\,(2Bt)]$$

$$= 2BT_s \sum_{n=-\infty}^{\infty} x[n] sinc\,(2B(t - nT_s))\,. \tag{A.11}$$

Since, from (A.9), this output is indeed the original analog waveform $x(t)$, we thus have

$$x(t) = \sum_{n=-\infty}^{\infty} x[n] sinc\left(\frac{t}{T_s} - n\right), \tag{A.12}$$

where we have assumed that the low-pass filter bandwidth $B = \frac{f_s}{2}$.

It is often the case that signals are not strictly bandlimited in practice. For example, the signals arriving at the antenna of a cognitive radio will not be bandlimited to a certain maximum frequency W as assumed in the Sampling theorem. However, the sampling rate of any practical A/D converter will be a finite value $f_s = 2W$ for some W. When input signal to the A/D converter has frequency content beyond this frequency W as in Figure A.3(a), then aliasing is unavoidable as implied by the sampling

(a)

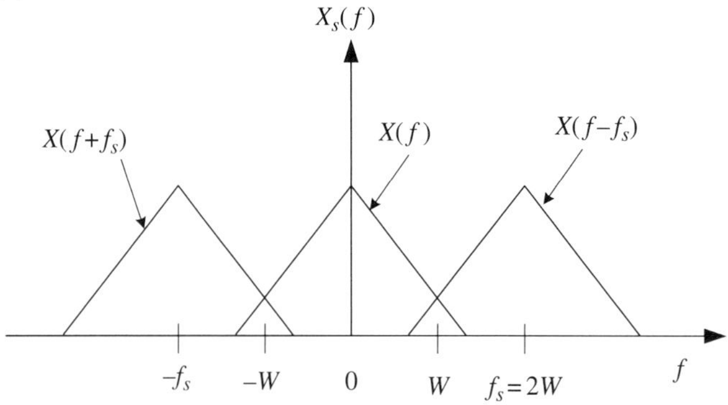

A signal $x(t)$ having frequency content beyond $W=\frac{f_s}{2}$ where f_s is the sampling rate.

(b)

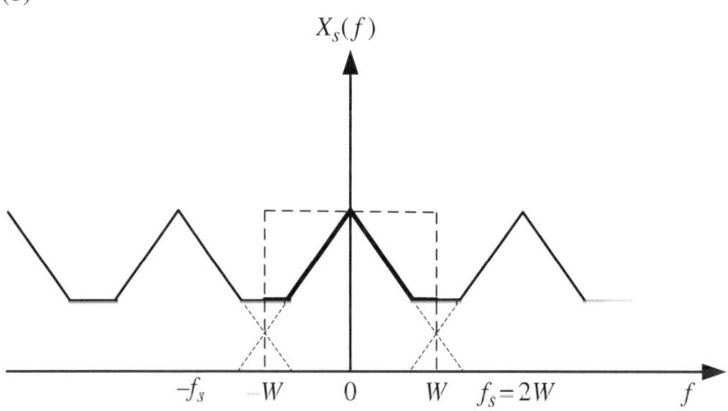

Sampled signal spectrum $X_s(f)$ is distorted even within $|f| \leq W$.

FIGURE A.3 When the signal $x(t)$ has frequency content beyond $\frac{f_s}{2}$, where f_s is the sampling rate used, the sampled signal spectrum will be distorted even within the frequency range $|f| \leq \frac{f_s}{2}$.

theorem. Although it is reasonable to expect no spectrum distortion for $|f| \leq W$ when we use a sampling rate of $f_s = 2W$, as can be seen from Figure A.3(b), in this situation, aliasing does distort the spectrum even for frequencies $|f| \leq W$. The reason is that the frequency content above $f = W$ in the original signal aliases into $|f| \leq W$ region.

In practice, it is desirable to avoid this effect and be able to obtain a distortionless spectrum for $|f| \leq W$ when using a sampling rate of $f_s = 2W$. This can be achieved by first filtering the original signal so that all frequencies above $f = W$ are removed. A low-pass filter that removes high-frequency content of a signal that may lead to aliasing and signal distortion is called an *antialiasing filter*. Figure A.4 shows an ideal antialiasing filter.

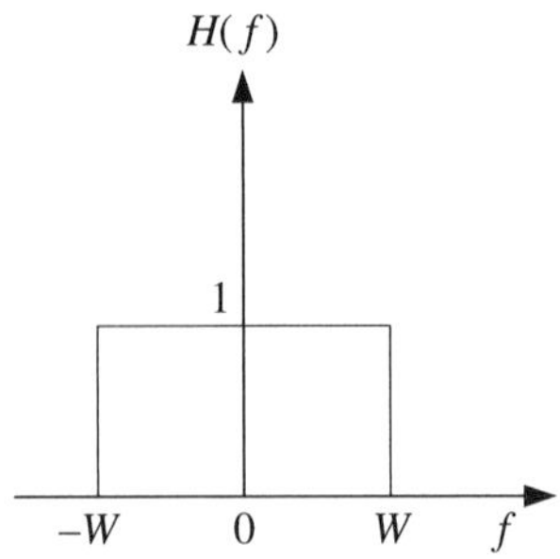

FIGURE A.4 An antialiasing filter.

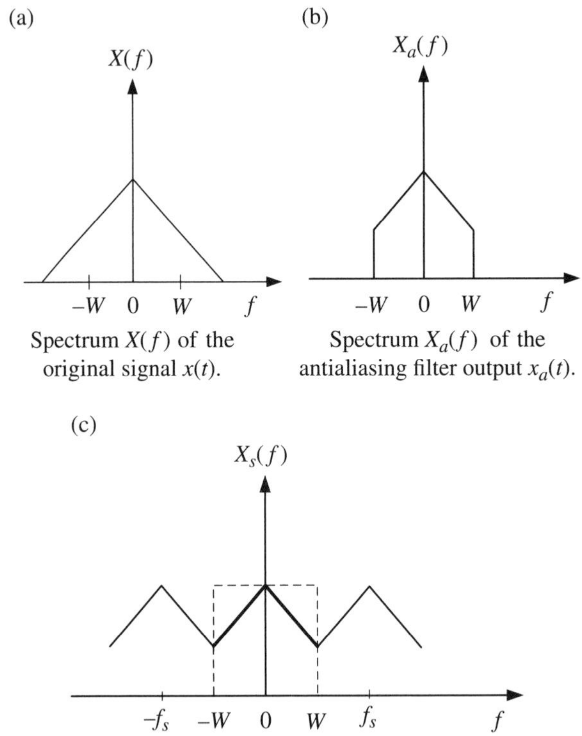

FIGURE A.5 The use of an antialiasing filter before the sampler can avoid signal distortion within the fundamental frequency interval $|f| \leq \frac{f_s}{2}$.

Let us denote by $x_a(t)$ the output of the antialiasing filter. Then $x_a(t)$ will have maximum frequency W, as shown in Figure A.5(b). Hence, the input to the sampler is a signal that has maximum frequency W. If the sampling rate is $f_s = 2W$, now, there is no aliasing for $|f| \leq W$. Hence, the sampled-signal spectrum $X_s(f)$ will be exactly equal to the original spectrum $X(f)$ for all $|f| \leq W$. This is shown in Figure A.5(c).

APPENDIX B

A COLLECTION OF USEFUL PROBABILITY DISTRIBUTIONS

B.1 UNIVARIATE DISTRIBUTIONS

- **Gaussian distribution**: X is a Gaussian random variable with mean μ and variance σ_X^2 if it has the following probability density function (pdf):

$$f_X(x) = \frac{1}{\sqrt{2\pi\sigma_X^2}} e^{\frac{-(x-\mu)^2}{2\sigma_X^2}}.$$

  When X is such a Gaussian random variable, we denote it as $X \sim \mathcal{N}(\mu, \sigma_X^2)$. Obviously in this case, $\mathbb{E}\{X\} = \mu$ and $\text{Var}(X) = \mathbb{E}\{(X-\mu)^2\} = \sigma_X^2$.
- **Standard Gaussian distribution**: X is a standard Gaussian (or standard normal) random variable if $X \sim \mathcal{N}(0, 1)$. The pdf of a standard Gaussian random variable X is

$$f_X(x) = \frac{e^{-\frac{x^2}{2}}}{\sqrt{2\pi}}.$$

  In other words, $\mathbb{E}\{X\} = 0$ and $\text{Var}(X) = 1$.

Signal Processing for Cognitive Radios, First Edition. Sudharman K. Jayaweera.

- **Lognormal distribution**: X is a lognormal random variable with parameter μ and σ^2 if it has the following pdf:

$$f_X(x)) = \begin{cases} \frac{1}{x\sqrt{2\pi\sigma^2}} e^{-\frac{(\log x-\mu)^2}{2\sigma^2}} & \text{if } x \geq 0 \\ 0 & \text{if } x < 0 \end{cases}.$$

If $X = e^Y$ where $Y \sim \mathcal{N}(\mu, \sigma^2)$, then X is lognormal distributed.

- **Exponential distribution**: X is an exponential random variable with parameter λ if it has the following pdf:

$$f_X(x)) = \begin{cases} \lambda e^{-\lambda x} & \text{if } x \geq 0 \\ 0 & \text{if } x < 0 \end{cases}.$$

Then $\mathbb{E}\{X\} = \frac{1}{\lambda}$ and $\mathrm{Var}(X) = \frac{1}{\lambda^2}$.

Sum of squares of two independent Gaussian random variables is an exponential random variable.

- **Rayleigh distribution**: X is a Rayleigh random variable if it has the pdf

$$f_X(x)) = \begin{cases} \frac{x}{\alpha^2} e^{-\frac{x^2}{2\alpha^2}} & \text{if } x \geq 0 \\ 0 & \text{if } x < 0 \end{cases}$$

where $\alpha > 0$.

Then $\mathbb{E}\{X\} = \alpha\sqrt{\frac{\pi}{2}}$ and $\mathrm{Var}(X) = (2 - \frac{\pi}{2})\alpha^2$.

Magnitude of sum of squares of two independent zero-mean Gaussian random variables is a Rayleigh random variable.

- **Chi-square distribution**: Z is a Chi-square random variable with n degrees of freedom, denoted as χ_n^2, if it has the pdf

$$f_Z(z) = \begin{cases} \frac{z^{\frac{n}{2}-1} e^{-\frac{z}{2}}}{2^{\frac{n}{2}} \Gamma\left(\frac{n}{2}\right)} & \text{if } z \geq 0 \\ 0 & \text{if } z < 0 \end{cases}$$

where $\Gamma(x)$ is the Gamma function defined as

$$\Gamma(x) = \int_0^{\infty} t^{x-1} e^{-t} dt.$$

Note that $\Gamma(x+1) = x\Gamma(x)$ for $x > 0$, $\Gamma(n) = (n-1)!$ for positive integer n and $\Gamma\left(\frac{1}{2}\right) = \sqrt{\pi}$.

Then $\mathbb{E}\{X\} = n$ and $\mathrm{Var}(X) = 2n$.

When $n = 2$, the Chi-square random variable reduces to an exponential random variable.

The sum of n-squared mutually independent standard Gaussian random variables is a the chi-square random variable with n degrees of freedom denoted as χ_n^2.

- **Gamma distribution**: X is a gamma random variable with a shape parameter $\alpha > 0$ and a scale parameter $\beta > 0$, denoted as $G(\alpha,\beta)$, if it has the pdf

$$f_X(x) = \begin{cases} \dfrac{x^{\alpha-1}}{\Gamma(\alpha)\beta^{\alpha}} e^{-\frac{x}{\beta}} & \text{if } x \geq 0 \\ 0 & \text{if } x < 0 \end{cases}.$$

Then $\mathbb{E}\{X\} = \alpha\beta$ and $\mathrm{Var}(X) = \alpha\beta^2$.

When $\beta = 2$ and $\alpha = \frac{n}{2}$ for an integer n, then $G(\alpha,\beta)$ becomes a chi-square random variable with n degrees of freedom χ_n^2.

Sometimes, instead of the scale parameter β, a rate parameter $\frac{1}{\beta}$ is used to specify the gamma distribution.

- **Beta distribution**: X is a beta random variable with parameters a and b if it has the pdf

$$f_X(x) = \begin{cases} \dfrac{x^{a-1}(1-x)^{b-1}}{B(a,b)} & \text{if } 0 < x < 1 \\ 0 & \text{otherwise,} \end{cases},$$

where $B(a,b)$ is the beta function defined as

$$B(a,b) = \int_0^1 x^{a-1}(1-x)^{b-1}dx,$$

which can be shown to be

$$B(a,b) = \frac{\Gamma(a)\Gamma(b)}{\Gamma(a+b)}.$$

Then $\mathbb{E}\{X\} = \frac{a}{a+b}$ and $\mathrm{Var}(X) = \frac{ab}{(a+b)^2(a+b+1)}$.

B.2 MULTIVARIATE DISTRIBUTIONS

- **Multivariate normal distribution (vector)**: If n-vector $\mathbf{X} = (X_1, X_2, \ldots, X_n)^T$ is multivariate Gaussian distributed, then its pdf is given by

$$f_{\mathbf{X}}(\mathbf{x}) = \frac{1}{(2\pi)^{\frac{n}{2}}|\Sigma|^{\frac{1}{2}}} e^{-\frac{1}{2}(\mathbf{x}-\boldsymbol{\mu})\Sigma^{-1}(\mathbf{x}-\boldsymbol{\mu})^T},$$

where $\mathbb{E}\{\mathbf{X}\} = \boldsymbol{\mu}$ is an n-vector, $Cov(\mathbf{X}) = \mathbb{E}\{(\mathbf{X}-\boldsymbol{\mu})(\mathbf{X}-\boldsymbol{\mu})^T\} = \Sigma$ is an $n \times n$ matrix, and $|\Sigma|$ denotes the determinant of the matrix Σ. We denote this by $\mathbf{X} \sim \mathcal{N}(\boldsymbol{\mu}, \Sigma)$.

In the special case of $n = 2$, if we denote $\mathbf{X} = (X_1, X_2)^T = (X, Y)^T$,

$$\begin{aligned} f_{\mathbf{X}}(\mathbf{x}) &= f_{XY}(x,y) \\ &= \frac{1}{2\pi\sigma_X\sigma_Y\sqrt{1-\rho^2}} e^{-\frac{1}{2(1-\rho^2)}\left[\frac{(x-\mu_X)^2}{\sigma_X^2} - 2\rho\left(\frac{x-\mu_X}{\sigma_X}\right)\left(\frac{y-\mu_Y}{\sigma_Y}\right) + \frac{(y-\mu_Y)^2}{\sigma_Y^2}\right]}, \end{aligned}$$

where X and Y are marginally Gaussian distributed as $X \sim \mathcal{N}(\mu_X, \sigma_X^2)$ and $Y \sim \mathcal{N}(\mu_Y, \sigma_Y^2)$ and ρ denotes the correlation coefficient of X and Y:

$$\rho = \frac{\mathbb{E}\{(X-\mu_X)(Y-\mu_Y)\}}{\sigma_X\sigma_Y}.$$

If $\rho = 0$, then X and Y are independent. When $\rho = 0$ for two arbitrarily distributed random variables, they are said to be uncorrelated.

- **Multivariate normal distribution (matrix)**: An $n \times d$ matrix Z is said to be multivariate Gaussian distributed with mean M and covariance matrix $C \otimes D$, where $\otimes$ denotes the Kronecker product and C and D are $n \times n$ and $d \times d$ positive-definite matrices, respectively, denoted as $Z \sim \mathcal{N}(M, C \otimes D)$, if it has the pdf [308]

$$f_Z(Z) = (2\pi)^{-\frac{nd}{2}} |C|^{-\frac{d}{2}} |D|^{-\frac{n}{2}} e^{-\frac{1}{2}\mathrm{tr}\left(C^{-1}(Z-M)D^{-1}(Z-M)^T\right)},$$

where $\mathrm{tr}(A)$ denotes the trace of matrix A and $\mathbb{E}\{\mathbf{Z}\} = M$ and the nd-vector $\mathbf{Z} = vec(Z^T)$ is multivariate Gaussian distributed such that $\mathbf{Z} \sim \mathcal{N}(\mathbf{m}, C \otimes D)$ with $\mathbf{m} = vec(M^T)$ and $vec(X) = (\mathbf{x}_1^T, \ldots, \mathbf{x}_n^T)^T$ if $X = [\mathbf{x}_1, \ldots, \mathbf{x}_n]$ is a $d \times n$ matrix with $\mathbf{x}_i$ denoting its ith column for $i = 1, \ldots, n$. Note that $Z \sim \mathcal{N}(M, C \otimes D)$ means that the nd-vector $\mathbf{Z} = vec(Z^T)$ is multivariate Gaussian distributed such that $\mathbf{Z} \sim \mathcal{N}(\mathbf{m}, C \otimes D)$ where $\mathbf{m} = vec(M^T)$ and $vec(X) = (\mathbf{x}_1^T, \ldots, \mathbf{x}_n^T)^T$ if $X = [\mathbf{x}_1, \ldots, \mathbf{x}_n]$ is a $d \times n$ matrix with $\mathbf{x}_i$ denoting its ith column for $i = 1, \ldots, n$.

In the special case of Z is $n \times d$ matrix such that

$$Z = \begin{bmatrix} \mathbf{Z}_1^T \\ \vdots \\ \mathbf{Z}_n^T \end{bmatrix}, \tag{B.1}$$

where d-vectors $\mathbf{Z}_i$, for $i = 1, \ldots, n$, are iid such that $\mathbf{Z}_i \sim \mathcal{N}(\mathbf{0}, \Sigma)$, for Σ a positive definite matrix, we have that $Z \sim \mathcal{N}(0, \mathbf{I}_n \otimes \Sigma)$.

- **Wishart distribution**: A $d \times d$ symmetric, positive definite matrix A is Wishart distributed with parameters $n \geq d$ and Σ, a $d \times d$ matrix, denoted as $A \sim W_d(\Sigma, n)$, if it has the pdf [308]

$$f_A(A) = \frac{1}{2^{\frac{nd}{2}} \Gamma_d\left(\frac{n}{2}\right) |\Sigma|^{\frac{n}{2}}} |A|^{\frac{n-d-1}{2}} e^{-\frac{1}{2}\mathrm{tr}(\Sigma^{-1}A)},$$

where $\mathrm{tr}(A)$ and $|A|$ denote, respectively, the trace and determinant of matrix A and $\Gamma_d(x)$ is the multivariate gamma function defined as

$$\Gamma_d(x) = \int_{A>0} |A|^{x \quad \frac{d+1}{2}} e^{\quad \mathrm{tr}(A)} (dA) \tag{B.2}$$

$$= \pi^{\frac{d(d-1)}{4}} \prod_{i=1}^{d} \Gamma\left(x - \frac{i-1}{2}\right), \tag{B.3}$$

for $\Re(x) > \frac{d-1}{2}$ and the integral in (B.2) is over the space of symmetric, positive definite $d \times d$ matrices.

Note that if $A = Z^T Z$ where Z is an $n \times d$ random matrix such that $Z \sim \mathcal{N}(0, \mathbf{I}_n \otimes \Sigma)$, then A is Wishart distributed such that $A \sim W_d(\Sigma, n)$.

APPENDIX C

CONJUGATE PRIORS

Signal Processing for Cognitive Radios, First Edition. Sudharman K. Jayaweera.

TABLE C.1 A collection of Conjugate Priors for Commonly Encountered Likelihoods [119]

Form of the Likelihood Function as a Function of the Parameter of Interest	Parameter	Conjugate prior	Posterior distribution of the parameter $f_{\theta\|\mathbf{x}}(\theta\|\underline{x})$
1. Bernoulli likelihood: $f_{\mathbf{x}\|\theta}(\mathbf{x}\|p) \propto p^{n\bar{x}}(1-p)^{n(1-\bar{x})}$ where $\bar{x} = \frac{1}{n}\sum_{i=1}^{n} x_i$	$\theta = p$	$\theta \sim beta(\alpha,\beta)$ with $\alpha > 0$ and $\beta > 0$ $g(\theta) = \begin{cases} \frac{\theta^{\alpha-1}(1-\theta)^{\beta-1}}{B(\alpha,\beta)} & \text{if } 0 < \theta < 1 \\ 0 & \text{otherwise} \end{cases}$	$\theta\|\mathbf{x} \sim beta(\alpha',\beta')$, where $\alpha' = \alpha + n\bar{x}$ $\beta' = \beta + n(1-\bar{x})$
2. Poisson likelihood: $f_{\mathbf{x}\|\theta}(\mathbf{x}\|\mu) \propto \begin{cases} \mu^{n\bar{x}}e^{-n\mu} & \begin{pmatrix} \text{if } x_i = 0,1,\ldots, \\ \text{for } i = 1,2,\ldots,n \end{pmatrix} \\ 0 & \text{otherwise,} \end{cases}$ where $\bar{x} = \frac{1}{n}\sum_{i=1}^{n} x_i$	$\theta = \mu$	$\theta \sim G(\alpha,\beta)$ with $\alpha > 0$ and $\beta > 0$ $g(\theta) = \begin{cases} \frac{\theta^{\alpha-1}e^{-\theta/\beta}}{\Gamma(\alpha)\beta^{\alpha}} & \text{if } \theta > 0 \\ 0 & \text{otherwise} \end{cases}$	$\theta\|\mathbf{x} \sim G(\alpha',\beta')$, where $\alpha' = \alpha + n\bar{x}$ $\beta' = \frac{\beta}{1+n}$
3. Negative binomial likelihood: $f_{\mathbf{x}\|\theta}(\mathbf{x}\|p) \propto \begin{cases} p^{nr\bar{x}}(1-p)^{n(\bar{x}-r)} & \begin{pmatrix} \text{if } x_i \geq r \\ \text{for } i = 1,2,\ldots,n \end{pmatrix} \\ 0 & \text{otherwise,} \end{cases}$ where $\bar{x} = \frac{1}{n}\sum_{i=1}^{n} x_i$	$\theta = p$	$\theta \sim beta(\alpha,\beta)$ with $\alpha > 0$ and $\beta > 0$ $g(\theta) = \begin{cases} \frac{\theta^{\alpha-1}(1-\theta)^{\beta-1}}{B(\alpha,\beta)} & \text{if } 0 < \theta < 1 \\ 0 & \text{otherwise} \end{cases}$	$\theta\|\mathbf{x} \sim beta(\alpha',\beta')$, where $\alpha' = \alpha + nr$ $\beta' = \beta + n(\bar{x}-r)$
4. Uniform likelihood: $f_{\mathbf{x}\|\theta}(\mathbf{x}\|L) \propto \begin{cases} \frac{1}{L^n} & \text{if } L > \max\{x_i\} \\ 0 & \text{otherwise} \end{cases}$	$\theta = L$	$\theta \sim Pareto(\alpha,\beta)$ with $\alpha > 0$ and $\beta > 0$ $g(\theta) = \begin{cases} \frac{\alpha\beta^{\alpha}}{\theta^{\alpha+1}} & \text{if } 0 > \beta \\ 0 & \text{otherwise} \end{cases}$	$\theta\|\mathbf{x} \sim Pareto(\alpha',\beta')$, where $\alpha' = \alpha + n$ $\beta' = \max\{x_1,\ldots,x_n,\beta\}$
5. Pareto likelihood: $\mathbf{x}\|b \sim Pareto(a,b)$ $f_{\mathbf{x}\|\theta}(\mathbf{x}\|b) \propto \begin{cases} b^{na} & \text{if } \min\{x_1,\ldots,x_n\} > b \\ 0 & \text{otherwise} \end{cases}$	$\theta = b$	$\theta \sim Pareto(\alpha,\beta)$ with $\alpha > 0$ and $\beta > 0$ $g(\theta) = \begin{cases} \frac{\alpha\beta^{\alpha}}{\theta^{\alpha+1}} & \text{if } 0 > \beta \\ 0 & \text{otherwise} \end{cases}$	$\theta\|\mathbf{x} \sim Pareto(\alpha',\beta')$, where $\alpha' = \alpha - an$, $\beta' = \beta$ with $\alpha > na$

(Continued)

TABLE C.1 Cont'd

Form of the Likelihood Function as a Function of the Parameter of Interest	Parameter	Conjugate prior	Posterior distribution of the parameter $f_{\theta\|\mathbf{x}}(\theta\|\underline{x})$
6. Pareto likelihood: $\mathbf{x}\|a \sim Pareto(a,b)$ $f_{\mathbf{x}\|\theta}(\mathbf{x}\|a) \propto \begin{cases} \frac{a^n b^{na}}{m^{a+1}} & \text{if } \min\{x_1,\ldots,x_n\} > b \\ 0 & \text{otherwise,} \end{cases}$ where $m = \prod_{i=1}^n x_i$	$\theta = a$	$\theta \sim G(\alpha,\beta)$ with $\alpha > 0$ and $\beta > 0$ $g(\theta) = \begin{cases} \frac{\theta^{\alpha-1}e^{-\theta/\beta}}{\Gamma(\alpha)\beta^\alpha} & \text{if } \theta > 0 \\ 0 & \text{otherwise} \end{cases}$	$\theta\|\mathbf{x} \sim G(\alpha^{'},\beta^{'})$, where $\alpha^{'} = \alpha + n,\ \beta^{'} = \frac{1}{\frac{1}{n}+\log(m)-n\log(b)}$ with $\beta < \frac{1}{n\log(b)-\log(m)}$
7. Exponential likelihood: $f_{\mathbf{x}\|\theta}(\mathbf{x}\|\lambda) \propto \begin{cases} \lambda^n e^{-n\bar{x}\lambda} & \begin{pmatrix} \text{if } x_i > 0, \\ \text{for } i = 1,\ldots,n \end{pmatrix} \\ 0 & \text{otherwise,} \end{cases}$ where $\bar{x} = \frac{1}{n}\sum_{i=1}^n x_i$	$\theta = \lambda$	$\theta \sim G(\alpha,\beta)$ with $\alpha > 0$ and $\beta > 0$ $g(\theta) = \begin{cases} \frac{\theta^{\alpha-1}e^{-\theta/\beta}}{\Gamma(\alpha)\beta^\alpha} & \text{if } \theta > 0 \\ 0 & \text{otherwise} \end{cases}$	$\theta\|\mathbf{x} \sim G(\alpha^{'},\beta^{'})$, where $\alpha^{'} = \alpha + n,\ \beta^{'} = \frac{\beta}{1+n\beta\bar{x}}$
8. Gamma likelihood: $\mathbf{x}\|\frac{1}{b} \sim G(a,b)$ $f_{\mathbf{x}\|\theta}(\mathbf{x}\|\frac{1}{b}) \propto \begin{cases} \frac{e^{-\frac{n\bar{x}}{b}}}{b^{na}} & \begin{pmatrix} \text{if } x_i > 0, \\ \text{for } i = 1,\ldots,n \end{pmatrix} \\ 0 & \text{otherwise,} \end{cases}$	$\theta = \frac{1}{b}$	$\theta \sim G(\alpha,\beta)$ with $\alpha > 0$ and $\beta > 0$ $g(\theta) = \begin{cases} \frac{\theta^{\alpha-1}e^{-\theta/\beta}}{\Gamma(\alpha)\beta^\alpha} & \text{if } \theta > 0 \\ 0 & \text{otherwise} \end{cases}$	$\theta\|\mathbf{x} \sim G(\alpha^{'},\beta^{'})$, where $\alpha^{'} = \alpha + na,\ \beta^{'} = \frac{\beta}{1+\beta n\bar{x}}$
9. Gamma likelihood: $\mathbf{x}\|a \sim G(a,b)$ $f_{\mathbf{x}\|\theta}(\mathbf{x}\|a) \propto \begin{cases} \frac{m^{a-1}}{b^{-na}\Gamma(a)^n} & \begin{pmatrix} \text{if } x_i > 0, \\ \text{for } i = 1,\ldots,n \end{pmatrix} \\ 0 & \text{otherwise,} \end{cases}$ where $m = \prod_{i=1}^n x_i$	$\theta = a$	$g(\theta) = \begin{cases} \frac{\alpha^{\theta-1}b^{\theta\gamma}}{K\Gamma(\theta)^\beta} & \text{if } \theta > 0 \\ 0 & \text{otherwise,} \end{cases}$ where $\alpha,\beta,\gamma > 0$ and $K = \int_0^\infty \frac{\alpha^{\theta-1}b^{\theta\gamma}}{\Gamma(\theta)^\beta}\mathrm{d}\theta$	$\theta\|\mathbf{x} \sim \begin{cases} \frac{(\alpha^{'})^{\theta-1}b^{\theta\gamma^{'}}}{K\Gamma(\theta)^{\beta^{'}}} & \text{if } \theta > 0 \\ 0 & \text{otherwise,} \end{cases}$, where $\alpha^{'} = \alpha m,\ \beta^{'} = \beta + n,\ \gamma^{'} = \gamma + n$

TABLE C.1 Cont'd

10. Gaussian likelihood: $\mathbf{x}\|a \sim \mathcal{N}(\mu,\sigma^2)$ $f_{\mathbf{x}\|\theta}(\mathbf{x}\|\mu) \propto e^{-\frac{n}{2\sigma^2}(\bar{x}-\mu)^2}$ where $\bar{x} = \frac{1}{n}\sum_{i=1}^{n} x_i$	$\theta = \mu$	$\theta \sim \mathcal{N}(m,\nu^2)$ $g(\theta) = \frac{1}{\sqrt{2\pi\nu^2}} e^{-\frac{(\theta-m)^2}{2\nu^2}}$	$\theta\|\mathbf{x} \sim \mathcal{N}(m',(\nu')^2)$, where $m' = \frac{\frac{m}{\nu^2}+\frac{n\bar{x}}{\sigma^2}}{\frac{1}{\nu^2}+\frac{n}{\sigma^2}}$, $(\nu')^2 = \nu^2 + \frac{n}{\sigma^2}$
11. Gaussian likelihood: $\mathbf{x}\|\sigma^2 \sim \mathcal{N}(\mu,\sigma^2)$ $f_{\mathbf{x}\|\theta}(\mathbf{x}\|\sigma^2) \propto \frac{e^{-\frac{1}{2\sigma^2}\sum_{i=1}^{n}(x_i-\mu)^2}}{(\sigma^2)^{n/2}}$	$\theta = \sigma^2$	$\partial \sim G(\alpha,\beta)$, with $\alpha > 0$ and $\beta > 0$ $g(\theta) = \begin{cases} \frac{\theta^{\alpha-1}e^{-\theta/\beta}}{\Gamma(\alpha)\beta^{\alpha}} & \text{if } \theta > 0 \\ 0 & \text{otherwise} \end{cases}$	$\theta\|\mathbf{x} \sim G(\alpha',\beta'$, where $\alpha' = \alpha + \frac{n}{2}$ $\beta' = \beta + \frac{\sum_{i=1}^{n}(x_i-\mu)^2}{2}$
12. Gaussian likelihood: $\mathbf{x}\|(\mu,\sigma^2) \sim \mathcal{N}(\mu,\sigma^2)$ $f_{\mathbf{x}\|\theta}(\mathbf{x}\|(\mu,\sigma^2)) \propto \frac{1}{\sigma^n} e^{-\frac{1}{2\sigma^2}\sum_{i=1}^{n}(x_i-\mu)^2}$	$\theta = (\theta_1,\theta_2)$ $\theta_1 = \mu$, $\theta_2 = \sigma^2$	$\theta = (\theta_1,\theta_2)$, $\theta_2 \sim G(\alpha,\beta)$, $(\theta_1\|\theta_2) \sim \mathcal{N}(m,\tau\theta_2^2)$, $g(\theta) = g(\theta_1\|\theta_2)g(\theta_2)$	$\theta\|\mathbf{x} \sim (\theta_1,\theta_2)\|\mathbf{x}$, where $\theta_2\|\mathbf{x} \sim G(\alpha',\beta')$, $\theta_1\|(\theta_2,\mathbf{x}) \sim \mathcal{N}(m',\tau'\theta_2^2)$, where $\alpha' = \alpha + \frac{n}{2}$, $\beta' = \left(\frac{1}{\beta} + \frac{\sum_{i=1}^{n}(x_i-\mu)^2}{2} + \frac{n\tau(\bar{x}-\mu)^2}{2(n+\tau)}\right)^{-1}$, $m' = \frac{m\tau+n\bar{x}}{n+\tau}$, $\tau' = n+\tau$
13. Lognormal likelihood: $\mathbf{x}\|\mu \sim lognormal(\mu,\sigma^2)$ $f_{\mathbf{x}\|\theta}(\mathbf{x}\|\mu) \propto e^{-\frac{n}{2\sigma^2}\left(\frac{\sum_{i=1}^{n}\log x_i-\mu}{n}\right)^2}$	$\theta = \mu$	$\theta \sim \mathcal{N}(m,\nu^2)$ $g(\theta) = \frac{1}{\sqrt{2\pi\nu^2}} e^{-\frac{(\theta-m)^2}{2\nu^2}}$	$\theta\|\mathbf{x} \sim \mathcal{N}(m',(\nu')^2$, where $m' = \frac{\frac{m}{\nu^2}+\frac{\sum_{i=1}^{n}x_i}{\sigma^2}}{\frac{1}{\nu^2}+\frac{n}{\sigma^2}}$, $(\nu')^2 = \nu^2 + \frac{n}{\sigma^2}$

(Continued)

TABLE C.1 Cont'd

Form of the Likelihood Function as a Function of the Parameter of Interest	Parameter	Conjugate prior	Posterior distribution of the parameter $f_{\theta\|\mathbf{x}}(\theta\|\underline{x})$
14. Lognormal likelihood: $\mathbf{x}\|\mu \sim lognormal(\mu, \sigma^2)$ $f_{\mathbf{x}\|\theta}(\mathbf{x}\|\sigma^2) \propto \frac{e^{-\frac{1}{2\sigma^2}\sum_{i=1}^{n}(\log x_i - \mu)^2}}{\sigma^n}$	$\theta = \sigma^2$	$\theta \sim G(\alpha, \beta)$ with $\alpha > 0$ and $\beta > 0$ $g(\theta) = \begin{cases} \frac{\theta^{\alpha-1}e^{-\theta/\beta}}{\Gamma(\alpha)\beta^{\alpha}} & \text{if } \theta > 0 \\ 0 & \text{otherwise} \end{cases}$	$\theta\|\mathbf{x} \sim G(\alpha', \beta')$, where $\alpha' = \alpha + \frac{n}{2}$ $\beta' = \beta + \frac{\sum_{i=1}^{n}(x_i - \mu)^2}{2}$
15. Lognormal likelihood: $\mathbf{x}\|(\mu, \sigma^2) \sim lognormal(\mu, \sigma^2)$ $f_{\mathbf{x}\|\theta}(\mathbf{x}\|(\mu, \sigma^2)) \propto \frac{1}{\sigma^n} e^{-\frac{1}{2\sigma^2}\sum_{i=1}^{n}(\log x_i - \mu)^2}$	$\theta = (\theta_1, \theta_2)$ $\theta_1 = \mu$, $\theta_2 = \sigma^2$	$\theta = (\theta_1, \theta_2)$, $\theta_2 \sim G(\alpha, \beta)$, with $\alpha > 0$ and $\beta > 0$ $g(\theta_1) = \sqrt{\frac{\tau\beta}{2\pi}} \frac{\Gamma(2\alpha+1)(1+\frac{\tau\beta}{2}(\theta_1 - m)^2)^{-\frac{2\alpha+1}{2}}}{\Gamma(\alpha)}$	$\theta_2\|\mathbf{x} \sim G(\alpha', \beta')$, $g(\theta_1\|\mathbf{x}) = \sqrt{\frac{\tau'\beta}{2\pi}} \frac{\Gamma(2\alpha+1)(1+\frac{\tau'\beta}{2}(\theta_1 - m')^2)^{-\frac{2\alpha+1}{2}}}{\Gamma(\alpha)}$, where $\alpha' = \alpha + \frac{n}{2}$, $\beta' = \left(\frac{1}{\beta} + \frac{\sum_{i=1}^{n}(\log x_i - \mu)^2}{2} + \frac{n\tau}{2}\frac{(\bar{x} - \mu)^2}{\tau + n}\right)^{-1}$, $m' = \frac{m\tau + n\bar{x}}{n + \tau}$, $\tau' = n + \tau$

REFERENCES

1. J. Mitola III and G. Q. Maguire, Jr., "Cognitive radio: Making software radios more personal," *IEEE Personal Communications*, vol. 6, no. 4, pp. 13–18, Aug. 1999.
2. R. J. Lackey and D. W. Upmal, "Speakeasy: The military software radio," *IEEE Communications Magazine*, vol. 33, no. 5, pp. 56–61, May 1995.
3. F. Torre, "Speakeasy—A new direction in tactical communications for the 21st century," in *Proceedings of the Tactical Communications Conference*, vol. 1, Fort Wayne, IN, Apr. 1992, pp. 139–142.
4. J. Mitola, "Software radios: Survey, critical evaluation and future directions," in *Proceedings of the National Telesystems Conference*, Washington, DC, May 1992.
5. P. Johnson, "New research lab leads to unique radio receiver," *E-Systems Team*, vol. 5, no. 4, pp. 6–7, May 1985.
6. H. Arslan and H. Celebi, "Software defined radio architectures for cognitive radios," in *Cognitive Radio, Software Defined Radio, and Adaptive Wireless Systems*, H. Arslan, Ed. New York: Springer, 2007.
7. P. G. Cook and W. Bonser, "Architectural overview of the SPEAKeasy system," *IEEE Journal on Selected Areas in Communications*, vol. 17, no. 4, pp. 650–661, Apr. 1999.
8. R. Vidano, "Speakeasy II—An IPT approach to software programmable radio development," in *Proceeding of the Military Communications Conference (MILCOM'97)*, vol. 3, Monterey, CA, 1997, pp. 1212–1215.
9. D. R. Stephens, B. Salisbury, and K. Richardson, "JTRS infrastructure architecture and standards," in *Proceedings of the 2006 IEEE Military Communications Conference (MILCOM 2006)*, Washington, DC, Oct. 2006, pp. 1–5.

Signal Processing for Cognitive Radios, First Edition. Sudharman K. Jayaweera.

10. Joint Tactical Radio System (JTRS), "Software communications architecture specification," Technical Report, Joint Program Executive Office (JPEO), San Diego, CA. 2006. Available at http://jtnc.mil/sca/Documents/SCAv2_2_2/SCA_version_2_2_2.pdf.
11. J. Place, D. Kerr, and D. Schaefer, "Joint tactical radio system," in *IEEE Military Communications Conference (MILCOM 2000)*, vol. 1, Los Angeles, CA, 2000, pp. 209–213.
12. A. M. Wyglinski, M. Nekovee, and Y. T. Hou, *Cognitive Radio Communications and Networks: Principles and Practice*. Burlington, MA: Academic Press, 2009.
13. S. Muir, J. Yedetore, L. Stich, and J. Chapin, "Integration of SDR cellular basestations into military telephone networks," in *IEEE Military Communications Conference (MILCOM 2007)*, vol. 1, Orlando, FL, 2007, pp. 1–7.
14. FCC, "Report of the spectrum efficiency working group," Technical Report, FCC Spectrum Policy Task Force, Nov. 2002. Available at http://transition.fcc.gov/sptf/files/SEWGFinalReport_1.pdf.
15. S. Haykin, "Cognitive radio: Brain-empowered wireless communications," *IEEE Journal on Selected Areas in Communications*, vol. 23, no. 2, pp. 201–220, Feb. 2005.
16. S. K. Jayaweera and C. G. Christodoulou, "Radiobots: Architecture, algorithms and realtime reconfigurable antenna designs for autonomous, self-learning future cognitive radios," Technical Report EECE-TR-11-0001. University of New Mexico, Albuquerque, NM, Mar. 2011. Available at http://repository.unm.edu/handle/1928/12306.
17. X. Wang and H. V. Poor, *Wireless Communications Systems: Advanced Techniques for Signal Reception*. Upper Saddle River, NJ: Prentice Hall, 2004.
18. A. F. Molisch, *Wireless Communications*. Hoboken, NJ: Wiley, 2005.
19. S. G. Glisic, *Advanced Wireless Communications: 4G Cognitive and Cooperative Broadband Technology*. Hoboken, NJ: Wiley, 2007.
20. J. Mitola III, "Cognitive radio: An integrated agent architecture for software defined radio," Ph.D. dissertation, KTH, Stockholm, Sweden, May 2000.
21. P. Michelon, "What are cognitive abilities and skills?", Dec. 2006. Available at http://www.sharpbrains.com/blog/2006/12/18/what-are-cognitive-abilities/. Accessed July 5, 2014.
22. S. S. Sparrow and S. M. Davis, "Recent advances in the assessment of intelligence and cognition," *Journal of Child Psychology Psychiatry*, vol. 41, no. 1, pp. 117–131, 2000.
23. R. S. Michalski, "Learning and cognition," in *2nd International World Conference on the Fundamentals of Artificial Intelligence*, Paris, France, July 1995.
24. H. Sato, "Two-user communication channels," *IEEE Transactions on Information Theory*, vol. IT-23, no. 3, pp. 295–304, May 1977.
25. T. S. Han and K. Kobayashi, "A new achievable rate region for the interference channel," *IEEE Transactions on Information Theory*, vol. IT-27, no. 1, pp. 49–60, Jan. 1981.
26. T. M. Cover and J. A. Thomas, *Elements of Information Theory*. Hoboken, NJ: John Wiley & Sons Inc., 2006.
27. Q. Zhao and B. M. Sadler, "A survey of dynamic spectrum access," *IEEE Signal Processing Magazine*, vol. 24, no. 3, pp. 79–89, May 2007.
28. S. K. Jayaweera and T. Li, "Dynamic spectrum leasing in cognitive radio networks via primary-secondary user power control games," *IEEE Transactions on Wireless Communications*, vol. 8, no. 6, pp. 3300–3310, June 2009.

29. S. K. Jayaweera and C. Mosquera, "A dynamic spectrum leasing (DSL) framework for spectrum sharing in cognitive radio networks," in *43rd Annual Asilomar Conference on Signals, Systems and Computers*, Pacific Grove, CA, Nov. 2009.
30. S. K. Jayaweera, G. Vazquez-Vilar, and C. Mosquera, "Dynamic spectrum leasing (DSL): A new paradigm for spectrum sharing in cognitive radio networks," *IEEE Transactions on Vehicular Technology*, vol. 59, no. 5, pp. 2328–2339, May 2010.
31. O. Simeone, I. Stanojev, S. Savazzi, Y. Bar-Ness, U. Spagnolini, and R. Pickholtz, "Spectrum leasing to cooperating secondary ad hoc networks," *IEEE Journal on Selected Areas in Communications*, vol. 26, pp. 203–213, Jan. 2008.
32. N. Devroye, P. Mitran, and V. Tarokh, "Limits on communications in a cognitive radio channel," *IEEE Communications Magazine*, vol. 44, no. 6, pp. 44–49, June 2006.
33. S. K. Jayaweera and M. Bkassiny. K. Avery, "Asymmetric cooperative communications based spectrum leasing via auctions in cognitive radio networks," *IEEE Transactions on Wireless Communications*, vol. 10, no. 8, pp. 2716–2724, Aug. 2011.
34. D. Pu, A. M. Wyglinski, and M. McLernon, "A frequency rendezvous approach for decentralized dynamic spectrum access networks," in *4th International Conference on Cognitive Radio Oriented Wireless Networks and Communications (CROWNCOM '09)*, Hannover, Germany, June 2009.
35. N. C. Theis and R. Thomas. L, "Rendezvous for cognitive radios," *IEEE Transactions on Mobile Computing*, vol. 10, no. 2, pp. 216–227, Mar. 2010.
36. President's Council of Advisors on Science and Technology (PCAST), "Realizing the full potential of government-held spectrum to spur economic growth," Technical Report. Executive Office of the President, Washington, DC, July 2012.
37. A. J. Johansson, "Wireless communication with medical implants: Antennas and propagation," Ph.D. dissertation, Lund University, Lund, Sweden, 2004.
38. P. Phunchongharn, E. Hossain, D. Niyato, and S. Camorlinga, "A cognitive radio system for E-health applications in a hospital environment," *IEEE Wireless Communications*, vol. 17, no. 1, pp. 20–28, Feb. 2010.
39. R. Doost-Mohammady and K. R. Chowdhury, "Transforming healthcare and medical telemetry through cognitive radio networks," *IEEE Wireless Communications*, vol. 19, no. 4, pp. 67–73, Aug. 2012.
40. S. C. Prabhakar, R. de Francisco, and G. Leus, "Performance evaluation of an IEEE 802.15.4 cognitive radio link in the 2360-2400 MHz band," in IEEE *Wireless Communications and Networking Conference (WCNC)*, Eindhoven, the Netherlands, Mar. 2011, pp. 2155–2160.
41. P. Conder, L. Linton, and M. Faulkner, "Cognitive radio developments for emergency communication systems," in *IEEE International Symposium on Dynamic Spectrum Access Networks (DySPAN 2011)*, Aachen, Germany, May 2011, pp. 658–659.
42. P. Pawelczak, R. V. Prasad, L. Xia, and I. G. M. M. Niemegeers, "Cognitive radio emergency networks—Requirements and design," in *First IEEE International Symposium on New Frontiers in Dynamic Spectrum Access Networks (DySPAN 2005)*, Baltimore, MD, Nov. 2005, pp. 601–606.
43. W. Wang, W. Gao, X. Bai, T. Peng, G. Chuai, and W. Wang, "A framework of wireless emergency communications based on relaying and cognitive radio," in *The 18th Annual IEEE International Symposium on Personal, Indoor and Mobile Radio Communications (PIMRC'07)*, Athens, Greece, 2007.

44. W. Wang, S. Bhattacharjee, M. Chatterjee, and K. Kwiat, "Collaborative jamming and collaborative defense in cognitive radio networks," *Pervasive and Mobile Computing*, vol. 9, no. 4, 572–587, 2013.
45. Y. Wu, B. Wang, and K. Liu, "Optimal defense against jamming attacks in cognitive radio networks using the Markov decision process approach," in *2010 IEEE Global Telecommunications Conference (GLOBECOM 2010)*, Miami, FL, Dec. 2010, pp. 1–5.
46. A. Sampath, H. Dai, H. Zheng, and B. Zhao, "Multi-channel jamming attacks using cognitive radios," in *Proceedings of 16th International Conference on Computer Communications and Networks (ICCCN 2007)*, Honolulu, HI, Aug. 2007, pp. 352–357.
47. Q. Zhu, H. Li, Z. Han, and T. Basar, "A stochastic game model for jamming in multi-channel cognitive radio systems," in *2010 IEEE International Conference on Communications (ICC 2010)*, Cape Town, South Africa, May 2010, pp. 1–6.
48. T. X. Brown and A. Sethi, "Potential cognitive radio denial of service attacks and remedies," in *Proceedings of the 2007 International Symposium on Advanced Radio Technologies (ISART 2007)*, Boulder, CO, Feb. 2007, pp. 26–28.
49. R. S. Prabhu, E. Valles, and P. Dafesh, "Software defined radio forum," in *Proceedings of Software Defined Radio (SDR) 2009 Technical Conference*, Washington, DC, Dec. 2009.
50. B. F. Lo and I. F. Akyildiz, "Multiagent jamming-resilient control channel game for cognitive radio Ad Hoc Networks," in *2012 IEEE International Conference on Communications (ICC 2012)*, Ottawa, ON, June 2012.
51. H. V. Poor, *An Introduction to Signal Detection and Estimation*. New York: Springer-Verlag, 1994.
52. S. A. Kassam, *Signal Detection in Non-Gaussian Noise*. New York: Springer-Verlag, 1988.
53. H. L. Van Trees, *Detection, Estimation, and Modulation Theory (Part I)*. Hoboken, NJ: Wiley-Interscience, 2001.
54. S. M. Kay, *Fundamentals of Statistical Signal Processing*, Volume 2: Detection Theory. Upper Saddle River, NJ: Prentice Hall, 1998.
55. S. Verdu, *Multiuser Detection*. Cambridge, UK: Cambridge University Press, 1998.
56. S. M. Kay, *Fundamentals of Statistical Signal Processing*, Volume 1: Estimation Theory. Upper Saddle River, NJ: Prentice Hall, 1993.
57. B. D. O. Anderson and J. B. Moore, *Optimal Filtering*. Englewood Cliffs, NJ: Prentice Hall, 1979.
58. E. L. Lehmann, *Testing Statistical Hypotheses*. New York: Wiley, 1986.
59. M. B. Priestley, *Spectral Analysis and Time Series*. London, UK: Academic Press, 1981.
60. W. A. Gardner, *Statistical Spectral Analysis*. Englewood Cliffs, NJ: Prentice Hall Inc., 1988.
61. P. Stoica and R. Moses, *Spectral Analysis of Signals*. Upper Saddle River, NJ: Pearson Prentice Hall, 2005.
62. A. Leon-Garcia, *Probability, Statistics and Random Processes for Electrical Engineering*. Upper Saddle River, NJ: Pearson Prentice Hall, 2008.
63. J. G. Proakis and D. G. Manolakis, *Digital Signal Processing: Principles, Algorithms and Applications*. Upper Saddle River, NJ: Prentice Hall International, 1996.

64. P. D. Welch, "The use of fast Fourier transform for the estimation of power spectra: A method based on time averaging over short modified periodograms," *IEEE Transactions on Audio and Electroacoustics*, vol. AU-15, pp. 70–73, June 1967.
65. W. Gardner, "Spectral correlation of modulated signals: Part I—Analog modulation," *IEEE Transactions on Communications*, vol. 35, no. 6, pp. 584–594, June 1987.
66. W. Gardner, W. Brown, and C.-K. Chen, "Spectral correlation of modulated signals: Part II—Digital modulation," *IEEE Transactions on Communications*, vol. 35, no. 6, pp. 595–601, June 1987.
67. W. Gardner, A. Napolitano, and L. Paura, "Cyclostationariy: Half a century of research," *Elsevier Signal Processing*, vol. 86, pp. 639–697, Sept. 2006.
68. M. L. Puterman, *Markov Decision Processes: Discrete Stochastic Dynamic Programming*. New York: John Wiley & Sons, 1994.
69. R. S. Sutton and A. G. Barto, *Reinforcement Learning: An Introduction*. Cambridge, MA: MIT Press, 1998.
70. A. R. Cassandra, "Exact and approximate algorithms for partially observable Markov decision processes," Ph.D. dissertation, Brown University, Providence, RI, 1998.
71. R. D. Smallwood and E. J. Sondik, "The optimal control of partially observable Markov processes over a finite horizon," *Operations Research*, vol. 21, pp. 1071–1088, 1973.
72. D. P. Bertsekas, *Dynamic Programming and Optimal Control*, 3rd ed., Volume 1. Belmont, MA: Athena Scientific, 2005.
73. C. Striebel, "Sufficient statistics in the control of stochastic systems," *Journal of Mathematical Analysis and Applications*, vol. 12, pp. 576–592, 1965.
74. G. E. Monahan, "A survey of partially observable Markov decision processes theory, models and algorithms," *Management Science*, vol. 28, no. 1, pp. 1–16, Jan. 1982.
75. E. J. Sondik, "The optimal control of partially observable Markov processes," Ph.D. dissertation, Stanford University, Stanford, CA, 1971.
76. L. P. Kaelbling, M. L. Littman, and A. R. Cassandra, "Planning and acting in partially observable stochastic domains," *Artificial Intelligence*, vol. 101, pp. 99–134, 1998.
77. K. Sawaki and A. Ichikawa, "Optimal control for partially observable Markov decision processes over an infinite horizon," *Journal of the Operational Research Society*, vol. 21, no. 1, pp. 1–4, 1978.
78. E. J. Sondik, "The optimal control of partially observable Markov processes over the infinite horizon: Discounted costs," *Operations Research*, vol. 26, no. 2, pp. 282–304, 1978.
79. R. O. Duda, P. E. Hart, and S. G. Stork, *Pattern Classification*. New York: John Wiley & Sons, Inc., 2001.
80. E. A. Parzen, "On estimation of a probability density function and mode," *Annals of Mathematical Statistics*, vol. 33, no. 3, pp. 1065–1076, 1962.
81. D. F. Specht, "Generation of polynomial discriminant functions for pattern recognition," *IEEE Transactions on Electronics Computers*, vol. EC-16, no. 3, pp. 308–319, 1967.
82. E. Fix and J. L. Hodges Jr., "Discriminatory analysis: Nonparametric discrimination: Consistency properties," *USAF School of Aviation Medicine*, vol. 4, pp. 261–279, 1951.
83. E. Fix and J. L. Hodges Jr., "Discriminatory analysis: Nonparametric discrimination: Small sample performance," *USAF School of Aviation Medicine*, vol. 11, pp. 280–322, 1952.

84. E. A. Patrick and F. P. Fischer, III, "A generalized k-nearest neighbor rule," *Information and Control*, vol. 16, no. 2, pp. 128–152, 1970.
85. C. M. Bishop, *Pattern Recognition and Machine Learning*. New York: Springer Science + Business Media, LLC., 2006.
86. K. Fukunaga, *Introduction to Statistical Pattern Recognition*. New York: Academic Press, 1972.
87. E. A. Patrick, *Fundamentals of Pattern Recognition*. Englewood Cliffs, NJ: Prentice-Hall, 1972.
88. J. B. MacQueen, "Some methods for classification and analysis of multivariate observations," in *Proceedings of 5th Berkeley Symposium on Mathematical Statistics and Probability*, University of California Press, Berkeley, CA, 1967, pp. 281–297.
89. S. Lloyd, "Least squares quantization in PCM," *IEEE Transactions on Information Theory*, vol. IT-28, no. 2, pp. 129–137, Mar. 1982.
90. J. Max, "Quantizing for minimum distortion," *IRE Transactions Information Theory*, vol. IT-6, pp. 7–12, 1960.
91. D. Pelleg and A. Moore, "X-means: Extending k-means with efficient estimation of the number of clusters," in *Seventh International Conference on Machine Learning (ICML '00)*, Stanford, CA, June–July 2000.
92. Y. W. Teh, M. I. Jordan, M. J. Beal, and D. M. Blei, "Hierarchical Dirichlet processes," *Journal of the American Statistical Association*, vol. 101, no. 476, pp. 1566–1581, 2006. Available at http://www.jstor.org/stable/27639773.
93. E. Fox, E. Sudderth, M. Jordan, and A. Willsky, "Bayesian nonparametric methods for learning Markov switching processes," *IEEE Signal Processing Magazine*, vol. 27, no. 6, pp. 43–54, Nov. 2010.
94. E. Fox, E. Sudderth, M. Jordan, and A. Willsky , "Bayesian nonparametric inference of switching dynamic linear models," *IEEE Transactions on Signal Processing*, vol. 59, no. 4, pp. 1569–1585, Apr. 2011.
95. N. Bouguila and D. Ziou, "A Dirichlet process mixture of Dirichlet distributions for classification and prediction," in *IEEE Workshop on Machine Learning for Signal Processing (MLSP '08)*, Cancún, Mexico, Oct. 2008, pp. 297–302.
96. M. D. Escobar, "Estimating normal means with a Dirichlet process prior," *Journal of the American Statistical Association*, vol. 89, no. 425, pp. 268–277, Mar. 1994. Available at http://www.jstor.org/stable/2291223.
97. M. D. Escobar and M. West, "Bayesian density estimation and inference using mixtures," *Journal of the American Statistical Association*, vol. 90, no. 430, pp. 577–588, 1995. Available at http://www.jstor.org/stable/2291069.
98. K. Ishiguro, T. Yamada, S. Araki, T. Nakatani, and H. Sawada, "Probabilistic speaker diarization with bag-of-words representations of speaker angle information," *IEEE Transactions on Audio, Speech, and Language Processing*, vol. 20, no. 2, pp. 447–460, Feb. 2012.
99. A. H. Harati Nejad Torbati, J. Picone, and M. Sobel, "Applications of Dirichlet process mixtures to speaker adaptation," in *IEEE International Conference on Acoustics, Speech and Signal Processing (ICASSP '12)*, Kyoto, Japan, Mar. 2012, pp. 4321–4324.
100. S. Li, Z. Yanning, M. Miao, and T. Guangjian, "SAR image segmentation method using DP mixture models," in *International Symposium on Computer Science and*

Computational Technology (ISCSCT '08), vol. 2, Shanghai, China, Dec. 2008, pp. 598–601.

101. M. Chen, J. Silva, J. Paisley, C. Wang, D. Dunson, and L. Carin, "Compressive sensing on manifolds using a nonparametric mixture of factor analyzers: Algorithm and performance bounds," *IEEE Transactions on Signal Processing*, vol. 58, no. 12, pp. 6140–6155, Dec. 2010.
102. M. Bkassiny, S. K. Jayaweera, and Y. Li, "Multidimensional Dirichlet process-based non-parametric signal classification for autonomous self-learning cognitive radios," *IEEE Transaction on Wireless Communications*, vol. 12, no. 11, pp. 5413–5423, Nov. 2013.
103. R. G. Gallager, *Principles of Digital Communications*. Cambridge, UK: Cambridge University Press, 2008.
104. J. E. Cavanaugh, "Lecture notes on model selection," Sept. 2012, unpublished.
105. G. Schwarz, "Estimating the dimension of a model," *The Annals of Statistics*, vol. 6, no. 2, pp. 461–464, 1978.
106. H. Akaike, "A new look at the statistical identification model," *IEEE Transactions on Automatic Control*, vol. 19, pp. 716–723, 1974.
107. A. R. Liddle, "Information criteria for astrophysical model selection," *Monthly Notices of the Royal Astronomical Society: Letters*, vol. 377, no. 1, pp. L74–L78, 2007.
108. T. S. Ferguson, "A Bayesian analysis of some nonparametric problems," *Annals of Statistics*, vol. 1, no. 2, pp. 209–230, 1973.
109. Y. W. Teh, "Dirichlet processes," in *Encyclopedia of Machine Learning*. Editors: C. Sammut and G. I. Webb, New York: Springer, 2010.
110. S. S. Wilks, *Mathematical Statistics*. New York: Wiley, 1962.
111. J. Sethuraman, "A constructive definition of Dirichlet priors," *Statistica Sinica*, vol. 4, pp. 639–650, 1994.
112. D. Blackwell and J. B. MacQueen, "Ferguson distributions via polya urn schemes," *The Annals of Statistics*, vol. 1, no. 2, pp. 353–355, 1973. Available at http://www.jstor.org/stable/2958020.
113. D. Aldous, "Exchangeability and related topics," *Ecole d'ete de Probabilites de Saint-Flour XIII-1983*, vol. 1117, pp. 1–198, 1985.
114. B. de Finetti, "Funzione caratteristica di un fenomeno aleatorio," in *Atti della R. Academia Nazionale dei Lincei, Serie 6. Memorie, Classe di Scienze Fisiche, Mathematice e Naturale*, vol. 4. 1931, pp. 251–299.
115. E. Hewitt and L. J. Savage, "Symmetric measures on Cartesian products," *Transactions of the American Mathematical Society*, vol. 80, pp. 470–501, 1955.
116. J. F. C. Kingman, "Uses of exchangeability," *Annals of Probability*, vol. 6, pp. 183–197, 1978.
117. P. Diaconis and D. Freedman, "Finite exchangeable sequences," *Annals of Probability*, vol. 8, pp. 745–764, 1980.
118. S. Geman and D. Geman, "Stochastic relaxation, Gibbs distributions, and the Bayesian restoration of images," *IEEE Transactions on Pattern Analysis and Machine Intelligence*, vol. 6, no. 6, pp. 721–741, Nov. 1984.
119. D. Fink, "A compendium of conjugate priors," DOE Progress Report, Montana State University, Bozeman, MT, May 1997.

120. M. West, P. Muller, and M. D. Escobar, “Hierarchical priors and mixture models, with applications in regression and density estimation.” In *Aspects of Uncertainty*. Editors: P. R. Freeman and A. F. M. Smith, New York: Wiley, 1994, pp. 363–386.

121. S. N. MacEachern and P. Muller, “Estimating mixture of Dirichlet process models,” *Journal of Computational and Graphical Statistics*, vol. 7, pp. 223–238, 1998.

122. S. N. MacEachern, “Computational methods for mixture of Dirichlet process models.” In D. Day, P. Muller, and D. Sinha (eds.), *Practical Nonparametric and Semiparametric Bayesian Statistics*. New York: Springer-Verlag, 1998, pp. 23–43.

123. S. Walker and P. Damien, “Sampling methods for Bayesian nonparametric inference involving stochastic processes.” In D. Day, P. Muller, and D. Sinha (eds.), *Practical Nonparametric and Semiparametric Bayesian Statistics*. New York: Springer-Verlag, 1998, pp. 243–254.

124. R. M. Neal, “Markov chain sampling methods for Dirichlet process mixture models,” *Journal of Computational and Graphical Statistics*, vol. 9, no. 2, pp. 249–265, June 2000.

125. M. West, “Hyperparameter estimation in Dirichlet process mixture models,” Technical Report, Duke University, Durham, NC, 1992.

126. A. Rabaoui, N. Viandier, J. Marais, and E. Duflos, “On selecting the hyperparameters of the DPM models for the density estimation of observation errors,” in *IEEE International Conference on Acoustics, Speech and Signal Processing (ICASSP ’11)*, Prague, Czech Republic, May 2011, pp. 4092–4095.

127. N. Devroye, M. Vu, and V. Tarokh, “Cognitive radio networks,” *IEEE Signal Processing Magazine*, vol. 25, pp. 12–23, Nov. 2008.

128. S. K. Jayaweera, Y. Li, M. Bkassiny, C. G. Christodoulou, and K. A. Avery, “Radiobots: The autonomous, self-learning future cognitive radios,” in *2011 International Symposium on Intelligent Signal Processing and Communications Systems (ISPACS)*, Chiang Mai, Thailand, Dec. 2011.

129. Q. Zhao, L. Tong, A. Swami, and Y. Chen, “Decentralized cognitive MAC for opportunistic spectrum access in ad hoc networks: A POMDP framework,” *IEEE Journal on Selected Areas in Communications*, vol. 25, no. 3, pp. 589–600, April 2007.

130. Y. Chen, Q. Zhao, and A. Swami, “Joint design and separation principle for opportunistic spectrum access in the presence of sensing errors,” *IEEE Transactions on Information Theory*, vol. 54, no. 5, pp. 2053–2071, May 2008.

131. J. Unnikrishnan and V. Veeravalli, “Cooperative sensing for primary detection in cognitive radio,” *IEEE Journal of Selected Topics in Signal Processing*, vol. 2, no. 1, pp. 18–27, Feb. 2008.

132. S. Ahmad, M. Liu, T. Javidi, Q. Zhao, and B. Krishnamachari, “Optimality of myopic sensing in multichannel opportunistic access,” *IEEE Transactions on Information Theory*, vol. 55, no. 9, pp. 4040–4050, Sept. 2009.

133. Y. Li, S. K. Jayaweera, M. Bkassiny, and K. A. Avery, “Optimal myopic sensing and dynamic spectrum access in cognitive radio networks with low-complexity implementations,” *IEEE Transactions on Wireless Communications*, vol. 11, no. 7, pp. 2412–2423, July 2012.

134. C. Ghosh, S. Roy, and M. B. Rao, “Modeling and validation of channel idleness and spectrum availability for cognitive networks,” *IEEE Journal on Selected Areas in Communications*, vol. 30, no. 10, pp. 2029–2039, Nov. 2012.

135. C. Watkins, "Learning from delayed rewards," Ph.D. dissertation, Cambridge University, Cambridge, UK, 1989.
136. C. Watkins and P. Dayan, "Q-learning," *Machine Learning*, vol. 8, no. 3, pp. 279–292, 1992.
137. T. Jaakkola, M. I. Jordan, and S. Singh, "On the convergence of stochastic iterative dynamic programming algorithms," *Neural Computation*, vol. 6, no. 6, 1185–1201, 1994.
138. A. Galindo-Serrano and L. Giupponi, "Distributed Q-learning for aggregated interference control in cognitive radio networks," *IEEE Transactions on Vehicular Technology*, vol. 59, no. 4, pp. 1823–1834, May 2010.
139. Y. Li, S. K. Jayaweera, C. Ghosh, and M. Bkassiny, "Learning-aided sensing scheduling for wideband cognitive radios," in *1st International Workshop on Wideband Mobile Cognitive Radios in 2013 IEEE Vehicular Technology Conference*, Las Vegas, NV, Sept. 2013.
140. Y. Li, S. K. Jayaweera, M. Bkassiny, and C. Ghosh, "Learning-aided sub-band selection algorithms for spectrum sensing in wide-band cognitive radios," *IEEE Transactions on Wireless Communications*, vol. 13, no. 4, pp. 2012–2024, April 2014.
141. Z. Tian and G. B. Giannakis, "A wavelet approach to wideband spectrum sensing for cognitive radios," in *1st International Conference on Cognitive Radio Oriented Wireless Networks and Communications*, Mykonos Island, Greece, June 2006, pp. 1–5.
142. S. Mallat and W. L. Hwang, "Singularity detection and processing with wavelets," *IEEE Transactions on Information Theory*, vol. 38, no. 2, pp. 617–643, Mar. 1992.
143. A. Grossman, "Wavelet transform and edge detection." In S. Albeverio, Ph. Blanchard, L. Streit and M. Hazewinkel (eds.), *Stochastic Processes in Physics and Engineering*. Dordrecht, the Netherlands: Reidel, 1986.
144. A. Grossman and J. Morlet, "Decomposition of hardy functions into square integrable wavelets of constant shape," *SIAM Journal of Mathematical Analysis*, vol. 15, pp. 723–736, 1984.
145. S. Zhong, "Edges representation from wavelet transform maxima," Ph.D. dissertation, New York University, New York, Sept. 1990.
146. B. M. Sadler and A. Swami, "Analysis of multiscale products for step detection and estimation," *IEEE Transactions on Information Theory*, vol. 45, no. 3, pp. 1043–1051, Apr. 1999.
147. D. Middleton, "Statistical physical models of electromagnetic interference," *IEEE Transactions Electromagnetic Compatibility*, vol. EMC-19, no. 1, pp. 106–127, Aug. 1977.
148. S. A. Kassam and H. V. Poor, "Robust techniques for signal processing: A survey," *Proceedings of the IEEE*, vol. 73, no. 3, pp. 433–481, Mar. 1985.
149. D. W. J. Stein, "Detection of random signals in Gaussian mixture noise," *IEEE Transactions on Information Theory*, vol. 41, no. 6, pp. 1788–1801, 1995.
150. J. Miller and J. Thomas, "The detection of signals in impulsive noise modeled as a mixture process," *IEEE Transactions on Communications*, vol. 24, no. 5, pp. 559–563, May 1976.
151. P. J. Huber and E. M. Ronchetti, *Robust Statistics*. Hoboken, NJ: John Wiley & Sons, Inc., 2009.
152. J. Miller and J. Thomas, "Robust detectors for signals in non-Gaussian noise," *IEEE Transactions on Communications*, vol. 25, no. 7, pp. 686–690, July 1977.

153. M. Bkassiny and S. K. Jayaweera, "Robust, non-Gaussian wideband spectrum sensing in cognitive radios," *IEEE Transaction in Wireless Communications*, Aug. 2014, Accepted for Publication.

154. T. Li, "A nonlinear method for robust spectral analysis," *IEEE Transactions on Signal Processing*, vol. 58, no. 5, pp. 2466–2474, 2010.

155. T. Li, "Laplace periodogram for time series analysis," *Journal of the American Statistical Association*, vol. 103, no. 482, pp. 757–768, 2008.

156. S. Banach, "Sur les operations dans les ensembles abstraits et leur application aux equations integrales," *Fundamenta Mathematicae*, vol. 3, pp. 133–181, 1922.

157. E. Candès, J. Romberg, and T. Tao, "Robust uncertainty principles: Exact signal reconstruction from highly incomplete frequency information," *IEEE Transactions on Information Theory*, vol. 52, no. 2, pp. 489–509, Feb. 2006.

158. D. Donoho, "Compressed sensing," *IEEE Transactions on Information Theory*, vol. 52, no. 4, pp. 1289–1306, Apr. 2006.

159. E. Candès and T. Tao, "Near optimal signal recovery from random projections: Universal encoding strategies?" *IEEE Transactions on Information Theory*, vol. 52, no. 12, pp. 5406–5425, Dec. 2006.

160. E. J. Candès and M. B. Wakin, "An introduction to compressive sampling," *IEEE Signal Processing Magazine*, pp. 21–30, Mar. 2008.

161. F. J. Harris, *Multirate Signal Processing for Communication Systems*. Upper Saddle River, NJ: Prentice Hall, 2004.

162. R. E. Ziemer and W. H. Tranter, *Principles of Communications: Systems, Modulation, and Noise*. Hoboken, NJ: John Wiley & Sons, Inc., 2002.

163. E. Candès and J. Romberg, "Sparsity and incoherence in compressive sampling," *Inverse Problems*, vol. 23, no. 3, pp. 969–985, 2007.

164. E. J. Candès, "Lectures on compressive sampling and frontiers in signal processing," in The Institute for Mathematics and its Applications, University of Minnesota, Minneapolis, MN, 2007.

165. A. Cohen, W. Dahmen, and R. DeVore, "Compressed sensing and best k-term approximation," *Journal of American Mathematical Society*, vol. 22, pp. 211–231, 2009.

166. Z. Tian and G. B. Giannakis, "Compressed sensing for wideband cognitive radios," in *IEEE International Conference on Acoustics, Speech and Signal Processing (ICASSP 2007)*, vol. 4, Honolulu, HI, Apr. 2007, pp. IV-1357–IV-1360.

167. M. Bkassiny, S. K. Jayaweera, Y. Li, and K. A. Avery, "Wideband spectrum sensing and non-parametric signal classification for autonomous self-learning cognitive radios," *IEEE Transactions on Wireless Communications*, vol. 11, no. 7, pp. 2596–2606, July 2012.

168. M. Bkassiny, S. K. Jayaweera, Y. Li, and K. A. Avery , "Blind cyclostationary feature detection based spectrum sensing for autonomous self-learning cognitive radios," in *IEEE International Conference on Communications (ICC '12)*, Ottawa, Canada, June 2012.

169. N. Shetty, S. Pollin, and P. Pawelczak, "Identifying spectrum usage by unknown systems using experiments in machine learning," in *IEEE Wireless Communications and Networking Conference (WCNC '09)*, Budapest, Hungary, Apr. 2009, pp. 1–6.

170. F. Millioz and N. Martin, "Estimation of a white Gaussian noise in the short time Fourier transform based on the spectral kurtosis of the minimal statistics: Application

to underwater noise," in *IEEE International Conference on Acoustics Speech and Signal Processing (ICASSP '10)*, Dallas, TX, Mar. 2010, pp. 5638–5641.

171. W. Gardner, "Measurement of spectral correlation," *IEEE Transactions on Acoustics, Speech and Signal Processing*, vol. 34, no. 5, pp. 1111–1123, Oct. 1986.
172. O. Setter, M. Sharir, and D. Halperin, "Constructing two-dimensional Voronoi diagrams via divide-and-conquer of envelopes in space," in *Proceedings of the Sixth International Symposium on Voronoi Diagrams (ISVD '09)*, Copenhagen, Denmark, June 2009, pp. 43–52.
173. W. Wei and J. M. Mendel, "A new maximum-likelihood method for modulation classification," in *Proceedings of Asilomar Conference*, Asilomar, CA, 1995, pp. 1132–1135.
174. A. Polydoros and K. Kim, "On the detection and classification of quadrature digital modulation in broad-band noise," *IEEE Transactions on Communications*, vol. 38, pp. 1199–1211, Aug. 1990.
175. C. Long, K. Chugg, and A. Polydoros, "Further results in maximum likelihood classification of QAM signals," in *Proceedings of the IEEE MILCOM Conference*, Long Branch, NJ, 1994, pp. 57–61.
176. C. Huang and A. Polydoros, "Likelihood methods for MPSK modulation classification," *IEEE Transactions on Wireless Communications*, vol. 43, pp. 1493–1504, Feb./Mar./Apr. 1995.
177. A. Swami and B. M. Sadler, "Hierarchical digital modulation classification using cumulants," *IEEE Transactions on Wireless Communications*, vol. 48, no. 3, pp. 416–429, Mar. 2000.
178. K. M. Chugg and C. Long. A, "Combined likelihood power estimation and multiple hypothesis modulation classification," in *Proceedings of the Asilomar Conference*, Asilomar, CA, 1995, pp. 1137–1141.
179. O. A. Dobre, A. Abdi, Y. Bar-Ness, and W. Su, "Survey of automatic modulation classification techniques: Classical approaches and new trends," *IET Communications*, vol. 1, no. 2, pp. 137–156, Apr. 2007.
180. Q. Zhu, M. Kam, and R. Yeager, "Non-parametric identification of QAM constellations in noise," in *IEEE International Conference on Acoustics, Speech, and Signal Processing (ICASSP '93)*, vol. 4, Minneapolis, MN, Apr. 1993, pp. 184–187.
181. F. Wang and X. Wang, "Fast and robust modulation classification via Kolmogorov-Smirnov test," *IEEE Transactions on Communications*, vol. 58, no. 8, pp. 2324–2332, Aug. 2010.
182. T. Clancy, A. Khawar, and T. Newman, "Robust signal classification using unsupervised learning," *IEEE Transactions on Wireless Communications*, vol. 10, no. 4, pp. 1289–1299, Apr. 2011.
183. H. Wu, M. Saquib, and Z. Yun, "Novel automatic modulation classification using cumulant features for communications via multipath channels," *IEEE Transactions on Wireless Communications*, vol. 7, no. 8, pp. 3098–3105, 2008.
184. K. Kim, I. Akbar, K. Bae, J. Urn, C. Spooner, and J. Reed, "Cyclostationary approaches to signal detection and classification in cognitive radio," in *2nd IEEE International Symposium on New Frontiers in Dynamic Spectrum Access Networks (DySPAN '07)*, Dublin, Ireland, Apr. 2007, pp. 212–215.
185. M. Ramon, T. Atwood, S. Barbin, and C. Christodoulou, "Signal classification with an SVM-FFT approach for feature extraction in cognitive radio," in *SBMO/IEEE MTT-S*

International Microwave and Optoelectronics Conference (IMOC '09), Belem, Brazil, Nov. 2009, pp. 286–289.

186. T. S. Rappaport, *Wireless Communications: Principles and Practice*. Upper Saddle River, NJ: Prentice Hall PTR, 2002.
187. W. C. Jakes, Ed., *Microwave Mobile Communications*. Piscataway, NJ: Wiley-IEEE Press, 1994.
188. P. K. Varshney, *Distributed Detection and Data Fusion*. New York: Springer-Verlag, 1996.
189. R. Viswanathan and P. K. Varshney, "Distributed detection with multiple sensors: Part I—Fundamentals," *Proceedings of the IEEE*, vol. 85, no. 1, pp. 54–63, Jan. 1997.
190. R. S. Blum, S. A. Kassam, and H. V. Poor, "Distributed detection with multiple sensors: Part II—Advanced topics," *Proceedings of the IEEE*, vol. 85, no. 1, pp. 64–79, Jan. 1997.
191. J. N. Tsistsiklis, "Decentralized detection." In H. V. Poor and J. B. Thomas (eds.), *Advances in Statistical Signal Processing, Signal Detection*. Greenwich, CT: JAI Press, 1993.
192. J. N. Tsistsiklis, "Decentralized detection by a large number of sensors," *Mathematics of Control, Signals, and Systems*, vol. 1, pp. 167–182, 1988.
193. R. R. Tenney and N. R. Sandell Jr., "Detection with distributed sensors," *IEEE Transactions on Aerospace and Electronic Systems*, vol. AES-17, no. 4, pp. 501–510, July 1981.
194. S. K. Jayaweera, "Bayesian fusion performance and system optimization in distributed stochastic Gaussian signal detection under communication constraints," *IEEE Transactions on Signal Processing*, vol. 55, pp. 1238–1250, Apr. 2007.
195. S. K. Jayaweera, and R. Viswanathan, Distributed signal processing for wireless sensor networks. In H. Hu, Y. Zhang, and J. Luo (eds.), *Distributed Antenna Systems: Open Architecture for Future Wireless Communications*. Boca Raton, FL: CRC Press, 2007.
196. S. K. Jayaweera and K. Altarazi, "Large system decision fusion performance in inhomogeneous sensor networks," in *IEEE International Conference on Advanced Video and Signal based Surveillance (AVSS '06)*, Sydney, Australia, Nov. 2006.
197. K. Altarazi, S. K. Jayaweera, and V. Aravinthan, "Performance of decentralized detection in a resource-constrained sensor network," in *39th Annual Asilomar Conference on Signals, Systems and Computers*, Pacific Grove, CA, Nov. 2005.
198. J. Chamberland and V. V. Veeravalli, "Decentralized detection in sensor networks," *IEEE Transactions on Signal Processing*, vol. 51, pp. 407–416, Feb. 2003.
199. J. Chamberland and V. V. Veeravalli, "Asymptotic results for decentralized detection in power constrained wireless sensor networks," *IEEE Journal on Selected Areas in Communications*, vol. 22, no. 6, pp. 1007–1015, Aug. 2004.
200. S. K. Jayaweera, "Large system decentralized detection performance under communication constraints," *IEEE Communications Letters*, vol. 9, pp. 769–771, Sept. 2005.
201. Q. Zhao, B. Krishnamachari, and K. Liu, "On myopic sensing for multi-channel opportunistic access: Structure, optimality, and performance," *IEEE Transactions on Wireless Communications*, vol. 7, no. 12, pp. 5431–5440, Dec. 2008.
202. K. Liu, Q. Zhao, and B. Krishnamachari, "Dynamic multichannel access with imperfect channel state detection," *IEEE Transactions on Signal Processing*, vol. 58, no. 5, pp. 2795–2808, May 2010.

203. J. Unnikrishnan and V. V. Veeravalli, "Algorithms for dynamic spectrum access with learning for cognitive radio," *IEEE Transactions on Signal Processing*, vol. 58, no. 2, pp. 750–760, Feb. 2010.

204. T. P. W. Wang, Y. Cui, and W. Wang, "Noncooperative power control game with exponential pricing for cognitive radio network," in *Vehicular Technology Conference (VTC2007-Spring)*, Dublin, Ireland, Apr. 2007.

205. J. O'Daniell, "Analysis and design of cognitive radio networks and distributed radio resource management algorithms," Ph.D. dissertation, Virginia Tech., Blacksburg, VA, 2006.

206. P. Zhou, W. Yuan, W. Liu, and W. Cheng, "Joint power and rate control in cognitive radio networks: A game-theoretical approach," in *Proceedings of the IEEE International Conference on Communications (ICC'08)*, Beijing, China, May 2008, pp. 3296–3301.

207. T. Li and S. K. Jayaweera, "Analysis of linear receivers in a target SINR game for wireless cognitive networks," in *Proceedings of the IEEE WICOM'08*, Dalian, China, Oct. 2008.

208. T. Li and S. K. Jayaweera, "A novel primary-secondary user power control game for cognitive radios," in *Proceedings of the 2008 International Symposium on Information Theory and Its Applications (ISITA2008)*, Auckland, New Zealand, Dec. 2008.

209. M. Dohler, E. Lefranc, and H. Aghvami, "Virtual antenna arrays for future wireless mobile communication systems," in *ICT2002*, Beijing, China, June 2002.

210. M. Dohler, E. Lefranc, and H. Aghvami, "Space-time block codes for virtual antenna arrays," in *PIMRC*, Lisbon, Portugal, Sept. 2002.

211. A. Sendonaris, E. Erkip, and B. Aazhang, "User cooperation diversity—Part I: System description," *IEEE Transactions on Communications*, vol. 51, no. 11, pp. 1927–1938, Nov. 2003.

212. A. Sendonaris, E. Erkip, and B. Aazhang, "User cooperation diversity—Part II: Implementation aspects and performance analysis," *IEEE Transactions on Communications*, vol. 51, no. 11, pp. 1939–1948, Nov. 2003.

213. S. K. Jayaweera, "Energy analysis of MIMO techniques in wireless sensor networks," in *38th Annual Conference on Information Sciences and Systems (CISS 04)*, Princeton, NJ, Mar. 2004.

214. J. N. Laneman, D. N. C. Tse, and G.W. Wornell, "Cooperative diversity in wireless networks: Efficient protocols and outage behavior," *IEEE Transactions on Information Theory*, vol. 50, no. 12, pp. 3062–3080, Dec. 2004.

215. S. K. Jayaweera, "An energy-efficient virtual MIMO communications architecture based on V-BLAST processing for distributed wireless sensor networks," in *First IEEE International Conference on Sensor and Ad-Hoc Communications and Networks (SECON 2004)*, Santa Clara, CA, Oct. 2004.

216. K. Azarian, H. El Gamal, and P. Schniter, "On the achievable diversity-multiplexing tradeoff in half-duplex cooperative channels," *IEEE Transactions on Information Theory*, vol. 51, no. 12, pp. 4152–4172, Dec. 2005.

217. S. K. Jayaweera, "Virtual MIMO-based cooperative communication for energy constrained wireless sensor networks," *IEEE Transactions on Wireless Communications*, vol. 5, no. 5, pp. 984–989, May 2006.

218. S. K. Jayaweera, "V-BLAST-based virtual MIMO architectures for energy-constrained wireless sensor networks," *IEEE Transactions on Communications*, vol. 55, no. 10, pp. 1867–1872, Oct. 2007.

219. G. El-Howayek and S. K. Jayaweera, "Distributed dynamic spectrum leasing (D-DSL) for spectrum sharing over multiple primary channels," *IEEE Transactions on Wireless Communications*, vol. 10, no. 1, pp. 55–60, Jan. 2011.

220. K. Hakim, S. K. Jayaweera, G. El-Howayek, and C. Mosquera, "Efficient dynamic spectrum sharing in cognitive radio networks: Centralized dynamic spectrum leasing (C-DSL)," *IEEE Transactions on Wireless Communications*, vol. 9, no. 9, pp. 2956–2967, Sept. 2010.

221. K. Hakim, S. K. Jayaweera, and C. Mosquera, "Analysis of linear receivers in a DSL game for interference spectrum sharing in cognitive radio network," in *Proceedings of the International Communications Conference (ICC 2010)*, Cape Town, South Africa, May 2010.

222. D. Fugenberg and J. Tirole, *Game Theory*. Cambridge, MA: MIT Press, 1991.

223. R. D. Yates, "A framework for uplink power control in cellular radio systems," *IEEE Journal on Selected Areas in Communications*, vol. 13, pp. 1341–1347, 1996.

224. S. K. Sarkar, T. G. Basavaraju, and C. Puttamadappa, *Ad Hoc Mobile Wireless Networks: Principles, Protocols, and Applications*. Boca Raton, FL: Auerbach Publications, 2007.

225. I. Stanojev, O. Simeone, Y. Bar-Ness, and T. Yu, "Spectrum leasing via distributed cooperation in cognitive radio," in *IEEE International Conference on Communications (ICC 2008)*, Beijing, China, May 2008, pp. 3427–3431.

226. A. V. Kariyapperuma and S. K. Jayaweera, "An asymptotically full rate cooperative communications scheme for DS-CDMA systems with non-orthogonal codes," in *Proceedings of the 2008 International Symposium on Information Theory and Its Applications (ISITA2008)*, Auckland, New Zealand, Dec. 2008.

227. H. W. Kuhn, "The Hungarian method for the assignment problem," *Naval Research Logistics Quarterly*, vol. 2, pp. 83–97, Mar. 1955.

228. C. H. Papadimitriou and K. Steiglitz, *Combinatorial Optimization: Algorithms and Complexity*. Mineola, NY: Dover Publications, 1998.

229. Y. Choi, J. Kim, and S. Bahk, "Downlink scheduling with fairness and optimal antenna assignment for MIMO cellular systems," in *IEEE Global Telecommunications Conference*, vol. 5, Dallas, TX, Nov. 2004, pp. 3165–3169.

230. G. Foschini and M. Gans, "On limits of wireless communications in a fading environment when using multiple antennas," *Wireless Personal Communications*, vol. 6, no. 3, pp. 311–335, Mar. 1998.

231. I. Telatar, "Capacity of multi-antenna Gaussian channels," *European Transactions on Telecommunications*, vol. 10, pp. 585–596, Nov.-Dec. 1999.

232. S. K. Jayaweera and H. V. Poor, "Capacity of multi-antenna systems with both receiver and transmitter channel state information," *IEEE Transactions on Information Theory*, vol. 49, no. 10, pp. 2697–2709, Oct. 2003.

233. S. M. Alamouti, "A simple transmit diversity technique for wireless communications," *IEEE Journal on Selected Areas in Communications*, vol. 16, no. 8, pp. 1451–1458, Oct. 1998.

234. V. Tarokh, N. Seshadri, and A. R. Calderbank, "Space-time codes for high rate wireless communication: Performance criterion and code construction," *IEEE Transactions on Information Theory*, vol. 44, no. 2, pp. 744–765, Mar. 1998.

235. V. Tarokh, H. Jafarkhani, and A. R. Calderbank, "Space-time block codes from orthogonal designs," *IEEE Transaction on Information Theory*, vol. 45, no. 5, pp. 1456–1467, July 1999.

236. G. J. Foschini, "Layered space-time architecture for wireless communication in a flat fading environment when using multi-element antennas," *Bell Labs Technical Journal*, vol. 1, no. 2, pp. 41–59, Autumn 1996.

237. R. Calderbank, A. Goldsmith, A. Pauraj, H. V. Poor, and E. Biglieri, Eds., *MIMO Wireless Communication*. Cambridge, UK: Cambridge University Press, 2008.

238. L. Zheng and D. N. C. Tse, "Diversity and multiplexing: A fundamental tradeoff in multiple antenna channels," *IEEE Transactions on Information Theory*, vol. 49, no. 5, pp. 1073–1096, May 2003.

239. S. Cui, A. J. Goldsmith, and A. Bahai, "Energy efficiency of MIMO and cooperative MIMO techniques in sensor networks," *IEEE Journal on Selected Areas in Communications*, vol. 22, no. 6, pp. 1089–1098, Oct. 2004.

240. J. N. Laneman and G. W. Wornell, "Distributed space-time-coded protocols for exploiting cooperative diversity in wireless networks," *IEEE Transaction on Information Theory*, vol. 49, no. 10, pp. 2415–2425, Oct. 2003.

241. G. Scutari and S. Barbarossa, "Distributed space-time coding for regenerative relay networks," *IEEE Transactions on Wireless Communications*, vol. 4, no. 5, pp. 2387–2399, Sept. 2005.

242. G. Kramer, M. Gastpar, and P. Gupta, "Cooperative strategies and capacity theorems for relay networks," *IEEE Transactions on Information Theory*, vol. 51, no. 9, pp. 3037–3063, Sept. 2005.

243. G. Scutari and S. Barbarossa, "Capacity bounds for cooperative diversity," *IEEE Transactions on Information Theory*, vol. 52, no. 4, pp. 1522–1544, Apr. 2006.

244. A. D. Murugan, P. K. Gopala, and H. E. Gamal, "Correlated sources over wireless channels: Cooperative source-channel coding," *IEEE Journal on Selected Areas in Communications*, vol. 22, no. 6, pp. 988–998, Aug. 2004.

245. F. H. P. Fitzek and M. D. Katz, Eds., *Cooperation in Wireless Networks: Principles and Applications*. Dordrecht, the Netherlands: Springer, 2006.

246. K. Liu, A. Sadek, W. Su, and A. Kwasinski, *Cooperative Communications and Networking*. Cambridge, UK: Cambridge University Press, 2008.

247. M. Dohler and Y. Li, *Cooperative Communications: Hardware, Channel & Phy*. Hoboken, NJ: Wiley, 2010.

248. K. B. Letaief and W. Zhang, "Cooperative communications for cognitive radio networks," *Proceedings of the IEEE*, vol. 97, no. 5, pp. 878–893, May 2009.

249. S. Chen, C. Ghosh, A. M. Alexander, and S. K. Jayaweera, "The impact of group cooperation over competitive secondary subnetworks," in *IEEE Pacific Rim Conference on Communications, Computers and Signal Processing*, Victoria, Canada, Aug. 2011.

250. M. Bkassiny and S. K. Jayaweera, "Optimal channel and power allocation for secondary users in cooperative cognitive radio networks," in *2nd International Conference on Mobile Lightweight Wireless Systems*, Barcelona, Spain, May 2010.

251. N. Devroye, M. Vu, and V. Tarokh, "Achievable rates in cognitive radio channels," *IEEE Transactions on Information Theory*, vol. 52, no. 5, pp. 1813–1827, May 2006.

252. Y. Chang, Z. Tao, J. Zhang, and C. Kuo, "A graph-based approach to multi-cell OFDMA downlink resource allocation," in *IEEE Global Telecommunications Conference*, New Orleans, LA, Dec. 2008, pp. 1–6.

253. N. J. Nilsson, "Introduction to machine learning," 1997, Draft of Incomplete Notes. Available at: http://robotics.stanford.edu/nilsson/mlbook.html.

254. S. Kulkarni and G. Harman, *An Elementary Introduction to Statistical Learning Theory*. Hoboken, NJ: John Wiley & Sons, 2011.

255. B. Widrow, "Generalization and information storage in networks of adaline 'neurons'." In M. Yovitz, G. Jacobi, and G. Goldstein (eds.), *Self-Organizing Systems*. Washington DC: Spartan, 1962, pp. 435–461.

256. M. Minsky and S. Papert, *Perceptrons*. Cambridge, MA: MIT Press, 1988.

257. F. Rosenblatt, "Perceptron simulation experiments," in *Proceedings of Institute of Radio Engineers*, vol. 48, pp. 301–309, 1960.

258. P. J. Werbos, "Beyond regression: New tools for prediction and analysis in the behavioral sciences," Ph.D. dissertation, Harvard University, Cambridge, MA, 1974.

259. D. E. Rumelhart, G. E. Hinton, and J. L. McClelland, "Learning representations by back-propagating errors," *Nature*, vol. 323, pp. 533–536, 1986.

260. M. Gandetto, M. Guainazzo, and C. S. Regazzoni, "Use of time-frequency analysis and neural networks for mode identification in a wireless software-defined radio approach," *EURASIP Journal on Applied Signal Processing*, vol. 12, pp. 1778–1790, 2004.

261. K. Tsagkaris, A. Katidiotis, and P. Demestichas, "Neural network-based learning schemes for cognitive radio systems," *Computer Communications*, vol. 31, pp. 3394–3404, 2008.

262. T. Kohonen, "The self-organizing map," *Proceedings of the IEEE*, vol. 78, no. 9, pp. 1464–1480, Sept. 1990.

263. V. Vapnik, *The Nature of Statistical Learning Theory*. New York: Springer-Verlag, 1995.

264. C. J. C. Burges, "A tutorial on support vector machines for pattern recognition," *Data Mining and Knowledge Discovery*, vol. 2, pp. 121–167, 1998.

265. M. Anthony and N. Biggs, Pac learning and neural networks. In M. A. Arbib (ed.), *Handbook of Brain Theory and Neural Networks*. Cambridge, MA: Bradford Books/MIT Press, 1995, pp. 694–697.

266. V. Vapnik, *Statistical Learning Theory*. New York: John Wiley & Sons, Inc, 1998.

267. B. E. Boser, I. M. Guyon, and V. Vapnik, "A training algorithm for optimal margin classifiers," in *Fifth Annual ACM Workshop on Computational Learning Theory*, Pittsburgh, PA, 1992, pp. 144–152.

268. A. Aizerman, E. Braverman, and L. Rozoner, "Theoretical foundations of the potential function method in pattern recognition learning," *Automation and Remote Control*, vol. 25, pp. 821–837, 1964.

269. W. V. Lovitt, *Linear Integral Functions*. New York: Dover, 1950.

270. H. Hu, J. Song, and Y. Wang, "Signal classification based on spectral correlation analysis and SVM in cognitive radio," in *22nd International Conference on Advanced Information Networking and Applications*, Okinawa, Japan, Mar. 2008.

271. G. Ding, Q. Wu, Y. Yao, J. Wang, and Y. Chen, "Kernel-based learning for statistical signal processing in cognitive radio networks," *IEEE Signal Processing Magazine*, vol. 30, pp. 123–136, July 2013.

272. P. A. Forero and V. Kekatos. G. Giannakis, "Robust clustering using outlier sparsity regularization," *IEEE Transactions on Signal Processing*, vol. 60, no. 8, pp. 4163–4177, Aug. 2012.
273. K. R. Müller, S. Mika, G. Rätsch, K. Tsuda, and B. Scholkopf, "An introduction to kernel-based learning algorithms," *IEEE Transactions on Neural Networks*, vol. 12, no. 2, pp. 181–201, Mar. 2001.
274. W. Liu, J. C. Principe, and S. Haykin, *Kernel Adaptive Filtering*. Hoboken, NJ: Wiley, 2010.
275. S. Mika, G. Ratsch, J. Weston, B. Scholkopf, and K. Muller, "Fisher discriminant analysis with kernels," in *Proceedings of the IEEE Signal Processing Society Workshop Neural Networks Signal Processing*, Madison, WI, 1999, pp. 41–48.
276. T. D. Atwood, "Rf channel characterization for cognitive radio using support vector machines," Ph.D. dissertation, University of New Mexico, Albuquerque, NM, 2009.
277. M. L. Littman, A. R. Cassandra, and L. P. Kaelbling, "Learning policies for partially observable environments: Scaling up," in *Proceedings of the 12th International Conference on Machine Learning*, Tahoe City, CA, 1995, pp. 362–370.
278. L. Chrisman, "Reinforcement learning with perceptual aliasing: The perceptual distinctions approach," in *Proceedings of the 10th National Conference on AI*, San Jose, CA, 1992.
279. R. A. McCallum, "First results with utile distinction memory for reinforcement learning," Technical Report 446, Department of Computer Science, University of Rochester, Rochester, NY, 1992.
280. B. F. Lo and I. F. Akyildiz, "Reinforcement learning for cooperative sensing gain in cognitive radio ad hoc networks," *Wireless Networks*, vol. 19, pp. 1237–1250, 2013.
281. G. D. Croon, M. F. V. Dartel, and E. O. Postma, "Evolutionary learning outperforms reinforcement learning on non-Markovian tasks," in *8th European Conference on Artificial Life Workshop on Memory and Learning Mechanisms in Autonomous Robots*, Kent, UK, 2005.
282. D. E. Moriarty, A. C. Schultz, and J. J. Grefenstette, "Evolutionary algorithms for reinforcement learning," *Journal of Artificial Intelligence Research*, vol. 11, pp. 241–276, 1999.
283. B. Wang, K. R. Liu, and T. Clancy, "Evolutionary cooperative spectrum sensing game: How to collaborate?" *IEEE Transactions on Wireless Communications*, vol. 58, no. 3, pp. 890–900, Mar. 2010.
284. F. Dandurand and T. Shultz, "Connectionist models of reinforcement, imitation, and instruction in learning to solve complex problems," *IEEE Transactions on Autonomous Mental Development*, vol. 1, no. 2, pp. 110–121, Aug. 2009.
285. J. Baxter and P. L. Bartlett, "Infinite-horizon policy-gradient estimation," *Journal of Artificial Intelligence Research*, vol. 15, pp. 319–350, 2001.
286. R. Sutton, D. Mcallester, S. Singh, and Y. Mansour, "Policy gradient methods for reinforcement learning with function approximation," in *Proceedings of the 12th Conference on Advances in Neural Information Processing Systems*, Denver, CO, 2001, pp. 1057–1063.
287. M. Bkassiny, Y. Li, and S. K. Jayaweera, "A survey on machine-learning techniques in cognitive radios," *IEEE Communications Surveys & Tutorials*, vol. 15, no. 3, pp. 1136–1159, 2013.

288. K. Leyton-Brown and Y. Shoham, *Essentials of Game Theory: A Concise, Multidisciplinary Introduction*. San Rafael, CA: Morgan & Claypool, 2008.
289. A. B. MacKenzie and L. A. DaSilva, *Game Theory for Wireless Engineers*. San Rafael, CA: Morgan & Claypool, 2006.
290. A. Blum and Y. Mansour, "Learning, regret minimization, and equilibria." In N. Nisan, T. Roughgarden, E. Tardos, and V. V. Vazirani (eds.), *Algorithmic Game Theory*. Cambridge, UK: Cambridge University Press, 2007, pp. 79–102.
291. J. Hannan, "Approximation to Bayes risk in repeated plays. In *Contributions to the Theory of Games*. Princeton, NJ: Princeton University Press, 1957, vol. 3, pp. 97–139.
292. N. Littlestone and M. K. Warmuth, "The weighted majority algorithm," *Information and Computation*, vol. 108, pp. 212–261, 1994.
293. P. Stone and M. M. Veloso, "Multiagent systems: A survey from a machine learning perspective," *Autonomous Robots*, vol. 8, pp. 345–383, 2000.
294. L. Panait and S. Luke, "Cooperative multi-agent learning: The state of the art," *Autonomous Agents and Multi-Agent Systems*, vol. 11, pp. 387–434, 2005.
295. M. Tan, Multiagent reinforcement learning: Independent vs. cooperative learning. *Readings in Agents*. San Francisco, CA: Morgan Kaufmann, 1993, pp. 487–494.
296. X. Zhang and J. Zhao, "Power control based on the asynchronous distributed pricing algorithm in cognitive radios," in *IEEE Youth Conference on Information Computing and Telecommunications (YC-ICT '10)*, Beijing, China, Nov. 2010, pp. 69–72.
297. J. Li, D. Chen, W. Li, and J. Ma, "Multiuser power and channel allocation algorithm in cognitive radio," in *International Conference on Parallel Processing (ICPP '07)*, Xi'an, China, Sep. 2007, p. 72.
298. L. Pillutla and V. Krishnamurthy, "Game theoretic rate adaptation for spectrum-overlay cognitive radio networks," in *IEEE Global Telecommunications Conference (GLOBECOM '08)*, New Orleans, LA, Dec. 2008, pp. 1–5.
299. H. Li, Y. Liu, and D. Zhang, "Dynamic spectrum access for cognitive radio systems with repeated games," in *IEEE International Conference on Wireless Communications, Networking and Information Security (WCNIS '10)*, Beijing, China, June 2010, pp. 59–62.
300. F. Fu and M. van der Schaar, "Stochastic game formulation for cognitive radio networks," in *3rd IEEE Symposium on New Frontiers in Dynamic Spectrum Access Networks (DySPAN '08)*, Chicago, IL, Oct. 2008, pp. 1–5.
301. L. Chen, S. Iellamo, M. Coupechoux, and P. Godlewski, "An auction framework for spectrum allocation with interference constraint in cognitive radio networks," in *IEEE INFOCOM '10*, San Diego, CA, Mar. 2010, pp. 1–9.
302. B. Latifa, Z. Gao, and S. Liu, "No-regret learning for simultaneous power control and channel allocation in cognitive radio networks," in *Computing, Communications and Applications Conference (ComComAp '12)*, Hong Kong, China, Jan. 2012, pp. 267–271.
303. Q. Zhu, Z. Han, and T. Basar, "No-regret learning in collaborative spectrum sensing with malicious nodes," in *IEEE International Conference on Communications (ICC '10)*, Cape Town, South Africa, May 2010, pp. 1–6.
304. Y. Xu, J. Wang, Q. Wu, A. Anpalagan, and Y.-D. Yao, "Opportunistic spectrum access in unknown dynamic environment: A game-theoretic stochastic learning solution," *IEEE Transactions on Wireless Communications*, vol. 11, no. 4, pp. 1380–1391, Apr. 2012.

305. M. Maskery, V. Krishnamurthy, and Q. Zhao, "Decentralized dynamic spectrum access for cognitive radios: Cooperative design of a non-cooperative game," *IEEE Transactions on Communications*, vol. 57, no. 2, pp. 459–469, Feb. 2009.
306. L. Giupponi, A. Galindo-Serrano, P. Blasco, and M. Dohler, "Docitive networks: An emerging paradigm for dynamic spectrum management," *IEEE Wireless Communications*, vol. 17, no. 4, pp. 47–54, Aug. 2010.
307. A. Galindo-Serrano, L. Giupponi, P. Blasco, and M. Dohler, "Learning from experts in cognitive radio networks: The docitive paradigm," in *Proceedings of the Fifth International Conference on Cognitive Radio Oriented Wireless Networks Communications*, Cannes, France, June 2010, pp. 1–6.
308. R. J. Muirhead, *Aspects of Multivariate Statistical Theory*. New York: John Wiley & Sons, 1982.

INDEX

Signal Processing for Cognitive Radios, First Edition. Sudharman K. Jayaweera.